FIFTH IFIP INTERNATIONAL CONFERENCE ON THEORETICAL COMPUTER SCIENCE – TCS 2008

IFIP – The International Federation for Information Processing

IFIP was founded in 1960 under the auspices of UNESCO, following the First World Computer Congress held in Paris the previous year. An umbrella organization for societies working in information processing, IFIP's aim is two-fold: to support information processing within its member countries and to encourage technology transfer to developing nations. As its mission statement clearly states,

> *IFIP's mission is to be the leading, truly international, apolitical organization which encourages and assists in the development, exploitation and application of information technology for the benefit of all people.*

IFIP is a non-profitmaking organization, run almost solely by 2500 volunteers. It operates through a number of technical committees, which organize events and publications. IFIP's events range from an international congress to local seminars, but the most important are:

- The IFIP World Computer Congress, held every second year;
- Open conferences;
- Working conferences.

The flagship event is the IFIP World Computer Congress, at which both invited and contributed papers are presented. Contributed papers are rigorously refereed and the rejection rate is high.

As with the Congress, participation in the open conferences is open to all and papers may be invited or submitted. Again, submitted papers are stringently refereed.

The working conferences are structured differently. They are usually run by a working group and attendance is small and by invitation only. Their purpose is to create an atmosphere conducive to innovation and development. Refereeing is less rigorous and papers are subjected to extensive group discussion.

Publications arising from IFIP events vary. The papers presented at the IFIP World Computer Congress and at open conferences are published as conference proceedings, while the results of the working conferences are often published as collections of selected and edited papers.

Any national society whose primary activity is in information may apply to become a full member of IFIP, although full membership is restricted to one society per country. Full members are entitled to vote at the annual General Assembly, National societies preferring a less committed involvement may apply for associate or corresponding membership. Associate members enjoy the same benefits as full members, but without voting rights. Corresponding members are not represented in IFIP bodies. Affiliated membership is open to non-national societies, and individual and honorary membership schemes are also offered.

FIFTH IFIP INTERNATIONAL CONFERENCE ON THEORETICAL COMPUTER SCIENCE – TCS 2008

IFIP 20th World Computer Congress, TC 1, Foundations of Computer Science, September 7-10, 2008, Milano, Italy

Edited by

Giorgio Ausiello
University of Rome
Italy

Juhani Karhumäki
University of Turku
Finland

Giancarlo Mauri
University of Milano
Italy

Luke Ong
Cambridge University
United Kingdom

Editors
Giorgio Ausiello
University of Rome
Italy

Juhani Karhumäki
University of Turku
Finland

Giancarlo Mauri
University of Milan
Italy

Luke Ong
Cambridge University
United Kingdom

p. cm. (IFIP International Federation for Information Processing, a Springer Series in Computer Science)

ISSN: 1571-5736/1861-2288 (Internet)
ISBN: 978-1-4419-3514-4 eISBN: 978-0-387-09680-3

Printed in acid-free paper

springer.com

IFIP 2008 World Computer Congress (WCC'08)

Message from the Chairs

Every two years, the International Federation for Information Processing hosts a major event which showcases the scientific endeavours of its over one hundred Technical Committees and Working Groups. 2008 sees the 20th World Computer Congress (WCC 2008) take place for the first time in Italy, in Milan from 7-10 September 2008, at the MIC - Milano Convention Centre. The Congress is hosted by the Italian Computer Society, AICA, under the chairmanship of Giulio Occhini.

The Congress runs as a federation of co-located conferences offered by the different IFIP bodies, under the chairmanship of the scientific chair, Judith Bishop. For this Congress, we have a larger than usual number of thirteen conferences, ranging from Theoretical Computer Science, to Open Source Systems, to Entertainment Computing. Some of these are established conferences that run each year and some represent new, breaking areas of computing. Each conference had a call for papers, an International Programme Committee of experts and a thorough peer reviewed process. The Congress received 661 papers for the thirteen conferences, and selected 375 from those representing an acceptance rate of 56% (averaged over all conferences).

An innovative feature of WCC 2008 is the setting aside of two hours each day for cross-sessions relating to the integration of business and research, featuring the use of IT in Italian industry, sport, fashion and so on. This part is organized by Ivo De Lotto. The Congress will be opened by representatives from government bodies and Societies associated with IT in Italy.

This volume is one of fourteen volumes associated with the scientific conferences and the industry sessions. Each covers a specific topic and separately or together they form a valuable record of the state of computing research in the world in 2008. Each volume was prepared for publication in the Springer IFIP Series by the conference's volume editors. The overall Chair for all the volumes published for the Congress is John Impagliazzo.

For full details on the Congress, refer to the webpage http://www.wcc2008.org.

Judith Bishop, South Africa, Co-Chair, International Program Committee
Ivo De Lotto, Italy, Co-Chair, International Program Committee
Giulio Occhini, Italy, Chair, Organizing Committee
John Impagliazzo, United States, Publications Chair

WCC 2008 Scientific Conferences

TC12	**AI**	Artificial Intelligence 2008
TC10	**BICC**	Biologically Inspired Cooperative Computing
WG 5.4	**CAI**	Computer-Aided Innovation (Topical Session)
WG 10.2	**DIPES**	Distributed and Parallel Embedded Systems
TC14	**ECS**	Entertainment Computing Symposium
TC3	**ED_L2L**	Learning to Live in the Knowledge Society
WG 9.7 **TC3**	**HCE3**	History of Computing and Education 3
TC13	**HCI**	Human Computer Interaction
TC8	**ISREP**	Information Systems Research, Education and Practice
WG 12.6	**KMIA**	Knowledge Management in Action
TC2 **WG 2.13**	**OSS**	Open Source Systems
TC11	**IFIP SEC**	Information Security Conference
TC1	**TCS**	Theoretical Computer Science

IFIP

- is the leading multinational, apolitical organization in Information and Communications Technologies and Sciences
- is recognized by United Nations and other world bodies
- represents IT Societies from 56 countries or regions, covering all 5 continents with a total membership of over half a million
- links more than 3500 scientists from Academia and Industry, organized in more than 101 Working Groups reporting to 13 Technical Committees
- sponsors 100 conferences yearly providing unparalleled coverage from theoretical informatics to the relationship between informatics and society including hardware and software technologies, and networked information systems

Details of the IFIP Technical Committees and Working Groups can be found on the website at http://www.ifip.org.

Conference chairs

Giorgio Ausiello (Rome)

Giancarlo Mauri (Milan)

Programme Committee

Track A: Algorithms, Complexity and Models of Computation

Ricardo Baeza-Yates (Santiago)
Marie-Pierre Béal (Paris)
Harry Buhrman (Amsterdam)
Xiaotie Deng (Hong Kong)
Josep Diaz (Barcelona)
Volker Diekert (Stuttgard)
Manfred Droste (Leipzig)
Ding-zhu Du (Dallas)
Juraj Hromkovic (Zurich)
Oscar Ibarra (Santa Barbara)
Pino Italiano (Rome)
Kazuo Iwama (Kyoto)
Juhani Karhumäki (Turku, chair)
Pekka Orponen (Helsinki)
George Paun (Bucharest)
Jiri Sgall (Prague)
Alexander Shen (Moscow)
Vijay Vazirani (Atlanta)
Mikhail Volkov (Ekaterinburg)

Track B: Logic, Semantics, Specification and Verification

Rajeev Alur (Pennsylvania)
Ulrich Berger (Swansea)
Andreas Blass (Ann Arbor)
Anuj Dawar (Cambridge)
Mariangiola Dezani-Ciancaglini (Turin)
Gilles Dowek (Paris)
Peter Dybjer (Gothenburg)
Masami Hagiya (Tokyo)
Martin Hofmann (Munich)
Leonid Libkin (Edinburgh)
Huimin Lin (Beijing)
Stephan Merz (Nancy)
Dale Miller (Paris)
Eugenio Moggi (Genova)
Anca Muscholl (Bordeaux)
Luke Ong (Oxford, chair)
Davide Sangiorgi (Bologna)
Thomas Schwentick (Dortmund)
Thomas Streicher (Darmstadt)
P. S. Thiagarajan (Singapore)
Wolfgang Thomas (Aachen)

Preface

The papers contained in this volume were presented at the *5th IFIP International Conference on Theoretical Computer Science* (IFIP TCS), 7-10 September 2008, Milan, Italy.

TCS is a bi-annual conference. The first conference of the series was held in Sendai (Japan, 2000), followed by Montreal (Canada, 2002), Toulouse (France, 2004) and Santiago (Chile, 2006). TCS is organized by IFIP TC1 (Technical Committee 1: Foundations of Computer Science) and Working Group 2.2 of IFIP TC2 (Technical Committee 2: Software: Theory and Practice). TCS 2008 was part of the 20th IFIP World Computer Congress (WCC 2008), constituting the TC1 Track of WCC 2008.

The contributed papers were selected from 36+45 submissions from altogether 30 countries. A total of 14+16 submissions were accepted as full papers. Papers in this volume are original contributions in two general areas: *Track A: Algorithms, Complexity and Models of Computation*; and *Track B: Logic, Semantics, Specification and Verification*. The conference also included seven invited presentations, from Luca Cardelli, Thomas Ehrhard, Javier Esparza, Antonio Restivo, Tim Roughgarden, Grzegorz Rozenberg and Avraham Trakhtman. These presentations are included (except one) in this volume. In particular, Luca Cardelli, Javier Esparza, Antonio Restivo, Tim Roughgarden and Avraham Trakhtman accepted our invitation to write full papers related to their talks.

We thank the local WCC organizers, IFIP TC1 and WG 2.2 for their support in the organization of IFIP TCS. We also thank the members of the Programme Committee and the additional reviewers for providing timely and detailed reviews. Finally, we want to thank William Blum and, in particular, Arto Lepistö for composing this proceedings.

Milan, Italy,
September, 2008

Juhani Karhumäki (PC Chair, Track A)
Luke Ong (PC Chair, Track B)

Contents

Part I
Track A

Invited talks

Ambiguity and Complementation in Recognizable Two-dimensional Languages

Dora Giammarresi[1] and Antonio Restivo[2]

[1] Dipartimento di Matematica, Università di Roma "Tor Vergata", via della Ricerca Scientifica, 00133 Roma, Italy.
giammarr@mat.uniroma2.it
[2] Dipartimento di Matematica e Applicazioni. Università di Palermo, via Archirafi, 34 - 90123 Palermo, Italy.
restivo@dipmat.math.unipa.it

1 Introduction

The theory of one-dimensional (word) languages is well founded and investigated since fifties. From several years, the increasing interest for pattern recognition and image processing motivated the research on *two-dimensional* or *picture* languages, and nowadays this is a research field of great interest. A first attempt to formalize the concept of finite state recognizability for two-dimensional languages can be attributed to Blum and Hewitt ([7]) who started in 1967 the study of finite state devices that can define two-dimensional languages, with the aim to finding a counterpart of what regular languages are in one dimension. Since then, many approaches have been presented in the literature following all classical ways to define regular languages: finite automata, grammars, logics and regular expressions.

In 1991, a unifying point of view was presented in [13] where the family of *tiling recognizable picture languages* is defined (see also [14]). The definition of recognizable picture language takes as starting point a well known characterization of recognizable word languages in terms of *local* languages and *projection*. Namely, any recognizable word language can be obtained as projection of a local word language defined over a larger alphabet. Such notion can be extended in a natural way to the two-dimensional case: more precisely, local picture languages are defined by means of a set of square arrays of side-length two (called *tiles*) that represents the only allowed blocks of that size in the pictures of the language (with special treatment for border symbols). Then, we say that a picture language is *tiling recognizable* if it can be obtained as a projection of a local picture language. The family of all tiling recognizable picture languages is called *REC*. Remark that, when we consider words as particular pictures (that is pictures in which one side has length one), this definition of recognizability coincides with the one for the words, i.e. the definition given in terms of finite automata.

The family *REC* can be characterized by several formalisms such as different variants of tiling systems, on-line tessellation automata, Wang systems, existential monadic second order logic, "special" regular expressions, etc.

Please use the following format when citing this chapter:

Giammarresi, D. and Restivo, A., 2008, in IFIP International Federation for Information Processing, Volume 273; *Fifth IFIP International Conference on Theoretical Computer Science*; Giorgio Ausiello, Juhani Karhumäki, Giancarlo Mauri, Luke Ong; (Boston: Springer), pp. 5–20.

(see [10, 14, 16, 19]). The number of different characterizations indicates that (tiling) recognizable picture languages form a robust a therefore somewhat natural class to study. Further this class inherits most of the important properties from the class of regular word languages (see also [16]). Moreover tiling recognizable picture languages have been considered and appreciated in the image processing and pattern recognition fields (see [9]).

On the other hand, recognizable picture languages do not share some properties that are fundamental in the theory of recognizable word languages. The first big difference regards the complement operation. It be proved (see [14]) that, contrary to the one-dimensional case, the family REC is not closed under complementation. As a consequence, it is interesting to consider the family $REC \cup co-REC$ of picture languages L such that either L itself or its complement ${}^cL$ is tiling recognizable. One has that REC is strictly included in $REC \cup co-REC$. An interesting problem (the *complement problem*) is to search for conditions on a picture language L such that both L and ${}^cL$ are tiling recognizable.

The non closure under complementation is related to the fact that the definition of recognizability in terms of tiling systems, i.e. in terms of local languages and projections, is implicitly non-deterministic. However, contrary to the one-dimensional case, does not exist a unique and clear notion of determinism in two dimensions (see [1]). A notion that indeed can be naturally expressed in terms of tiling systems is the notion of *ambiguity*. Informally, a tiling system is *unambiguous* if every picture has a unique counter-image in its corresponding local language. Observe that an unambiguous tiling system can be viewed as a generalization in two dimensions of the definition of unambiguous automaton that recognizes a word language. A recognizable two-dimensional language is *unambiguous* if it is recognized by a unambiguous tiling system.

We denote by $UREC$ the family of all unambiguous recognizable picture languages. Obviously it holds true that $UREC \subseteq REC$. Remark that, in the one dimensional case, $UREC$ is equal to REC. In [3], it is shown that it is *undecidable* whether a given tiling system is unambiguous. Furthermore some closure properties of $UREC$ are proved. The main result in [3] is that, for pictures, $UREC$ is strictly included in REC. In other words, there exist picture languages in REC that are *inherently ambiguous.*

The aim of this paper is to shed new light on the relations between the complement problem and the unambiguity in the family of recognizable picture languages. Remark that the interest for such relations was also raised by W. Thomas in [24].

Following some ideas in [15], we present a novel general framework to study properties of recognizable picture languages and then use it to study the relations between classes $REC \cup co-REC$, REC and $UREC$. The strict inclusions among these classes have been proved in [8], [20], [3], respectively, using ad-hoc techniques. Here we present again those results in a unified formalism and proof method with the major intent of establishing relations between the complement problem and unambiguity in the family of recognizable picture languages.

We consider some complexity functions on picture languages and combine two main techniques. First, following the approach of O. Matz in [20], we consider, for each positive integer m, the set $L(m)$ of pictures of a language L having one dimension (say the vertical one) of size m. Language $L(m)$ can be viewed as a word language over the alphabet (of the columns) $\Sigma^{m,1}$. The idea is then to measure the complexity of the picture language L by evaluating the grow rate, with respect to m, of some numerical parameters of $L(m)$. In order to specify such numerical parameters we make use, as a second technique, of the Hankel matrix of a word language. The parameters are indeed expressed in terms of some elementary matrix-theoretic notions of the Hankel matrices of the word languages $L(m)$. In particular, we consider here three parameters: the *number of different rows*, the *rank*, and the *maximal size of a permutation sub-matrix*.

We state a main theorem that establishes some bounds on corresponding complexity functions based on those three parameters, respectively. Then, as applications for those bounds we analyze the complexity functions of some examples of picture languages in the case of unary alphabet. By means of those languages we re-prove the strict inclusions of families $REC \cup co-REC$, REC and $UREC$ even in the case of unary alphabet.

Moreover we show an example of a language in REC that does not belong to $UREC$ and whose complement is not in REC. This language introduces further discussions on relations between unambiguity and non-closure under complement.

2 Recognizable Two-dimensional languages

In this section we introduce some definitions about two-dimensional languages and their operations. Then we recall definitions and basic properties of tiling recognizable two-dimensional languages firstly introduced in 1992 in [13] that correspond to family REC. Furthermore, we give the definition of unambiguous recognizable picture languages and of class $UREC$. The notations used together with all the results and proofs mentioned here can be found in [14].

Let Σ be a finite alphabet. A *picture* (or *two-dimensional word*) over Σ is a two-dimensional rectangular array of elements of Σ. Given a picture p, let $p(i,j)$ denote the symbol in p with coordinates (i,j), moreover the *size* of p is given by a pair (m,n) where m and n are the number of rows and columns of p, respectively. The set of all pictures over Σ of size (x,y) for all $x,y \geq 1$ is denoted by Σ^{++} and a *two-dimensional language* over Σ is a subset of Σ^{++}. Very often we will refer to two-dimensional languages as *picture languages*. Remark that in this paper we do not consider the case of empty pictures (i.e. pictures where the number of rows and/or columns can be zero). The set of all pictures over Σ of fixed size (m,n), with $m,n \geq 1$ is denoted by $\Sigma^{m,n}$. We give a first example of a picture language.

Example 1. Let L be the language of square pictures over an alphabet Σ:

$$L = \{ p \mid p \text{ has size } (n, n),\ n > 0 \}.$$

Between pictures and picture languages there are defined two different concatenation operations along the horizontal and vertical directions called *column concatenation* and *row concatenation*, respectively. Notice that they are partial operations because they are defined between pictures with same number of rows (for the column concatenation) or same number of columns (for row concatenation). Furthermore, by iterating the concatenation operations, we obtain the column and row *closure* or *star*.

In order to describe recognizing strategies for pictures, it is needed to identify the symbols on the boundary. Then, for any picture p of size (m, n), we consider picture $\widehat{p}$ of size $(m+2, n+2)$ obtained by surrounding p with a special *boundary symbol* $\# \notin \Sigma$. We call *tile* a square picture of dimension $(2, 2)$ and given a picture p we denote by $B_{2,2}(p)$ the set of all blocks of p of size $(2, 2)$.

Let Γ be a finite alphabet. A two-dimensional language $L \subseteq \Gamma^{++}$ is *local* if there exists a finite set Θ of tiles over the alphabet $\Gamma \cup \{\#\}$ such that $L = \{x \in \Gamma^{++} | B_{2,2}(\widehat{x}) \subseteq \Theta\}$. We will write $L = L(\Theta)$. Therefore tiles in Θ represent all the *allowed blocks* of size $(2, 2)$ for the pictures in L. The family of local picture languages will be denoted by LOC. We now give an example of a local two-dimensional language.

Example 2. Let $\Gamma = \{0, 1\}$ be an alphabet and let Θ be the following set of tiles over Γ.

$$\Theta = \left\{ \begin{array}{cccccc}
\begin{array}{|c|c|}\hline 0 & \# \\ \hline 1 & \# \\ \hline\end{array} &
\begin{array}{|c|c|}\hline 0 & \# \\ \hline 0 & \# \\ \hline\end{array} &
\begin{array}{|c|c|}\hline \# & \# \\ \hline 0 & 0 \\ \hline\end{array} &
\begin{array}{|c|c|}\hline \# & \# \\ \hline 0 & 1 \\ \hline\end{array} &
\begin{array}{|c|c|}\hline \# & \# \\ \hline \# & 1 \\ \hline\end{array} &
\begin{array}{|c|c|}\hline \# & \# \\ \hline 0 & \# \\ \hline\end{array} \\
\\
\begin{array}{|c|c|}\hline \# & 1 \\ \hline \# & 0 \\ \hline\end{array} &
\begin{array}{|c|c|}\hline \# & 0 \\ \hline \# & 0 \\ \hline\end{array} &
\begin{array}{|c|c|}\hline 0 & 0 \\ \hline \# & \# \\ \hline\end{array} &
\begin{array}{|c|c|}\hline 0 & 1 \\ \hline \# & \# \\ \hline\end{array} &
\begin{array}{|c|c|}\hline \# & 0 \\ \hline \# & \# \\ \hline\end{array} &
\begin{array}{|c|c|}\hline 1 & \# \\ \hline \# & \# \\ \hline\end{array} \\
\\
\begin{array}{|c|c|}\hline 1 & 0 \\ \hline 0 & 1 \\ \hline\end{array} &
\begin{array}{|c|c|}\hline 0 & 0 \\ \hline 0 & 1 \\ \hline\end{array} &
\begin{array}{|c|c|}\hline 0 & 1 \\ \hline 0 & 0 \\ \hline\end{array} &
\begin{array}{|c|c|}\hline 0 & 0 \\ \hline 0 & 0 \\ \hline\end{array} & &
\end{array} \right\}$$

The language $L(\Theta)$ is the language of squares pictures (i.e. pictures of size (n, n) with $n \geq 2$) in which all diagonal positions (i.e. those of the form (i, i)) carry symbol 1, whereas the remaining positions carry symbol 0. That is, pictures as the following:

1	0	0	0	0	0
0	1	0	0	0	0
0	0	1	0	0	0
0	0	0	1	0	0
0	0	0	0	1	0
0	0	0	0	0	1

Notice that the language of squares over a one-letter alphabet is not a local language because there is no "local strategy" to compare the number of rows and columns using only one symbol.

Let Γ and Σ be two finite alphabets. A mapping $\pi : \Gamma \to \Sigma$ will be in the sequel called *projection*. The projection $\pi(p)$ of $p \in \Gamma^{++}$ of size (m,n) is the picture $p' \in \Sigma^{++}$ such that $p'(i,j) = \pi(p(i,j))$ for all $1 \leq i \leq m, 1 \leq j \leq n$. Similarly, if $L \subseteq \Gamma^{++}$ is a picture language over Γ, we indicate by $\pi(L)$ the projection of language L, i.e. $\pi(L) = \{p' | p' = \pi(p), p \in L\} \subseteq \Sigma^{++}$.

A quadruple $\mathcal{T} = (\Sigma, \Gamma, \Theta, \pi)$ is called *tiling system* if Σ and Γ are finite alphabets, Θ is a finite set of tiles over $\Gamma \cup \{\#\}$ and $\pi : \Gamma \to \Sigma$ is a projection. Therefore, a tiling system is composed by a local language over Γ (defined by the set Θ) and a projection $\pi : \Gamma \longrightarrow \Sigma$. A two-dimensional language $L \subseteq \Sigma^{++}$ is *tiling recognizable* if there exists a tiling system $\mathcal{T} = (\Sigma, \Gamma, \Theta, \pi)$ such that $L = \pi(L(\Theta))$. Moreover, we will refer to $L' = L(\Theta)$ as an *underling local language* for L and to Γ as a *local alphabet* for L. Let $p \in L$, if $p' \in L'$ is such that $\pi(p') = p$, we refer to p' as a *counter-image* of p in the underling local language L'.

The family of all two-dimensional languages that are *tiling recognizable* is denoted by REC. As first example consider the following.

Example 3. Let L be the language of square pictures (i.e. pictures of size (n,n)) over one-letter alphabet $\Sigma = \{a\}$. Language L is in REC because it can be obtained as projection of local language in Example 2 by mean of projection $\pi(0) = \pi(1) = a$.

We remark that a tiling system $\mathcal{T} = (\Sigma, \Gamma, \Theta, \pi)$ for a picture language is in some sense a generalization to the two-dimensional case of an automaton that recognizes a word language. Indeed, in one-dimensional case, the quadruple $(\Sigma, \Gamma, \Theta, \pi)$ corresponds exactly to the state-graph of the automaton: the alphabet Γ is in a one-to-one correspondence with the edges, the set Θ describes the edges adjacency, the mapping π gives the labelling of the edges in the automaton. Then, the set of words of the underlying local language defined by set Θ corresponds to all accepting paths in the state-graph and its projection by π gives the language recognized by the automaton. As consequence, when rectangles degenerate in strings the definition of recognizability coincides with the classical one for strings (cf. [11]).

The family REC is closed with respect to different types of operations. In particular: the family REC is closed under alphabetic projection, under row and column concatenation, under row and column stars and under union and intersection operations (see [14]).

2.1 Examples of recognizable languages

First family of examples of recognizable picture languages can be obtained as immediate application of closure properties. In fact, as we do in the word case, we can define sort of *picture regular expressions* starting from finite languages and using operations of union, intersection, row and column concatenations and closures and projection.

In this way we can list the following as recognizable two-dimensional languages: languages of pictures with odd number of rows, of pictures with even numbers of as, of pictures with first row equal to the last row, of pictures that contains to equal columns and so on.

In some sense we can consider all the properties of recognizable word languages and "make" the corresponding two-dimensional ones and get a recognizable two-dimensional language. But this does not exhausts the family of all recognizable two-dimensional languages! In fact going from one to two dimensions, such generalization of finite automata can recognize much more properties.

As first example, consider the set of pictures over $\Sigma = \{\{a, b\}$ of size $(n, 2n)$ where the first row is the word $a^n b^n$. The tiling system for this language is quite straightforward. Furthermore, in [26] it is proved that even the language of pictures over $\Sigma = \{\{a, b\}$ where the number of as is equal to the number of bs (providing that the size (m, n) of the pictures is such that $m \leq 2^n$ and $n \leq 2^m$). Therefore in two dimensions we can "count" within a recognizable setting.

Another way to interpret a picture over a two-letters alphabet $\Sigma = \{\{a, b\}$, more in the spirit of pattern recognition, is to consider, for example, the as as background and the bs as the "figure". In [25] it is exhibited a tiling system for the language of connected figures.

Very interesting is the examples of *Chinese boxes* in [9]. Pictures are defined on $\{\{0, 1\}$ alphabet and contain rectangular frames or boxes, placed anywhere. Frames may be nested one inside the other but they may not overlap, touch each other, or touch the border. The perimeter of a frame are encoded by 1 and the background by 0 symbols. It is proved that Chinese boxes are recognizable. Remark that Chinese boxes can be viewed as the two-dimensional version of the "well-formed parenthesis languages" that is not regular in one-dimension.

A family of recognizable two-dimensional languages that is worthwhile to consider are the languages of *pictures on one-letter alphabet*. This corresponds also to consider the *shapes* of the pictures without looking to the inside contents.

Remark that, in this case, a picture is defined by a pair of positive numbers corresponding to its size (m, n) and then a picture language is a set of pairs of natural numbers. Furthermore, given a function f defined on the set of natural numbers, one can consider the set of pictures of sizes $(n, f(n))$ for each n. It can be proved that several families of functions are tiling recognizable like polynomial and exponential functions (see [12] or [14]).

Alternatively, given a set of natural numbers, one can consider the set of square pictures of corresponding sizes. There are some surprising sets of recognizable numbers. One for all, the set of *primes* is proved to be tiling recognizable in [5] where it is also given a characterization involving the Turing Machine.

2.2 Ambiguity and complementation

The examples in previous section indicate that tiling systems are devices having a strong expressive power. Let us observe that, in the one-dimensional case, "well-formed parenthesis" and "counting" are some kind of prototype concepts for non recognizability. On the contrary, examples in the previous section show that the natural extensions of such concepts to two-dimensions define picture languages that are tiling recognizable. So the notion of (tiling) recognizability appears to have, in two dimensions, a stronger expressive power with respect to the one-dimensional case.

At the same time, recognizable picture languages do not share some properties that are fundamental in the theory of recognizable word languages. The first big difference regards the complement operation. In [14], using a combinatorial argument, it is showed that language in Example 6 is not tiling recognizable while it is not difficult to write a picture regular expressions for its complement. This proves the following theorem.

Theorem 1. *REC is* not *closed under complement.*

As consequence of this theorem, it is interesting to consider the family $REC \cup co-REC$ of picture languages L such that either L itself or its complement ${}^cL$ is tiling recognizable. Previous theorem states that REC is strictly included in $REC \cup co-REC$.

Closure by complement for a family of languages is usually related to the existence of a *deterministic* computational model recognizing the languages in the family. Remark that the definition of recognizability in terms of tiling systems, i.e. in terms of local languages and projections, is implicitly non-deterministic. This can be easily understood if we refer to the one-dimensional case: if no particular constraints are given for the tiling system, this corresponds in general to a non-deterministic automaton.

Contrary to the one-dimensional case, there are however some difficulties to define determinism in two dimensions, since tiling systems are not computational models in strict sense. As remarked in [1], they are not effective devices for recognition unless a *scanning strategy* for pictures is fixed (for a word the natural scanning strategy is to read it from left to right). So in [1] is introduced a notion of *tiling automaton* as a tiling system equipped with a scanning strategy and, in this framework, some definitions of determinism are proposed.

Actually, a notion that can be naturally expressed in terms of tiling systems is the notion of *ambiguity*. Informally, a tiling system is *unambiguous* if every picture has a unique counter-image in its corresponding local language. In a more formal way, a tiling system $\mathcal{T} = (\Sigma, \Gamma, \Theta, \pi)$ is *unambiguous* if for any picture $x \in L(\mathcal{T})$ there exists a *unique* local picture $y \in L(\Theta)$ such that $x = \pi(y)$.

An alternative definition for *unambiguous tiling system* is that function π extended to $\Gamma^{++} \to \Sigma^{++}$ is injective. Observe that an unambiguous tiling system can be viewed as a generalization in two dimensions of the definition of unambiguous automaton that recognizes a word language.

A recognizable two-dimensional language $L \subseteq \Sigma^{++}$ is *unambiguous* if it is recognized by an unambiguous tiling system $\mathcal{T} = (\Sigma, \Gamma, \Theta, \pi)$. We denote by $UREC$ the family of all unambiguous recognizable two-dimensional languages. Obviously it holds true that $UREC \subseteq REC$.

In [3], it is shown that it *undecidable* whether a given tiling system is unambiguous. Furthermore some closure properties of $UREC$ are proved. The main result in [3] is the following theorem.

Theorem 2. *$UREC$ is strictly included in REC.*

This theorem shows that there exist languages in REC that are inherently ambiguous.

In the sequel we will focus on possible relationships between Theorem 1 and Theorem 2, i.e. on the relations between the complement problem and the ambiguity of a picture language. In next section we present a novel general framework to study such a problem, by introducing some complexity functions on picture languages.

3 Hankel matrices and complexity functions

In this section we introduce a novel tool to study picture languages based on combining two main techniques: the Matz's technique (that associates to a given picture language L an infinite sequence $(L(m))_{m\geq 1}$ of word languages) and the technique that describes a word language by means of its Hankel matrix. As results there will be the definitions of some complexity functions for picture languages that will be used to state some necessary conditions on recognizable picture languages.

We first describe a technique, introduced by O. Matz in [20]. Let $L \subseteq \Sigma^{++}$ be a picture language. For any $m \geq 1$, we consider the subset $L(m) \subseteq L$ containing all pictures with exactly m rows. Such language $L(m)$ can be viewed as a word language over the alphabet $\Sigma^{m,1}$ of the columns, i.e. words in $L(m)$ have a "fixed height m". For example, if

$$p = \begin{array}{|ccccc|} a & b & b & a & a \\ a & a & b & b & a \\ b & b & a & b & a \\ a & a & a & a & b \end{array} \in L$$

then the word

$$w = \begin{bmatrix} a \\ a \\ b \\ a \end{bmatrix} \begin{bmatrix} b \\ a \\ b \\ a \end{bmatrix} \begin{bmatrix} b \\ b \\ a \\ a \end{bmatrix} \begin{bmatrix} a \\ b \\ b \\ a \end{bmatrix} \begin{bmatrix} a \\ b \\ b \\ a \end{bmatrix} \begin{bmatrix} a \\ a \\ a \\ b \end{bmatrix}$$

belongs to the word language $L(4)$ over the alphabet of columns

$$\Sigma^{4,1} = \left\{ \begin{bmatrix} x \\ y \\ s \\ t \end{bmatrix} \mid x, y, s, t \in \Sigma \right\}.$$

Observe that studying the sequence $(L(m))_{m \geq 1}$ of word languages corresponding to a picture languages L does not capture the whole structure of L because in some sense it takes into account only its horizontal dimension. Nevertheless it will be very useful to state some conditions for the recognizability of the picture language L.

We first report a lemma given in [20]. Let L be a recognizable picture languages and let $\mathcal{T} = (\Sigma, \Gamma, \Theta, \pi)$ a tiling system recognizing L.

Lemma 1. *For all $m > 1$ there exists a finite automaton $\mathcal{A}(m)$ with γ^m states that recognizes word language $L(m)$, where $\gamma = |\Gamma \cup \{\#\}|$.*

The proof of the above lemma constructs explicitly such non-deterministic finite automaton $\mathcal{A}(m) = (\Sigma^{1,m}, Q_m, I_m, F_m, \delta_m)$ where $\Sigma^{1,m}$ is the alphabet of the columns of height m over Σ; the set of states Q_m is the set of all possible columns of m. The transitions from a given state p to state q are defined by using the adjacency allowed by the set of local tiles. This construction implies directly the following corollary.

Corollary 1. *If $L \in UREC$, then $\mathcal{A}(m)$ is unambiguous.*

Hankel matrices were firstly introduced in [28] in the context of formal power series (see also [6] and [27]). Moreover they are used under different name in communication complexity (see [18]).

Definition 1. Let $S \subseteq A^*$ be a string language. The Hankel matrix of S is the infinite boolean matrix $H_S = [h_{xy}]_{x \in A^*, y \in A^*}$ where

$$h_{xy} = \begin{cases} 1 & \text{if } xy \in S \\ 0 & \text{if } xy \notin S. \end{cases}$$

Therefore both the rows and the columns of H_S are indexed by the set of strings in A^* and the 1s in the matrix gives the description of language S in the way described above.

Let us observe that, in the case of one letter alphabet, the Hankel matrix of a (string) language is a Hankel matrix in the classical sense, i.e. a matrix, with rows and columns indexed by non negative integers, with constant skew diagonals. In other words it is a matrix in which the (i, j)th entry depends only on the sum $i+j$. Such matrices are sometimes known as persymmetric matrices or, in older literature, orthosymmetric matrices.

Given an Hankel matrix H_S, we call *submatrix of* H_S a matrix K_S specified by a pair of languages (U, V), with $U, V \subseteq A^*$, that is obtained by intersecting all rows and all columns of H_S that are indexed by the strings in U and V, respectively. Moreover, given two Hankel submatrices K_S^1 and K_S^2, their intersection is the submatrix specified by the intersections of the corresponding index sets respectively.

Moreover we recall some further notations on matrices. A *permutation* matrix is a boolean matrix that has exactly one 1 in each row and in each column. Usually when dealing with permutation matrices, one makes a correspondence between a permutation matrix $D = [d_{ij}]$ of size n with a permutation function $\sigma = \mathbb{N} \longrightarrow \mathbb{N}$ by assuming that $d_{ij} = 1 \Leftrightarrow j = \sigma(i)$.

Finally we recall that the *rank* of a matrix is the size of the biggest submatrix with non-null determinant (with respect to field $\mathbb{Z}$). Alternatively, the rank is defined as the maximum number of row or columns that are linearly independent. Then, observe that, by definition, the rank of a permutation matrix coincides with its size.

Given a picture language L over the alphabet Σ, we can associate to L an infinite sequence $(H_L(m))_{m\geq 1}$ of matrices, where each $H_L(m)$ is the Hankel matrix of string language $L(m)$ associated to L.

We can define the following functions from the set of natural numbers $\mathbb{N}$ to $\mathbb{N} \cup \infty$.

Definition 2. Let L be a picture language.
i) The *row complexity function* $R_L(m)$ gives the number of distinct rows of the matrix $H_L(m)$;
ii) The *permutation complexity function* $P_L(m)$ gives the size of the maximal permutation matrix that is a submatrix of $H_L(m)$;
iii) The *rank complexity function* $K_L(m)$ gives the rank of the matrix $H_L(m)$.

Notice the all the functions $R_L(m)$, $P_L(m)$ and $K_L(m)$ defined above are independent from the order of the rows (columns, resp.) of the Hankel matrix $H_L(m)$. In the sequel we will use any convenient order for the set of strings that index the rows and the columns. We can immediately state the following lemma.

Lemma 2. *Given a picture language L, for each $m \in \mathbb{N}$:*

$$P_L(m) \leq K_L(m) \leq R_L(m).$$

Example 4. Consider the language L of squares over a two-letters alphabet $\Sigma = \{a, b\}$ described in Example 1. Observe that, for each $m \geq 0$, $L(m)$ is the finite language of all possible strings of length m over the alphabet of the columns $\Sigma^{m,1}$. Then consider the Hankel matrix of $L(m)$: it has all its 1s in the positions indexed by pairs (x, y) of strings such that $|x| + |y| = m$. Now assume that the strings that index the rows and the columns of the Hankel matrix are ordered by length: we can have some non-zero positions only in the upper-right portion of $H_L(m)$ that indexed by all possible strings of length $\leq m$ on the alphabet $\Sigma^{m,1}$, included the empty word. More specifically, in this portion the matrix $H_L(m)$ has all 0s with the exception of a chain of rectangles of all 1s from the top-right to the bottom left corner. This is represented in the following figure where the numbers $0, 1, \ldots, m-1, m$ indicate the length of the index words.

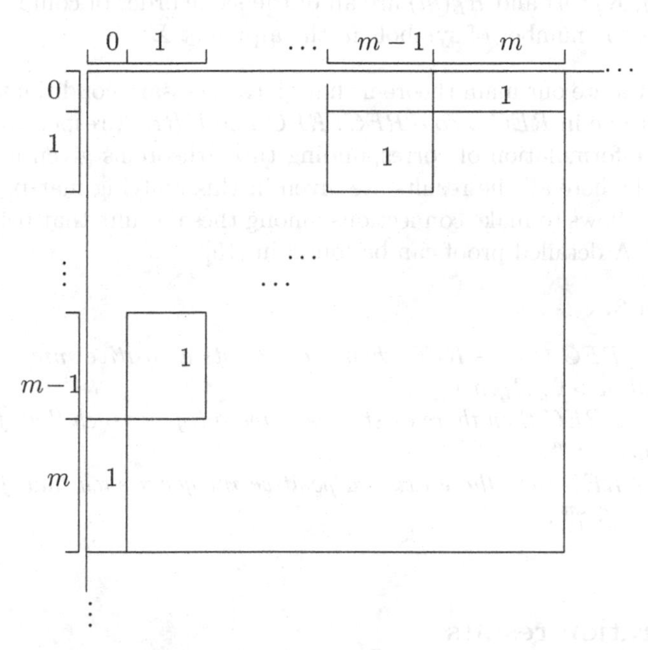

It is easy to verify that the number of different rows in $H_L(m)$ is equal to $m+1$ and this is also the number of rows of a permutation submatrix and this is also the rank of $H_L(m)$.

Then for this language it holds that for all positive m:

$$P_L(m) = K_L(m) = R_L(m) = m + 1.$$

Example 5. As generalization of the above Example 4, consider the language L of pictures over an alphabet Σ of size $(n, f(n))$ where $f(n)$ is a non-negative function defined on the set of natural numbers, that is:

$$L = \{p \mid p \text{ is of size } (n, f(n)\}.$$

Similar arguments as in the above example show that, for each $m \geq 0$, language $L(m)$ is a finite language (it contains all strings of length $f(m)$ over the alphabet of the columns $\Sigma^{m,1}$) and then, for all positive m: $P_L(m) = K_L(m) = R_L(m) = f(m) + 1$.

Example 6. Consider the language L of pictures over an alphabet Σ of size $(n, 2n)$ such that the two square halves are equal, that is:

$$L = \{p \mathbin{\varobar} p \mid p \text{ is a square}\}.$$

Again, as in the Example 4, for each $m \geq 0$, language $L(m)$ is a finite language (it contains all strings of length $2m$ over the alphabet of the columns $\Sigma^{m,1}$ of the form ww). Then, doing all the calculations, one obtains that, for all positive m, $P_L(m)$, $K_L(m)$ and $R_L(m)$ are all of the same order of complexity $O(\sigma^{m^2})$, where σ is the number of symbols in the alphabet Σ.

We now state our main theorem that gives necessary conditions for a picture language to be in $REC \cup co-REC$, REC and $UREC$, respectively. Although this is a re-formulation of corresponding three theorems given in [8], [20], [3], respectively, here all the results are given in this unifying matrix-based framework that allows to make connections among these results that before appeared unrelated. A detailed proof can be found in [15].

Theorem 3.

i) If $L \in REC \cup co-REC$ then there exists a positive integer γ such that, for all $m > 0$, $R_L(m) \leq 2^{\gamma^m}$.
ii) If $L \in UREC$ then there exists a positive integer γ such that, for all $m > 0$, $K_L(m) \leq \gamma^m$.
iii) If $L \in REC$ then there exists a positive integer γ such that, for all $m > 0$, $P_L(m) \leq \gamma^m$.

4 Separation results

In this section we state some separation results for the classes of recognizable picture languages here considered. We start by showing that there exist languages L such that are neither L nor ${}^cL$ are recognizable.

Let L_f be a picture language over Σ with $|\Sigma| = \sigma$ of pictures of size $(n, f(n))$ where f is a non-negative function over $\mathbb{N}$. In Example 5 it is remarked that $R_{L_f}(m) = f(m) + 1$. Then, if we choose a function "greater" than the bound in Theorem 3 - *i)*, we obtain the following.

Corollary 2. *Let $f(n)$ be a function that has asymptotical growth rate greater than 2^{γ^n}, then $L_f \notin REC \cup co-REC$.*

We now consider an example of picture language L over one letter alphabet that, together with its complement ${}^{c}L$, will be checked for the inequalities of the Theorem 3. In such a way we show that, even in the case of one letter alphabet, classes $REC \cup co-REC$, REC and $UREC$ are strictly separated.

The proofs of the following results are based on some arithmetic properties of the function $F(n)$ that is introduced below (cf. [21]). Denote by $lcm(x_1, x_2, ..., x_h)$ the lowest common multiple of the integers $x_1, x_2, ..., x_h$. Consider the function

$$G(m) = lcm(m+1, m+2, ..., 2m).$$

It holds the following.

Lemma 3. $G(m) = 2^{\Omega(m)}$.

Consider now the function $F(n) = G(2^n)$ and the language

$$L = \{(n, m) \mid \ m \text{ is not multiple of } F(n)\}.$$

Theorem 4. $L \in REC$.

We now calculate our complexity functions for language L. It is not difficult to verify that, for all $n > 0$, the Hankel matrix $H_L(n)$ is such that in its submatrix composed by the first $F(n)$ rows and the first $F(n)$ columns (i.e.the rows and the columns indexed $0, 1, ..., F(n)$) every element in the main skew diagonal is equal to 0 and all other elements are equal to 1. We represent it below.

	0	1	2	3	...	...	...	$F(n)$
1	1	1	1	1	1	1	1	0
1	1	1	1	1	1	1	0	1
2	1	1	1	1	1	0	1	1
3	1	1	1	1	0	1	1	1
...	1	1	1	0	1	1	1	1
...	1	1	0	1	1	1	1	1
...	1	0	1	1	1	1	1	1
$F(n)$	0	1	1	1	1	1	1	1

On can easily check that, for all $n > 0$:

$$R_L(n) = K_L(n) > F(n)$$

and

$$P_L(n) = 2.$$

Since $L \in REC$ and $F(n) = G(2^n) = 2^{\Omega(2^n)}$, from the inequality $R_L(n) > F(n)$ it holds the following proposition.

Proposition 1. *The bound given in Theorem 3 -* i) *is tight.*

From the inequality $K_L(n) > F(n)$ and Theorem 3 - *iii)*, one derives the following result.

Theorem 5. *UREC is strictly included in REC.*

This result was firstly proved in [3] and in the unary case in [4].

Consider now the language ${}^cL$. For all $n > 0$, the Hankel matrix $H_{^cL}(n)$ is obtained from the matrix $H_L(n)$ by interchanging the zero's and the one's. It follows that, for all $n > 0$,

$$R_{^cL}(n) = K_{^cL}(n) > F(n)$$

and

$$P_{^cL}(n) > F(n).$$

By the previous inequality and Theorem 3 - *ii)* it follows that ${}^cL \notin REC$ and then one has the following theorem.

Theorem 6. *REC is strictly included in* $REC \cup co - REC$.

Therefore we can conclude that also in the unary case it holds the following hierarchy:

$$UREC \subsetneq REC \subsetneq REC \cup co - REC.$$

5 Final remarks and open questions

We presented an unifying framework based on Hankel matrices to deal with recognizable picture languages. As result, we stated three necessary conditions for the classes $REC \cup co - REC$, REC and $UREC$. The first natural question that arises regards the non-sufficiency of such statements, more specifically the possibility of refining them to get sufficient conditions. Observe that the technique we used of reducing a picture language L in a sequence of string languages $(L(m))_{m>0}$ on the columns alphabets $\Sigma^{m,1}$ allows to take into account the "complexity" of a picture language along only the horizontal dimension. Then the question is whether by combining conditions that use such both techniques along the two dimensions we could get strong conditions for the recognizability of the given picture language.

The novelty of these matrix-based complexity functions gives a common denominator to study relations between the *complement problem* and *unambiguity* in this family of recognizable picture languages. In 1994, in the more general context of graphs, Wolfgang Thomas et. al. had pointed the close relations between these two concepts. In particular, paper [24] ends with the following question formulated specifically for grids graphs and a similar notion of recognizability (here, we report it in our terminology and context).

Question 1. Let $L \subseteq \Sigma^{++}$ be a language in REC such that also ${}^cL \in REC$. Does this imply that $L \in UREC$?

As far as we know, there are no negative examples for this question. On the other hand, we have seen a language L that belongs to REC such that its complement does not and L itself is not in $UREC$. Then we can formulate another question.

Question 2. Let $L \subseteq \Sigma^{++}$ be a language in REC such that ${}^cL \notin REC$. Does this imply that $L \notin UREC$?

Remark that, since our language is on unary alphabet, the above questions are meaningful also in this special case.

As further work we believe that this matrix-based complexity function technique to discriminate class of languages could be refined to study relations between closure under complement and unambiguity. Notice that a positive answer to any of a single question above does not imply that $UREC$ is closed under complement. Moreover observe that the two problems can be rewritten as whether $REC \cap co\text{–}REC \subseteq UREC$ and whether $UREC \subseteq REC \cap co\text{–}REC$, respectively, i.e. they correspond to verify two inverse inclusions. As consequence, if both conjectures were true then we would conclude not only that $UREC$ is closed under complement but also that it the *largest* subset of REC closed under complement.

References

1. M. Anselmo, D. Giammarresi, M. Madonia. Tiling Automaton: A Computational Model for Recognizable Two-Dimensional Languages. *Procs. CIAA07*, LNCS 44783, 290-302, Springer Verlag, 2007.
2. M. Anselmo, D. Giammarresi, M. Madonia. From Determinism to Non-determinism in Recognizable Two-dimensional Languages.*Procs. DLT07*, LNCS 4588, 36-47, Springer Verlag, 2007.
3. M. Anselmo, D. Giammarresi, M. Madonia, A. Restivo. Unambiguous Recognizable two-dimensional languages. *RAIRO - Inf. Theor. Appl.* 40 (2006) 277-293
4. M. Anselmo,M. Madonia Deterministic and Unambiguous Two-dimensional Languages over One-letter Alphabet.*Theoretical Computer Science*, to appear.
5. A. Bertoni, M. Goldwurm, V. Lonati. On the Complexity of Unary Tiling-Recognizable Picture Languages. *Procs. STACS'07*, LNCS 4393, 381-392, Springer Verlag, 2007.
6. J. Berstel, C. Reutenauer. *Rational series and their languages.* EATCS monographs on Theoretical Computer Science. Springer 1988.
7. M. Blum and C. Hewitt. Automata on a two-dimensional tape. *IEEE Symposium on Switching and Automata Theory*, pages 155–160, 1967.
8. J. Cervelle Langages de Figures. *Rapport de Stage. Ecole Normale Supériure de Lyon, Dept de Mathématiques et Informatique*, 1997.
9. S. Crespi Reghizzi, M. Pradella. A SAT-based parser and completer for pictures specified by tiling. *Pattern Recognition* 41(2): 555-566 (2008).
10. De Prophetis, L., Varricchio, S.: Recognizability of rectangular pictures by wang systems. Journal of Automata, Languages, Combinatorics. **2** (1997) 269-288

11. S. Eilenberg. *Automata, Languages and Machines.* Vol. A, Academic Press, 1974.
12. D. Giammarresi. Two-dimensional languages and recognizable functions. In *Proc. Developments in language theory*, Finland, 1993. G.Rozenberg and A. Salomaa (Eds). *World Scientific Publishing Co.* , 1994.
13. D. Giammarresi, A. Restivo. Recognizable picture languages. Int. Journal Pattern Recognition and Artificial Intelligence. Vol. 6, No. 2& 3, pages 241 –256, 1992.
14. D. Giammarresi, A. Restivo. Two-dimensional languages. *Handbook of Formal Languages*, G. Rozenberg, *et al.* Eds, Vol. III, pag. 215–268. Springer Verlag, 1997.
15. D. Giammarresi, A. Restivo. Matrix-based complexity functions and recognizable picture languages. In *Logic and Automata: History and Perspectives.* E. Grader, J. Flum, T. Wilke Eds. , pag 315-337. Texts in Logic and Games 2. Amsterdam University Press, 2007.
16. D. Giammarresi, A. Restivo, S. Seibert, W. Thomas. Monadic second order logic over pictures and recognizability by tiling systems. *Information and Computation*, Vol 125, 1, 32–45, 1996.
17. I. Glaister, J. Shallit. A lower bound technique for the size of nondeterministic finite automata. *Information Processing Letters*, pages 125:32, vol 45, 1996.
18. J. Hromkovic, J. Karumäki, H. Klauck, G. Schnitger, S. Seibert. Communication Complexity Method for Measuring Nondeterminism in Finite Automata. *Information and Computation*, Vol. 172, pag 202-217, 2002.
19. K. Inoue and I. Takanami. A Characterization of recognizable picture languages. In *Proc. Second International Colloquium on Parallel Image Processing*, A. Nakamura et al. (Eds.), Lecture Notes in Computer Science 654, Springer-Verlag, Berlin 1993.
20. O. Matz. On piecewise testable, starfree, and recognizable picture languages. In *Foundations of Software Science and Computation Structures*, M. Nivat, Ed., vol. 1378, Springer-Verlag, Berlin, 1998.
21. O. Matz. Dot-Depth and Monadic Quantifier Alternation over Pictures Technical Report 99-08, Rheinisch-Westflischen Technischen Hochschule Aachen, 1999. `http://automata.rwth-aachen.de/m̃atz/bibliography.html#MatzDiss.`
22. O. Matz, W. Thomas. The Monadic Quantifier Alternation Hierarchy over Graphs is Infinite. In *IEEE Symposium on Logic in Computer Science, LICS.* IEEE 1997: 236-244
23. I. Mäurer. Weighted Picture Automata and Weighted Logics. In B. Durand, W. Thomas (Eds.)*Procs. STACS 2006*, LNCS 3885, pp. 313-324, Springer-Verlag, 2006.
24. A. Potthoff, S. Seibert, W. Thomas. Nondeterminism versus determinism of finite automata over directed acyclic graphs. *Bull. Belgian Math. Soc.* 1, pages 285–298, 1994.
25. K. Reinhardt. On Some Recognizable Picture-Languages. In *Proc. 23rd MFCS 1998*, pages 760-770, LNCS 1450, Springer-Verlag, Berlin, 1998.
26. K. Reinhardt. The #a = #b Pictures Are Recognizable. In *Proc. 18th STACS 2001*, pages 527-538, LNCS 2010, Springer-Verlag, Berlin, 2001.
27. A. Salomaa, M. Soittola. Authomata-Theoretic Aspect of Formal Power Series. Springer, 1978.
28. M. P. Schützenberger. On the definition of a family of automata. *Information and Control.* vol.4, pp. 245–270, 1961.
29. W. Thomas. On Logics, Tilings, and Automata. In *Proc. 18th ICALP*, pages 441–453, LNCS 510, Springer-Verlag, Berlin, 1991.

Algorithmic Game Theory: Some Greatest Hits and Future Directions

Tim Roughgarden*

Department of Computer Science, Stanford University, 353 Serra Mall, Stanford, CA 94305

Abstract. We give a brief and biased survey of the past, present, and future of research on the interface of theoretical computer science and game theory.

1 Introduction

By the end of the 20th century, the widespread adoption of the Internet and the emergence of the Web had changed fundamentally society's relationship with computers. The primary role of a computer evolved from a stand-alone, well-understood machine for executing software to a conduit for global communication, content-dissemination, and commerce. Two aftershocks of this phase transition were inevitable: theoretical computer science would respond by formulating novel problems, goals, and design and analysis techniques relevant for Internet applications; and game theory, with its deep and beautiful study of interaction between competing or cooperating individuals, would play a crucial role. Research on the interface of theoretical computer science and game theory, an area now known as *algorithmic game theory (AGT)*, has exploded phenomenally over the past ten years.

The central research themes in AGT differ from those in classical microeconomics and game theory in important, albeit predictable, ways. Firstly in application areas: Internet-like networks and non-traditional auctions (such as digital goods and search auctions) motivate much of the work in AGT. Secondly in its quantitative engineering approach: AGT research typically models applications via concrete optimization problems and seeks optimal solutions, impossibility results, upper and lower bounds on feasible approximation guarantees, and so on. Finally, AGT usually adopts reasonable (e.g., polynomial-time) computational complexity as a binding constraint on the feasible behavior of system designers and participants. These themes, which have played only a peripheral role in traditional game theory, give AGT its distinct character and relevance.

Sections 2–4 touch on the current dominant research trends in AGT, loosely following the organization of the first book in the field [94]; Section 5 highlights a number of prominent open questions. We discuss only (a subset of the) topics

* Supported in part by NSF CAREER Award CCF-0448664, an ONR Young Investigator Award, and an Alfred P. Sloan Fellowship. Email: tim@cs.stanford.edu.

Please use the following format when citing this chapter:

Roughgarden, T., 2008, in IFIP International Federation for Information Processing, Volume 273; *Fifth IFIP International Conference on Theoretical Computer Science*; Giorgio Ausiello, Juhani Karhumäki, Giancarlo Mauri, Luke Ong; (Boston: Springer), pp. 21–42.

studied by "the STOC/FOCS community"; see [4, 54, 79, 123] for alternative perspectives on computer science and game theory.

2 Algorithmic Mechanism Design

Algorithmic mechanism design studies optimization problems where the underlying data (such as a value of a good or a cost of performing a task) is *a priori unknown* to the algorithm designer, and must be implicitly or explicitly elicited from self-interested participants (e.g., via a bid). The high-level goal is to design a protocol, or "mechanism", that interacts with participants so that *self-interested behavior yields a desirable outcome.*

There is a complex interaction between the way an algorithm employs elicited data and participant behavior. For example, in a "first-price" sealed-bid auction (where the winner pays its bid), bidders typically shade their bids below their maximum willingness to pay, by an amount that depends on knowledge or beliefs about the other bids. In the "second-price" or "Vickrey" variant [130], where the winner pays only the value of the second-highest bid, each participant may as well bid its true value for the good. (Do you see why?)

Nisan and Ronen [93] proposed the systematic study of what can and cannot be efficiently computed or approximated when the problem data is held by selfish agents, and also coined the term "algorithmic mechanism design (AMD)". (See [76, 101, 119] for related contemporaneous work on combinatorial auctions in the AI literature.) Auction design is the most obvious motivation for this subfield, but there are many others. See [92] for a list of traditional economic applications, together with [71] and [37] for overviews of two modern "killer applications" — keyword search auctions and spectrum auctions, respectively. The economic literature on mechanism design is very rich (e.g., [60]), but AMD has contributed in several ways. We concentrate here on its emphasis on complexity bounds and worst-case approximation guarantees, but mention additional aspects of AMD at the end of the section.

The technical core of AMD is the following deep question:

(Q1) *to what extent is "incentive-compatible" efficient computation fundamentally less powerful than "classical" efficient computation?*

To translate question (Q1) into mathematics, reconsider the Vickrey auction for selling a single good. Each bidder i has a private (true) willingness-to-pay v_i and submits to the auctioneer a bid b_i. The auction comprises two algorithms: an *allocation algorithm*, which picks a winner, namely the highest bidder; and a *payment algorithm*, which uses the bids to charge payments, namely 0 for the losers and the second-highest bid for the winner. One easily checks that this auction is *truthful* in the following sense: for every bidder i and every set of bids by the other players, player i maximizes its "net value" (value for the good, if received, minus its payment, if any) by submitting its true private value: $b_i = v_i$.

Moreover, no false bid is competitive with truthful bidding for all possible bids by the other players. Assuming all players bid truthfully (as they should), the Vickrey auction solves the *social welfare maximization* problem, in the sense that the good is allocated to the participant with the highest value for it.

More generally, consider a feasible region Ω, n participants each with a real-valued private objective function $t_i(\cdot)$ defined on Ω, and a designer objective function $f(t_1, \ldots, t_n)$. In the Vickrey auction, Ω has one outcome per participant (indicating the winner), $t_i(\omega)$ is v_i if i wins in ω and 0 otherwise, and f is $\sum_i t_i(\omega)$. Classical optimization would ask: *given the t_i's*, optimize the objective function f over Ω. The AMD analog is only harder: simultaneously *determine the (private) t_i's* and optimize the corresponding f over Ω. Sometimes the latter problem is no more difficult that the former (as with the Vickrey auction) — when is it strictly more difficult?

Characterizations and the Limits of Approximation. Question (Q1) is the subject of intense study by the AGT community. We confine our discussion here to mechanisms M that share the following properties with the Vickrey auction: M first asks each participant i for a "bid function" $b_i(\cdot)$, hopefully identical to the private objective function $t_i(\cdot)$; M then invokes an allocation algorithm $x(b_1, \ldots, b_n)$ and a payment algorithm $\pi(b_1, \ldots, b_n)$ to determine an outcome ω and payments $p_1, \ldots, p_n$, respectively; and truthful reporting always maximizes the resulting "utility" $t_i(\omega) - p_i$ of a player, no matter what other players do. We call such mechanisms *simple.*[2] The allocation algorithm of a simple mechanism is essentially solving the classical optimization version of the problem with known t_i's (assuming all players bid truthfully, as they should).

Call an allocation algorithm *implementable* if, for some cleverly chosen payment algorithm π, coupling x with π yields a (truthful) simple mechanism. For a single-good auction, if x is the "highest-bidder" allocation algorithm, then defining π as in the Vickrey auction shows that x is implementable. If x is the "second-highest bidder" allocation algorithm, then it is not implementable: *no* payment algorithm can be matched with x to yield a truthful mechanism. (This is not obvious but not hard to prove.) Thus some but not all algorithms are implementable. We can mathematically phrase the question (Q1) as follows: *are implementable algorithms less powerful than arbitrary algorithms for solving fundamental optimization problems?*

This question is interesting for both polynomial-time and computationally unbounded algorithms. There is a strong positive result in the latter scenario, achieved by a far-reaching generalization of the Vickrey auction known as the "VCG mechanism" (see e.g. [92]): for every mechanism design problem with a sum objective ($\sum_i t_i(\omega)$, and weighted variants), the optimal (not necessarily polynomial-time) allocation algorithm is implementable. This is not generally the case for non-sum objectives [10, 93].

---

[2] The usual term is "truthful, direct-revelation". Our restriction to simple mechanisms is partially but not fully without loss of generality; see Section 5.

Far less is known about polynomial-time implementability. Most intriguing are the many mechanism design problems that are derived from an NP-complete problem and for which the optimal allocation algorithm is implementable. For these, *any separation between implementable and non-implementable polynomial-time algorithms must be conditional on* $P \neq NP$, *and no such separation is known.* Any resolution of this issue would be conceptually and technically remarkable: either incentive-compatibility imposes no additional difficulty for a massive class of important mechanism design problems, or else there is a non-trivial way of amplifying (conditional) complexity-theoretic approximation lower bounds using information-theoretic strategic requirements.

Understanding the reach of implementable algorithms generally involves two interrelated goals: characterization theorems and approximation bounds (see also [72]).

(G1) Usefully characterize the implementable allocation algorithms x for the problem.

(G2) Prove upper and lower bounds on the best-achievable approximation ratio of an implementable algorithm (subject to polynomial running time, if desired).

The second goal quantifies the limitations of implementable algorithms using a worst-case approximation measure. The first goal aims to reformulate the unwieldy definition of implementability into a form more amenable to (both upper and lower) approximation bounds. Versions of the second goal pervade modern algorithmic research: for a given "constrained computational model", where the constraint can be either computational (as for polynomial-time approximation algorithms) or information-theoretic (as for online algorithms), quantify its limitations for optimization and approximation. Goal (G1) reflects the additional difficulty in AMD that even the "computational model" (of implementable algorithms) induced by strategic constraints is poorly understood — for example, determining whether or not a given algorithm is online is intuitively far easier than checking if one is implementable.

Single-Parameter Mechanism Design. This two-step approach is vividly illustrated by the important special case of *single-parameter problems*, where goal (G1) has been completely resolved. A mechanism design problem is *single-parameter* if all outcomes are real n-vectors and participants' private objective functions have the form $t_i(\omega) = v_i\omega_i$ for a private real number v_i (the "single parameter"); ω_i and v_i can be thought of as the quantity received and the value-per-unit of a good, respectively. (A single-item auction is the special case in which each ω is a standard basis vector.) An algorithm for a single-parameter problem is *monotone* if a greater value begets a greater allocation: increasing the value of a v_i (keeping other v_j's fixed) can only increase the ith component of the computed solution. For example, the "highest bidder" allocation algorithm for a single-good auction is monotone, while the "second-highest bidder" allocation algorithm is not. More generally, monotonicity characterizes implementability for single-parameter problems.

Theorem 1 ([10, 90, 105]). *An algorithm for a single-parameter mechanism design problem is implementable if and only if it is monotone.*

Theorem 1 should be viewed as a useful solution to the first goal (G1), and it reduces implementable algorithm design to monotone algorithm design. An analogous characterization applies to randomized algorithms, where the monotonicity and truthfulness conditions concern expected allocations and expected participant utilities, respectively [10].

Archer and Tardos [10] were the first to systematically study approximation in single-parameter mechanism design problems. Among other contributions, they identified a natural candidate problem for a conditional separation between implementable and non-implementable polynomial-time approximation algorithms: minimizing the makespan of parallel related machines with private machine speeds. (In a scheduling context, each player is a machine with a private speed $s_i = -1/v_i$, allocations describe the sum of job processing times assigned to each machine, and monotonicity dictates that declaring a slower speed can only decrease the amount of work received.) The problem admits an (exponential-time) implementable optimal algorithm, but all classical polynomial-time approximation algorithms for it (e.g., the PTASes in [43, 58]) are not monotone and hence not implementable [10]. Archer and Tardos [7, 10] devised a randomized monotone 2-approximation algorithm for the problem, and several subsequent papers gave monotone deterministic approximation algorithms (see Kovács [70] for the best bound of 2.8 and references). Very recently, Dhangwatnotai et al. [40] proved that, allowing randomization, monotone polynomial-time algorithms are competitive with arbitrary polynomial-time algorithms for makespan minimization.

Theorem 2 ([40]). *There is a monotone randomized PTAS, and a corresponding truthful in expectation mechanism, for makespan minimization on parallel related machines.*

Whether or not there is a conditional separation between implementable and arbitrary polynomial-time algorithms remains open. In light of Theorem 2, the most likely candidate problems for obtaining such a separation are multi-parameter; we discuss these next.

Multi-Parameter Mechanism Design. Many important mechanism design problems are not single-parameter. *Combinatorial auctions*, in which each participant aims to acquire a heterogeneous set of goods and has unrelated values for different sets, are a practical and basic example. (See [24, 38] for much more on the topic.) Multi-parameter mechanism design is complex and our current understanding of goals (G1) and (G2) is fairly primitive for most problems of interest. Because of its importance and bounty of open questions, the subject has been a hotbed of activity over the past few years; we briefly indicate the primary research threads next.

New characterizations of implementable algorithms are useful (and possibly essential) for understanding their approximation capabilities, and are interesting in their own right. Rochet's Theorem [107] is a classical characterization of implementable algorithms in terms of a certain shortest-path condition known as *cycle monotonicity* (see [132]) that is general but difficult to use to prove upper or lower approximation bounds (see [74] for an exception). Archer and Kleinberg [8] give a promising reformulation of Rochet's Theorem that could lend itself to new approximation bounds. Saks and Yu [118] show that in the common special case where the t_i's are drawn from convex sets, implementability is equivalent to a simpler 2-cycle condition known as *weak monotonicity*; see also [8, 87] for new alternative proofs and [85] for a recent analog in discrete domains.

But what kinds of algorithms meet these technical conditions? The answer depends on the "richness" of the domain in which the private information (the t_i's) lie — richer domains possess more potentially profitable false declarations, making the space of implementable algorithms more highly constrained. For the extreme case of "unrestricted domains", where Ω is an abstract outcome set and the t_i's are arbitrary real-valued functions on Ω, *Robert's Theorem* [106] states that there are almost no implementable algorithms: only the VCG-like "affine maximizers", all minor variants on the algorithm that always chooses the outcome maximizing $\sum_i t_i(\omega)$. This should be viewed as a negative result, since affine maximizers have limited polynomial-time approximation capabilities in most important problems (see e.g. [41]). However, applications usually involve more structured domains. This point motivates an important research agenda, still in its embryonic stages, to identify the types of domains for which Robert's Theorem holds (see [100] for a surprising new example) and characterize the additional implementable mechanisms for domains in which Robert's Theorem breaks down (see [20, 73] and [42, 33] for partial but highly non-trivial results on combinatorial auctions and machine scheduling, respectively).

The design and analysis of good truthful multi-parameter mechanisms has proceeded apace despite our limited understanding of implementability. Much of this research has coalesced around welfare maximization in combinatorial auctions (see [24]), where Ω is the ordered partitions $(S_1, \ldots, S_n)$ of a set of m goods among the n players, the private information t_i describes player i's valuation (willingness to pay) $v_i(S)$ for each of the 2^m possible subsets S of goods, and the optimization problem is to choose an allocation maximizing $\sum_i v_i(S_i)$.[3] While the aforementioned VCG mechanism truthfully solves this optimization problem in exponential time, its polynomial-time approximability varies with the degree of structure imposed on valuations. General valuations exhibit both "complements", where goods are useful only when purchased in tandem (as with a pair of tennis shoes), and "substitutes", where goods are redundant (as with a pair of tennis rackets). Early research focused on valuations with complements but no substitutes and largely succeeded in designing implementable polynomial-time algorithms with approximation ratios matching the best-possible ones for arbitrary polynomial-time algorithms (assuming $P \neq NP$) [76, 89]. Some of these

[3] Valuations are typically modeled either as a "black box" that can be queried or implicitly via a compact representation of size polynomial in m; an "efficient algorithm" in this context has running time polynomial in both n and m.

guarantees have been extended to general valuations (see [24]). Unfortunately, with complements, the underlying welfare maximization problem includes the Maximum Independent Set problem as a special case and thus reasonable approximation guarantees are possible only under strong additional assumptions (as in [9, 17]).

Recent work has focused on classes of valuations with substitutes but no complements, including subadditive valuations (satisfying $v(S \cup T) \leq v(S) + v(T)$ for all S, T) and submodular valuations (satisfying the stronger condition that $v(S \cup \{j\}) - v(S) \leq v(T \cup \{j\}) - v(T)$ for all $T \subseteq S$ and $j \notin S$). Here, excellent (constant-factor) approximation guarantees appear possible, though challenging to obtain. Beginning in [75], a number of papers have proved constant-factor upper and lower bounds for polynomial-time approximation of welfare maximization with complement-free valuations by *non-implementable* algorithms; see [47] and [135] for two recent gems. Remarkably, no constant-factor *implementable* algorithm is known for any such problem. For problems with a sum objective, welfare maximization with complement-free bidders appears to be the most likely candidate to separate the power of implementable and non-implementable algorithms. See [100] for a very recent communication complexity-based separation, a significant research breakthrough.

Further Aspects of AMD. This section focused on the design of computationally efficient truthful mechanisms with provable approximation guarantees for three reasons: it comprises a large portion of AMD research; there remain numerous deep open questions on the topic; and appreciating its motivating questions and key results requires minimal economics background. We emphasize that AMD has several other thriving aspects, including: revenue-maximization with worst-case guarantees, and related algorithmic pricing problems (surveyed in [56]); revenue guarantees and cost-sharing mechanism design (see [61, 83]); online mechanism design, in which participants arrive and depart over time (surveyed in [102]); and new models and goals for Internet-suitable mechanism design, such as distributed mechanisms (see [48]) and mechanisms restricted to use little [84] or no [57, 78, 122] payments.

3 Quantifying Inefficiency and the Price of Anarchy

The truthful mechanisms studied in Section 2 are strategically degenerate in that the best course of action of a player (i.e., truthtelling) does not depend on the actions taken by the others. This was possible because a designer (like a search engine owner) had tremendous control over the game being played. Strategic games that occur "in the wild" are rarely so well behaved. Even in a design context, when the designer cannot directly dictate the allocation of resources (such as traffic rates or routing paths in a large network), dependencies between different players' optimal courses of action are generally unavoidable, and these dependencies usually preclude exact optimization of standard objective functions. This motivates adopting an *equilibrium concept* — a rigorous proposal for the expected outcome(s) of a game with self-interested participants — and an *approximation measure* that quantifies the inefficiency of a game's equilibria, in order to address the following basic question:

(Q2) *when, and in what senses, are game-theoretic equilibria guaranteed to approximately optimize natural objective functions?*

Such a guarantee implies that imposing additional control over the system is relatively small, and is particularly reassuring when implementing an optimal solution is infeasible (as in a typical Internet application).

We only address this question for the most popular modeling choices (Nash equilibria and the price of anarchy, respectively) and the most well-studied application area (routing games). The end of the section provides pointers to some of the many other results in the area.

Routing with Congestion. General tight bounds on the inefficiency of equilibria were first proved in a model of "selfish routing" [115]. The model is originally from [18, 136] and is thoroughly discussed in [110]; the price of anarchy was originally suggested in [69] for a scheduling model, results on which are surveyed in [131].

Consider a directed multicommodity network — a directed graph with fixed flow rates between given source-sink vertex pairs — in which selfish users choose paths to minimize individual cost. Edge costs are *congestion-dependent*, with $c_e(f_e)$ denoting the cost incurred by flow on edge e when there are f_e units of such flow. In an *equilibrium*, each selfish user with source s_i and sink t_i chooses an s_i-t_i path P that minimizes $\sum_{e \in P} c_e(f_e)$, given the routing selections of the other users. Such games are strategically non-trivial in that the routing decision of one user can alter the optimal path for another.

To keep things simple, assume that each selfish user controls a negligible fraction of the overall traffic, and that all edge cost functions are continuous and non-decreasing. Equilibrium flows are then, by definition, those on which all flow is routed on shortest paths, given the congestion: $f_P > 0$ for a path P implies $\sum_{e \in P} c_e(f_e)$ is minimum over all paths with the same source and destination (if not, some selfish users using this path would switch to a cheaper one). All equilibrium flows are interchangeable in that they have equal cost — $\sum_e c_e(f_e) f_e$, as in classical minimum-cost flow — and one is guaranteed to exist [18].

For example, in a "Pigou-like network" (named after [103]), r units of selfish users decide between two parallel edges e_1 and e_2 connecting a source s to a sink t. Suppose the second edge has some cost function $c_2(\cdot)$, and the first edge has a constant cost function c_1 everywhere equal to $c_2(r)$. Such networks are strategically trivial, just like the simple mechanisms of Section 2: the second edge always has no larger cost than the first, even in the worst case when it is fully congested. For this reason, routing all flow on the second edge is an equilibrium. This equilibrium in generally suboptimal, in that it fails to minimize the cost $\sum_{e \in P} c_e(f_e)$ over all feasible flows. For example, if $r = 1$ and $c_2(x) = x$, the equilibrium flow has cost 1, while splitting the traffic equally between the two edges yields an (optimal) flow with cost 3/4. The latter flow is not an equilibrium because of a "congestion externality": a selfish network user routed on the first edge would switch to the second edge, indifferent to the fact that this switch (slightly) increases the cost incurred by a large portion of the population.

The *price of anarchy (POA)* of such a selfish routing network is the ratio of costs of an equilibrium and an optimal flow — 4/3 in the example above. The closer the POA is to 1, the lesser the consequences of selfish behavior. Simple exploration of Pigou-like networks suggests that, at least in this simple family of examples, the POA is governed by the "degree of nonlinearity" of the cost function c_2; in particular, the POA can be arbitrarily large in Pigou-like networks with unrestricted cost functions. A key result formalizes and extends this intuition to *arbitrary* multicommodity networks: among all multicommodity networks with cost functions lying in a set $\mathcal{C}$ (e.g., bounded-degree polynomials with nonnegative coefficients), the largest-possible POA is already achieved in Pigou-like networks [109]. Conceptually, *complex topologies do not amplify the worst-case POA*. Technically, this reduction permits the easy calculation of tight bounds on the worst-case POA in most interesting cases. For example, the POA of every multicommodity selfish routing network with affine cost functions (of the form $c_e(f_e) = a_e f_e + b_e$ for $a_e, b_e \geq 0$) is at most 4/3, matching the lower bound noted above. See [113, 112] for recent surveys detailing these and related results.

While there is no explicit design aspect to these POA bounds, they nicely justify a common rule of thumb used in real-life network design and management: *overprovisioning networks with extra capacity ensures good performance.* This postulate was first formalized mathematically and proved in [115]. Here we provide a conceptually similar but technically different result, which is a special case of the POA bounds in [109] (see also [110, §3.6]). Suppose every edge e of a network has a *capacity* u_e and a corresponding cost function $c_e(f_e) = 1/(u_e - f_e)$. (If $f_e \geq u_e$, we interpret the cost as infinite.) This is the standard M/M/1 queueing delay function with service rate u_e, a common model in the network literature (e.g. [19]). We say the network is *β-overprovisioned* for $\beta \in (0, 1)$ if, at equilibrium, at least a β fraction of each edge's capacity remains unused. The following tight bound on the POA holds for such networks.

Theorem 3 (Consequence of [109]). *The POA of a β-overprovisioned network is at most $\frac{1}{2}(1 + \frac{1}{\sqrt{\beta}})$.*

Thus even 10% extra capacity reduces the price of anarchy of selfish routing to roughly 2.

Designing for Good Equilibria. In the same spirit as mechanism design and our prescriptive interpretation of Theorem 3, we would like to use inefficiency measures such as the POA to inform how to *design* systems to have good equilibria.

Two variants of this idea have been explored in a number of different models: improving the POA of a given game (see [113] for a survey of selfish routing examples), and designing a *family* of games to minimize the worst-case POA. We focus on the latter idea, first proposed in [32], where a number open issues remain. See [62, 117] for surveys of other work on this important topic.

We follow the network cost-allocation example in [27], which was motivated by the network formation games of [6] (see [111, 127] for relevant surveys). As in a selfish routing network, each player selects a path in a multicommodity network to minimize its incurred cost. For technical convenience, we now assume that each player controls a single (non-negligible) unit of flow and uses a single path to route it. The key difference between the two models is the cost structure. If f_e units of flow use an edge e of a selfish routing network, this creates total cost $f_e \cdot c_e(f_e)$ which is distributed evenly among the edges' users (for a per-unit cost of $c_e(f_e)$). In a network cost-allocation game, each edge e has a fixed price p_e for being used by one or more players — for installing infrastructure or leasing a large fixed amount of bandwidth, say — to be somehow distributed among the edges' users. The average per-player cost of an edge is thus *decreasing* with the number of users, giving players an incentive to cooperate via shared paths. Our benchmark is the minimum-cost way of connected all of the players' source-sink pairs, a Steiner connectivity problem (equivalent to the minimum-cost Steiner tree problem if all players share a common sink vertex). An obvious question is: *how should we distribute costs to minimize the worst-case equilibrium efficiency loss over all networks?* This cost-allocation design decision does not affect the underlying optimization problem, but it fundamentally determines the incentives, and hence the Nash equilibria, in the resulting path selection game.

For example, *Shapley cost-sharing* dictates sharing each edge cost equally among its users. So if k players choose paths $P_1, \ldots, P_k$, the cost incurred by the ith player is $\sum_{e \in P_i} p_e / f_e$, where f_e is the number of players choosing a path including e. At a (pure-strategy) *Nash equilibrium*, no player can switch paths to strictly decrease its cost. Shapley cost-sharing always leads to at least one equilibrium [6], and generally to multiple equilibria. For example, in a network of parallel links, all with costs strictly between 1 and k, every link corresponds to a different Nash equilibrium (if all players use a link with price p, each player pays only $p/k < 1$, and a unilateral deviation to a different link would cost more than this). The POA is traditionally defined by the worst equilibrium [69], and this example yields a linear lower bound for the worst-case POA of Shapley cost-sharing (there is an easy matching upper bound). Can we do better?

The answer is different for undirected and directed networks. An alternative to Shapley cost-sharing is *ordered* cost-sharing, a simple priority scheme: order the players arbitrarily, with the first user of an edge (according to this order) paying its full cost. Up to tie-breaking, there is a unique Nash equilibrium under ordered cost-sharing: the first player chooses a shortest path between its source and sink, the second player chooses a shortest path given the edges already paid for by the first player, and so on. Indeed, the equilibria are in one-to-one correspondence to the possible outputs of well-studied greedy online algorithms for Steiner connectivity problems [13, 59]. This correspondence implies that, in undirected networks, *ordered cost-sharing has exponentially better worst-case POA than Shapley cost-sharing.* There is also a matching lower bound.

Theorem 4 ([27]). *In undirected cost-allocation games, ordered cost-sharing attains the minimum-possible worst-case POA (up to constant factors).*

The proof of Theorem 4 is highly non-trivial, and hinges on a complete classification of the cost-sharing methods that are guaranteed to induce at least one Nash equilibrium in all networks. These turn out to be precisely the finite "concatenations" of weighted Shapley values (in the sense of [65]); Shapley cost-sharing is the special case of uniform weights and no concatenation, while ordered cost-sharing arises from the concatenation of k different one-player (trivial) Shapley values. No method of this type can outperform ordered cost-sharing by more than a constant factor [27].

In directed networks, it is easy to show that all cost-sharing methods, including ordered ones, have linear worst-case POA (like Shapley cost-sharing). We can obtain a more refined comparison by analyzing the ratio of the best (instead of the worst) Nash equilibrium and a minimum-cost solution, a quantity known as the price of stability (POS). The worst-case POS of Shapley cost-sharing in directed networks is precisely the kth Harmonic number $\mathcal{H}_k \approx \ln k$ [6]. A consequence of the classification above is that no other method has superior worst-case POS (or POA).

Theorem 5 ([27]). *In directed cost-allocation games, Shapley cost-sharing attains the minimum-possible worst-case POS and POA.*

Further Aspects of Quantifying Inefficiency. We have barely scratched the surface of recent work on equilibrium efficiency analyses. Many different models of routing games have studied from this perspective — following [108, 116], often in the more abstract guise of "congestion games" — see [68, 112] for an incomplete survey. See [94, Chapters 19-21] and [113] for overviews of efficiency analyses in some other application domains. See [5, 31] for efficiency analyses of equilibrium concepts other than Nash equilibria. See [15, 28, 66] for recent efficiency guarantees in models that allow altruistic and/or malicious participants, rather than only self-interested ones.

In addition to the aforementioned work on designing games with efficient equilibria, a second current and important trend in the area is to prove POA-type bounds under increasingly weak assumptions on the rationality of participants. Recall that in Section 2, our only assumption was that participants will make use of a "foolproof" strategy (one that dominates all others), should one be available. This section implicitly assumed that selfish participants can reach a Nash equilibrium of a game without such foolproof strategies, presumably through repeated experimentation. This much stronger assumption has been addressed in two different ways in the recent literature. The first is to formally justify this assumption by positing natural experimentation strategies (or "dynamics") and proving that they quickly reach a (possibly approximate) equilibrium; see [14, 21, 30, 44, 50] for a sampling of examples. The second is to prove POA-like guarantees on system performance that apply even if such experimentation strategies fail to converge to an equilibrium. Remarkably, such bounds exist in, for example, the selfish routing networks discussed in this section; see [53, 86] and [22] for two different formalizations of this approach.

4 Complexity of Equilibrium Computation

Equilibrium concepts such as the Nash equilibrium obviously play a starring role in game theory and microeconomics. If nothing else, a notion of equilibrium describes outcomes that, once reached, persist under some model of individual behavior. In engineering applications we generally demand a stronger interpretation of an equilibrium, as a credible *prediction* of the long-run state of the system. But none the standard equilibrium notions or the corresponding proofs of existence suggest how to arrive at an equilibrium with a reasonable amount of effort. The Pavlovian response of any theoretical computer scientist would be to pose the following queries.

(Q3) *When can the participants of a game quickly converge to an equilibrium? More modestly, when can a centralized algorithm quickly compute an equilibrium outcome?*

These questions are important for two reasons. Algorithms for equilibrium computation can be useful practically, for example in game-playing (e.g. [52]) and for multi-agent reasoning (see [124] for an introduction). Second, resolving the computational complexity of an equilibrium concept has economic implications: a polynomial-time algorithm for computing an equilibrium is a crucial step toward establishing its credibility, while an intractability result casts doubt on its predictive power (a type of critique dating back at least 50 years [104]).

There has been a frenzy of recent work on these questions, for many different fundamental equilibrium concepts. Perhaps the most celebrated results in the area concern the $PPAD$-completeness of computing mixed-strategy Nash equilibria in general games with two or more players [29, 39]. To briefly convey the spirit of the area with a minimum of technical fuss, we instead discuss the complexity of converging to and computing pure-strategy Nash equilibria in variants of the routing games studied in Section 3. The end of the section mentions the key differences between the two settings, as well as surveys of other central equilibrium computation problems (such as market and correlated equilibria).

Pure Equilibria in Network Congestion Games. Recall the selfish routing networks of Section 3. The *atomic* variant is similar to the cost allocation games of the section, in that each of k players controls a non-negligible fraction of the overall traffic (say one unit each) and routes it on a single path. Each edge cost function $c_e : \{1, 2, \ldots, k\} \to \mathcal{R}+$, describing the per-player cost along an edge as a function of its number of users, is non-decreasing. Similarly to the cost allocation games in Section 3, in a (pure-strategy) Nash equilibrium (PNE) $P_1, \ldots, P_k$, each player simultaneously chooses a *best response*: a path with minimum-possible cost $\sum_e c_e(f_e)$, given the choices of others.

Best-response dynamics (BRD) is a simple model of experimentation by players over time: while the current outcome is not a PNE, choose an arbitrarily player that is not using a best response, and update its path to a best response. The update of one player usually changes the best responses of the others; for this reason, BRD cycles forever in many games. In an atomic selfish routing network, however, every iteration of BRD strictly decreases the *potential function* $\Phi(P_1, \ldots, P_k) = \sum_{e \in E} \sum_{i=1}^{f_e} c_e(i)$, and thus BRD is guaranteed to terminate, necessarily at a PNE [88, 108]. The number of distinct outcomes is generally exponential in the size of the network and the number of players; does convergence require polynomial or exponential time? Can we compute a PNE of such a game by other means in polynomial time?

Computing a PNE of an atomic selfish routing game is a member of $TFNP$ ("total functional NP), an intriguing class of search problems for which all instances have a (short and efficiently verifiable) witness [82]. Intuitively, all (well-formed) instances have a solution (in our case, a PNE); the only issue is finding one in polynomial time.

Assume for the moment that the problem lies outside P; how would we amass evidence for this fact? We can't expect to prove that a $TFNP$ problem is NP-hard in a meaningful sense; a short argument shows that such a reduction would imply $NP = coNP$ [82]. We also can't expect to show that it is $TFNP$-complete, since $TFNP$ is a "semantic class" — informally, there is no apparent way to efficiently check membership in $TFNP$ given (say) a Turing machine description of a NP search problem — and thus unlikely to contain complete problems (see [63, 125]). Our best option is therefore to define a "syntactic subclass" of $TFNP$ that contains as many problems as possible (including computing PNE) while admitting complete problems.

We follow [114] in motivating the appropriate subclass. View the definition of NP (existence of short witnesses and an efficient verifier) as a minimal constraint ensuring that a problem is solvable by brute-force search (enumerating all possible witnesses) using polynomial time per iteration. Computing a PNE of an atomic selfish routing games appears to be easier because there is a *guided search* algorithm (namely BRD) that is guaranteed to find a legitimate witness. What are the minimal ingredients that guarantee that a problem admits an analogous guided search procedure? This question was answered twenty years ago in the context of local search algorithms, by the definition of the class PLS, for "polynomial local search" [64]. A PLS problem is abstractly described by *three* polynomial-time algorithms: one to accept an instance and output an initial candidate solution; one to evaluate the objective function value of a candidate solution; and one that either verifies local optimality (for some local neighborhood) or else returns a new candidate solution with strictly better objective function value. PLS can be phrased as a syntactic class and it therefore admits a generic complete problem [64]. The analog of Cook's Theorem (a reduction from the generic complete problem to a concrete one), proved in [64], states that a particular local search problem for Boolean circuits called "Circuit Flip" is PLS-hard. Circuit Flip has been used to establish the PLS-completeness of many other problems (e.g. [121, 137]).

Solutions of a PLS problem correspond to local optima, and one can obviously be found (generally in exponential time) via local search. Computing a PNE of an atomic selfish routing game can be cast as a PLS problem by adopting the potential function as an objective, and define two outcomes to be neighbors if they differ in the path of only one player. Local minima then correspond to the PNE of the game.

Solving a PLS problem means computing a locally optimal solution by whatever means (not necessarily by local search). For example, in *single-commodity* atomic selfish routing games, where all players have the same source and sink, a PNE can be computed in polynomial time using minimum-cost flow [46] despite the fact that BRD (i.e., local search) can require an exponential number of iterations [1]. If $P = PLS$, then given only an abstract description of a PLS problem in terms of the three algorithms above, there is a generic, problem-independent way of finding a "shortcut" to a locally optimal solution, exponentially faster than rote traversal of the path suggested by the guided search algorithm. For both this conceptual reason and its inclusion of many well-known and apparent difficult problems, it is generally believed that $P \neq PLS$. PLS-hardness should therefore viewed as strong evidence that a $TFNP$ search problem is not solvable in polynomial time. Computing a PNE of a (multicommodity) atomic selfish routing network is hard in this sense.

Theorem 6 ([46]). *The problem of computing a PNE of an atomic selfish routing game is PLS-complete.*

See also [1] for an alternative proof, and [1, 2, 46, 126] for further PLS-completeness results on PNE.

The reductions in PLS-completeness results such as Theorem 6 nearly always give unconditional exponential lower bounds on the worst-case running time of the generic local search algorithm (or BRD, in the present context). Even if $P = PLS$, the following corollary holds.

Corollary 1. *There is a constant $c > 0$ such that for arbitrarily large n, there is an n-player atomic selfish routing network and an initial outcome from which BRD requires 2^{cn} iterations to converge to a PNE, no matter how players are selected in each step of BRD.*

Mixed-Strategy Nash Equilibria and PPAD. A *mixed strategy* is a probability distribution over the pure strategies of a player. A collection of mixed-strategies is a *(mixed-strategy) Nash equilibrium (MNE)* if every player simultaneously chooses a mixed strategy maximizing its expected utility, given the mixed strategies chosen by the others. Resorting to mixed strategies is necessary to establish the existence of Nash equilibria in arbitrary finite games with two or more players [91], but they are not without conceptual controversy (see e.g. [96, §3.2]). Regardless, computing an MNE of a finite game is clearly a central equilibrium computation problem.

First consider the two-player ("bimatrix") case, where the input is two $m \times n$ payoff matrices (one for each player) with integer entries. There is a non-obvious exponential-time algorithm for computing an MNE in bimatrix games, which enumerates over all possible pairs of supports for the two players and solves a linear system for each to check for a feasible solution (see e.g. [98, 114, 124]). There is a still less obvious "guided search" algorithm, the *Lemke-Howson (LH) algorithm* [77]; see [133] for a careful exposition. Its worst-case running time is exponential [120]. The LH algorithm is a path-following algorithm in the spirit of local search, but is not guided by an objective or potential function and thus does not obviously prove that computing a MNE of a bimatrix game is in PLS. A related but apparently different subclass of $TFNP$, called $PPAD$ (for "polynomial parity argument, directed version"), was defined in [97] to capture the complexity of this and related problems (mostly from combinatorial topology, such as computing approximate Brouwer fixed points). Its formal definition parallels that of PLS, with a $PPAD$ problem consisting of the minimal ingredients (again easily phrased as three polynomial-time algorithms) necessary to execute a LH-like search procedure. $PPAD$-hardness is viewed as a comparable negative result to PLS-hardness (for the same reasons). Computing an MNE of a bimatrix game is hard in this sense.

Theorem 7 ([29, 39]). *The problem of computing an MNE of a bimatrix game is $PPAD$-complete.*

This hardness result trivially applies to games with any constant number of players. It extends to computing a natural notion of an "ϵ-approximate MNE" for values of ϵ as large as inverse polynomial [29], thus ruling out an FPTAS for computing ϵ-approximate MNE (unless $P = PPAD$). Unlike PLS-completeness results, $PPAD$-completeness results are not known to have immediate unconditional consequences in the spirit of Corollary 1. However, a lower bound on the convergence time of certain dynamics to an MNE was recently proved in [55] (without relying on Theorem 7).

The proof of Theorem 7 is necessarily intricate because in the result is a "Cook's Theorem for $PPAD$" — while several $PPAD$-complete problems were previously known [97], all of them have the flavor of "generic" complete problems, in which an instance includes a description of an arbitrary polynomial-time algorithm. For example, instances of $PPAD$-complete fixed-point problems included an encoding of a polynomial-time algorithm that computes the values of some continuous function restricted to a subdivided simplex. The proof of Theorem 7 effectively encodes arbitrary computation in terms of a bimatrix game, so its sophistication should come as no surprise. Many of the first "non-generic" PLS-complete problems required similarly intricate reductions (e.g. [121]). See [98] for a nice high-level survey of the proof of Theorem 7 and the sequence of results that led to it.

Further Aspects of Equilibrium Computation. Another genre of equilibrium computation problems bustling with activity is *market* or *price equilibria* — prices for goods at which decentralized and selfish exchange "clears the market", yielding a Pareto efficient allocation of the goods. As with mixed Nash equilibria, such equilibria exist under weak conditions [11] but their efficient computation is largely open. The last five years have seen a number of new polynomial-time algorithm (surveyed in [129] and [34]) and a few scattered hardness results (see [34]), but many basic questions remain open (see [129]).

Back in finite games, equilibrium computation in extensive-form games — specified by a game tree in which paths represent sequences of actions by the various players and by nature, see e.g. [134] — was studied early on by the AI community (surveyed in [67]) and more recently in the theoretical computer science literature (e.g. [85]). Special classes of extensive-form games defined in [36] are, along with some number-theoretic problems like factoring, among the most prominent candidates for problems in $(NP \cap coNP) \setminus P$ (see [63]). Other equilibrium concepts in finite games have also been studied recently. For correlated equilibria [12], an equilibrium concept with fundamental connections to no-regret learning algorithms (see [23]), sweeping positive algorithmic results are possible [99]. In repeated games, computing a Nash equilibrium is polynomial-time solvable in two-player games [81] but $PPAD$-hard with three or more players [25], despite the overwhelming number of equilibria guaranteed by the "folk theorem" for such games.

5 Future Directions

The astonishing and accelerating rate of progress in algorithmic game theory, nourished by deep connections with other areas of theoretical computer science and a consistent infusion of new motivating applications, leaves no doubt that it will continue to flourish for many years to come. There is presently a surplus of challenging open questions across all three of the areas surveyed in Sections 2–4; we record a small handful to prove the point.

We first mention some concrete problems that are well known in the AGT community. A few in AMD include: prove better upper or lower bounds on the achievable approximation guarantees of implementable algorithms for combinatorial auctions (see [24] for a reasonably current survey); characterize the multi-parameter domains for which affine maximizers are the only implementable algorithms (see [100] for the latest developments); and develop some understanding of the power of randomization in polynomial-time implementability (see [3] for an entry point). Some personal favorites involving equilibrium efficiency analyses are: determine the POA in atomic selfish routing networks with fractional routing and the POS in Shapley cost allocation games (see [35] and [49], respectively, for partial results); develop a general analytical technique to extract tight efficiency loss bounds from potential functions and/or variational inequalities (see [111]); and, in the spirit of [27], identify how to distribute delays (via an appropriate queuing policy) to minimize the worst-case POA in selfish routing networks. Central open questions in equilibrium computation include the complexity of computing approximate mixed-strategy Nash equilibria (see [26, 80, 128] for the state-of-the-art), the complexity of computing market equilibria with reasonably general (concave) participant utility functions (see [129]), and the complexity of the stochastic games in $NP \cap coNP$ defined in [36] (see also [63]).

Speaking more informally and long-term, we expect that all areas of AGT will (and should) grapple with appropriate models of agent behavior over the next several years. Some type of non-worst-case behavioral assumption is inevitable for systems with independent participants: all of the results described in this survey, even the welfare guarantee of the simple Vickrey auction, depend on such assumptions. AGT has minimized controversy thus far by adopting well-known notions from traditional game theory, such as the Nash equilibrium. But if traditional game theory applied "off the shelf" to modern computer science applications, there would be no need for AGT at all. See [51] for a compelling argument — made over a decade ago but more appropriate than ever — about why models of rationality and equilibrium concepts should be completely rethought given the characteristics of an Internet-like strategic environment.

Behavioral assumptions are essential to address modern computer applications, yet are largely foreign to the mainstream "STOC/FOCS" mentality and its emphasis on minimal assumptions and worst-case analysis. Can we retain this unquestionably useful and well-motivated bias while expanding our field's reach? *Of course:* shining examples of worst-case guarantees coupled with novel behavioral models have already begun to sprout in the AGT literature. For example: mechanism implementation in undominated strategies [16] and in ex post collusion-proof Nash equilibrium [95]; the price of total anarchy in [22]; and the complexity of unit-recall games [45]. If history is any guide, these represent only the vanguard of what promises to be a rich and relevant theory.

References

1. H. Ackermann, H. Röglin, and B. Vöcking. On the impact of combinatorial structure on congestion games. In *FOCS '06*, pages 613–622.
2. H. Ackermann and A. Skopalik. On the complexity of pure Nash equilibria in player-specific network congestion games. In *WINE '07*, pages 419–430.
3. G. Aggarwal, A. Fiat, A. V. Goldberg, J. D. Hartline, N. Immorlica, and M. Sudan. Derandomization of auctions. In *STOC '05*, pages 619–625.
4. E. Altman, T. Boulogne, R. El Azouzi, T. Jiménez, and L. Wynter. A survey on networking games in telecommunications. *Computers & Operations Research*, 33(2):286–311, 2006.

5. N. Andelman, M. Feldman, and Y. Mansour. Strong price of anarchy. In *SODA '07*, pages 189–198.
6. E. Anshelevich, A. Dasgupta, J. Kleinberg, É. Tardos, T. Wexler, and T. Roughgarden. The price of stability for network design with fair cost allocation. In *FOCS '04*, pages 295–304.
7. A. Archer. *Mechanisms for Discrete Optimization with Rational Agents*. PhD thesis, Cornell University, 2004.
8. A. Archer and R. D. Kleinberg. Truthful germs are contagious: a local-to-global characterization of truthfulness. In *EC '08*.
9. A. Archer, C. H. Papadimitriou, K. Talwar, and É. Tardos. An approximate truthful mechanism for combinatorial auctions with single parameter agents. In *SODA '03*, pages 205–214.
10. A. Archer and É. Tardos. Truthful mechanisms for one-parameter agents. In *FOCS '01*, pages 482–491.
11. K. Arrow and G. Debreu. Existence of an equilibrium for a competitive economy. *Econometrica*, 22:265–290, 1954.
12. R. J. Aumann. Subjectivity and correlation in randomized strategies. *Journal of Mathematical Economics*, 1(1):67–96, 1974.
13. B. Awerbuch, Y. Azar, and Y. Bartal. On-line generalized Steiner problem. In *SODA '96*, pages 68–74.
14. B. Awerbuch, Y. Azar, A. Epstein, V. S. Mirrokni, and A. Skopalik. Fast convergence to nearly optimal solutions in potential games. In *EC '08*.
15. M. Babaioff, R. D. Kleinberg, and C. H. Papadimitriou. Congestion games with malicious players. In *EC '07*, pages 103–112.
16. M. Babaioff, R. Lavi, and E. Pavlov. Single-value combinatorial auctions and implementation in undominated strategies. In *SODA '06*, pages 1054–1063.
17. Y. Bartal, R. Gonen, and N. Nisan. Incentive compatible multi unit combinatorial auctions. In *TARK '03*, pages 72–87.
18. M. J. Beckmann, C. B. McGuire, and C. B. Winsten. *Studies in the Economics of Transportation*. Yale University Press, 1956.
19. D. P. Bertsekas and R. G. Gallager. *Data Networks*. Prentice-Hall, 1987. Second Edition, 1991.
20. S. Bikhchandani, S. Chatterji, R. Lavi, A. Mu'alem, N. Nisan, and A. Sen. Weak monotonicity characterizes dominant strategy implementation. *Econometrica*, 74(4): 1109–1132, 2006.
21. A. Blum, E. Even-Dar, and K. Ligett. Routing without regret: On convergence to Nash equilibria of regret-minimizing algorithms in routing games. In *PODC '06*, pages 45–52.
22. A. Blum, M. Hajiaghayi K. Ligett, and A. Roth. Regret minimization and the price of total anarchy. In *STOC '08*.
23. A. Blum and Y. Mansour. Learning, regret minimization, and equilibria. In Nisan et al. [94], chapter 4, pages 79–101.
24. L. Blumrosen and N. Nisan. Combinatorial auctions. In Nisan et al. [94], chapter 11, pages 267–299.
25. C. Borgs, J. Chayes, N. Immorlica, A. T. Kalai, V. S. Mirrokni, and C. H. Papadimitriou. The myth of the folk theorem. In *STOC '08*.
26. H. Bosse, J. Byrka, and E. Markakis. New algorithms for approximate Nash equilibria in bimatrix games. In *WINE '07*, pages 17–29.
27. H. Chen, T. Roughgarden, and G. Valiant. Designing networks with good equilibria. In *SODA '08*, pages 854–863.
28. P.-A. Chen and D. Kempe. Altruism, selfishness, and spite in traffic routing. In *EC '08*.
29. X. Chen, X. Deng, and S.-H. Teng. Settling the complexity of two-player Nash equilibria. *Journal of the ACM*, 2008.
30. S. Chien and A. Sinclair. Convergence to approximate Nash equilibria in congestion games. In *SODA '07*, pages 169–178.

31. G. Christodoulou and E. Koutsoupias. On the price of anarchy and stability of correlated equilibria of linear congestion games. In *EC '05*, pages 59–70.
32. G. Christodoulou, E. Koutsoupias, and A. Nanavati. Coordination mechanisms. In *ICALP '04*, pages 345–357.
33. G. Christodoulou, E. Koutsoupias, and A. Vidali. A characterization of 2-player mechanisms for scheduling. Submitted, 2008.
34. B. Codenotti and K. Varadarajan. Computation of market equilibria by convex programming. In Nisan et al. [94], chapter 6, pages 135–158.
35. R. Cominetti, J. R. Correa, and N. E. Stier Moses. The impact of oligopolistic competition in networks. 2008.
36. A. Condon. The complexity of stochastic games. *Information and Computation*, 96:203–224, 1992.
37. P. Cramton. Spectrum auctions. In *Handbook of Telecommunications Economics*, chapter 14, pages 605–639. 2002.
38. P. Cramton, Y. Shoham, and R. Steinberg, editors. *Combinatorial Auctions*. MIT Press, 2006.
39. C. Daskalakis, P. W. Goldberg, and C. H. Papadimitriou. The complexity of comuting a Nash equilibria. *SIAM Journal on Computing*, 2008.
40. P. Dhangwatnotai, S. Dobzinski, S. Dughmi, and T. Roughgarden. Truthful approximation schemes for single-parameter agents. Submitted, 2008.
41. S. Dobzinski and N. Nisan. Limitations of VCG-based mechanisms. In *STOC '07*, pages 338–344.
42. S. Dobzinski and M. Sundararajan. On characterizations of truthful mechanisms for combinatorial auctions and scheduling. In *EC '08*.
43. L. Epstein and J. Sgall. Approximation schemes for scheduling on uniformly related and identical parallel machines. *Algorithmica*, 39(1):43–57, 2004.
44. E. Even-Dar, A. Kesselman, and Y. Mansour. Convergence time to Nash equilibria. In *ICALP '03*, pages 502–513.
45. A. Fabrikant and C. H. Papadimitriou. The complexity of game dynamics: BGP oscillations, sink equlibria, and beyond. In *SODA '08*, pages 844–853.
46. A. Fabrikant, C. H. Papadimitriou, and K. Talwar. The complexity of pure Nash equilibria. In *STOC '04*, pages 604–612.
47. U. Feige. On maximizing welfare when utility functions are subadditive. In *STOC '06*, pages 41–50.
48. J. Feigenbaum, M. Schapira, and S. Shenker. Distributed algorithmic mechanism design. In Nisan et al. [94], chapter 14, pages 363–384.
49. A. Fiat, H. Kaplan, M. Levy, S. Olonetsky, and R. Shabo. On the price of stability for designing undirected networks with fair cost allocations. In *ICALP '06*, pages 608–618.
50. S. Fischer and B. Vöcking. On the evolution of selfish routing. In *ESA '04*, pages 323–334.
51. E. J. Friedman and S. J. Shenker. Learning and implementation on the Internet. Working paper, 1997.
52. A. Gilpin, T. Sandholm, and T. B. Sorensen. Potential-aware automated abstraction of sequential games, and holistic equilibrium analysis of Texas Hold'em poker. In *AAAI '07*.
53. M. X. Goemans, V. Mirrokni, and A. Vetta. Sink equilibria and convergence. In *FOCS '05*, pages 142–151.
54. J. Y. Halpern. Computer science and game theory: A brief survey. In S. N. Durlauf and L. E. Blume, editors, *Palgrave Dictionary of Economics*. 2008.
55. S. Hart and Y. Mansour. The communication complexity of uncoupled Nash equilibrium procedures. *Games and Economic Behavior*, 2008.
56. J. Hartline and A. Karlin. Profit maximization in mechanism design. In Nisan et al. [94], chapter 13, pages 331–362.
57. J. D. Hartline and T. Roughgarden. Optimal mechanism design and money burning. In *STOC '08*.

58. D. Hochbaum and D. B. Shmoys. A polynomial approximation scheme for scheduling on uniform processors: Using the dual approximation approach. *SIAM J. Comput.*, 17(3):539–551, 1988.
59. M. Imase and B. M. Waxman. Dynamic Steiner tree problem. *SIAM Journal on Discrete Mathematics*, 4(3), 1991.
60. M. O. Jackson. A crash course in implementation theory. *Social Choice and Welfare*, 18(4):655–708, 2001.
61. K. Jain and M. Mahdian. Cost sharing. In Nisan et al. [94], chapter 15, pages 385–410.
62. R. Johari. The price of anarchy and the design of scalable resource allocation mechanisms. In Nisan et al. [94], chapter 21, pages 543–568.
63. D. S. Johnson. The NP-completeness column: Finding needles in haystacks. *ACM Transactions on Algorithms*, 3(2), 2007. Article 24.
64. D. S. Johnson, C. H. Papadimitriou, and M. Yannakakis. How easy is local search? *Journal of Computer and System Sciences*, 37(1):79–100, 1988.
65. E. Kalai and D. Samet. On weighted Shapley values. *International Journal of Game Theory*, 16(3):205–222, 1987.
66. G. Karakostas and A. Viglas. Equilibria for networks with malicious users. *Mathematical Programming*, 110(3):591–613, 2007.
67. D. Koller and A. Pfeffer. Representations and solutions for game-theoretic problems. *Artificial Intelligence*, 94(1-2):167–215, 1997.
68. S. C. Kontogiannis and P. G. Spirakis. Atomic selfish routing in networks: A survey. In *WINE '05*, pages 989–1002.
69. E. Koutsoupias and C. H. Papadimitriou. Worst-case equilibria. In *STACS '99*, pages 404–413.
70. A. Kovács. Tighter approximation bounds for LPT scheduling in two special cases. In *CIAC*, pages 187–198, 2006.
71. S. Lahaie, D. Pennock, A. Saberi, and R. Vohra. Sponsored search auctions. In Nisan et al. [94], chapter 28.
72. R. Lavi. Computationally efficient approximation mechanisms. In Nisan et al. [94], chapter 12, pages 301–329.
73. R. Lavi, A. Mu'alem, and N. Nisan. Towards a characterization of truthful combinatorial auctions. In *FOCS '03*, pages 574–583.
74. R. Lavi and C. Swamy. Truthful mechanism design for multi-dimensional scheduling via cycle monotonicity. In *EC '07*, pages 252–261.
75. B. Lehmann, D. J. Lehmann, and N. Nisan. Combinatorial auctions with decreasing marginal utilities. In *EC '01*, pages 18–28.
76. D. Lehmann, L. I. O'Callaghan, and Y. Shoham. Truth revelation in approximately efficient combinatorial auctions. *Journal of the ACM*, 49(5):577–602, 2002.
77. C. E. Lemke and J. T. Howson, Jr. Equilibrium points of bimatrix games. *SIAM Journal*, 12(2):413–423, 1964.
78. H. Levin, M. Schapira, and A. Zohar. Interdomain routing and games. In *STOC '08*, 2008.
79. N. Linial. Game-theoretic aspects of computing. In R. J. Aumann and S. Hart, editors, *Handbook of Game Theory with Economic Applications*, volume 2, chapter 38, pages 1339–1395. 1994.
80. R. J. Lipton, E. Markakis, and A. Mehta. Playing large games using simple strategies. In *EC '03*, pages 36–41.
81. M. L. Littman and P. Stone. A polynomial-time Nash equilibrium algorithm for repeated games. *Decision Support Systems*, 39(1):55–66, 2005.
82. N. Megiddo and C. H. Papadimitriou. On total functions, existence theorems and computational complexity. *Theoretical Computer Science*, 81(2):317–324, 1991.
83. A. Mehta, T. Roughgarden, and M. Sundararajan. Beyond Moulin mechanisms. In *Proceedings of the 8th ACM Conference on Electronic Commerce (EC)*, pages 1–10, 2007.

84. M. Mihail, C. H. Papadimitriou, and A. Saberi. On certain connectivity properties of the Internet topology. *Journal of Computer and System Sciences*, 72(2):239–251, 2006.
85. P. B. Miltersen and T. B. Sørensen. Fast algorithms for finding proper strategies in game trees. In *SODA '08*, pages 874–883.
86. V. S. Mirrokni and A. Vetta. Convergence issues in competitive games. In *APPROX '04*, pages 183–194.
87. D. Monderer. Monotonicity and implementability. In *EC '08*.
88. D. Monderer and L. S. Shapley. Potential games. *Games and Economic Behavior*, 14(1):124–143, 1996.
89. A. Mu'alem and N. Nisan. Truthful approximation mechanisms for restricted combinatorial auctions. In *AAAI '02*, pages 379–384.
90. R. Myerson. Optimal auction design. *Mathematics of Operations Research*, 6:58–73, 1981.
91. J. F. Nash. Equilibrium points in N-person games. *Proceedings of the National Academy of Science*, 36(1):48–49, 1950.
92. N. Nisan. Introduction to mechanism design (for computer scientists). In Nisan et al. [94], chapter 9, pages 209–241.
93. N. Nisan and A. Ronen. Algorithmic mechanism design. *Games and Economic Behavior*, 35(1/2):166–196, 2001.
94. N. Nisan, T. Roughgarden, É. Tardos, and V. Vazirani, editors. *Algorithmic Game Theory*. Cambridge University Press, 2007.
95. N. Nisan, M. Schapira, and A. Zohar. Best-reply mechanisms. Working paper, 2008.
96. M. J. Osborne and A. Rubinstein. *A Course in Game Theory*. MIT Press, 1994.
97. C. H. Papadimitriou. On the complexity of the parity argument and other inefficient proofs of existence. *Journal of Computer and System Sciences*, 48:498–532, 1994.
98. C. H. Papadimitriou. The complexity of finding nash equilibria. In Nisan et al. [94], chapter 2, pages 29–51.
99. C. H. Papadimitriou and T. Roughgarden. Computing correlated equilibria in multiplayer games. *Journal of the ACM*, 2008.
100. C. H. Papadimitriou, M. Schapira, and Y. Singer. On the hardness of being truthful. Manuscript, 2008.
101. D. Parkes. *Iterative Combinatorial Auctions: Achieving Economic and Computational Efficiency*. PhD thesis, University of Pennsylvania, 2001.
102. D. C. Parkes. Online mechanisms. In Nisan et al. [94], chapter 16, pages 411–439.
103. A. C. Pigou. *The Economics of Welfare*. Macmillan, 1920.
104. M. O. Rabin. Effective computability of winning strategies. In M. Dresher, A. W. Tucker, and P. Wolfe, editors, *Contributions to the Theory Games*, volume 3. Princeton University Press, 1957.
105. J. Riley and W. Samuelson. Optimal auctions. *American Economic Review*, 71:381–92, 1981.
106. K. Roberts. The characterization of implementable choice rules. In J.-J. Laffont, editor, *Aggregation and Revelation of Preferences*, pages 321–349. 1979.
107. J. C. Rochet. A necessary and sufficient condition for rationalizability in a quasilinear context. *Journal of Mathematical Economics*, 16:191–200, 1987.
108. R. W. Rosenthal. A class of games possessing pure-strategy Nash equilibria. *International Journal of Game Theory*, 2(1):65–67, 1973.
109. T. Roughgarden. The price of anarchy is independent of the network topology. *Journal of Computer and System Sciences*, 67(2):341–364, 2003.
110. T. Roughgarden. *Selfish Routing and the Price of Anarchy*. MIT Press, 2005.
111. T. Roughgarden. Potential functions and the inefficiency of equilibria. In *Proceedings of the International Congress of Mathematicians*, volume III, pages 1071–1094, 2006.
112. T. Roughgarden. Routing games. In Nisan et al. [94], chapter 18, pages 461–486.
113. T. Roughgarden. Selfish routing and the price of anarchy. *OPTIMA*, 74:1–15, 2007.
114. T. Roughgarden. Computing equilibria: A computational complexity perspective. *Economic Theory*, 2008.

115. T. Roughgarden and É. Tardos. How bad is selfish routing? *Journal of the ACM*, 49(2):236–259, 2002.
116. T. Roughgarden and É. Tardos. Bounding the inefficiency of equilibria in nonatomic congestion games. *Games and Economic Behavior*, 49(2):389–403, 2004.
117. T. Roughgarden and É. Tardos. Introduction to the inefficiency of equilibria. In Nisan et al. [94], chapter 17, pages 443–459.
118. M. E. Saks and L. Yu. Weak monotonicity suffices for truthfulness on convex domains. In *EC '05*, pages 286–293.
119. T. Sandholm. Algorithm for optimal winner determination in combinatorial auctions. *Artificial Intelligence*, 135(1):1–54, 2002.
120. R. Savani and B. von Stengel. Hard-to-solve bimatrix games. *Econometrica*, 74(2):397–429, 2006.
121. A. A. Schäffer and M. Yannakakis. Simple local search problems that are hard to solve. *SIAM Journal on Computing*, 20(1):56–87, 1991.
122. J. Schummer and R. V. Vohra. Mechanism design without money. In Nisan et al. [94], chapter 10, pages 243–265.
123. Y. Shoham. Computer science and game theory. *Communications of the ACM*, 2008.
124. Y. Shoham and K. Leyton-Brown. *Multiagent Systems: Algorithmic, Game Theoretic and Logical Foundations*. Cambridge University Press, 2008.
125. M. Sipser. On relativization and the existence of complete sets. In *ICALP '82*, pages 523–531.
126. A. Skopalik and B. Vöcking. Inapproximability of pure Nash equilibria. In *STOC '08*.
127. É. Tardos and T. Wexler. Network formation games and the potential function method. In Nisan et al. [94], chapter 19, pages 487–516.
128. H. Tsaknakis and P. G. Spirakis. An optimization approach for approximate Nash equilibria. In *WINE '07*, pages 42–56.
129. V. V. Vazirani. Combinatorial algorithms for market equilibria. In Nisan et al. [94], chapter 5, pages 103–134.
130. W. Vickrey. Counterspeculation, auctions, and competitive sealed tenders. *Journal of Finance*, 16(1):8–37, 1961.
131. B. Vöcking. Selfish load balancing. In Nisan et al. [94], chapter 20, pages 517–542.
132. R. Vohra. Paths, cycles and mechanism design. Working paper, 2007.
133. B. von Stengel. Computing equilibria for two-person games. In R. J. Aumann and S. Hart, editors, *Handbook of Game Theory with Economic Applications*, volume 3, chapter 45, pages 1723–1759. North-Holland, 2002.
134. B. von Stengel. Equilibrium computation for two-player games in strategic and extensive form. In Nisan et al. [94], chapter 3, pages 53–78.
135. J. Vondrak. Optimal approximation for the submodular welfare problem in the value oracle model. In *STOC '08*.
136. J. G. Wardrop. Some theoretical aspects of road traffic research. In *Proceedings of the Institute of Civil Engineers, Pt. II*, volume 1, pages 325–378, 1952.
137. M. Yannakakis. Computational complexity. In E. Aarts and J. K. Lenstra, editors, *Local Search in Combinatorial Optimization*, chapter 2, pages 19–55. 1997.

Synchronizing Road Coloring

A.N. Trahtman* †

Bar-Ilan University, Dep. of Math., 52900, Ramat Gan, Israel

Abstract. The synchronizing word of a deterministic automaton is a word in the alphabet of colors (considered as letters) of its edges that maps the automaton to a single state. A coloring of edges of a directed graph is synchronizing if the coloring turns the graph into a deterministic finite automaton possessing a synchronizing word.
The road coloring problem is the problem of synchronizing coloring of a directed finite strongly connected graph with constant outdegree of all its vertices if the greatest common divisor of lengths of all its cycles is one. The problem was posed by Adler, Goodwyn and Weiss over 30 years ago and evoked noticeable interest among the specialists in the theory of graphs, finite automata, coding and symbolic dynamics. Many partial solutions of the problem have been found and different generalizations were considered.
The positive solution of the road coloring problem is presented below. We reproduce from the literature also the statements used in our proof. The necessary and sufficient conditions of synchronizing road coloring of directed graph with constant outdegree of a vertex are presented.

Key words: road coloring problem, graph, deterministic finite automaton, synchronization.

Introduction

The road coloring problem originates in [2] and was stated explicitly in [1] for a strongly connected directed finite graph with constant outdegree of all its vertices where the greatest common divisor (gcd) of lengths of all its cycles is one. The edges of the graph are unlabelled. The task is to find a labelling of the edges that turns the graph into a deterministic finite automaton possessing a synchronizing word. So the road coloring problem is connected with the problem of existence of synchronizing word for deterministic complete finite automaton. The condition on gcd is necessary [1], [5]. It can be replaced by the equivalent property that there does not exist a partition of the set of vertices on subsets V_1, V_2, ..., $V_k = V_1$ $(k > 1)$ such that every edge which begins in V_i has its end in V_{i+1} [5], [14]. The outdegree of the vertex can be considered also as the size of an alphabet where the letters denote colors.
The road coloring problem is important in automata theory: a synchronizing

* Email: trakht@macs.biu.ac.il

† http://www.cs.biu.ac.il/~trakht/syn.html

Please use the following format when citing this chapter:

Trahtman, A.N., 2008, in IFIP International Federation for Information Processing, Volume 273; *Fifth IFIP International Conference on Theoretical Computer Science*; Giorgio Ausiello, Juhani Karhumäki, Giancarlo Mauri, Luke Ong; (Boston: Springer), pp. 43–53.

coloring makes the behavior of an automaton resistant against input errors since, after detection of an error, a synchronizing word can reset the automaton back to its original state, as if no error had occurred. The problem appeared first in the context of symbolic dynamics and is important also in this area.
Together with the Černy conjecture, the road coloring problem belongs to the most fascinating problems in the theory of finite automata [13], [16], [17]. The problem is discussed even in "Wikipedia" - the popular Internet Encyclopedia. However, at the same time it proved to be hard and was considered as a "notorious open problem" [12], [5] and "unfeasible" [8].
Several partial solutions in this area have been found within last thirty years. In [14] it is shown that a graph with no multiple edges (i.e. no distinct edges in G have the same source and the same target) and with a simple cycle of prime length has a synchronizing coloring. In [6] it is shown that a graph of outdegree two with a simple cycle of length relatively prime to the weight of the graph (i.e. the sum of the components of an integer Perron left eigenvector chosen with relatively prime components) has a synchronizing coloring. The conjecture is true for Eulerian digraphs [10] (i.e. the indegree of any vertex is equal to the outdegree). In [5] the problem is solved for the class of automata having always stable synchronizing pair of states. The class is closed under some kind of homomorphism. The conjecture has positive solution also if the outdegree of vertices is relatively great [7]. Another special case, proven in [15], is that a graph with all vertices of indegree 1 except one (these graphs are trees where all leaves merge with the root), has a synchronizing coloring. In [9] it is shown that a graph of outdegree k which is decomposable in k disjoint monochromatic subgraphs containing exactly one cycle, has a synchronizing coloring if the greater common divisor of the lengths of the monochromatic cycles equals 1. The last result was strengthened in [4] for strongly disjoint set of cycles. The structure theory of the minimal ideal of a finite semigroup plays an essential role in [3].
The concept from [6] of the weight of a vertex supposed by Friedman and the concept of a stable pair of states of Culik, Karhumaki and Kari [5], [10] with corresponding results and consequences are essentially used in our proof. We also reproduce from the literature the proofs of some related statements for to complete the picture.
The road coloring conjecture is settled in the affirmative: a finite directed strongly connected graph with constant outdegree of all vertices has a synchronizing coloring iff the great common divisor of the lengths of all its cycles is equal to one.
The necessary and sufficient conditions of synchronizing road coloring of directed graph with constant outdegree of a vertex are presented.

Preliminaries

A finite directed strongly connected graph with constant outdegree of all its vertices where the gcd of lengths of all its cycles is one will be called *AGW graph* as aroused by Adler, Goodwyn and Weiss.
If there exists a path in an automaton from the state $\mathbf{p}$ to the state $\mathbf{q}$ and the edges of the path are consecutively labeled by $\sigma_1, ..., \sigma_k$, then for $s = \sigma_1...\sigma_k \in \Sigma^+$ let us write $\mathbf{q} = \mathbf{p}s$.
Let Ps be the map of the subset P of states of an automaton by help of $s \in \Sigma^+$ and let Ps^{-1} be the maximal set of states Q such that $Qs \subseteq P$. For the transition graph Γ of an automaton let Γs denote the map of the set of states of the automaton.
$|P|$ - the size of the subset P of states from an automaton (of vertices from a graph).
A word $s \in \Sigma^+$ is called a *synchronizing* word of the automaton with transition graph Γ if $|\Gamma s| = 1$.
A coloring of a directed finite graph is *synchronizing* if the coloring turns the graph into a deterministic finite automaton possessing a synchronizing word.
A pair of distinct states $\mathbf{p}, \mathbf{q}$ of an automaton (of vertices of the transition graph) will be called *synchronizing* if $\mathbf{p}s = \mathbf{q}s$ for some $s \in \Sigma^+$. In the opposite case, if for any s $\mathbf{p}s \neq \mathbf{q}s$, we call the pair *deadlock*.
A synchronizing pair of states $\mathbf{p}$, $\mathbf{q}$ of an automaton is called *stable* if for any word u the pair $\mathbf{p}u, \mathbf{q}u$ is also synchronizing [5], [10].
We call the set of all outgoing edges of a vertex a *bunch* if all these edges are incoming edges of only one vertex.
Let u be a left eigenvector with positive components having no common divisor of adjacency matrix of a graph with vertices $\mathbf{p}_1$, ..., $\mathbf{p}_n$. The i-th component u_i of the vector u is called *the weight* of the vertex $\mathbf{p}_i$ and denoted by $w(\mathbf{p}_i)$. The sum of the weights of the vertices from a set D is denoted by $w(D)$ and is called *the weight* of D [6].
The subset D of states of an automaton (of vertices of the transition graph Γ of the automaton) such that $w(D)$ is maximal and $|Ds| = 1$ for some word $s \in \Sigma^+$ let us call *F-maximal* as introduced by Friedman [6].
The subset Γs of states (of vertices of the transition graph Γ) for some word s such that every pair of states from the set is deadlock will be called an *F-clique*.

1 Some properties of *F*-clique and of coloring free of stable pairs

The road coloring problem was formulated for *AGW* graphs [1], [2] and only such graphs are considered below. The primitive cases of graphs with loops and of only one color can be also omitted [1], [14]. Let us formulate some important results from [6], [5] and [10] together with some useful statements [1], [14] in the following form:

Theorem 1 *[6] There exists a partition of Γ on F-maximal sets (of the same weight).*

Proof. Let Γ have outdegree d everywhere. The vector $e = (1, ..., 1)$ is a right eigenvector with eigenvalue d of the adjacency matrix [11] A of Γ. Since Γ is strongly connected, by Perron-Frobenius Theorem [11] the matrix A with non-negative elements has a positive left eigenvector $w = (w_1, ..., w_n)$ of integer components with the same eigenvalue d, i.e. $wA = dw$. The component w_i of the vector w is defined as the weight $w(\mathbf{p}_i) > 0$ of the state $\mathbf{p}_i$.
Let $\mathbf{q}$ be arbitrary state and Q be the set of states $q\sigma^{-1}$ for all $\sigma \in \Sigma$. Then $\sum_{\mathbf{r} \in Q} w(\mathbf{r}) = dw(\mathbf{q})$ because $wA = dw$. Consequently for any set R of states $dw(R) = \sum_{\sigma \in \Sigma} R\sigma^{-1}$. It implies, in particular, that from $w(R) > w(R\sigma^{-1})$ for some $\sigma \in \Sigma$ follows $w(R) < w(R\alpha^{-1})$ for some another $\alpha \in \Sigma$. Therefore for F-maximal set R holds $w(R) = w(R\sigma^{-1})$ for any σ and $w(R) = w(Rs^{-1})$ for any $s \in \Sigma^+$.
For F-maximal set R and some word s $|Rs| = 1$. Since Γ is strongly connected, for any state $\mathbf{p}$ there exists a word $t = t_1 s$ such that $Rt = \mathbf{p}$. So for any state $\mathbf{p}$ and some word t the set of states $\mathbf{p}t^{-1}$ is F-maximal. For any state $\mathbf{r} \notin \mathbf{p}t^{-1}$ and some word u is also F-maximal. The set $\mathbf{p}t^{-1}u^{-1}$ is F-maximal, too. Both obtained F-maximal sets $\mathbf{p}t^{-1}u^{-1}$ and $\mathbf{r}u^{-1}$ are disjoint. The continuation of this process for states outside obtained F-maximal sets gives us a partition of Γ on F-maximal sets.

Lemma 1 *[1], [5], [14]. Let Γ be directed graph. Then the greatest common divisor of lengths of all its cycles is k if and only if there exists a partition of the set of vertices on subsets V_1, V_2, ..., $V_{k+1} = V_1$ such that every edge which begins in V_i has its end in V_{i+1}.*

Proof. Indeed, in the case of such partition of size $k > 1$ the length of any cycle of the graph is divided by k.
In the case k is a common divisor of length of all cycles of the graph let us enumerate the vertices of the graph. We begin from an arbitrary vertex and suppose $n(\mathbf{q}) = n(\mathbf{p}) + 1$ (modulo k) if there exist an edge $\mathbf{p} \to \mathbf{q}$. The contradiction in the enumeration is impossible because the difference between length of cycles is divided by k. Then suppose $\mathbf{q} \in V_m$ if $n(\mathbf{q}) = m$ (modulo k). Therefore every edge which begins in V_i ends in V_{i+1}. So the desired partition exists.

Let us recall that a binary relation ρ on the set of the states of an automaton is called congruence if ρ is equivalence and for any word u from $\mathbf{p} \rho \mathbf{q}$ follows $\mathbf{p}u \rho \mathbf{q}u$.

Theorem 2 *[5], [10] Let us consider a coloring of AGW graph Γ. Stability of states is a binary relation on the set of states of the obtained automaton; denote this relation by ρ. Then ρ is a congruence relation, Γ/ρ presents an AGW graph and synchronizing coloring of Γ/ρ implies synchronizing recoloring of Γ.*

Proof. Suppose $\mathbf{p}\ \rho\ \mathbf{q}$ and $\mathbf{q}\ \rho\ \mathbf{r}$. Then for any word u there exists a word s such that $\mathbf{p}us = \mathbf{q}us$. The couple of states $\mathbf{q}, \mathbf{r}$ is stable, whence there exists a word t such that for arbitrary u $\mathbf{q}ust = \mathbf{r}ust$. So for any u there exists a word st such that $\mathbf{p}ust = \mathbf{r}ust$. Hence $\mathbf{p}\ \rho\ \mathbf{r}$ and the relation ρ is transitive and stable. It implies the equivalence of ρ. From $\mathbf{p}\ \rho\ \mathbf{q}$ follows $\mathbf{p}s\ \rho\ \mathbf{q}s$ for any s (because the pair $\mathbf{p}s$, $\mathbf{q}s$ is also stable) and therefore the relation ρ is a congruence.
The outdegree of a state in the quotient automaton Γ/ρ is equal to the same number of colors as in Γ, Γ/ρ is strongly connected just as Γ.
The condition on *gcd* can be replaced by the equivalent property that there does not exist a partition of the set of vertices on subsets $V_1, V_2, ..., V_k = V_1$ such that every edge which begins in V_i has its end in V_{i+1} (Lemma 1).
The non-trivial such partition of Γ/ρ exists only if Γ has also such partition. Every edge with beginning in image of V_i has its end in image of V_{i+1}. Therefore the condition on *gcd* is stable, whence Γ/ρ is *AGW* graph.
Suppose now that Γ/ρ has a synchronizing coloring. The synchronizing coloring of Γ/ρ induces a synchronizing coloring of the original automaton as follows: we color all the preimages of an edge of Γ/ρ by the same color. For any pair of states from Γ the synchronizing word of the images of the states in Γ/ρ takes both states into one equivalence class of the relation ρ on Γ. Any couple of states from this class is stable and therefore synchronizing. So via such coloring any pair of states from Γ is synchronizing.

The last theorem shows that if every *AGW* graph has a coloring with a stable pair, then every *AGW* graph has a synchronizing coloring. So the problem is reduced to the search of a coloring with stable pair.

Lemma 2 *Let w be the weight of F-maximal set of the AGW graph Γ via some coloring. Then the size of every F-clique of the coloring is the same and equal to $w(\Gamma)/w$ (the size of partition of Γ on F-maximal sets).*

Proof. Two states from an F-clique could not belong to one F-maximal set because this pair is not synchronizing. By Theorem 1 there exists a partition of Γ on F-maximal sets of weight w. So the partition consists from $w(\Gamma)/w$ F-maximal sets and to every F-maximal set belongs at most one state from F-clique. Consequently, the size of any F-clique is not greater than $w(\Gamma)/w$.
Let Γs be an F-clique. The sum of the weights $\mathbf{q}s^{-1}$ for all $\mathbf{q} \in \Gamma s$ is the weight of Γ. So

$$w(\Gamma) = \sum_{q \in \Gamma s} w(\mathbf{q}s^{-1})$$

The number of addends (the size of the F-clique) is not greater than $w(\Gamma)/w$. The weight of the set $\mathbf{q}s^{-1}$ for every $\mathbf{q} \in \Gamma s$ is not greater than w. Therefore $\mathbf{q}s^{-1}$ is an F-maximal set of weight w for every $\mathbf{q} \in \Gamma s$ and the size of any F-clique is $w(\Gamma)/w$, the number of F-maximal sets in the corresponding partition of Γ.

Lemma 3 *Let F be F-clique via some coloring of AGW graph Γ. For any word s the set Fs is also an F-clique and any state [vertex]* $\mathbf{p}$ *belongs to some F-clique.*

Proof. Any pair $\mathbf{p}$, $\mathbf{q}$ from an F-clique F is a deadlock. To be deadlock is a stable binary relation, therefore for any word s the pair $\mathbf{p}s$, $\mathbf{q}s$ from Fs also is a deadlock. So all pairs from Fs are deadlocks.
For the F-clique F there exists a word t such that $\Gamma t = F$. Thus $\Gamma ts = Fs$, whence Fs is an F-clique.
For any $\mathbf{r}$ from a strongly connected graph Γ, there exists a word u such that $\mathbf{r} = \mathbf{p}u$ for $\mathbf{p}$ from the F-clique F, whence $\mathbf{r}$ belongs to the F-clique Fu.

Lemma 4 *Let A and B ($|A| > 1$) be distinct F-cliques via some coloring without stable pairs of the AGW graph Γ. Then $|A|-|A\cap B| = |B|-|A\cap B| > 1$.*

Proof. Let us assume the contrary: $|A| - |A\cap B| = 1$. By Lemma 2, $|A| = |B|$. So $|B| - |A \cap B| = 1$, too. The pair of states $\mathbf{p} \in A \setminus B$ and $\mathbf{q} \in B \setminus A$ is not stable. Therefore for some word s the pair $(\mathbf{p}s, \mathbf{q}s)$ is a deadlock. Any pair of states from the F-clique A and from the F-clique B as well as from F-cliques As and Bs is a deadlock. So any pair of states from the set $(A \cup B)s$ is a deadlock. One has $|(A \cup B)s| = |A| + 1 > |A|$.
In view of Theorem 1, there exists a partition of size $|A|$ (Lemma 2) of Γ on F-maximal sets. To every F-maximal set belongs at most one state from $(A\cup B)s$ because every pair of states from this set is a deadlock and no deadlock could belong to an F-maximal set. This contradicts the fact that the size of $(A\cup B)s$ is greater than $|A|$.

Lemma 5 *Let some vertex of AGW graph Γ have two incoming bunches. Then any coloring of Γ has a stable couple.*

Proof. If a vertex $\mathbf{p}$ has two incoming bunches from vertices $\mathbf{q}$ and $\mathbf{r}$, then the couple $\mathbf{q}$, $\mathbf{r}$ is stable for any coloring because $\mathbf{q}\alpha = \mathbf{r}\alpha = \mathbf{p}$ for any letter (color) $\alpha \in \Sigma$.

2 The spanning subgraph of cycles and trees with maximal number of edges in the cycles

Définition 1 *Let us call a subgraph S of the AGW graph Γ a* spanning subgraph *of Γ if to S belong all vertices of Γ and exactly one outgoing edge of every vertex.*
A maximal subtree of the spanning subgraph S with root on a cycle from S and having no common edges with cycles from S is called a tree *of S.*
The length of path from a vertex $\mathbf{p}$ *through the edges of the tree of the spanning set S to the root of the tree is called the* level *of* $\mathbf{p}$ *in S.*

Remark 1 *Any spanning subgraph S consists of disjoint cycles and trees with roots on cycles; any tree and cycle of S is defined identically, the level of the vertex from cycle is zero, the vertices of trees except root have positive level, the vertex of maximal positive level has no incoming edge from S. The edges of every given color by any coloring form a spanning subgraph and for any spanning subgraph there exists a corresponding coloring.*

Lemma 6 *Let N be a set of vertices of level n from some tree of the spanning subgraph S of AGW graph Γ. Then in a coloring of Γ where all edges of S have the same color α, any F-clique F satisfies $|F \cap N| \leq 1$.*

Proof. Some power of α synchronizes all states of given level of the tree and maps them into the root. Any couple of states from an F-clique could not be synchronized and therefore could not belong to N.

Lemma 7 *Let AGW graph Γ have a spanning subgraph R of only disjoint cycles (without trees). Then Γ also has another spanning subgraph with exactly one vertex of maximal positive level.*

Proof. The spanning subgraph R has only cycles and therefore the levels of all vertices are equal to zero. In view of $gcd = 1$ in the strongly connected graph Γ, not all edges belong to a bunch. Therefore there exist two edges $u = \mathbf{p} \to \mathbf{q} \notin R$ and $v = \mathbf{p} \to \mathbf{s} \in R$ with common first vertex $\mathbf{p}$ but such that $\mathbf{q} \neq \mathbf{s}$. Let us replace the edge $v = \mathbf{p} \to \mathbf{s}$ from R by u. Then only the vertex $\mathbf{s}$ has maximal level $L > 0$ in the new spanning subgraph.

Lemma 8 *Let any vertex of an AGW graph Γ have no two incoming bunches. Then Γ has a spanning subgraph such that all its vertices of maximal positive level belong to one non-trivial tree.*

Let us consider a spanning subgraph R with a maximal number of vertices [edges] in its cycles. In view of Lemma 7, suppose that R has non-trivial trees and let $L > 0$ be the maximal value of the level of a vertex.
Further consideration is necessary only if at least two vertices of level L belong to distinct trees of R with distinct roots.
Let us consider a tree T from R with vertex $\mathbf{p}$ of maximal level L and edge $\bar{b}$ from vertex $\mathbf{b}$ to the tree root $\mathbf{r} \in T$ on the path of length L from $\mathbf{p}$. Let the root $\mathbf{r}$ belong to the cycle H of R with the edge $\bar{c} = \mathbf{c} \to \mathbf{r} \in H$. There exists also an edge $\bar{a} = \mathbf{a} \to \mathbf{p}$ that does not belong to R because Γ is strongly connected and $\mathbf{p}$ has no incoming edge from R.

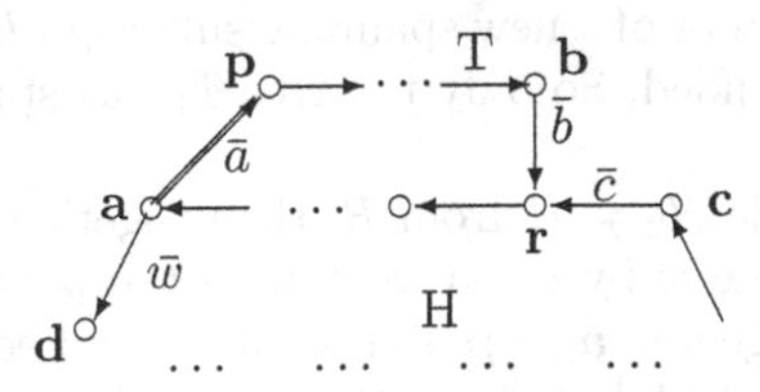

We consider the path in T from $\mathbf{p}$ to $\mathbf{r}$ of maximal length L. Our aim is to

extend the maximal level of the vertex on the extension of the tree T much more than the maximal level of vertex of other trees from R. We plan to use the following three changes:
1) replace the edge $\bar{w}$ from R with first vertex $\mathbf{a}$ by the edge $\bar{a} = \mathbf{a} \to \mathbf{p}$,
2) replace the edge $\bar{b}$ from R by some other outgoing edge of the vertex $\mathbf{b}$,
3) replace the edge $\bar{c}$ from R by some other outgoing edge of the vertex $\mathbf{c}$.
If one of the ways does not succeed let us go to the next assuming the situation in which the previous way fails and excluding the successfully studied cases. So we diminish the considered domain. We can use sometimes two changes together. Let us begin with
1) Suppose first $\mathbf{a} \notin H$. If $\mathbf{a}$ belongs to a path in T from $\mathbf{p}$ to $\mathbf{r}$ then a new cycle with part of the path and edge $\mathbf{a} \to \mathbf{p}$ is added to R extending the number of vertices in its cycles in spite of the choice of R. In opposite case the level of $\mathbf{a}$ in the new spanning subgraph is $L+1$ and the vertex $\mathbf{r}$ is a root of the new tree containing all vertices of maximal level (in particular, the vertex $\mathbf{a}$ or its ancestors in R).
So let us assume $\mathbf{a} \in H$ and suppose $\bar{w} = \mathbf{a} \to \mathbf{d} \in H$. In this case the vertices $\mathbf{p}$, $\mathbf{r}$ and $\mathbf{a}$ belong to a cycle H_1 with new edge $\bar{a}$ of a new spanning subgraph R_1. So we have the cycle $H_1 \in R_1$ instead of $H \in R$. If the length of path from $\mathbf{r}$ to $\mathbf{a}$ in H is r_1 then H_1 has length $L+r_1+1$. A path to $\mathbf{r}$ from the vertex $\mathbf{d}$ of the cycle H remains in R_1. Suppose its length is r_2. So the length of the cycle H is r_1+r_2+1. The length of the cycle H_1 is not greater than the length of H because the spanning subgraph R has maximal number of edges in its cycles. So $r_1+r_2+1 \geq L+r_1+1$, whence $r_2 \geq L$. If $r_2 > L$, then the length r_2 of the path from $\mathbf{d}$ to $\mathbf{r}$ in a tree of R_1 (and the level of $\mathbf{d}$) is greater than L and the level of $\mathbf{d}$ (or of some other ancestor of $\mathbf{r}$ in a tree from R) is the desired unique maximal level.
So assume for further consideration $L = r_2$ and $\mathbf{a} \in H$. Analogously, for any vertex of maximal level L with root in the cycle H and incoming edge from a vertex $\mathbf{a}_1$ the proof can be reduced to the case $\mathbf{a}_1 \in H$ and $L = r_2$ for the corresponding new value of r_2.
2) Suppose the set of outgoing edges of the vertex $\mathbf{b}$ is not a bunch. So one can replace in R the edge $\bar{b}$ from the vertex $\mathbf{b}$ by an edge $\bar{v}$ from $\mathbf{b}$ to a vertex $\mathbf{v} \neq \mathbf{r}$.
The vertex $\mathbf{v}$ could not belong to T because in this case a new cycle is added to R and therefore a new spanning subgraph has a number of vertices in the cycles greater than in R.
If the vertex $\mathbf{v}$ belongs to another tree of R but not to cycle, then T is a part of a new tree T_1 with a new root of a new spanning subgraph R_1 and the path from $\mathbf{p}$ to the new root is extended. So only the tree T_1 has states of new maximal level.
If $\mathbf{v}$ belongs to some cycle $H_2 \neq H$ from R, then together with replacing $\bar{b}$ by $\bar{v}$, we replace also the edge $\bar{w}$ by $\bar{a}$. So we extend the path from $\mathbf{p}$ to the new root $\mathbf{v}$ at least by the edge $\bar{a} = \mathbf{a} \to \mathbf{p}$ and by almost all edges of H. Therefore the new maximal level $L_1 > L$ has either the vertex $\mathbf{d}$ or its ancestors from the

old spanning subgraph R.
Now there remains only the case when $\mathbf{v}$ belongs to the cycle H. The vertex $\mathbf{p}$ also has level L in new tree T_1 with root $\mathbf{v}$. The only difference between T and T_1 (just as between R and R_1) is the root and the incoming edge of the root. The new spanning subgraph R_1 has also a maximal number of vertices in cycles just as R. Let r_3 be the length of the path from $\mathbf{d}$ to the new root $\mathbf{v} \in H$.
For the spanning subgraph R_1, one can obtain $L = r_3$ just as it was done on the step 1) for R. From $\mathbf{v} \neq \mathbf{r}$ follows $r_3 \neq r_2$, though $L = r_3$ and $L = r_2$.
So for further consideration suppose that the set of outgoing edges of the vertex $\mathbf{b}$ is a bunch to $\mathbf{r}$.
3) The set of outgoing edges of the vertex $\mathbf{c}$ is not a bunch to $\mathbf{r}$ because $\mathbf{r}$ has another bunch from $\mathbf{b}$.
Let us replace in R the edge $\bar{c}$ by an edge $\bar{u} = \mathbf{c} \to \mathbf{u}$ such that $\mathbf{u} \neq \mathbf{r}$. The vertex $\mathbf{u}$ could not belong to the tree T because in this case the cycle H is replaced by a cycle with all vertices from H and some vertices of T whence its length is greater than $|H|$. Therefore the new spanning subgraph has a number of vertices in its cycles greater than in spanning subgraph R in spite of the choice of R.

So remains the case $\mathbf{u} \notin T$. Then the tree T is a part of a new tree with a new root and the path from $\mathbf{p}$ to the new root is extended at least by a part of H from the former root $\mathbf{r}$. The new level of $\mathbf{p}$ therefore is maximal and greater than the level of any vertex in some another tree.
Thus anyway there exists a spanning subgraph with vertices of maximal level in one non-trivial tree.

Theorem 3 *Any AGW graph Γ has coloring with stable couples.*

Proof. By Lemma 5, in the case of vertex with two incoming bunches Γ has a coloring with stable couples. In opposite case, by Lemma 8, Γ has a spanning subgraph R such that the vertices of maximal positive level L belong to one tree of R.
Let us give to the edges of R the color α and denote by C the set of all vertices from the cycles of R. Then let us color the remaining edges of Γ by other colors arbitrarily.
By Lemma 3, in a strongly connected graph Γ for every word s and F-clique F of size $|F| > 1$, the set Fs also is an F-clique (of the same size by Lemma 2) and for any state $\mathbf{p}$ there exists an F-clique F such that $\mathbf{p} \in F$.
In particular, some F has non-empty intersection with the set N of vertices of maximal level L. The set N belongs to one tree, whence by Lemma 6 this intersection has only one vertex. The word α^{L-1} maps F on an F-clique F_1 of size $|F|$. One has $|F_1 \setminus C| = 1$ because the sequence of edges of color α from any tree of R leads to the root of the tree, the root belongs to a cycle colored by α from C and only for the set N with vertices of maximal level holds $N\alpha^{L-1} \not\subseteq C$.
So $|N\alpha^{L-1} \cap F_1| = |F_1 \setminus C| = 1$ and $|C \cap F_1| = |F_1| - 1$.
Let the integer m be a common multiple of the lengths of all considered cycles from C colored by α. So for any $\mathbf{p}$ from C as well as from $F_1 \cap C$ holds $\mathbf{p}\alpha^m = \mathbf{p}$.

Therefore for an F-clique $F_2 = F_1\alpha^m$ holds $F_2 \subseteq C$ and $C \cap F_1 = F_1 \cap F_2$. Thus two F-cliques F_1 and F_2 of size $|F_1| > 1$ have $|F_1| - 1$ common vertices. So $|F_1 \setminus (F_1 \cap F_2)| = 1$. Consequently, in view of Lemma 4, there exists a stable couple in the considered coloring.

Theorem 4 *Every AGW graph Γ has synchronizing coloring.*

The proof follows from Theorems 3 and 2.

3 The necessary and sufficient conditions of synchronizing coloring of an arbitrary graph

Theorem 5 *Let every vertex of strongly connected directed finite graph Γ have the same number of outgoing edges. Then Γ has synchronizing coloring if and only if the greatest common divisor of lengths of all its cycles is one.*

In view of Theorem 4, we must prove only the necessity of the condition on *gcd*.
Proof [1], [5].
Suppose $d > 1$ is the greatest common divisor of lengths of all cycles of Γ. Let us consider a tree T with root $\mathbf{p}$ and with all vertices of the graph. Suppose $t(\mathbf{p}) = 0$ and for every edge $\mathbf{r} \to \mathbf{q}$ of the tree suppose $t(\mathbf{r}) = t(\mathbf{q}) + 1$ (modulo d). So $t(\mathbf{q}) < d$ for every vertex $\mathbf{q}$.
Let the edge $\mathbf{u} \to \mathbf{v}$ be outside of T. If $t(\mathbf{u}) \neq t(\mathbf{v}) + 1$ (modulo d) then two paths from $\mathbf{p}$ to $\mathbf{v}$ through the edge $\mathbf{u} \to \mathbf{v}$ and the edges of T and through only the edges of T have not equal (modulo d) lengths. Therefore in strongly connected graph Γ there are two cycles having not equal lengths (modulo d). It contradicts to the choice of d as *gcd* of lengths of all cycles.
So for any edge $\mathbf{u} \to \mathbf{v}$ one has $t(\mathbf{u}) = t(\mathbf{v}) + 1$ (modulo d). Consequently by whatever coloring for any word s of the colors one has $t(\mathbf{u}s) = t(\mathbf{v}s) + 1$. So any word s could not unite $\mathbf{v}$ and $\mathbf{u}$, whence Γ has no synchronizing coloring.

Let us recall that the vertex $\mathbf{q}$ of the graph Γ is called a *sink* if there exists a way on Γ from any vertex to $\mathbf{q}$.

Theorem 6 *A finite directed graph Γ with constant outdegree of all its vertices has synchronizing coloring if and only if Γ has a sink and in the strongly connected component H of the sink the greatest common divisor of lengths of all cycles is one.*

Proof. The necessity of a sink is obvious, the necessity of conditions on H follows from Theorem 5 because any subgraph of Γ has synchronizing coloring.
Let us go to the sufficiency. There exists in Γ a tree T with root in sink. Let us give all edges from $T \setminus H$ common color α. Therefore the word α^i for some i maps Γ on H. So the proof is reduced to the conditions of Theorem 5.

References

1. R.L. Adler, L.W. Goodwyn, B. Weiss. *Equivalence of topological Markov shifts*, Israel J. of Math. 27, 49-63, 1977.
2. R.L. Adler, B. Weiss. *Similarity of automorphisms of the torus*, Memoirs of the Amer. Math. Soc. 98, Providence, RI, 1970.
3. G. Budzban, A. Mukherjea. *A semigroup approach to the Road Coloring Problem*, Probability on Algebraic Structures. Contemporary Mathematics, 261, 195-207, 2000.
4. A. Carbone. *Cycles of relatively prime length and the road coloring problem*, Israel J. of Math., 123, 303-316, 2001.
5. K. Culik II, J. Karhumaki, J. Kari. *A note on synchronized automata and Road Coloring Problem*, Developments in Language Theory (5th Int. Conf., Vienna, 2001), Lecture Notes in Computer Science, 2295, 175-185, 2002.
6. J. Friedman. *On the road coloring problem*, Proc. of the Amer. Math. Soc. 110, 1133-1135, 1990.
7. E. Gocka, W. Kirchherr, E. Schmeichel, *A note on the road-coloring conjecture*. Ars Combin. 49, 265-270, 1998.
8. R. Hegde, K. Jain, *Min-Max theorem about the Road Coloring Conjecture* EuroComb 2005, DMTCS proc., AE, 279-284, 2005.
9. N. Jonoska, S. Suen. *Monocyclic decomposition of graphs and the road coloring problem*, Congressum numerantium, 110, 201-209, 1995.
10. J. Kari. *Synchronizing finite automata on Eulerian digraphs*, Springer, Lect. Notes in Comp. Sci., 2136, 432-438, 2001.
11. P. Lankaster. *Theory of Matrices*, Acad. Press, NY - London, 1969.
12. D. Lind, B. Marcus. *An Introduction of Symbolic Dynamics and Coding*, Cambridge Univ. Press, 1995.
13. A. Mateescu, A. Salomaa, *Many-Valued Truth Functions, Černy's Conjecture and Road Coloring*, Bull. of European Ass. for TCS, 68, 134-148,1999.
14. G.L. O'Brien. *The road coloring problem*, Israel J. of Math., 39, 145-154, 1981.
15. D. Perrin, M.P. Schützenberger. *Synchronizing prefix codes and automata, and the road coloring problem*, In Symbolic Dynamics and Appl., Contemp. Math., 135, 295-318, 1992.
16. J.E. Pin. *On two combinatorial problems arising from automata theory*, Annals of Discrete Math., 17, 535-548, 1983.
17. A.N. Trahtman. *Notable trends concerning the synchronization of graphs and automata*, CTW06, El. Notes in Discrete Math., 25, 173-175, 2006.

Contributed talks

Leader Election in Anonymous Rings: Franklin Goes Probabilistic

Rena Bakhshi[1], Wan Fokkink[1,2], Jun Pang[3], and Jaco van de Pol[4]

[1] Vrije Universiteit Amsterdam, Department of Computer Science
rbakhshi@few.vu.nl, wanf@cs.vu.nl
[2] CWI, Amsterdam
[3] Université du Luxembourg, Faculté des Sciences, de la Technologie et de la Communication
jun.pang@uni.lu
[4] University of Twente, Department of Computer Science
j.c.vandepol@ewi.utwente.nl

Abstract. We present a probabilistic leader election algorithm for anonymous, bidirectional, asynchronous rings. It is based on an algorithm from Franklin [22], augmented with random identity selection, hop counters to detect identity clashes, and round numbers modulo 2. As a result, the algorithm is finite-state, so that various model checking techniques can be employed to verify its correctness, that is, eventually a unique leader is elected with probability one. We also sketch a formal correctness proof of the algorithm for rings with arbitrary size.

1 Introduction

Leader election is the problem of electing a unique leader in a distributed network. It is required that all processes execute the same local algorithm.[1] Leader election is a fundamental problem in distributed computing and has numerous applications. For example, it is an important tool for breaking symmetry in a distributed system. Moreover, by choosing a process as the leader, it is possible to execute centralized algorithms in a decentralized environment. Leader election can also be used to recover from token loss for token-based algorithms, by making the leader responsible for generating a new token when the current one is lost.

There is a broad range of leader election algorithms. These algorithms vary in communication mechanism (asynchronous vs. synchronous), process names (unique identities vs. an anonymous network), network topology (e.g. ring, acyclic graph, complete graph). Here we focus on asynchronous communication with reliable channels but no message order preservation, and a bidirectional ring topology.

A classic leader election algorithm for unidirectional rings was given by Chang and Roberts [12]. It requires that each process has a unique identity,

[1] Else, the problem would be trivial: let one process perform the event "leader", while all other processes perform the event "not leader".

Please use the following format when citing this chapter:

Bakhshi, R., et al., 2008, in IFIP International Federation for Information Processing, Volume 273; *Fifth IFIP International Conference on Theoretical Computer Science*; Giorgio Ausiello, Juhani Karhumäki, Giancarlo Mauri, Luke Ong; (Boston: Springer), pp. 57–72.

with a total ordering on identities; the process with the largest identity becomes the leader. The basic idea is that each process sends a message around the ring bearing its identity, where only the message with the largest identity completes the round trip. This algorithm requires $\mathcal{O}(n^2)$ messages in the worst case, but $\mathcal{O}(n \log n)$ on average. Franklin [22] developed a leader election algorithm for bidirectional rings with a worst-case message complexity of $\mathcal{O}(n \log n)$. The algorithm proceeds in election rounds, and each process is either active or passive. At the start of an election round, each active process sends its identity to its nearest active neighbours, and in return it receives such messages from these neighbours. An active process only progresses to the next election round if its own identity is larger than the two incoming identities. Peterson [42] and Dolev, Klawe and Rodeh [15] independently adapted Franklin's algorithm for unidirectional rings.

Sometimes the processes in a network cannot be distinguished by means of unique identities. Firstly, there is no concept of identity, e.g. Lego MindStorms robots. Secondly, as the number of processes in a network increases, it may become difficult to keep the identities of all processes distinct; or a network may accidentally assign the same identity to different processes. Thirdly, identities cannot always be sent around the network, for instance for reasons of efficiency; this is for instance the case in the leader election algorithm used within the IEEE 1394 (FireWire) standard, see [38]. In a so-called anonymous (or uniform) network, processes do not carry an identity. Angluin [1] showed that there does not exist a terminating deterministic algorithm for electing a leader in an anonymous, asynchronous network.

Itai and Rodeh [32, 33] studied how to break the symmetry in anonymous networks using probabilistic algorithms. They presented a probabilistic algorithm, based on the Chang-Roberts algorithm, to elect a leader in an anonymous unidirectional ring, under the assumption that all processes know the ring size.[2] At the start of an election round, active processes select a random identity from a finite domain, which they send around the ring. Active processes with the largest identity start a new election round if they detect a name clash, meaning that another process selected the same identity in the current round. Since the size of the ring is known, each process can recognise its own message by means of a hop counter, included in each message. The Itai-Rodeh algorithm terminates with probability one, and all its terminal states are correct, meaning that exactly one leader is elected. The average-case message complexity is $\mathcal{O}(n \log n)$.

In the Itai-Rodeh algorithm, an old message that has been overtaken by other messages in the ring, could in principle result in a situation where no leader is elected. To overcome this problem, successive election rounds are numbered, and each process and message is supplied with a round number. Thus an old message can be recognized and ignored. Fokkink and Pang [20, 21] showed that in case of FIFO channels, round numbers can be omitted from the Itai-Rodeh

---

[2] The latter assumption is essential; see e.g. [46, Sect. 9.4.1].

algorithm. They analysed the resulting algorithm using the probabilistic model checker PRISM [29].

In Sect. 2, we present a probabilistic leader election algorithm for anonymous bidirectional rings, based on Franklin's algorithm. As in the Itai-Rodeh algorithm, it is assumed that all processes know the ring size, and at the start of an election round, active processes select a random identity from a finite domain. We do not impose any assumption on the channel behaviour, i.e. the order of messages is not necessarily preserved between any pair of processes. Once again, each process can recognise its own message by means of a hop counter that is included in each message. However, instead of an infinite range of round numbers, we only need to keep track of round numbers modulo 2. This means that our probabilistic leader election algorithm is finite-state, and thus can be verified using explicit state space exploration (see Sect. 4). Furthermore, it implies that infinite executions, in which no leader is ever elected, violate "global fairness " (i.e., if in an infinite execution a transition from one global state of the system to another one $\gamma \rightarrow \gamma'$ can be taken infinitely often, then it is taken infinitely often); see Sect. 7.

We modelled our probabilistic version of Franklin's algorithm in the process algebraic language μCRL [8], and analysed for up to ring size six that a unique leader is elected. For ring size five, in case of a domain of three process identities, and for ring size six, in case of a domain of two process identities, we used the distributed version of the μCRL toolset [7] to store the generated state space over a cluster of computers. Moreover, we sketched a formal correctness proof for the algorithm in Sect. 4.

The model checker CADP [17] provided counter-examples to show that: (1) round numbers cannot be omitted from the probabilistic Franklin algorithm altogether (see Sect. 3), and (2) in case of a probabilistic version of the Dolev-Klawe-Rodeh algorithm, round numbers modulo 2 do not suffice (see Section 8). We used several optimizations, described in Sect. 5, to increase the efficiency of model checking of the probabilistic Franklin algorithm, notably confluence reduction. Moreover, using the probabilistic model checker PRISM, we made a performance comparison of two versions of the probabilistic Franklin algorithm: one in which fresh identities are chosen at the start of each election round, and one in which fresh identities are only chosen at the detection of an identity clash (see Sect. 6).

Related Work

Higham and Myers [28] present a leader election algorithm for anonymous, unidirectional rings of known size; their algorithm is similar to the algorithm of Itai and Rodeh, augmented with a time-out mechanism.

Fischer and Jiang [19] give a self-stabilizing leader election algorithm for anonymous, unidirectional rings, based on a *leader oracle* Ω?, which for some point onwards is guaranteed to return the same leader to all processes (see also Sect. 7).

Several papers [11, 30, 34, 18] present leader election algorithms for anonymous rings of prime size, in the presence of a central demon, which acts as a scheduler.[3]

Leader election is related to token circulation for solving mutual exclusion problem, where having a token is interpreted as a permission to enter the critical section. A self-stabilizing token circulation algorithm guarantees eventual circulation of a unique token, even if the system is started from a global state where several tokens are present. Israeli and Jalfon [31] propose a self-stabilizing token circulation algorithm in an anonymous, bidirectional, asynchronous ring, in the presence of a centralized demon. In their algorithm, tokens move to the left or to the right with probability $\frac{1}{2}$, and merge when they meet, eventually reducing the number of tokens to one. However, without knowledge of ring size, the processes can never be sure whether a single token is left.

Mayer, Ofek, Ostrovsky and Yung [40] show that on an anonymous ring, leader election is equivalent to providing a self-stabilizing round-robin token management scheme. Angluin, Aspnes, Fischer and Jiang [2] construct a self-stabilizing leader election algorithm for anonymous, unidirectional, asynchronous rings of odd size in the framework of their model of population protocols. Beauquier, Gradinariu and Johnen [5] present a randomized self-stabilizing leader election algorithm under an arbitrary scheduler (no fairness assumption is required) on anonymous, unidirectional rings of known size, in the shared variables model. Both algorithms are based on token circulation.

Several other papers present self-stabilizing token circulation algorithms for anonymous, unidirectional rings: the algorithm of Herman [27] works on synchronous rings of odd size; Duchon, Hanusse and Tixeuil [16] present algorithms for synchronous rings of arbitrary size; Beauquier, Gradinariu and Johnen [4] and Datta, Gradinariu and Tixeuil [13] use several types of tokens and assume the synchronous communication model of shared variables, while the algorithm of Rosaz [44] uses the same idea in asynchronous message passing systems; the algorithms of Kakugawa and Yamashita [36] for asynchronous rings and Johnen [35] for shared memory settings run under unfair distributed schedulers. Of these papers, [4, 36, 35] require knowledge of ring size.

Mayer, Ostrovsky and Yung [41] give a randomized compiler for anonymous rings that transforms a self-stabilizing algorithm based on bidirectional communication to one that requires unidirectional, synchronous communication.

[3] Dijkstra [14] noted that such a leader election algorithm cannot exist if the ring size is a composite number.

2 Franklin's Algorithm for Anonymous Rings

We consider a ring consisting of processes $p_0, \ldots, p_{n-1}$ for $n \geq 2$. Processes are anonymous, meaning that they do not carry a unique identity. Message-passing communication between processes is asynchronous, message order is not preserved between any pair of processes. Channels are bidirectional, so that a process p_i can send messages to its neighbours $p_{(i+1) \bmod n}$ and $p_{(i-1) \bmod n}$; a sent message is included in the message queue of its destination. It is assumed that receiving a message, processing it, and possibly sending a subsequent message take zero time. Channels are reliable, and the message queues are guided by a fair scheduler, meaning that every sent message will eventually be processed at its destination.

Each process is either active or passive. In our probabilistic version of Franklin's algorithm, an active process p_i maintains three parameters:

- $id_i \in \{1, \ldots, k\}$, for some $k \geq 2$, is its identity, not necessarily unique;
- $state_i$ ranges over $\{active, leader\}$;
- $bit_i \in \{T, F\}$ represents the number of the current election round modulo 2.

Passive processes simply pass on messages (increasing their hop counter by one).

All messages are of the form (*id*, *hop*, *bit*), travelling in both clockwise and counter-clockwise direction, where:

- *id* stores the identity of the process that originally sent the message;
- *bit* is a bit that represents the election round of this process modulo 2 (at the time that it sent the message);
- *hop* $\in \{1, \ldots, n\}$ is a counter, which initially has the value 1, and which is increased by one every time it is passed on by a process.

At the start of an election round, each active process p_i randomly selects an identity $id \in \{1, \ldots, k\}$, and sends a message with its identity to each of its two neighbours; initially, this message is of the form (id_i, 1, bit_i). Next, p_i receives such messages that originate from its two nearest active neighbours. Upon receipt of these messages, p_i determines whether it stays active for the next election round, by comparing three identities. If either of the messages it received has a larger identity than its own identity, then it becomes passive. Otherwise, it starts a new election round with a new identity. If a process gets a message with the hop counter equal to the network size n, the process becomes the leader ($state_i := leader$).

We now provide a more precise description of the algorithm. Initially, all processes p_i are active ($state_i = active$), and their bit bit_i is set to T.

- At the start of an election round with round number *bit*, an active process selects an identity $id \in \{1, \ldots, k\}$ and sends the message $(id, 1, bit)$ in both directions.
- Upon receipt of a message (id, hop, bit), a *passive* process passes on the message in the same direction, increasing the hop counter by one, i.e., $(id, hop + 1, bit)$.

- Upon receipt of a message (id, hop, bit) with $bit_i = bit$, an *active* process p_i executes the following steps:
  - if $hop = n$, then p_i becomes the leader ($state_i := leader$);
  - if $hop < n$, then p_i stores the message, and waits for a message with the bit bit_i from the opposite direction.
- An active process p_i stores messages that carry a bit $\neg bit_i$, to process them in the round with the appropriate bit.
- Upon receipt of messages with a bit bit_i from both directions, p_i checks whether either of these messages carries an identity larger than its own identity. If this is the case, then p_i becomes passive; otherwise, p_i starts a new election round, with an inverted bit as round number ($bit_i := \neg bit_i$) and a new identity.

3 Round Numbers Modulo 2 are Needed

Initially, we thought that our probabilistic version of the Franklin algorithm could maybe do without round numbers altogether. However, a model checking verification using the μCRL toolset [8] showed us that this is not true.[4] Fig. 3 shows a scenario where no leader is elected for a ring of size three and three identities. In this figure, black processes are active and white processes are passive.

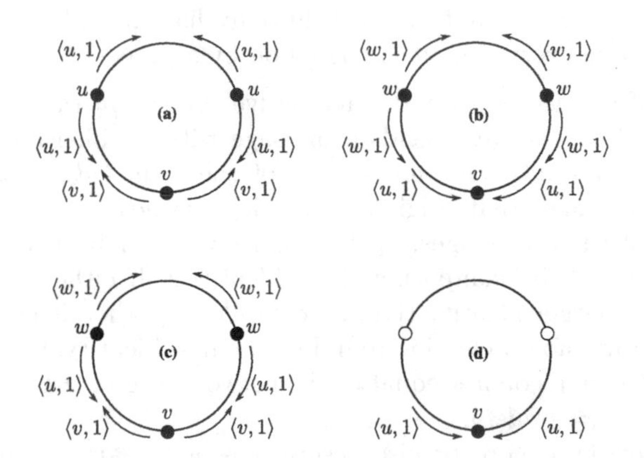

Fig. 1 Probabilistic Franklin algorithm without round numbers is flawed.

Initially, all processes are active; two processes select the same identity u, and one selects an identity $v < u$; all processes send a message with their

[4] Lamport [37] actually advocates that all distributed algorithms should be model checked before publication.

identity in both directions (Fig. 3(a)). At the receipt of a message from both neighbours, the processes with identity u select a new identity $w < v$, and send messages carrying this new identity (Fig. 3(b)). The two messages $(u, 1)$ are overtaken by two messages with identity w. As $w < v$, process v proceeds to a next election round, in which it selects the identity v again, and sends messages $(v, 1)$ in both direction (Fig. 3(c)). Upon the receipt of messages $(v, 1)$ and $(w, 1)$, the processes with identity w become passive (Fig. 3(d)). Finally, the outdated messages $(u, 1)$ make the process with identity v passive as well; all processes have become passive now.

4 Correctness Analysis

We say that an execution of the algorithm has *terminated* if each process is either passive or elected as the leader, and there are no remaining messages in the channels. We argue that the probabilistic Franklin algorithm for anonymous bidirectional rings terminates with probability one, and upon termination a unique leader has been elected.

For a start, we modelled the probabilistic Franklin algorithm with round numbers modulo 2 in the μCRL framework [8], with channels that have an unbounded capacity, each implemented as a buffer, and its correctness has been verified for rings with a size up to six processes. The input language for μCRL is based on process algebra and abstract data types. Our μCRL specification is available at [3]. Tables 1(a) and 1(b) provide state space generation results for domains of two and three identities, respectively. To carry out the verification for six processes (in case of two identities) and five processes (in case of three identities), a distributed version of the μCRL toolset was used. The resulting state space was reduced using branching bisimulation equivalence [23, 26], which eliminates internal and communication transitions (i.e., only "leader" transitions are not abstracted away), while maintaining the branching structure of the state space. In case of the distributed version of the μCRL toolset, we applied a distributed reduction algorithm from [9].

Table 1 State space generation statistics

(a) State space for two identities

# Procs	States	Transitions
2	657	1,368
3	15,445	43,968
4	380,609	1,396,512
5	9,819,065	44,242,920
6	260,753,105	1,393,967,976

(b) State space for three identities

# Procs	States	Transitions
2	1,525	3,564
3	55,009	168,102
4	2,095,777	8,182,092
5	84,381,157	401,681,445

The μCRL specification language does not allow to express probabilities. Still we could verify that although there are infinite executions, with probability one, eventually always a leader is elected. This is because branching bisimulation equivalence abstracts away from infinite executions that violate global fairness. That is, after minimization modulo this equivalence, such executions have been eliminated. The minimized state space of our algorithm consisted of only two states s_1 and s_2, where the initial state s_1 can perform a leader action to s_2, which is a terminated state.

We now sketch a formal correctness proof of the probabilistic Franklin algorithm.

Proposition 1. *If channels are FIFO, the probabilistic Franklin algorithm behaves correctly, even if processes and messages do not keep track of round numbers at all. That is, upon termination, exactly one leader has been elected.*

Proof. In case of FIFO channels, it is guaranteed that in each election round, an active process always receives messages from the left and the right that were created in this election round (cf. [20, 21]). Therefore round numbers are redundant.

In each election round, active processes with the largest identity in that round do not become passive. And an active process can only become the leader if all other processes have become passive. From this it follows that upon termination there is a unique leader. □

We now focus on showing that in the probabilistic Franklin algorithm, round numbers modulo 2 suffice to enforce FIFO behaviour of these channels.

Lemma 1. *After initialization, and before a leader is elected, the following invariant holds for the algorithm. Between each pair of active processes p, p' there are exactly two messages m, m'.*

- *If m, m' travel in opposite directions, p, p', m, m' all carry the same bit as round number.*
- *If m, m' travel in the same direction, p, p' have opposite bits, and m, m' have opposite bits.*

Proof. Fig. 2 depicts three cases (a symmetric variant of Fig. 2(a) is omitted) consisting of a triple of adjacent active processes, wherein the middle process with the bit $b \in \{T, F\}$ receives two incoming messages from its neighbours (Figs. 2(a), 2(d), 2(g)). If neither of the messages it received has a larger identity than its own identity, then it starts a new election round with the bit $\neg b$ (Figs. 2(b), 2(e), 2(h)). Otherwise, it becomes passive (Figs. 2(c), 2(f), 2(i)). In all six cases, the invariant holds. □

Theorem 1. *In the probabilistic Franklin algorithm (with reliable, but not necessarily FIFO channels), upon termination, exactly one leader has been elected.*

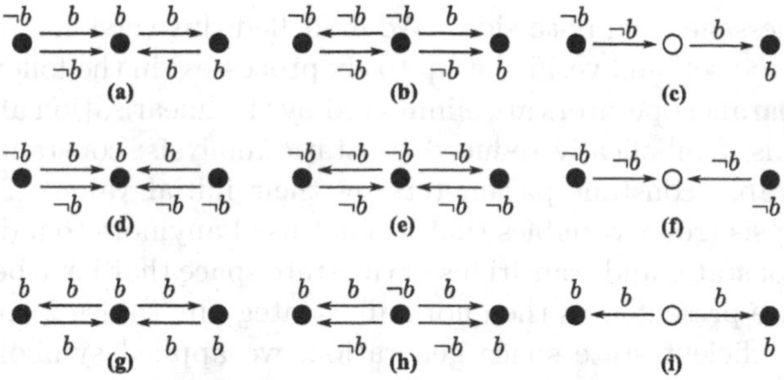

Fig. 2 Illustration of the invariant

Proof. From the invariant in Lemma 1 it follows that in the probabilistic Franklin algorithm with round numbers modulo 2, channels behave as FIFO queues. Namely, if there are two messages travelling to an active process in the same direction, they have opposite bits. So the active process can recognize which of these two messages was created in its current election round. Hence, the theorem follows from Prop. 1. □

Theorem 2. *The probabilistic Franklin algorithm terminates with probability one.*

Proof. When there are $\ell \geq 2$ active processes in the ring, these processes all remain active only if they all the time choose the same identity. Otherwise, at least one active process will become passive. The probability that all active processes select the same identity in one election round is $(\frac{1}{k})^{\ell-1}$, where k is the number of possible identities. Thus, the probability for all ℓ active processes to choose the same identity m times in a row is $(\frac{1}{k})^{m(\ell-1)}$. As $k \geq 2$, the probability that the number of active processes eventually decreases is one.

On average, the probabilistic Franklin algorithm takes $\mathcal{O}(n \log n)$ messages to terminate. (On average, in each election round about $\frac{3}{8}$ of the active processes become passive, so there are in the order of $\log n$ rounds; and each election round takes $2n - 2$ messages.)

5 Optimisation Techniques for Generating the State Space

To obtain a smaller state space, we simplified the algorithm described in Sect. 2: every round a node could decide to read always first a message from the left neighbour, and then from the right neighbour. This does not really influence the behaviour of the algorithm, because a node only take visible actions after receiving a message from both sides. We modelled this version of the algorithm in the μCRL toolset, and verified it up to six processes[5] in the following manner.

First the parallel operators are eliminated by the linearisation algorithm from [25]. Next it is symbolically reduced by static analysis: constant propagation (replace provably constant parameters by their initial value) [24], and dead variable analysis (reset variables that are not used anymore to a default value). The number of states and transitions of the state space that have been generated in this way are presented as the "normal" strategy in Tables 2 and 3.

For more efficient state space generation, we applied symbolic confluence reduction [10]. To this end, a theorem prover can be used to automatically detect and mark confluent τ's, i.e. internal transitions and hidden communications that are not causally related, for instance, because they occur at different parallel components. Confluence can then be exploited on a symbolic level by giving priority to confluent τ's, marked by the theorem prover. This reduction keeps only the confluent τ's going out of a state, and all the other transitions going out of the state are removed. This symbolic prioritization is implemented in the Confelm tool [6] from μCRL. We used it to remove confluent τ-summands, marked by the theorem prover Confcheck [43] from μCRL.

We also experimented with an on-the-fly τ-reduction [10, 39]. It is based on Tarjan's algorithm for decomposition of a graph into its strongly connected components [45]. In this reduction, for each state a representative state is computed, which it can reach by means of confluent τ-transitions. To compute the representative of a state, a depth-first search traversal via the confluent τ-transitions is made, until a state with a known representative is encountered, or a 'terminal' strongly connected component of confluent τ-transitions is found. (Terminal means that there are no outgoing confluent τ-transitions.) In the former case the known representative is returned, and in the latter case the state where the terminal strongly connected component was entered is returned. In the state space generation algorithm from [6], only representatives of states are generated.

After generation of the (partially reduced) state space, we performed a full reduction modulo branching bisimulation. The resulting state space consisted of only two states s_1 and s_2, where the initial state s_1 can perform a leader action to s_2, which is a terminated state.

Tables 2 and 3 show state space generation results of the simplified algorithm (states and transitions) for domains of two and three identities, respectively, with the different reduction strategies. For the on-the-fly τ-reduction, the number of states generated in the end (external states) differs from the number of states that are internally computed (internal states).

[5] A distributed version of the μCRL toolset [7] was used for six processes (in case of two identities) and five processes (in case of three identities) with the "normal" state space generation strategy.

Strategy	# Proc.	2	3	4	5	6
normal	s.	385	7,613	152,065	3,162,337	67,758,817
	t.	664	17,880	459,488	11,736,100	298,484,184
confelm	s.	205	2,875	40,881	606,783	9,280,633
	t.	340	6,342	114,384	2,069,040	37,381,488
confelm	ext. s.	165	1,819	21,409	263,963	3,348,345
+	int. s.	181	2,343	30,039	395,723	5,350,021
on-the-fly	t.	276	4,086	60,576	902,820	13,449,324

Table 2 State space for the probabilistic Franklin algorithm with 2 identities.

Strategy	# Proc.	2	3	4	5
normal	s.	877	26,299	802,489	25,919,965
	t.	1,680	65,853	2,560,848	100,868,445
confelm	s.	469	9,874	214,957	4,952,449
	t.	876	23,310	637,884	17,778,660
confelm	ext. s.	385	6,400	116,785	2,242,609
+	int. s.	433	8,518	170,131	3,524,305
on-the-fly	t.	732	15,570	353,508	8,137,080

Table 3 State space for the probabilistic Franklin algorithm with 3 identities

6 Performance Comparison with PRISM

In Sect. 2, we presented the probabilistic Franklin algorithm in which an active process chooses a fresh identity at the start of each election round (Algorithm A). There is one variant of this algorithm (Algorithm B) in which an active process only chooses a fresh identity at the start of a new election round if either of the two messages it received in the previous election round carried an identity equal to its own identity.

The probabilistic model checker PRISM [29] has the ability to automatically compute precise quantitative results based on exhaustive analysis of a formal model. For both versions, we used PRISM version 3.1.1 to calculate the probabilities of electing a unique leader within t "discrete time steps" (up to 150), where each such step corresponds to one transition in the algorithm. The experimental results presented in Fig. 3 indicate that Algorithm A has a much better performance than Algorithm B. Note that when t moves to infinity, both algorithms elect a unique leader with probability one.

7 Global Fairness

Fischer and Jiang [19] give a leader election algorithm for anonymous, unidirectional rings, without requiring global knowledge of ring size. Instead, they require a *leader oracle* $\Omega?$, which can by each process be asked who is the leader, and which for some point onwards is guaranteed to return the same answer to all processes. Under the assumption of what they call global fairness (i.e., if in an infinite execution a transition from one global state of the system to another

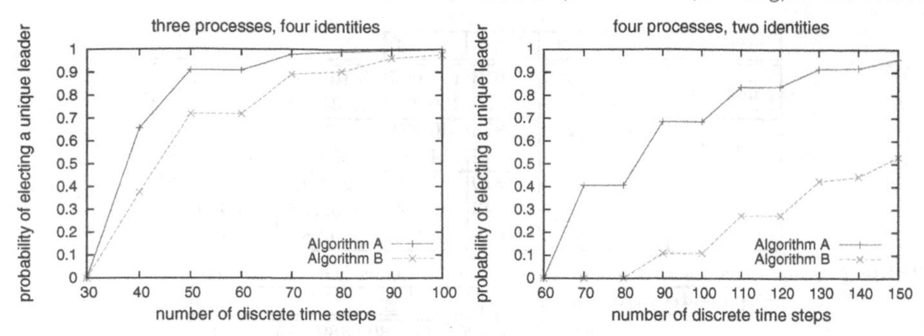

Fig. 3 The probability of electing a unique leader with deadlines.

one $\gamma \to \gamma'$ can be taken infinitely often, then it is taken infinitely often), they prove that their algorithm always terminates successfully.

Fischer and Jiang [19, p403] write: "We leave open the question of whether such an algorithm exists without the help of Ω?." Actually, in the absence of global knowledge of ring size, it is straightforward to provide a negative answer to this question. Namely, if such a leader election algorithm existed, then upon successful termination, the leader could start a traversal to determine the correct ring size. However, for anonymous rings, each ring size computation algorithm has a positive probability of computing the wrong ring size (see [46, Sect. 9.4.1]).

On the other hand, under the assumption of global knowledge of ring size, our probabilistic version of Franklin's algorithm provides a positive answer to the question of Fischer and Jiang, in the case of bidirectional rings. Namely, owing to the fact that our algorithm is finite-state, each globally fair infinite execution should at some point reach a configuration in which one active process selects a larger identity than all other active processes, meaning that the execution will terminate with this process as leader. But this contradicts with the fact that the execution is infinite. In other words, in our algorithm each infinite execution is not globally fair.

We note that this argumentation does not apply to the Itai-Rodeh algorithm, due to the presence of an infinite range of round numbers. As a consequence, in that algorithm no infinite execution visits a configuration infinitely often.

8 Probabilistic Dolev-Klawe-Rodeh Algorithm

The Dolev-Klawe-Rodeh algorithm is an adaptation of Franklin's algorithm to unidirectional rings. In an election round, an active process p compares its identity with the identities of the two closest active processes on its left. The process p proceeds to the next election round only if the identity of the closest active process on its left is the largest of these three identities. A natural question is whether the idea of round numbers modulo 2 would also apply to that algorithm. We therefore modelled a probabilistic version of the Dolev-Klawe-

Rodeh algorithm with round numbers modulo 2 in μCRL. We detected by a model checking analysis using μCRL and CADP toolsets that this algorithm is flawed, in the sense that no leader may be elected. This is due to the fact that in the probabilistic Dolev-Klawe-Rodeh algorithm, round numbers modulo 2 do not enforce FIFO behaviour of channels. This is depicted in Fig. 4.

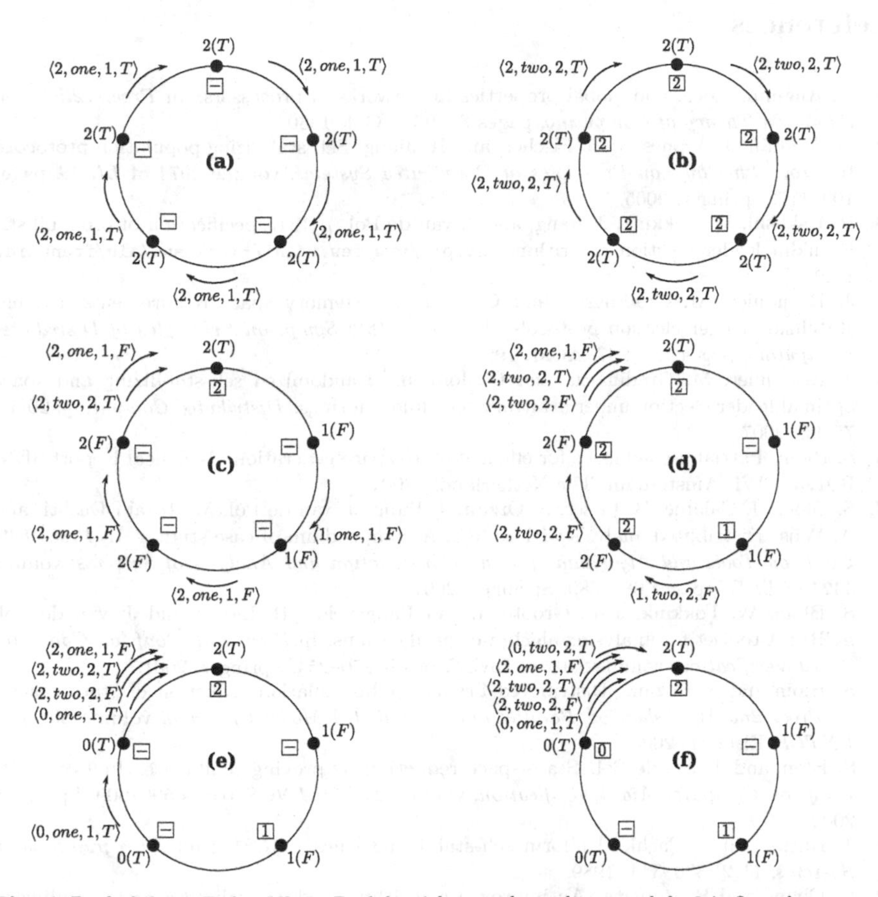

Fig. 4 Probabilistic Dolev-Klawe-Rodeh with round numbers modulo 2 is flawed.

In Fig. 4, a scenario is depicted in which a message $\langle 2, two, 2, T\rangle$ is overtaken by a newer message $\langle 0, two, 2, T\rangle$. In this picture, processes carry a round number modulo 2 (T or F). Moreover, messages carry a value (the first parameter), a hop counter (the third parameter), and a round number modulo 2 (the fourth parameter). The second parameter in a message, *one* or *two*, keeps track whether a message is travelling from its originator to the next active process, or has been forwarded by an active process, respectively.

Acknowledgement

We thank Bert Lisser for performing the experiments on the distributed version of the μCRL toolset.

References

1. D. Angluin. Local and global properties in networks of processors. In *Proc. 12th ACM Symp. on Theory of Computing*, pages 82–93. ACM, 1980.
2. D. Angluin, J. Aspnes, M. J. Fischer, and H. Jiang. Self-stabilizing population protocols. In *Proc. 9th Conf. on Principles of Distributed Systems*, volume 3974 of *LNCS*, pages 103–117. Springer, 2005.
3. R. Bakhshi, W. Fokkink, J. Pang, and J. van de Pol. μCRL specification of probabilistic Franklin leader election algorithm. `http://www.few.vu.nl/~rbakhshi/alg/franklin.mcrl`.
4. J. Beauquier, M. Gradinariu, and C. Johnen. Memory space requirements for self-stabilizing leader election protocols. In *Proc. 18th Symp. on Principles of Distributed Computing*, pages 199–207. ACM, 1999.
5. J. Beauquier, M. Gradinariu, and C. Johnen. Randomized self-stabilizing and space optimal leader election under arbitrary scheduler on rings. *Distributed Computing*, 20(1): 75–93, 2007.
6. S. Blom. Partial τ-confluence for efficient state space generation. Technical Report SEN-R0123, CWI, Amsterdam, The Netherlands, 2001.
7. S. Blom, J. Calamé, B. Lisser, S. Orzan, J. Pang, J. van de Pol, M. Torabi Dashti, and A. Wijs. Distributed analysis with μCRL: A compendium of case studies. In *Proc. 13th Conf. on Tools and Algorithms for the Construction and Analysis of Systems*, volume 4424 of *LNCS*, pages 683–689. Springer, 2007.
8. S. Blom, W. Fokkink, J.-F. Groote, I. van Langevelde, B. Lisser, and J. van de Pol. μCRL: A toolset for analysing algebraic specifications. In *Proc. 13th Conf. on Computer Aided Verification*, volume 2102 of *LNCS*, pages 250–254. Springer, 2001.
9. S. Blom and S. Orzan. Distributed branching bisimulation reduction of state spaces. In *Proc. 2nd Workshop on Parallel and Distributed Model Checking*, volume 89(1) of *ENTCS*. Elsevier, 2003.
10. S. Blom and J. van de Pol. State space reduction by proving confluence. In *Proc. 14th Conf. on Computer Aided Verification*, volume 2404 of *LNCS*, pages 596–609. Springer, 2002.
11. J. Burns and J. Pachl. Uniform self-stabilizing rings. *ACM Trans. Program. Lang. Systems*, 11(2):330–344, 1989.
12. E. Chang and R. Roberts. An improved algorithm for decentralized extrema-finding in circular configurations of processes. *Commun. ACM*, 22(5):281–283, 1979.
13. A. Kumar Datta, M. Gradinariu, and S. Tixeuil. Self-stabilizing mutual exclusion using unfair distributed scheduler. In *Proc. 14th Int. Parallel & Distributed Processing Symp.*, pages 465–470. IEEE Computer Society, 2000.
14. E. Dijkstra. Self-stabilizing systems in spite of distributed control. *Commun. ACM*, 17(11):643–644, 1974.
15. D. Dolev, M. Klawe, and M. Rodeh. An $O(n \log n)$ unidirectional algorithm for extrema finding in a circle. *J. of Algorithms*, 3(3):245–260, 1982.
16. P. Duchon, N. Hanusse, and S. Tixeuil. Optimal randomized self-stabilizing mutual exclusion in synchronous rings. In *Proc. 18th Symp. on Distributed Computing*, volume 3274 of *LNCS*, pages 216–229. Springer Verlag, 2004.
17. J.-C. Fernandez, H. Garavel, A. Kerbrat, L. Mounier, R. Mateescu, and M. Sighireanu. CADP - a protocol validation and verification toolbox. In *Proc. 8th Conf. on Computer Aided Verification*, volume 1102 of *LNCS*, pages 437–440. Springer, 1996.

18. F. Fich and C. Johnen. A space optimal, deterministic, self-stabilizing, leader election algorithm for unidirectional rings. In *Proc. 15th Conf. on Distributed Computing*, volume 2180 of *LNCS*, pages 224–239. Springer, 2001.
19. M. Fischer and H. Jiang. Self-stabilizing leader election in networks of finite-state anonymous agents. In *Proc. 10th Conf. on Principles of Distributed Systems*, volume 4305 of *LNCS*, pages 395–409. Springer, 2006.
20. W. Fokkink and J. Pang. Simplifying Itai-Rodeh leader election for anonymous rings. In *Proc. 4th Workshop on Automated Verification of Critical Systems*, volume 128(6) of *ENTCS*, pages 53–68. Elsevier, 2005.
21. W. Fokkink and J. Pang. Variations on Itai-Rodeh leader election for anonymous rings and their analysis in PRISM. *J. of Universal Computer Science*, 12(8):981–1006, 2006.
22. R. Franklin. On an improved algorithm for decentralized extrema finding in circular configurations of processors. *Commun. ACM*, 25(5):336–337, 1982.
23. R. van Glabbeek and P. Weijland. Branching time and abstraction in bisimulation semantics. *J. of the ACM*, 43(3):555–600, 1996.
24. J. F. Groote and B. Lisser. Computer assisted manipulation of algebraic process specifications. *SIGPLAN Notices*, 37(12):98–107, 2002.
25. J.-F. Groote, A. Ponse, and Y. Usenko. Linearization in parallel pcrl. *J. Log. Algebr. Program.*, 48(1-2):39–70, 2001.
26. J.-F. Groote and F. Vaandrager. An efficient algorithm for branching bisimulation and stuttering equivalence. In *Proc. 17th Colloq. on Automata, Languages and Programming*, volume 443 of *LNCS*, pages 626–638. Springer, 1990.
27. T. Herman. Probabilistic self-stabilization. *Inf. Process. Lett.*, 35(2):63–67, 1990.
28. L. Higham and S. Myers. Self-stabilizing token circulation on anonymous message passing. In *Proc. 2nd Conf. on Principles of Distributed Systems*, pages 115–128. Hermes, 1998.
29. A. Hinton, M. Kwiatkowska, G. Norman, and D. Parker. PRISM: A tool for automatic verification of probabilistic systems. In *Proc. 12th Conf. on Tools and Algorithms for the Construction and Analysis of Systems*, volume 3920 of *LNCS*, pages 441–444. Springer, 2006.
30. S.-T. Huang. Leader election in uniform rings. *ACM Trans. Program. Lang. Systems*, 15(3):563–573, 1993.
31. A. Israeli and M. Jalfon. Token management schemes and random walks yield self-stabilizing mutual exclusion. In *Proc. 9th ACM Symp. on Principles of Distributed Computing*, pages 119–131. ACM, 1990.
32. A. Itai and M. Rodeh. Symmetry breaking in distributive networks. In *Proc. 22nd Symp. on Foundations of Computer Science*, pages 150–158. IEEE, 1981.
33. A. Itai and M. Rodeh. Symmetry breaking in distributed networks. *Inf. Comput.*, 88(1):60–87, 1990.
34. G. Itkis, C. Lin, and J. Simon. Deterministic, constant space, self-stabilizing leader election on uniform rings. In *Proc. 9th Workshop on Distributed Algorithms*, volume 972 of *LNCS*, pages 288–302. Springer, 1995.
35. C. Johnen. Service time optimal self-stabilizing token circulation protocol on anonymous unidrectional rings. In *Proc. 21st Symp. on Reliable Distributed Systems*, pages 80–89. IEEE Computer Society, 2002.
36. H. Kakugawa and M. Yamashita. Uniform and self-stabilizing fair mutual exclusion on unidirectional rings under unfair distributed daemon. *J. Parallel Distrib. Comput.*, 62(5):885–898, 2002.
37. L. Lamport. Checking a multithreaded algorithm with $^{+}$CAL. In *Proc. 20th Symp. on Distributed Computing*, volume 4167 of *LNCS*, pages 151–163. Springer, 2006.
38. S. Maharaj and C. Shankland. A survey of formal methods applied to leader election in IEEE 1394. *J. of Universal Computer Science*, 6(11):1145–1163, 2000.
39. R. Mateescu. On-the-fly state space reductions for weak equivalences. In *Proc. 10th Workshop on Formal Methods for Industrial Critical Systems*, pages 80–89. ACM, 2005.

40. A. Mayer, Y. Ofek, R. Ostrovsky, and M. Yung. Self-stabilizing symmetry breaking in constant-space. In *Proc. 24th ACM Symp. on Theory of Computing*, pages 667–678. ACM, 1992.
41. A. Mayer, R. Ostrovsky, and M. Yung. Self-stabilizing algorithms for synchronous unidirectional rings. In *Proc. 7th ACM-SIAM Symp. on Discrete Algorithms*, pages 564–573. Society for Industrial and Applied Mathematics, 1996.
42. G. Peterson. An $O(n \log n)$ unidirectional algorithm for the circular extrema problem. *ACM Trans. Program. Lang. Systems*, 4(4):758–762, 1982.
43. J. van de Pol. A prover for the μCRL toolset with applications, version 0.1. Technical Report SEN-R0106, CWI, Amsterdam, The Netherlands, 2001.
44. L. Rosaz. Self-stabilizing token circulation on asynchronous uniform unidirectional rings. In *Proc. 19th ACM Symp. on Principles of Distributed Computing*, pages 249–258. ACM, 2000.
45. R. Tarjan. Depth-First Search and Linear Graph Algorithms. *SIAM J. on Computing*, 1(2):146–160, 1972.
46. G. Tel. *Introduction to Distributed Algorithms.* Cambridge University Press, 2000. 2nd edition.

Inverse Problems Have Inverse Complexity

Tobias Berg and Harald Hempel
Fakultät für Mathematik und Informatik
Friedrich-Schiller-Universität Jena
07740 Jena, Germany
tberg@minet.uni-jena.de
hempel@uni-jena.de

Abstract. In this paper we show that inverting problems of higher complexity is easier than inverting problems of lower complexity. While inverting Σ_i^p3CNFSAT is known to be coNP-complete [6] for $i = 1$ we prove that it remains coNP-complete for $i = 2$ and is in P for all $i \geq 3$. Relatedly, we show that inverting Σ_i^p3DNFSAT is in P for all $i \geq 1$.

1 Introduction

Do problems of higher complexity also always have inverse problems of higher complexity? We answer this question to the negative by showing that within the polynomial hierarchy complete problems from higher levels have easier inverse problems than those from lower levels. More precisely, we prove that while inverting Σ_i^p3CNFSAT is coNP-complete for $i = 1$ [6] it is also coNP-complete for $i = 2$, yet is in P for all $i \geq 3$. In contrast, inverting Σ_i^p3DNFSAT is easy, i.e., in P, for all $i \geq 1$.

Standard NP decision problems A are of the nature given an object x find out if there exists a proof for the membership of x in A. The inverse problem would then be given a set of proofs for membership in A does there exist an object x such the proofs for membership of x in A are exactly the given ones? For example, while the well known satisfiability problem SAT asks if a given Boolean formula has a satisfying assignment, the computational problem INVERSE SAT is defined as follows: Given a set of assignments does there exist a Boolean formula F such that the given assignments are exactly the satisfying assignments of F. While INVERSE SAT is (trivially) in P it has been shown that INVERSE 3SAT is coNP-complete [6]. Note that our proofs showing that inverting Σ_i^p3CNFSAT as well as inverting Σ_j^p3DNFSAT is in P for $i \geq 3$ and $j \geq 1$, respectively, are constructive. Hence, not only the decision if a formula F such that the given assignments are exactly the satisfying assignments of F exists but also actually finding F can be done in polynomial time.

In general, the study of inverse problems contributes to the field of identifying meaningful structures in data and efficient knowledge representation. Finding a computationally appealing representation for a given set of data can only be an easy problem if the corresponding inverse problem is easy. Furthermore, the

Please use the following format when citing this chapter:

Berg, T. and Hempel, H., 2008, in IFIP International Federation for Information Processing, Volume 273; *Fifth IFIP International Conference on Theoretical Computer Science*; Giorgio Ausiello, Juhani Karhumäki, Giancarlo Mauri, Luke Ong; (Boston: Springer), pp. 73–86.

study of inverse NP-problems may be helpful in gaining more insight into the nature of NP-completeness and may also be helpful in characterizing "natural" verifiers.

It has been shown that for many NP-complete problems inverting their natural verifier is coNP-complete [6, 2, 7]. However, the complexity of inverse problems in general heavily depends on the underlying verifier [2]. Formally, NP is the set of all languages A such that there exists a polynomial time computable 2-ary predicate V (also called a polynomial-time-verifier or NP-verifier) such that for all $x \in \Sigma^*$ we have $x \in A$ if and only if there exists a polynomial size bounded string π such that $(x, \pi) \in V$. The inverse NP-problem of A relative to V, INVS_V, is given a set of strings $\{\pi_1, \pi_2, \ldots, \pi_k\}$ does there exist a string x such that $\{\pi_1, \pi_2, \ldots, \pi_k\} = \{\pi : (x, \pi) \in V\}$? There are NP-complete problems that have NP-verifiers that can be inverted in P while the inversion of other of their NP-verifiers is Σ_2^p-complete [2]. Despite these results we feel that studying the inverse problems relative to the canonical (natural) NP-verifiers will give the true answer concerning the complexity of the inverse problems.

In this paper we study inverse problems from the classes Σ_i^p from the polynomial hierarchy thereby giving answers to some open questions posed in [2]. We introduce the notion of a verifier for the classes Σ_i^p and define the inverse problem for such verifiers. After giving upper bounds for the complexity of these inverse problems based on Σ_i^p-verifiers we study the inverse problem for some specific Σ_i^p-complete satisfiability problems such as Σ_i^p3DNFSAT and Σ_i^p3CNFSAT, yielding the above mentioned results.

We mention in passing that inverse NP-problems in a slightly different setting, namely in a setting where the solutions are not given explicitely as a list but implicitely in form of a boolean circuit accepting exactly those solutions have been studied in [4]. Also, lower and upper bounds for the inversion of RE problems have been found by the authors [1].

This paper is organized as follows: After formally introducing some notation and giving some remarks on previous results in Section 2, we will translate these concepts to problems from Σ_i^p in Section 3. In Section 3 we will also give an upper bound for inverting a reasonable restricted subset of verifiers for Σ_i^p-languages. We will furthermore examine the inverse complexity of some natural verifiers for specific Σ_i^p-complete satisfiability problems yielding the above mentioned results that are interesting beyond the scope of inverse problems.

2 Preliminaries

We assume the reader to be familiar with the basic concepts and notations of complexity theory (see [9, 5]).

Our alphabet will be $\Sigma = \{0, 1\}$. For a string $\alpha \in \Sigma^*$ let α^i denote the ith letter of α, i.e., $\alpha = \alpha^1\alpha^2\alpha^3 \ldots \alpha^{|\alpha|}$. As is standard in complexity theory an assignment for a Boolean formula F with variables $x_1, x_2, \ldots, x_n$ is a length n

string $\alpha = \alpha^1\alpha^2\alpha^3 \ldots \alpha^n$ which, for all $1 \leq i \leq n$, assigns the Boolean value α^i to the variable x_i. Recall that a 3CNF (3DNF) formula is a Boolean formula in conjunctive (disjunctive) normal form having exactly 3 literals per clause.

Verifiers, the language associated with a verifier, sets of proofs, and inverse problems relative to a given verifier can in general be defined as follows:

Definition 1. 1. A relation V is called a verifier if and only if $V \subseteq \Sigma^* \times \Sigma^*$.
2. For any verifier V and any string $x \in \Sigma^*$, the set of proofs for x with respect to V, short $V(x)$, is defined as

$$V(x) = \{\pi \in \Sigma^* : (x, \pi) \in V\}.$$

3. The language associated with V, $L(V)$, is defined as

$$L(V) = \{x \in \Sigma^* : V(x) \neq \emptyset\}.$$

4. The inverse problem relative to a verifier V, INVS_V, is defined as

$$\mathrm{INVS}_V = \{\Pi \subseteq \Sigma^* : (\exists x \in L(V))[V(x) = \Pi]\}.$$

One could also define the inverse problem as $\mathrm{INVS}_V = \{\Pi \subseteq \Sigma^* : (\exists x \in \Sigma^*)[V(x) = \Pi]\}$. However, this marginal change in definition – adding the empty set to the inverse problem – should not result in any differences regarding the complexity of both types of inverse problems. We take the freedom to sometimes write $V(x, \pi)$ instead of $(x, \pi) \in V$ for verifiers V and strings x and π.

The class NP can be viewed as the class of languages having polynomial-time verifiers.

Definition 2. A verifier V is called a polynomial-time verifier if and only if
1. $V \in \mathrm{P}$ and
2. there is a polynomial p such that for all $x, \pi \in \Sigma^*$, $(x, \pi) \in V \rightarrow |\pi| \leq p(|x|)$.

In this paper polynomial-time verifiers will also be called NP-verifiers. It is well-known that a language A is in NP if and only if there exists an NP-verifier V such that $A = L(V)$.

Inverse NP-problems are exactly the inverse problems relative to NP-verifiers and have been introduced in [2]. Clearly, it does not make sense to speak of inverting NP-problems without specifying the verifier. And in fact, inverting different NP-verifiers for one and the same NP-problem has different complexity. In [2] it has been shown that for every problem $A \in \mathrm{NP}$ there exists an NP-verifier V such that $L(V) = A$ and $\mathrm{INVS}_V \in P$. Here one can even show that there exists such a verifier that is fair [2].

Definition 3. [2] An NP-verifier V is called fair if and only if there exists a polynomial q such that for all $x \in L(V)$ there exists a string $x' \in L(V)$ such that $V(x) = V(x')$ and $|x'| \leq q(||V(x)||)$, where $||V(x)||$ denotes the length of the encoding of the set $V(x)$.

In contrast, it has been shown that several NP-problems have NP-verifiers such that the inverse problem relative to those verifiers is coNP-complete [6, 2, 7]. And it is also known that there is a tight Σ_2^p upper bound for inverting fair NP-verifiers [2], where Σ_2^p denotes the second level of the polynomial hierarchy.

Recall that for a complexity class $\mathcal{C}$ the classes $\mathrm{P}^{\mathcal{C}}$ and $\mathrm{NP}^{\mathcal{C}}$ are defined as the classes of languages that can be accepted by polynomial-time deterministic and nondeterministic, respectively, oracle Turing machines that make queries to a language from $\mathcal{C}$. Based on this concept the Σ_i^p levels of the polynomial-time hierarchy are defined as follows.

Definition 4. [8, 10] The complexity classes Σ_i^p are inductively defined via

1. $\Sigma_0^p = \mathrm{P}$ and
2. $\Sigma_{i+1}^p = \mathrm{NP}^{\Sigma_i^p}$ for all $i \geq 1$.

A useful characterization of the classes Σ_i^p was proven in [8].

Theorem 1. [8] *A language $A \subseteq \Sigma^*$ belongs to Σ_i^p if and only if there exists a predicate $V \in P$ and polynomials $p_1, ..., p_i$ such that for all $x \in \Sigma^*$ the following holds:*

$$x \in A \leftrightarrow (\exists y_1 \in \Sigma^*)(\forall y_2 \in \Sigma^*)(\exists y_3 \in \Sigma^*) \ldots (Qy_i \in \Sigma^*)[|y_1| \leq p_1(|x|) \wedge ... \wedge |y_i| \leq p_i(|x|) \wedge (x, y_1, ..., y_i) \in V].$$

If i is even then $Q = \forall$ and if i is odd then $Q = \exists$.

As we have pointed out earlier the complexity of inverse problems heavily depends on the underlying verifier. In order to study inverse NP-problems researchers have focused on inverting "natural" NP-verifiers, i.e., NP-verifiers that have proofs that closely reflect the canonic statement of the original NP-problem. For instance, in the case of SATISFIABILITY the most natural proof would be an assignment and a natural NP-verifier for SATISFIABILITY would be

$$V_{\mathrm{SAT}} = \{(F, \alpha) : F \text{ is a Boolean formula and } \alpha \text{ satisfies } F\}.$$

The first "natural" NP-verifiers have been studied in [6], where the complexity of inverting various syntactically constrained satisfiability problems has been studied. Following this line of research the coNP-completeness of the inverse problem (with respect to some "natural" NP-verifier) for some more NP-complete problems has been shown :

- 3SAT [6]
- CLIQUE, EXACT COVER, VERTEX COVER, SUBSET SUM (=KNAPSACK), STEINER TREE IN GRAPHS, PARTITION [2]
- HAMILTONIAN CIRCUIT, 3-D MATCHING [7]

For formal definition of these problems see [3].

3 The Inverse Problem for Σ_i^p

It has been suggested in [2] to examine the inverse problems for classes different than NP. In this section we will lay the ground for studying inverse Σ_i^p problems.

The class Σ_i^p is defined as the class of all languages that can be decided by a nondeterministic polynomial-time oracle Turing machine with queries to a Σ_{i-1}^p oracle, $\Sigma_i^p = \mathrm{NP}^{\Sigma_{i-1}^p}$. This leads to the following definition of a Σ_i^p-verifier.

Definition 5. A verifier V is called a Σ_i^p-verifier if and only if

1. $V \in \mathrm{P}^{\Sigma_{i-1}^p}$,
2. there exists a polynomial p such that for all $x, \pi \in \Sigma^*$, $(x, \pi) \in V \rightarrow |\pi| \leq p(|x|)$.

Observation 1 *For every language $A \subseteq \Sigma^*$, A is in Σ_i^p if and only if there exists a Σ_i^p-verifier V such that $L(V) = A$.*

Fair Σ_i^p-verifiers can be defined in analogy to Definition 3.

Definition 6. A Σ_i^p-verifier V is called a fair Σ_i^p-verifier if and only if there exists a polynomial q such that $(\forall x \in L(V))(\exists x' \in L(V))[V(x) = V(x') \wedge |x'| \leq q(||V(x)||)]$, where $||V(x)||$ denotes the length of the encoding of the set $V(x)$.

Informally, a Σ_i^p-verifier is called fair if for any set of proofs Π either

- there exists a polynomially length-bounded string (theorem) x' with exactly the proofs from Π or
- there exists no theorem with the set of proofs Π.

With this definitions in mind, what is an upper complexity bound for inverting a fair Σ_i^p-verifier?

Theorem 2. *If V is a fair Σ_i^p-verifier ($i \geq 1$), then* $\mathrm{INVS}_V \in \Sigma_{i+1}^p$.

Proof. The case $i = 1$ has been shown in [2]. So let $i \geq 2$ and let V be a fair Σ_i^p-verifier, i.e., $V \in \mathrm{P}^{\Sigma_{i-1}^p}$ and there exist two polynomials p and q such that

1. for all $x, \pi \in \Sigma^*$, $(x, \pi) \in V \rightarrow |\pi| \leq p(|x|)$
2. for all $x \in L(V)$ there exists $x' \in L(V)$ such that both $V(x) = V(x')$ and $|x'| \leq q(||V(x)||)$.

We define the following set A:

$$A = \{(\Pi, x) : \Pi \subseteq \Sigma^* \wedge x \in \Sigma^* \ \wedge \\ (\forall \pi \in \Sigma^* : \pi \leq p(|x|))[\pi \in V(x) \iff \pi \in \Pi]\}.$$

It is not hard to see that $A \in \Pi_i^p$ since $V \in \mathrm{P}^{\Sigma_{i-1}^p}$. Observe that the set A can be also written as $A = \{(\Pi, x) : \Pi \subseteq \Sigma^* \wedge x \in \Sigma^* \ \wedge \ V(x) = \Pi\}$. It follows that

$$\begin{aligned}\mathrm{INVS}_V &= \{\Pi \subseteq \Sigma^* : (\exists x \in \Sigma^*)[V(x) = \Pi]\} \\ &= \{\Pi \subseteq \Sigma^* : (\exists x \in \Sigma^* : |x| \leq q(||\Pi||))[V(x) = \Pi]\} \\ &= \{\Pi \subseteq \Sigma^* : (\exists x \in \Sigma^* : |x| \leq q(||\Pi||))[(\Pi, x) \in A]\}\end{aligned}$$

and thus $\mathrm{INVS}_V \in \Sigma^p_{i+1}$.

Even though inverting fair Σ^p_i-verifiers has, in general, an upper complexity bound of Σ^p_{i+1}, inversion of fair Σ^p_i-verifiers can be very easy in special cases.

Lemma 1. *For all $i \geq 1$ and all $B \in \Sigma^p_i$ there exists a fair Σ^p_i-verifier S such that* $\mathrm{INVS}_S \equiv^{\log}_m B$.

Proof. The proof is based on a proof given in [2]. Let B be a set from Σ^p_i and let R be a Σ^p_i-verifier such that $L(R) = B$. Consider the verifier S that is defined by $((x, \pi), x) \in S \leftrightarrow (x, \pi) \in R$ for all $x, \pi \in \Sigma^*$. Clearly, S is a Σ^p_i-verifier. It is straightforward to verify that S is also a fair Σ^p_i-verifier. Note that for all (x, π), the set $S((x, \pi))$ contains at most one proof, namely x itself.

We will now show that $x \in B \leftrightarrow \{x\} \in \mathrm{INVS}_S$ which yields the claim. First assume that $x \in B$ and thus there exists a certificate π such that $(x, \pi) \in R$ and thus $((x, \pi), x) \in S$. Since $S((x, \pi)) \subseteq \{x\}$ it follows that $S((x, \pi)) = \{x\}$. We conclude that $\{x\} \in \mathrm{INVS}_S$.

For the other direction assume that $x \notin B$ and hence for all $\pi \in \Sigma^*$ it holds that $(x, \pi) \notin R$ and thus $((x, \pi), x) \notin S$ for all certificates π. It follows that $S((x, \pi)) = \emptyset$ for all $\pi \in \Sigma^*$ which implies $\{x\} \notin \mathrm{INVS}_S$.

3.1 The Inverse Problem for Σ^p_i3CNFSAT

In the next two subsection we would like to examine the inverse complexity of natural verifiers for some selected complete problems in Σ^p_i. In particular we will look at the quantified versions of 3CNF-SAT and 3DNF-SAT and their natural verifiers.

Definition 7. An $i+1$-tuple $(F, X, Y_1, Y_2, \ldots, Y_{i-1})$ is called a type-i-formula if and only if F is a Boolean formula with variables from the set $X \cup Y_1 \cup \ldots \cup Y_{i-1}$ and $X, Y_1, \ldots, Y_{i-1}$ are pairwise disjoint sets. The set Σ^p_iSAT is defined as

$$\begin{aligned}\Sigma^p_i\mathrm{SAT} = \{\ (F, X, Y_1, \ldots, Y_{i-1}) : &\ (F, X, Y_1, \ldots, Y_{i-1}) \text{ is a type-}i\text{-formula} \quad \wedge \\ &(\exists \alpha \in \{0,1\}^{|X|})(\forall \beta_1 \in \{0,1\}^{|Y_1|}) \ldots \\ &(Q\, \beta_{i-1} \in \{0,1\}^{|Y_{i-1}|})[F(\alpha, \beta_1, \ldots, \beta_{i-1}) = 1]\}\end{aligned}$$

where $Q = \forall$ if i is even and $Q = \exists$ if i is odd. Here $F(\alpha, \beta_1, \ldots, \beta_{i-1})$ denotes the truth value of F when using α as a truth assignment for the variables of X and for all $1 \leq j \leq i-1$ using β_j as a truth assignment for the variables of Y_j.

It is well know that for all $i \geq 1$, the language Σ_i^pSAT is Σ_i^p-complete [10]. When restricting the formulas in Σ_i^pSAT to 3CNF or 3DNF formulas the set

$$\Sigma_i^p\text{3CNFSAT} = \{F : F \in \Sigma_i^p\text{SAT} \wedge F \text{ is a 3CNF-formula}\}$$

is Σ_i^p-complete for odd i's [10]. If i is even then the set

$$\Sigma_i^p\text{3DNFSAT}= \{F : F \in \Sigma_i^p\text{SAT} \wedge F \text{ is a 3DNF-formula}\}$$

is Σ_i^p-complete [10].

Let $i \geq 1$ be a natural number. The natural choice for a Σ_i^p-verifier for Σ_i^pSAT is certainly S_i, where

$$S_i(F,\alpha) \leftrightarrow (F, X, Y_1, \ldots, Y_{i-1}) \text{ is a type-}i\text{-formula} \wedge (\forall \beta_1 \in \{0,1\}^{|Y_1|}) \\ (\exists \beta_2 \in \{0,1\}^{|Y_2|}) \ldots (Q\, \beta_{i-1} \in \{0,1\}^{|Y_{i-1}|})[F(\alpha, \beta_1, ..., \beta_{i-1}) = 1]$$

and $Q = \forall$ if i is even and $Q = \exists$ if i is odd. Analogously, the natural verifiers for Σ_i^p3CNFSAT and Σ_i^p3DNFSAT are C_i and D_i, where

$$(F,\alpha) \in C_i \leftrightarrow F \text{ is a type-}i\text{-formula in 3CNF} \wedge (F,\alpha) \in S_i,$$
$$(F,\alpha) \in D_i \leftrightarrow F \text{ is a type-}i\text{-formula in 3DNF} \wedge (F,\alpha) \in S_i,$$

and $Q = \forall$ if i is even and $Q = \exists$ if i is odd.

In this subsection we will concentrate on the inverse problem for the verifier C_i. In the next subsection we will proof results for the verifier D_i.

Lemma 2. *For all $i \geq 1$ it holds that* $\text{INVS}_{C_i} \subseteq \text{INVS}_{C_{i+1}}$.

Proof. The proof is obvious, since every type-i-formula $(F, X, Y_1, ..., Y_{i-1})$ in 3CNF has exactly the same satisfying assignments as the type-$i+1$-formula $(F, X, Y_1, ..., Y_{i-1}, Y_i)$ in 3CNF, where $Y_i = \emptyset$.

Next we will show a partial converse to Lemma 2, i.e., that $\text{INVS}_{C_1} = \text{INVS}_{C_2}$ (Theorem 3). Before formally stating and proving the theorem we will recall some helpful concepts and prove some lemmata.

Definition 8. [6]

1. Let $\Pi \subseteq \{0,1\}^n$ be a set of Boolean vectors, $\Pi = \{\pi_1, \pi_2, \ldots, \pi_k\}$. We call a Boolean vector $m \in \{0,1\}^n$ 3-compatible with Π if for any triple of indices (i_1, i_2, i_3), $1 \leq i_1 \leq i_2 \leq i_3 \leq n$, there exists a vector $\pi_j \in \Pi$ such that $m^{i_1} = \pi_j^{i_1}$ and $m^{i_2} = \pi_j^{i_2}$ and $m^{i_3} = \pi_j^{i_3}$.
2. A set $\Pi \subseteq \{0,1\}^n$ of Boolean vectors is called a 3CNF-set if and only if there is a 3CNF formula F such that the set of satisfying assignments of F is equal to Π.

Informally put, a vector $m \in \{0,1\}^n$ is 3-compatible with a set of Boolean vectors Π if and only for any sequence of three bit positions there exists a string in Π that agrees with m in these three positions.

The following Lemma from [6] gives a very tight connection between the notions of 3-compatibility and 3CNF-sets, namely, a set of Boolean vectors is a 3CNF-set if and only if it is closed under 3-compatibility.

Lemma 3. [6] *Let $\Pi \subseteq \{0,1\}^n$ be a set of assignments. Then Π is a 3CNF-set if and only if for all $m \in \{0,1\}^n$ that are 3-compatible with Π we have $m \in \Pi$.*

As an easy example consider the set $\Pi := \{0111, 1011, 1101, 1110\}$. The Boolean vector 1111 is 3-compatible with Π. But since $1111 \notin \Pi$ we conclude by Lemma 3 that there can not exist a 3CNF-formula F with exactly the satisfying assignments from Π.

Lemma 4. *Let $\Pi \subseteq \{0,1\}^n$ be a 3CNF-set. For all i, $1 \leq i \leq n$, and all $c \in \{0,1\}$ the set*

$$Cut_c^i(\Pi) := \{ \alpha : \alpha \in \Pi \ \wedge \ \alpha^i = c \}$$

is a 3CNF-set.

Proof. Let $\Pi \subseteq \{0,1\}^n$ be a 3CNF-set, let $1 \leq i \leq n$, and $c \in \{0,1\}$. In order to show that $Cut_c^i(\Pi)$ is a 3CNF-set we use Lemma 3. We need to show that for all assignments α it holds that whenever α is 3-compatible with $Cut_c^i(\Pi)$ it also is an element of $Cut_c^i(\Pi)$. We will give a proof by contradiction.

So assume that $\alpha \in \{0,1\}^n$ is 3-compatible with $Cut_c^i(\Pi)$ yet $\alpha \notin Cut_c^i(\Pi)$. Hence $\alpha \notin \Pi$ or $\alpha^i \neq c$. We now argue that in both cases we have a contradiction. So assume that $\alpha \notin \Pi$. Since α is 3-compatible with $Cut_c^i(\Pi)$ it is also 3-compatible with any superset of $Cut_c^i(\Pi)$ and thus also 3-compatible with Π. However, by Lemma 3 we have that Π contains every assignment that is 3-compatible with Π, a contradiction. In case $\alpha^i \neq c$ we have an outright contradiction with the fact that α is 3-compatible with $Cut_c^i(\Pi)$. By definition for any three positions $1 \leq i_1 \leq i_2 \leq i_3 \leq n$ there exists a vector in $Cut_c^i(\Pi)$ that agrees with α in these three positions yet $\alpha^i \neq c$ and all vectors $\beta \in Cut_c^i(\Pi)$ satisfy $\beta^i = c$.

Lemma 5. *Let $\Pi \subseteq \{0,1\}^n$ be a 3CNF-set. For all $1 \leq i, j \leq n$ and all $c_1, c_2 \in \{0,1\}$ the set*

$$Cut_{c_1,c_2}^{i,j}(\Pi) := \{ \alpha : \alpha \in \Pi \ \wedge \ (\alpha^i = c_1 \ \vee \ \alpha^j = c_2) \}$$

is a 3CNF-set.

The proof is quite similar to the proof of Lemma 4 and thus omitted. We are now prepared to state and prove the main results of this section.

Theorem 3. $\mathrm{INVS}_{C_1} = \mathrm{INVS}_{C_2}$.

Proof. Due to Lemma 2 it suffices to show $\mathrm{INVS}_{C_2} \subseteq \mathrm{INVS}_{C_1}$.

Let $\Pi \in \mathrm{INVS}_{C_2}$. By definition of INVS_{C_2} we have $\Pi \neq \emptyset$ and there exists a type-2-formula (F, X, Y) in 3CNF over the variable set $X \cup Y$ of the form $F =$

$K_1 \wedge ... \wedge K_p$ where each K_i is a clause of the form $(z_1 \vee z_2 \vee z_3)$, $z_1, z_2, z_3 \in X \cup Y$, such that $C_2(F) = \Pi$. Recall that by definition of C_2 it holds that $(F, \alpha) \in C_2$ if and only if F is a type-2-formula in 3CNF and $(\forall \beta_1 \in \{0,1\}^{|Y|})F(\alpha, \beta) = 1$. In the remainder of this proof an assignment for a type-2-formula (F', X', Y') in 3CNF will be denoted by $\alpha\beta$, where α is the part of the assignment that assigns truth values to the variables from X' whereas β is the part of the assignment that assigns truth values to the variables from Y'.

We will now show that the set $C_2(F) = \Pi$ is itself a 3CNF-set and thus $\Pi \in \text{INVS}_{C_1}$. Observe that F does not contain a clause consisting solely of literals from the variable set Y since otherwise $C_2(F) = \Pi = \emptyset$, a contradiction. Hence, each clause of F contains at least one literal from the variable set X. We will construct a sequence of type-2-formulas in 3CNF $(F_0, X, Y), (F_1, X, Y), \ldots, (F_{n_1}, X, Y), (F_{n_1+1}, X, Y), \ldots, (F_{n_1+n_2}, X, Y)$ over the variable set $X \cup Y$ such that $C_2(F_{n_1+n_2}) = \Pi$. Indeed, we will prove by induction that for each $0 \leq i \leq n_1 + n_2$ the set $C_2(F_i)$ is a 3CNF-set.

Define F_0 to be the 3CNF formula that consists of all clauses from F that contain no literal from the variable set Y. Note that $C_2(F_0)$ is a 3CNF-set since F_0 is satisfied independent of assignments to the variables from Y. Let n_1 be the number of clauses in F that contain exactly one literal from the variable set Y. For each i, $0 \leq i \leq n_1 - 1$, let F_{i+1} be a type-2-formula in 3CNF such that $F_{i+1} = F_i \wedge K$ where K is a clause from F that contains exactly one literal from the variable set Y and K is not part of F_i. We will now argue that for all i, $1 \leq i \leq n_1$, $C_2(F_i)$ is a 3CNF-set. We will do this inductively. Recall that $C_2(F_0)$ is a 3CNF-set and assume that for some q, $1 \leq q \leq n_1$, $C_2(F_{q-1})$ is a 3CNF-set. Consider F_q. Let $F_q = F_{q-1} \wedge (\ell_i \vee \ell_j \vee \ell_k)$ where ℓ_i and ℓ_j are literals of the variables x_i and x_j, respectively, from X and ℓ_k is a literal of the variable y_k from Y. Observe that those assignments $\alpha\beta$ for F_q that (implicitly) assign the truth value 0 to ℓ_i, ℓ_j and ℓ_k can not satisfy F_q. It follows that no assignment α that assigns the truth value 0 to ℓ_i and ℓ_j can be in $C_2(F_q)$. On the other hand, any assignment from $C_2(F_{q-1})$ that assigns 1 to ℓ_i or ℓ_j or both is also in $C_2(F_q)$. Since trivially $C_2(F_{q-1}) \supseteq C_2(F_q)$ we have that $C_2(F_q) = Cut^{i,j}_{a,b}(C_2(F_{q-1}))$ where $a = 1$ if $\ell_i = x_i$ and $a = 0$ if $\ell_i = \overline{x_i}$ and similarly $b = 1$ if $\ell_j = x_j$ and $b = 0$ if $\ell_j = \overline{x_j}$. By Lemma 5 and the induction hypothesis we conclude that $C_2(F_q)$ is a 3CNF-set.

Let n_2 be the number of clauses in F that contain exactly two literals from the variable set Y. Similar to the above inductive argument related to clauses that contain exactly one literal from X one can easily show that $C_2(F_q) = Cut^i_a(C_2(F_{q-1}))$ for an appropriately chosen $a \in \{0,1\}$. By Lemma 4 and the induction hypothesis we have that $C_2(F_q)$ is a 3CNF-set.

To complete the proof observe that $F_q = F$ and thus $C_2(F) = \Pi$ is a 3CNF-set and hence $\Pi \in \text{INVS}_{C_1}$.

It has been shown in [6] that INVS_{C_1} is coNP-complete. Using the last theorem we have the following corollary.

Corollary 1. INVS_{C_2} *is* coNP-*complete.*

Next we will show that except the two coNP-complete problems INVS_{C_1} and INVS_{C_2} all other problems INVS_{C_i}, $i \geq 3$, are in P. We will do so by showing that for every syntactically correct set of assignments Π there exists a type-3-3CNF-formula with exactly the satisfying assignments from Π.

Theorem 4. *For all n and all $\Pi \subseteq \Sigma^n$ it holds that $\Pi \in \mathrm{INVS}_{C_3}$.*

Proof. Let $\Pi = \{\alpha_1, ..., \alpha_p\} \subseteq \Sigma^n$ for some $n \in \mathbb{N}$. In order to show $\Pi \in \mathrm{INVS}_{C_3}$ we will construct a type-3-3CNF-formula (F, X, Y_1, Y_2) over the variable sets X, Y_1, and Y_2 where $|X| = n$, $|Y_1| = 1$, and $|Y_2| = 2p + 1$ such that $C_3(F) = \Pi$.

We define an auxiliary set of assignments $\Pi' \subseteq \{0,1\}^{|X|+|Y_1|+|Y_2|}$ as follows:

$$\Pi' = \left\{ \begin{array}{l} \alpha_1\ 0\ 000...00001, \\ \alpha_1\ 1\ 000...00011, \\ \alpha_2\ 0\ 000...00111, \\ \alpha_2\ 1\ 000...01111, \\ \quad\vdots \\ \alpha_p\ 0\ 001...11111, \\ \alpha_p\ 1\ 011...11111 \end{array} \right\} .$$

Claim: Π' is a 3CNF-set.

Proof of Claim: According to Lemma 3 it suffices to show that every assignment $\gamma \in \Sigma^{n+2p+2}$ that is 3-compatible with Π' is also an element of Π'.

So let $\gamma \in \Sigma^{n+2p+2}$ be an assignment that is 3-compatible with Π'. Hence it holds for any three positions k_1, k_2, and k_3, $1 \leq k_1 \leq k_2 \leq k_3 \leq n + 2p + 2$, that γ agrees with some $\gamma' \in \Pi'$ at these three positions. Since all assignments in Π' have a 0 at position $n+2$ and 1 at position $n+2p+2$ and since γ due to its 3-compatibility with Π' agrees with some assignment from Π' in particular at positions $n + 2$, $n + 2p + 2$ and 1 it follows that $\gamma^{n+2} = 0$ and $\gamma^{n+2p+2} = 1$. Hence, there exists a position k, $n + 2 \leq k \leq n + 2p + 1$, such that $\gamma^k = 0$ and $\gamma^{k+1} = 1$. Furthermore, for all positions k', $1 \leq k' \leq n + 2p + 2$, there exists an assignment $\gamma' \in \Pi'$ such that γ and γ' are equal at the positions k, $k + 1$ and k'. However, there is only one assignment $\widehat{\gamma} \in \Pi'$ that has a 0 at position k and a 1 at position $k + 1$. Hence $\widehat{\gamma}$ and γ have to agree at all positions k', $1 \leq k' \leq n + 2p + 2$. It follows that $\gamma = \widehat{\gamma}$ and thus $\gamma \in \Pi'$. This concludes the proof of the claim.

By the claim there exists a 3CNF-formula F' for which Π' is exactly the set of its satisfying assignments, $C_1(F') = \Pi'$. Let (F, X, Y_1, Y_2) denote a type-3-3CNF-formula where $F = F'$ and X, Y_1, and Y_2 are the sets of variables that correspond to the first n, the $n + 1$st, and the last $2p + 1$ truth values in each assignment in Π'. Now it is immediate that $C_3(F) = \Pi$.

Corollary 2. *For all n and all $\Pi \subseteq \Sigma^n$ it holds that a type-3-3CNF-formula (F, X, Y_1, Y_2) such that $\Pi = C_3(F)$ can be constructed in time polynomial in the size of Π.*

The proof is immediate from the proof of Theorem 4 and the fact that given a set of assignments a so called candidate formula for that set of assignments can be constructed in polynomial time [2].

Note that INVS_{C_3} already contains all possible syntactically correct proof sets Π for C_3, i.e., all proof sets where all certificates have the same length. To decide if Π belongs to INVS_{C_3} one therefore simply has to test if all certificates of Π have the same length, which can be tested in polynomial time in the size of Π. By Lemma 2 we furthermore have that for all $i \geq 3$ it holds that $\mathrm{INVS}_{C_i} = \mathrm{INVS}_{C_3}$.

Corollary 3. *For all $i \in \mathbb{N}$, $i \geq 3$, it holds*

1. $\mathrm{INVS}_{C_i} = \mathrm{INVS}_{C_3} = \{\Pi \subseteq \{0,1\}^* : (\exists n \in \mathbb{N})[\Pi \subseteq \{0,1\}^n]\}$.
2. $\mathrm{INVS}_{C_i} \in \mathrm{P}$.

Summarizing the results from this section, we can state that the inverse problems for the languages Σ_i^p3CNFSAT (based on their natural verifiers C_i) become easier with growing i.

3.2 The Inverse Problem for Σ_i^p3DNFSAT

In this subsection we will focus on the inverse problems related to Σ_i^p3DNFSAT as defined in Subsection 3.1.

We start by examining the problem Σ_1^p3DNFSAT. Note that Σ_1^p3DNFSAT belongs to P. Despite the fact that members of languages from P do not need any certificate, we feel that the verifier D_1 as defined in Section 3.1 is a natural verifier for Σ_1^p3DNFSAT. However, it is not immediately clear, that INVS_{D_1} is in P as well.

Theorem 5. $\mathrm{INVS}_{D_1} \in P$.

Proof. Let us first take a look at the structure of the proof set Π for a 3DNF-formula F. Let $F = \mathcal{M}_1 \vee ... \vee \mathcal{M}_m$ be a 3DNF-formula over the variable set $X = \{x_1, ..., x_n\}$ consisting of 3-monomials $\mathcal{M}_1, \ldots, \mathcal{M}_m$. If $\mathcal{M} = (\ell_i \wedge \ell_j \wedge \ell_k)$, where $1 \leq i < j < k \leq n$ and either $\ell_t = x_t$ or $\ell_t = \overline{x_t}$ for all $t \in \{i, j, k\}$, is a monomial of the formula F then all assignments $\alpha \in \{0,1\}^n$ that assign the truth value 1 to ℓ_i, ℓ_j, and ℓ_k are satisfying assignments for the monomial $\mathcal{M}$. We denote the set of assignments for the formula F that satisfy the monomial $\mathcal{M}$ by $\Pi_{\mathcal{M}}$, i.e.,

$$\Pi_{\mathcal{M}} = \{\alpha \in \{0,1\}^n : \alpha \text{ as an assignment for } F \text{ satisfies } \mathcal{M}\}.$$

It is obvious that for the set of satisfying assignments Π of the formula F we have $\Pi = \Pi_{\mathcal{M}_1} \cup ... \cup \Pi_{\mathcal{M}_m}$.

In order to decide if a given set of assignments Π is contained in INVS_{D_1} we have to test if there exist 3-monomials $\mathcal{M}_1, ..., \mathcal{M}_m$ such that $\Pi = \Pi_{\mathcal{M}_1} \cup$

$\ldots \cup \Pi_{\mathcal{M}_m}$. A deterministic polynomial-time algorithm for this decision problem works as follows: On input $\Pi \subseteq \{0,1\}^*$ test if there exists a natural number n such that $\Pi \subseteq \{0,1\}^n$. If so continue and otherwise reject the input Π. Next, test for each of the $8\binom{n}{3}$ possible 3-monomials $\mathcal{M}$ over the variable set $X = \{x_1, ..., x_n\}$ if $\Pi_{\mathcal{M}} \subseteq \Pi$. In case $\Pi_{\mathcal{M}} \subseteq \Pi$ mark all those assignments α in Π that are contained in $\Pi_{\mathcal{M}}$, otherwise continue with the next monomial. As a final step, check if there are unmarked assignments in Π and accept if this is not the case and reject otherwise.

Note that this algorithm runs in polynomial time in the size of Π. If all assignments are marked in the final stage of the algorithm it is immediate that the formula F, consisting of all 3-monomials $\mathcal{M}$ satisfying $\Pi_{\mathcal{M}} \subseteq \Pi$, has exactly the satisfying assignments from Π. If there is an unmarked assignment in Π then there exists no 3DNF-formula F such that $D_1(F) = \Pi$. This is since any unmarked assignment in Π has to be the satisfying assignment for a 3-monomial $\mathcal{M}$ that has additional assignments not contained in Π. This procedure can be accomplished in polynomial time in the size of Π.

A close look at the proof of Theorem 5 reveals that following corollary holds.

Corollary 4. *For all n and all $\Pi \subseteq \Sigma^n$ it holds that a 3DNF-formula F over n variables such that $\Pi = D_1(F)$ can be constructed in time polynomial in the size of Π.*

Regarding INVS_{D_i} for $i \geq 1$ we can in analogy to Lemma 2 state the following.

Lemma 6. *For all $i \geq 1$ it holds that* $\mathrm{INVS}_{D_i} \subseteq \mathrm{INVS}_{D_{i+1}}$.

Next we will introduce the main idea used in the proof of Theorem 6 at an easy example, otherwise the proof of Theorem 6 would become slightly intricate.

Lemma 7. *For all $\Pi \subseteq \{0,1\}^*$ with $|\Pi| = 1$ it holds that $\Pi \in \mathrm{INVS}_{D_2}$.*

Proof. Let $\Pi \subseteq \{0,1\}^*$ such that $|\Pi| = 1$ and let $\alpha \in \{0,1\}^n$ denote the single string contained in Π, i.e., $\Pi = \{\alpha\}$.

We will define a Σ_2^p-3DNF-formula (F, X, Y) with $D_2(F) = \{\alpha\}$, $X = \{x_1, x_2, \ldots, x_n\}$, and $Y = \{y_1, y_2, \ldots, y_{n-3}\}$. The formula F is defined as

$$
\begin{aligned}
F = (x_1^{\alpha^1} \wedge x_2^{\alpha^2} \wedge y_1) \vee &\\
& (\overline{y_1} \wedge x_3^{\alpha^3} \wedge y_2) \vee \\
& \quad (\overline{y_2} \wedge x_4^{\alpha^4} \wedge y_3) \vee \\
& \quad\quad \ldots \vee \\
& \quad\quad\quad (\overline{y_{n-4}} \wedge x_{n-2}^{\alpha^{n-2}} \wedge y_{n-3}) \vee \\
& \quad\quad\quad\quad (\overline{y_{n-3}} \wedge x_{n-1}^{\alpha^{n-1}} \wedge x_n^{\alpha^n}),
\end{aligned}
$$

where for any variable z, $z^0 = \overline{z}$ and $z^1 = z$. It remains to show that $D_2(F) = \{\alpha\}$.

First, observe that when assigning α to the variables from X the formula F is satisfied independent of the assignment of the variables from Y. Second, let $\alpha' \in \{0,1\}^n$, $\alpha \neq \alpha'$, be an assignment for the variables from X. Since $\alpha \neq \alpha'$ there exists $1 \leq i \leq n$ such that $\alpha^i \neq \alpha'^i$. However, it follows that the assignment $\alpha'\beta$, where $\beta^j = 0$ for all j smaller than $i-2$ and $\beta^j = 1$ for all other j, does not satisfy F.

This shows that the only assignment for the variables of X such that for all assignments of the variables from Y the formula F is satisfied is indeed α.

The main idea of the proof of Lemma 7 can be also used to prove the main result of this section. Similar to Theorem 4 we have that all syntactically correct set of proof are contained in INVS_{D_2}.

Theorem 6. *For all n and all $\Pi \subseteq \{0,1\}^n$ it holds that $\Pi \in \text{INVS}_{D_2}$.*

Proof. Let $\Pi \subseteq \{0,1\}^n$, $\Pi = \{\alpha_1, \alpha_2, \ldots, \alpha_k\}$. Just as in the proof of Lemma 7 we will construct a formula F such that $D_2(F) = \Pi$. Informally, the formula F will consist of k subformulas in 3DNF $F_1, F_2, \ldots, F_k$ such that for all $1 \leq i \leq k$, $D_2(F_i) = \{\alpha_i\}$.

Let $X = \{x_1, x_2, \ldots, x_n\}$ and $Y = \{y_1, y_2, \ldots, y_{n-3}\}$ be disjoint sets. For each i, $1 \leq i \leq k$ we define a Σ_2^p-3DNF-formula (F_i, X, Y) as follows:

$$F_i = (x_1^{\alpha_i^1} \wedge x_2^{\alpha_i^2} \wedge y_1) \vee (\overline{y_1} \wedge x_3^{\alpha_i^3} \wedge y_2) \vee \ldots \vee \overline{y_{n-3}} \wedge x_{n-1}^{\alpha_i^{n-1}} \wedge x_n^{\alpha_i^n}).$$

The Σ_2^p-3DNF-formula (F, X, Y) is defined via $F = F_1 \vee F_2 \vee \ldots \vee F_k$. It follows from the proof of Lemma 7 that for each i, $1 \leq i \leq k$, there is exactly one assignment α for the variables of X, namely α_i, such that for all assignments β for the variables of Y, we have that $F_i(\alpha, \beta)$ is satisfied. It follows that $D_2(F) = \Pi$.

The proof of Theorem 6 contains an algorithm that given a set of assignments $\Pi \subseteq \Sigma^n$ constructs a type-2-3DNF-formula (F, X, Y) such that $D_2(F) = \Pi$. We hence have the following corollary.

Corollary 5. *For all n and all $\Pi \subseteq \Sigma^n$ it holds that a type-2-3DNF-formula (F, X, Y) such that $\Pi = D_2(F)$ can be constructed in time polynomial in the size of Π.*

In light of Lemma 6 we also have

Corollary 6. $\text{INVS}_{D_i} \in \text{P}$ *for all $i \geq 1$.*

Rounding off this section, we recall the verifier S_i ($i \in \mathbb{N}$) defined as

$$S_i(F,\alpha) \leftrightarrow (F,X,Y_1,\ldots,Y_{i-1}) \text{ is a type-}i\text{-formula} \wedge (\forall \beta_1 \in \{0,1\}^{|Y_1|}) \\ (\exists \beta_2 \in \{0,1\}^{|Y_2|}) \ldots (Q\,\beta_{i-1} \in \{0,1\}^{|Y_{i-1}|})[F(\alpha,\beta_1,...,\beta_{i-1}) = 1]$$

where $Q = \forall$ if i is even and $Q = \exists$ if i is odd. It can be seen as the natural verifier for the language Σ_i^pSAT.

Corollary 7. $\mathrm{INVS}_{S_i} \in \mathrm{P}$ *for all* $i \geq 1$.

The corollary again is obvious since every type-i-3DNF-formula is a type-i-formula.

Acknowledgments The authors would like to thank the anonymous referees for their very helpful comments.

References

1. Tobias Berg and Harald Hempel. Inverse problems have inverse complexity, 2008. Technical Report, FSU Jena.
2. Hubie Chen. Inverse NP problems. In *Mathematical foundations of computer science 2003*, volume 2747 of *Lecture Notes in Comput. Sci.*, pages 338–347. Springer, Berlin, 2003.
3. Michael R. Garey and David S. Johnson. *Computers and intractability.* W. H. Freeman and Co., San Francisco, Calif., 1979. A guide to the theory of NP-completeness, A Series of Books in the Mathematical Sciences.
4. Edith Hemaspaandra, Lane A. Hemaspaandra, and Harald Hempel. All superlinear inverse schemes are coNP-hard. *Theoret. Comput. Sci.*, 345(2-3):345–358, 2005.
5. Lane A. Hemaspaandra and Mitsunori Ogihara. *The complexity theory companion.* Texts in Theoretical Computer Science. An EATCS Series. Springer-Verlag, Berlin, 2002.
6. Dimitris Kavvadias and Martha Sideri. The inverse satisfiability problem. *SIAM J. Comput.*, 28(1):152–163 (electronic), 1999.
7. Michael Krüger and Harald Hempel. Inverse Hamiltonian cycle and inverse 3-D matching are coNP-complete. In *Algorithms and computation*, volume 4288 of *Lecture Notes in Comput. Sci.*, pages 243–252. Springer, Berlin, 2006.
8. Albert R. Meyer and Larry J. Stockmeyer. The equivalence problem for regular expressions with squaring requires exponential space. In *FOCS*, pages 125–129. IEEE, 1972.
9. Christos H. Papadimitriou. *Computational complexity.* Addison-Wesley Publishing Company, Reading, MA, 1994.
10. Larry J. Stockmeyer. The polynomial-time hierarchy. *Theoret. Comput. Sci.*, 3(1):1–22 (1977), 1976.

Literal Shuffle of Compressed Words

Alberto Bertoni[1], Christian Choffrut[2], and Roberto Radicioni[1]

[1] Dip. di Scienze dell'Informazione, Università degli Studi di Milano,
Via Comelico 39/41, 20135 Milano - Italy
{bertoni,radicioni}@dsi.unimi.it

[2] L.I.A.F.A. (Laboratoire d'Informatique Algorithmique,
Fondements et Applications),
Université Paris VII, 2 Place Jussieu, 75221 Paris - France
Christian.Choffrut@liafa.jussieu.fr

Abstract. Straight-Line Programs (SLP) are widely used compressed representations of words. In this work we study the rational transformations and the literal shuffle of words compressed via SLP, proving that the first preserves the compression rate, while the second does not. As a consequence, we prove a tight bound for the descriptional complexity of 2D texts compressed via SLP. Finally, we observe that the Pattern Matching Problem for texts expressed by the literal shuffle of compressed words is NP-complete. However, we present a parameter-tractable algorithm for this problem, working in polynomial time whenever the length of the pattern is polynomially related to that of the text.

1 Introduction

Straight-line programs (SLP) are a widely accepted representation of compressed texts (see, for instance, [16, 13, 12, 11]). A SLP is a grammar in Chomsky Normal Form generating only one word; the grammar can be seen as a compressed representation of the word. Such a representation suggests a natural measure of descriptional complexity for a word, consisting of the SLP of smallest size that generates it. The compression rate of SLPs is comparable to that of Lempel-Ziv factorization. Indeed, given the LZ-encoding of a word, it is possible to obtain a SLP of the same compressed size, up to a log factor, that generates the same word ([17]).

Since the output size of these compression techniques could be logarithmic with respect to the length of the generated word, it is useful to design algorithms for problems on compressed texts without full unpacking. Generally, in this context, grammar compression is more convenient than LZ-factorization. For some problems, such as Equality and Pattern Matching with grammar compressed words as input, polynomial time algorithms have been found ([15, 9]); for other problems, the compressed version becomes NP-hard (for instance, computing Hamming distance, as proved in [9]).

In this work, we consider some operations on strings and study the problem of implementing such operations in compressed representations. In particular, we consider the rational transformations and the literal shuffle. The literal shuffle

---

Please use the following format when citing this chapter:

Bertoni, A., Choffrut, C. and Radicioni, R., 2008, in IFIP International Federation for Information Processing, Volume 273; *Fifth IFIP International Conference on Theoretical Computer Science*; Giorgio Ausiello, Juhani Karhumäki, Giancarlo Mauri, Luke Ong; (Boston: Springer), pp. 87–100.

consists of merging two words of equal length from left to right alternating exactly one symbol of the first word and one of the second (for example the literal shuffle of “lug” and “one” is “lounge”). The “inverse operation” R (L) consists of selecting the subword composed by the symbols in odd (even, respectively) position in the input word.

These operations play an important role in Cooley-Tukey Algorithm for the fast computation of the Discrete Fourier Transform [3]. This technique is based on a Divide and Conquer strategy which recursively breaks up a string by using R and L operations, while the merging phase consists of applying the literal shuffle to the partial solutions. A natural question is whether it is possible to apply this technique in a compressed context, that is, to execute efficiently R, L and the literal shuffle on grammar compressed strings.

First of all, we prove that rational transformations preserve the compression rate, while, in general, this does not hold for literal shuffle. This fact is proved by exploiting a construction that relates the circuital complexity of boolean functions with the descriptional complexity defined in terms of SLPs.

This result is then applied to compressed pictures. 2D-texts can be compressed by using a 2D version of SLPs. The structure of 2D SLPs is more complex than that of SLPs. Indeed, it is known that, while factors of logarithmically compressible words are still logarithmically compressible, this does not hold for 2D texts. In particular, there exists an infinite number of logarithmically compressible pictures having at least one section (row or column) not logarithmically compressible ([2]). We obtain a bound for the descriptional complexity of the sections of a compressed picture which depends on their position in the picture. Such a bound is proved to be tight, in some sense.

Finally, we study the problem of deciding whether a word is a factor of a text, where both the word and the text are represented by the literal shuffle of compressed words given as input. We prove that the problem is NP-complete also if the word to be searched for is 11. However, we present an algorithm working in polynomial time whenever the length of the pattern is polynomially related to that of the text.

2 Preliminaries

Given a word $w \in \Sigma^*$, we denote by $w[i]$ the i-th symbol of w and by $w[i,j]$ the factor $w[i] \cdots w[j]$ of w, where $1 \leq i \leq j \leq |w|$. We call $\mathrm{Fact}(w)$ the set of the factors of w.

For the sake of simplicity, in the following we consider $\Sigma = \{0,1\}$; given a word $x \in \{0,1\}^n$, $b(x)$ is the base-2 integer whose binary representation is x and, with an abuse of notation, we intend $\underline{x}$ as the vector of components $b(x[1]), \dots, b(x[n])$. By $\underline{0}$ ($\underline{1}$), we denote a vector whose components are all 0 (1, respectively); its dimension is specified only in the case of ambiguity. Given two vectors $\underline{a} = (a_1, \ldots, a_n)$ and $\underline{b} = (b_1, \ldots, b_m)$, we denote by $\underline{a} \odot \underline{b}$ the concatenation $(a_1, \ldots, a_n, b_1, \ldots, b_m)$ and, if $n = m$, by $\underline{a} \cdot \underline{b}$ the sum $\sum_i a_i b_i$.

2.1 Straight-Line Programs

A *straight-line program* (*SLP*) is a sequence of labelled instructions of the form

$$X_1 = 0, \quad X_2 = 1, \quad X_k = X_i X_j \qquad 0 < i, j < k, \quad k = 3, \ldots, n.$$

The output of a SLP Φ is the word generated by performing all the concatenations from X_3 to X_n and is denoted by eval(Φ), while we write $\text{eval}_\Phi(X_k)$ for the word obtained by performing the first k concatenations in Φ. The number n of instructions in Φ is called its *size* and is denoted by $|\Phi|$. For every $w \in \{0,1\}^*$, as *descriptional complexity* of w we consider the size $g(w)$ of the smallest SLP generating w.

Since the computational complexity of a word can be logarithmic with respect of its size, many classical problems on words are studied in their compressed version, that is, considering SLPs as input instead of words.

For instance, the input of the compressed version of Equality is a pair (Φ, Ψ) of straight-line programs and the question is to decide whether eval(Φ) = eval(Ψ). Analogously, the question of the compressed version of Pattern Matching is to decide whether eval(Ψ) is a factor of eval(Φ). The first result in this direction is in [15] where a polynomial time algorithm for Equality is shown, while the best algorithm for Compressed Pattern Matching is presented in [9].

2.2 Lempel-Ziv Factorization

The *LZ-factorization* of a word w is a decomposition $f_1 \cdots f_k = w$, where $f_1 = w[1]$ and f_{i+1} is the shortest factor not appearing in $f_1 \cdots f_i$. We call *LZ-factors* of w the factors appearing in its LZ-factorization. The LZ-encoding of w is the sequence $LZ(w) = (f_1, \ldots, f_k)$, where every LZ-factor $f_i = w[a, b]$ is exclusively expressed by a and b. LZ-encoding gives a very efficient lossless compression technique, used in several compression standards ([8, 7]).

The size of $LZ(w)$ is the number of its LZ-factors and is denoted by $|LZ(w)|$. In [17] it is shown that $g(w) \geq |LZ(w)|$ and $g(w) = O(|LZ(w)| \times \log |w|)$ for every w. Moreover, we give a simple lower bound for the size of a LZ-factorization:

Lemma 1. *For every $w \in \{0,1\}^*$, $|LZ(w)| \geq |Fact(w) \cap 10^*1|$.*

Proof. The LZ-factorization of w contains a LZ-factor for each first occurrence of 10^t1 with different t. Indeed, if we scan w from left to right and run into a factor 10^t1 for the first time, then two cases are possible: either we already ran into a sequence of zeros of length $s > t$, and then a new LZ-factor necessarily starts immediately after 10^t1 in w, or the longest sequence of zeros has length $s < t$, then a new LZ-factor starts from the $(s+1)$th zero of 10^t1. □

3 Rational Transformations and SLPs

In this section we study rational transformations of compressed words. First, we recall the notion of deterministic rational transducer, defining from word to word rational transformations. Then we prove that rational transformations on compressed words preserve the compression rate.

A *deterministic rational transducer* is a 5-tuple $A = (\Sigma, \Gamma, Q, q_0, \delta)$, where Σ is the input alphabet, Γ is the output alphabet, Q is the set of states, $q_0 \in Q$ is the initial state and $\delta = (\delta_Q, \delta_\Gamma)$ is the transition function, with $\delta_Q : Q \times \Sigma \to Q$ and $\delta_\Gamma : Q \times \Sigma \to \Gamma^*$. We denote $(\delta_Q(q,\sigma), \delta_\Gamma(q,\sigma))$ as $\delta(q,\sigma)$.

The extension of δ_Q to Σ^* is similar to the case of finite state automata, we set $\delta_Q^*(q,\epsilon) = q$ and $\delta_Q^*(q, w\sigma) = \delta_Q(\delta_Q^*(q,w),\sigma)$ for every $w \in \Sigma^*$. The extension of δ_Γ is different: we set $\delta_\Gamma^*(q,\epsilon) = \epsilon$, and

$$\delta_\Gamma^*(q, w\sigma) = \delta_\Gamma^*(q,w)\delta_\Gamma(\delta_Q^*(q,w),\sigma),$$

where $w \in \Sigma^*$. The rational transformation applied by A to a word $w \in \Sigma^*$ is the word $A(w) = \delta_\Gamma^*(q_0, w)$.

To our aim, we consider the set $S = \{w \mid \delta_\Gamma(q,\sigma) = w, q \in Q, \sigma \in \Sigma\}$ and define the size of A as $|A| = |Q| + \sum_{w \in S} |w|$. Hence, in the context of compressed words, we introduce the following problem:

PROBLEM: *Compressed Rational Transformation (CRT)*
INSTANCE: A deterministic rational transducer A and a SLP Φ;
QUESTION: A SLP generating $A(\text{eval}(\Phi))$.

The CRT problem can be solved in polynomial time, as stated in the following

Theorem 1. *Given a rational transducer A and a SLP Φ, there is a $O(|A| \times |\Phi|)$ algorithm for the CRT problem with input A and Φ.*

Proof. Let $A = (\Sigma, \Gamma, Q, q_0, \delta)$ and $n = |\Phi|$. We first compute a table T with entries (X_k, q) with $X_k \in \Phi$ and $q \in Q$, such that $T(X_k, q) = \delta_Q^*(q, \text{eval}_\Phi(X_k))$. The table T can be computed in time $O(|Q| \times |\Phi|)$ by giving an order to the pairs (X_k, q_i) which preserves the order in Φ and then, for every instruction $X_k = X_i X_j$ and every $q \in Q$, computing the entry $T(X_k, q)$ as $T(X_j, T(X_i, q))$. If $X_k = \sigma$, then $T(X_k, q) = \delta_Q(q, \sigma)$.

We obtain a new SLP Ψ by translating each variable X_k of Φ in $|Q|$ variables $X(k, q)$ of the form

$$X(k,q) = \begin{cases} X(i,q)\ X(j,T(X_i,q)) & \text{if } X_k = X_i X_j, \\ \Psi(q,\sigma) & \text{if } X_k = \sigma,\ \sigma \in \Sigma, \end{cases}$$

where $\Psi(q,\sigma)$ is a SLP such that $\text{eval}(\Psi(q,\sigma)) = \delta_\Gamma(q,\sigma)$. By setting $X(n,q_0)$ as the last variable of Ψ, we have $\text{eval}_\Psi(X(n,q_0)) = A(\text{eval}(\Phi))$ by construction. Every $\Psi(q,\sigma)$ has size at most $|\delta_\Gamma(q,\sigma)|$, hence $|\Psi| = O(|A| \times |\Phi|)$. □

A straightforward consequence of Theorem 1 is that the compression properties of SLPs are preserved by rational transformations.

Corollary 1. *Let A be a fixed rational transducer. Then, $g(A(w)) = O(g(w))$ for every $w \in \Sigma^*$.*

Example 1. Consider the following rational transducer

$$A = (\{0,1\}, \{0,1\}, \{q_0, q_1, q_2\}, q_0, \delta),$$

where $\delta(q_0, 0) = (q_1, \epsilon)$, $\delta(q_1, 0) = \delta(q_1, 1) = \delta(q_2, 0) = (q_0, 0)$, $\delta(q_0, 1) = (q_2, \epsilon)$ and $\delta(q_2, 1) = (q_0, 1)$.

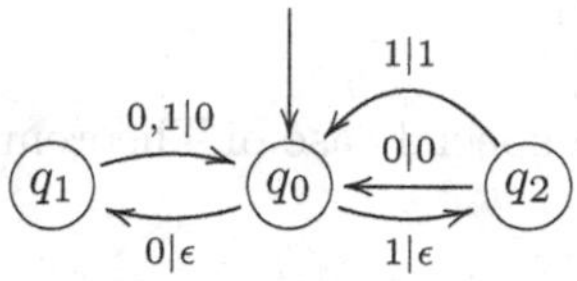

Such a transducer reads the symbols of a word in $\{0,1\}^*$ two by two, and writes 1 for 11 and 0 for 00, 01 and 10.

Let $\Phi = (X_1 = 0, X_1 = 1, [X_k = X_{k-1}X_{k-2}]_{k\in[2..6]})$ be the SLP that generates the 6th Fibonacci word $\text{eval}(\Phi) = 10110101$. Applying the algorithm of Th. 1, we obtain the (opportunely simplified) straight-line program $\Psi = (X_1 = 0, X_2 = 1, X_3 = X_1X_1, X_4 = X_1X_2, X_5 = X_4X_3)$, which generates the word 0100.

4 Lohrey Strings

In this section we recall a construction due to Lohrey ([10]), useful for study the computational complexity of some compressed word problems. The SUBSETSUM problem consists in deciding, given as input a vector $\underline{w}$ of integers and a target integer t, if there is at least one selection of entries in $\underline{w}$ whose sum is t. It can be formally defined as

PROBLEM: *Subset Sum (*SUBSETSUM*)*
INSTANCE: $\underline{w} \in \mathbf{N}^n$, $t \in \mathbf{N}$;
QUESTION: does there exist $\underline{x} \in \{0,1\}^n$ such that $\underline{x} \cdot \underline{w} = t$?

It is a well known NP-complete problem and its counting version, consisting in defining the cardinality of the set $\{\underline{x} \in \{0,1\}^n \mid \underline{x} \cdot \underline{w} = t\}$, is $\sharp P$-complete. In the context of straight-line programs, it has been used to prove that computing the Hamming distance of two compressed words is a $\sharp P$-complete problem ([9]). The proof makes use of the so called Lohrey strings [10], couples of words representing instances of SubsetSum problem that have an exponential compression rate.

Let $I = (\underline{w}, t)$ be an instance of SubsetSum, with $\underline{w} \in \mathbf{N}^n$ and define $s = \underline{1} \cdot \underline{w}$. The *Lohrey strings* of I are the two words

$$\xi(I) = (0^t 1 0^{s-t})^{2^n} \qquad \xi'(I) = \prod_{\substack{x \in \{0,1\}^n \\ b(x)=0..2^n-1}} (0^{\underline{x}\cdot\underline{w}} 1 0^{s-\underline{x}\cdot\underline{w}})$$

of length $(s+1)2^n$. Informally, $\xi(I)$ encodes t by 2^n blocks of length $s+1$ made of zeros in all places except in the $(t+1)$-th. On the other hand, $\xi'(I)$ encodes the sums of all the possible subsets of $\underline{w}$ by setting to 1 the only bit in position $\underline{x} \cdot \underline{w}$ in the x-th block, for every $x \in \{0,1\}^n$.

The relevance of Lohrey strings is depicted by the following

Lemma 2. *Let $I = (\underline{w}, t)$ be an instance of* SubsetSum *with $\underline{w} \in \mathbf{N}^n$. Then, $g(\xi(I)), g(\xi'(I)) = n^{O(1)}$.*

Proof. This lemma is a special case of Theorem 6 in [10]. □

5 Literal Shuffle of Compressed Words

In this section we consider the operations of bitwise AND and literal shuffle between words. Let $x, y \in \{0,1\}^n$, with $n > 0$; the bitwise AND $x \wedge y$ is $(x \wedge y)[i] = x[i] \wedge y[i]$ for $i = 1, \ldots, n$, while the *literal shuffle* ([1]) of x and y is defined as

$$x \sqcup\!\sqcup y = x[1]y[1]x[2]y[2] \cdots x[n]y[n],$$

Its "inverse operations" are L and R, where, for a word $w \in \{0,1\}^{2n}$,

$$L(w) = w[1]w[3] \cdots w[2n-1], \qquad R(w) = w[2]w[4] \cdots w[2n].$$

Operations L, R and $\sqcup\!\sqcup$ play an important role in many algorithms (such as Fast Fourier Transform for analysis and compression of digital signals) and it would be interesting to work using these operations in a compressed representation. L and R preserve the compression rate, since it is easy to construct the deterministic rational transducers implementing such operations. Unfortunately, this does not hold for the literal shuffle, as proved in this section.

First of all, we prove a technical lemma that allows to transform constructions using boolean circuits into constructions using SLPs.

Lemma 3. *Let C be a circuit computing the boolean function $f(\underline{x})$ in the variables $x_1, \ldots, x_n$. Then, there exist two SLP Φ and Ψ such that*

- $|\Phi|, |\Psi| = |C|^{O(1)}$;
- $|eval(\Phi)| = |eval(\Psi)| = 2^{n+m}$, *with* $m = n^{O(1)}$;
- $f(\underline{x}) = 1 \implies \exists! z \in \{0,1\}^m \mid eval(\Phi)[b(xz)] = eval(\Psi)[b(xz)] = 1$;
- $f(\underline{x}) = 0 \implies \forall z \in \{0,1\}^m \mid eval(\Phi)[b(xz)] \wedge eval(\Psi)[b(xz)] = 0$;

Proof. Without loss of generality, suppose C is built using NAND gates. Then, it is easy to construct a 3-CNF formula ϕ for f with $O(|C| + n)$ variables and clauses by adding the boolean variables $y_1, \ldots, y_{|C|}$ to the initial $x_1, \ldots, x_n$. Let y_k represent the output of a gate k in C and let a and b be its inputs. Then, ϕ contains the clauses defining $y_k = \overline{a \wedge b}$. Moreover, it contains further clauses for $y_{|C|} = 1$, being $y_{|C|}$ the output of the circuit.

In this construction, if $f(\underline{x}) = 1$ then there exists a unique $\underline{y}$ such that $\phi(\underline{x}, \underline{y}) = 1$, whereas if $f(\underline{x}) = 0$ then $\phi(\underline{x}, \underline{y}) = 0$ for all $\underline{y}$.

By using a minor variant of the reduction from 3-SAT to SUBSETSUM (see, for example, [6] and [4, pag. 223]), we can reduce $\phi(\underline{x}, \underline{y})$ to an instance $I_n(\underline{\alpha}, \underline{\beta}, \underline{\gamma}; t)$ of SUBSETSUM, with $|\underline{\gamma}| = n^{O(1)}$, such that

- $\phi(\underline{x}, \underline{y}) = 1$ implies $\exists! \underline{w}$ such that $\underline{x} \cdot \underline{\alpha} + \underline{y} \cdot \underline{\beta} + \underline{w} \cdot \underline{\gamma} = t$;
- $\phi(\underline{x}, \underline{y}) = 0$ implies $\underline{x} \cdot \underline{\alpha} + \underline{y} \cdot \underline{\beta} + \underline{w} \cdot \underline{\gamma} \neq t$ for every $\underline{w}$.

Let now $\xi(I_n)$ and $\xi'(I_n)$ be the Lohrey strings associated with the instance $I_n(\underline{\alpha}, \underline{\beta}, \underline{\gamma}; t)$. Then, we have two words of length $2^{n+m+|\gamma|}$ such that

- $f(\underline{x}) = 1 \implies \exists! z \in \{0,1\}^{m+|\gamma|} \mid \xi(I_n)[b(xz)] = \xi'(I_n)[b(xz)] = 1$;
- $f(\underline{x}) = 0 \implies \forall z \in \{0,1\}^{m+|\gamma|} \mid \xi(I_n)[b(xz)] \wedge \xi'(I_n)[b(xz)] = 0$;

By Lemma 2, $g(\xi(I_n)), g(\xi'(I_n)) = n^{O(1)}$. □

Now, we are able to prove the main result of this section:

Theorem 2. *For all* $n > 0$, *there exist two words* w_n *and* w'_n *of equal length such that* $g(w_n), g(w'_n) = n^{O(1)}$ *and* $g(w_n \wedge w'_n) = \Omega(2^{n/2})$.

Proof. Let C be the circuit computing the boolean function

$$f(\underline{x}, \underline{y}) = \begin{cases} 1 & \text{if } b(x) = b(y)^2 \\ 0 & \text{if } b(x) \neq b(y)^2 \end{cases}$$

A circuit C for f can be realized with $O(|x|^2)$ variables and $O(|x|^2)$ 3-clauses by iterated sums.

Consider the instance $I_n(\underline{\alpha}, \underline{\beta}, \underline{\gamma}; t)$ of SUBSETSUM defined for C as in Lemma 3, fix the representation $q(s)$ of the sth perfect square s^2 and let $z(s)$ be the unique string such that $\underline{q(s)} \cdot \underline{\alpha} + \underline{z(s)} \cdot (\underline{\beta} \odot \underline{\gamma}) = t$. Let now $\xi(n)$ and $\xi'(n)$ be the Lohrey strings associated with I_n and let be $\xi = \xi(n) \wedge \xi'(n)$. The position t_s of the sth 1 in ξ is equal to $b(q(s)z(s))$. Since $z(s)$ is unique, we have

$$s^2 2^M \leq t_s < s^2 2^M + 2^M,$$

where $M = |z(s)|$. It follows that

$$s2^{M+1} \leq t_{s+1} - t_s < (s+1)2^{M+1}.$$

This implies that $t_{s+1} - t_s \neq t_{j+1} - t_j$ whenever $s \neq j$. As a consequence, fixing $\hat{s} = \max\{s \mid s^2 < 2^n\}$, it holds

$$|\text{Fact}(\xi) \cap 10^*1| = \left|\{10^{t_{s+1}-t_s}1 \mid 1 \leq s \leq \hat{s}\}\right| = \hat{s} \geq 2^{n/2} - 1.$$

By Lemma 1, we have $|LZ(\xi)| \geq |\text{Fact}(\xi) \cap 10^*1|$. Hence $g(\xi) = \Omega(2^{n/2})$, while $g(\xi(n)), g(\xi'(n)) = n^{O(1)}$ by Lemma 2. □

In Example 1, we have a rational transducer A such that $A(x \sqcup\!\sqcup y) = x \wedge y$. Hence, by exploiting Theorem 1, the previous result can be extended to the literal shuffle of words.

Corollary 2. *For all $n > 0$, there exist two words w_n and w'_n of equal length such that $g(w_n), g(w'_n) = n^{O(1)}$ and $g(w_n \sqcup\!\sqcup w'_n) = \Omega(2^{n/2})$.*

5.1 Picture Straight-Line Programs

A natural representation of 2D texts can be obtained by using a 2D extension of SLPs. Informally, a binary picture of width M and height N is a matrix $T \in \{0,1\}^{N \times M}$. We refer to the rows and columns of a picture as its sections. A 2D-SLP of size n is a sequence of labelled instructions of the form

$$X_k = 1 \mid 0 \mid X_i \oslash X_j \mid X_i \ominus X_j, \quad 0 < i,j < k \quad k = 1,\dots,n.$$

The operator $\oslash$ is the horizontal concatenation between two pictures of equal height, while $\ominus$ is the vertical concatenation of two pictures of equal width. The output of a 2D-SLP Φ is a binary picture $T = \text{eval}(\Phi)$, obtained performing all the concatenations in Φ; the descriptional complexity $g(T)$ of a picture T is the size of the smallest 2D-SLP generating T.

The structure of 2D-SLPs is more complex than that of SLPs. In particular, while the factors of logarithmically compressible words are still logarithmically compressible, an analogous property does not hold for subpictures of pictures. For instance, in [2] it is proved that, for each n, there exists a picture T_n with $g(T_n) = n$ having at least one section w_n with $g(w_n) = 2^{\Omega(n)}$.

Here, we prove a stronger result on the sections of compressed pictures.

Theorem 3. *Let T be a $N \times M$ picture and let c_i be its i-th column and r_j its j-th row. Then, $g(c_i) \leq g(T) \times \min\{i, M-i\}$ and $g(r_j) \leq g(T) \times \min\{j, N-j\}$.*

Proof. Let $\Phi = (X_1, \dots, X_n)$ be a 2D-SLP such that $\text{eval}(\Phi) = T$. Then, we construct a SLP Φ_s for the sth column of T in the following way. Without loss of generality, suppose that $s < \lfloor M/2 \rfloor$ and, for each variable X_k of Φ, define s variables of the form

$$X_k(s) = \begin{cases} \sigma & \text{if } X_k = \sigma; \\ X_i(s)X_j(s) & \text{if } X_k = X_i \ominus X_j; \\ X_i(s) & \text{if } X_k = X_i \oslash X_j \text{ and } s \le |X_i|; \\ X_j(s - |X_i|) & \text{if } X_k = X_i \oslash X_j \text{ and } s > |X_i|. \end{cases}$$

Then, $|\Phi_s| = s \times |\Phi|$ and $\text{eval}_{\Phi_s}(X_n(s))$ is the sth column of T. The technique can be easily adapted for the columns of position grater that $\lceil M/2 \rceil$ and for all the rows of T. □

The previous result gives an upper bound for the compression rate of the sections of a picture. However, we would ask how much strict is this bound. Next theorem proves that the bound is, in some sense, optimal.

Theorem 4. *There exists an infinite number of pictures* $\{T_n\}_{n \in \mathbf{N}}$ *such that, for each column* c_i *of* T_n *in position* i*, it holds* $g(c_i) = (g(T_n) \times \min\{i, M - i\})^{\Omega(1)}$.

Proof. Consider the Lohrey strings $\xi(n), \xi'(n)$ related with the instance of SUBSETSUM in the proof of Th. 2 and let $l = |\xi(n)| = |\xi'(n)|$. By Lemma 2, we have two polynomial size SLPs Φ and Φ' such that $\text{eval}(\Phi) = \xi(n)$ and $\text{eval}(\Phi') = \xi'(n)$. Moreover, let Z_i be the $2^i \times 2^{i-1}$ picture containing all zeros. For $i = 1, \ldots, k$, all the Z_i can be compressed by a SLP of size $O(k)$. Then, we can construct a polynomial size SLP for the picture T described recursively by

$$X_1 = \xi(n) \ominus \xi'(n);$$
$$X_{i+1} = (Z_i \oslash X_i) \ominus (X_i \oslash Z_i), \quad i = 1, \ldots, 2n.$$

In this way, we obtain a picture T of size $2^{2n+1} \times (l + 2^{2n} - 1)$ of the form

$$\begin{bmatrix} 0 & \cdots & \cdots & \cdots & \cdots & 0 & a_1 & a_2 & \cdots & a_l \\ \vdots & \cdots & \cdots & \cdots & \cdots & 0 & b_1 & b_2 & \cdots & b_l \\ \vdots & \cdots & \cdots & \cdots & 0 & a_1 & a_2 & \cdots & a_l & 0 \\ \vdots & \cdots & \cdots & \cdots & 0 & b_1 & b_2 & \cdots & b_l & 0 \\ \vdots & \cdots & \cdots & 0 & a_1 & a_2 & \cdots & a_l & 0 & \vdots \\ \vdots & \cdots & \cdots & 0 & b_1 & b_2 & \cdots & b_l & 0 & \vdots \\ 0 & \cdots & \cdot^{\cdot^{\cdot}} & \cdot^{\cdot^{\cdot}} & \cdot^{\cdot^{\cdot}} & \cdot^{\cdot^{\cdot}} & \cdot^{\cdot^{\cdot}} & \cdot^{\cdot^{\cdot}} & \vdots & \vdots \\ a_1 & a_2 & \cdots & a_l & 0 & \cdots & \cdots & \cdots & \cdots & 0 \\ b_1 & b_2 & \cdots & b_l & 0 & \cdots & \cdots & \cdots & \cdots & 0 \end{bmatrix}$$

with $\xi(n) = a_1 \cdots a_l$ and $\xi'(n) = b_1 \cdots b_l$. Clearly, $g(T) = n^{\Theta(1)}$. Note that every column c_i of T in position i such that $l \le i < 2^{2n}$ contains all zeros except a factor $\xi(n) \sqcup\!\sqcup \xi'(n)$. So, $g(c_i) = n^{\Theta(1)} \, g(\xi(n) \sqcup\!\sqcup \xi'(n)) = (n2^n)^{\Theta(1)}$.

The cases $1 \le i < l$ and $2^{2n} \le i < 2^{2n} + l$ are symmetric, so we focus on the former. In this case, only the prefix $p_i = (\xi(n) \sqcup\!\sqcup \xi'(n))[1, 2i]$ appears at the end of c_i, while the rest of the column is all zeros. By Corollary 1, $O(g(p_i)) = g(\xi[1, 2i])$, where $\xi = \xi(n) \wedge \xi'(n)$. The structure of ξ is such that

$$g(\xi[1, 2i]) \ \ge \ |\text{Fact}\,(\xi[1, 2i]) \cap 10^*1| \ = \ \Omega(\sqrt{i}).$$

Hence, again $g(c_i) = (ni)^{\Omega(1)}$. □

Obviously, the same result holds for the case of rows of pictures.

6 Pattern Matching and Literal Shuffle

In this section, we represent a word $w = \text{eval}(\Phi) \sqcup\!\sqcup \text{eval}(\Psi)$ by means of the pair (Φ, Ψ) with size $|\Phi| + |\Psi|$ and study the compressed pattern matching problem in this representation. More formally, the problem can be stated as

PROBLEM: *Compressed Pattern Matching with* $\sqcup\!\sqcup$ *(*$CPM_{\sqcup\!\sqcup}$*)*
INSTANCE: Four SLPs Φ, Ψ, Φ', Ψ', such that
$|\text{eval}(\Phi)| = |\text{eval}(\Phi')|$ and $|\text{eval}(\Psi)| = |\text{eval}(\Psi')|$;
QUESTION: is $\text{eval}(\Psi) \sqcup\!\sqcup \text{eval}(\Psi')$ a factor of $\text{eval}(\Phi) \sqcup\!\sqcup \text{eval}(\Phi')$?

It can be easily observed that $\text{CPM}_{\sqcup\!\sqcup}$ is reducible to the Compressed Pattern Matching for pictures composed by two lines. So, it appears more difficult than Compressed Pattern Matching for words, solvable in polynomial time ([9]), but easier than Compressed Pattern Matching for pictures, which is Σ_2^P-complete ([2]).

$\text{CPM}_{\sqcup\!\sqcup}$ is clearly in NP. Moreover, it is NP-complete even if the pattern is the string 11.

Theorem 5. *The problem of deciding, given two SLPs Φ and Φ', whether* 11 *is a factor of eval(Φ)* $\sqcup\!\sqcup$ *eval(Φ') is NP-hard.*

Proof. By reduction to SUBSETSUM. Let $\xi(I)$ and $\xi'(I)$ be the Lohrey strings associated with an instance I of SUBSETSUM and let Φ and Φ' be the associated SLPs of smallest size. Then, define a deterministic rational transducer A such that $A(x) = 0x[1]0x[2]\cdots 0x[|x|]$, for every word x. By Th. 1, $g(A(x)) = O(g(x))$. Moreover, $A(\text{eval}(\Phi)) \sqcup\!\sqcup A(\text{eval}(\Phi'))$ contains the factor 11 if and only if the instance I of SUBSETSUM admits a solution. □

Despite this hardness result, we exhibit an algorithm for $\text{CPM}_{\sqcup\!\sqcup}$ working in polynomial time if the length of the pattern is polynomially related with that of the text.

We recall some notation about SLPs used in [9]. The *positions* 0 and n in a word $w \in \Sigma^n$ are the points immediately before $w[1]$ and after $w[n]$, respectively, while a position i, with $1 \leq i < n$ is the point between $w[i]$ and $w[i+1]$. A factor $w[i, j]$ touches a position k in w if $i - 1 \leq k \leq j$. Given a nonterminal symbol X_k in a SLP Φ such that $X_k = X_i X_j$, its *cut position* is $|\text{eval}_\Phi(X_i)|$.

Finally, by the triple of nonnegative integers (p, d, r) we codify the *arithmetical progression* $\{p, p+d, p+2d, \ldots, p+rd\}$ and recall the following result.

Lemma 4. *Let (p, d, r) and (p', d', r') be two arithmetical progressions, where p, d, r, p', d', r' are n-bits integers. Then, deciding if their intersection is empty requires $O(n^2)$ time.*

Proof. The problem of deciding if $(p, d, r) \cap (p', d', r')$ is the empty set consists of verifying the existence of two integers x, y such that

1. $p + dx = p' + d'y$;
2. $0 \leq x \leq r$ and $0 \leq y \leq r'$.

Such equations are equivalent to the diophantine equation $Ax - By = C$, where $c = MCD(d, d', p' - p)$ and $A = d/c$, $B = d'/c$, $C = (p' - p)/c$.

If $MCD(A, B) > 1$, then the previous equation has no solution and $(p, d, r) \cap (p', d', r') = \emptyset$. Otherwise, one solution (x_0, y_0) can be obtained by computing the $(h - 1)$th convergent, where h is the number of terms in the continued fraction for A/B ([14]).

All the other solutions are of the form $x = c(x_0 + kB)$, $y = c(y_0 + kA)$. By setting

$$\begin{aligned} \underline{k} &= \min\{k \mid 0 \leq c(x_0 + kB)\}, \\ \overline{k} &= \max\{k \mid c(x_0 + kB) \leq r\}, \\ \underline{k}' &= \min\{k \mid 0 \leq c(y_0 + kA)\}, \\ \overline{k}' &= \max\{k \mid c(y_0 + kA) \leq r\}, \end{aligned}$$

one can conclude that $(p, d, r) \cap (p', d', r') \neq \emptyset$ if and only if $[\underline{k}, \overline{k}] \cap [\underline{k}', \overline{k}'] \neq \emptyset$.

The most expensive task in this process is the computation of the convergent of a fraction of two n-bits integers, which takes $O(n^2)$ time. □

Many compressed pattern matching algorithms are based on the following ([5])

Lemma 5. *Given two words w and v, all the occurrences of v in w touching a fixed position form a single arithmetical progression.*

Some compressed pattern matching algorithms use as a data structure the so called *AP*-table, which we recall in a simplified version. Given two SLPs Φ and Ψ, the *AP*-table for Φ and Ψ is a vector where, for every symbol X_k in Φ, the k-th entry is the (possibly empty) arithmetical progression $(p[X_k], d[X_k], r[X_k])$ identifying the starting positions of the occurrences of eval(Ψ) that touch the cut position of X_k. The *AP*-table for Φ and Ψ is computable in time $O(|\Phi|^3 \times |\Psi|)$ ([9]).

Given a SLP Φ having variables $X_1, \ldots, X_n$, consider the following partial function $t : \{X_1, \ldots, X_n\} \longrightarrow \{X_1, \ldots, X_n\}$, such that

$$t(X_k) = \begin{cases} X_i & \text{if } X_k = X_i X_j \text{ and } |\text{eval}_\Phi(X_i)| \geq |\text{eval}_\Phi(X_k)|/2, \\ X_j & \text{if } X_k = X_i X_j \text{ and } |\text{eval}_\Phi(X_i)| < |\text{eval}_\Phi(X_k)|/2, \\ \perp & \text{otherwise.} \end{cases}$$

Now, select the path $z_1, \ldots, z_j$ in the derivation tree of eval(Φ), defined as $z_1 = X_n$, $z_{i+1} = t(z_i)$ for $1 \leq i < j$, where z_j is the first occurrence of $\perp$. Then, this path identifies a sequence of factors $f_1, \ldots, f_j$ of eval(Φ), each of which is a prefix or a suffix of its predecessor. Hence, their starting positions $g_1, \ldots, g_j$ can be computed in time $O(|\Phi|^2)$.

Now, suppose that $|\text{eval}(\Psi)| > |\text{eval}(\Phi)|/2$. Then, the possible starting positions of eval(Ψ) in eval(Φ) are the elements defined in the arithmetical progressions

$$ar_i(\Phi, \Psi) = (g_i + p[z_i], d[z_i], r[z_i]), \quad \text{for } i = 1, \ldots, j.$$

Theorem 6. *Let Φ, Ψ, Φ' and Ψ' be four SLPs such that $|eval(\Phi)| = |eval(\Phi')| = N$ and $|eval(\Psi)| = |eval(\Psi')| = M$. Then, the CPM$_\sqcup$ problem can be solved in time $O(Nn^4/M)$, where $n = |\Phi| + |\Psi| + |\Phi'| + |\Psi'|$.*

Proof. Suppose that $N < 2M$. Then,

$$\text{eval}(\Psi) \sqcup \text{eval}(\Psi') \in \text{Fact}\,(\text{eval}(\Phi) \sqcup \text{eval}(\Phi'))$$

if and only if, for some integers i, s, at least one of the following facts hold:

- $ar_i(\Phi, \Psi) \cap ar_s(\Phi', \Psi') \neq \emptyset$;
- $ar_i(\Phi', \Psi) \cap ar'_s(\Phi, \Psi') \neq \emptyset$;

where $ar'_s(\Phi, \Psi')$ is $ar_s(\Phi, \Psi')$ left-shifted by one, i.e., if $ar_s(\Phi, \Psi') = (g+p, d, r)$, then $ar'_s(\Phi, \Psi') = (g+p-1, d, r)$. By Lemma 4, non-emptiness of each intersection can be verified in $O(n^2)$ time and, since i and s range over $[1, n]$, we can solve the problem in time $O(n^4)$.

If, on the contrary, $N \geq 2M$, then, for $0 \leq i \leq 2N/M - 2$, we construct the SLPs Φ_i, such that

$$\text{eval}(\Phi_i) = \text{eval}(\Phi)[i\lceil M/2\rceil + 1, \min\{N, (i+3)\lceil M/2\rceil\}]$$

and we do the same for Φ'. This construction requires $O(n^2 N/M)$ time and guarantees that the following sentences are equivalent:

1. The word eval(Ψ) ⊔ eval(Ψ') is a factor of eval(Φ) ⊔ eval(Φ').
2. There exists k ($0 \leq k \leq N - M$) such that at least one of these facts hold:
   - eval(Ψ) is a factor of eval(Φ) and eval(Ψ') is a factor of eval(Φ'), both starting in position k;
   - eval(Ψ) is the factor of eval(Φ') starting in position k and eval(Ψ') is the factor of eval(Φ) starting in position $k + 1$.
3. There exists i ($0 \leq i \leq 2N/M - 2$) such that eval(Ψ) ⊔ eval(Ψ') is a factor of eval(Φ_i) ⊔ eval(Φ'_i).

Since $|\Phi_i| = |\Phi'_i| < 2M$, for every i, we can verify whether eval(Ψ) ⊔ eval(Ψ') is a factor of eval(Φ_i) ⊔ eval(Φ'_i) in time $O(n^4)$. Hence, the problem is solvable in time $O(n^4 N/M)$. □

7 Conclusions

We investigated the possibility of performing rational transformations and the literal shuffle of words compressed via SLPs, without full unpacking. We proved that the last operation does not preserve the compression rate; hence, some techniques like Cooley-Tukey algorithm for FFT can not be applied in a compressed context. On the other hand, rational transformations can be performed without fully uncompressing the SLPs in input.

These results lead to a deeper insight into the relations between SLPs for words and SLPs for pictures. Indeed we showed that the descriptional complexity of the sections of a picture can strongly depend on their distance from the borders.

The literal shuffle has been finally exploited as a compressed representation of pictures having two lines. The associated compressed pattern matching problem lies in the half way between the same problems for compressed words and for compressed pictures. We proposed a parameter-tractable algorithm working in polynomial time, where the parameter is the ratio between the length of the text and that of the pattern.

Acknowledgements We would like to thank Antonio Restivo for some useful discussions.

References

1. Béatrice Bérard. Literal shuffle. *Theoret. Comput. Sci.*, 51(3):281–299, 1987.
2. Piotr Berman, Marek Karpinski, Lawrence L. Larmore, Wojciech Plandowski, and Wojciech Rytter. On the complexity of pattern matching for highly compressed two-dimensional texts. In *Combinatorial pattern matching (Aarhus, 1997)*, volume 1264 of *Lecture Notes in Comput. Sci.*, pages 40–51. Springer, Berlin, 1997.
3. James W. Cooley and John W. Tukey. An algorithm for the machine calculation of complex Fourier series. *Math. Comp.*, 19:297–301, 1965.
4. Michael R. Garey and David S. Johnson. *Computers and intractability.* W. H. Freeman and Co., San Francisco, Calif., 1979. A guide to the theory of NP-completeness, A Series of Books in the Mathematical Sciences.
5. Leszek Gąsieniec, Marek Karpinski, Wojciech Plandowski, and Wojciech Rytter. Randomized efficient algorithms for compressed strings: the finger-print approach (extended abstract). In *Combinatorial pattern matching (Laguna Beach, CA, 1996)*, volume 1075 of *Lecture Notes in Comput. Sci.*, pages 39–49. Springer, Berlin, 1996.
6. Richard M. Karp. Reducibility among combinatorial problems. In *Complexity of computer computations (Proc. Sympos., IBM Thomas J. Watson Res. Center, Yorktown Heights, N.Y., 1972)*, pages 85–103. Plenum, New York, 1972.
7. Abraham Lempel and Jacob Ziv. On the complexity of finite sequences. *IEEE Trans. Information Theory*, IT-22(1):75–81, 1976.
8. Abraham Lempel and Jacob Ziv. A universal algorithm for sequential data compression. *IEEE Trans. Information Theory*, IT-23(3):337–343, 1977.
9. Yury Lifshits. Processing compressed texts: A tractability border. In *Combinatorial Pattern Matching*, volume 4580 of *Lecture Notes in Comput. Sci.*, pages 228–240. Springer, Berlin, 2007.

10. Markus Lohrey. Word problems on compressed words. In *Automata, languages and programming*, volume 3142 of *Lecture Notes in Comput. Sci.*, pages 906–918. Springer, Berlin, 2004.
11. Markus Lohrey. Word problems and membership problems on compressed words. *SIAM J. Comput.*, 35(5):1210–1240 (electronic), 2006.
12. N. Markey and Ph. Schnoebelen. A PTIME-complete matching problem for SLP-compressed words. *Inform. Process. Lett.*, 90(1):3–6, 2004.
13. Masamichi Miyazaki, Ayumi Shinohara, and Masayuki Takeda. An improved pattern matching algorithm for strings in terms of straight-line programs. *J. Discrete Algorithms (Oxf.)*, 1(1):187–204, 2000.
14. C. D. Olds. *Continued fractions.* Random House, New York, 1963.
15. Wojciech Plandowski. Testing equivalence of morphisms on context-free languages. In *Algorithms—ESA '94 (Utrecht)*, volume 855 of *Lecture Notes in Comput. Sci.*, pages 460–470. Springer, Berlin, 1994.
16. Wojciech Plandowski and Wojciech Rytter. Complexity of language recognition problems for compressed words. In *Jewels are forever*, pages 262–272. Springer, Berlin, 1999.
17. Wojciech Rytter. Application of Lempel-Ziv factorization to the approximation of grammar-based compression. *Theoret. Comput. Sci.*, 302(1-3):211–222, 2003.

Reconstructing words from a fixed palindromic length sequence*

Alexandre Blondin Massé[1], Srečko Brlek[1], Andrea Frosini[2], Sébastien Labbé[1] and Simone Rinaldi[2]

[1] Laboratoire de Combinatoire et d'Informatique Mathématique,
Université du Québec à Montréal,
C. P. 8888 Succursale "Centre-Ville", Montréal (QC), CANADA H3C 3P8,
Brlek.Srecko@uqam.ca, [blondin_masse.alexandre, Labbe.Sebastien]@courrier.uqam.ca

[2] Dip. di Scienze Matematiche ed Informatiche Roberto Magari,
Università degli Studi di Siena
Pian dei Mantellini 44, 53100 Siena, Italy
[frosini, rinaldi]@unisi.it

Abstract. To every word w is associated a sequence G_w built by computing at each position i the length of its longest palindromic suffix. This sequence is then used to compute the palindromic defect of a finite word w defined by $D(w) = |w| + 1 - |\mathrm{Pal}(w)|$ where $\mathrm{Pal}(w)$ is the set of its palindromic factors. In this paper we exhibit some properties of this sequence and introduce the problem of reconstructing a word from G_w. In particular we show that up to a relabelling the solution is unique for 2-letter alphabets.

Key words: Palindromic complexity, defect, lacunas, reconstruction.

1 Introduction

Among the many ways of measuring the information content of a finite word, counting the number of its distinct factors or subwords of given length has been widely used and known as its complexity. A refinement of this notion amounts to restrict the factors to palindromes. The motivations for the study of palindromic complexity comes from many areas ranging from the study of Schrödinger operators in physics [4, 7, 20] to number theory [6] and combinatorics on words where it appears as a powerful tool for understanding the local structure of words. It has been recently studied in various classes of infinite words, an account of which may be found in the survey provided by Allouche et al. [5].

In particular, the palindromic factors give an insight on the intrinsic structure, due to its connection with the usual complexity, of many classes of words. For instance, they completely characterize Sturmian words [23], and for the class of smooth words they provide a connection with the notion of recurrence [12, 13].

* with the support of NSERC (Canada)

Please use the following format when citing this chapter:

Blondin Massé, A., et al., 2008, in IFIP International Federation for Information Processing, Volume 273; *Fifth IFIP International Conference on Theoretical Computer Science*; Giorgio Ausiello, Juhani Karhumäki, Giancarlo Mauri, Luke Ong; (Boston: Springer), pp. 101–114.

The problem of reconstructing words from partial information arise naturally. We mention a few of them that have been solved for a fixed alphabet Σ:

Some set A of factors is fixed. Find the shortest words containing the set A of all factors of given length k. This leads to the De Bruijn sequences [15, 17, 19] whose construction uses a graph G_k where vertices are the given words of length k, and where edges model the scanning of the word by a window of size k. The solution is then obtained by computing all Eulerian cycles in the graph. It is worth noting that finding the lexicographically smallest such word is much easier: it is given by the lexicographic concatenation of Lyndon words on Σ whose lengths are divisible by k (See Fredericksen et al. [18]).

Some set A is fixed along with some suitable hypothesis. Construct all words w such that the set of its factors $A = F(w)$. The technique used for this problem is based on constructing a set of minimal forbidden words, that is the extensions of words in A that do not belong to A [9]. That technique was also used in [11] to construct words whose language of palindromes is a fixed set P. It turns out that it is a rational language. Concerning multisets of subsequences, instead of factors we mention a general result. If the set A contains sufficiently many subsequences of length k, then the solution is unique [26]: indeed, for a word w of lengh $n > 7$ and $k \geq \lceil n/2 \rceil$ the subsequences uniquely determine w, and for $k < \log_2 n$ they do not. See also an interesting combinatorial approach depending on the Burrows-Wheeler transform (See Mantaci et al. [25]).

Fixed complexity. The most famous example is that of Sturmian words (see Lothaire [22] for a substantial review) which are characterized by the complexity $P(n) = n + 1$ established by M. Morse [28]. Sturmian words are the discretization of lines with irrational slopes, and they are easily constructed from the continued fraction expansion corresponding to the irrational slope. The complexity is therefore not enough to characterize completely a word. However, in the case of the Thue-Morse complexity [10, 24], there are essentially only two such words [1, 2].

In this paper we introduce the problem of reconstructing a word from sequences describing its palindromic complexity. Droubay, Justin and Pirillo [16] noted that the palindrome complexity $|\text{Pal}(w)|$ of a word w is bounded by $|w| + 1$, and observed that it is computed by a sequential algorithm listing the first occurrences of longest palindromic suffixes, called *unioccurrent* in [16]. For our study we need the following two auxiliary functions on words. Given a word of length n, $w : [0..(n-1)] \longrightarrow \Sigma$, we define two functions $G_w, H_w : \mathbb{N} \longrightarrow \mathbb{N}$ by $G_w(i) = |\text{LPS}(w[0..i])|$ and

$$H_w(i) = \begin{cases} G_w(i) & \text{if it is the first occurrence of LPS}(w[0..i]) \\ 0 & \text{otherwise} \end{cases} \tag{1}$$

We first exhibit some combinatorial properties of the palindromic factors in words (Section 3) and use them in order to obtain properties of the sequences

G and H (Section 4). Finally we study the problem of reconstructing words from given sequences, and establish conditions for unicity on 2-letter alphabets.

2 Preliminaries

In what follows, Σ is a finite *alphabet* whose elements are called *letters*. By *word* we mean a finite sequence of letters $w : [0..(n-1)] \longrightarrow \Sigma$, where $n \in \mathbb{N}$. The length of w is $|w| = n$ and $w[i]$ or w_i denote its i-th letter. The set of n-length words over Σ is denoted Σ^n. By convention, the *empty* word is denoted ε and its length is 0. The free monoid generated by Σ is defined by $\Sigma^* = \bigcup_{n \geq 0} \Sigma^n$.

The set of right infinite words is denoted by Σ^ω and we set $\Sigma^\infty = \Sigma^* \cup \Sigma^\omega$. Given a word $w \in \Sigma^\infty$, a *factor* f of w is a word $f \in \Sigma^*$ satisfying

$$\exists x \in \Sigma^*, \exists y \in \Sigma^\infty, w = xfy.$$

If $x = \varepsilon$ (resp. $y = \varepsilon$) then f is called *prefix* (resp. *suffix*). The set of all factors of w is denoted by $\text{Fact}(w)$, those of length n is $\text{Fact}_n(w) = \text{Fact}(w) \cap \Sigma^n$, and $\text{Pref}(w)$ is the set of all prefixes of w. The number of occurrences of a factor f in w is denoted $|w|_f$. A *period* of a word w is an integer $p < |w|$ such that $w[i] = w[i+p]$, for all $i < |w| - p$. If $w = pu$, with $|w| = n$ and $|p| = k$, then $p^{-1}w = w[k..(n-1)] = u$ is the word obtained by erasing p. A word is said to be *primitive* if it is not a power of another word. Two words u and v are *conjugate* when there are words x, y such that $u = xy$ and $v = yx$. The conjugacy class of a word w is denoted by $[w]$; note that the length is invariant under conjugacy. For a given word w of length n, any of its conjugates is obtained by cyclic permutation, that is $\sigma^i(w) = w[i..(n-1)]w[0..(i-1)]$.

The *reversal* of $u = u_0u_1 \cdots u_{n-1} \in \Sigma^n$ is the word $\widetilde{u} = u_{n-1}u_{n-2} \cdots u_0$, and a *palindrome* is a word p such that $p = \widetilde{p}$. Since every word contains palindromes, the letters and ε being necessarily part of them, the set of its palindromic factors is $\text{Pal}(w)$, and its *palindromic complexity* is denoted by $|\text{Pal}(w)|$. Conjugacy is an equivalence relation having numerous properties and for our purpose we need the following one easily obtained by induction: let p and q be two palindromes, then $\sigma^i(pq) = p'q'$, for some palindromes p' and q'. We start by quoting Lemma 1 of [8] in order to establish a useful combinatorial property.

Lemma 1 (Blondin Massé et al. [8]) *Assume that $w = xy = yz$. Then for some u, v, and some $i \geq 0$ we have from* [21]

$$x = uv, y = (uv)^i u, z = vu; \tag{2}$$

and the following conditions are equivalent :

(i) $x = \widetilde{z}$;
(ii) *u and v are palindromes;*

(iii) *w is a palindrome;*
(iv) *xyz is a palindrome.*

Moreover, if one of the equivalent conditions above holds then

(v) *y is a palindrome.*

As a consequence we have the following proposition.

Proposition 1 *Assume that $w = xp = qz$ where p and q are palindromes such that $|q| > |x|$. Then w has period $|x| + |z|$, and $x\widetilde{z}$ is a product of two palindromes.*

Proof. Since $|q| > |x|$, there exists a non-empty word y such that $q = xy$ and $p = yz$. It follows that

$$w\,\widetilde{x} = q\,z\,\widetilde{x} = x\,y\,z\,\widetilde{x} = x\,p\,\widetilde{x} = x\,\widetilde{p}\,\widetilde{x} = x\,\widetilde{z}\,\widetilde{y}\,\widetilde{x} = x\,\widetilde{z}\,\widetilde{q} = x\,\widetilde{z}\,q.$$

Considering $qz\widetilde{x} = x\widetilde{z}q$, we obtain from Equation (2) that $|x\widetilde{z}|$ is a period of $w\widetilde{x}$. From Lemma 1 (iii), there exist palindromes u, v such that $x\widetilde{z} = uv$. □

In order to compute the palindromic complexity we need the function LPS : $\Sigma^* \longrightarrow \Sigma^*$ which associates to any word w its longest palindromic suffix LPS(w).

Droubay, Justin and Pirillo [16] noted that the palindrome complexity $|\mathrm{Pal}(w)|$ of a word w is bounded by $|w| + 1$, and that finite Sturmian (and even episturmian) words realize the upper bound. Moreover they implicitly show that the palindrome complexity is computed by an algorithm listing the longest palindromic suffixes which amounts to compute for a word w the functions $G_w, H_w : \mathbb{N} \longrightarrow \mathbb{N}$ defined by

$$G_w(i) = |\mathrm{LPS}(w[0..i])|;$$

$$H_w(i) = \begin{cases} G_w(i) & \text{if it is the first occurrence of LPS}(w[0..i]); \\ 0 & \text{otherwise.} \end{cases}$$

We often omit the subscript w in G_w and H_w when the context is clear. As an example let $w = aababbaababaaabaab$. Then we have the following table :

i	0	1	2	3	4	5	6	7	8	9	10	11	12	13	14	15	16	17
w	a	a	b	a	b	b	a	a	b	a	b	a	a	a	b	a	a	b
G	1	2	1	3	3	2	4	2	4	3	3	5	7	3	5	7	5	4
H	1	2	1	3	3	2	4	0	4	0	0	5	7	3	5	7	5	0

A position in the word w where H vanishes is called a *lacuna* in [8]. For instance the set of lacunas for w in the example above is $\{7, 9, 10, 17\}$. Equivalent words, that is words obtained by relabelling of the alphabet, have obviously the same functions G and H. For instance, on the 2-letter alphabet $\{a, b\}$, we have $G_w = G_{\overline{w}}$ and $H_w = H_{\overline{w}}$, where $\overline{(\ )}$ is the morphism defined by $\overline{a} = b, \overline{b} = a$.

The palindromic defect of a finite word w is defined in Brlek et al. [11] by $D(w) = |w| + 1 - |\mathrm{Pal}(w)|$, and words for which $D(w) = 0$, that is, such that H does not vanish for any index are called *full*. In that paper it is also shown that there exist periodic full words, and an optimal algorithm is provided to check if an infinite periodic word is full or not. Moreover, a characterization by means of a rational language is given for the language L_P of words whose palindromic factors belong to a fixed and finite set P of palindromes.

3 Properties of the functions G and H

First observe that a word w is full if and only if $G_w = H_w$. Now we describe the shortest words having a fixed defect value d. For instance, on a 2-letter alphabet, the shortest words having one lacuna, i.e. when $d = 1$, are

$$w_1 = aababbaa, w_2 = aabbabaa, w_3 = bbabaabb \text{ and } w_4 = bbaababb.$$

Observe that this set is closed under reversal ($w_1 = \widetilde{w_2}; w_3 = \widetilde{w_4}$) and complementation ($w_1 = \overline{w_3}; w_2 = \overline{w_4}$). On the other hand, one of the shortest words having two lacunas is the following.

i	0	1	2	3	4	5	6	7	8	9
w	b	a	a	b	a	b	b	a	a	b
G	1	1	2	4	3	3	2	4	2	4
H	1	1	2	4	3	3	2	4	0	0

The example above extends to the infinite periodic word $W = (baab.ab)^\omega$

i	0	1	2	3	4	5	6	7	8	9	10	11	12	13	14	15	16	17	...
W	b	a	a	b	a	b	b	a	a	b	a	b	b	a	a	b	a	b	...
G	1	1	2	4	3	3	2	4	2	4	3	3	2	4	2	4	3	3	...
H	1	1	2	4	3	3	2	4	0	0	0	0	0	0	0	0	0	0	...

where *baab.ab* is not the product of two palindromes, so that $|\mathrm{Pal}(W)|$ is finite by virtue of a previous result (see Theorems 4 and 6 in [11]). More generally, we have the following result.

Proposition 2 *Let $M(k, d)$ be the length of a shortest word on a k-letter alphabet Σ having defect d, we have :*

$$M(k,d) = \begin{cases} 8 & \text{if } k = 2, d = 1, \\ d+8 & \text{if } k = 2, d \geq 2, \\ d+k & \text{if } k \geq 3. \end{cases}$$

Proof. The first two cases follow from the observations above. For $k \geq 3$, let w be a word such that $|w| = M(k, d)$. Since every letter occurs in w and w

has defect value d, we have $M(k,d) \geq d+k$. Now, consider the infinite periodic word $w = (\alpha_1\alpha_2 \cdots \alpha_k)^\omega$, where α_i is a letter. Observe that $\text{Pal}(w) = \Sigma$, so that each prefix of length $n \geq k+1$ has defect value $n-k$. Hence $M(k,d) = d+k$ for $k \geq 3$. □

Lemma 2 *Let w be a nonempty word, and let $W = w^\omega$. Then we have*

(i) *$G_w(0) = 1$, and if $H_w(i) = 1$ then $w[i]$ is the first occurrence of a letter;*
(ii) *if $w = pq$ is primitive with $p, q \in \text{Pal}(\Sigma^*)$ then $\lim_{n \longrightarrow \infty} G_W(n) = \infty$;*
(iii) *if w is not the product of two palindromes then G_W is eventually periodic.*

Proof. (i) Obvious. (ii) In this case by Theorem 4 of [11] the palindromic language of W is infinite. Since for all $k \geq 0, (pq)^k p$ is a palindromic prefix of W, there are infinitely many palindromic prefixes of W. Moreover, we have $G_W(i) = G_W(i - |w|) + |w|$ for $i \geq 2|w|$.

(iii) Here again by Theorem 4 of [11], the palindromic language of W is finite. Therefore, let u be the shortest prefix of W containing all the palindromes, and let k be the smallest integer such that $u \in \text{Pref}(w^k)$ then we have $G_W(i) = G_W(i + k|w|)$. □

Examples. Let $W = (abc)^\omega$, whose palindromic language is $P = \{a, b, c\}$ taken from [11] (Section 3). Then we have the following values for G and H:

i	0	1	2	3	4	5	6	7	8	
W	a	b	c	a	b	c	a	b	c	...
G	1	1	1	1	1	1	1	1	1	...
H	1	1	1	0	0	0	0	0	0	...

Here are some typical periodic words with their characteristic functions:

W	G_W
a^n	$[1, 2, 3, 4, 5, \cdots]$
$a.b^n$	$[1, 1, 2, 3, 4, 5, \cdots]$
$(ab)^n$	$[1, 1, 3, 3, 5, 5, 7, 7, 9, 9, \cdots, (2n+1), (2n+1), \cdots]$

Moreover they are all full, since G and H coincide.

Another periodic example illustrating Lemma 2 (ii) is $W = (aba.cbc)^\omega$. Its palindromic language is infinite and W has infinitely many palindromic prefixes, and we have

i	0	1	2	3	4	5	6	7	8	9	10	11	12	13	14	15	16	17	18	19	20	
W	a	b	a	c	b	c	a	b	a	a	b	a	c	b	c	a	b	a	c	b	c	...
H	1	1	3	1	0	3	5	7	9	5	7	9	11	13	15	11	13	15	17	19	21	...

Observe also that there are non periodic words U such that G_U is periodic. Indeed, take any nonperiodic word, for instance the Fibonacci word defined as

$$F = \varphi^{\omega}(a) = abaababaabaabab\cdots, \quad \text{where} \quad \varphi(a) = ab; \varphi(b) = a.$$

Define the morphism $\theta : \{a,b\} \longrightarrow \{a,b,c,d\}^*$ by $a \mapsto abcd; b \mapsto acbd$. Then the word $W = \theta(F)$ is nonperiodic, but $G_W = (1111)^{\omega}$. Nevertheless we have the following result showing a local periodical behaviour.

Lemma 3 *Let $w \in \Sigma^*$. If there exists i such that $G(i) = G(i+k) = l$, with $l \geq k$, then the factor $f = w[(i-l+1)..(i+k)]$ has period $2k$, and any factor of length $2k$ of f is the product of two palindromes.*

Proof. Assume that q and p are the longest palindromic suffixes of length l at positions respectively i and $i+k$. Then there exist x and z such that $f = qz = xp$. In the case $l = k$, we have $f = qp$ and the claim is true. If $l > k$ there exists a non-empty word y such that $q = xy$ and $p = yz$. It follows from Proposition 1 that $2|x|$ is a period of f, and $x\tilde{z}$ is a product of two palindromes. Therefore, any factor of length $2k$ is the product of two palindromes since it is a conjugate of $x\tilde{z}$. □

The function G satisfies the following properties

Proposition 3 *For any finite word $w \in \Sigma^*$, the following properties hold :*

(i) $G(i) \leq \max\{|p| : p \in \mathrm{Pal}(w[0..i])\} \leq i+1$
(ii) $G(j) \leq G(i) + 2(j-i)$, *for all* $j \geq i$;
(iii) $G(i{+}1) = G(i) \implies G(i)$ *and* $G(i{+}1)$ *are odd, and* $G(i{+}2) \in \{G(i){+}2, 2\}$;
(iv) $G(i+1) = G(i) + 1 \implies \mathrm{LPS}(w[0..i]) = \alpha^{G(i)+1}$, *for some* $\alpha \in \Sigma$;

Proof. (i) is obvious. (ii) First, note that $G(i+1) \leq G(i) + 2$ since the longest palindromic suffix at position $i+1$ contains a palindrome of length $G(i+1) - 2$ ending at position i. The result follows by induction.

(iii) Follows from Lemma 3. (iv) Let p and q be the respective palindromes at positions i and $(i+1)$. Then we have $q = p\alpha$ for some $\alpha \in \Sigma$, and we conclude by using Proposition 1. □

Lemma 4 *Let $i \leq k$. If $G(k) = G(i) + 2(k-i)$, then $G(j) = G(i) + 2(j-i)$ for all $i \leq j \leq k$.*

Proof. $G(k) - 2(k-j) \leq G(j) \leq G(i) + 2(j-i)$ and the left term is equal to

$$G(i) + 2(k-i) - 2(k-j) = G(i) + 2(j-i). \quad \square$$

The next proposition is obtained by adapting the proof of Proposition 3.

Proposition 4 *For any finite word $w \in \{a,b\}^*$, the function H satisfies*

(i) $H(i+1) - H(i) \leq 2$;
(ii) $H(i+1) = H(i) \implies H(i+1)$ *and* $H(i)$ *are both odd;*
(iii) $H(i) \leq \max\{|p| : p \in \mathrm{Pal}(w)\}$;
(iv) *if* $H([i..(i+k+2)]) = [n, 0, \cdots, 0, m]$ *for some* i, *then* $m < n + 2k$.

4 Reverse engineering the functions G and H

Here we tackle the following problems. Given a (finite or infinite) sequence s of integers, does there exist a word w such that $H_w = s$ or $G_w = s$? If such a word w exists, under which conditions is it unique up to permutation of the letters ?

We say that a finite/infinite sequence s is *G-consistent* (resp. *H-consistent*) on Σ if there is at least one nonempty word $w \in \Sigma^\infty$ such that for all i, $G_w(i) = s[i]$ (resp. $H_w(i) = s[i]$). If there is only one such word (up to permutation of the letters) then s is said to be *unambiguous.* A first simple result follows:

Proposition 5 *Let Σ be an alphabet of at most 3 letters. Then any G-consistent sequence on Σ is unambiguous.*

Proof. Let s be a G-consistent sequence. We proceed by induction on the length of s. Then $s[0] = 1$ so that the base of the induction is trivially satisfied by choosing one letter in Σ. Assume that $s[0..i]$ is unambiguous. Then there exist a word w, such that $G_w[0..i] = s[0..i]$. Two cases arise:

(a) $s[i+1] > 1$: we set $w[i+1] = w[i+2-s[i+1]]$.
(b) $s[i+1] = 1$: if $|s|_1 = 2$ then $|\Sigma| = 2$, so that $w[i+1] \in \Sigma \setminus \{w[0]\}$.
If $|s|_1 > 2$ then $|\Sigma| = 3$ and we have to consider two cases:
- if $|s[0..i]|_1 = 2$, then we set $w[i+1]$ to the remaining letter;
- if $|s[0..i]|_1 > 2$, then $w[0..i] = p\gamma\beta^k\alpha^l$ where $\Sigma = \{\alpha, \beta, \gamma\}$, $p \in \Sigma^*$, and $k, l \geq 1$, and we set $w[i+1] = \gamma$. □

Observe that for larger alphabets, that is when $|\Sigma| > 3$, G-consistent sequences are not necessarily unambiguous, as shown in the following examples.

Example. Let $\Sigma = \{a, b, c, d\}$ and consider the sequence $s = [1, 1, 1, 3, 2, 1, 3, 5]$. There is a unique word $w[0..4]$ which is G-consistent with $s[0..4]$:

i	0	1	2	3	4	5	6	7
s	1	1	1	3	2	1	3	5
w	a	b	c	b	b	**a**	b	b
w'	a	b	c	b	b	**d**	b	b

while two different words are consistent with $s[0..5]$, a fact that follows from Lemma 2(i).

One can easily see that the previous ambiguity is related with the presence of more than three 1's in the sequence s. However here is a word w having four occurrences of 1, but uniquely determined by G as well.

Example. Let $\Sigma = \{a, b, c, d\}$ and let $s = [1, 1, 1, 3, 1, 3, 5, 7, 9]$. There is a unique word which is G-consistent with s:

i	0	1	2	3	4	5	6	7	8
s	1	1	1	3	1	3	5	7	9
w	a	b	c	b	d	b	c	b	a

The situation is clearly explained by the following statement

Proposition 6 *Let s be a G-consistent sequence. If there exist two distinct words w, w' consistent with s, then there exists i such that $G_w(i) = G_{w'}(i) = 1$ and $H_w(i) = 0$ or $H_{w'}(i) = 0$.*

Proof. Indeed, if $s[i] = 1$, then $w[i]$ is either a new letter or, a previously encountered letter such that the longest palindromic $\text{LPS}(w[0..i])$ is the letter itself. □

Consider now the same problem for the function H. Since the functions G and H coincide for full words, we have immediately the next result.

Corollary 1 *Any full word (thus any Sturmian word) is uniquely determined by the function H.*

So, the function G_w encodes all the information on w, but this is no longer true for the function H_w. Indeed, there exist H-consistent sequences that are not unambiguous as shown in the following example: consider the word $w = abbabbbabaabb$. Then, we have

$$H_{wa} = H_{wb} = (1, 1, 2, 4, 3, 5, 3, 5, 7, 3, 2, 4, 0, 0) \tag{3}$$

but $wa \neq wb$.

Observe that the counterpart of Proposition 6 does not hold for the function H. Indeed, every 1 in the sequence $s = H_w$ corresponds necessarily to a new letter in w. Consequently the presence of 1's does not cause ambiguity, and $|s|_1 = |\Sigma|$, as shown below for a 5-letter alphabet $\{a, b, c, d, e\}$.

i	0	1	2	3	4	5	6	7	8
s	1	1	2	1	3	5	1	3	5
w	a	b	b	**c**	b	b	**e**	b	b
w'	a	b	b	**d**	b	b	**c**	b	b

We point out that w and w' are the same word up to a relabelling of the letters.

For words which are not full, the study of the H function is more complex. However, there are some special conditions ensuring that an H-consistent sequence s on a given Σ is also unambiguous.

Proposition 7 *Let s be an H-consistent sequence such that $s = s_1\, 0^k\, m\, s_2$ with $m \neq 0$, and s_1 does not contain any 0. If $m > 2k+1$ then there is a unique word $w[\,0..(|s_1| + k)\,]$ such that $H_w = s[\,0..(|s_1| + k)\,]$.*

The proof is similar to that of proof of Proposition 5. As an example, consider the following H-consistent sequence on $\Sigma = \{a, b\}$

i	0	1	2	3	4	5	6	7	8	9	10	11
s	1	2	1	2	4	6	4	3	3	0	0	6
w	a	a	b	b	a	a	b	a	b	x	y	z

where $s_1 = [1, 2, 1, 2, 4, 6, 4, 3, 3]$, $m = 6$, and $k = 2$. The last three elements of w can be uniquely determined since the factor $b\,a\,b\,x\,y\,z$ has to be a palindrome, that is $x = z = b$, and $y = a$.

Note that the bound k on the length of a subsequence of 0's in s given in Proposition 5 does not depend on the cardinality of Σ. On the other hand, observe that if $|\,\Sigma\,| = 2$ and $k = 1$ the sequence s is still uniquely determined. For instance, consider the sequence $s[0..12] = [1, 2, 1, 3, 3, 2, 4, 6, 5, 3, 5, 0, 3]$, with $\Sigma = \{a, b\}$, and therefore $m = 3$:

i	0	1	2	3	4	5	6	7	8	9	10	11	12
s	1	2	1	3	3	2	4	6	5	3	5	0	3
w	a	a	b	a	b	b	a	b	b	b	a	a	a

Here the cardinality of the alphabet allows only one possible choice of the letter consistent with $H_w(11) = 0$, consequently $m \not> 2k + 1 = 3$, but the word w is uniquely determined as well.

4.1 Infinite words

In the case of infinite words, the situation is similar and ambiguous H can also occur. We start by recalling some facts. From [8, 11], we know that, when analyzing the defect and the lacunas of an infinite word, it can present

(a) an infinite palindromic complexity with a finite number of lacunas;
(b) a finite palindromic complexity with an infinite number of lacunas;
(c) both infinite palindromic complexity and number of lacunas.

In general, in none of the three cases the function H is unambiguous, as it is shown in the following examples.

Case (a): consider two words U and V having the same prefix of length 23

$$u_1 = a\,b\,b\,a\,a\,b\,b\,a\,b\,a\,b\,a\,a\,a\,a\,b\,a\,b\,b\,b\,b\,a\,a,$$

and such that $U = u_1\,a\,(a\,b)^\omega$ and $V = u_1\,b\,(b\,a)^\omega$. The two sequences $H_U([0..22])$ and $H_V([0..22])$ are equal since they share a common prefix. Now, since the suffix parts of U and V starting at position 23 satisfy $\overline{U[\geq 23]} = V[\geq 23]$, we have $H_U = H_V$. Since, both u and v are eventually periodic, and since their period is the product of two palindromes, the palindromic complexity of both u and v is infinite. Finally, an easy check reveals that the suffix sequence of the function H, for $n > 2$, is

$$H_U([\geq 23]) = (0,0,0,0,0,0,0,7,7,9,9,11,11,\ldots,(2n+3),(2n+3),\ldots).$$

Case (b): let $w = abbabbbabaabb$, already used in Equation (3), and consider the words U and V defined as follows, by means of the word w:

$$U = w \cdot ab \cdot bbaaa \cdot (baabba)^{\omega},$$

$$V = w \cdot ba \cdot bbaaa \cdot (baabba)^{\omega}.$$

The sequences H_U and H_V coincide and are

$$H_U = H_V = (1,1,2,4,3,5,3,5,7,3,2,4,0,0,0,0,0,0,0,3,5,0,5,0,0,0,\\ 0,0,0,0,0,0,0,0,0,\cdots)$$

All the terms after position 22 are equal to 0 since the words U and V are eventually periodic, with a period which is not the product of two palindromes.

Case (c): finally, consider the words U and V defined as follows (using again $w = abbabbbabaabb$):

$$U = w \cdot ab \cdot bbaaa \cdot baab \cdot baabba \cdot (baab)^2 \cdot baabba \ldots (baab)^n \cdot baabba \ldots$$

$$V = w \cdot ba \cdot bbaaa \cdot baab \cdot baabba \cdot (baab)^2 \cdot baabba \ldots (baab)^n \cdot baabba \ldots.$$

The sequences H_U and H_V coincide and their first terms are

$$H_U = H_V = (1,1,2,4,3,5,3,5,7,3,2,4,0,0,0,0,0,0,0,3,5,0,5,0,0,0,\\ 6,8,6,8,0,0,0,0,0,0,0,0,0,0,10,12,10,12,0,0,0,0,0,0,\\ 0,0,0,0,0,0,0,0,14,16,14,16,0,0,\ldots).$$

The two sequences have an infinite number of new palindromes since the palindromic factor $baab$ is repeated an increasing number of times at each step. At the same time the set of lacunas is infinite since the factor $baabba$, which is not the product of two palindromes, occurs infinitely many times.

5 Further work

The problem of reconstructing words from the functions G and H leads to many interesting developments, some of them requiring a deeper analysis in order to produce efficient decision algorithms.

Consistency. Deciding if a given finite sequence s of numbers is G-consistent (resp. H-consistent) may be easily achieved. Indeed, let $k = |s|_1$. This implies that the smallest alphabet Σ we have to consider contains at most k letters (exactly k for H-consistency, by virtue of Lemma 2 (i)). Taking an order on the letters of Σ permits to restrict the study to classes of words equivalent under permutations of letters. A close look to the proof of Proposition 5 reveals all the information in order to construct sequentially all words consistent with s: indeed, it suffices to check at each position i, if $\mathrm{LPS}(w[0..i]) = s[i]$.

Random and exhaustive generation. The algorithms described above may be used for constructing trees of words. Indeed, at each step i one constructs a trie of words having height $i+1$ and satisfying $G[i] = s[i+1]$ (resp. $H[i] = s[i+1]$). The process stops if either it is impossible to construct the next step, or ends successfully if $i = |s|$. In case of a successful termination it is easy to check if every non leaf node has a unique son, solving the unambiguity problem of the sequence s. These are the basic tools for constructing randomly or exhaustively many classes of words, for instance all full words of length n.

Enumeration. Counting classes of G-consistent or H-consistent sequences follows naturally. For instance, given a fixed length n, it amounts to count for a fixed alphabet Σ the set $\{ H_w : w \in \Sigma^n \}$. Indeed, a greedy algorithm can be implemented to obtain the first values: it suffices to generate all words in Σ^n, and to compute H for each such word.

The enumeration formula of the finite Sturmian words is known [27]. Since they are full, a closely related counting problem is that of determining a formula for the number of non-Sturmian full words on the alphabet $\{a, b\}^*$. Determining the number of words having a fixed number of lacunas is also challenging.

Characterization of special classes of G or H functions. For instance, given a 2-letter alphabet $\Sigma = \{a, b\}$ one might look for a description of the following sets of functions:

$$\mathcal{G} = \{G_w : w \in \{a,b\}^*\} \qquad \mathcal{H} = \{H_w : w \in \{a,b\}^* \text{ such that } w \text{ is full}\}.$$

In another direction it would be interesting to describe infinite words on fixed alphabets whose G (or H) sequence is automatic.

Constrained reconstruction. Given a finite set of palindromes P, how can we determine the shortest full word containing all the palindromes of P and only those palindromes? The answer is based on Theorem 1 of [11]. Indeed, the language of words having exactly P as palindromic factors is rational. Therefore there exists a deterministic minimal automaton recognizing all these words. For each palindrome q in P, there is a unique path starting from the initial state whose trace is q. Collecting the target states $\mathrm{T}(P)$ of all paths computing P, it suffices then to compute the shortest path starting from the initial state and containing all states in $\mathrm{T}(P)$. It may or may not exist, and if it does not, one might relax the conditions by allowing some extra palindromes in order to find a solution.

Structure of full words. Let w be a finite full word on a 2-letter alphabet Σ. One can easily prove that H_w and $H_{\widetilde{w}}$ have the same elements, while $H_w = H_{\widetilde{w}}$ if and only if $\widetilde{w} = \overline{w}$ or $\widetilde{w} = w$. The two sets of longest unioccurrent palindromic suffixes of w and $\widetilde{w}$ naturally define a permutation on the set $\{1, 2, \ldots, |w|\}$. More precisely, let p_1, p_2, ..., $p_{|w|}$ be the longest palindromic suffixes of w in order of their first occurrence in w and let x_i be the position of the last occurrence of p_i in w. We define the permutation π_w on $\{1, 2, \ldots, |w|\}$ by

$$\pi_w(i) = |w| + |p_i| - |x_i|.$$

Now, let q_1, q_2, ..., $q_{|w|}$ be the longest palindromic suffixes of $\widetilde{w}$ in order of their first occurrence in $\widetilde{w}$. Then $p_i = q_{\pi_w(i)}$, for $i = 1, 2, \ldots, |w|$. We illustrate this fact by an example: let $w = ababbabab$, so that $\widetilde{w} = bababbaba$. Then we have the following table showing that $H_w \neq H_{\widetilde{w}}$,

i	1	2	3	4	5	6	7	8	9
LPSU(w)	*a*	*b*	*aba*	*bab*	*bb*	*abba*	*babbab*	*ababbaba*	*babab*
LPSU($\widetilde{w}$)	*b*	*a*	*bab*	*aba*	*babab*	*bb*	*abba*	*babbab*	*ababbaba*

and the permutation π_w is $(2, 1, 4, 3, 6, 7, 8, 9, 5)$.
We would like to study the combinatorial properties of π_w in relation with those of the word w. In particular we are interested in characterizing the permutations π_w associated with full words. A similar study can also be performed on arbitrary alphabets provide one replaces the $\overline{(\,)}$ operation by an arbitrary permutation of the alphabet Σ.

Acknowledgements

We are grateful to the anonymous referees for their careful reading and useful comments.

References

1. Aberkane, A. and Brlek, S. (2002) Suites de même complexité que celle de Thue-Morse, *Actes des Journées Montoises d'informatique théorique (9-11 septembre 2002, Montpellier, France)* 85–89.
2. Aberkane, A., Brlek, S. and Glen, A. (2007) Sequences having the Thue-Morse complexity, *Disc. Math.* 15p. (submitted)
3. Allouche, J.-P. (1994) Sur la complexité des suites infinies, *Bull. Belg. Math. Soc.* 1:133–143.
4. Allouche, J.P. (1997) Schrödinger operators with Rudin-Shapiro potentials are not palindromic, *J. Math. Phys.* 38:1843–1848.
5. Allouche, J.P., Baake, M., Cassaigne, J., and Damanik, D. (2003) Palindrome complexity, *Theoret. Comput. Sci.* 292:9–31.

6. Allouche, J.P., and Shallit, J. (2000) Sums of digits, overlaps, and palindromes, *Disc. Math. and Theoret. Comput. Sci.* 4:1–10.
7. Baake, M. (1999) A note on palindromicity, *Lett. Math. Phys.* 49:217–227.
8. Blondin Massé, A., Brlek, S., and Labbé, S. (2008) Palindromic lacunas of the Thue-Morse word, GASCOM 2008 (To appear)
9. Fici, G., Mignosi, F., Restivo, A. and Sciortino, M. (2006) Word assembly through minimal forbidden words, *Theoret. Comput. Sci.*, 359/1-3: 214–230.
10. Brlek, S. (1989) Enumeration of factors in the Thue-Morse word, *Disc. Appl. Math.* 24:83–96.
11. Brlek, S., Hamel, S., Nivat, M., and Reutenauer, C. (2004) On the Palindromic Complexity of Infinite Words, in J. Berstel, J. Karhumäki, D. Perrin, Eds, Combinatorics on Words with Applications, Int. J. of Found. of Comput. Sci., 15/2:293–306
12. Brlek, S., and Ladouceur, A. (2003) A note on differentiable palindromes, *Theoret. Comput. Sci.* 302:167–178.
13. Brlek, S., Jamet, D., and Paquin, G., (2008) Smooth Words on 2-letter alphabets having same parity, *Theoret. Comput. Sci.* 393/1-3:166181.
14. Brlek, S., Dulucq, S., Ladouceur, A. and Vuillon L. (2006) Combinatorial properties of smooth infinite words, *Theoret. Comput. Sci.* 352/1-3:306–317.
15. de Bruijn N. G. (1946) A Combinatorial Problem, *Koninklijke Nederlandse Akademie v. Wetenschappen* 49: 758764.
16. Droubay, X., Justin, J., and Pirillo, G. (2001) Episturmian words and some constructions of de Luca and Rauzy, *Theoret. Comput. Sci.* 255:539–553.
17. Flye Sainte-Marie, C. (1894). Question 48, *L'Intermdiaire Math.* 1: 107110.
18. Fredericksen, H. and Maiorana, J. (1978) Necklaces of beads in k colors and k-ary de Bruijn sequences *Disc. Math.* 23/3, 207–210
19. Good, I. J. (1946) Normal recurring decimals, *J. London Math. Soc.* 21 (3): 167–169.
20. Hof, A., Knill, O., and Simon, B. (1995) Singular continuous spectrum for palindromic Schrödinger operators, *Commun. Math. Phys.* 174:149–159.
21. Lothaire M. (1983) Combinatorics on words, Addison-Wesley.
22. Lothaire, M. (2002) Algebraic Combinatorics on words, Cambridge University Press.
23. de Luca, A. (1997) Sturmian words: structure, combinatorics, and their arithmetics, *Theoret. Comput. Sci.* 183:45–82.
24. de Luca, A. , and Varricchio, S. (1989) Some combinatorial properties of the Thue-Morse sequence, *Theoret. Comput. Sci.* 63:333–348.
25. Mantaci, S., Restivo, A., Rosone, G. and Sciortino, M. (2008) A New Combinatorial Approach to Sequence Comparison *Theory Comput. Syst.* 42/3:411–429.
26. Manvel, B., Meyerowitz, A., Schwenk*, A., Smith, K. and Stockmeyer, P. (1991) Reconstruction of sequences, *Disc. Math.* 94/3: 209–219.
27. Mignosi, F. (1991) On the number of factors of Sturmian words, *Theoret. Comput. Sci.* 82/1: 71–84.
28. Morse, M. and Hedlund, G. (1940) Symbolic Dynamics II. Sturmian trajectories, *Amer. J. Math.* 62:1–42.

The mv-decomposition: definition and application to the distance-2 broadcast problem in multi-hops radio networks

Olivier Cogis, Benoît Darties, Sylvain Durand, Jean-Claude König, and Geneviève Simonet

LIRMM UMR5506 - Université Montpellier II - 161, rue Ada, 34392 Montpellier Cedex 05, France - firstname.lastname@lirmm.fr

Abstract We present a new tool called the "mv-decomposition", and we describe some interesting algorithmic properties about it. We propose an algorithm with a complexity of $O(m)$ to build a mv-decomposition for each bipartite graph. We use this mv-decomposition to propose a solution to the distance-2 broadcast problem in a synchronous multi-hops radio networks where adjacent transmissions are subject to interferences. More precisely, we propose two algorithms of resolution: the first one guarantees a complete distance-2 broadcast scheme using $O((\log n)^2)$ slots for a time complexity of $O(m(\log n)^2)$, while the second builds a solution with a minimal number of transmissions for a time complexity of $O(m)$.

1 Introduction

In a multi-hops radio network, nodes communicate with each other via multi-hops wireless links. The use of the radio medium implies some restrictions and properties: whenever a node transmits, all the nodes in its communication range may receive the transmission. Incoming messages have to be forwarded to reach nodes which are located at more than one hop from the source. Since all nodes share the same frequency channel, a collision may occurs if two or more neighbors transmit simultaneously, preventing correct reception of the message. This paper deals with the broadcast problem which refers to the sending of a message from a source node to all the other nodes of the network. We consider the simplified communication model used in [4, 5]: nodes send messages in synchronous slots. In each slot each node acts either as a transmitter or as a receiver. A node acting as a receiver in a given slot gets a message if and only if exactly one of its neighbors transmits in this slot. In addition, the topology of the network is assumed to be known by all the nodes. This model has been widely considered to analyze the complexity of the broadcast problem. According to this model, a valid broadcast strategy consists of finding a schedule scheme, i.e. a particular schedule of transmissions among the network nodes.

Many research have focused on producing schedule-based broadcasting schemes in known radio networks. Chlamtac and Kutten have proved that finding a scheme with a minimum number of slots is a NP-Hard problem [4]. Authors from [5] have first proposed a polynomial algorithm in $O(nm(\log n)^2)$ for constructing a schedule which achieves a broadcast in $O(D.(\log n)^2)$ slots, where D is the source eccentricity,

---

Please use the following format when citing this chapter:

Cogis, O., et al., 2008, in IFIP International Federation for Information Processing, Volume 273; *Fifth IFIP International Conference on Theoretical Computer Science*; Giorgio Ausiello, Juhani Karhumäki, Giancarlo Mauri, Luke Ong; (Boston: Springer), pp. 115–126.

n the number of nodes and m the number of links. Other results have progressively reduced this bound to $O(D.\log n + (\log n)^2)$ in [2, 11], then $O(D + (\log n)^5)$ [9], to end with $O(D + (\log n)^4)$ [10]. In this last paper authors announce a schedule scheme for broadcasting which requires $O(D + (\log n)^3)$ slots when the network graph is planar. In [1], the authors present a class of 2-diameter graphs which require $\Omega((\log n)^2)$ slots to complete a broadcast.

The broadcast problem has also been studied under the assumption that the topology is unknown: a first scheme using $O(n^{11/6})$ slots has been proposed in [6]. This bound has been decreased in multiple works [12, 7, 13] to reach $O(n(\log n)^2)$ slots in [8]. Actual lower bounds for the broadcast problem without knowledge of the topology are in $\Omega(n \log n)$ [6, 3].

This paper is organized as follows: in section 2 we present a new tool: the mv-decomposition, and describe some of its algorithmic properties. We use the mv-decomposition in section 3 to propose strategies with performance guarantees for the distance-2 broadcast problem: this problem is a restricted version of the broadcast problem in which the objective consists of informing nodes located at two hops away from the source node. We conclude this section by giving an algorithm which constructs a distance-2 broadcast strategy requiring $O((\log n)^2)$ slots. The quality of the strategy returned by our algorithm is the same as the solution proposed by [5], but the computation time complexity is improved from $O(nm(\log n)^2)$ to $O(m(\log n)^2)$.

2 A new tool: the mv-decomposition

This section is organized as follows: in a first step, we propose a common graph model for radio networks, and we introduce some useful definitions and notations. Then we propose a new tool, which we call the mv-decomposition, and pose some algorithmic properties. We also propose an algorithm with a complexity of $O(m)$ to compute an mv-decomposition for each bipartite graph.

2.1 Model description and definitions

A radio network is commonly modelized by an undirected graph $G = (V, E)$, where V represents the network nodes, and E contains pairs of nodes which can directly communicate. The source node is noted s.

Let $G = (X, Y, E)$ be a bipartite graph. A *cover* of a subset $Y' \subseteq Y$ in G, is a subset $X' \subseteq X$ such that $Y' \subseteq N_G(X')$, where $N_G(X')$ is the union of neighborhoods of vertices of X' in G.

We say that X' is a *minimal cover* (for the inclusion) of Y' in G when X' is a cover of Y, but none of its subsets is.

For a given cover X' of Y in G, we note $mv_G(X')$ the set of neighbors of X' which are adjacent to exactly one element of X'.

Lemma 1. *let $G = (X,Y,E)$ be a bipartite graph and $X' \subseteq X$ a minimal cover of $Y' \subseteq Y$.*

Then each vertex of X' has an adjacent vertex in Y which is not adjacent to any other vertex of X'. In other words, $mv_g(X') \geq |X'|$.

Proof. Let $X' \subseteq X$ be a cover of $Y' \subseteq Y$ in G_B, and x be a vertex of X'. If each neighbor of x in Y is also adjacent to another vertex in X', then $X' - \{x\}$ is still a cover of Y'.

Let $G = (X,Y,E)$ be a bipartite graph such that X covers Y. We say that a collection $(X_i)_{i\in I}$ of subsets of X *saturates* Y in G_B when $Y = \bigcup_{i\in I} mv_G(X_i)$. Then the saturation cost of G is the minimal cardinal of a collection of subsets of X which saturates Y in G. We note it $\sigma(G)$.

2.2 The mv-decomposition: definition and properties

In the following sections, let us define $X_0 = X$ and $Y_0 = Y$. A *mv-decomposition* of a bipartite graph G consists of the data of an integer K, a collection $(X_i)_{1\leq i\leq K}$ of K subsets of X which saturates Y in G, and two others collections $(Y_i)_{1\leq i\leq K}$ and $(Z_i)_{1\leq i\leq K}$, such that for each i with $X_i \neq \emptyset$ we have :

- $X_{i+1} \subseteq X_i$ is a minimal cover of Y_i,
- Z_i is defined such that the subgraph of G induced by $X_i \cup Z_i$ is a perfect matching: each vertex has degree 1,
- $Y_{i+1} = Y_i - Z_{i+1}$.

The *depth* of an mv-decomposition is the smallest value K, for which $Y_K = \emptyset$. Let us note that, for any collection $(X_i)_{1\leq i\leq K}$ which saturates Y in G, one can deduce an mv-decomposition of depth K, by computing the sets Y_i and Z_i from the knowledge of X_i. That is why, in the following, an mv-decompostion is sometime described as the collection $(X_i)_{1\leq i\leq K}$.

Property 1 *Let $G = (X,Y,E)$ be a bipartite graph such that X covers Y. Then for each mv-decomposition we have:*

1. *$\{X_i\}_{0\leq i\leq K}$ et $\{Y_i\}_{0\leq i\leq K}$ are two sequences such that $X_i \subseteq X_{i-1}$ and $Y_i \subseteq Y_{i-1}$, with $X_K \neq \emptyset$ and $Y_K = \emptyset$. In addition X_i covers Y_i for $0 \leq i \leq K$.*
2. *$\{Z_j\}_{i\leq j\leq K}$ is a partition of Y_{i-1}. In particular $\{Z_i\}_{1\leq i\leq K}$ is a partition of Y.*
3. *For each i such that $1 \leq i \leq K$, we have $|Z_i| = |X_i| \neq \emptyset$, and $Z_i \subseteq mv_G(X_i)$.*
4. *For each i such that $1 \leq i \leq K$, each vertex x of X_i has, for each j such that $1 \leq j \leq i$, exactly one neighbor in Z_j which is not adjacent to any other vertex of X_i.*

Proof. Let us consider $X_i \subseteq X$ and $Y_i \subseteq Y$, $Y_i \neq \emptyset$, such that X_i covers Y_i (true for $i = 0$). Then Y_i has some minimal cover $X_{i+1} \subseteq X_i$. Lemma 1 allows to affirm that for each $X_{i+1} \neq \emptyset$, Z_{i+1} is defined and not empty, and then that Y_{i+1} is strictly included in Y_i. This also guarantees that X_{i+1} is a cover of Y_{i+1}. This proves points (1), (2) and (3).

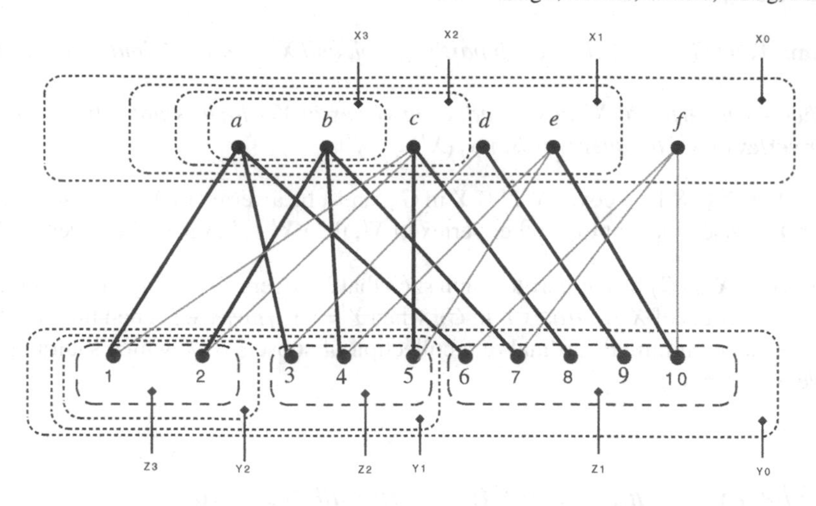

Fig. 1 an mv-decomposition of a bipartite graph

For each i and j such that $1 \leq j \leq i \leq K$, we have $X_i \subseteq X_j$. Then any vertex x of X_i is also a vertex of X_j. Since the subgraph of G induced by vertices $X_j \cup Z_j$ is a perfect matching, then there exists a vertex $z_j \in Z_j$ adjacent to x but not with any other vertex of X_j.

For any mv-decomposition of G with a depth K, we have $K \leq \Delta_G(X)$, where $\Delta_G(X)$ is the maximum degree of a node of X in G.

Property 2 *Let $G = (X,Y,E)$ be a bipartite graph such that X covers Y. Then for any mv-decomposition of G with a depth K, we have:*

$$K \leq \Delta_G(X) \tag{1}$$

Where $\Delta_G(X)$ is the maximum degree of a vertex of X in G.

Proof. According to points 2 and 4 of property 1, we have $d_G(x) \geq K$. This allows us to conclude.

We propose the algorithm "mv-decomposition" which computes an mv-decomposition from a given bipartite graph $G = (X,Y,E)$.

Algorithm 1: mv-decomposition

```
Data: A bipartite graph G = (X,Y,E)
Result: A collection (X_i)_{1<=i<=K} of subsets of X which saturates Y in G
   // Variables declaration :
1  Stack P[x]: stack of vertices of Y that are adjacent only to x, ∀x ∈ X.
2  int L: number of vertices of Y which have been saturated.
3  int i: actual depth
   // variables initialization :
4  L = 0 ; X[0] = X; i = 1
5  Initialize P[x], ∀x ∈ X.
6  while L < |Y| do
7      X[i] = ∅
       // Computing a minimal cover X[i] :
8      foreach x ∈ X[i−1] do
9          if |P[x]| = 0 then
               // Suppress x from the neighborhood of its
               neighbors :
10             foreach y ∈ N(x) do
11                 N(y) = N(y) − {x}
12                 if |N(y)| = 1 then
                       // If y has only one neighbor z, it is
                       added to P[z]
13                     P[N(y)] = P[N(y)] ∪ {y}
14                 end
15             end
16         else
               // x is selected in the current cover.
17             X[i] = X[i] ∪ {x}
               // a vertex y becomes the receiver of the
               transmission of x :
18             Let y ∈ P[x]. P[x] = P[x] − {y}.
19             L++
20         end
21     end
22 end
```

Theorem 1 *The algorithm mv-decomposition has a complexity of $O(m)$.*

Proof. The initialization phase (line 5) runs in $O(m)$ and consists of filling the stacks $P[x], \forall x \in X$.

Thereafter, for a given $x \in X$:

- The part of code between lines 9 and 16 is executed at most once, and consists of suppressing the vertex x from G.
- The part of code between lines 17 and 21 is executed at most $d_G(x)$ times.

The part of code between lines 9 and 15 has a complexity of $O(d_G(x))$ (by using advanced implementation techniques). The part of code between lines 17 and 21 has a complexity of $O(1)$.

It is concluded that the overall complexity of the algorithm is of the order of :

$$O(\sum_{x \in X} d_G(x)) = O(m)$$

3 Using the mv-decomposition to solve the distance-2 broadcast problem

We employ the mv-decomposition to define solutions with performance guarantees for the distance-2 broadcast problem in multi-hops synchronous radio networks. This problem is a particular case of the broadcast-problem and can be described as follows: let us consider a single source broadcast problem. After the first slot is completed, all the nodes which are adjacent to the source node have a knowledge of the broadcasted information. Their transmissions must be scheduled in order to inform all the nodes that are two hops away from the source. A recursive approach of this process, depending on the distance of nodes from the source, allows to broadcast the message on the whole network.

The data can be restricted to a bipartite graph $G = (X, Y, E_B)$ where X and Y respectively denotes the set of vertices at distance 1 and 2 of s in G, and E the set of possible direct communications: $E_B = \{\{x,y\} | x \in X, y \in Y, \{x,y\} \in E\}$. We say that finding a distance-2 broadcast strategy consists of broadcasting a single message from nodes of X to nodes of Y. In a synchronous model, two important criterias are the number of required slots, and the number of realized transmissions.

In the first sub-section, we use the mv-decomposition to propose a distance-2 broadcast strategy with a minimal (not minimum, which is an NP-hard problem) number of transmissions, and a number of slots bounded by the maximum degree of the graph. In the second sub-section, we propose an algorithm to compute a distance-2 broadcast strategy with $O(\log n)^2)$ slots, for a time complexity of $O(m(\log n)^2)$.

3.1 *Minimizing both the number of slots and the number of transmissions*

Let I be an instance of the distance-2 broadcast problem composed of a bipartite graph $G = (X, Y, E)$ such that X covers Y.

The following theorem establishes a link between a cover of Y in a bipartite graph, and the number of required transmissions for the distance-2 broadcast problem on the same graph.

Theorem 2 *Let $G = (X, Y, E)$ be a bipartite graph such that X covers Y. Then we have:*

1. *If C is a minimal cover of Y, then there exists a broadcast strategy from X to Y with a minimal number of transmissions equal to the cardinality of C.*
2. *Finding a broadcast strategy with a minimum number of transmissions is tantamount to finding a minimum cover.*

Proof. Let $C \subseteq X$ be a minimal cover of Y of cardinality k, with $C = \{c_i\}_{1 \le i \le k}$. Let us consider the following strategy : during each slot exactly one node of C is transmitting. All the nodes of C have transmitted the information after k slots, and no interference has occurred. Thus all the nodes of Y have successfully received the information, and we infer points 1 and 2. The number of transmissions is clearly equal to the cardinality of the cover C. Let us note that as C is minimal, each element of C has to transmit at least once. Q.E.D.

Let us consider a collection $(X_i)_{1 \le i \le K}$ of subsets of X resulting from the mv-decomposition of G. From this mv-decomposition we can propose a distance-2 broadcast strategy S_1: At slot i, all the nodes of X_{K+1-i} are transmitting the message. Since the collection $(X_i)_{1 \le i \le K}$ saturates Y, each node of Y can receive the information. The number of transmissions is equal to $\sum_{i=1}^{K} |X_i|$. This number is not minimal, since X_1 is already a minimal cover of Y.

We define a second strategy S_2 as follows :

- During the first slot, all the nodes of X_K transmit the message.
- During the slot i with $2 \le i \le K$, all the nodes of $X_{K+1-i} - X_{K+2-i}$ transmit the message.

This second strategy differs from the previous one in the fact that when a node transmits at slot i, it does not transmit anymore. We propose the following property :

Property 3 *The strategy S_2 produces a complete broadcast from X to Y.*

sketch of proof: We recall that $X_{i+1} \subseteq X_i$ for all i such that $1 \le i \le k-1$. Each node of X_1 transmits exactly once. The validity of this strategy can be deduced if we compare it with S_1 . □

The number of used slots by strategy S_2 is K. Its cost in number of transmissions is equal to:

$$|X_K| + \sum_{i=2}^{K} |X_{K+1-i} - X_{K+2-i}| = |X_1|$$

If the set X_1 is a minimal cover of Y, then we obtain a valid broadcast strategy (all the nodes will receive the information), where the number of transmissions is minimal, in accordance with Theorem 2.

With both strategies S_1 and S_2, the number of used slots is less or equal than $\Delta_G(X)$, in agreement with property 2. We show in the following sub-section that we can obtain a strategy with a better cost in term of number of slots.

3.2 Minimizing the number of slots

In this sub-section, we propose another strategy to solve the distance-2 broadcast problem. The objective is here to minimize the number of used slots, regardless the number of effective transmissions.

Our approach consists of showing that one can ensure that enough number of nodes of Y could receive the message in exactly one slot. By generalizing this property, we obtain a valid broadcast strategy, and we evaluate its cost in number of slots. First we establish some properties concerning the receptivity, ie the maximum number of nodes that can receive a transmission correctly in one slot.

Property 4 *Let $G = (X,Y,E)$ be a bipartite graph such that X covers Y. Then the receptivity $\rho(G)$ satisfies the following inequation:*

$$\max_{X' \subseteq X} |mv_G(X')| = \rho(G) \geq \max(\Delta_G(X), \frac{|Y|}{\Delta_G(X)})$$

Proof. Let x be a vertex of X having degree $\Delta_G(X)$. Then $|mv_G(\{x\})| = \Delta_G(X)$. The inequality $\rho(G) \geq \Delta_G(X)$ is deduced from the definition of $\rho(G)$.

Let $X' \subseteq X$ be a minimal cover of Y in G. While $Y = \bigcup_{x \in X'} N_G(x)$, then we have :

$$|Y| \leq \sum_{x \in X'} |N_G(x)| = \sum_{x \in X'} d_G(x) \leq |X'| . \Delta_G(X) \leq mv_G(X') . \Delta_G(X)$$

The second inequality $\rho(G) \geq \frac{|Y|}{\Delta_G(X)}$ is deduced again from the definition of $\rho(G)$.

As an immediate corollary of property 4, we have :

$$\rho(G) \geq \sqrt{|Y|}$$

In fact, we are going to improve this bound to show that :

$$\rho(G) \geq \frac{|Y|}{1 + \ln|Y|}$$

Property 5 *Let $G = (X,Y,E)$ be a bipartite graph such that X covers Y. Then for each mv-decomposition for G, we have :*

$$\forall i | 1 \leq i \leq K, |mv_G(X_i)| \geq i \times |X_i| \tag{2}$$

Proof. According to point 4 of property 1, each node x of X_i has in each Z_j, with $1 \leq j \leq i$, a neighbor which is not adjacent to any other vertex of X_i. According to point 2 of the same property, these i neighbors are pairwise distinct.

Theorem 3 *Let $G = (X,Y,E)$ be a bipartite graph such that X covers Y. Then for each mv-decomposition of G, we have :*

$$\rho(G) \geq \max_{1 \leq i \leq K} |mv_G(X_i)| \geq \frac{|Y|}{H_K} \tag{3}$$

$$\sigma(G) \leq K \tag{4}$$

where H_n is the harmonic number $H_n = 1 + \frac{1}{2} + \frac{1}{3} + \cdots + \frac{1}{n}$.

Proof. The first inequality of (3) stems from the definition of $\rho(G)$. The second is deduced from the followings:

$$\begin{aligned}
|Y| &= \sum_{i=1}^{K} |Z_i| \\
&= \sum_{i=1}^{K} |X_i| \\
&\leq \sum_{i=1}^{K} \frac{|mv_G(X_i)|}{i} \\
&\leq \sum_{i=1}^{K} \frac{\max_{1 \leq i \leq K} |mv_G(X_i)|}{i} = \max_{1 \leq i \leq K} |mv_G(X_i)| \times H_K
\end{aligned}$$

Now we prove the inequality 4. Let y be any vertex of Y. According to point 2 of property 1, $y \in Z_i$ for one i such that $1 \leq i \leq K$. According to point 3 of the same property, $y \in mv_G(X_i)$. Then $Y = \bigcup_{i=1}^{K} mv_G(X_i)$, and allows us to conclude.

Theorem 4 *Let $G = (X, Y, E)$ be a bipartite graph such that X covers Y. Then :*

$$\rho(G) \geq \frac{|Y|}{1 + ln\Delta_G(X)} \tag{5}$$

$$\sigma(G) \leq \Delta_G(X) \tag{6}$$

(Let us remind that $\sigma(G)$ is the saturation cost of G, i.e. the minimal cardinality of a collection of subsets of X which saturates Y in G). We note it $\sigma(G)$.

Proof. Can be deduced from theorem 3 and property 2, bearing in mind that the harmonic number H_n is an increasing function of n which satisfies $H_n \leq 1 + \ln n$.

We propose the following algorithm :

Algorithm 2: the algorithm "Saturation"

Data: A bipartite graph $G = (X,Y,E)$ such that X covers Y
Result: A collection $(W_t)_{1 \leq t \leq L}$ which saturates Y in G

1 $R = Y$;
2 $t = 0$;
3 **while** $R \neq \emptyset$ **do**
4 $\quad t = t+1$;
5 $\quad$ compute an mv-decomposition of G[X,R];
6 $\quad$ Let K_t be its depth, and let $(X_i^t)_{0 \leq i \leq K_t}$ be the resulting sequence.;
7 $\quad$ choose i in $\{1,\dots,K_t\}$ so that the cardinality of $mv_{G_{[X,R]}}(X_i^t)$ is maximum;
8 $\quad R = R - mv_{G_{[X,R]}}$;
9 $\quad W_t = X_i^t$;
10 **end**
11 L = t;
12 Return $\{W_t\}_{1 \leq t \leq L}$;

Clearly, $\{W_t\}_{1 \leq t \leq L}$ is a collection of subsets of X and saturates Y.

A valid broadcast strategy can be logically deduced from $\{W_t\}_{1 \leq t \leq L}$, if the vertices of W_i emit at slot i. The number of slots is the cardinality of $\{W_t\}_{1 \leq t \leq L}$, ie the number of iterations of the algorithm.

Theorem 5 *The algorithm "'Saturation" runs in $O((\ln |Y|)^2)$ iterations. In other words, a broadcast strategy constructed from the collection $(W_t)_{1 \leq t \leq L}$ requires $O((\ln |Y|)^2)$ slots.*

Proof. In agreement with property 2, $K_t \leq \Delta_{G_{[X,R]}} \leq |R|$. During one iteration we have, in accordance with theorem 3 :

$$\left| mv_{G_{[X,R]}}(X_i^t) \right| \geq \frac{|R|}{H_k} \geq \frac{|R|}{1+\ln |R|}$$

Let us note u_n the cardinality of the set R after the nth iteration. Then we have :

$$\begin{aligned} u_0 &= |Y| \\ u_{n+1} &\leq u_n \left(1 - \frac{1}{1+\ln u_n}\right), 0 \leq n \leq L \\ u_L &= 0 \end{aligned}$$

Let $(v_n)_{n \in N}$ be the geometric sequence defined as:

$$v_n = |Y| \left(1 - \frac{1}{1+\ln |Y|}\right)^n$$

We have :

$$v_{\theta((\ln|Y|)^2)} = 0$$

Indeed :

$$\begin{aligned} v_n < 1 &\Leftrightarrow \ln v_n < 0 \\ &\Leftrightarrow \ln|Y| + n\ln\left(1 - \frac{1}{1+\ln|Y|}\right) < 0 \end{aligned}$$

While $\ln(1+x) \leq x$, in order that $v_n < 1$, it requires that $\ln|Y| - \frac{n}{1+\ln|Y|} < 0$, soit $n > \ln|Y| \times (1+\ln|Y|)$.

Clearly we have $v_n \geq u_n, \forall n$ and then $L \leq \ln|Y| \times (1+\ln|Y|)$.

Theorem 6 *The algorithm "Saturation" has a time complexity of* $O(m \times (\ln|Y|)^2)$.

Proof. During each iteration, the algorithm computes an mv-decomposition. In agreement with theorem 1, any mv-decomposition can be computed in $O(m)$. Let us recall that the number of iterations of this algorithm is in $O((\ln|Y|)^2)$, in accordance with theorem 5.

Thus, we have proposed an algorithm to compute a strategy using $O((\log n)^2)$ slots. The quality of the solution returned by our algorithm is the same as the algorithm of [5], but we have improved the complexity from $O(mn(\log n)^2)$ to $O(m(\log n)^2)$.

4 Conclusion

We have proposed the mv-decomposition as a new theoretical tool with interesting algorithmic properties. These properties have been used to develop different algorithms for the distance-2 broadcast problem in multi-hops synchronous radio networks. The mv-decomposition allows to create broadcast solutions where the number of transmissions is minimal, ensuring a number of slots below the maximum degree of the graph. The algorithm which computes this solution has a complexity of $O(m)$.

We have also proposed an algorithm which builds a distance-2 broadcast strategy of $O((\ln|Y|)^2)$ slots for a time complexity $O(m(\log n)^2)$. This improves the result of [5] which announces a broadcast strategy with the same number of slots for a time complexity $O(mn(\log n)^2)$. An interesting perspective would be to adapt the mv-decomposition for the distance-3 broadcast problem, by including a weight function on the elements of Y, and to generalize this approach for the broadcast problem on arbitrary graphs.

References

1. N. Alon, A. Bar-Noy, N. Linial, and D. Peleg. A lower bound for radio broadcast. *J. Comput. Syst. Sci.*, 43(2):290–298, 1991.

2. R. Bar-Yehuda, O. Goldreich, and A. Itai. On the time-complexity of broadcast in multi-hop radio networks: An exponential gap between determinism and randomization. *J. Comput. Syst. Sci.*, 45(1):104–126, 1992.
3. D. Bruschi and M. Del Pinto. Lower bounds for the broadcast problem in mobile radio networks. *Distrib. Comput.*, 10(3):129–135, 1997.
4. I. Chlamtac and S. Kutten. On broadcasting in radio networks - Problem analysis and protocol design. *IEEE Transactions on Communications*, 33:1240–1246, December 1985.
5. I. Chlamtac and O. Weinstein. The wave expansion approach to broadcasting in multihop radio network. *IEEE Transaction Communication*, (39):426–433, 1991.
6. B. Chlebus, L. Gąsieniec, A. Gibbons, A. Pelc, and W. Rytter. Deterministic broadcasting in unknown radio networks. In *SODA '00: Proceedings of the eleventh annual ACM-SIAM symposium on Discrete algorithms*, pages 861–870, Philadelphia, PA, USA, 2000. Society for Industrial and Applied Mathematics.
7. B. Chlebus, L. Gąsieniec, A. Östlin, and J.M. Robson. Deterministic radio broadcasting. In *ICALP '00: Proceedings of the 27th International Colloquium on Automata, Languages and Programming*, pages 717–728, London, UK, 2000. Springer-Verlag.
8. M. Chrobak, L. Gasieniec, and W. Rytter. Fast broadcasting and gossiping in radio networks. *J. Algorithms*, 43(2):177–189, 2002.
9. I. Gaber and Y. Mansour. Broadcast in radio networks. In *SODA '95: Proceedings of the sixth annual ACM-SIAM symposium on Discrete algorithms*, pages 577–585, Philadelphia, PA, USA, 1995. Society for Industrial and Applied Mathematics.
10. G. Kortsarz and M. Elkin. An improved algorithm for radio broadcast (submitted), 2005.
11. D. R. Kowalski and A. Pelc. Centralized deterministic broadcasting in undirected multi-hop radio networks. In *APPROX-RANDOM*, pages 171–182, 2004.
12. G. De Marco and A. Pelc. Faster broadcasting in unknown radio networks. *Inf. Process. Lett.*, 79(2):53–56, 2001.
13. D. Peleg. Deterministic radio broadcast with no topological knowledge, 2000.

Partitioning Random Graphs with General Degree Distributions

Amin Coja-Oghlan[1] and André Lanka[2]

[1] Carnegie Mellon University, Department of Mathematical Sciences,
Pittsburgh, PA 15213, USA
amincoja@andrew.cmu.edu
[2] Fakultät für Informatik, Technische Universität Chemnitz
Straße der Nationen 62, 09107 Chemnitz, Germany
lanka@informatik.tu-chemnitz.de

Abstract. We consider the problem of recovering a planted partition (e.g., a small bisection or a large cut) from a random graph. During the last 30 years many algorithms for this problem have been developed that work provably well on models resembling the Erdős-Rényi model $G_{n,m}$. Since in these random graph models edges are distributed very uniformly, the recent theory of large networks provides convincing evidence that real-world networks, albeit looking random in some sense, cannot sensibly be described by these models. Therefore, a variety of new types of random graphs have been introduced. One of the most popular of these new models is characterized by a prescribed expected degree sequence. We study a natural variant of this model that features a planted partition, the main result being that there is a polynomial time algorithm for recovering (a large share of) the planted partition efficiently. In contrast to prior work, the algorithm's input *only* consists of the graph, i.e., no further parameters of the distribution (such as the expected degree sequence) are required.

1 Introduction

To solve various types of graph partitioning problems, *spectral heuristics* are in common use. Such heuristics represent the input graph by a suitable matrix and exploit the eigenvectors of that matrix in order to solve the combinatorial problem of interest. Spectral techniques have been used to either cope with "classical" NP-hard graph partitioning problems such as GRAPH COLORING or MAX CUT, or to solve less well defined problems such as recovering a "latent" clustering of the vertices of a graph. Examples of such clustering problems occur in information retrieval [4], scientific simulation [18], or bioinformatics [10]. Furthermore, an important advantage of spectral methods is their efficiency, as there are very fast algorithms for computing eigenvectors, in particular in the case of sparse graphs/matrices.

Despite their success in applications (e.g., [17, 18]), for most of the known spectral heuristics there are counterexamples known showing that these algorithms perform badly in the "worst case". Thus, understanding the conditions that cause spectral heuristics to succeed (as well as their limitations) is an im-

Please use the following format when citing this chapter:

Coja-Oghlan, A. and Lanka, A., 2008, in IFIP International Federation for Information Processing, Volume 273; *Fifth IFIP International Conference on Theoretical Computer Science*; Giorgio Ausiello, Juhani Karhumäki, Giancarlo Mauri, Luke Ong; (Boston: Springer), pp. 127–141.

portant research problem. To address this problem, quite a few authors have performed rigorous analyses of spectral techniques on suitable models of *random graphs.* Examples include Alon and Kahale [3] (Graph Coloring), Boppana [5] (Minimum Bisection), and McSherry [15] (recovering a latent partition).

Since the random graph models studied in the aforementioned papers are closely related to the simple models $G_{n,p}$ and $G_{n,m}$ pioneered by Erdős and Rényi, the resulting graphs have a very simple degree distribution. In fact, the vertex degrees are concentrated about a constant number of values. By contrast, the recent theory of complex networks shows that in many cases real-world instances of partitioning problems have a considerably more involved degree distribution [1]. Since most spectral heuristics are very sensitive to fluctuations of the degree distribution, this means that most of the previous spectral methods do not apply to such real-world inputs. Indeed, none of the algorithms from [3, 5, 15] can cope with heavily-tailed degree distributions such as those resulting from the ubiquitous "power law".

Therefore, in the present paper we present and analyze a spectral heuristic for partitioning random graphs with a general degree distribution (including, but not limited to "power laws"). In fact, the result comprises *sparse* graphs, i.e., the case that the average degree remains bounded as the number of vertices grows. This case is of particular practical interest, as many real-world networks turn out to be sparse [1].

The present work is an extension of our prior paper [9] on the same subject. The crucial improvement that we achieve in the present work is that the algorithm *only* requires the graph as an input. By contrast, the algorithm in [9] requires further inputs (namely, parameters of the random graph model such as the expected degree of each vertex), which generally will not be available in practice. Hence, the present work is a step towards spectral methods that apply to graphs with general degree distributions – and in fact to *sparse* graphs.

In Section 2 we describe the random graph model and state the main result. Then, in Section 3 we discuss related work, and Section 4 contains the algorithm and its analysis.

2 The random graph model and the main result

We consider random graphs with a planted partition and a given expected degree sequence. The model coincides with the one studied in [9] and resembles the model investigated in Dasgupta, Hopcroft, and McSherry [11]. Moreover, it is based on the "given expected degrees" model of Chung and Lu [7], which we modify in order to incorporate a planted partition.

Let $V = \{1, \ldots, n\}$ be the set of nodes. The first parameter of the model is a symmetric 2×2-matrix $\Phi = (\phi_{ij})$ of full rank with non-negative constants as entries. Furthermore, for each vertex u there is a weight $w_u > 0$; let $\overline{w} =$

$\sum_{u \in V} w_u/n$ be the average weight. In addition, let V_1, V_2 be a partition of V into two subsets; this is going to be *planted partition* that the algorithm is supposed to recover. For each $u \in V$ we let $\psi(u) \in \{1,2\}$ denote the index of the subset u belongs to, that is $u \in V_{\psi(u)}$.

Now, the random graph $G = G(V_1, V_2, \Phi, w_1, \ldots, w_n) = (V, E)$ is obtained by inserting each possible edge $\{u, v\}$ with $u, v \in V$ independently with probability

$$\phi_{\psi(u),\psi(v)} \cdot \frac{w_u \cdot w_v}{\overline{w} \cdot n}. \tag{1}$$

Of course, we insist on the parameters Φ and w_u being chosen such that each of the above terms is bounded above by 1. Let d_u signify the degree of $u \in V$, and let w'_u be the expected degree. Then (1) yields

$$w'_u = \mathbf{E}[d_u] = \frac{w_u}{\overline{w} \cdot n} \cdot \sum_{v \in V} w_v \cdot \phi_{\psi(u),\psi(v)}. \tag{2}$$

We say that the random graph $G = G(V_1, V_2, \Phi, w_1, \ldots, w_n)$ has some property $\mathcal{P}$ *with high probability* ("w.h.p.") if the probability that $\mathcal{P}$ holds tends to 1 as $n \to \infty$, uniformly for any feasible choice of V_1, V_2, Φ and $w_1, \ldots, w_n$.

Let us briefly discuss the meaning of the model's parameters. As (2) shows, the expected degree of $u \in V$ is proportional to w_u. Thus, the purpose of the weights w_u is to model the desired degree sequence (e.g., a power law). Furthermore, the matrix Φ rules the edge density inside the classes V_1, V_2 and the density of the bipartite graph consisting of the V_1-V_2 edges; for by (2) the edge density of V_1 (resp. V_2) is proportional to ϕ_{11} (resp. ϕ_{22}), and the V_1-V_2-edge density is proportional to $\phi_{12} = \phi_{21}$. Thus, the weight w_u influences the degree of u, while the matrix Φ yields what proportion of u's neighbors belong to V_1 or V_2.

For instance, to model a graph with a small bisection, we could set $\phi_{11} = \phi_{22} = 0.51$ and $\phi_{12} = 0.49$. Moreover, we let $V_1, V_2 \subset V$ be two randomly chosen disjoint sets of size $n/2$. Finally, setting $w_u = d \cdot u^{\frac{1}{2}}$, we obtain a graph with a power law degree distribution (with average degree about $2d$) and a "planted bisection" containing about 49% of all edges. Other examples include graphs with planted independent sets, planted dense spots etc.

Theorem 1. *There is a polynomial time algorithm $\mathcal{A}$ such that the following holds. Let $\varepsilon, \delta > 0$ be arbitrarily small but fixed, and let $C = C(\varepsilon, \delta)$ be a sufficiently large constant. Moreover, assume that*

1. $|V_1|, |V_2| \geq \delta n$,
2. for all $u \in V$ the weight w_u satisfies $\varepsilon \overline{w} \leq w_u \leq n^{1-\varepsilon}$, and
3. the average weight satisfies $\overline{w} \geq C$.

Then w.h.p. $\mathcal{A}$ applied to $G = G(V_1, V_2, \Phi, w_1, \ldots, w_n)$ outputs a partition V'_1, V'_2 that differs from the planted partition V_1, V_2 on at most $n \cdot \ln \overline{w} / \overline{w}^{0.98}$ vertices; that is, $\min\{|V_1 \triangle V'_1| + |V_2 \triangle V'_2|, |V_1 \triangle V'_2| + |V_2 \triangle V'_1|\} \leq n \cdot \ln \overline{w} / \overline{w}^{0.98}$.

Note that the number of vertices that $\mathcal{A}$ may not classify correctly decreases as $\overline{w}$ grows. Indeed, if $\overline{w} = O(1)$, i.e., if G is a *sparse* graph with average degree $O(1)$, then it is *impossible* to recover the partition V_1, V_2 perfectly. A simple reason for this is that w.h.p. both V_1 and V_2 will contain a linear number $\Omega(n)$ of isolated vertices. Nevertheless, a large share of the vertices gets partitioned correctly w.h.p. Moreover, we emphasize that the input of the algorithm *only* consists of the graph G; no further parameters of the model are revealed to $\mathcal{A}$.

Although we have stated Thereom 1 only for a planted partition V_1, V_2 with two classes, the techniques generalize to the case of an arbitrarily large but bounded number k of classes. We omit the details to simplify the exposition.

3 Related work

The general relationship between spectral properties of the adjacency matrix of a graph and clustering problems has been investigated thoroughly [2]. Usually this relationship is based on some separation between the few largest eigenvalues in absolute value (which then represent the clusters) and the remaining eigenvalues. Along these lines theoretically rigorous analyses of spectral methods have been conducted, mainly stating that a certain algorithm performs well on a certain random graph model. Indeed, this has lead to provably efficient algorithms for clustering problems in situations where purely combinatorial algorithms do not seem to work; examples include Alon and Kahale [3] (3-coloring), Boppana [5] (graph bisection), and McSherry [15] (recovering a "latent" partition). In particular [3] has inspired further results (e.g., Flaxman's work on 3-SAT [12]).

However, the aforementioned results do not yield spectral algorithms for clustering graphs whose degree distribution features a heavy upper tail, e.g., a power law degree distribution. Nonetheless, these degree distributions occur prominently in large real world networks [1]. In fact, Mihail and Papadimitriou [16] proved that in the case of a power-law the spectrum of the adjacency matrix merely reflects the upper tail of the degree distribution, but provides no clue on global graph properties (such as the presence of dense clusters or a large cut). Furthermore, in the case of a heavily-tailed degree distribution it is not an option to just remove high degree vertices, because significant parts of the graph may just be ignored in this way. Thus, the adjacency matrix is inappropriate to represent graphs with heavy-tailed degree distributions.

To cope with a heavily-tailed degree distribution, the Laplacian matrix has been used in both theoretical (e.g. [6]) and practically oriented work [17]. However, for randomly generated graphs the Laplacian is significantly more difficult to study than the adjacency matrix (because the entries are heavily dependent). Nonetheless, Dasgupta, Hopcroft, and McSherry [11] showed that clustering problems on sufficiently dense random graphs with a general degree distribution (say, average degree $\gg \ln^6(n)$, where n is the number of vertices) can be solved efficiently using the Laplacian. More precisely, [11] deals with essentially the same model as considered in the present paper (though they additionally deal with the case $k > 2$). However, the assumption that the average degree is $\gg \ln^6 n$ turns out to be crucial in [11] (because the paper employs the "trace method" from Füredi and Komlós [13] for analyzing the Laplacian spectrum). Hence, in comparison to [11] the new aspect of the present work is that our result covers *sparse* graphs (of average degree $O(1)$), which seem most appropriate to model real networks [1]. In fact, the case of sparse graphs is posed as an open problem in [11].

In a prior paper [9] we studied the same random graph model and presented an algorithm for recovering (a large part of) the planted partition efficiently, provided that the *expected* degree distribution $(\mathbf{E}[d_v])_{v \in V}$ is given as a further input parameter to the algorithm. This assumption is crucial in that paper, because the algorithm exploits the spectrum of the matrix $\mathbf{M} = (\mathbf{m}_{uv})_{u,v \in V}$ with entries

$$\mathbf{m}_{uv} = \begin{cases} (\mathbf{E}[d_u]\,\mathbf{E}[d_v])^{-1} & \text{if } u, v \text{ are adjacent,} \\ 0 & \text{otherwise.} \end{cases} \tag{3}$$

In fact, in the sparse case (average degree $O(1)$), the vertex degrees d_v are *not* tightly concentrated about their means (as there tails of Poisson type), so that it is impossible to recover/approximate the expected degree distribution $(\mathbf{E}[d_v])_{v \in V}$ sufficiently well in terms of the actual degree distribution $(d_v)_{v \in V}$. Therefore, the assumption that the algorithm is given the expected degree sequence is inevitable in order to set up the matrix (3). Of course, this assumption is rather impractical, because it reduces the applicability of the algorithm to artificially generated instances.

To avoid the assumption that the expected degree sequence is given to the algorithm, we fix (3) by instead considering the matrix $M = (m_{uv})_{u,v \in V}$ with entries

$$m_{uv} = \begin{cases} (d_u d_v)^{-1} & \text{if } u, v \text{ are adjacent,} \\ 0 & \text{otherwise.} \end{cases} \tag{4}$$

Hence, we replace the expected degrees by the actual vertex degrees of the input graph. In effect, while the entries of (3) are mutually independent (up to the trivial dependence due to symmetry), the entries of (4) are *mutually dependent.* This issue complicates the *analysis* of the algorithm – in particular, the analysis of the spectrum of M – significantly; to cope with these new issues, we build upon methods that we developed recently in [8]. Furthermore, the algorithm needs to proceed more carefully, as the actual vertex degrees may deviate from their means considerably. Thus, in comparison to [9] the contribution of the present work is that we obtain a much more practical algorithm, and present significantly more sophisticated techniques for analyzing its performance on random graphs.

4 The algorithm and its analysis

Throughout this section we keep the notation and the assumptions of Theorem 1.

4.1 Notation and preliminaries

If ξ is a vector, then $\|\xi\|$ denotes its ℓ_2-norm. Moreover, for a $m \times n$ matrix B we let $\|B\| = \max_{\xi \in \mathbf{R}^n,\ \|\xi\|=1} \|B\xi\|$ denote the operator norm. The transpose of B is written as B^t. Furthermore, $\mathbf{1}$ signifies the vector with all entries equal to 1 (in any dimension). If $\xi \in \mathbf{R}^S$ and $U \subseteq S$, then $\xi_{|U} \in \mathbf{R}^S$ signifies the vector obtained by replacing the i'th component of ξ by 0 if $i \notin U$, whereas $\xi_U \in \mathbf{R}^U$ is obtained from ξ by deleting all entries ξ_v with $v \notin U$. In addition, if B is a $m \times n$ matrix and $X \subseteq \{1, \ldots, m\}$, $Y \subseteq \{1, \ldots, n\}$, then $B_{X \times Y}$ denotes the minor of B induced on $X \times Y$. Further, if $M = (m_{uv})$ is a matrix and X (resp. Y) is a set of rows (columns), then we set

$$s_M(X, Y) = \sum_{x \in X} \sum_{y \in Y} m_{xy}.$$

If u is a vertex of a graph $G = G(V_1, V_2, \Phi, w_1, \ldots, w_n)$, then $N(u) = \{v : \{u, v\} \in E\}$ denotes the neighborhood of u. Moreover, for two sets U_1, U_2 of vertices we define the volume of (U_1, U_2) to be

$$\mathrm{Vol}(U_1, U_2) = \sum_{u \in U_1} \sum_{v \in U_2} \phi_{\psi(u), \psi(v)} \cdot \frac{w_u \cdot w_v}{\overline{w} \cdot n};$$

if U_1 and U_2 are disjoint, then $\mathrm{Vol}(U_1, U_2)$ equals the *expected* number of U_1-U_2-edges. In other words, if $A = A(G)$ is the adjacency matrix, then $\mathrm{Vol}(U_1, U_2) = \mathbf{E}\left[s_A(U_1, U_2)\right]$.

The following Chernoff bounds will prove useful in several places (cf. [14, Theorems 2.1 and 2.8]).

Fact 2. *Let X be the sum of independent 0–1 random variables. Then*

1. $\mathbf{Pr}\left[X \geq \mathbf{E}[X] + t\right] \leq \exp\left(-\frac{t^2}{2 \cdot (\mathbf{E}[X] + t/3)}\right)$

2. $\mathbf{Pr}\left[X \leq \mathbf{E}[X] - t\right] \leq \exp\left(-\frac{t^2}{2 \cdot \mathbf{E}[X]}\right)$

for all $t \geq 0$.

Finally, we collect a few simple observations concerning the random graph model.

Lemma 3. *Suppose that $G = G(V_1, V_2, \Phi, w_1, \ldots, w_n)$ is a random graph.*

1. *Let u_1, u_2 be two vertices belonging to the same set of the planted partition. Then* $w_{u_1}/w'_{u_1} = w_{u_2}/w'_{u_2}$.
2. *There exists a constant $C = C(\Phi, \varepsilon, \delta)$ such that $1/C \le w'_u/w_u \le C$ for all $u \in V$.*
3. *The expected average degree of G equals $\overline{w}' = \sum_{u\in V} w'_u/n = \Theta(\overline{w})$.*

Since by Lemma 3 the quotient w_u/w'_u coincides for all $u \in V_i$, we abbreviate

$$W_i = w_u/w'_u = \Theta(1), \qquad \text{and} \qquad W = \overline{w}/\overline{w}' = \Theta(1). \tag{5}$$

4.2 The algorithm

The algorithm $\mathcal{A}$ for Theorem 1 reads as follows.

Algorithm 4.
Input: A graph $G = (V, E)$.
Output: A partition V'_1, V'_2 of V.

1. Calculate the average degree $\overline{d} = \sum_{u=1}^{n} d_u/n$ of G and set $d_{\mathrm{m}} = \overline{d}/\ln \overline{d}$.
2. Construct the matrix $M = (m_{uv})_{u,v\in V}$ as described in (4).
3. Let $U = \{u \in V : d_u \ge d_{\mathrm{m}}\}$ be the set of all vertices whose degree is "not too small".
4. Obtain M^* from M by replacing any entry m_{uv} with $(u, v) \notin U \times U$ by 0.
5. Let s_1, s_2 be the eigenvectors of M^* with the two largest eigenvalues in absolute value. Scale s_i such that $\|s_i\| = \sqrt{n}$.
6. If at least one of s_1, s_2 enjoys the following property:

   There are $c_1, c_2 \in \mathbb{R}$ with $|c_1 - c_2| > 1/4$ such that more than $n/\sqrt{d_{\mathrm{m}}}$ vertices $v \in U$ satisfy $|s_i(v) - c_1| \le 1/32$ and more than $n/\sqrt{d_{\mathrm{m}}}$ vertices satisfy $|s_i(v) - c_2| \le 1/32$, (6)

   then let $s \in \{s_1, s_2\}$ be such an eigenvector. Furthermore, let V'_1 be the vertices whose corresponding entries in s are closer to c_1 than to c_2 and set $V'_2 = V \setminus V'_1$. Otherwise, if neither s_1 nor s_2 enjoys (6), let $V'_1 = V$ and $V'_2 = \emptyset$ (in this case, the algorithm fails).

Before we sketch the analysis of the algorithm, let us briefly discuss the basic ideas that it is based on. In its first step, $\mathcal{A}$ just computes the average degree and the value d_{m}. This value is assumed to be a lower bound on the degree that a vertex should typically have; that is, all vertices with degree $< d_{\mathrm{m}}$ are considered exceptional. Note that this is consistent with assumption 2. of Theorem 1, which entails that $\mathbf{E}[d_u] \ge \delta\varepsilon^2 \cdot \min_{\phi_{ij}>0}(\phi_{ij}) \cdot \overline{d} > d_{\mathrm{m}}$ for all $u \in V$.

Step 2 of the algorithm then sets up the matrix M, whose eigenvectors we are going to use in order to partition G. Note that the entry corresponding to an edge $\{u, v\}$ is normalized by the product $d_u d_v$ of the vertex degrees; this normalization is crucial as it ensures that the upper tail of the degree distribution does not dominate the spectrum of M (in contrast to the case of the adjacency matrix, cf. Section 3).

While the normalization of the entries of M ensures that the upper tail of the degree distribution does not dominate the spectrum of M, vertices of atypically small degree may induce large eigenvalues (cf. [8]). Therefore, before computing the dominant eigenvectors s_1, s_2 in Step 5, Steps 3 and 4 remove all entries of M that involve low degree vertices. By the Chernoff bound (Fact 2), in this way we just remove a tiny (though linear) fraction of the vertices.

Finally, Step 6 exploits the entries of s_1 and s_2 to compute a partition. The basic insight is that the entries of s_1 and s_2 are essentially constant on the two classes V_1, V_2, and that indeed the entries of s_1 and s_2 differ on each class significantly; this second fact follows from our assumption that the density matrix D has full rank. However, if s_1 and s_2 do not have these properties, then the algorithm will fail to partition the graph correctly and just output a trivial partition.

In order to analyze the algorithm (and thus to prove Theorem 1), we basically need to study the eigenvalues and -vectors of M^*. The main ingredient of the analysis is the following result on the spectrum of the minors $M^*_{V_i \times V_j}$, i.e., the sub-matrices of M^* consisting of the rows V_i and the columns V_j.

Theorem 5. *With high probability the following holds for any two indices* $1 \leq i, j \leq 2$.

1. $\dfrac{\mathbf{1}^t}{\|\mathbf{1}^t\|} \cdot {M^*}_{V_i \times V_j} \cdot \dfrac{\mathbf{1}}{\|\mathbf{1}\|} = \phi_{ij} \cdot W_i \cdot W_j \cdot \dfrac{\sqrt{|V_i| \cdot |V_j|}}{\overline{w} \cdot n} \cdot \left(1 \pm O\left({d_{\mathrm{m}}}^{-0.49}\right)\right).$

2. For any u, v *with* $\|u\| = \|v\| = 1$ *and* $u \perp \mathbf{1}$ *or* $v \perp \mathbf{1}$ *we have the bound*

$$\left|u^t \cdot {M^*}_{V_i \times V_j} \cdot v\right| = O\left({\overline{w}}^{-1.49} + {d_{\mathrm{m}}}^{-1.5}\right) = O(1/(\overline{w} \cdot {d_{\mathrm{m}}}^{0.49})).$$

The assumptions of Theorem 1 ensure that the expression on the r.h.s. of 1. is of order $1/\overline{w}$, whereas the expression in 2. is of order $1/(\overline{w} \cdot {d_{\mathrm{m}}}^{0.49})$. Thus, the intuitive meaning of Theorem 5 is that the dominant singular value of $M^*_{V_i \times V_j}$ corresponds approximately to the singular vectors $\mathbf{1}_{V_i}$ and $\mathbf{1}_{V_j}$. By combining the estimates from Theorem 5 for all index pairs $1 \leq i, j \leq 2$, we obtain the following result concerning the eigenvectors of M^*.

Corollary 6. *W.h.p.* M^* *has exactly two eigenvalues whose absolute value is* $\Theta(1/\overline{w})$, *whereas all the other eigenvalues are* $O\left(1/(\overline{w} \cdot {d_{\mathrm{m}}}^{0.49})\right)$ *in absolute value. Moreover, if* s_1, s_2 *are orthogonal eigenvectors of norm* $\sqrt{n}$ *with the largest two eigenvalues in absolute value, then there is an index* $j \in \{1, 2\}$ *such that*

$$s_j = \alpha \mathbf{1}_{|V_1} + \beta \mathbf{1}_{|V_2} + \gamma u, \quad \text{where } u \perp \mathbf{1}_{|V_1}, \mathbf{1}_{|V_2}, \|u\| = \sqrt{n}$$

and $|\alpha - \beta| > \frac{1}{4}$ *and* $\gamma = O({d_{\mathrm{m}}}^{-0.49})$.

Corollary 6 implies that w.h.p. step 6 of $\mathcal{A}$ will succeed in finding a vector that satisfies (6). Moreover, a simple calculation based on the above eigenvalue bounds shows that the number of falsely classified vertices (i.e., the symmetric difference of the partitions (V_1', V_2') and (V_1, V_2)) is at most $O(n/{d_m}^{0.98})$, whence Theorem 1 follows.

The values of α and β correspond to the c_i in the algorithm. If some vertex classified falsely, its entry in s_j is twisted due to its value in $\gamma \cdot u$. Because of the large distance between α and β, such entries are bounded below by some constant. As $|\gamma| = O({d_m}^{-0.49})$ the value in u has to be $\Omega({d_m}^{0.49})$. Since $\|u\| = \sqrt{n}$ we have at most $O(n/{d_m}^{0.98})$ such entries.

4.3 Proof of Corollary 6

At first we show that M^* has the exactly two eigenvalues whose absolute value is $\Theta(1/\overline{w})$, whereas all the other eigenvalues are $O\left(1/(\overline{w} \cdot d_m^{0.49})\right)$ in absolute value. Let g, h be two vectors from the space spanned by $\mathbf{1}_{|V_1}$ and $\mathbf{1}_{|V_2}$. Namely, $g = a_1 \cdot \mathbf{1}_{|V_1}/\|\mathbf{1}_{|V_1}\| + a_2 \cdot \mathbf{1}_{|V_2}/\|\mathbf{1}_{|V_2}\|$ with $a_1^2 + a_2^2 = 1$ and $h = b_1 \cdot \mathbf{1}_{|V_1}/\|\mathbf{1}_{|V_1}\| + b_2 \cdot \mathbf{1}_{|V_2}/\|\mathbf{1}_{|V_2}\|$ with $b_1^2 + b_2^2 = 1$. Note, $\|g\| = \|h\| = 1$. By Theorem 5 we have with probability $1 - o(1)$ that

$$
\begin{aligned}
h^t M^* g &= \sum_{i,j=1}^{2} b_i \cdot \frac{\mathbf{1}_{|V_i}}{\|\mathbf{1}_{|V_i}\|} \cdot M^* \cdot a_j \cdot \frac{\mathbf{1}_{|V_j}}{\|\mathbf{1}_{|V_j}\|} = \sum_{i,j=1}^{2} b_i \cdot a_j \cdot \frac{\mathbf{1}^t \cdot {M^*}_{V_i \times V_j} \cdot \mathbf{1}}{\sqrt{|V_i| \cdot |V_j|}} \\
&= \sum_{i,j=1}^{2} b_i \cdot a_j \cdot \phi_{ij} \cdot W_i \cdot W_j \cdot \frac{\sqrt{|V_i| \cdot |V_j|}}{\overline{w} \cdot n} \cdot \left(1 \pm O\left({d_m}^{-0.49}\right)\right) \\
&= \sum_{i,j=1}^{2} \left(b_i \cdot a_j \cdot \phi_{ij} \cdot W_i \cdot W_j \cdot \frac{\sqrt{|V_i| \cdot |V_j|}}{\overline{w} \cdot n} \right) \pm O\left(1/(\overline{w} \cdot d_m^{0.49})\right) \\
&= \frac{1}{\overline{w}} \cdot (b_1\ b_2) \cdot P \cdot \begin{pmatrix} a_1 \\ a_2 \end{pmatrix} \pm O\left(1/(\overline{w} \cdot d_m^{0.49})\right)
\end{aligned}
$$

with

$$
P = \begin{pmatrix} W_1 \cdot \sqrt{\frac{|V_1|}{n}} & 0 \\ 0 & W_2 \cdot \sqrt{\frac{|V_2|}{n}} \end{pmatrix} \cdot \begin{pmatrix} \phi_{11} & \phi_{12} \\ \phi_{12} & \phi_{22} \end{pmatrix} \cdot \begin{pmatrix} W_1 \cdot \sqrt{\frac{|V_1|}{n}} & 0 \\ 0 & W_2 \cdot \sqrt{\frac{|V_2|}{n}} \end{pmatrix}.
$$

Remember, Φ has full rank as well as both remaining factors of P. We conclude that the matrix P has full rank. The W_i are $\Theta(1)$ as $|V_i|/n$, too. This shows that the spectral properties of P are determined only by Φ, ε and δ and do not rely on $w_1, \ldots, w_n$ or n. P has two eigenvectors with *constant* nonzero eigenvalues. Let $(e_1\ e_2)^t$ and $(f_1\ f_2)^t$ be two orthonormal eigenvectors of P to the eigenvalues λ_1 and λ_2. Set

$$g_1 = e_1 \cdot \frac{\mathbf{1}_{|V_1}}{\|\mathbf{1}_{|V_1}\|} + e_2 \cdot \frac{\mathbf{1}_{|V_2}}{\|\mathbf{1}_{|V_2}\|} \quad \text{and} \quad g_2 = f_1 \cdot \frac{\mathbf{1}_{|V_1}}{\|\mathbf{1}_{|V_1}\|} + f_2 \cdot \frac{\mathbf{1}_{|V_2}}{\|\mathbf{1}_{|V_2}\|}.$$

By the calculation above get

$$\begin{aligned} \left|g_1^t \cdot M^* \cdot g_1\right| &= \left|\frac{1}{\overline{w}} \cdot (e_1\ e_2) \cdot P \cdot \begin{pmatrix} e_1 \\ e_2 \end{pmatrix} \pm O\left(1/(\overline{w} \cdot {d_{\mathrm{m}}}^{0.49})\right)\right| \\ &= \left|\frac{1}{\overline{w}} \cdot \lambda_1 \pm O\left(1/(\overline{w} \cdot {d_{\mathrm{m}}}^{0.49})\right)\right| = \Theta(1/\overline{w}) \end{aligned}$$

whereas

$$\begin{aligned} \left|g_1^t \cdot M^* \cdot g_2\right| &= \left|\frac{1}{\overline{w}} \cdot (e_1\ e_2) \cdot P \cdot \begin{pmatrix} f_1 \\ f_2 \end{pmatrix} \pm O\left(1/(\overline{w} \cdot {d_{\mathrm{m}}}^{0.49})\right)\right| \\ &= \left|\frac{1}{\overline{w}} \cdot 0 \pm O\left(1/(\overline{w} \cdot {d_{\mathrm{m}}}^{0.49})\right)\right| \end{aligned}$$

Thus for $1 \leq i, j \leq 2$ we have

$$\left|g_i^t \cdot M^* \cdot g_j\right| = \begin{cases} \Theta(1/\overline{w}) & \text{for } i = j \\ O\left(1/(\overline{w} \cdot {d_{\mathrm{m}}}^{0.49})\right) & \text{for } i \neq j \end{cases}. \tag{7}$$

For any unit-vector $u \perp g_1, g_2$ (what equals $u \perp \mathbf{1}_{|V_1}, \mathbf{1}_{|V_2}$) we have by Theorem 5 for all unit-vectors v

$$\left|u^t \cdot M^* \cdot v\right| \leq \sum_{i,j=1}^{2} \left|u_{V_i}^t \cdot {M^*}_{V_i \times V_j} \cdot v_{V_j}\right| = O\left(1/(\overline{w} \cdot {d_{\mathrm{m}}}^{0.49})\right)$$

and analogously

$$\left|v^t \cdot M^* \cdot u\right| = O\left(1/(\overline{w} \cdot {d_{\mathrm{m}}}^{0.49})\right).$$

Both bounds and (7) together with the Courant-Fischer-characterization of eigenvalues yield the first part of the claim.

We are left to show that M^* w.h.p. has an eigenvector s_j as desired. Let e be an eigenvector of M^* with norm $\|e\| = \sqrt{n}$ to the eigenvalue $\Theta(1/\overline{w})$ (in absolute value). We can decompose e such that $e = \alpha \cdot \mathbf{1}_{|V_1} + \beta \cdot \mathbf{1}_{|V_2} + \gamma \cdot u$ for some $u \perp \mathbf{1}_{|V_1}, \mathbf{1}_{|V_2}$ with $\|u\| = \sqrt{n}$. By Theorem 5 we conclude on the one hand

$$\left|e^t \cdot M^* \cdot u\right| = \|e\| \cdot \|u\| \cdot O\left(1/(\overline{w} \cdot {d_{\mathrm{m}}}^{0.49})\right) = O\left(n/\left(\overline{w} \cdot {d_{\mathrm{m}}}^{0.49}\right)\right)$$

as $u \perp \mathbf{1}_{|V_1}, \mathbf{1}_{|V_2}$. Because of $e^t \cdot M^* = \Theta(1/\overline{w}) \cdot e^t$ we have on the other hand

$$\left|e^t \cdot M^* \cdot u\right| = \Theta(1/\overline{w}) \cdot \left|e^t \cdot u\right| = \Theta(1/\overline{w}) \cdot |\gamma| \cdot u^t u = \Theta(1/\overline{w}) \cdot |\gamma| \cdot n,$$

so that $|\gamma| = O(d_{\mathrm{m}}{}^{-0.49})$. Let s_1, s_2 be as in the lemma and

$$s_j = \alpha_j \cdot \mathbf{1}_{|V_1} + \beta_j \cdot \mathbf{1}_{|V_2} + \gamma_j \cdot u_j$$

the decomposition with $u_j \perp \mathbf{1}_{|V_1}, \mathbf{1}_{|V_2}$ and $\|u_j\| = \sqrt{n}$ as described. Assume for a contradiction that we have $|\alpha_j - \beta_j| \leq 1/4$ for both $j = 1, 2$. As

$$n = s_j^t \cdot s_j = \alpha_j^2 \cdot |V_1| + \beta_j^2 \cdot |V_2| + \gamma_j^2 \cdot n$$

we get

$$\alpha_j^2 + \beta_j^2 \geq \alpha_j^2 \cdot \frac{|V_1|}{n} + \beta_j^2 \cdot \frac{|V_2|}{n} = 1 - \gamma_j^2 \geq 1 - O(d_{\mathrm{m}}{}^{-0.98}).$$

Clearly, for both $j = 1, 2$ we have $|\alpha_j| > 1/2$ or $|\beta_j| > 1/2$, yielding that the sign of α_j equals the sign of β_j for both $j = 1, 2$. We get

$$|\alpha_1 \cdot \alpha_2 + \beta_1 \cdot \beta_2| = |\alpha_1 \cdot \alpha_2| + |\beta_1 \cdot \beta_2| \geq \frac{1}{2} \cdot \frac{1}{4} + \frac{1}{4} \cdot \frac{1}{2} = \frac{1}{4}$$

and

$$\begin{aligned} 0 = s_1^t \cdot s_2 &= |\alpha_1 \cdot \alpha_2 \cdot |V_1| + \beta_1 \cdot \beta_2 \cdot |V_2| + \gamma_1 \cdot \gamma_2 \cdot u_1^t \cdot u_2| \\ &\geq \delta n \cdot |\alpha_1 \cdot \alpha_2 + \beta_1 \cdot \beta_2| - |\gamma_1 \cdot \gamma_2| \cdot n \geq \delta n/4 - O\left(n/d_{\mathrm{m}}{}^{0.98}\right). \end{aligned}$$

This is a contradiction since $\delta > 0$ is constant and d_{m} is large. So at least one s_j has $|\alpha_j - \beta_j| > 1/4$. □

4.4 Proof of Theorem 5: The spectrum of $M^*{}_{V_i \times V_j}$

The main difficulty in the (rather involved) proof of Theorem 5 is the fact that the entries of M^* are mutually dependent, because we normalize by the actual vertex degrees (cf. Step 2 of the algorithm and (4)). Furthermore, in case of sparse graphs (which is included in Theorem 1), it is possible that all (or most) weights w_u remain bounded as $n \to \infty$. In this case the expected degrees are bounded as well. In effect, the actual degrees of the vertices are *not* concentrated about their expectations, but may deviate by up to $\Omega(\log n/\log\log n)$. Hence, we need to cope with the dependence of the matrix entries as well as with deviations of the vertex degrees from their expectations.

To this end, we mark vertices $u \in V_i$ as "bad" if the number of u's neighbors in V_j is far from its expectation (of course, this is just a part of the *analysis* – the algorithm cannot identify these "bad" vertices). Similarly, we mark vertices from V_j as "bad". Now, it is possible that some "good" vertices inside V_i and/or V_j have many "bad" neighbors. We mark such vertices as "bad", too. Repeating this process, we obtain a subset $R_{ij} \subseteq V_i$ of "good" vertices, which firstly have about as many neighbors in V_j as expected and secondly have only a few "bad" neighbors in V_j. Analogously we obtain "good" vertices $C_{ij} \subseteq V_j$. Then, we shall analyze the sub-matrix induced on $R_{ij} \times C_{ij}$ separately from the rest.

More precisely, the sets $R_{ij} \subseteq V_i$ and $C_{ij} \subseteq V_j$ are the outcome of the following process. Let c be a sufficiently large constant (the value gets determined later), and let $A = A(G)$ be the adjacency matrix of G.

1. Let $R' = \{u \in V : \forall j' : |s_A(u, V_{j'}) - \mathrm{Vol}(u, V_{j'})| \leq \mathrm{Vol}(u, V_{j'})^{0.51}\}$.
2. Let $C' = \{v \in V : \forall i' : |s_A(V_{i'}, v) - \mathrm{Vol}(V_{i'}, v)| \leq \mathrm{Vol}(V_{i'}, v)^{0.51}\}$.
3. Set $R'_{ij} := R' \cap V_i$ and $C'_{ij} := C' \cap V_j$.
4. While there is some $u \in R'_{ij}$ with
$$s_A(u, V_j \setminus C'_{ij}) \geq \mathrm{Vol}(u, V_j) \cdot c/d_{\mathrm{m}} \qquad \text{then} \quad R'_{ij} := R'_{ij} \setminus \{u\}.$$
5. While there is some $v \in C'_{ij}$ with
$$s_A(V_i \setminus R'_{ij}, v) \geq \mathrm{Vol}(V_i, v) \cdot c/d_{\mathrm{m}} \qquad \text{then} \quad C'_{ij} := C'_{ij} \setminus \{v\}.$$
6. Repeat Steps 4 – 5 until R'_{ij} and C'_{ij} remain unchanged.
7. $R_{ij} := R'_{ij}$. $C_{ij} := C'_{ij}$.

We abbreviate R_{ij} by $\mathcal{R}$ and C_{ij} by $\mathcal{C}$, $V_i \setminus R_{ij}$ by $\overline{\mathcal{R}}$, and $V_j \setminus C_{ij}$ by $\overline{\mathcal{C}}$. Due to the first step of the above process all $u \in \mathcal{R}$ and $v \in \mathcal{C}$ satisfy

$$\begin{aligned} |s_A(u, V) - \mathrm{Vol}(u, V)| &\leq 2 \cdot \mathrm{Vol}(u, V)^{0.51}, \\ |s_A(V, v) - \mathrm{Vol}(V, v)| &\leq 2 \cdot \mathrm{Vol}(V, v)^{0.51}. \end{aligned} \tag{8}$$

Let us briefly discuss the above process. For a vertex $u \in V_1$ the standard deviation of the number $s_A(u, V_j)$ of neighbors of u in V_j from its expectation $\mathrm{Vol}(u, V_j)$ is of order $O(\mathrm{Vol}(u, V_j)^{0.5})$ (because $s_A(u, V_j)$ is a sum of independent 0/1-random variables). Therefore, the Chernoff bound (Fact 2) entails that w.h.p. "most" of the vertices in V_i belong to R'. Moreover, the larger $\mathrm{Vol}(u, V_j)$, the more likely it is that $u \in R'$. Hence, we expect $\mathrm{Vol}(V_i \setminus R', V_j)$ (as well as $\mathrm{Vol}(V_i, V_j \setminus C')$) to be fairly small. Consequently, as a vertex removed from R'_{ij} in Step 4 has relatively many neighbors inside the set $V_j \setminus C'_{ij}$ of small volume, we expect that Step 4 will remove only a small number of vertices. Thus, the final sets $\mathcal{R}$ and $\mathcal{C}$ should constitute the dominant fraction of the volume of G. The following lemma, whose proof is omitted, shows that this is actually the case.

Lemma 7. *W.h.p. we have* $\mathrm{Vol}(\overline{\mathcal{R}}, V_j) \leq n/{d_{\mathrm{m}}}^4$, $\mathrm{Vol}(V_i, \overline{\mathcal{C}}) \leq n/{d_{\mathrm{m}}}^4$, *and* $\mathrm{Vol}(\overline{\mathcal{R}}, \overline{\mathcal{C}}) \leq n/{d_{\mathrm{m}}}^8$.

A consequence of Lemma 7 is that both $\overline{\mathcal{R}}$ and $\overline{\mathcal{C}}$ contain only a few vertices. For by the choice of d_{m} (cf. Step 1 of $\mathcal{A}$) for all $u \in V_i$ and all $v \in V_j$ we have

$$d_{\mathrm{m}} \leq \mathrm{Vol}(u, V_j) \leq \mathrm{Vol}(u, V) = w'_u \qquad \text{and} \qquad d_{\mathrm{m}} \leq \mathrm{Vol}(V_i, v) \leq w'_v . \tag{9}$$

Thus, $d_{\mathrm{m}} \cdot |\overline{\mathcal{R}}| \le \mathrm{Vol}(\overline{\mathcal{R}}, V_j) \le n/{d_{\mathrm{m}}}^4$, which yields $|\overline{\mathcal{R}}| \le n/{d_{\mathrm{m}}}^5$. As $\delta \cdot n \le |V_i|$, we get

$$|\overline{\mathcal{R}}| \le \frac{|V_i|}{\delta \cdot {d_{\mathrm{m}}}^5} \le \frac{|V_i|}{{d_{\mathrm{m}}}^4} \qquad \text{and} \qquad |\mathcal{R}| = |V_i| - |\overline{\mathcal{R}}| \ge |V_i| \cdot \left(1 - \frac{1}{{d_{\mathrm{m}}}^4}\right), \tag{10}$$

(provided that $\overline{w} > 1/\delta^2$ is sufficiently large). Analogously,

$$|\overline{\mathcal{C}}| \le |V_i| / {d_{\mathrm{m}}}^4 \qquad \text{and} \qquad |\mathcal{C}| \ge |V_i| \cdot \left(1 - 1/{d_{\mathrm{m}}}^4\right) . \tag{11}$$

To proceed, we subdivide $M^*_{V_i \times V_j}$ into four parts $M^*_{\overline{\mathcal{R}} \times \mathcal{C}}$, $M^*_{\mathcal{R} \times \overline{\mathcal{C}}}$, $M^*_{\mathcal{R} \times \mathcal{C}}$, and $M^*_{\overline{\mathcal{R}} \times \overline{\mathcal{C}}}$, which we shall analyze separately. With respect to $M^*_{\mathcal{R} \times \mathcal{C}}$, we have the following.

Lemma 8. *With high probability we have*

1. $\mathbf{1}^t \cdot M^*_{\mathcal{R} \times \mathcal{C}} \cdot \mathbf{1} = \phi_{ij} \cdot W_i \cdot W_j \cdot \dfrac{|\mathcal{R}| \cdot |\mathcal{C}|}{\overline{w} \cdot n} \cdot \left(1 \pm O(1/{d_{\mathrm{m}}}^{0.49})\right) = \Theta(n/\overline{w})$,
2. $|u^t \cdot M^*_{\mathcal{R} \times \mathcal{C}} \cdot v| = O\left(1/\overline{w}^{1.49}\right)$ *for any* u, v *with* $\|u\| = \|v\| = 1$ *and* $u \perp \mathbf{1}$ *or* $v \perp \mathbf{1}$, *and*
3. $\|M^*_{\mathcal{R} \times \mathcal{C}}\| = \Theta\left(1/\overline{w}\right)$.

The proof of Lemma 8 is based on the fact that on $\mathcal{R} \times \mathcal{C}$ the vertex degrees behave at least roughly as expected. Therefore, we can relate the spectrum of $M^*_{\mathcal{R} \times \mathcal{C}}$ to the spectrum of $\mathbf{M}_{\mathcal{R} \times \mathcal{C}}$, where $\mathbf{M}$ is the matrix from (3). Since the entries of $\mathbf{M}$ are mutually independent (up to the trivial dependence resulting from symmetry), the analysis of its spectrum is significantly simpler than the analysis of M; in fact, this analysis has been carried out in [9]. Nonetheless, in order to relate $\mathbf{M}_{\mathcal{R} \times \mathcal{C}}$ and $M^*_{\mathcal{R} \times \mathcal{C}}$, we need to analyze the degree distribution of G thoroughly, which requires considerable technical work (omitted).

As a next step, we analyze the three "small" blocks $M^*_{\overline{\mathcal{R}} \times \mathcal{C}}$, $M^*_{\mathcal{R} \times \overline{\mathcal{C}}}$ and $M^*_{\overline{\mathcal{R}} \times \overline{\mathcal{C}}}$.

Lemma 9. *With high probability we have that* $\|M^*_{\mathcal{R} \times \overline{\mathcal{C}}}\|$, $\|M^*_{\overline{\mathcal{R}} \times \mathcal{C}}\|$ *and* $\|M^*_{\overline{\mathcal{R}} \times \overline{\mathcal{C}}}\|$ *are* $O({d_{\mathrm{m}}}^{-1.5})$.

The proof of Lemma 9 is based on combinatorial ideas, and, in particular, the fact that the volumes of $\overline{\mathcal{R}}$ and $\overline{\mathcal{C}}$ are relatively small (cf. Lemma 7). Therefore, for instance the subgraph induced on $\overline{\mathcal{R}} \times \overline{\mathcal{C}}$ has a very simple combinatorial structure (it is essentially forest-like), which allows a direct analysis of $M^*_{\overline{\mathcal{R}} \times \overline{\mathcal{C}}}$. Details are omitted.

4.4.1 Proof of Theorem 5. With respect to the first statement, we have

$$\mathbf{1}^t \cdot M^*_{V_i \times V_j} \cdot \mathbf{1} = \mathbf{1}^t \cdot M^*_{\mathcal{R} \times \mathcal{C}} \cdot \mathbf{1} + \mathbf{1}^t \cdot M^*_{\mathcal{R} \times \overline{\mathcal{C}}} \cdot \mathbf{1} + \mathbf{1}^t \cdot M^*_{\overline{\mathcal{R}} \times \mathcal{C}} \cdot \mathbf{1} + \mathbf{1}^t \cdot M^*_{\overline{\mathcal{R}} \times \overline{\mathcal{C}}} \cdot \mathbf{1} . \tag{12}$$

Item 1. of Lemma 8 gives for the first term

$$\begin{aligned}\mathbf{1}^t \cdot {M^*}_{\mathcal{R}\times\mathcal{C}} \cdot \mathbf{1} &= \phi_{ij} \cdot W_i \cdot W_j \cdot \frac{|\mathcal{R}| \cdot |\mathcal{C}|}{\overline{w} \cdot n} \cdot (1 \pm O(1/{d_{\mathrm{m}}}^{0.49})) \\ &\overset{(10),(11)}{=} \phi_{ij} \cdot W_i \cdot W_j \cdot \frac{|V_i| \cdot |V_j|}{\overline{w} \cdot n} \cdot (1 \pm O(1/{d_{\mathrm{m}}}^{0.49})).\end{aligned}$$

Lemma 9 shows that the second summand in (12) is bounded by

$$\begin{aligned}\left|\mathbf{1}^t \cdot {M^*}_{\mathcal{R}\times\overline{\mathcal{C}}} \cdot \mathbf{1}\right| &\le \sqrt{|\mathcal{R}| \cdot |\overline{\mathcal{C}}|} \cdot \|{M^*}_{\mathcal{R}\times\overline{\mathcal{C}}}\| \overset{(11)}{\le} \sqrt{|V_i| \cdot |V_j| / {d_{\mathrm{m}}}^4} \cdot O({d_{\mathrm{m}}}^{-1.5}) \\ &= \sqrt{|V_i| \cdot |V_j|} \cdot O\left({d_{\mathrm{m}}}^{-3.5}\right) = \sqrt{|V_i| \cdot |V_j|} \cdot O(1/(\overline{w} \cdot {d_{\mathrm{m}}}^{0.49})).\end{aligned}$$

The same bound holds for both $\left|\mathbf{1}^t \cdot {M^*}_{\overline{\mathcal{R}}\times\mathcal{C}} \cdot \mathbf{1}\right|$ and $\left|\mathbf{1}^t \cdot {M^*}_{\overline{\mathcal{R}}\times\overline{\mathcal{C}}} \cdot \mathbf{1}\right|$. Dividing each summand for (12) by $\sqrt{|V_i| \cdot |V_j|}$ we get the desired bound on $\frac{\mathbf{1}^t}{\|\mathbf{1}^t\|} \cdot {M^*}_{V_i\times V_j} \cdot \frac{\mathbf{1}}{\|\mathbf{1}\|}$.

For the second item of Theorem 5 we assume that $u \perp \mathbf{1}$, yielding $u^t \cdot (\mathbf{1}_{|\mathcal{R}} + \mathbf{1}_{|\overline{\mathcal{R}}}) = 0$, so that

$$\left|u^t \cdot \mathbf{1}_{|\mathcal{R}}\right| = \left|u^t \cdot \mathbf{1}_{|\overline{\mathcal{R}}}\right| \le \|u\| \cdot \|\mathbf{1}_{|\overline{\mathcal{R}}}\| \le \sqrt{|\overline{\mathcal{R}}|}. \tag{13}$$

We decompose u as $u = a \cdot \mathbf{1}_{|\mathcal{R}} / \|\mathbf{1}_{|\mathcal{R}}\| + b \cdot u_l$ with $\|u_l\| = 1$ and $u_l \perp \mathbf{1}_{|\mathcal{R}}$. Clearly $u_{l|\mathcal{R}} \perp \mathbf{1}_{|\mathcal{R}}$, too, and $a^2 + b^2 = 1$. A straightforward computation yields

$$|a| = \left|u^t \cdot \frac{\mathbf{1}_{|\mathcal{R}}}{\|\mathbf{1}_{|\mathcal{R}}\|}\right| \overset{(13)}{\le} \frac{\sqrt{|\overline{\mathcal{R}}|}}{\|\mathbf{1}_{|\mathcal{R}}\|} \overset{(10)}{<} 2/{d_{\mathrm{m}}}^2. \tag{14}$$

Let v be some arbitrary unit-vector. Then we can rewrite $\left|u^t \cdot {M^*}_{V_i\times V_j} \cdot v\right|$ as

$$\left|u^t \cdot {M^*}_{V_i\times V_j} \cdot \left(v_{|\mathcal{C}} + v_{|\overline{\mathcal{C}}}\right)\right| \le \left|u^t \cdot {M^*}_{V_i\times V_j} \cdot v_{|\mathcal{C}}\right| + \|{M^*}_{\mathcal{R}\times\overline{\mathcal{C}}}\| + \|{M^*}_{\overline{\mathcal{R}}\times\overline{\mathcal{C}}}\|.$$

The second and the third summand are $O({d_{\mathrm{m}}}^{-1.5})$ by Lemma 9. The first one we bound as follows

$$\begin{aligned}\left|u^t \cdot {M^*}_{V_i\times V_j} \cdot v_{|\mathcal{C}}\right| &= \left|\left(a \cdot \frac{{\mathbf{1}^t}_{|\mathcal{R}}}{\|{\mathbf{1}^t}_{|\mathcal{R}}\|} + b \cdot u_l\right) \cdot {M^*}_{V_i\times\mathcal{C}} \cdot v_{\mathcal{C}}\right| \\ &\le |a| \cdot \|{M^*}_{\mathcal{R}\times\mathcal{C}}\| + |(b \cdot u_l) \cdot {M^*}_{V_i\times\mathcal{C}} \cdot v_{\mathcal{C}}| \\ &\overset{(14)}{<} 2/{d_{\mathrm{m}}}^2 \cdot O(1/\overline{w}) + \left|b \cdot \left(u_{l|\mathcal{R}} + u_{l|\overline{\mathcal{R}}}\right) \cdot {M^*}_{V_i\times\mathcal{C}} \cdot v_{\mathcal{C}}\right| \\ &\le O\left({d_{\mathrm{m}}}^{-1.5}\right) + |u_{l\mathcal{R}} \cdot {M^*}_{\mathcal{R}\times\mathcal{C}} \cdot v_{\mathcal{C}}| + \|{M^*}_{\overline{\mathcal{R}}\times\mathcal{C}}\| \\ &= O\left({d_{\mathrm{m}}}^{-1.5}\right) + O\left(\overline{w}^{-1.49}\right) + O\left({d_{\mathrm{m}}}^{-1.5}\right).\end{aligned}$$

We got the last step because of $u_{l|\mathcal{R}} \perp \mathbf{1}_{|\mathcal{R}}$ and Lemma 8. So, $\left|u^t \cdot {M^*}_{V_i \times V_j} \cdot v\right|$ is $O\left(\overline{w}^{\,-1.49}\right) + O\left({d_{\mathrm{m}}}^{-1.5}\right)$ as desired. The case $v \perp \mathbf{1}$ and u arbitrary can be handled analogously. □

References

1. Aiello, W., Chung, F., Lu, L.: A random graph model for massive graphs. Proc. 33rd. SToC (2001), 171–180.
2. Alon, N. Spectral techniques in graph algorithms. Proc. LATIN (1998), LNCS 1380, Springer, 206–215.
3. Alon, N., Kahale, N.: A spectral technique for coloring random 3-colorable graphs. SIAM J. Comput. **26** (1997) 1733–1748.
4. Azar, Y., Fiat, A., Karlin, A.R., McSherry, F., Saia, J.: Spectral analysis of data. Proc. 33rd STOC (2001) 619–626
5. Boppana, R.B.: Eigenvalues and graph bisection: An average case analysis. Proc. 28th FoCS (1987), 280–285.
6. Chung, F.K.R.: Spectral Graph Theory. American Mathematical Society (1997).
7. Chung, F.K.R., Lu, L.: Connected components in random graphs with given expected degree sequences. Annals of Combinatorics **6** (2002) 125–145.
8. Coja-Oghlan, A., Lanka, A.: The Spectral Gap of Random Graphs with Given Expected Degrees. Proc. ICALP (2006), LNCS 4051, Springer, 15–26.
9. Coja-Oghlan, A., Goerdt, A., Lanka, A.: Spectral Partitioning of Random Graphs with Given Expected Degrees. Preprint, submitted for publication. Available at http://www.tu-chemnitz.de/informatik/TI/publications/spj10.pdf
10. Ding. C.H.Q.: Analysis of gene expression profiles: class discovery and leaf ordering. Proc. 6th International Conference on Computational biology (2002) 127–136
11. Dasgupta, A., Hopcroft, J.E., McSherry, F.: Spectral Analysis of Random Graphs with Skewed Degree Distributions. Proc. 45th FOCS (2004) 602–610.
12. Flaxman, A.: A spectral technique for random satisfiable 3CNF formulas. Proc. 14th SODA (2003) 357–363.
13. Füredi, Z., Komloś, J.: The eigenvalues of random symmetric matrices. Combinatorica **1** (1981) 233–241.
14. Janson, S., Łuczak, T., Ruciński, A.: Random graphs. John Wiley and Sons 2000.
15. McSherry, F.: Spectral Partitioning of Random Graphs. Proc. 42nd FoCS (2001) 529–537.
16. Mihail, M., Papadimitriou, C.H.: On the Eigenvalue Power Law. Proc. 6th RANDOM (2002) 254–262.
17. Pothen, A., Simon, H.D., Liou, K.-P.: Partitioning sparse matrices with eigenvectors of graphs. SIAM J. Matrix Anal. Appl. **11** (1990) 430–452
18. Schloegel, K., Karypis, G., Kumar, V.: Graph partitioning for high performance scientific simulations. in: Dongarra, J., Foster, I., Fox, G., Kennedy, K., White, A. (eds.): CRPC parallel computation handbook. Morgan Kaufmann (2000)

On the Longest Common Factor Problem

Maxime Crochemore[1], Alessandra Gabriele[2], Filippo Mignosi[3], and Mauriana Pesaresi[4]

[1] King's College London, U.K. and Université Paris-Est, Institut Gaspard-Monge, France
maxime.crochemore@kcl.ac.uk
[2] Dipartimento di Matematica e Applicazioni, Università di Palermo, Italy
sandra@math.unipa.it
[3] Dipartimento di Informatica, Università dell'Aquila, Italy mignosi@ns.di.univaq.it
[4] Dipartimento di Informatica, Università di Pisa, Italy
pesaresi@di.unipi.it

Abstract The Longest Common Factor (LCF) of a set of strings is a well studied problem having a wide range of applications in Bioinformatics: from microarrays to DNA sequences analysis. This problem has been solved by Hui (2000) who uses a famous constant-time solution to the Lowest Common Ancestor (LCA) problem in trees coupled with use of suffix trees. A data structure for the LCA problem, although linear in space and construction time, introduces a multiplicative constant in both space and time that reduces the range of applications in many biological applications.
In this article we present a new method for solving the LCF problem using the suffix tree structure with an auxiliary array that take space $O(n)$. Our algorithm works in time $O(n \log a)$, where n is the total input size and a is the size of the alphabet.
We also consider a different version of our algorithm that applies to DAWGs. In this case, we prove that the algorithm works in both time and space proportional to data DAWG's size.

1 Introduction

In *1976* E.M.McCreight settled a Kunt's open problem by introducing a new data structure on string: the Suffix Tree. Since then, many other problems have been settled by using suffix trees or similar structures such as Patricia trees, DAWG, CDAWG and suffix array (cf. for instance [2, 8, 7, 5, 14] and references therein). Some other applications can be retrieved by exploring the "Pattern Matching Pointers" maintained by S. Lonardi (cf. [13]).

The most commonly used data structures are Suffix Trees, Suffix Arrays, DAWGs and CDAWGs. Usually any problem that can be settled by the aid of one of such data structure can also be settled by using any of the other ones. Despite this fact the passage from one data structure to another is not automatic nor always easy and, in some rare cases, not yet proved (see [1] for example). Each of these structures has some advantage and some disadvantage. Some relation among the data structures and their size is reported in [3]. The size of an implementation of the above data structures is often evaluated by the

Please use the following format when citing this chapter:

Crochemore, M., et al., 2008, in IFIP International Federation for Information Processing, Volume 273; *Fifth IFIP International Conference on Theoretical Computer Science*; Giorgio Ausiello, Juhani Karhumäki, Giancarlo Mauri, Luke Ong; (Boston: Springer), pp. 143–155.

average number of bytes necessary to store one letter of the original text. It is commonly admitted that these ratios are 4 for suffix arrays, 9 to 11 for suffix trees, and 5 for CDAWGs (cf. [3] for further information).

This paper deals with particular data structures: DAWGs.

The problem we consider is reported by D. Gusfield, [8, 8,Sec. 7.6, 9.4]. In the exact case, it is the following: given a set of m strings, for any $k = 2, .., m$ find the longest factors that are common to at least k strings. The word *common* in the exact case means *occurring with equality.* The first solution to this problem has been given by Hui, ([10, 11]), who uses a famous constant-time solution to the Lowest Common Ancestor (LCA) problem in trees coupled with the use of suffix trees (see [9, 16, 6]). A data structure for the LCA, although it is linear in space and time, introduces a multiplicative constant in both space and time that reduces the range of applications in many biological applications.

Since DAWGs and CDAWGs are not trees, this solution cannot be used for the structures we are interested in. Therefore we look for a totally new solution. So, our solution turns out to be simpler and more efficient than Hui's one of about one order of magnitude. This solution is an extension from that of suffix trees to DAWGs.

This paper is organized as follows. In the next section we describe our solution for the problem based on the use of suffix trees, while in the Section 3 we extend our solution to DAWGs. The fourth section contains our conclusions and some conjectures on the approximate case of the problem. Hence in the Appendixes A and B, we report the specialized pseudo-code related to the procedures used in our algorithm.

2 A Simpler Solution

We assume the reader familiar with suffix trees and Generalized Suffix Trees.

Let S be a set of input strings S_i, $1 \leq i \leq m$, on the alphabet $\{0, 1\}$. Let u be the word composed of the concatenated labels of transitions along the unique path from the root to the node p in the Generalized Suffix Tree.

We want to compute a table ℓ having $m-1$ entries: where entry $\ell[k]$ provides the length of the longest factor common to at least k of the input strings and also points to one of the common factors having that length.

Our preprocessing is as follows. We build the Generalized Suffix Tree for the m strings. Then perform a depth-traversal of the tree and put all nodes in a stack in the order they appear. Define s to be an array of pointers representing the input strings useful to increase the algorithm's performances.

Each node stores the following information:

- i represents the string identifier whose suffix is the node path-label. If this is not a suffix, this field is empty.
- *num* is the number of distinct string identifiers that appear at the leaves in the subtree rooted in p. Observe that this approach is the same as the one

used by Gusfield in [8, Sec.7.6]. The difference lays in how to compute these values, that he calls $C[p]$.

So we must first compute the *num* values and then use them to update the table.

2.1 Computing the num values

For each node p, we create an auxiliary node *size* that stores the values $num(p)$ and points to the strings it represents in s.

When for the vertex p we have $num(p) = b$, this means that in its subtree there are nodes representing suffixes from b different input strings. In other words, p is the common factors of exactly b different input strings. In the algorithm we call these nodes representative in the operation of $Union$ that plays an important role in the computing of our values.

The operation $Union$ is the union between disjoint sets of elements that, in our case, are nodes *size* linked to visited nodes. All pointers to auxiliary node of smaller size must point to the other node *size* and, naturally, we must also update the sizes of the involved nodes, i.e. the field *num*.

$Union$ operates as follows. Let a be a node with $num(a) = 2$ and let b with $num(b) = 3$. When we visit the a and b's father p, we execute a $Union$ of his children. The result is that $num(p) = 5$ and the p's label becomes a common factor of 5 input strings.

We keep the disjoint strings sets as follows. We use an array s of m pointers that represent the input strings and for which $s[i]$ points permanently to the last met factor of the string s_i. Since the last factor of a string is unique, the sets to merge are always disjoints.

In the algorithm we use three procedures called $NodeSize\ Test$, $String\ Test$ and $Union$ (that implements the union operation). Now we explain how they work, while in the Appendix A we show the code of them.

- $NodeSize\ Test$ procedure: we check if the node *size* is already created. If not, we create it.
- $String\ Test$ procedure: when we visit a new node, we must update the information about the last visited factor of some string. Note that after this test and related "cut-append" of pointers, node *size* stores the current *num* value, while the internal node stores the real one. Because nodes *size* are representatives in the $Union$, then they must be updated in every time. In Figure 1 we show the effect of this test.
- $Union$ procedure: after we have found the smallest son its pointer is redirected to the largest one, the *num* value is updated, and the new node *size* resulting from the merging is merged with the father node *size*.

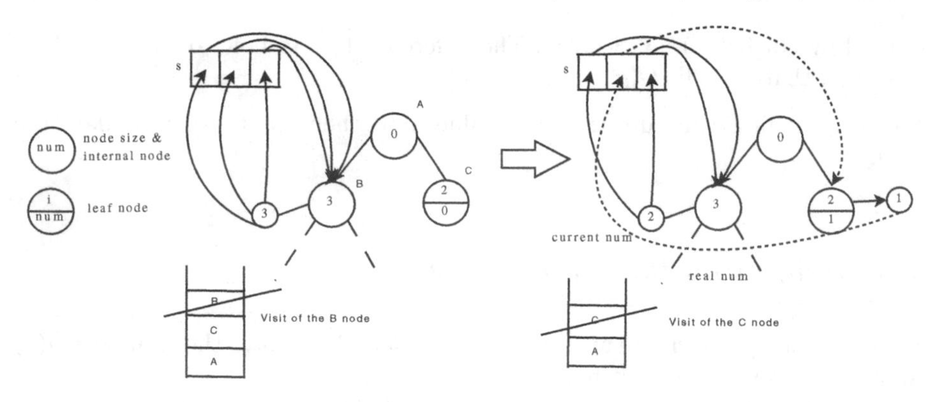

Fig. 1: In the left figure, is shown the situation after the visit of the B node, at the top of the stack. It's the common factor of all input strings. Then the node C is traversed. Therefore it's a leaf node representing the string s_2, then the algorithm "cut" the pointer from B's node *size* to the element $s[2]$, appending it to C's node *size*. In fact, the last factor of s_2 is the path-label of C. Observe that the values about the size of the nodes are updated during this test.

Now we describe the algorithm to compute in an efficient way the *num* information.

```
CountNum (stack, s)
 1. while (stack isNotEmpty) do
 2.       p = pop(stack);
 3.       NodeSize test;
 4.       String test;
 5.       if p has sons then
 6.             Union operation;
 7. End CountNum.
```

2.2 The method

Once the *num* values are known and the string-depth of every node is known, the desired $\ell(k)$ values can be easily found with a linear-time traversal of $GST(S)$.

When encountering a node p with $num(p) = k$, we compare the string-depth of p to the current value of $\ell(k)$. If the first value is greater than the second, we change $\ell(k)$ to the depth of p and update its pointer to the node representing the factor with the current value of $\ell(k)$.

Eventually, the resulting table holds the desired $\ell(k)$ values.

2.3 Time and space analysis

Building $GST(S)$ requires linear space and time in the size of input [8, Sec. 6.4], i.e. $O(n)$ with $n = |S_1| + \cdots + |S_m|$. Algorithm *CountNum* executes a single post-order traversal of $GST(S)$ and its main operation is the *Union*. Since the operation "cut-append" of a pointer from a node to another is done in constant time, then we have to know how many pointers could be involved during it.

Theorem 1. ***[CountingNum]*** *During the execution of* $CountNum(S, s)$ *algorithm, the number of "cut-append" operations is less than* n*, with* $n = |S_1| + \cdots + |S_m| = |S|$.

Proof. The statement is proved by induction on the total size n of the representatives u of the nodes p. Recall that u is p's representative if it is the concatenated labels of transitions of the unique path from the root to the node p in $GST(S)$.

Basic step: **let m be the minimal size of S.** The root points directly to the m leaves. Therefore the number of "cut-append" operations is equal to $m - 1$: we append all auxiliary leaves to the root. Since the input's size is equal to m then our thesis is proved.

Inductive step: by induction we suppose that our thesis is true for every tree with representatives' size equal to $n - 1$.
We prove the thesis is true at the level n.
Let our visit be stopped in a node with two sons. The first subtree has NL_1 leaves and representatives' size equal to n_1, while the second has NL_2 leaves and size equal to n_2.
When we get to the bottom level n, we add a character to every representative for each leaf. So the total representatives' size of the level n is equal to $n_1 + NL_1 + n_2 + NL_2$.
The *Union* simply appends all leaves of one subtree to the other subtree.

$$\begin{aligned} cut - append &= NL_1 + NL_2 < \\ &< n_1 + NL_1 + n_2 + NL_2 = n. \end{aligned} \tag{1}$$

When we visit the $GST(S)$'s root, the total representative's size is equal to the input length, n. Hence, the total number of *Union* operations is linear in the input length. ■

During a run of the algorithm $O(n)$ "cut-append" operations are executed, each of which takes constant time, so the overall *Union* takes $O(n)$ time.

Hence only $O(n)$ time is needed to execute the algorithm and to compute all *num* numbers. Once these are known, only $O(n)$ additional time is needed to build the output table.

Hui's solution take $O(mn)$ time because it uses an array of k elements for each node of the tree to calculate the *num* values. We solve the original problem simply using $O(m+n)$ space, because the algorithm makes use of a unique array.

Theorem 2. *The Lowest Common Factor Problem on a set of m input's strings, represented by a Generalized Suffix Tree, can be solved in $O(n)$ time, with $n = |s_1| + \cdots + |s_m|$, and $O(m+n)$ space.*

3 An Optimal Solution

In this section we deal with the data structures that plays an important role in this paper, the Generalized Directed Acylic Word Graph (Generalized DAWG). We assume the reader familiar with DAWGs.

Now we recall the definition of DAWG.

Definition 1. The DAWG for a set of strings $s_1, \cdots, s_m$ is a directed acyclic graph, with a node marked as initial and m distinct nodes $F_1, \cdots, F_m$ marked as final. Edges are labeled with non empty factors of at least one of the strings. Labels of two edges leaving the same node cannot begin with the same character. For every string s_i in the set, all suffixes of s_i are spelled by patterns starting at the initial node and ending at node F_i. Paths ending at non final nodes correspond to strict classes of factors of the congruence relations $\equiv_{Suf(S)}$.

Let S be our input set of strings.

We want to analyse the meaning of the state u in terms of "representative". In $DAWG(S)$ there are more edges entering the same state than in the corresponding tree, according to Def.1. So we define the *representative* of a state as the longest path from the initial state to it.

Like for Suffix Trees, we want to compute a table that gives for entry k the length of a longest factor common to at least k strings and also points to it.

Now our preprocessing is less easy than in the previous section because more paths are not distinct. We build a Generalized DAWG for the m input strings. Each final state represents an input string (e.g., s_i) and is marked with a non null identifier (e.g., i).

Observe that in a DAWG two or more outgoing edges from the same state could finish in the same state and so we would like that the path from an internal state to other one is unique. Hence we keep only the representative of a factor's class. Since to solve the LCF Problem we need the longest labels of the paths, we keep only the transitions with the longest labels and we delete all other ones that have the same origin and target states. In this way the number of transition is drastically reduced and we obtain a pruned DAWG, denoted by D, having a deterministic transition function between adjacent states.

Now we are ready to perform a particular breadth-traversal of the new structure to store all states in a stack, in a way that is similar to the procedure done on suffix trees. We put nodes in the stack in the order they appear. Our problem is that we traverse some nodes more times and we must store them only once. Hence, if a node is already stored, we delete its previous occurrences, we put its new occurrence and we increase a counter related to the node. Define s to be an array of pointers representing the input strings like above. In our data structure each state stores the following informations:

- i is the string identifier whose suffix is the state path-label,
- num is the number of distinct string identifiers that appear in the subgraph rooted in the state,
- $count$ is the counter mentioned above.

As in the previous section, we first compute the num values and then we use them to update the output's table.

3.1 How to compute desired values?

The algorithm to calculate num is almost the same as for suffix trees. The only difference is in the *String Test* procedure, because here there is another parameter to check, the *count* value. In the algorithm we use three procedures called *NodeSize Test*, *StringD Test* and *UnionD*. *NodeSize Test* procedure has already been described in the previous section.

Now we explain how the *StringD* test works, while in the Appendix B we show its code and the *UnionD* one. First we perform the following test on the field *count* of the sons of the current state:

- if *count* is not null for some son, we decrease the value of *count* and we "cut" only the pointer from array s to the previous state to link it to the actual node, because this one represents the last factor of the interesting string. Observe that we delete a node *size* when *count* become null. So, for *count* times we must replace the node *size*. This fact causes an additional extra-space but it permits to perform the execution in linear time;
- otherwise we call the classical *String* test.

The complete algorithm is the following:

```
COUNTNUMBIS (stack, s)
 1. while (stack isNotEmpty) do
 2.       p = pop(stack);
 3.       NodeSize test;
 4.       StringD test;
 5.       if p has sons then
 6.             UnionD operation;
 7. End CountNum.
```

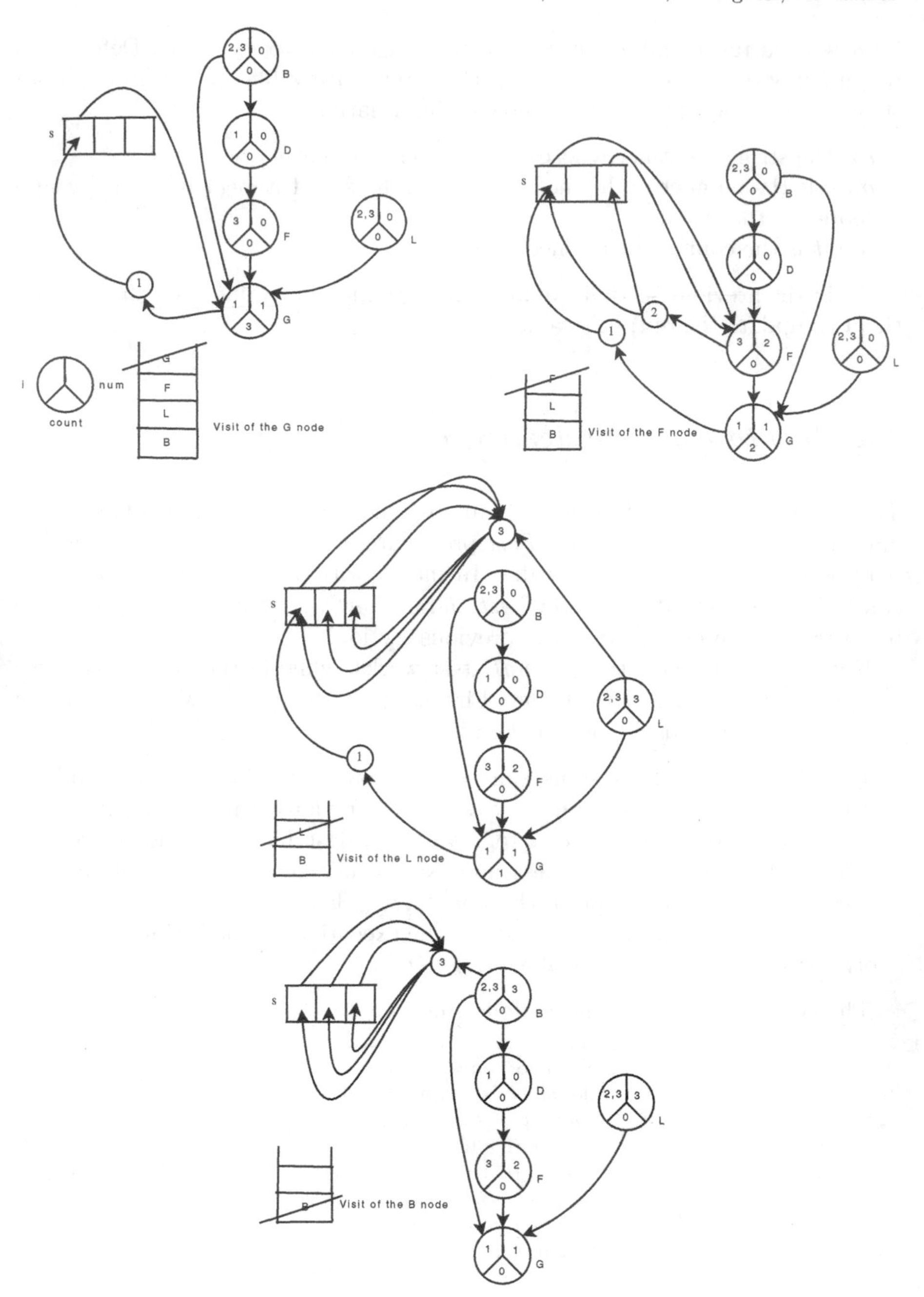

Fig. 2: How *StringD* works. Visiting the state G is the same as visiting the corresponding tree node. When we traverse state F, we create a duplicate pointer to $s[1]$ not to lose the information related to state G: in fact, two other edges arrive in this state and they need to know that G is a suffix of $s[1]$. Note that the field *count* of state G is decreased. Therefore F is also a suffix of $s[2]$, then we perform a traditional *Union*. The same happens when traversing states L and B. In last case, since the field *count* of state G is null, then we can delete its node *size* because we have visited all its neighbors.

3.2 Building the output table

Once the *num* value and the string-depth of every state are known, the desired $\ell(k)$ values can be easily found with a linear-time traversal of D.

When encountering a state p with $num(p) = k$, the string-depth of p is compared to the current value of $\ell(k)$ and if the first one is greater than the second, $\ell(k)$ is changed to the depth of p and its pointer is updates to the node representing the factor with the current value of $\ell(k)$.

Finally the resulting table holds the desired $\ell(k)$ values.

3.3 Time analysis

Let n be the input size, with $n = |S_1| + \cdots + |S_m|$. Building $DAWG(S)$ requires linear time in the input size as described in [12].

Algorithm *CountNumBIS* executes a single traversal of D and its main operation is *Union*. Since the operation "cut-append" of a pointer from a node to another is constant, then we would like to know how many pointers could be involved during it.

Let D be the Generalized *DAWG* over S. We can use a breadth-first visit of D to re-create the original Suffix Tree. Each path from initial state to a final state in *DAWG* is used to build a path from the root to a leaf in the Suffix Tree. Note that the technique is the same as McCreight's one (cf. [15]) to create suffix trees directly from input's strings.

After this traversal, we have created a suffix tree with a number ns of nodes that is larger than the number nc of *DAWG* states, with same edges and related labels. Hence representatives of suffix tree states are the same as that of *DAWG*s.

Since $nc \leq ns$, from Theorem 1, we have the following result:

Theorem 3. ***[LCSS Counting Bis]*** *During the execution of algorithm* $CountNumBIS(S, s)$*, the number of "cut-append" operations is less than* n*, with* $n = |S_1| + \cdots + |S_m| = |S|$.

During the run of the algorithm there are $O(n)$ "cut-append" operations executed, each of which takes constant time, so all *Union* executions take $O(n)$ time in total.

Hence only $O(n)$ time is needed to execute the algorithm and to compute all num_S numbers. Once these are known, only $O(n)$ additional time is needed to build the output table.

Finally, we can state:

Theorem 4. *The Lowest Common Factor Problem on a set of* m *input strings, represented by a Generalized Directed Acyclic Graph, can be solved in* $O(n)$ *time, with* $n = |s_1| + \cdots + |s_m|$*, and* $O(m + n)$ *space.*

4 Conclusions

In this paper we introduced an algorithm that we show to require less space than the previous Hui's solution, when we use a data structure like Suffix Trees. We obtain a solution that requires a unique k-array, where k is the number of input's strings, to store all information, instead of using a k-array for each node as in the Hui's solution. Both algorithms run in linear time.
Another advantage of our algorithm is about the size of the implementation of the data structure used that is often evaluated by the average number of bytes necessary to store one letter of the original text. It is commonly admitted that these ratios are 9 to 11 for suffix trees and 5 for DAWGs (cf. [3] for further information). Moreover a data structure for the LCA problem, although linear in space and construction time, introduces a multiplicative constant (from 2 to 4) in both space and time. While Hui's implementation introduce a factor of 40 to solve the problem, our implementation with the DAWGs reduces this multiplicative constant to nearly 5.
Recent experiments [4] have showed that DAWGs are space thrifty not only in exact problems, but also in the approximate cases, where some "errors" or "faults" are allowed. To build the approximate DAWG of a word in optimal time remains an open problem. Now, we think that our solution of the exact problem can be applied to these data structures to solve the approximate case. If the conjecture reported in [4] is true and if it is possible to build approximate DAWGs in optimal time, then our solution will drastically outperform previous solutions.

References

1. M.I. Abouelhoda, S. Kurtz, and E. Ohlebusch. Replacing suffix trees with enhanced suffix arrays. *Journal of Discrete Algorithms*, 2(1):53–86, 2004.
2. A. Apostolico. The myriad virtues of suffix trees. In A.Apostolico and Z.Galil, editors, *Combinatorial Algorithms on Words*, volume 12 of *F*, pages 85–96. 1985.
3. M. Crochemore. Reducing space for index implementation. *Theoretical Computer Science*, 292(1):185–197, 2003.
4. M. Crochemore, C. Epifanio, A. Gabriele, and F. Mignosi. On the suffix automaton with mismatches. In *To appear in Lecture Notes in Computer Science.* CIAA'07, 2007.
5. M. Crochemore, C. Hancart, and T. Lecroq. *Algorithmique du texte.* Vuibert Informatique, 2001.
6. J. Fischer and V. Heun. Theoretical and practical improvements on the rmq-problem, with applications to lca and lce. In *Springer LNCS*, volume 4009 of *Proceedings of the 17th Annual Symposium on Combinatorial Pattern Matching(CPM'06)*, pages 36–48, 2006.
7. R. Grossi and G.F. Italiano. Suffix trees and their applications in string algorithms. Proceedings of the 1st South American Workshop on String Processing, pages 57–76, 1993.
8. D. Gusfield. *Algorithms on Strings, Trees, and Sequences: Computer Science and Computational Biology.* Cambridge University Press, 1997.

9. D. Harel and R.E. Tarjan. Fast algorithms for finding nearest common ancestors. *SIAM Journal of Computing*, 13:338–355, 1984.
10. L.C.K. Hui. Color set size problem with applications to string matching. In *Proceedings 3rd Symposium on Combinatorial Pattern Matching*, volume 644 of *Springer LNCS*, pages 227–240, 1992.
11. L.C.K. Hui. A practical algorithm to find longest common substring in linear time. *International Journal of Computer Systems Science & Engineering*, 15(2):73–76, 2000.
12. S. Inenaga, H. Hoshino, A. Shinohara, M. Takeda, S. Arikawa, G. Mauri, and G. Pavesi. On-line construction of compact directed acyclic word graphs. In *Discrete Applied Mathematics*, volume 146 of *12th Annual Symposium on Combinatorial Pattern Matching*, pages 156–179, 2005.
13. Stefano Lonardi. Pattern Matching Pointers. http://www.cs.ucr.edu/ stelo/pattern.html, 2008.
14. M.G. Maas. Matching statistics: efficient computation and a new practical algorithm for the multiple common substring problem. *Software Practice and Experience*, 36:305–331, 2006.
15. E.M. McCreight. A space-economical suffix tree construction algorithm. *Journal of the ACM*, 23(2):262–272, 1976.
16. B. Schieber and U. Vishkin. On finding lowest common ancestors: simplifications and parallelizzation. *SIAM Journal on Computing*, 17:1253–1262, 1988.

5 Appendix

Now we detail the procedures used by the algorithm for Suffix Tree. Recall the data structures in a formally way.
The auxiliary structure s is an m-array of pointers.
The node of $GST(S)$ are formed by three fields (and not two):

- the fields i and num;
- the field ns is a pointer to the node $size$ related to our node.

The node size has two fields (and not one):

- the field num;
- the field ns is a set of pointers to the structures s, one for each string that the node representing.

```
NodeSize test (GST(S), s)
  1. if p.ns = Nil then
  2.       p.ns =new node.
  3. Return(GST(S), s).
```

```
STRING TEST (GST(S), s)
 1. for every i of p do
 2.        if p.i! = Nil then
 3.               s[p.i].ns.num − −
 4.               if s[p.i].ns.num = 0 then
 5.                      delete s[p.i].ns;
 6.        s[p.i] = p;
 7.        p.ns.ns = s[p.i];
 8.        p.ns.num + +;
 9.        p.num + +.
10. Return(GST(S), s).
```

```
UNION (GST(S), s)
 1. Merge between nodes size
 2.        merge between pointers to s;
 3.        sum between the fiels num;
 4.        have created a new node size m;
 5. p.ns = merge(p.ns, m);
 6. p.num = p.ns.num.
 7. End Union operation.
```

6 Appendix

Now we detail the procedures used by the algorithm for DAWG. Recall the data structures in a formally way.
The auxiliary structure s is an m-array of pointers.
The node of $DAWG(S)$ are formed by four fields (and not three):

- the fields i, num and $count$;
- the field ns is a pointer to the node *size* related to our node.

The node size has two fields (and not one):

- the field num;
- the field ns is a set of pointers to the structures s, one for each string that the node representing.

```
NODESIZE TEST (D, s)
 1. if p.ns = Nil then
 2.        p.ns =new node.
 3. Return(D, s).
```

```
STRINGD TEST (D, s)
 1. for every i of p do
 3.        if s[p.i].count! = 0 then
 4.              s[p.i].count − −;
 5.              if s[p.i].count = 0 then
 6.                    delete s[p.i].ns;
 7.              s[p.i] = p
 8.        else
 9.              s[p.i].ns.num − −
 10.             if s[p.i].ns.num = 0 then
 11.                   delete s[p.i].ns;
 12.             s[p.i] = p;
 13.             p.ns.ns = s[p.i];
 14.       p.ns.num + +;
 15.       p.num + +.
 16. Return(D, s).
```

```
UNIOND (D, s)
 1. Merge between nodes size of p's sons with the field count null
 2.       merge between pointers to s;
 3.       sum between the fields num;
 4.       have created a new node size m;
 5. if q.count! = 0 and q.ns.ns! = p.ns.ns with q son of p then
 6.                   q.count − −;
 7.                   if q.count = 0 then
 8.                         delete q.ns;
 9.                   duplicate q.ns pointers and append to m;
 10.                  m.num = m.num + +
 11. p.ns = merge(p.ns, m);
 12. p.num = p.ns.num.
 13. End UnionD operation.
```

Stable Dynamics of Sand Automata*

Alberto Dennunzio[1], Pierre Guillon[2], and Benoît Masson[3]

[1] Università degli studi di Milano-Bicocca, Dipartimento di Informatica Sistemistica e Comunicazione, viale Sarca 336, 20126 Milano, Italy
dennunzio@disco.unimib.it

[2] Université Paris-Est, Laboratoire d'Informatique de l'Institut Gaspard Monge, UMR CNRS 8049, 5 bd Descartes, 77454 Marne la Vallée Cedex 2, France
pierre.guillon@univ-mlv.fr

[3] Laboratoire d'Informatique Fondamentale de Marseille (LIF)-CNRS, Aix-Marseille Université, 39 rue Joliot-Curie, 13453 Marseille Cedex 13, France
benoit.masson@lif.univ-mrs.fr

Abstract. In this paper, we study different notions of stability for sand automata, dynamical systems inspired by sandpile models and cellular automata. First, we study the topological stability properties of equicontinuity and ultimate periodicity, proving that they are equivalent. Then, we deal with nilpotency. The classical definition for cellular automata being meaningless in that setting, we define a more suitable one. Finally, we prove that this dynamical behavior is undecidable.

1 Introduction

Self-organized criticality (SOC [2]) is a common phenomenon observed in a huge variety of processes in physics, biology and computer science. A SOC system evolves to a "critical state" after some finite transient. Examples of SOC systems are: sandpiles, snow avalanches, star clusters in the outer space, earthquakes, forest fires, load balance in operating systems [1]. Among them, sandpile models are a paradigmatic formal model for SOC systems [10].

In [3], the authors introduced sand automata as a generalization of sandpile models and transposed them in the setting of discrete dynamical systems. A key-point of [3] was to introduce a (locally compact) metric topology to study the dynamical behavior of sand automata. A first and important result was a fundamental representation theorem similar to the well-known theorem of Hedlund for cellular automata [11, 3]. In [4, 5], the authors investigate sand automata by dealing with some basic set properties and decidability issues. Then, in [8], a new compact topology is introduced, inspired by the strong relation between sand automata and cellular automata. It is proved that with

* This work has been supported by the Interlink/MIUR project "Cellular Automata: Topological Properties, Chaos and Associated Formal Languages", by the ANR Blanc "Projet Sycomore" and by the PRIN/MIUR project "Formal Languages and Automata: Mathematical and Applicative Aspects".

Please use the following format when citing this chapter:

Dennunzio, A., Guillon, P. and Masson, B., 2008, in IFIP International Federation for Information Processing, Volume 273; *Fifth IFIP International Conference on Theoretical Computer Science*; Giorgio Ausiello, Juhani Karhumäki, Giancarlo Mauri, Luke Ong; (Boston: Springer), pp. 157–169.

this new topology, the representation theorem still holds, while the compactness provides new opportunities for further topological studies of the model.

In this paper we continue the study of sand automata dynamics of [4, 5], using the topological framework from [8]. We focus on stability, which is a major issue for isolating the realistic sandpile models satisfying the SOC principles. More precisely, we study different types of stability. First, we deal with the topological stability, i.e., the equicontinuity and ultimate periodicity properties. We prove that they are equivalent. We also show the insignificance of expansivity, a form of strong instability. This fact suggests that the topological classification for cellular automata from [13] cannot be easily generalized to sand automata.

Then, we study nilpotency, a very strong form of dynamical stability. The classical definition of nilpotency for cellular automata [7, 12] is no more meaningful here, since it would prevent any sand automaton from being nilpotent. Therefore, we introduce a new definition which captures the intuitive idea that a nilpotent automaton destroys all configurations: a sand automaton is nilpotent if all configurations get closer and closer to a uniform configuration, not necessarily reaching it. Finally, we prove that this behavior is undecidable, using the undecidability of the nilpotency of spreading cellular automata.

The paper is structured as follows. Section 2 recalls basic definitions and results on cellular automata and sand automata. In Section 3, results on the topological stability of sand automata are proved and discussed. Nilpotency of sand automata is then defined and proved undecidable in Section 4.

2 Definitions

For all $a, b \in \mathbb{Z}$ with $a \leq b$, let $[a, b] = \{a, a+1, \ldots, b\}$ and $\widetilde{[a, b]} = [a, b] \cup \{+\infty, -\infty\}$. Let $\mathbb{N}_+$ be the set of positive integers.

For a vector $i \in \mathbb{Z}^d$, denote by $|i|$ the infinite norm of i. Let A a (possibly infinite) alphabet, and $r \in \mathbb{N}, d \in \mathbb{N}_+$. Denote by $\mathcal{M}_r^d$ the set of all the d-dimensional matrices with values in A and entry vectors in the hyper-rectangle $[-r, r]^d$.

2.1 Cellular Automata

Let A be a finite alphabet. A *CA configuration* of dimension d is a function from $\mathbb{Z}^d$ to A. The set $A^{\mathbb{Z}^d}$ of all the CA configurations is called the *CA configuration space.* This space is usually equipped with the Tychonoff metric d_T defined by

$$\forall x, y \in A^{\mathbb{Z}^d}, \quad \mathsf{d}_T(x, y) = 2^{-k} \quad \text{where} \quad k = \min\left\{|j| : j \in \mathbb{Z}^d, x_j \neq y_j\right\} \ .$$

The topology induced by d_T coincides with the product topology induced by the discrete topology on A. It makes the CA configuration space is a Cantor space: it is compact, perfect (i.e., it has no isolated points) and totally disconnected.

A *cellular automaton* (CA) is a quadruple $\langle A, d, r, g\rangle$, where A is the alphabet, also called the *state set*, $d \in \mathbb{N}_+$ is the dimension, $r \in \mathbb{N}$ is the *radius* and $g : \mathcal{M}_r^d \to A$ is the *local rule* of the automaton. The local rule g induces a *global rule* $G : A^{\mathbb{Z}^d} \to A^{\mathbb{Z}^d}$ defined as follows,

$$\forall x \in A^{\mathbb{Z}^d}, \forall i \in \mathbb{Z}^d, \quad G(x)_i = g\big(M_r^i(x)\big) \ ,$$

where $M_r^i(x) \in \mathcal{M}_r^d$ is the *finite portion* of x of reference position $i \in \mathbb{Z}^d$ and radius r defined by $\forall k \in [-r,r]^d$, $M_r^i(x)_k = x_{i+k}$.

For any $k \in \mathbb{Z}^d$ the *shift map* $\sigma^k : A^{\mathbb{Z}^d} \to A^{\mathbb{Z}^d}$ is defined by $\forall x \in A^{\mathbb{Z}^d}, \forall i \in \mathbb{Z}^d$, $\sigma^k(x)_i = x_{i+k}$. A function $F : A^{\mathbb{Z}^d} \to A^{\mathbb{Z}^d}$ is said to be *shift-commuting* if $\forall k \in \mathbb{Z}^d$, $F \circ \sigma^k = \sigma^k \circ F$. Note that CA are exactly the class of all shift-commuting functions which are (uniformly) continuous with respect to the Tychonoff metric (Hedlund's theorem from [11]). For the sake of simplicity, we will make no distinction between a CA and its global rule G.

For a given CA, a state $s \in A$ is *quiescent* (resp., *spreading*) if for all matrices $U \in \mathcal{M}_r^d$ such that $\forall k \in [-r,r]^d$, (resp., $\exists k \in [-r,r]^d$) $U_k = s$, it holds that $g(U) = s$. Remark that a spreading state is also quiescent. A CA is said to be spreading if it has a spreading state. In the sequel, the spreading state of any spreading CA will be denoted $0 \in A$.

2.2 SA Configurations

A *SA configuration* (or simply *configuration*) is a set of sand grains organized in piles and distributed all over the d-dimensional lattice $\mathbb{Z}^d$. A *pile* is an element of $\widetilde{\mathbb{Z}} = \mathbb{Z} \cup \{-\infty, +\infty\}$ which represents a number of grains. One pile is positioned in each point of the lattice $\mathbb{Z}^d$. Formally, a configuration x is a function from $\mathbb{Z}^d$ to $\widetilde{\mathbb{Z}}$ which associates any vector $i = (i_1, \ldots, i_d) \in \mathbb{Z}^d$ with the number $x_i \in \widetilde{\mathbb{Z}}$ of grains in the pile of position i. When the dimension d id known without ambiguity, we note 0 the null vector of Z^d. Denote by $\mathcal{C} = \widetilde{\mathbb{Z}}^{\mathbb{Z}^d}$ the set of all configurations.

A configuration $x \in \mathcal{C}$ is said to be *constant* if there is an integer $c \in \mathbb{Z}$ such that for any vector $i \in \mathbb{Z}^d$, $x_i = c$. In that case we write $x = \underline{c}$. A configuration $x \in \mathcal{C}$ is said to be *bounded* if there exist two integers $m_1, m_2 \in \mathbb{Z}$ such that for all vectors $i \in \mathbb{Z}^d$, $m_1 \le x_i \le m_2$. Denote by $\mathcal{B}$ the set of all bounded configurations.

A *measuring device* β_r^m of precision $r \in \mathbb{N}$ and reference height $m \in \mathbb{Z}$ is a function from $\widetilde{\mathbb{Z}}$ to $\widetilde{[-r,r]}$ defined as follows

$$\forall n \in \widetilde{\mathbb{Z}}, \quad \beta_r^m(n) = \begin{cases} +\infty & \text{if } n > m + r \ , \\ -\infty & \text{if } n < m - r \ , \\ n - m & \text{otherwise.} \end{cases}$$

A measuring device is used to evaluate the relative height of two piles, with a bounded precision. This is the technical basis of the definition of cylinders, distance and ranges which are used all along this article.

In [8], a topology, inspired by the topology on CA configurations, is defined as follows.

Definition 1 (cylinder). For any configuration $x \in \mathcal{C}$, for any $r \in \mathbb{N}$, and for any $i \in \mathbb{Z}^d$, the *cylinder* of x centered on i and of radius r is the d-dimensional matrix $C_r^i(x) \in \mathcal{M}_r^d$ defined on the finite alphabet $\widetilde{[-r, r]}$ by

$$\forall k \in [-r, r]^d, \quad \left(C_r^i(x)\right)_k = \beta_r^0(x_{i+k}) \ .$$

Definition 2. For any pair of configurations $x, y \in \mathcal{C}$, we define

$$\mathsf{d}(x, y) = 2^{-k} \quad \text{where} \quad k = \min\left\{r \in \mathbb{N} : C_r^0(x) \neq C_r^0(y)\right\} \ .$$

As a consequence, two configurations x, y are compared by putting boxes (the cylinders) at height 0 around the corresponding piles indexed by 0. The integer k is the size of the smallest cylinders in which a difference appears between x and y.

With the topology induced by d, the SA configuration space is perfect, totally disconnected, and, unlike the original topology used in [11, 3], compact (see [8]).

2.3 Sand Automata

For any integer $r \in \mathbb{N}$, for any configuration $x \in \mathcal{C}$ and any index $i \in \mathbb{Z}^d$ with $x_i \neq \pm\infty$, the *range* of center i and radius r is the d-dimensional matrix $R_r^i(x) \in \mathcal{M}_r^d$ on the finite alphabet $A = \widetilde{[-r, r]} \cup \bot$ such that

$$\forall k \in [-r, r]^d, \quad \left(R_r^i(x)\right)_k = \begin{cases} \bot & \text{if } k = 0 \ , \\ \beta_r^{x_i}(x_{i+k}) & \text{otherwise.} \end{cases}$$

The range is used to define a sand automaton. It is a kind of cylinder, where the observer is always located on the top of the pile x_i (called the *reference*). It represents what the automaton is able to see at position i. Sometimes the central $\bot$ symbol may be omitted for simplicity sake. The set of all possible ranges of radius r, in dimension d, is denoted by $\mathcal{R}_r^d$.

A *sand automaton* (SA) is a deterministic finite automaton working on configurations. Each pile is updated synchronously, according to a local rule which computes the variation of the pile by means of the range. Formally, a SA is a triple $\langle d, r, f \rangle$, where d is the dimension, r is the *radius* and $f : \mathcal{R}_r^d \to [-r, r]$ is the *local rule* of the automaton. The *global rule* $F : \mathcal{C} \to \mathcal{C}$ is defined by

$$\forall x \in \mathcal{C}, \forall i \in \mathbb{Z}^d, \quad F(x)_i = \begin{cases} x_i & \text{if } x_i = \pm\infty \ , \\ x_i + f(R_r^i(x)) & \text{otherwise.} \end{cases}$$

The following example illustrates a sand automaton whose behavior will be studied in Section 4. For more examples, we refer to [5].

Example 1 (the automaton $\mathcal{N}$). This automaton destroys a configuration by collapsing all piles towards the lowest one. It decreases a pile when there is a lower pile in the neighborhood. Let $\mathcal{N} = \langle 1, 1, f_{\mathcal{N}} \rangle$ of global rule $F_{\mathcal{N}}$ where

$$\forall a, b \in \widetilde{[-1, 1]}, \quad f_{\mathcal{N}}(a, b) = \begin{cases} -1 & \text{if } a < 0 \text{ or } b < 0 \ , \\ 0 & \text{otherwise.} \end{cases}$$

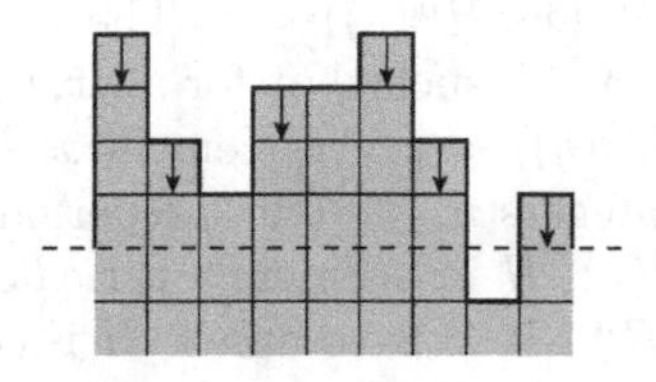

Fig. 1 Illustration of the behavior of $\mathcal{N}$.

When no misunderstanding is possible, we identify a SA with its global rule F. For any $k \in \mathbb{Z}^d$, we extend the definition of the *shift map* to $\mathcal{C}$, $\sigma^k : \mathcal{C} \to \mathcal{C}$ is defined by $\forall x \in \mathcal{C}, \forall i \in \mathbb{Z}^d$, $\sigma^k(x)_i = x_{i+k}$. The *raising map* $\rho : \mathcal{C} \to \mathcal{C}$ is defined by $\forall x \in \mathcal{C}, \forall i \in \mathbb{Z}^d$, $\rho(x)_i = x_i + 1$. A function $F : \mathcal{C} \to \mathcal{C}$ is said to be *vertical-commuting* if $F \circ \rho = \rho \circ F$. A function $F : \mathcal{C} \to \mathcal{C}$ is *infinity-preserving* if for any configuration $x \in \mathcal{C}$ and any vector $i \in \mathbb{Z}^d$, $F(x)_i = +\infty$ if and only if $x_i = +\infty$ and $F(x)_i = -\infty$ if and only if $x_i = -\infty$.

With the topology from [8], the Hedlund-like representation theorem for SA from [3] remains valid.

Theorem 1 ([8]). *A mapping $F : \mathcal{C} \to \mathcal{C}$ is a SA if and only if F is (uniformly) continuous, shift-commuting, vertical-commuting and infinity-preserving.*

3 Some Dynamical Behaviors

SA are very interesting models, whose complexity lies somewhere between d-dimensional and $d+1$-dimensional CA. Indeed, the latter can simulate d-dimensional SA, which can, in turn, simulate the former [5, 8]. We are interested in studying the SA complexity from the stability point of view. The concepts that first come to mind to formalize the notion of stability are inspired by the topological classifications given in [9, 13] for CA. In [13], one-dimensional CA are classified into four classes, from the most stable to the most unstable behavior: equicontinuous CA, non-equicontinuous CA admitting an equicontinuity configuration, sensitive but not positively expansive CA, positively expansive CA. Things are very different as soon as we get into dimension $d = 2$, as noted in [15, 14]. The question is now whether the complexity of the SA model is closer to that of the lower or the higher-dimensional CA. In this section we consider the above mentioned concepts in the SA settings, and we introduce the notion of ultimate periodicity, useful for the characterization of SOC systems. We prove that there exist no positively expansive SA and characterize equicontinuous SA as the ultimately periodic SA.

First, recall basic definitions. Let (X, m) be a metric space and let $H : X \to X$ be a continuous application. An element $x \in X$ is an *equicontinuity point* for H if for any $\varepsilon > 0$, there exists $\delta > 0$ such that for all $y \in X$, $m(x, y) < \delta$ implies that $\forall n \in \mathbb{N}$, $m(H^n(x), H^n(y)) < \varepsilon$. The map H is *equicontinuous* if for any $\varepsilon > 0$, there exists $\delta > 0$ such that for all $x, y \in X$, $m(x, y) < \delta$ implies that $\forall n \in \mathbb{N}$, $m(H^n(x), H^n(y)) < \varepsilon$. An element $x \in X$ is *ultimately periodic* for H if there exist two integers $n \geq 0$ (the preperiod) and $p > 0$ (the period) such that $H^{n+p}(x) = H^n(x)$. H is *ultimately periodic* if there exist $n \geq 0$ and $p > 0$ such that $H^{n+p} = H^n$. If X is compact, H is equicontinuous (resp. ultimately periodic) iff all elements of X are equicontinuity points (resp. ultimately periodic). H is *sensitive* (to the initial conditions) if there is a constant $\varepsilon > 0$ such that for all points $x \in X$ and all $\delta > 0$, there is a point $y \in X$ and an integer $n \in \mathbb{N}$ such that $m(x, y) < \delta$ but $m(F^n(x), F^n(y)) > \varepsilon$. H is *positively expansive* if there is a constant $\varepsilon > 0$ such that for all distinct points $x, y \in X$, there exists $n \in \mathbb{N}$ such that $m(H^n(x), H^n(y)) > \varepsilon$.

The classification from [13] is no more relevant in the SA context since the class of positively expansive SA is empty. This result can be related to the absence of positively expansive two-dimensional CA (see [15]), though the proof is much different.

Proposition 1. *There are no positively expansive SA.*

Proof. Let F be a SA and for any $k \in \mathbb{N}$, let $\delta = 2^{-k}$. Take two distinct configurations $x, y \in \mathcal{C}$ such that $\forall i \in [-k, k]^d, x_i = y_i = +\infty$. By infinity-preservingness, we get $\forall n \in \mathbb{N}, \forall i \in [-k, k]^d, F^n(x)_i = F^n(y)_i = +\infty$, hence $d(F^n(x), F^n(y)) < \delta$. □

We now prove that two different notions of stability, such as equicontinuity and ultimate periodicity, are equivalent. We need the following lemma, which allows a better understanding of equicontinuity for SA.

Lemma 1. *If F is an equicontinuous SA, then the variation of a pile is bounded by the differences in an initial neighborhood, i.e., there exists an integer $l \in \mathbb{N}$ such that all configurations $x \in \mathcal{C}$ with $x_0 = 0$ satisfy*

$$\forall n \in \mathbb{N}, \quad |F^n(x)_0| \leq \max_{\substack{|i| \leq l \\ |x_i| < \infty}} \{|x_i|\} .$$

Proof. If F is equicontinuous, in particular, for $\varepsilon = 2^0$, there exists $\delta = 2^{-l}$ such that for all $x, y \in \mathcal{C}$, if $C_l^0(x) = C_l^0(y)$, then $\forall n \in \mathbb{N}, C_0^0(F^n(x)) = C_0^0(F^n(y))$. First, consider a configuration y which has *infinite l-neighborhood*, i.e., $\forall i \in [-l,l]^d, y_i \notin [-l,l]$. Let z defined by $z_i = +\infty$ if $y_i \geq 0$ and $z_i = -\infty$ if $y_i < 0$, in such a way that $C_l^0(y) = C_l^0(z)$. Then $\forall n \in \mathbb{N}, C_0^0(F^n(y)) = C_0^0(F^n(z)) = C_0^0(z)$, i.e., $F^n(y)_0 < -l \Leftrightarrow y_0 < -l$ and $F^n(y)_0 > l \Leftrightarrow y_0 > l$.

Now, let $x \in \mathcal{C}$ such that $x_0 = 0$ and $m = \max_{|i| \leq l, |x_i| < \infty}\{|x_i|\}$. Notice that $\rho^{l+m+1}(x)$ has infinite l-neighborhood, since $x_i \leq m$ or $x_i = +\infty$ for $|i| \leq l$. Hence, as seen before, $\forall n \in \mathbb{N}, F^n(x)_0 \leq m$. A symmetrical reasoning on $\rho^{-l-m-1}(x)$ gives $\forall n \in \mathbb{N}, |F^n(x)_0| \leq m$. □

Proposition 2. *A SA is equicontinuous if and only if it is ultimately periodic.*

Proof. Let F be a SA such that with $F^{n+p} = F^n$ for some $n \geq 0$, $p > 0$. Since $F, F^2, \ldots, F^{n+p-1}$ are uniformly continuous maps, for any $\varepsilon > 0$ there exists $\delta > 0$ such that for all $x, y \in \mathcal{C}$ with $\mathsf{d}(x,y) < \delta$, it holds that $\forall q \in \mathbb{N}$, $q < n+p$, $\mathsf{d}(F^q(x), F^q(y)) < \varepsilon$. Since for any $t \in \mathbb{N}$, F^t is equal to some F^q with $q < n+p$, the map F is equicontinuous.
Let F be an equicontinuous SA and l, as in Lemma 1, such that for all $x, y \in \mathcal{C}$, if $C_l^0(x) = C_l^0(y)$, then $\forall n \in \mathbb{N}, C_0^0(F^n(x)) = C_0^0(F^n(y))$. Let $x \in \mathcal{C}$ such that x_0 is finite. Should we vertically shift it, we can assume $x_0 = 0$. Let $y \in \mathcal{C}$ defined by $y_i = \max\{\min\{x_i, l+1\}, -l-1\}$ if $|i| \leq l$ and $y_i = +\infty$ otherwise, in such a way that $C_l^0(x) = C_l^0(y)$. By Lemma 1, $\forall i \in [-l,l]^d, \forall n \in \mathbb{N}, |F^n(y)_i| \leq 2l+2$. So we can find some preperiod q_y and some period p_y such that $\forall i \in [-l,l]^d, F^{p_y+q_y}(y)_i = F^{q_y}(y)_i$. Since the other piles are infinite, and then invariant, we get $F^{p_y+q_y}(y) = F^{q_y}(y)$. As a consequence, $C_0^0(F^{p_y+q_y}(x)) = C_0^0(F^{q_y}(x))$. Define p (resp., q) as the least common multiple (resp., maximum) of all p_y (resp., q_y) for $y \in \mathcal{C}$ such that $|y_i| \leq l+1$ if $|i| \leq l$ and $y_i = +\infty$ otherwise. Then, for any $x \in \mathcal{C}$, $C_0^0(F^{p+q}(x)) = C_0^0(F^q(x))$; in particular for vertical and horizontal shifts of x, which gives $F^{p+q}(x) = F^q(x)$. □

An important open question in the dynamical behavior of SA is the existence of non-sensitive SA without any equicontinuity configuration. An example for two-dimensional CA is given in [14], but the involved method can hardly be adapted for SA. However, we conjecture that such SA exist, which would lead to a classification of SA into four classes: equicontinuous, admitting an equicontinuity configuration (but not equicontinuous), non-sensitive without equicontinuity configurations, sensitive.

4 Nilpotency

In this section we give a definition of nilpotency, the most stable dynamics of a dynamical system, adapted to SA. Then, we prove that this nilpotency behavior is undecidable (Theorem 3).

4.1 Nilpotency of CA

Here we recall the basic definitions and properties of nilpotent CA. Nilpotency is among the simplest dynamical behavior that an automaton may exhibit. Intuitively, a system is nilpotent if it destroys every piece of information in any initial configuration, reaching a common constant configuration after a while. For CA, this is formalized as follows.

Definition 3 (CA nilpotency [7, 12]). A CA G is nilpotent if

$$\exists c \in A, \quad \exists N \in \mathbb{N} \quad \forall x \in A^{\mathbb{Z}^d}, \quad \forall n \geq N, \quad G^n(x) = \underline{c} \ .$$

Remark that, because of the compactness of the CA configuration space, a CA is nilpotent if and only if it is nilpotent for all initial configurations (i.e., all configurations eventually reach the same configuration).

Spreading CA have the following stronger characterization.

Proposition 3 ([6]). *A CA G, with spreading state 0, is nilpotent if and only if for all configurations $x \in A^{\mathbb{Z}^d}$, $\lim_{n \to \infty} \mathsf{d}_T(G^n(x), \underline{0}) = 0$.*

This equivalence is very useful since the CA nilpotency has been proved undecidable in [12], even for the restricted class of spreading CA.

Theorem 2 ([12]). *For a given state s, it is undecidable to know whether a cellular automaton with spreading state s is nilpotent.*

4.2 Nilpotency of SA

A direct adaptation of Definition 3 to SA is vain. Indeed, assume F is a SA of radius r. For any $k \in \mathbb{Z}^d$, consider the configuration $x^k \in \mathcal{B}$ defined by $x_0^k = k$ and $x_i^k = 0$ for any $i \in \mathbb{Z}^d \backslash \{0\}$. Since the pile of height k may decrease at most by r during one step of evolution of the SA, and the other piles may increase at most by r, x^k requires at least $\lceil k/2r \rceil$ steps to reach a constant configuration. Thus, there exists no common integer n such that all configurations x^k reach a constant configuration in time n. This is a major difference with CA, which is essentially due to the unbounded set of states and to the infinity-preserving property.

Thus, we propose to label as nilpotent the SA which make every pile approach a constant value, but not necessarily reaching it ultimately. This nilpotency notion, inspired by Proposition 3, is formalized as follows for a SA F:

$$\exists c \in \mathbb{Z}, \quad \forall x \in \mathcal{C}, \quad \lim_{n\to\infty} \mathsf{d}(F^n(x), \underline{c}) = 0 \ .$$

Remark that c shall not be taken in the full state set $\widetilde{\mathbb{Z}}$, because allowing infinite values for c would not correspond to the intuitive idea that a nilpotent SA "destroys" a configuration (otherwise, the raising map would be nilpotent). Anyway, this definition is not satisfying because of the vertical commutativity: two configurations which differ by a vertical shift reach two different configurations, and then no nilpotent SA may exist. A possible way to work around this issue is to make the limit configuration depend on the initial one:

$$\forall x \in \mathcal{C}, \quad \exists c \in \mathbb{Z}, \qquad \lim_{n\to\infty} \mathsf{d}(F^n(x), \underline{c}) = 0 \ .$$

Again, since SA are infinity-preserving, an infinite pile cannot be destroyed (nor, for the same reason, can an infinite pile be built from a finite one). Therefore nilpotency has to involve the configurations of $\mathbb{Z}^{\mathbb{Z}^d}$, i.e., the ones without infinite piles. Moreover, every configuration $x \in \mathbb{Z}^{\mathbb{Z}^d}$ made of regular steps (i.e., in dimension 1, for all $i \in \mathbb{Z}$, $x_i - x_{i-1} = x_{i+1} - x_i$) is invariant by the SA rule (possibly composing it with the vertical shift). So it cannot reach nor approach a constant configuration. Thus, the larger reasonable set on which nilpotency might be defined is the set of bounded configurations $\mathcal{B}$. This leads to the following formal definition of nilpotency for SA.

Definition 4 (SA nilpotency).

$$\forall x \in \mathcal{B}, \quad \exists c \in \mathbb{Z}, \qquad \lim_{n\to\infty} \mathsf{d}(F^n(x), \underline{c}) = 0 \ .$$

The following proposition shows that the class of nilpotent SA is nonempty. Remark that similar nilpotent SA can be constructed with any radius and in any dimension.

Proposition 4. *The SA $\mathcal{N}$ from Example 1 is nilpotent.*

Proof. Let $x \in \mathcal{B}$, let $i \in \mathbb{Z}$ such that for all $j \in \mathbb{Z}$, $x_j \geq x_i$. Clearly, after $x_{i+1} - x_i$ steps, $F_{\mathcal{N}}^{x_{i+1}-x_i}(x)_{i+1} = F_{\mathcal{N}}^{x_{i+1}-x_i}(x)_i = x_i$. By immediate induction, we obtain that for all $j \in \mathbb{Z}$ there exists $n_j \in \mathbb{N}$ such that $F_{\mathcal{N}}^{n_j}(x)_j = x_i$, hence $\lim_{n\to\infty} \mathsf{d}(F_{\mathcal{N}}^n(x), \underline{x_i}) = 0$. □

4.3 Undecidability

The main result of this section is that SA nilpotency is undecidable (Theorem 3), by reducing to it the nilpotency of spreading CA. This emphasizes the fact that the dynamical behavior of SA is very difficult to predict. We think that this result might be used as the reference undecidable problem for further questions on SA.

Problem Nil
INSTANCE: a SA $\mathcal{A} = \langle d, r, \lambda \rangle$;
QUESTION: is $\mathcal{A}$ nilpotent?

Theorem 3. *The problem* ***Nil*** *is undecidable.*

Proof. This is proved by reducing **Nil** to the nilpotency of spreading cellular automata. Remark that it is sufficient to show the result in dimension 1. Let $\mathcal{S}$ be a spreading cellular automaton $\mathcal{S} = \langle A, 1, s, g \rangle$ of global rule G, with finite set of integer states $A \subset \mathbb{N}$ containing the spreading state 0. We simulate $\mathcal{S}$ with the sand automaton $\mathcal{A} = \langle 1, r = \max(2s, \max A), f \rangle$ of global rule F using the following technique, also developed in [5]. Let $\xi : A^{\mathbb{Z}} \to \mathcal{B}$ be a function which inserts markers every two cells in the CA configuration to obtain a bounded SA configuration. These markers allow the local rule of the SA to know the absolute state of each pile and behave as the local rule of the CA. To simplify the proof, the markers are put at height 0 (see Figure 2):

$$\forall y \in A^{\mathbb{Z}}, \forall i \in \mathbb{Z}, \quad \xi(y)_i = \begin{cases} 0 \text{ (marker)} & \text{if } i \text{ is odd} , \\ y_{i/2} & \text{otherwise.} \end{cases}$$

This can lead to an ambiguity when all the states in the neighborhood of size $4s + 1$ are at state 0, as shown in the picture. But as in this special case the state 0 is quiescent for g, this is not a problem: the state 0 is preserved, and markers are preserved.

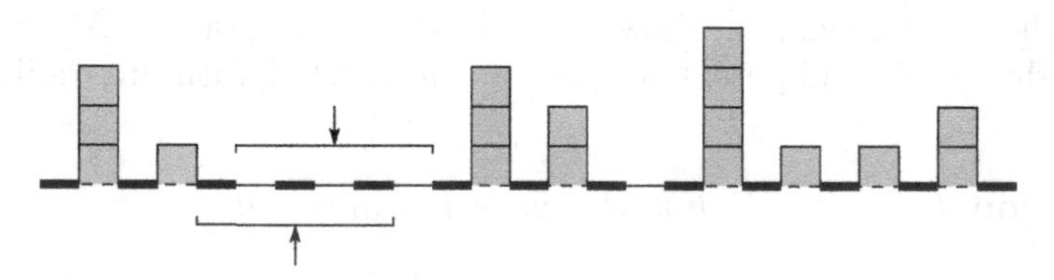

Fig. 2 Illustration of the function ξ used in the simulation of the spreading CA $\mathcal{S}$ by $\mathcal{A}$. The thick segments are the markers used to distinguish the states of the CA, put at height 0. There is an ambiguity for the two piles indicated by the arrows: with a radius 2, the neighborhoods are the same, although one of the piles is a marker and the other the state 0.

The local rule f is defined as follows, for all ranges $R \in \mathcal{R}_r^1$,

$$f(R) = \begin{cases} 0 & \text{if } R_{-2s+1}, R_{-2s+3}, \ldots, R_{-1}, R_1, \ldots, R_{2s-1} \in A \ , \\ g(R_{-2s}+a, R_{-2s+2}+a, \ldots, R_{-2}+a, a, R_2+a, \ldots, R_{2s}+a) - a & \\ \quad \text{if } R_{-2s+1} = R_{-2s+3} = \cdots = R_{2s-1} = a < 0 \text{ and } -a \in A \ . \end{cases} \tag{1}$$

The first case is for the markers (and state 0) which remain unchanged, the second case is the simulation of g in the even piles. As proved in [5], for any $y \in A^{\mathbb{Z}}$ it holds that $\xi(G(y)) = F(\xi(y))$. The images by f of the remaining ranges will be defined later on, first a few new notions need to be introduced.

A sequence of consecutive piles $(x_i, \ldots, x_j)$ from a configuration $x \in \mathcal{B}$ is said to be *valid* if it is part of an encoding of a CA configuration, i.e., $x_i = x_{i+2} = \cdots = x_j$ (these piles are markers) and for all $k \in \mathbb{N}$ such that $0 \le k < (j-i)/2$, $x_{i+2k+1} - x_i \in A$ (this is a valid state). We extend this definition to configurations, when $i = -\infty$ and $j = +\infty$, i.e., $x \in \rho^c \circ \xi(A^{\mathbb{Z}})$ for a given $c \in \mathbb{Z}$ ($x \in \mathcal{B}$ is valid if it is the raised image of a CA configuration). A sequence (or a configuration) in *invalid* if it is not valid.

First we show that starting from a valid configuration, the SA $\mathcal{A}$ is nilpotent if and only if $\mathcal{S}$ is nilpotent. This is due to the fact that we chose to put the markers at height 0, hence for any valid encoding of the CA $x = \rho^c \circ \xi(y)$, with $y \in A^{\mathbb{Z}}$ and $c \in \mathbb{Z}$,

$$\lim_{n\to\infty} \mathsf{d}_T(G^n(y), \underline{0}) = 0 \quad \text{if and only if} \quad \lim_{n\to\infty} \mathsf{d}(F^n(x), \underline{c}) = 0 \ .$$

It remains to prove that for any invalid configuration, $\mathcal{A}$ is also nilpotent. In order to have this behavior, we add to the local rule f the rules of the nilpotent automaton $\mathcal{N}$ for every invalid neighborhood of width $4s+1$. For all ranges $R \in \mathcal{R}_r^1$ not considered in Equation (1),

$$f(R) = \begin{cases} -1 & \text{if } R_{-r} < 0 \text{ or } R_{-r+1} < 0 \text{ or } \cdots \text{ or } R_r < 0 \ , \\ 0 & \text{otherwise.} \end{cases} \tag{2}$$

Let $x \in \mathcal{B}$ be an invalid configuration. Let $k \in \mathbb{Z}$ be any index such that $\forall l \in \mathbb{Z}, x_l \ge x_k$. Let $i, j \in \mathbb{Z}$ be respectively the lowest and greatest indices such that $i \le k \le j$ and $(x_i, \ldots, x_j)$ is valid (i may equal j). Remark that for all $n \in \mathbb{N}$, $(F^n(x)_i, \ldots, F^n(x)_j)$ remains valid. Indeed, the markers are by construction the lowest piles and Equations (1) and (2) do not modify them. The piles coding for non-zero states can change their state by Equation (1), or decrease it by 1 by Equation (2), which in both cases is a valid encoding. Moreover, the piles x_{i-1} and x_{j+1} will reach a valid value after a finite number of steps: as long as they are invalid, they decrease by 1 until they reach a value which codes for a valid state. Hence, by induction, for any indices $a, b \in \mathbb{Z}$, there exists $N_{a,b}$ such that for all $n \ge N_{a,b}$ the sequence $(F^n(x)_a, \ldots, F^n(x)_b)$ is valid.

In particular, after $N_{-2Nr-1,2Nr+1}$ step, there is a valid sequence of length $4Nr+3$ centered on the origin (here, N is the number of steps needed by $\mathcal{S}$ to reach the configuration $\underline{0}$, given by Definition 3). Hence, after $N_{-2Nr,2Nr} + N$

steps, the local rule of the CA $\mathcal{S}$ applied on this valid sequence leads to 3 consecutive zeros at positions $-1, 0, 1$. All these steps are illustrated on Figure 3.

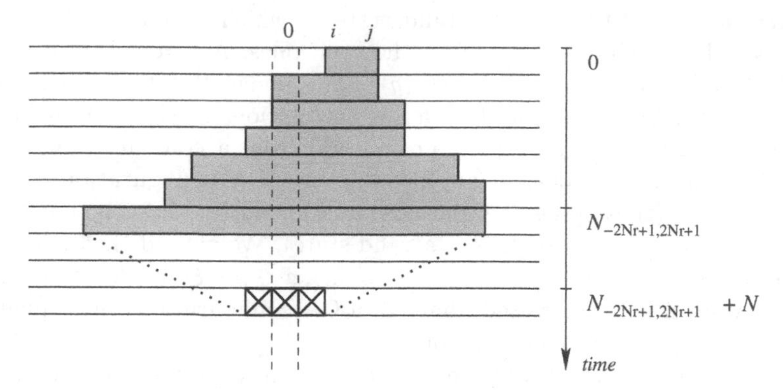

Fig. 3 Destruction of the invalid parts. The lowest valid sequence (in gray) extends until it is large enough. Then after N other steps the 3 central piles (hatched) are destroyed because the rule of the CA is applied correctly.

In a similar way, we prove that for all $n \geq N_{-2Nr-k,2Nr+k} + N$, the sequence $(F^n(x)_{-k}, \ldots, F^n(x)_k)$ is constant and does not evolve as n grows. Therefore, there exists $c \in \mathbb{Z}$ such that $\lim_{n\to\infty} \mathsf{d}(F^n(x), \underline{c}) = 0$. We just proved that $\mathcal{A}$ is nilpotent, i.e., $\lim_{n\to\infty} \mathsf{d}(F^n(x), \underline{c}) = 0$ for all $x \in \mathcal{B}$, if and only if $\mathcal{S}$ is nilpotent (because of the equivalence of definitions given by Proposition 3), so **Nil** is undecidable (Theorem 2). □

5 Conclusion

In this article we have continued the study of sand automata, using the compact topology on the SA configuration space introduced in [8]. This topology, inspired by the topology on CA, may facilitate studies about dynamical and topological properties of SA, as for the proof of the equivalence between equicontinuity and ultimate periodicity (Proposition 2).

Then, we have given a definition of nilpotency. Although it differs from the standard one for CA, it captures the intuitive idea that a nilpotent automaton "destroys" configurations. Finally, we have proved that SA nilpotency is undecidable (Theorem 3). This fact enhances the idea that the behavior of a SA is hard to predict. We also think that this result might be used as a fundamental undecidability result, which could be reduced to other SA properties.

Besides, in the context of CA, nilpotency clearly implies ultimate periodicity. It appears that with our definitions, nilpotency of SA is not necessarily a particular case of ultimate periodicity ($\mathcal{N}$ is not ultimately periodic). However, it would be interesting to see if it could be linked to other weaker stability notions.

Moreover, the study of global properties such as injectivity and surjectivity and their corresponding dimension-dependent decidability problems could help to understand if d-dimensional SA look more like d-dimensional or $d+1$-dimensional CA. Unfortunately, deciding these dynamical properties remains a major problem. Similarly, it would be interesting to solve the open question of the dichotomy between sensitive SA and those with equicontinuous configurations. A potential counter-example would give a more precise idea of the dynamical behaviors represented by SA.

References

1. P. Bak. *How nature works - The science of SOC.* Oxford University Press, 1997.
2. P. Bak, C. Tang, and K. Wiesenfeld. Self-organized criticality. *Physical Review A*, 38(1):364–374, 1988.
3. J. Cervelle and E. Formenti. On sand automata. In 20th *Symposium on Theoretical Aspects of Computer Science (STACS'03)*, volume 2607 of *Lecture Notes in Computer Science*, pages 642–653. Springer, 2003.
4. J. Cervelle, E. Formenti, and B. Masson. Basic properties for sand automata. In 30th *Conference on Mathematical Foundations of Computer Science (MFCS'05)*, volume 3618 of *Lecture Notes in Computer Science*, pages 192–211. Springer, 2005.
5. J. Cervelle, E. Formenti, and B. Masson. From sandpiles to sand automata. *Theoretical Computer Science*, 381:1–28, 2007.
6. J. Cervelle and P. Guillon. Towards a Rice theorem on traces of cellular automata. In 32nd *Conference on Mathematical Foundations of Computer Science (MFCS'07)*, volume 4708 of *Lecture Notes in Computer Science*, pages 310–319. Springer, 2007.
7. K. Čulik, J. Pachl, and S. Yu. On the limit sets of cellular automata. *SIAM Journal of Computing*, 18(4):831–842, 1989.
8. A. Dennunzio, P. Guillon, and B. Masson. Topological properties of sand automata as cellular automata. In 1st *Symposium on Cellular Automata (JAC'08)*, 2008. To appear.
9. Robert H. Gilman. Classes of linear automata. *Ergodic Theory & Dynamical Systems*, 7:105–118, 1988.
10. E. Goles and M. A. Kiwi. Game on line graphs and sandpile automata. *Theoretical Computer Science*, 115:321–349, 1993.
11. G. A. Hedlund. Endomorphisms and automorphisms of the shift dynamical system. *Mathematical Systems Theory*, 3:320–375, 1969.
12. J. Kari. The nilpotency problem of one-dimensional cellular automata. *SIAM Journal of Computing*, 21(3):571–586, 1992.
13. P. Kůrka. Languages, equicontinuity and attractors in cellular automata. *Ergodic Theory & Dynamical Systems*, 17:417–433, 1997.
14. M. Sablik and G. Theyssier. Topological dynamics of 2D cellular automata. In 4th *Conference on Computability in Europe (CiE'08)*, volume 5028 of *Lecture Notes in Computer Science.* Springer, 2008. To appear.
15. M. A. Shereshevsky. Expansiveness, entropy and polynomial growth for groups acting on subshifts by automorphisms. *Indagationes Mathematicæ*, 4(2):203–210, 1993.

On tractability of Cops and Robbers game

Fedor V. Fomin[1], Petr A. Golovach[1], and Jan Kratochvíl[2]

[1] Department of Informatics, University of Bergen, PB 7803, 5020 Bergen, Norway, fedor.fomin|petr.golovach@ii.uib.no*

[2] Department of Applied Mathematics, and Institute for Theoretical Computer Science, Charles University, Malostranské nám. 25, 118 00 Praha 1, Czech Republic honza@kam.mff.cuni.cz†

Abstract. The Cops and Robbers game is played on undirected graphs where a group of cops tries to catch a robber. The game was defined independently by Winkler-Nowakowski and Quilliot in the 1980s and since that time has been studied intensively. Despite of that, its computation complexity is still an open question. In this paper we prove that computing the minimum number of cops that can catch a robber on a given graph is NP-hard. Also we show that the parameterized version of the problem is W[2]-hard. Our proof can be extended to the variant of the game where the robber can move s times faster than cops. We also provide a number of algorithmic and complexity results on classes of chordal graphs and on graphs of bounded cliquewidth. For example, we show that when the velocity of the robber is twice cop's velocity, the problem is NP-hard on split graphs, while it is polynomial time solvable on split graphs when players posses the same speed. Finally, we establish that on graphs of bounded cliquewidth (this class of graphs contains, for example, graphs of bounded treewidth), the problem is solvable in polynomial time in the case the robber's speed is at most twice the speed of cops.

Key words: Pursuit-evasion games on graphs, complexity, parameterized complexity, algorithms, cliquewidth

1 Introduction

Cops and Robbers is a pursuit-evasion game with two players cop $\mathcal{C}$ and robber $\mathcal{R}$ which play alternately on a finite connected undirected graph G. Player $\mathcal{C}$ has a team of cops who attempt to capture the robber. At the beginning of the game $\mathcal{C}$ selects vertices and put cops on these vertices. Then $\mathcal{R}$ put the robber on a vertex. The players take turns starting with $\mathcal{C}$. At every move each of the cops can be either moved to an adjacent vertex or kept on the same vertex. (Several cops can occupy the same vertex at some move.) $\mathcal{R}$ responds by moving the robber to some vertex along some path of length at most s, which does not contain vertices occupied by cops. (In other words, cops are moving with a unit speed and the speed of robber is s, and robber cannot run through a vertex

* Supported by the Norwegian Research Council.

† Supported by Czech research grant 1M0545.

Please use the following format when citing this chapter:

Fomin, F. V., Golovach, P. A. and Kratochvíl, J., 2008, in IFIP International Federation for Information Processing, Volume 273; *Fifth IFIP International Conference on Theoretical Computer Science*; Giorgio Ausiello, Juhani Karhumäki, Giancarlo Mauri, Luke Ong; (Boston: Springer), pp. 171–185.

occupied by a cop.) We say that a cop *catches* the robber at some move if at that move they occupy the same vertex. Player $\mathcal{C}$ wins if in a finite number of moves one of his cops catches the robber. Player $\mathcal{R}$ wins if he can avoid such a situation. For an integer s and a graph G, we denote by $c_s(G)$ the minimum number of cops sufficient for $\mathcal{C}$ to win on graph G against the robber moving at the speed of s.

The variant of the game with $s = 1$, i.e. when cops and robber have the same speed, was studied intensively. The game was defined (for one cop) by Winkler and Nowakowski [25] and Quilliot [28] who also characterized graphs with the cop number one. Aigner and Fromme [2] initiated the combinatorial study of the problem with several cops and obtained a number of important results. In particular, they observed that if a girth of G (the minimum length of a cycle) is at least 5, then $c_1(G)$ is at least the minimum vertex degree of G. Another interesting result proved in [2] is that on planar graphs 3 cops can always catch the robber. This result can be generalized on graphs of bounded genus [27, 31]. Andreae [5] extended the result of Aigner and Fromme to graphs containing no fixed graph H as a minor. Different combinatorial (lower and upper) bounds on the cop number for different graph classes are discussed in [4, 13, 15, 16, 20, 22, 23] (see also the survey [3]).

There is a resemblance of Cops and Robbers game, at least for large values of $s \to \infty$, to the helicopter search game defined by Seymour and Thomas [32], which is the game-theoretic interpretation of the well known treewidth parameter. In Seymour-Thomas game the robber can move arbitrarily fast, but players make their moves simultaneously. See the survey of Bodlaender for an overview of pursuit-evasion games related to treewidth [7].

Despite of such an intensive study of the combinatorial properties of the game almost no algorithmic results on this game are known. Perhaps the only algorithmic result known about Cops and Robbers game (for $s = 1$) is the observation that determining whether the cop number of a graph on n vertices is at most k can be done by a backtracking algorithm which runs in time $O(n^{O(k)})$ (thus polynomial for fixed k) [6, 17, 19].

Similar result holds for every $s \geq 1$. Given an integer k and a graph G on n vertices, the question if $c_s(G) \leq k$ can be answered (and the corresponding winning strategy of k cops can be computed) by constructing the game graph on $2\binom{n+k-1}{k}n$ nodes (every node of the game graph corresponds to a possible position in G of k cops and one robber, taking into account two possibilities for the turn), and then by making use of backtracking find if some cop-winning position can be obtained from an initial position. While the proof of the following proposition is standard and easy (and we omit it here), it serves as the main tool for obtaining all polynomial time algorithms in this work.

Proposition 1. *For a given integer $k \geq 1$ and a graph G on n vertices, the question if $c_s(G) \leq k$ can be answered in time $\binom{n+k-1}{k}^2 \cdot n^{O(1)} = n^{O(k)}$.*

Thus for every fixed k, one can decide in polynomial time if k cops can catch the robber on a given graph G. There are several natural questions around

Proposition 1. The first is, what is the complexity of the problem when k is part of the input? Another question, is the problem fixed parameter tractable? There are many search and pursuit-evasion problems which are fixed parameter tractable, i.e. for which deciding if k searchers (cops) can catch evader (robber) on an n-vertex graph can be done in time $O(f(k) \cdot n^{O(1)})$ (we refer to Bodlaender's survey [7] for examples of such problems).

There are several variants of similar games like the k-pebbles game, or the cat and k-mouse game, which solutions require $n^{\Omega(k)}$ steps (see e.g. Adachi et al. [1]). However, all these games are played on directed graphs or the games should either start, or end in specified positions (holes or cheese for mouses), and the proofs are strongly based on these specific properties. Following this line of research, Goldstein and Reingold [17] proved that the version of the Cops and Robbers game on *directed* graphs is EXPTIME-complete. Also they have shown that the version of the game on undirected graphs when cops and robber are given their initial positions is also EXPTIME-complete. They also conjectured that the game on undirected graphs (for $s = 1$) is also EXPTIME-complete. Again, their proofs strongly relies on the specific settings (adding directions or fixing initial positions) and cannot be transferred to the standard Cops and Robbers game on undirected graphs, and their conjecture is still open.

Our results. We prove that for every $s \geq 1$, deciding if $c_s(G) \leq k$ is NP-hard. We also show that the parameterized version of the problem is $W[2]$-hard. Loosely speaking, this means that the existence of a $O(f(k) \cdot n^{O(1)})$-time algorithm deciding if $c_s(G) \leq k$, where f is a function only of the parameter k and G is a graph on n vertices, would imply that $FPT = W[2]$, which is considered to be very unlikely in parameterized complexity. (We refer to the books [12, 14, 24] for an information on parameterized complexity.) We also show that for $s \geq 2$, the problem remains NP-hard and $W[2]$-hard even when input is restricted to split graphs. We find it a bit surprising, especially for $s = \infty$, i.e. when the speed of robber is not bounded, because all known search and pursuit-evasion problems on undirected graphs which look quite similar to this case, are polynomially solvable or at least fixed parameter tractable for chordal graphs. For example, for helicopter search game [32] the minimum number of cops equals treewidth plus one and can be easily calculated for chordal graphs. For node searching (see [18]) the corresponding problem can be solved in polynomial time for split graphs but remains NP-complete on chordal graphs. See also [26] for related results. Note also that for $s = 1$ one cop always can capture robber on the chordal graph [28]. By continuing investigating the complexity of the problem on classes of chordal graphs, we show that for every fixed s, the computation of $c_s(G)$ on interval graphs can be done in polynomial time. Finally, we investigate the complexity of the problem on graphs of bounded cliquewidth. We prove that on graphs of bounded cliquewidth the computation of numbers $c_s(G)$ can be done in polynomial time for $s = 1, 2$. While most of polynomial time algorithms on graphs of bounded cliquewidth (and treewidth) are based on dynamic programming approach [11], this is not the case for the

Cops and Robbers problem. Our proof is based on combinatorial bounds and Proposition 1.

2 Cops and Robbers is NP hard

All this section is devoted to the proof of the following result

Theorem 1. *For every $s \geq 1$, the following problem is NP-hard*

INSTANCE: A graph G and a positive integer k.
QUESTION: Is $c_s(G) \leq k$?

Moreover, the parameterized version

INSTANCE: A graph G.
PARAMETER: A positive integer k.
QUESTION: Is $c_s(G) \leq k$?

of the Cops and Robbers problem is $W[2]$-hard for every $s \geq 1$.

2.1 Bipartite graphs with large girth and degrees of vertices

Let us start with auxiliary results. We want to construct a bipartite graph with girth at least six and large minimum vertex degree with some additional properties. (Let us remind that the girth of a graph G is the minimum cycle length in G.) The study of such graphs has a long history (see e.g. [8]). There are different approaches for obtaining such graphs. Most of them are geometrical or algebraic. For our reduction we use algorithmic construction which is based on the construction of Krishnan et al. [21].

For positive integers n, m and r we construct a bipartite graph $H(n,m,r)$ with rmn^2 edges and bipartition (X,Y), $|X| = |Y| = nm$. Set X is partitioned into sets $U_1, U_2, \ldots, U_n$, and set Y is partitioned into sets $W_1, W_2, \ldots, W_n$, $|U_i| = |W_i| = m$ for $i = 1, 2, \ldots, n$. We denote by $H_{i,j}$ the subgraph of $H(n,m,r)$ induced by $U_i \cup W_j$, and by $\deg_{i,j}(z)$ the degree of vertex z in $H_{i,j}$. We also denote by E the set of edges in $H(n,m,r)$ and by $\mathrm{dist}(x,y)$ the distance between vertices x and y in $H(n,m,r)$.

The graph $H(n,m,r)$ is constructed by the following procedure which starts from empty graph on vertices $X \cup Y$ and add edges according the following rules:

```
for k := 1 to rm do
    let t := ⌈k/m⌉;
    if k is odd then
        for i := 1 to n do
            for j := 1 to n do
                choose a vertex x ∈ U_i of minimum degree in H_{i,j};
                let S := {z ∈ W_j : dist(x, z) > 1 and deg_{i,j}(z) < t + 1};
                select a vertex y ∈ W_j such that
                dist(x, y) = max_{z∈S} dist(x, z); add (x, y) to E;
    else
        for j := 1 to n do
            for i := 1 to n do
                choose a vertex y ∈ W_j of minimum degree in H_{i,j};
                let S := {z ∈ U_i: dist(y, z) > 1 and deg_{i,j}(z) < t + 1};
                select a vertex x ∈ U_i such that
                dist(x, y) = max_{z∈S} dist(x, z); add (x, y) to E;
```

Value of t is called the phase number of the algorithm. Clearly, the algorithm has to complete r phases. If k is odd then we say that n^2 edges, added by the algorithm for this value of k, are added during the odd phase t. Correspondingly, if k is even then we say that n^2 edges, added by the algorithm for this value of k, are added during the even phase t.

The following lemma, which is the direct analog of Lemma 1 from [21], establishes the key invariants maintained by the algorithm. We omit the proof of this lemma here.

Lemma 1. *For every $1 \le t \le r$ the following holds:*

1. *When the algorithm completes an odd phase t, the average degree of vertices of U_i in $H_{i,j}$ is r and $t-1 \le \deg_{i,j}(x) \le t+1$ for $x \in U_i$ and $i, j \in \{1, 2, \ldots, n\}$;*
2. *When the algorithm completes an even phase t, the average degree of vertices of W_j in $H_{i,j}$ is r and $t-1 \le \deg_{i,j}(y) \le t+1$ for $y \in W_j$ and $i, j \in \{1, 2, \ldots, n\}$.*

It can be easily seen that if set S is empty then the algorithm cannot add an edge. Next lemma gives sufficient condition, which makes such situation impossible.

Lemma 2. *If $r < \frac{m+3}{6}$ then the algorithm completes all r phases.*

This lemma is a simplified version of the lemma 2 of [21] and we omit its proof here.

Now we can summarize properties of the algorithm and of the graph $H(n, m, r)$ which will be used in our reduction.

Lemma 3. *Let* $m \geq 2n(r+1)\frac{(n(r+1)-1)^6-1}{(n(r+1)-1)^2-1}$. *Then*

1. *The algorithm constructs graph* $H(n,m,r)$ *in time* $O(r \cdot m \cdot n^2)$*;*
2. *For every vertex* $z \in V(H_{i,j})$ *and every* $i,j \in \{1,2,\dots,n\}$*, we have* $r-1 \leq \deg_{i,j}(z) \leq r+1$*;*
3. *For every vertex* z*,* $\deg(z) \leq n(r+1)$*.*
4. *The girth of* $H(n,m,r)$ *is at least six.*

Proof. The first three items are immediate corollaries of Lemmata 1 and 2.

In order to prove 4, let us assume that a cycle of length $g = 2p$, $p \geq 1$, where g is the girth of $H(n,m,r)$, was created during the phase t of the algorithm. Without loss of generality, we can assume that the last edge (x,y) of this cycle was added during odd phase t, and $x \in U_i$, $y \in W_j$. Let $D = \{z \in W_j \colon \mathrm{dist}(x,z) \geq g\}$. Since vertex x had no neighbors in D, we have that for every $z \in D$ $\deg_{i,j}(z) = t+1$ during the even phase t. By Lemma 1, $|D| \leq \frac{m}{2}$. Thus $|W_j \setminus D| \geq \frac{m}{2}$. Clearly $\mathrm{dist}(x,z) \leq g-1 = 2p-1$ for every $z \in W_j \setminus D$. Let us estimate the number of vertices at distance at most $2p-1$ from x in $H(n,m,r)$. Since the maximum vertex degree in $H(n,m,r)$ is at most $n(r+1)$, we have that the number of vertices at distance at most $2p-1$ from x is at most

$$n(r+1) + n(r+1)(n(r+1)-1)^2 + \cdots + n(r+1)(n(r+1)-1)^{2(p-1)} = n(r+1)\frac{(n(r+1)-1)^{2p}-1}{(n(r+1)-1)^2-1}.$$

Thus

$$n(r+1)\frac{(n(r+1)-1)^6-1}{(n(r+1)-1)^2-1} \leq \frac{m}{2} \leq n(r+1)\frac{(n(r+1)-1)^{2p}-1}{(n(r+1)-1)^2-1},$$

which yields $g = 2p \geq 6$. □

2.2 Proof of Theorem 1

Now we are ready to proceed with the proof of the main result of this section. We use reduction from the well known NP-complete Minimum Dominating set problem

INSTANCE: A graph G and a nonnegative integer k.
QUESTION: Does G contain a dominating set (i.e. a set of vertices D such that every vertex of G is either in D, or is adjacent to a vertex of D) of cardinality at most k?

Let G be a graph with the vertex set $V(G) = \{v_1, v_2, \dots, v_n\}$. Let $r = k+2$ and

$$m = \left\lceil 2n(r+1)\frac{(n(r+1)-1)^6-1}{(n(r+1)-1)^2-1}\right\rceil.$$

For every vertex $v_i \in V(G)$ we add $2m$ new vertices and make each new vertex adjacent to vertices from $N[v_i]$ (in G). We use m of the new vertices to compose the set U_i, and the other m vertices to compose the set W_i. Then we apply the algorithm from the previous section to construct the bipartite graph $H(n, m, r)$ on the vertex set

$$(U_1 \cup U_2 \cup \cdots \cup U_n) \cup (W_1 \cup W_2 \cup \cdots \cup W_n).$$

Denote the resulting graph by G'. By Lemma 3, G' is constructed in time polynomial in n and k.

Now we prove that graph G has a dominating set of size at most k if and only if $c_s(G') \le k$.

We say that vertex is dominated by the cop if this vertex is occupied by the cop or some adjacent vertex is occupied by the cop.

Let $S \subseteq V(G)$ be a dominating set in G of size $\le k$. Since cops placed in vertices of S dominate all vertices of G', for every vertex choice of robber he will be caught after the first move of cops.

In opposite direction, let us assume that G has no dominating set of size k and describe the strategy of the robber avoiding cops. Let S be the set of vertices chosen by cops for their initial position. Since this set is not a dominating set in G, we have that there is a vertex $v_i \in V(G)$ which is not dominated by cops. Degree of every vertex of $H(n, m, r)$ is at most $n(r+1)$ and thus k cops dominate at most $kn(r+1)$ vertices in U_i. The set U_i contains m vertices, therefore,

$$m = \left\lceil 2n(r+1)\frac{(n(r+1)-1)^6-1}{(n(r+1)-1)^2-1}\right\rceil > kn(r+1).$$

So there is a vertex $u \in U_i$ which is not dominated by cops. The robber chooses this vertex as his initial position. Suppose now that after some robber's move the robber occupies vertex $u \in U_i$ which is not dominated by cops. If after the next move of cops this vertex is still not dominated then the robber stays there. If it it becomes dominated, then the robber do the following. Let S be the set of vertices of G occupied by cops. Since this set is not a dominating set in G, there is vertex $v_j \in V(G)$ which is not dominated by cops standing at S. The vertex u has at least $r-1 = k+1$ neighbors in W_j. Since graph $H(n, m, r)$ has the girth at least six, we have that at least one of these neighbors is not dominated by cops. Then the robber moves into this vertex (note that he moves along the path of length 1). Clearly, this strategy of the robber gives him possibility to avoid cops. This completes the NP-hardness part of the proof.

To prove $W[2]$-hardness, it is sufficient to observe that our reduction from dominating set (which is $W[2]$-hard) is an FPT reduction.

3 Complexity on Split and Interval graphs

A graph G is a *split graph* if the vertex set of G can be partitioned into sets C and I, such that C is a clique, and I is an independent set. It is well known that the treewidth of a split graph can be computed in linear time (actually it is true for a larger class of chordal graphs). It is also well known that $c_1(G) = 1$ on a superclass of chordal graphs and can be computed in polynomial time [25]. Also the treewidth of a chordal graph can be computed in polynomial time, and thus the search game of Seymour-Thomas is tractable on chordal graphs. However, for $s \geq 2$ problem of computing of $c_s(G)$ becomes difficult even for split graphs.

Theorem 2. *For every $s \geq 2$ the following problem is NP-hard:*

INSTANCE: A split graph G, and a nonnegative integer k.
QUESTION: Is $c_s(G) \leq k$?

Moreover, for every $s \geq 2$ the parameterized version of the problem is $W[2]$-hard on split graphs.

Proof. The proof of this theorem uses the constructions from the proof of Theorem 1. It is known that the Minimum Dominating set problem is NP-complete (and its parameterized version is W[2]-hard) even when the input is restricted to split graphs [29].

Let G be a split graph with clique C and independent set $I = \{v_1, v_2, \ldots, v_p\}$. Let also $r = k + 2$ and $m = \lceil 2(r+1)\frac{r^6-1}{r^2-1} \rceil$. Each vertex $v_i \in I$ is replaced by new m vertices, which form set V_i. Let $N(v_i)$ be the set of neighbors of v_i in the original graph G. We make every new vertex from V_i be adjacent to all vertices from $N(v_i)$. Then we add m vertices forming a set W to the clique (i.e. these vertices are joined by edges with each other and vertices of C). Now we construct p copies of the graph $H(1, m, r)$ with vertex sets $V_1 \cup W, V_2 \cup W, \ldots, V_p \cup W$ ($V_i = X$ and $W = Y$ for each copy of $H(1, m, r)$). The resulting graph is denoted by G'. Clearly, this graph is a split graph, and can be constructed in polynomial time.

Now we prove that for any $s \geq 2$, graph G has a dominating set of size at most k if and only if $c_s(G') \leq k$.

Suppose that $S \subseteq V(G)$ is a dominating set in G and $|S| \leq k$. Clearly we can assume that $S \subseteq C$. It can be easily seen that S is a dominating set in G'. We place cops in vertices of S, and for every possible choice of an initial position, the robber would be captured after the first move of cops.

Assume now that for every $S \subseteq V(G)$, $|S| \leq k$, S is not a dominating set of G, and describe the strategy of the robber. Suppose that cops have chosen initial positions, and S is the set of vertices of G occupied by cops. Since this set is not a dominating set in G, there is $i \in \{1, 2, \ldots, p\}$ such that vertices of V_i are not dominated by cops standing on vertices of S. Since each vertex $u \in W$ is adjacent to no more than $k + 3$ vertices of V_i and $k(k+3) + 1 \leq m$, we have that there is vertex $x \in V_i$ which is not dominated by cops standing on vertices of W. The robber chooses this vertex as his initial position. Suppose now that after some moves the robber occupies vertex $x \in V_i$ which is not dominated by cops. If after next move of cops this vertex is still not dominated, then the robber stays there. Suppose that it became dominated. Let S be the set of vertices of G occupied by cops. Since this set is not a dominating set in G, there is $j \in \{1, 2, \ldots, p\}$ such that vertices of V_j are not dominated by cops standing on vertices of S. Vertex x has at least $k+1$ adjacent vertices in W. So there is vertex $y \in W$ which is adjacent to x and is not occupied by cops. Now vertex y has at least $k+1$ neighbors in V_j. Since graph $H(1, m, r)$ has the girth at least six, at least one vertex $z \in V_j$ in the neighborhood of y is not dominated by cops. Then the robber can move from x to y and then to z. Such a strategy provides the robber an opportunity to avoid capture.

To establish the parameterized complexity on split graph we observe, that the parameterized version of the dominating set problem remains to be $W[2]$-hard on split graphs and that the described reduction from dominating set is an FPT reduction. □

Another well known class of chordal graphs are interval graphs. An *interval graph* is the intersection graph of a set of intervals on the real line, i.e. every vertex corresponds to an interval and two vertices are adjacent if and only if the corresponding intervals intersect. We show that for every interval graph G and integer s, $c_s(G)$ can be computed in polynomial time. Actually the only property of interval graphs we need is the existence in interval graphs dominating pairs. A *dominating pair* in a connected graph G is a pair of two (not necessary different) vertices u and v such that the vertex set of every u,v-path in G is a dominating set. A *caterpillar* is a tree which consists of a path, called *backbone*, and leaves adjacent to vertices of the backbone. For a graph G and integer p, the p-th power of G, G^p is the graph on vertex set $V(G)$, and vertices u, v are adjacent in G^p if and only if the distance between them is at most p in G.

Lemma 4. *Let T be a spanning caterpillar of a graph G, and let p be an integer such that G is a subgraph of T^p. Then $c_s(G) \leq \max\{1, ps - 1\}$.*

Proof. We describe a winning strategy for $k = \max\{1, ps - 1\}$ cops. Suppose that $P = (v_1, v_2, \ldots, v_r)$ is a backbone of T. Cops occupy first k vertices of the backbone. Then they move along P simultaneously. If after some robber's move he is standing on the vertex adjacent to the vertex occupied by a cop, then this cop makes capturing move.

For a vertex v we use $N[v]$ to denote the closed neighborhood of v, i.e. the set of all vertices adjacent or equal to v. We use induction to prove that if at some step cops occupy vertices $v_i, v_{i+1}, \ldots, v_{i+k-1}$ then the robber cannot move to any vertex of set $\bigcup_{j=1}^{i+k-1} N[v_j]$ without being captured after the next move of cops. Clearly, this holds after the first move of cops. Let us consider the i-th move. By the induction assumption, before this move of cops the robber is at some vertex $x \notin \bigcup_{j=1}^{i+k-2} N[v_j]$. If he is going to move to vertex $y \in \bigcup_{j=1}^{i+k-1} N[v_j]$ he has to go along some path of length at most s which does not contain cops. Since $G \subseteq T^p$, the distance between x and y in T is at most ps. Then $y \in \bigcup_{j=i}^{i+k-1} N[v_j]$, i.e that y is adjacent to a vertex occupied by some cop and thus the robber is caught at the next move of cops. □

Lemma 5. *Let G be a connected graph with dominating pair. Then $c_s(G) \leq 5s-1$.*

Proof. Let u and v be a dominating pair, and P be a shortest u,v-path in G. Then P is the backbone of a spanning caterpillar T in G. Since P is a shortest path, $G \subseteq T^5$. Now we apply Lemma 4. □

Combining Proposition 1 with Lemma 5, we obtain the following result.

Corollary 1. *For every positive integer s, $c_s(G)$ can be computed in time $n^{O(s)}$ on graphs with a dominating pair.*

Corollary 1 yields polynomial time algorithms on many graph classes containing a dominating pair. This include not only interval graphs and cocomparability graphs, but more general class of AT-free graphs. (See [9, 10] for definition and properties of AT-free graphs.)

4 Graphs of bounded cliquewidth

Cliquewidth is a graph parameter that measures in a certain sense the complexity of a graph. This parameter was introduced by Courcelle, Engelfriet, and Rozenberg [11].

Let G be a graph, and k be a positive integer. A k-graph is a graph whose vertices are labeled by integers from $\{1, 2, \ldots, k\}$. We call the k-graph consisting of exactly one vertex labeled by some integer from $\{1, 2, \ldots, k\}$ an initial k-graph. The cliquewidth is the smallest integer k such that G can be constructed from initial k-graphs by means of repeated application of the following three operations:

- Disjoint union (denoted by $\oplus$).
- Relabeling: changing all labels i to j (denoted by $\rho_{i \to j}$).
- Join: connecting all vertices labeled by i with all vertices labeled by j (denoted by $\eta_{i,j}$).

If graph G has cliquewidth k it is possible to construct the expression tree for G. The expression tree is a rooted tree T of the following form:

- The nodes of T are of four types i, $\oplus$, η and ρ.
- Introduce nodes $i(v)$ are leaves of T, corresponding to initial k-graphs with vertices v, which are labeled i.
- A union node $\oplus$ stands for a disjoint union of graphs associated with children.
- A join node $\eta_{i,j}$ with one child is associated with the k-graph, which is the result of join operation for the graph corresponding to the child.
- A relabel node $\rho_{i\to j}$ also with one child is associated with the k-graph, which is the result of relabeling operation for the graph corresponding to the child.
- The graph G is isomorphic to the graph associated with the root of T (with all labels removed).

For node v of T we denote by T_v the subtree of T induced by v and it's descendants, and by G_v is denoted k-graph associated with this node. Clearly, T_v is the expression tree for G_v.

Theorem 3. *Let G be a connected graph with cliquewidth k. Then $c_1(G) \leq k$ and $c_2(G) \leq 2k$.*

Proof. If our graph has one vertex then the statement is trivial. So assume that G contains at least two vertices.

We start with the first bound. Let T be an expression tree for G. We describe a cops strategy, which is constructed by tracing of T starting from the root. The key idea of the cop's strategy is to force the robber stay in vertices of graph G_v, where v is a child of considered node of T.

It is assumed that at the beginning cops occupy some vertices of G. We say that a cop moves to vertex z if he is moved to this vertex by a sequence of moves. In the process of the pursuit cops are assigned to sets of vertices of the graph. Correspondingly, these cops (sets) are called *assigned*, and other cops are called *free*.

Let u be a vertex of T. It is assumed inductively that the robber occupies some vertex of G_u, and that all vertices of $V(G_u)$, which are adjacent to vertices of $V(G) \setminus V(G_u)$, are dominated by assigned cops. Suppose that $S_1, S_2, \ldots, S_r$ are disjoint sets of vertices of G_u, to which cops are assigned. The cop assigned to the set S_i occupies some vertex, which is adjacent to all vertices of this set, and every set has exactly one assigned cop. If u is the root, then $r = 0$. Now we consider different cases.

Case 1. *u is an introduce node.* Since this vertex is dominated by some cop, this case is trivial.

Case 2. *u is a union node.* Let $v_1, v_2, \ldots, v_t$ be the children of u. Since G_u is a disjoint union of $G_{v_1}, G_{v_2}, \ldots, G_{v_r}$, we have that the robber can stay only in vertices of the graph G_{v_i} for some $1 \leq i \leq r$. If for some $j \in \{1, 2, \ldots, r\}$ $S_j \cap V(G_{v_i}) = \emptyset$, then the cop assigned to this set is declared free. For other sets we put $S_j = S_j \cap V(G_{v_i})$. Finally, we put $u = v_i$ and cops proceed with the new list of assigned sets.

Case 3. *u is a join node $\eta_{i,j}$ with the child v.* Let $X \subseteq V(G_u)$ be the set of vertices labeled by i, and $Y \subset V(G_u)$ be the set of vertices labeled by j. If X is not included in the list of assigned sets, then vertex $z \in Y$ is chosen, some free cop is moved to this vertex, and this cop is assigned to X. Similarly, if Y is not included to the list of assigned sets then vertex $z \in X$ is chosen, some free cop is moved to this vertex and is assigned to Y. The game proceeds with the new list of assigned sets for $u = v$.

Case 4. *u is a relabel node $\rho_{i \to j}$ with the child v.* Let $X \subset V(G_u)$ be the set of relabeled vertices. If for some $t \in \{1, 2, \ldots, r\}$, $X \subset S_t$, then set S_t is partitioned into X and $S_t \setminus X$, and one additional free cop is moved to a vertex dominating X. This cop is assigned to X and the one that was assigned to S_t is assigned to $S_t \setminus X$. Then cops proceed further with the new list of assigned sets for $u = v$.

By following this strategy, Cop player is guaranteed that at some moment he reaches a position in the game when it is his turn to make a move and that the robber occupies a vertex of some assigned set. Since each of the assigned vertices is dominated by a cop, it follows that at some moment Cop player can win the game by catching the robber.

Let us prove that k cops are sufficient to perform this strategy. We use here the following property: For every $u \in V(T)$ with assigned sets $S_1, S_2, \ldots, S_r$, no label is used on vertices from two different sets. This property can be shown by inductive arguments. By definition, it holds when u is the root of T. Suppose that after some step of the pursuit two different sets S_i and S_j have vertices with same label. But it means that in the process of construction of G from G_u these sets have to be subjected to relabeling and join operations simultaneously. Then all vertices of these sets should be included into one assigned set after some join operation. Thus $r \leq k$, which yields that $c_1(G) \leq k$.

The second bound is proved similarly. Main difference is that we assign not one but two cops to a set. Let u be a vertex of T. For the case $s = 1$ cops were able to succeed by dominating all vertices of $V(G_u)$, which are adjacent to vertices of $V(G) \setminus V(G_u)$. In the case $s = 2$, this is not sufficient and cops also have to control all vertices of $V(G) \setminus V(G_u)$, which are adjacent to vertices of $V(G_u)$. Except this, the proof of this bound is almost identical to the case of $s = 1$ and we omit it here. □

In combination with Proposition 1, Theorem 3 implies that

Corollary 2. *For every graph G of bounded cliquewidth the numbers $c_1(G)$ and $c_2(G)$ can be computed in polynomial time.*

Let us remark that the results of this section cannot be extended for $s \geq 3$ because $c_s(G)$ is not bounded by the cliquewidth of a graph. Consider, for example, complete n-partite graph with partition sets $V_1, V_2, \ldots, V_n$, $|V_i| = n$ for every $i \in \{1, 2, \ldots, n\}$. Then we add n vertices $v_1, v_2, \ldots, v_n$ and for every $i \in \{1, 2, \ldots, n\}$ make v_i adjacent to all vertices from V_i. Let G_n be the resulting graph. It is easy to see that this graph has cliquewidth at most 3 and that $c_s(G_n) = n$ for $s \geq 3$.

5 Open problems

Many interesting algorithmic question around Cops and Robbers game remain open and we conclude with asking some of them.

- The most challenging question is due to Goldstein and Reingold in [17]: Is the testing of $c_1(G) \leq k$ EXPTIME-complete? If the answer is "yes", is the problem EXPTIME-complete for every fixed s? Can it be so that for large s, say for $s \geq \sqrt{n}$, the problem is in NP?
- We have shown that for every graph G of bounded cliquewidth and $s \leq 2$, the number $c_s(G)$ can be computed in polynomial time. What is the computational complexity of the problem on graphs of bounded cliquewidth for $s = 3$ or for $s = \infty$?
- For a graph G of treewidth k, for every $s \geq 1$, it is possible to prove that $c_s(G) \leq k + 1$, which implies that $c_s(G)$ can be computed in time $n^{O(k)}$. What is the parameterized complexity of computing c_s with the treewidth (or the cliquewidth) of a graph as a parameter?
- In the proof of Theorem 1, for a given graph G on n vertices, we construct a graph G' on $O(n^{10})$ vertices such that $\gamma(G) = c_s(G')$, where $\gamma(G)$ is the domination number of G. Combined with the non-approximability for dominating set problem [30], this implies the following

  Corollary 3. *There is a constant $c > 0$ such that there is no polynomial time algorithm to approximate $c_s(G)$ within a multiplicative factor $c \log n$, unless $P = NP$.*

  An interesting question here is if there is an $n^{1-\varepsilon}$-approximation algorithm for the Cops and Robbers game.
- We have shown that for every fixed s, the solution of the Cops and Robbers game can be solved in polynomial time on interval graphs. Can $c_\infty(G)$ be computed in polynomial time on interval graphs?

References

1. A. Adachi, S. Iwata, and T. Kasai, *Some combinatorial game problems require $\Omega(n^k)$ time*, J. ACM, 31 (1984), pp. 361–376.
2. M. Aigner and M. Fromme, *A game of cops and robbers*, Discrete Appl. Math., 8 (1984), pp. 1–11.
3. B. Alspach, *Searching and sweeping graphs: a brief survey*, Matematiche (Catania), 59 (2006), pp. 5–37.
4. T. Andreae, *Note on a pursuit game played on graphs*, Discrete Appl. Math., 9 (1984), pp. 111–115.
5. T. Andreae, *On a pursuit game played on graphs for which a minor is excluded*, J. Combin. Theory Ser. B, 41 (1986), pp. 37–47.
6. A. Berarducci and B. Intrigila, *On the cop number of a graph*, Adv. in Appl. Math., 14 (1993), pp. 389–403.

7. H. L. Bodlaender, *A partial k-arboretum of graphs with bounded treewidth*, Theoretical Computer Science, 209 (1998), pp. 1–45.
8. B. Bollobás, *Extremal graph theory*, vol. 11 of London Mathematical Society Monographs, Academic Press Inc. [Harcourt Brace Jovanovich Publishers], London, 1978.
9. A. Brandstädt, V. B. Le, and J. P. Spinrad, *Graph classes: a survey*, SIAM Monographs on Discrete Mathematics and Applications, Society for Industrial and Applied Mathematics (SIAM), Philadelphia, PA, 1999.
10. D. G. Corneil, S. Olariu, and L. Stewart, *Asteroidal triple-free graphs*, SIAM J. Discrete Math., 10 (1997), pp. 399–430.
11. B. Courcelle, J. Engelfriet, and G. Rozenberg, *Context-free handle-rewriting hypergraph grammars.*, in Graph-Grammars and Their Application to Computer Science, H. Ehrig, H.-J. Kreowski, and G. Rozenberg, eds., vol. 532 of Lecture Notes in Computer Science, Springer, 1990, pp. 253–268.
12. R. G. Downey and M. R. Fellows, *Parameterized complexity*, Springer-Verlag, New York, 1999.
13. S. L. Fitzpatrick and R. J. Nowakowski, *Copnumber of graphs with strong isometric dimension two*, Ars Combin., 59 (2001), pp. 65–73.
14. J. Flum and M. Grohe, *Parameterized Complexity Theory*, Texts in Theoretical Computer Science. An EATCS Series, Springer-Verlag, Berlin, 2006.
15. P. Frankl, *Cops and robbers in graphs with large girth and Cayley graphs*, Discrete Appl. Math., 17 (1987), pp. 301–305.
16. P. Frankl, *On a pursuit game on Cayley graphs*, Combinatorica, 7 (1987), pp. 67–70.
17. A. S. Goldstein and E. M. Reingold, *The complexity of pursuit on a graph*, Theoret. Comput. Sci., 143 (1995), pp. 93–112.
18. J. Gustedt, *On the pathwidth of chordal graphs*, Discrete Appl. Math., 45 (1993), pp. 233–248.
19. G. Hahn and G. MacGillivray, *A note on k-cop, l-robber games on graphs*, Discrete Math., 306 (2006), pp. 2492–2497.
20. Y. O. Hamidoune, *On a pursuit game on Cayley digraphs*, European J. Combin., 8 (1987), pp. 289–295.
21. K. M. Krishnan, R. Singh, L. S. Chandran, and P. Shankar, *A combinatorial family of near regular LDPC codes*, ArXiv Computer Science e-prints, cs/0609146, (2006).
22. M. Maamoun and H. Meyniel, *On a game of policemen and robber*, Discrete Appl. Math., 17 (1987), pp. 307–309.
23. S. Neufeld and R. J. Nowakowski, *A vertex-to-vertex pursuit game played with disjoint sets of edges*, in Finite and infinite combinatorics in sets and logic (Banff, AB, 1991), vol. 411 of NATO Adv. Sci. Inst. Ser. C Math. Phys. Sci., Kluwer Acad. Publ., Dordrecht, 1993, pp. 299–312.
24. R. Niedermeier, *Invitation to fixed-parameter algorithms*, vol. 31 of Oxford Lecture Series in Mathematics and its Applications, Oxford University Press, Oxford, 2006.
25. R. Nowakowski and P. Winkler, *Vertex-to-vertex pursuit in a graph*, Discrete Math., 43 (1983), pp. 235–239.
26. S.-L. Peng, M.-T. Ko, C.-W. Ho, T.-s. Hsu, and C. Y. Tang, *Graph searching on some subclasses of chordal graphs*, Algorithmica, 27 (2000), pp. 395–426.
27. A. Quilliot, *A short note about pursuit games played on a graph with a given genus*, J. Combin. Theory Ser. B, 38 (1985), pp. 89–92.
28. A. Quilliot, *Some results about pursuit games on metric spaces obtained through graph theory techniques*, European J. Combin., 7 (1986), pp. 55–66.
29. V. Raman and S. Saurabh, *Short cycles make w-hard problems hard: Fpt algorithms for w-hard problems in graphs with no short cycles*, Accepted for publication in Algorithmica.

30. R. RAZ AND S. SAFRA, *A sub-constant error-probability low-degree test, and a sub-constant error-probability PCP characterization of NP*, in STOC, 1997, pp. 475–484.
31. B. S. W. SCHROEDER, *The copnumber of a graph is bounded by* $\lfloor \frac{3}{2}$ *genus* $(G)\rfloor + 3$, in Categorical perspectives (Kent, OH, 1998), Trends Math., Birkhäuser Boston, Boston, MA, 2001, pp. 243–263.
32. P. D. SEYMOUR AND R. THOMAS, *Graph searching and a min-max theorem for tree-width*, J. Combin. Theory Ser. B, 58 (1993), pp. 22–33.

Computability of Tilings

Grégory Lafitte[1] and Michael Weiss[2]

[1] Laboratoire d'Informatique Fondamentale de Marseille (LIF), CNRS – Aix-Marseille Université, 39, rue Joliot-Curie, F-13453 Marseille Cedex 13, France
[2] Centre Universitaire d'Informatique, Université de Genève, Battelle bâtiment A, 7 route de Drize, 1227 Carouge, Switzerland

Abstract Wang tiles are unit size squares with colored edges. To know whether a given finite set of Wang tiles can tile the plane while respecting colors on edges is undecidable. Robinson's tiling is an auto-similar tiling in which the computation of a Turing machine can be carried out. By using this construction and by considering a strong notion of simulation between tilings, we prove computability results for tilings. In particular, we prove theorems on tilings that are similar to Kleene's recursion theorems. Then we define and show how to construct reductions between sets of tile sets. We generalize this construction to be able to transform a tile set with a given recursively enumerable property into a tile set with another property. These reductions lead naturally to a Rice-like theorem for tilings.

Introduction

In [17], Wang introduced the study of tilings with colored tiles. A tile is a unit size square with colored edges. Two tiles can be assembled if their common edge has the same color. To tile consists in assembling tiles from a tile set (a finite set of different tiles) on the grid $\mathbb{Z}^2$. The tiles can be repeated as many time as needed, but cannot be turned.

Two questions arose from these definitions. The first one, conjectured true by Wang, was to know whether any tile set that can tile the whole plane can also tile it in a periodic way, *i.e.*, there exists two linearly independant vector u and $v \in \mathbb{Z}^2$ such that for any position $z \in \mathbb{Z}^2$, the tiles at position z, $z+u$ and $z+v$ in the tiling are the same. The second one, known as the *domino problem*, is to know if one can decide whether a given tile set can generate a tiling of the plane.

Both of the questions were answered by Berger in [3]. In his thesis, Berger constructed for any Turing machine M and any input w, a tile set $\tau_{M,w}$ such that this tile set can generate a tiling of the plane if and only if the computation of M stops on the input w. This construction proved the undecidability of the domino problem, and also proved that there exist aperiodic tile sets, *i.e.*, tile set that produces only aperiodic tiling (similarly, a tile set is said to be periodic if it generates at least one periodic tiling). This technical construction was improved later, and simplified constructions of aperiodic tile sets can be found in [16] and [1].

Since the main argument of Berger's proof was to simulate the behavior of a given Turing machine with a tile set, then one of the most important fact concerning tilings

---

Please use the following format when citing this chapter:

Lafitte, G. and Weiss, M., 2008, in IFIP International Federation for Information Processing, Volume 273; *Fifth IFIP International Conference on Theoretical Computer Science*; Giorgio Ausiello, Juhani Karhumäki, Giancarlo Mauri, Luke Ong; (Boston: Springer), pp. 187–201.

is that tilings can constitute a Turing equivalent computation model. This computation model is particularly relevant as a model of computation on the plane.

The study of tilings has made possible the resolution of mathematical logical problems ([1]). Then researchers have been interested in studying the kinds of tilings that one tile set can produce ([16] and more recently [5, 8, 12]). Others have defined tools to quantify the regular structure of a tiling ([6, 2, 13]). Recently, notions of simulation between tilings have been defined to obtain a first approach to computability results on tilings ([12, 14]).

In this paper, we aim at proving computability results for tilings. To reach this goal, we use the construction most used nowadays: Robinson's tiling. In [16], Robinson has built a tile set that generates only auto-similar aperiodic tilings. The construction is based on a hierarchy of squares of ever increasing sizes. In each of these squares, some zone can be used to simulate the behavior of a Turing machine. In [12], notions of simulation and reduction between tilings and tile sets have lead to notions of universality for tilings and completeness for tile sets. Finer notions of simulation have been defined in [14]. These notions rely on Robinson's construction to study the computability of problems related to simulation. In this paper, we make a heavy usage of this construction to prove classical computability results for tilings.

In classical computability (recursion theory) all theorems derive from the enumeration and *s-m-n* theorems. Kleene's recursion (or fixed point) theorem is a direct application of *s-m-n*. With tilings, an *s-m-n* approach would be unnatural because of the particular geometrical nature of computation in this model. Nevertheless, Kleene's theorem is a tool that seems to be more naturally fitted to be transposed on tilings. Our goal in this paper is to show how a computability can be shaped on the geometrical computation model of tilings, and not merely to use classical computability to obtain tools on tilings. In traditional computability, Kleene's theorem states that for any recursive modification of programs M, there exists a program p which is a fixed point for M, *i.e.*, p and $M(p)$ compute the same function. So two Turing machines can be seen as equivalent if they compute the same function. To obtain a Kleene-like theorem for tilings, we need notions of comparison of tile sets: one such notion is the exact simulation. The general idea is to say that a tile set τ exactly simulates a tile set τ' if τ generates a set of rectangles of equal sizes which are isomorphic to the tiles of τ'. From this, we can obtain Kleene-like theorems for tilings.

Beyond Kleene-like theorems, we show how to construct reductions between sets of tile sets. Reductions are fundamental notions in computability theory. Natural notions of reductions between sets of tile sets are also fundamental for tilings. In fact, the idea behind the construction of these reductions lies in Kleene's recursion theorem with parameters: to inject some property in the fixed point being constructed. The reduction constructed is not only interesting for applications but also in itself: it shows how to transform a tile set with a certain property into another tile set with another property. A generalization of this construction leads to another main computability result: Rice's theorem. This theorem states that for any property P on the set of partial recursive functions, if there exist at least one function which satisfies P and one which does not then it is not decidable to know if a given Turing machine computes a function satisfying the property P. Again with the exact simulation, we can state this theorem for tilings as follows: if A is a set of tile sets, then it is not decidable to know whether a given tile set τ exactly simulates a tile set of A. We note that in [4], a first and different approach to a Rice-like theorem for the local constraints has been done, where local constraints are a tiling equivalent model. In this paper, the authors show that it is not decidable to know whether two local constraints can produce the same set of tilings. Our approach is different since we consider the exact simulation as the way to compare tile sets. With the exact simulation, we show how to build reductions between tile sets which lead naturally to a Rice-like theorem.

The main result of this paper is to obtain different Kleene-like theorems using Robinson's construction. We also show that some of these results can be proved with another natural construction introduced in [9] to construct an aperiodic self-similar tiling using Kleene's theorem.

From there, we show how to construct reductions between sets of tile sets and obtain a Rice-like theorem for tilings. The striking aspect of this work holds primarily in the fact that these reductions exist and in the detailed description of their construction.

In Sec. 1, we recall the basic notions of tilings and simulation between tile sets and recall the two main definitions of simulation, the total and the exact ones introduced in [14]. In Sec. 2, we recall the construction of Robinson's tiling and how it can carry out the simulation of a Turing machine. In Sec. 3, we improve this construction to obtain a famous result proved in [10]: the set of periodic tile sets is Σ_1-complete. In Sec. 4, we prove three Kleene-like theorems for tilings. In the last section, we define how to construct reductions between sets of tile sets and prove a Rice-like theorem for tilings.

1 Notions of simulation

We begin with the basic notions of tilings. A tile is an oriented unit size square with colored edges from C, where C is a finite set of colors. A tile set is a finite set of tiles. To tile consists in placing the tiles of a given tile set on the grid $\mathbb{Z}^2$ such that two adjacent tiles share the same color on their common edge. Since a tile set can be described with a finite set of integers, then we can enumerate the tile sets, and τ_i designates the ith tile set.

Let τ be a tile set. A tiling P generated by τ is called a τ-tiling. It is associated to a tiling function f_P where $f_P(x,y)$ gives the tile at position (x,y) in P. When we say that we superimpose the tiles of a tile set τ on the tiles of a tile set τ', we mean that for any tile $t \in \tau$ and any tile $t' \in \tau'$, we build a tile $u = t \times t'$ where the colors of the sides of u are the cartesian product of the colors of the sides of t and t'. Then two tiles $u_1 = t_1 \times t'_1$ and $u_2 = t_2 \times t'_2$ match if and only if t_1 and t_2 match and t'_1 and t'_2 match.

Different notions of reduction have been introduced in [12] and in [14]. We recall some of the notions relative to these reductions and we refer the reader to these papers for detailed explanations and properties.

A pattern is a finite tiling. If it is generated by τ, we call it a τ-pattern. A finite set of rectangular τ-patterns of even size is a τ-pattern set. By analogy with tilings, to tile with a pattern set consists in placing the patterns on a regular subgrid of $\mathbb{Z}^2$ in such a way that the connection between two patterns respects the local constraint of color matching. We call a tiling P generated by a pattern set M, an M-tiling. If M is a set of τ-patterns, then for any M-tiling P, there exists a τ-tiling Q which is a representation of P at the unit tile level.

From this remark we obtain notions of simulation. We say that a pattern tiling P simulates a tiling P' if there exists a function R from the patterns of P to the tiles of P' such that if we replace the patterns of P by their corresponding tiles given by R, then we obtain P'. In such a case, we write $P' \trianglelefteq^R P$ and say that P' reduces to P. If R is not determined, we denote the fact that P' reduces to P by $P' \trianglelefteq P$. The main thing in this reduction is that R is not necessarily a one-to-one function. Different patterns of P can represent the same tile of P'.

This is the least restrictive notion of simulation that we have. We require of a tile set to be able to simulate the behavior of another tile set with patterns. This can be done by any tile set that can produce rectangle patterns whose sides can encode colors. From this simulation, we can define notions of universality for tilings and completeness for tile sets: a tiling P is strongly universal if for any tile set τ, there exists a τ-tiling Q such that $Q \trianglelefteq P$ and a tile set τ is complete if for any tile set τ' and any τ'-tiling Q there exists a τ-tiling P such that $Q \trianglelefteq P$. Therefore, universality is a property of tilings. A tiling is universal if it can simulate the behavior of at least one tiling for any tile set. Completeness is a property of tile sets. A tile set τ is complete if for any tiling P it can generate a tiling having the behavior of P.

In [14], two finer notions have been introduced:

Definition 1. Let τ and τ' be two tile sets. We say that τ *totally simulates* τ' if there exist $a, b \in \mathbb{Z}$ and a reduction R from the $a \times b$ patterns of τ to the tiles of τ' such that the two following conditions are respected:

1. for any τ'-tiling Q, there exists a τ-tiling P such that $Q \trianglelefteq^R P$,
2. for any τ-tiling P, there exists a τ'-tiling Q such that $Q \trianglelefteq^R P$.

We denote it by $\tau' \trianglelefteq_t \tau$ (or $\tau' \trianglelefteq_t^R \tau$ to specify the reduction R).

If $\tau' \trianglelefteq_t \tau$, then there exists a reduction R such that any τ-tiling can be cut in rectangle patterns of size $a \times b$ such that if one replaces these patterns by their corresponding tiles given by R then one obtains a τ'-tiling. And the set of all τ'-tilings that reduce to a τ-tiling is exactly the set of all τ'-tiling. The total simulation is thus more specific than the simulation introduced in [12]. In this way, τ can be seen as a tile set which *computes* in a same way than τ'.

A tile set τ exactly simulates a tile set τ' if τ totally simulates τ' and if the reduction R between τ and τ' is one-to-one. In the total simulation, different patterns can represent the same tile; in the exact one, any tile is represented by only one pattern. It is this simulation that we use to prove our computability theorems for tilings.

To be able to study these notions of simulation, we now recall the classical Robinson construction and some of its specific aspects that we will use later on.

2 Basic notions of simulation of a tile set

Since Berger's proof of the domino problem, we know that we can simulate a Turing machine with a tiling. To any Turing machine M and any input w, we can associate a tile set which simulates the behavior of the computation of M on w. Nowadays, the most used construction to simulate a Turing machine is based on Robinson's tile set (Fig. 1). In [16], Robinson built an aperiodic tiling. This tiling is based on a hierarchy of squares of ever-increasing sizes (Fig. 1.1) shows this hierarchy for the first three levels. These squares are of sizes $2^n + 1$. The idea is to dedicate spaces (the white spaces in Fig. 1.2) in each square of size $2^{2n} + 1$ to simulate a Turing machine by forcing the lowest southwest tile of any of these squares to have the tile representing the initial state of M on the input w. For more details and explanations of this construction, we refer the reader to [1].

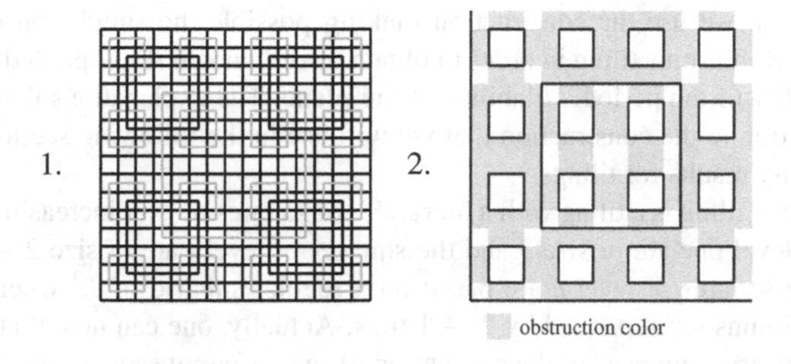

Fig. 1 The hierarchical structure and the obstruction zone in Robinson's tiling

In [12], a tile set is simulated by a Turing machine, in the sense that for any tile set τ, we build a Turing machine M_τ that produces space×time diagrams of same size which are isomorphic to the tiles of τ, where the size of the space×time diagrams are the length and width of the diagrams , *i.e.*, the time and space needed to reach a final state.. This can be done with a Turing machine that takes as input two integers: i, the code of the index of a tile set, and j, the code of a color of τ_i. The Turing machine checks if j is the code of a color of the south side of τ_i. If yes, it computes in a non-deterministic way a tile of τ_i with south color j, as shown in Fig. 2. Then we can simulate this Turing machine in Robinson's tiling and obtain a tile set which simulates totally or exactly, depending on the conditions used, another tile set. For a detailed explanation we refer the reader to [12] and also [14] where constructions of particular tile sets with simulation conditions are built.

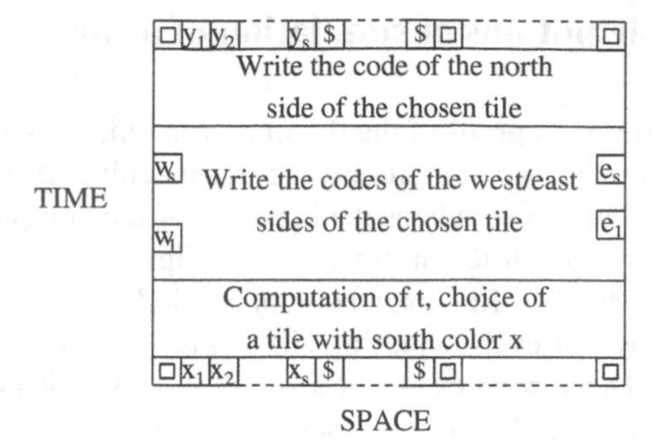

Fig. 2 The space×time diagram of a Turing machine representing the simulation of a tile

3 Periodicity if and only if a Turing machine stops

In this section we use the construction making possible the simulation of a Turing machine in Robinson's tiling in order to obtain a well known result proved in [10]: the undecidability of the periodic tilability of the plane. The explanations that follow are an introduction to the construction that we will use in the following sections to prove computability results for tilings.

Robinson's tiling is a tiling with a hierarchy of squares of ever increasing sizes. The squares of level one are of size 3 and the squares of level i are of size $2^i + 1$. We can see that the squares of level n are based on a regular subgrid of $\mathbb{Z}^2$ where two lines and two columns are separated by $2^n - 1$ tiles. Actually, one can note that these lines and columns are composed of the alternation of two different sequences of $2^n - 1$ tiles separated by corner tiles, one of these sequences representing the side of a square of the nth level. We call this subgrid on which is based the squares of level n, the nth grid. Therefore, the sides of any squares of level n is part of the nth grid.

We can tile Robinson's tiling in a sequence of stages: at stage one, we tile the first grid on $\mathbb{Z}^2$. At stage n, we tile the nth grid and modify, if needed, the tiles of the lowest grids with which the nth grid intersects. This can be done without changing the structure of squares made until this stage. We can proceed like that until the end of the process and we will obtain Robinson's tiling. But we can see that after having tiled the nth level, if we choose to add to our tiling a simple grid, *i.e.*, a grid that does not contains square of the Robinson hierarchy, of same size than the nth grid, and translated in such a way that its corner tiles are in the middle of the squares of the nth grid, then we complete the tiling and make it periodic since we have stopped the self-similarity. Fig. 3 shows the black grid which is inserted in the tiling.

We add to Robinson's tile set special tiles that can generate squares of Robinson's tiling marked with a special color. Thus, at a certain level n, we can decide to tile the nth grid either with the tiles of Robinson's tile set or with the special marked tiles. The special colored tiles have the particularity to not allow squares of higher level to intersect it. Therefore, when one has decided to tile a level with these special tiles, then the self-similarity of Robinson's tiling stops. The only way to complete the tiling,

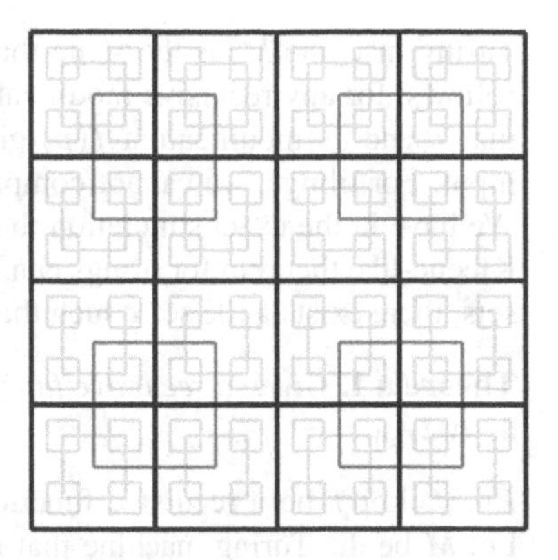

Fig. 3 The blocking color (dark gray) forces the completion of the tiling by adding a regular subgrid (black) that stops the self-similarity of Robinson's tiling (clear gray)

is to do as said in the previous paragraph: we tile a simple grid and, by stopping the self-similarity, we obtain a periodic tiling.

Fig. 3 shows what happens when one decides to tile the squares of level n with the blocking color (here, in black gray). Since no other square of higher level can be added to the tiling, the only way to complete the tiling is to add a simple grid formed of squares of sizes $2^n + 1$ (here, in black).

We now have to add a condition to force to tile with the special colored tiles. Let M be a Turing machine. We build the tile set τ_M which simulates M on the empty input. On the lowest southwest tile of any square of level $2n$, we begin the simulation of M with τ_M with the condition that if a final state is reached before reaching the perimeter of the square, then a special color is sent to the north side of the square that forces the perimeter of the square of level $2n$ - and thus the whole $(2n)^{th}$ grid - to be tiled with the special colored tiles. Then the self-similarity is stopped and the tiling is periodic if and only if M stops on the empty input.

To be more precise, we can compute the exact period of this tiling. If we choose to stop the self-similarity of Robinson's tiling at the level $2n$, then the squares of the hierarchy are of size $2^{2n} + 1$ and at least $2^{2n} - 1$ tiles separate two sides of two squares of level $2n$. Therefore, the smallest period is a square of size 2^{2n+1}. In Fig. 3, the period is represented by a square composed of four blue squares.

In the following sections, we used these different constructions to obtain computability results for tilings.

4 Kleene-like theorems for tilings

The first result we want to obtain is a theorem like Kleene's fixed point theorem but for tilings. Kleene's theorem, in classical computability, states that for any recursive function f, there exists a Turing machine M_e[1] such that the function computed by the

[1] Where M_e denote the eth Turing machine according to an acceptable enumeration of Turing Machines

Turing machine M_e is the same than the one computed by $M_{f(e)}$. We can state it as follows: for any recursive modification of programs f, there exists a program p such that p and its modification $f(p)$ give the same result when computing on the same input. For tilings, we cannot compare functions but we can compare their behavior. We have in the exact simulation the notion of comparison that we need. Therefore, a Kleene-like theorem for tilings can be stated as follows: for any modification f of tile sets, there exists a tile set τ such that τ exactly simulates the modification of τ by f.

Theorem 1. *Given a recursive function f, there exists an e such that τ_e simulates exactly $\tau_{f(e)}$.*

Proof. Let f be a recursive function and M_f a Turing machine which computes f. Let M be the Turing machine that has the following behavior: when the input is the empty word, M computes an integer i. After having computed i, M simulates M_f on the input i. We consider Robinson's tiling where the lowest southwest corner of each square of level n, and thus of size $2^{2n}+1$, of the hierarchy of Robinson's tiling is a tile representing the initial state of M. The simulation of the computation of M is made in this square until it has computed the value $f(i)$. When this value has been computed, a special color is sent to the north board of the square that colors the whole perimeter of this square with this special color (Fig. 4.1). This special color is also a blocking color, *i.e.*, the self-similarity of Robinson's tiling is stopped. Then we send the bits composing $f(i)$ to the south board of the square. This can be done by superimposing the bits of $f(i)$ on the computation tiles.

Therefore, the first line of the square is marked with the bits of $f(i)$ and with the special color, as well as the whole perimeter of the square. When the square is marked with the special color, the computation of a new Turing machine, say N, can begin. N is a Turing machine which takes as inputs an integer x, the index of a tile set, and an integer y, the index of a color of τ_x and computes a tile of the tile set τ_x with south color y, *i.e.*, the space$\times$time diagram of the computation of N on x and y is isomorphic to a tile of τ_x with south color y. In our tiling, we want to simulate a tile of the tile set $\tau_{f(i)}$. Since we already have the bits of $f(i)$ on the first line, we just need to add an integer y, following $f(i)$, which represents the index of a color of the tile set $\tau_{f(i)}$, and then begin the computation of N on $f(i)$ and y (Fig. 4.2).

If y is not a south color of a tile of $\tau_{f(i)}$, then the computation enters an error state, and the tiling cannot be completed. Therefore, the tiling process keeps going on if and only if we have chosen a valid color y. Then N computes the simulation of a tile with south color y. Thus, there exists a level $2n$ such that any square of this level carries out the computation of a tile of $\tau_{f(i)}$.

The last thing that has to be done, to guarantee that two neighboring squares of level $2n$ carry out the simulation of two tiles that match, is to send the codes of the colors on the sides of the squares of level $2n$ outside the square. This guarantees that the zone between two neighboring squares contains the code of a common color.

Those squares of level $2n$ are the biggest of the tiling, since the self-similarity has been stopped. Two squares, carrying out the simulation of the same tile, are composed exactly of the same tiles. There exists only one way for a square to carry out the simulation of a given tile. Therefore, the reduction is an isomorphism and the tile set

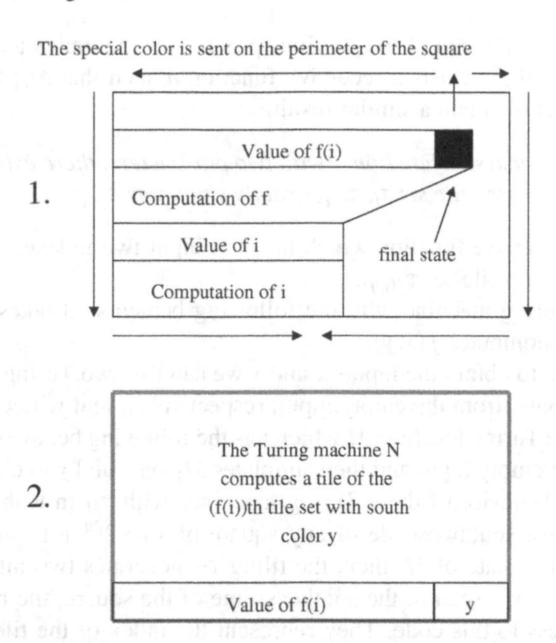

Fig. 4 The computation of M and N in a square of computation of Robinson's tiling

can simulate any tiling generated by $\tau_{f(i)}$ and does not generate a tiling that does not simulate a $\tau_{f(i)}$-tiling. Therefore, the simulation is exact.

We would like for our tile set to have access to its own index to be able to simulate itself but modified by f. This is not an all natural fact, since each time that we add tiles to our tile set to try to encode the code of the tile set, we change the code of the tile set. To prove this, we need Kleene's theorem with parameters which states that for any recursive function g with two parameters, there exists a recursive function n such that for any index of Turing machine e, $M_{n(e)}$ and $M_{g(n(e),e)}$ compute the same function. We consider a recursive function g which takes as inputs a tile set that generates Robinson's tiling, or a Turing machine able to simulate this tile set, and a Turing machine M, and outputs the code $g(\text{Robinson's tile set}, M)$ of a Turing machine which has the following behavior: it computes the index of the tile set which is the simulation of M in Robinson's tiling. By Kleene's theorem, there exists a function n such that $M_{n(M)} = M_{g(n(M),M)}$. Here, n(Robinson's tile set) is our fixed point and represents a Robinson tiling which has access to its own code. This proves that when we simulate a Turing machine in a tiling, we can always suppose that we can do it by having access to the code of this tile set written somewhere in the tilings that it generates.

Therefore, we can suppose that there exists M which gives the index i of its own tile set and thus, the tile set simulated exactly itself modified by f. This proves that this tile set τ_i exactly simulates $\tau_{f(i)}$. □

We now show another version of Kleene's theorem on tilings: Kleene's theorem with parameters. This theorem in a classical computability setting is of great usefulness, as shown at the end of the previous proof. This theorem states that for any recursive function f, there exists a recursive function n such that $M_{n(y)} = M_{f(n(y),y)}$. For tilings, we expect to obtain a similar result.

Theorem 2. *For any recursive function f with two parameters, there exists a recursive function n such that for any tile set τ_i, $\tau_{n(i)}$ exactly simulates $\tau_{f(n(i),i)}$.*

Proof. Let f be a recursive function which takes as input two indexes i, j of tile sets and transforms them in a tile set $\tau_{f(i,j)}$.

Let M_f be the Turing machine with the following behavior: it takes as input two integers x and y and computes $f(x,y)$.

As we did before, to obtain the inputs x and y we can use two Turing machines M^x and M^y which compute, from the empty input, respectively x and y. Let τ_M be the tile set that simulates the Turing machine M which has the following behavior: it simulates M^x and M^y from the empty input and then simulates M_f on x and y to obtain $f(x,y)$.

We simulate the behavior of these Turing machines with τ_M in Robinson's tiling. To do that, the lowest southwest tile of any square of size $2^{2n}+1$ contains the tile representing the initial state of M: then, the tiling τ_M generates two integers x and y and computes $f(x,y)$. We send to the southeast line of the square, the bits of $f(x,y)$, to have a plain access to this code. They represent the index of the tile set we want to simulate. As we did before, the final state of M sends a special color to the north side of the square that forces the perimeter of the square to be colored with this special color. This special color triggers the computation of a new Turing machine, say N, that simulates the tiles of the tile set $f(x,y)$. If the square is big enough to carry out the computation of the tiles of the tile set of index $f(x,y)$, then a blocking color is sent to the north side of the square of computation which forces the whole perimeter of the square to be colored with this blocking color and stops the self-similarity of Robinson's tiling. As we have seen in the previous proof, stopping the self-similarity allows the simulation to be exact.

Therefore, we have a tile set τ^{M^x,M^y}, depending on M^x and M^y, which simulates exactly the tile set $\tau_{f(x,y)}$. For any tile set τ_i, and any Turing machine M^i which computes i when given the empty input, by using Kleene's theorem with parameters, we have seen that we can find a Turing machine $\mathscr{M}^x$ such that $\mathscr{M}^x$ outputs the index of the tile set $\tau^{\mathscr{M}^x,M^i}$, *i.e.*, the tile set that has the following behavior: it simulates $\mathscr{M}^x$ on the empty input, which gives the code of the tile set, say k; then it simulates M^i which outputs i and computes $f(k,i)$. Finally, it simulates the tile set with index $f(k,i)$. Let n be the recursive function that transforms the index i into the index of the tile set $\tau^{\mathscr{M}^x,M^i}$, *i.e.*, k. Therefore, $n(i)$ is a fixed point. Indeed, $\tau_{n(i)} = \tau^{\mathscr{M}^x,M^i}$ exactly simulates the tile set $\tau_{f(\mathscr{M}^x(\varepsilon),M^i(\varepsilon))} = \tau_{f(n(i),i)}$. □

The two previous theorems can be proved without using Robinson's construction. To do that, we can use the construction introduced in the paper [9]. In this paper, the authors use Kleene's recursion theorem to build an aperiodic tiling. The idea is to cut $\mathbb{Z}^2$ with rectangular equal patterns, where each tile of the rectangle knows its position in this rectangle. This can be done by using a special tile for any position of these rectangles. Then one superimposes on each rectangle the computation of a Turing

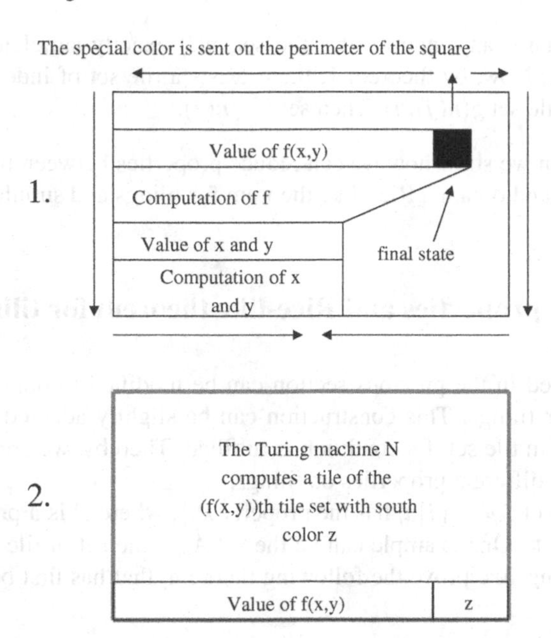

Fig. 5 The computation of M and N in a square of computation of Robinson's tiling

machine simulating a tile of a tile set. One can modify this tile set, say τ, in such a way that each rectangle simulates a tile of τ. By using Kleene's recursion theorem, one obtains a tile set that simulates itself and thus, cannot be periodic.

We can also use this construction to prove our theorem. Since we can know the time needed to compute x, y and $f(x,y)$ then we can apply the same argument and simulate M_x, M_y, the computation of $f(x,y)$ and the simulation of the tiles of $f(x,y)$ in a determined rectangle. The conclusion is the same than in the previous proof. We just have to simulate, as before, the tile set which simulates itself modified by f. Therefore, the recursive function n, that takes as input the code i of a tile set, and outputs the code $n(i)$ of a tile set which computes: $M^{n(i)}$, M^i, and the tiles of the tile set with index $f(M^{n(i)}(\varepsilon), M^i(\varepsilon)) = f(n(i), i)$, is a fixed point and $\tau_{n(i)}$ exactly simulates the tiles of the tile set $\tau_{f(n(i),i)}$.

Another version of Kleene's theorem that we prove is the doubled-fixed point theorem: if f and g are two recursive functions of two variables, then there exist a and b such that: $M_a = M_{f(a,b)}$ and $M_b = M_{g(a,b)}$. In the context of tilings, we obtain the following theorem:

Corollary 1. *Let f and g be two recursive functions of two variables. Then there exist two indexes k and j of tile sets such that τ_k exactly simulates $f(k, j)$ and τ_j exactly simulates $g(k, j)$.*

Proof. We use the two Kleene-like theorems we have just introduced. Since f is a recursive function with two variables, then, by theorem 2, there exists a recursive function n such that for any index i of a tile set, $n(i)$ exactly simulates the tile set with index $f(n(i),i)$. Now, by theorem 1, there exists a tile set of index j which exactly simulates the tile set $g(n(j),j)$. Then set $k = n(j)$. □

In the next section we show how we can reduce properties between tilings to study their computability, and obtain a Rice-like theorem for tilings and simulation.

5 Reductions of properties and Rice-like theorem for tilings

The construction used in the previous section can be modified to obtain other computability results for tilings. This construction can be slightly adapted to obtain the simulation of a certain tile set if a condition is fulfilled. Thereby, we are able to study the computability of different properties on tilings.

We consider the set $A_P = \{ i \mid \tau_i \text{ has the property } P \}$, where P is a property on the tilings generated by τ_i. One example can be the set A_{per}, the set of tile sets that generates a periodic tiling. We prove the following theorem, that has first been proved in [10]:

Theorem 3. *$A_{per} \equiv K_0$, where K_0 is the set of pairs $\langle i,w \rangle$ such that the Turing machine M_i stops on the input w, and thus is Σ_1-complete.*

Proof. In Sec. 3, we have shown that $K_0 \leq_1 A_{per}$. It suffices to show that A_{per} is in Σ_1. The property "*τ is periodic*" can be defined as follows: there exists an n such that τ generates a pattern of size n which is a periodic pattern. Thus, A_{per} can be defined with an $\exists$ arithmetical property. □

To prove the previous theorem, we have reduced the halting problem to the problem to know whether a tile set generates periodicity, by forcing a tile set to have a property if a Turing machine halts on a given input. This kind of argument can be generalized to tile sets to obtain reduction between sets of tile sets. We have the following definition:

Definition 2. Let A and B be two sets of tile sets. A reduces to B (noted $A \leq B$) if there exists a recursive function f such that $i \in A \Leftrightarrow f(i) \in B$.

We show a first kind of reduction between sets of tile sets by proving that the set of periodic tile sets reduces to non-recursive tile sets, *i.e.*, tile sets that produces only non recursive tilings of the plane.

Theorem 4. *Let B_{nr} be the set of non recursive tile sets, i.e., tile sets that produce only tilings of the plane which cannot be defined by a recursive function. Then $A_{per} \leq B_{nr}$ and thus, B_{nr} is not a recursive set.*

Proof. Let τ be a periodic tile set and ρ be a non-recursive tile set. Since [11] and [15], we know that such tile sets exist. Let M be the Turing machine that enumerates the rectangle patterns generated by τ and which stops if and only if τ generates a periodic pattern. As we did before, we simulate M in Robinson's tiling and we block the self-similarity of Robinson's tiling if a final state is reached. Thus, if a period exists, then there exists a level of squares which is all tiled with the blocking color.

We want for our property of generating only non-recursive tilings to appear if and only if τ generates a periodic tiling. We have shown that we can simulate a tile set τ with another tile set, by inserting in Robinson's tiling the simulation of a Turing machine that has the particularity to produce space$\times$time diagrams which are isomorphic to the tiles of τ. Therefore, if a square is marked with the blocking color, it allows the beginning of the computation of a new Turing machine, say N, which has the particularity to produce space$\times$time diagrams which are isomorphic to the tiles of ρ. Without loss of generality, we can consider that N takes always less time and space than M to reach a final state, and thus, if a square can carry out the computation of M, it can also carry out the one of N. Let τ' be this tile set. A simulation of a tile of ρ by τ' is made in a square if and only if the computation of M stops in this square. By adding the condition that the color of the sides of the squares of level n are sent outside the square to force the matching with the neighboring squares, then we obtain the simulation of a ρ-tiling. This tiling cannot be recursive, since it would imply that the tiling it simulates is recursive too. If τ does not generate a periodic tiling, then the squares never carry out the simulation of tiles of ρ and thus, the tile set τ' can generate recursive tilings.

By construction, we have that the self-similarity is stopped and the simulation of the tiles of ρ is made if and only if τ is periodic. As seen before, τ' exactly simulates ρ and thus τ' cannot be recursive. Therefore, τ is periodic if and only if τ' is not recursive.

The reduction that associates to any tile set τ, the tile set τ' shows that B_{nr} is not a recursive set. □

In the previous proof, we have reduced the property of being periodic to the property of being non recursive. This construction can be generalized to obtain other reductions. The main argument of the proof is that, as for Kleene's theorem with parameters, we can inject in a tiling the computation of a program who checks if a property is satisfied in order to obtain a tiling with another property if the previous one is satisfied. The property that we want to verify can be any property P such that it is recursively enumerable to know whether a tile set satisfies it or not. Therefore, we can reduce tile sets satisfying a recursively enumerable property to tile sets with another property. Such recursively enumerable property can be, for example: τ does not tile the plane, τ simulates exactly ρ (where ρ is fixed), τ generates patterns using all its tiles.... Then, if the property is satisfied, we can trigger the start of an exact simulation of a tile set satisfying another property.

By generalizing this kind of construction, we can obtain a Rice-like theorem for exact simulation of sets of tile sets. The only thing we need, is to have a set of tile sets such that if a tile set τ satisfies the property, then any tile set simulating exactly τ has the property too. We define formally this property:

Definition 3. Let A be a set of tile sets. A is an *exact index set* if for any index $i \in A$ of a tile set, if a tile set τ_j exactly simulates τ_i then $j \in A$.

Rice's theorem for Turing machines states that to know whether a Turing machine accepts a language which is in a set A of recursively enumerable languages is not decidable except if A is trivial (empty or if it contains all enumerable languages). We can compare Turing machines by the functions they accept. For tile sets, we do not have a notion of function to compare them. Therefore, if we want a Rice-like theorem for tile sets, the set of tile sets has to be an exact index set and contains the tile sets which "compute" in a same way.

Theorem 5. *Let A be an exact index set. Then the set A is recursive if and only if A is trivial, i.e., $A \neq \mathbb{N}$ and $A \neq \emptyset$.*

Proof. Let A be an exact index set. Since A is not trivial, thus there exist at least one index $i \in A$ and one index $j \notin A$. We first suppose that Robinson's tile set is not in A.

We will reduce L_{per} to L_A as we did in the previous proof. For that, we just have to build from a tile set τ_k, a tile set $\tau_{f(k)}$ such that $\tau_{f(k)}$ simulates τ_i - whose index is in A - if τ_k is periodic, and does not simulate it if τ_k is not periodic. Therefore, this tile set is in A since A is an exact index set.

If τ_k is not periodic, then the only tile set that $\tau_{f(k)}$ exactly simulates is Robinson's tile set.

Therefore, $\tau_k \in L_{per} \Leftrightarrow \tau_{f(k)} \in L_A$.

If Robinson's tile set is in A, then we just have to consider $\overline{L_A}$ instead of $L(A)$. □

To have a better intuitive understanding of this theorem, we can state it as follows: let P be a property on the tilings generated by a tile set satisfying the following statement: if τ satisfies P, then any τ', that exactly simulates τ, satisfies P. Then to know whether a given tile set satisfies P or not is undecidable except if any or no tile set satisfies P.

Acknowledgements

We warmly thank Bruno Durand, Andrei Romashchenko and Alexander Shen for giving us a preprint of their paper [9] and for the discussions on their exciting use of Kleene's fixed point theorem for constructing an aperiodic tile set.

References

1. ALLAUZEN (C.) and DURAND (B.), *The Classical Decision Problem*, appendix A: "Tiling problems", p. 407–420. Springer, 1996.
2. BALLIER (A.), DURAND (B.) and JEANDEL (E.), « Structural Aspects of Tilings », to appear in *Proceedings of the Symposium on Theoretical Aspects of Computer Science*, 2008.
3. BERGER (R.), « The undecidability of the domino problem », *Memoirs of the American Mathematical Society*, vol. **66**, 1966, p. 1–72.
4. CERVELLE (J.) and DURAND (B.), « Tilings: recursivity and regularity », *Theoretical Computer Science*, vol. **310**, n° 1-3, 2004, p. 469–477.
5. CULIK II (K.) and KARI (J.), « On aperiodic sets of Wang tiles », in *Foundations of Computer Science: Potential - Theory - Cognition*, p. 153–162, 1997.

6. DURAND (B.), « Tilings and quasiperiodicity », *Theoretical Computer Science*, vol. **221**, n° 1-2, 1999, p. 61–75.
7. DURAND (B.), « De la logique aux pavages », *Theoretical Computer Science*, vol. **281**, n° 1-2, 2002, p. 311–324.
8. DURAND (B.), LEVIN (L. A.) and SHEN (A.), « Complex tilings », in *Proceedings of the Symposium on Theory of Computing*, p. 732–739, 2001.
9. DURAND (B.), ROMASHCHENKO (A.) and SHEN (A.), « Fixed point and aperiodic tilings », preprint.
10. GUREVICH (Y.) and KORIAKOV (I.), « A remark on Berger's paper on the domino problem », in *Siberian Journal of Mathematics*, **13**:459–463, 1972. (In Russian).
11. HANF (W. P.), « Non-recursive tilings of the plane. I », *Journal of Symbolic Logic*, vol. **39**, n° 2, 1974, p. 283–285.
12. LAFITTE (G.) and WEISS (M.), « Universal Tilings », in *Proceedings of the Symposium on Theoretical Aspects of Computer Science*, Lecture Notes in Computer Science n° **4393**, p. 367–380, 2007.
13. LAFITTE (G.) and WEISS (M.), « A topological study of tilings », to appear in *Proceedings of the conference on Theory and Aspects of Models of Computation* , TAMC'08, 2008.
14. LAFITTE (G.) and WEISS (M.), « Simulation between tilings », submitted to Computability in Europe, CIE'08, 2008.
15. MYERS (D.), « Non-recursive tilings of the plane. II », *Journal of Symbolic Logic*, vol. **39**, n° 2, 1974, p. 286–294.
16. ROBINSON (R.), « Undecidability and nonperiodicity for tilings of the plane », *Inventiones Mathematicae*, vol. **12**, 1971, p. 177–209.
17. WANG (H.), « Proving theorems by pattern recognition II », *Bell System Technical Journal*, vol. **40**, 1961, p. 1–41.
18. WANG (H.), « Dominoes and the $\forall\exists\forall$-case of the decision problem », in *Proceedings of the Symposium on Mathematical Theory of Automata*, p. 23–55, 1962.

A Classification of Degenerate Loop Agreement

Xingwu Liu[1], Juhua Pu[2], and Jianzhong Pan[3]

[1] Institute of Computing Technology, Chinese Academy of Sciences, Beijing, China, xingwuliu@gmail.com

[2] School of Computer Science and Engineering, BeiHang University, Beijing, China, pujh@buaa.edu.cn

[3] School of Mathematics and System Sciences, Chinese Academy of Sciences, Beijing, China, pjz@amss.ac.cn

Abstract. Loop agreement is a type of distributed decision tasks including many well-known tasks such as set agreement, simplex agreement, and approximation agreement. Because of its elegant combinatorial structure and its important role in the decidability problem of distributed decision tasks, loop agreement has been thoroughly investigated. A classification of loop agreement tasks has been proposed, based on their relative computational power: tasks are in the same class if and only if they can implement each other. However, the classification does not cover such important tasks as consensus, because any loop agreement task allows up to three distinct output values in an execution. So, this paper considers classifying a variation of loop agreement, called degenerate loop agreement, which includes consensus. A degenerate loop agreement task is defined in terms of its decision space and two distinguished vertices in the space. It is shown that there are exactly two equivalence classes of degenerate loop agreement tasks: one represented by the trivial task, and the other by consensus. The classification is totally determined by connectivity of the decision space of a task; if the distinguished points are connected in the space, the task is equivalent to the trivial task, otherwise to consensus.

Key words: distributed computing, loop agreement, computability, classification

1 Introduction

A distributed computing system consists of finitely many sequential processes communicating via accessing shared read/write registers and other mechanisms [10]. The mechanisms include communication channels, synchronizing primitives, and general services [1, 6]. The processes are asynchronous and may fail by stopping, so it is indistinguishable whether an irresponsive process has failed or is only running slowly. A protocol is a distributed program in such a system. A task is a distributed coordination problem where each process starts with a private input value and decides an output value such that the decisions of all processes meet some specification [7]. Well-known examples of tasks include consensus[5], set consensus[4], and renaming [2]. A protocol is said to solve a task if starting with any legal input assignment, the outputs produced in any execution of the protocol meet the task specification.

Loop agreement [8] is an interesting type of tasks in the theory of distributed computing. A loop agreement task is defined in terms of an edge loop in a 2-complex, with

Please use the following format when citing this chapter:

Liu, X., Pu, J. and Pan, J., 2008, in IFIP International Federation for Information Processing, Volume 273; *Fifth IFIP International Conference on Theoretical Computer Science*; Giorgio Ausiello, Juhani Karhumäki, Giancarlo Mauri, Luke Ong; (Boston: Springer), pp. 203–213.

three distinguished points on the loop. It stands for a task with the distinguished points as input values and the vertices of the 2-complex as output values. In an execution, if the inputs are the same, the outputs all coincide with the input; if the inputs have two distinct values, the outputs span a simplex along the segment of the loop connecting the two points; otherwise, the outputs span an arbitrary simplex in the complex. Loop agreement is attractive for the following reasons. 1. It has elegant combinatorial structure. 2. It plays a critical role in proving the undecidability of a variety of distributed tasks [7]. 3. It is so general as to include many well-known tasks such as set agreement and approximation agreement.

There are two very influential pieces of work on the computability issue of loop agreement [7, 8]. Ref. [7] showed that a loop agreement task is solvable in certain models if and only if the loop is contractible in the 2-complex, so the solvability of loop agreement tasks in these models is undecidable.

In [8], a classification of loop agreement tasks was presented based on their relative computational power. It considered whether a task T_1 can implement T_2, i.e. T_2 can be solved by calling an instance of a solution to T_1, followed by a protocol using shared read/write registers. Loop agreement tasks can be classified according to the equivalence relation induced by implementation. [8] assigned an algebraic signature to each loop agreement task, which is a pair of the fundamental group of the 2-complex and the path class represented by the loop. It was shown that T_1 can implement T_2 if and only if there is a homomorphism from the signature of T_1 to that of T_2. As a result, the signature completely characterizes the computability of a loop agreement task.

The above work is so elegant. However, its significance is a little weakened in that loop agreement does not include consensus. Consensus is a task whose set of input values is $\{0,1\}$, and in any execution, all the processes agree on the input to some process. Consensus is among the most important tasks in distributed computing, due to its universality [6]. As a result, this paper choose to study an variation of loop agreement, called degenerate loop agreement, which includes consensus. The aim is to adapt the classification of loop agreement tasks in [8] to degenerate loop agreement tasks.

The main contribution of this paper is a complete classification of degenerate loop agreement tasks. Based on the equivalence relation induced by mutual implementation, degenerate loop agreement tasks are divided into two classes: one represented by consensus, the other by the trivial task. The classification is topologically determined; any disconnected task is equivalent to consensus, while connected ones are equivalent to the trivial task.

The rest of this paper is organized as follows. In Section 2, preliminaries on complexes and distributed tasks are presented. In Section 3, degenerate loop agreement tasks are defined. Section 4 proves that there are exactly two classes of degenerate loop agreement tasks, up to the equivalence induced by implementation. Section 5 concludes this paper.

2 Preliminaries

This section will introduce our distributed computing model and formalize the notion of a task. Necessary material from combinatorial topology is also presented, since degenerate loop agreement will be specified using simplicial complexes. Simplicial complexes and their topological properties have long been utilized in distributed computability theory [3, 9, 12]. This paper will exploit connectivity of 1-complexes.

2.1 System model and task formalization

The computing model and task formalization coincide with those in [8], so we will present very briefly. Interested readers please refer to Subsection 3.1 of [8].

We adopt the shared-memory model [10] for distributed computing, where a system consists of a finite set of asynchronous sequential processes, which communicate through accessing shared memory. The shared memory includes read/write registers and possibly more powerful objects and services. A process may delay indefinitely, or fail by stopping.

A task is a distributed coordination problem in which each process starts with a private input value, communicates with others via shared memory, produces an output value, and halts.

Formally, an n-process task T is specified by a triple $(\mathscr{I},\mathscr{O},\Delta)$, where $\mathscr{I} \subseteq (D_I \bigcup\{\bot\})^n \setminus \{(\bot,\cdots,\bot)\}$ is the set of input vectors, $\mathscr{O} \subseteq (D_O \bigcup\{\bot\})^n \setminus \{(\bot,\cdots\bot)\}$ is the set of output vectors, and $\Delta \subseteq \mathscr{I} \times \mathscr{O}$ is the task specification. D_I and D_O are respectively the input and output data types. $\mathscr{I}$ and $\mathscr{O}$ are both prefix-closed [8]. An element $I \in \mathscr{I}$ represents an assignment of input values in an execution: if $I_i \neq \bot$, the ith process starts with input I_i, otherwise it does not participate in that execution. The meaning of output vectors can be likewise understood. Δ carries an input vector to a set of matching output vectors, specifying the *legal* outputs for that input assignment. Here, vectors $I \in \mathscr{I}$ and $O \in \mathscr{O}$ are said to match, when for any i, $I_i = \bot$ if and only if $O_i = \bot$.

An n-process protocol is said to t-resiliently solve a task $(\mathscr{I},\mathscr{O},\Delta)$, if for every execution where the input vector is I and at least $n-t$ processes decide, the decision vector is a prefix of some output vector in $\Delta(I)$. When $t=n-1$, the protocol is said to be wait-free.

We also borrow the notion of implementation from [8]. A task T_1 is said to be implementable from task T_2, if T_1 can be solved by calling an instance of a protocol that solves T_2, possibly followed by access to shared read/write registers. Implementation naturally induces an equivalence relation where two tasks are equivalent if and only if they are mutually implementable. This relation partitions tasks into equivalence classes, which is the very idea of the classification in this paper.

2.2 Simplicial Complexes

We recall the notion of simplicial complexes and simplicial maps. Readers can also refer to the standard textbook [13, 11] for more information.

Arbitrarily choose a finite set of points $\{v_0, v_1, \cdots, v_m\}$ in the $n-$dimensional Euclidean space R^n. If they are affinely independent, the convex closure $s = \left\{ \sum_{i=0}^{m} \lambda_i v_i \in R^n | \sum_{i=0}^{m} \lambda_i = 1 \text{ and } \lambda_i \geq 0 \text{ for } 0 \leq i \leq m \right\}$ is called the simplex spanned by $\{v_0, v_1, \cdots, v_m\}$, denoted by $\overline{\{v_0, v_1, \cdots, v_m\}}$, and m is called the dimension of s. The simplex spanned by any subset of $\{v_0, v_1, \cdots, v_m\}$ is called a face of s. Each v_i is called a vertex of s.

Two simplices are said to well-positioned, if the intersection of them is either empty or a face of each of them. A finite set of pairwise well-positioned simplices, together with all their faces, is called a (simplicial) complex. A complex is said to be an $n-$complex, if all the simplexes are of dimension no more than n. A complexes C' is said to be a subcomplex of C, if $C' \subseteq C$. Vertices A, B in a complex C are said to connected, if there is a sequence of vertices $v_0 = A, v_1, \cdots, v_n = B$, such that for each $0 \leq i \leq n-1$, $\{v_i, v_{i+1}\}$ spans a simplex in C; such a sequence of vertices is called a path connecting A and B.

A map f from complex C to C' is simplicial, if for each vertex v of C, $f(v)$ is also a vertex of C', and for each simplex $s = \overline{\{v_0, v_1, \cdots, v_m\}} \in C$, $f(s)$ is spanned by the set $\{f(v_i) | 0 \leq i \leq m\}$. Obviously, to define a simplicial map, one only has to define its behavior on vertices.

3 Degenerate Loop Agreement

Definition 1. A 1-complex K, together with two distinct vertices $A, B \in K$, determines a task $(\mathscr{I}, \mathscr{O}, \Delta)$ where $\mathscr{I} = (D_I \bigcup \{\perp\})^n \setminus \{(\perp, \cdots, \perp)\}$, $\mathscr{O} = (D_O \bigcup \{\perp\})^n \setminus \{(\perp, \cdots, \perp)\}$, $D_I = \{0, 1\}$, D_O is the set of vertices of K, and

$$\Delta(I) = \begin{cases} \{O | O \text{ matches } I, \text{ and } val(O) = \{A\}\} \; if \; val(I) = \{0\} \\ \{O | O \text{ matches } I, \text{ and } val(O) = \{B\}\} \; if \; val(I) = \{1\} \\ \{O | O \; matches \; I, \; and \; \overline{val(O)} \in K\} \quad otherwise \end{cases} \tag{1}$$

The task is called a degenerate loop agreement task and denoted by $T = (K, A, B)$. K is called the decision space of T.

Intuitively, the input values of $T = (K, A, B)$ are 0 and 1, and the output ones are the vertices of K. When all the inputs are 0 (or 1, respectively), all processes decide A (or B, respectively); otherwise, the decided values spans a simplex in K.

Hereunder, a degenerate loop agreement task $T = (K, A, B)$ will be illustrated by the complex K marked with A and B. See Figure 1 as an example.

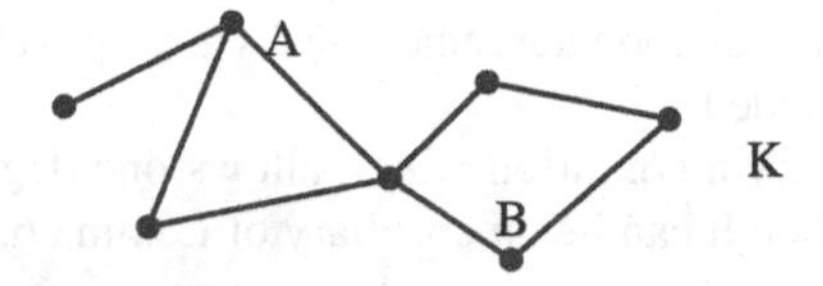

Fig. 1 The illustration of a task $T = (K, A, B)$

Example 1. A famous example of degenerate loop agreement task is consensus, which intuitively means that all processes must agree on a value from their inputs. Formally, $consensus = (\{A, B\}, A, B)$, as illustrated in Figure 2.

Fig. 2 consensus

There is a canonical fact on consensus.

Lemma 1. (Theorem 12.6, [10]) *Consensus can't be solved using read/write registers.*

Example 2. Another example of degenerate loop agreement task is $T = (K, A, B)$, where K consists of the simplex $\overline{\{A, B\}}$ and its faces. See Figure 3 as an illustration of T. Since T can be solved by the protocol where each process trivially outputs its input, it is called the trivial task in this paper.

Fig. 3 The trivial task

There is an obvious fact on the trivial task. The proof is omitted here.

Lemma 2. *A degenerate loop agreement task can be solved using read/write registers if and only if it can be implemented by the trivial task.*

4 A Classification of Degenerate Loop Agreement

The main result is that degenerate loop agreement is divided into two classes, as stated in Theorem 1 at the end of this section. To prove this theorem, this section is organized as follows. First, Corollary 1 *normalizes* degenerate loop agreement tasks by removing *redundant* components from their decision spaces. Second, Lemma 6 shows that all disconnected degenerate loop agreement tasks are equivalent. Third, Lemma 8 shows that all connected degenerate loop agreement tasks are equivalent. Some other technical lemmas are also included.

First of all, we identify a condition which allows one degenerate loop agreement task to implement another. It can be an corollary of Lemma 6.2 in [8], but we provide a much simpler proof.

Lemma 3. *Given two degenerate loop agreement tasks T=(K,A,B) and* $T' = (K', A', B')$*, if there is a simplicial map* $f : K \to K'$ *such that* $f(A) = A'$ *and* $f(B) = B'$*, then T implements* T'*.*

Proof: Choose an arbitrary protocol P for T, and construct a protocol P_f as follows. Each process of P_f first runs protocol P, resulting in a temporary decision value v. Then it outputs $f(v)$ as its final decision. We show that P_f solves T'.

Consider an arbitrary execution of P_f, with S_I/S_O as its set of input/output values, respectively. Assume S'_O to be the set of output values of P in this execution. The following is a case analysis.

Case 1: $S_I = \{0\}$. Then $S'_O = \{A\}$ since P solves T. Because $f(A) = A'$, we have $S_O = \{A'\}$. Likewise, if $S_I = \{1\}$, then $S_O = \{B'\}$.

Case 2: $S_I = \{0,1\}$. Then S'_O spans a simplex in K. Because $f : K \to K'$ is a simplicial map, $S_O = \{f(v) | v \in S'_O\}$ spans a simplex in K'.

As a result, P_f solves T', and hence T implements T'. □

Then we show that a task gets stronger if some part of its decision space is removed, as shown in the following lemma.

Lemma 4. *Given two 1-complexes K and* K'*, if K is a subcomplex of* K' *and A,B are vertices of K, then the degenerate loop agreement task* $T = (K,A,B)$ *implements* $T' = (K',A,B)$*.*

Proof: The inclusion $i : K \to K'$, $v \mapsto v$ is a simplicial map. By Lemma 3, $T = (K,A,B)$ implements $T' = (K',A,B)$. □

Definition 2. Given a degenerate loop agreement task $T = (K,A,B)$, a connected component C of K is called an idle component of T, if C contains neither A nor B.

Lemma 5. *Let C be an idle component of a degenerate loop agreement task T=(K,A,B). Then T is equivalent to* $T' = (K \setminus C, A, B)$*.*

Proof: On the one hand, T' implement T, by Lemma 4.

On the other hand, define a simplicial map $f : K \to K'$,

$$f(v) = \begin{cases} A \; if \; v \; is \; a \; vertex \; in \; C \\ v \; otherwise \end{cases} \tag{2}$$

See Figure 4 for an illustration of f. By Lemma 3, T' implement T.

To sum up, T is equivalent to T'. □

According to Lemma 5, a task can be equivalently transformed by eliminating all its idle components, so we immediately have the following corollary.

Corollary 1. *Any degenerate loop agreement task is equivalent to one without idle components.*

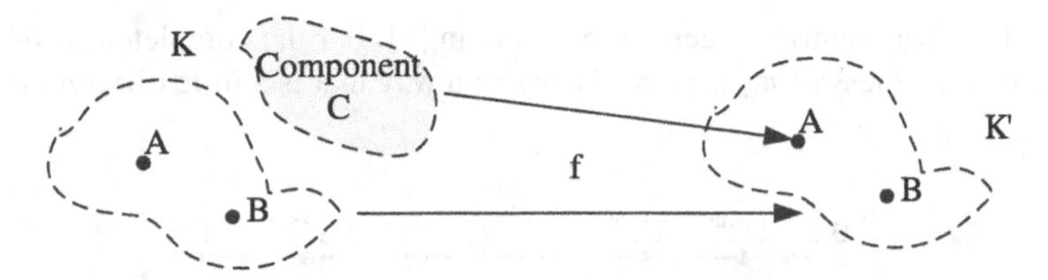

Fig. 4 The map f in Lemma 5

As a result, all the tasks hereunder are assumed to have no idle components, without loss of generality.

Definition 3. A degenerate loop agreement task $T = (K, A, B)$ is said to be connected if A and B are connected in K. Otherwise it is said to be disconnected.

Connectivity is a topological property. The following lemmas show that it plays a critical role in classifying degenerate loop agreement tasks.

Lemma 6. *Any two disconnected degenerate loop agreement tasks are equivalent.*

proof: The basic idea is to show that any disconnected degenerate loop agreement task $T = (K, A, B)$ is equivalent to consensus. Without loss of generality, assume the decision space of consensus is $K' = \{A, B\}$.

First, K' is a subcomplex of K. By Lemma 4, consensus implements T.

Second, define a simplicial map $f : K \to K'$,

$$f(v) = \begin{cases} A \ if \ v \ is \ in \ the \ component \ containing \ A \\ B \ if \ v \ is \ in \ the \ component \ containing \ B \end{cases} \tag{3}$$

See Figure 5 for an illustration of f. By Lemma 3, T implements consensus.

Altogether, T is equivalent to consensus, and the lemma holds. □

To show that that all connected degenerate loop agreement tasks are also equiva-

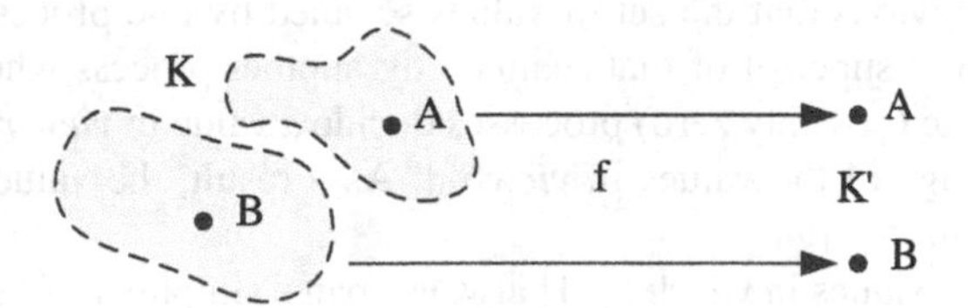

Fig. 5 The map f in Lemma 6

lent, we have to construct a protocol π_m for a special task $\tau_m = (\kappa_m, 0, 1)$, where m is a positive integer. The decision space κ_m of τ_m is a 1-complex in R^1, consisting of the simplices $\{\frac{i}{2^m}, \frac{i+1}{2^m}\}$, $0 \leq i \leq 2^m - 1$, as well as their faces. κ_m is illustrated in Figure 6. The $n-$process protocol π_m is illustrated in Figure 7. It is actually the 1-dimensional version of the barycentric agreement protocol in [8]. For the completeness of presentation, the correctness of π_m is proved here, in a way that is a little different from that in [8].

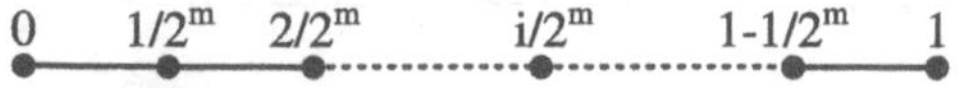

0 1/2^m 2/2^m i/2^m 1-1/2^m 1

Fig. 6 Decision space κ_m of the task τ_m

```
shared array view[m-1][n];
protocol (input l)
  for r=0..m-1 do
      view[r][P]:=l;
      l:=average of the values in scan(view[r]);
      end for
  decide l;
end protocol
```

Fig. 7 The protocol π_m (for process P)

Lemma 7. *The protocol π_m solves τ_m.*

Proof: First, π_m is wait-free, since each process does not wait for others to progress and it only executes a bounded number of steps before terminating.

Second, We claim that for each r, the values in $view[r]$ always span a simplex in κ_r. The proof is by induction.

Step 1. When $r = 0$, $view[r]$ contains either 0, 1, or 0 and 1, so it spans a simplex in κ_0. The claim holds in this case.

Step 2. Hypothesize that the claim holds for $r_0 < m-1$.

Step 3. It is obvious that the set of values scanned by one process when $r = r_0$ is either a subset or a superset of that scanned by another process when $r = r_0$. Hence when $r = r_0$, some (possibly zero) processes decide a value in $view[r_0]$, and the others decides the average of the values in $view[r_0]$. As a result, the values in $view[r_0+1]$ spans a simplex in κ_{r_0+1}.

To sum up, the values in $view[m-1]$ always span a simplex in κ_{m-1}. Following the argument in step 3, we have that the final values decided by the protocol π_m spans a simplex in κ_m. Furthermore, it is clear that when all the inputs are A, the processes only decides A, likewise for the case of B. So, π_m solves τ_m. □

Now we are ready to adapt Lemma 6 to the case of connected tasks.

Lemma 8. *Any two connected degenerate loop agreement tasks are equivalent.*

Proof: Our idea is to show that any connected degenerate loop agreement task $T = (K, A, B)$ is equivalent to the trivial task. The proof proceeds in two steps. Without loss of generality, assume that the trivial task is $T' = (K', 0, 1)$, where $K' = \{0, 1, \overline{\{0,1\}}\}$.

Step 1: to prove that T implement the trivial task. Define a simplicial map $f : K \rightarrow K'$,

$$f(v) = \begin{cases} 0 \; if \; v = A \\ 1 \; otherwise \end{cases} \tag{4}$$

See Figure 8 for an illustration of f. By Lemma 3, T implements the trivial task.

Step 2: to prove that the trivial task implements T. Because K is connected, there

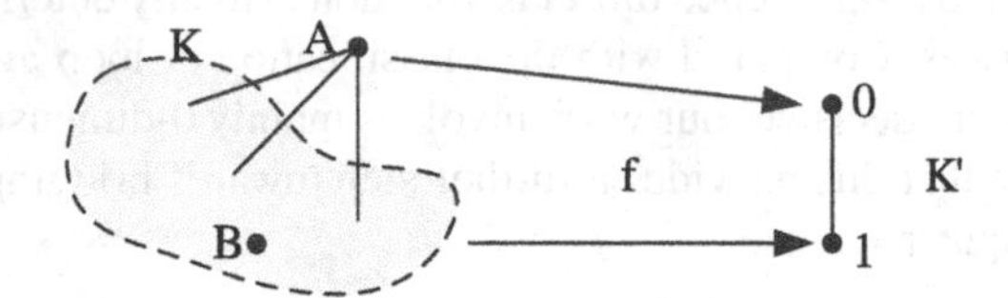

Fig. 8 The map f in Lemma 8

is a path in K connecting A and B. Fix one such path $u_0, u_1, u_2, \cdots, u_n$, where $u_0 = A$ and $u_n = B$. Let $m = \lceil log_2 n \rceil$. By Lemmas 7 and 2, the trivial task implements τ_m.

We now have to show that τ_m implements T. Define a simplicial map $g : \kappa_m \rightarrow K$,

$$g(\frac{i}{2^m}) = \begin{cases} u_i \; 0 \leq i \leq n \\ B \; n \leq i \leq 2^m \end{cases} \tag{5}$$

See Figure 9 for an illustration of f. By Lemma 3, τ_m implements T, so the trivial task implements T.

To sum up, every connected degenerate loop agreement task is equivalent to the trivial task, and the lemma holds. □

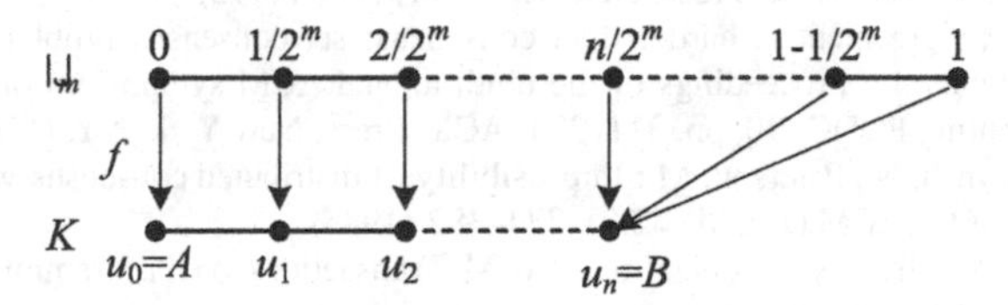

Fig. 9 The map g in Lemma 8

Theorem 1. *There are two equivalence classes of degenerate loop agreement tasks.*

Proof: By Lemma 6 and Lemma 8, degenerate loop agreement tasks can be divided into at most two equivalence classes: one represented by consensus, and the other by the trivial task. By Lemma 1, consensus can not be implemented from the trivial task. As a result, there are exactly two equivalence classes of degenerate loop agreement tasks. □

5 Conclusion

Loop agreement is an interesting type of distributed decision tasks and has been thoroughly studied. However, it does not include the important task of consensus, so this paper considers one of its variation, called degenerate loop agreement, which includes consensus. Classifying degenerate loop agreement tasks is explored to characterize their computational power: two tasks are in the same class if and only if they can implement each other. It turn out that there are exactly two classes: one represented by consensus, including all disconnected tasks, and the other by the trivial task, including all connected tasks. Hence this classification is totally determined by topology of the decision spaces. Compared with the classification of loop agreement where 1-dimensional holes are decisive, our work involves mainly 0-dimension holes, i.e. connectivity. We hope that this provides a further step towards bridging the gap between topology and computer science.

Acknowledgements The work is supported by China's Natural Science Foundation (60603004).

References

1. Attie, P., Guerraoui, R., Kouznetsov, P., Lynch, N., Rajsbaum, S.: The impossibility of boosting distributed service resilience. In: Proceedings of the 25th IEEE International Conference on Distributed Computing Systems, pp. 39–48. IEEE Computer Society, Washington DC, USA (2005)
2. Attiya, H., Bar-Noy, A., Dolev, D., Koller, D., Peleg, D., Reischuk, R.: Achievable cases in an asynchronous environment. In: Proceedings of the 28th Annual Symposium on Foundations of Computer Science, pp. 337–346. IEEE Computer Society (1987)
3. Borowsky, E., Gafni, E.: Generalized flp impossibility result for t-resilient asynchronous computations. In: Proceedings of the twenty-fifth annual ACM symposium on Theory of computing STOC '93, pp. 91–100. ACM Press, New York, NY, USA (1993)
4. Chaudhuri, S.: Agreement is harder than consensus: set consensus problems in totally asynchronous systems. In: Proceedings of the ninth annual ACM symposium on Principles of distributed computing PODC '90, pp. 311–324. ACM Press, New York, NY, USA (1990)
5. Fischer, M., Lynch, N., Paterson, M.: Impossibility of distributed consensus with one faulty process. Journal of the ACM (JACM) **32**(2), 374–382 (1985)
6. Herlihy, M.: Wait-free synchronization. ACM Transactions on Programming Languages and Systems **13**(1), 124–149 (1991)
7. Herlihy, M., Rajsbaum, S.: The decidability of distributed decision tasks (extended abstract). In: Proceedings of the twenty-ninth annual ACM symposium on theory of computing, pp. 589–98. ACM Press, New York, NY, USA (1997)
8. Herlihy, M., Rajsbaum, S.: A classification of wait-free loop agreement tasks. Theoretical Computer Science **291**(1), 55–77 (2003)
9. Hoest, G., Shavit, N.: Towards a topological characterization of asynchronous complexity. In: Proceedings of the sixteenth annual ACM symposium on Principles of distributed computing PODC '97, pp. 199–208. ACM Press, New York, NY, USA (1997)
10. Lynch, N.: Distributed Algorithms. Morgan Kaufmann Publishers (1996)
11. Munkres, J.: Elements of Algebraic Topology. Perseus Press (1993)

12. Saks, M., Zaharoglou, F.: Wait-free k-set agreement is impossible: The topology of public knowledge. In: Proceedings of the twenty-fifth annual ACM symposium on Theory of computing STOC '93, pp. 91–100. ACM Press, New York, NY, USA (1993)
13. Spanier, E.: Algebraic Topology. Springer (1994)

On the expressive power of univariate equations over sets of natural numbers*

Alexander Okhotin[1,2], and Panos Rondogiannis[3]

[1] Academy of Finland
[2] Dept. of Mathematics, University of Turku, Finland
[3] Dept. of Informatics and Telecommunications, University of Athens, Greece

Abstract. Equations of the form $X = \varphi(X)$ are considered, where the unknown X is a set of natural numbers. The expression $\varphi(X)$ may contain the operations of set addition, defined as $S+T = \{m+n \mid m \in S, n \in T\}$, union and intersection, as well as ultimately periodic constants. An equation with a non-periodic solution of exponential growth is constructed. At the same time it is demonstrated that no sets with super-exponential growth can be represented. It is also shown that a restricted class of these equations cannot represent sets with super-linearly growing complements. The results have direct implications on the power of conjunctive grammars with one nonterminal symbol.

1 Introduction

Language equations, in which the unknowns are formal languages, have recently become an active topic of study [5]. Formal languages are typically considered over an alphabet containing at least two letters. For a unary alphabet $\Sigma = \{a\}$, they can be regarded as sets of natural numbers. Then the operation of concatenating such languages turns into pairwise addition of sets: $S + T = \{m+n \mid m \in S,\ n \in T\}$. Language equations accordingly become equations over sets of numbers. Even in this seemingly simple case they already have quite surprising properties.

Consider systems of equations of the form

$$X_i = \varphi_i(X_1, \ldots, X_n) \quad (1 \leqslant i \leqslant n), \qquad (*)$$

where the unknowns X_i are subsets of $\mathbb{N}_0 = \{0, 1, 2, \ldots\}$, while the right-hand sides φ_i contain union, addition and singleton constants. These systems are equivalent to language equations of the same form (*) over a unary alphabet using the operations of union and concatenation, and accordingly represent context-free grammars. As it is well-known that all unary context-free languages

* Research supported by the Academy of Finland under grant 118540, and by the Greek General Secretariat for Research and Technology under the program $\Pi ENE\Delta$ (grant number $03E\Delta$ 330).

Please use the following format when citing this chapter:

Okhotin, A. and Rondogiannis, P., 2008, in IFIP International Federation for Information Processing, Volume 273; *Fifth IFIP International Conference on Theoretical Computer Science*; Giorgio Ausiello, Juhani Karhumäki, Giancarlo Mauri, Luke Ong; (Boston: Springer), pp. 215–227.

are regular, least solutions of systems (*) over sets of numbers are vectors of ultimately periodic sets.

Another kind of equations are systems of the form (*) with *addition and complementation.* An example of such an equation with a non-periodic solution was given by Leiss [6]. Later Okhotin and Yakimova [8] established the main properties of systems of such equations (in the more general case of language equations) and gave a direct proof that a certain rather simple non-periodic set is not representable.

Consider systems of the same general form (*), in which the allowed operations are *union, intersection and addition.* These systems correspond to an extension of the context-free grammars, the *conjunctive grammars* [7], which are again considered over a unary alphabet. The question of whether conjunctive grammars can generate any non-regular unary languages has been an open problem for some years [7], until recently solved by Jeż [3], who constructed a grammar for the language $\{ a^{4^n} \mid n \geqslant 0 \}$. This grammar can be regarded as a system (*) of four equations over sets of numbers using union, intersection and addition, such that one of the four components of its least solution is $\{ 4^n \mid n \geqslant 0 \}$.

The set $\{ 4^n \mid n \geqslant 0 \}$ grows exponentially, so this example left a question of whether any super-exponentially growing sets are representable. A strong answer was given by Jeż and Okhotin [4], who showed that for every given recursive function it is possible to represent a set that grows faster.

Despite these extensive positive results (and maybe to some extent *due* to these positive results), no results saying that some particular set *cannot* be represented by such equations could so far be obtained. The $DTIME(n^2) \cap DSPACE(n)$ complexity upper bound for conjunctive grammars over a unary alphabet is the only known restriction. Otherwise, no techniques of proving non-representability of sets by equations with union, intersection and addition are known.

This paper considers a particular case of systems (*) with $n = 1$: these are equations of the form $X = \varphi(X)$, where X is a unique variable and φ is an expression containing arbitrarily nested union, intersection, sum and ultimately periodic constants. Every such equation has a least solution given by $\bigcup_{n=0}^{\infty} \varphi^n(\varnothing)$. It is shown that these equations can represent a certain non-periodic set of an exponential growth rate: namely, the example of Jeż [3] is reconstructed using one variable instead of four. At the same time it is proved that no sets that grow asymptotically faster than exponential can be represented. Another class of sets is shown to be non-representable by a restricted class of such equations: these are *dense sets*, that is, sets with super-linearly growing complements. In overall, it is demonstrated that one-variable equations are weaker in power than systems of multiple equations. This also demonstrates that conjunctive grammars with a single nonterminal cannot generate all conjunctive languages.

2 Conjunctive grammars and systems of equations

Conjunctive grammars form a natural extension of the context-free grammars that supports intersection in the right-hand sides of rules:

Definition 1 ([7]). A conjunctive grammar is a quadruple $G = (\Sigma, N, P, S)$, where Σ and N are disjoint finite nonempty sets of terminal and nonterminal symbols respectively, P is a finite set of rules, each of the form

$$A \to \alpha_1 \& \ldots \& \alpha_n \qquad (n \geqslant 1,\ A \in N,\ \alpha_i \in (\Sigma \cup N)^*) \tag{1}$$

and $S \in N$ is the start symbol. A grammar is said to be linear conjunctive if furthermore each α_i in each rule (1) is in $\Sigma^* N \Sigma^*$ or in Σ^*.

One way to define the semantics of conjunctive grammars is by term rewriting. Consider terms over concatenation and conjunction. Then a subterm A can be rewritten with $(\alpha_1 \& \ldots \& \alpha_n)$ for every rule (1), and any subterm of the form $(w \& \ldots \& w)$, with $w \in \Sigma^*$, can be rewritten with w. Then $L(G)$ is defined as the set of all strings $w \in \Sigma^*$ that are derivable from the term S.

An equivalent definition can be given using language equations.

Definition 2. For every conjunctive grammar $G = (\Sigma, N, P, S)$, the associated system of language equations is a system of equations in variables N, in which each variable assumes a value of a language over Σ, and which contains the following equation for every variable A:

$$A = \bigcup_{A \to \alpha_1 \& \ldots \& \alpha_m \in P} \ \bigcap_{i=1}^{m} \alpha_i \quad (\text{for all } A \in N)\,. \tag{2}$$

Each instance of a symbol $a \in \Sigma$ in such a system defines a constant language $\{a\}$, while each empty string denotes a constant language $\{\varepsilon\}$. A solution of such a system is a vector of languages $(\ldots, L_C, \ldots)_{C \in N}$, such that the substitution of L_C for C, for all $C \in N$, turns each equation (2) into an equality.

Let $(\ldots, L_C, \ldots)$ be the least solution of the system and define $L_G(C) = L_C$ for all $C \in N$ and $L(G) = L_G(S)$.

Consider conjunctive grammars over a one-symbol alphabet, with $\Sigma = \{a\}$. A formal language $L \subseteq a^*$ can be regarded as a set of numbers $\{\, n \mid a^n \in L \,\}$. The operation of concatenation of languages is replaced with pairwise addition of sets: for all $S, T \subseteq \mathbb{N}$, define

$$S + T = \{\, m + n \mid m \in S, \text{ and } n \in T \,\}$$

Thus a system of language equations (2) corresponding to a conjunctive grammar over $\{a\}$ can be regarded as a system of equations over sets of natural numbers.

For unary languages, being regular means to be ultimately periodic as a set of numbers. A set S is *ultimately periodic* if there exist numbers $d, p \geqslant 0$, such that for any $n \geqslant d$, the number n is in S if and only if $n+p$ is in S. Such a set is also said to be periodic starting from d with period p.

The first example of a system of equations with union, intersection and addition representing a non-periodic set (originally presented in the form of a conjunctive grammar) is as follows:

Example 1 (Jeż [3]). The system of equations

$$\begin{cases} X_1 = \big((X_1+X_3) \cap (X_2+X_2)\big) \cup \{1\} \\ X_2 = \big((X_1+X_1) \cap (X_6+X_2)\big) \cup \{2\} \\ X_3 = \big((X_1+X_2) \cap (X_6+X_6)\big) \cup \{3\} \\ X_6 = \big((X_1+X_2) \cap (X_3+X_3)\big) \end{cases}$$

has the least solution $X_k = \{\, k \cdot 4^n \mid n \geqslant 0 \,\}$, for $k = 1, 2, 3, 6$.

The idea of this construction is best understood in terms of *positional notation* of numbers. Let $\Sigma_k = \{0, 1, \ldots, k-1\}$ be digits in base-k notation. For every $w \in \Sigma_k^*$, let $(w)_k$ be the number defined by this string of digits. Define $(L)_k = \{\, (w)_k \mid w \in L \,\}$. Now the solution of the above system can be represented in base-4 notation as the vector $\big((10^*)_4, (20^*)_4, (30^*)_4, (120^*)_4\big)$. Let us substitute this vector into the right-hand side of the first equation:

$$\begin{aligned} &\big((10^*)_4 + (30^*)_4\big) \cap \big((20^*)_4 + (20^*)_4\big) = \\ &\quad = \big((10^*30^*)_4 \cup (10^+)_4 \cup (30^*10^*)_4\big) \cap \big((20^*20^*)_4 \cup (10^+)_4\big) = (10^+)_4 \end{aligned}$$

Taking the singleton $\{1\}$ into account, the set $(10^*)_4$ is obtained.

In order to minimize the number of brackets, the subsequent examples will assume the following default precedence of operations: addition has the highest precedence, intersection has intermediate precedence, and the precedence of union is the lowest. Also, singleton constants $\{n\}$ will sometimes be written as n.

Let us define the notion of a growth rate of a set. Every infinite set of numbers $L = \{i_1, i_2, \ldots, i_n, \ldots\}$, with $0 \leqslant i_1 < i_2 < \ldots < i_n < \ldots$, can be regarded as an increasing integer sequence. The *growth rate* of such sequences is represented by a function $g(n) = i_n$. The set from Example 1 has exponential growth rate.

The method of manipulating positional notations of numbers using addition of sets has been further extended in the following way. Consider a linear conjunctive grammar generating base-k positional notations of some numbers. Then the set of these numbers can be specified by a system of equations over sets of numbers.

Theorem 1 (Jeż, Okhotin [4]). *For every $k \geqslant 2$ and for every linear conjunctive grammar G over Σ_k there exists a system of equations $X = \varphi_i(X_1, \ldots, X_n)$ over sets of natural numbers with the least solution $X_i = S_i$, in which $S_1 = (L(G))_k$.*

This theorem has several important implications. One of them is that the growth rate of representable sets is not bounded by any fixed recursive function.

Theorem 2 (Jeż, Okhotin [4]). *For every recursively enumerable set of natural numbers S there exists a system $X_i = \varphi_i(X_1, \ldots, X_n)$ over sets of natural numbers with the least solution $X_i = S_i$, such that the growth function of S_1 is greater than that of S at any point.*

There are four variables in the system in Example 1, while Theorems 1–2 use quite many variables. The purpose of this paper is to investigate the expressibility of univariate equations.

3 Equations with one variable

Consider an equation

$$X = \varphi(X),$$

where the unknown X is a set of natural numbers, while φ uses union, intersection and addition, as well as ultimately periodic constants. These operations can, in general, be arbitrarily nested. It is known from the fixed point theory that $\bigcup_{i \geqslant 0} \varphi^i(\varnothing)$ is the least (wrt set inclusion) among all the solutions of the equation.

A particular case of such equations are those corresponding to one-nonterminal conjunctive grammars, where φ must be a union of intersections of sums, and it is interesting to note that already in this case every ultimately periodic set can be represented using singleton constants.

Lemma 1 (Alhazov [1]). *Every unary regular language is generated by a one-nonterminal conjunctive grammar.*

Proof. Let $K \cup (a^p)^+ L$ be the given language, where $K, L \subseteq \{\varepsilon, a, \ldots, a^{p-1}\}$. Then the required grammar is

$$S \to a^i \quad (a^i \in K \cup a^p L \cup a^{2p} L)$$
$$S \to a^p S \& a^{2p} S$$

□

The question is, whether any non-periodic sets can be represented using univariate equations. As the following lemma demonstrates, this is indeed the case:

Lemma 2. *The following one-variable equation has the unique solution $\{\, 4^n - 8 \mid n \geqslant 3 \,\} \cup \{\, 2 \cdot 4^n - 15 \mid n \geqslant 3 \,\} \cup \{\, 3 \cdot 4^n - 11 \mid n \geqslant 3 \,\} \cup \{\, 6 \cdot 4^n - 9 \mid n \geqslant 3 \,\}$:*

$$X = \big(11 + X + X \cap 22 + X + X\big) \cup \big(1 + X + X \cap 9 + X + X\big) \cup$$
$$\cup \big(7 + X + X \cap 12 + X + X\big) \cup \big(13 + X + X \cap 14 + X + X\big) \cup \{56, 113, 181\}$$

Here addition is assumed to have higher precedence than intersection.

The idea behind this construction is to encode four variables from Example 1 into a single variable. The unique solution of the constructed equation is a union of four disjoint sets:

$$\begin{aligned} L_1 &= \{\, 4^n - 8 \mid n \geqslant 3 \,\} \\ L_2 &= \{\, 2 \cdot 4^n - 15 \mid n \geqslant 3 \,\} \\ L_3 &= \{\, 3 \cdot 4^n - 11 \mid n \geqslant 3 \,\} \\ L_6 &= \{\, 6 \cdot 4^n - 9 \mid n \geqslant 3 \,\} \end{aligned}$$

Each of them represents the corresponding component of the solution of the system from Example 1. These components are represented with an *offset*: the numbers in L_1, L_2, L_3 and L_6 are smaller by $d_1 = 8$, $d_2 = 15$, $d_3 = 11$ and $d_6 = 9$, respectively.

Consider first the following system:

$$\begin{cases} Y_1 = \big(11{+}Y_1{+}Y_3 \cap 22{+}Y_2{+}Y_2\big) \cup \{56\} \\ Y_2 = \big(1{+}Y_1{+}Y_1 \cap 9{+}Y_6{+}Y_2\big) \cup \{113\} \\ Y_3 = \big(7{+}Y_6{+}Y_6 \cap 12{+}Y_1{+}Y_2\big) \cup \{181\} \\ Y_6 = 13{+}Y_3{+}Y_3 \cap 14{+}Y_1{+}Y_2 \end{cases} \tag{3}$$

This system is obtained from the system in Example 1 as follows. First, the constant sets $\{1\}$, $\{2\}$ and $\{3\}$ are replaced with $\{64\}$, $\{128\}$ and $\{192\}$, so that the values of n in the solution start from 3. Then the substitution $X_1 = Y_1 + 8$, $X_2 = Y_2 + 15$, $X_3 = Y_3 + 11$, $X_6 = Y_6 + 9$ is applied. It is easy to see that the solution of system (3) is the vector (L_1, L_2, L_3, L_6).

Note that each set L_i is a subset of a periodic set $\{\, 64m - d_i \mid m \geqslant 1 \,\}$. Let us call every such periodic superset *a track*. The sum of any two of these sets, $L_i + L_j$, is a subset of $\{\, 64m - d_i - d_j \mid m \geqslant 2 \,\}$, which is a track as well. The numbers 8, 15, 11 and 9 have been chosen so that the sums of all pairs of these numbers are pairwise distinct: $d_i + d_j = d_k + d_\ell$ with $i \leqslant j$ and $k \leqslant \ell$ implies $i = k$ and $j = \ell$. In other words, the tracks are pairwise disjoint, and the calculations in the right-hand sides of different equations occur in different tracks.

This property is used to ensure that if the same set $L_1 \cup L_2 \cup L_3 \cup L_6$ is substituted for *every variable* in the right-hand sides of (3), then every right-hand side still evaluates to L_1, L_2, L_3 and L_6, respectively. Now the equation in Lemma 2 is obtained from the system (3) by identifying all four variables into one.

It must be admitted that these ideas do not work in general, and Lemma 2 is not proved by a formal transformation. However, they happen to work for the given example and with the given assignment of offsets to variables. The lemma can actually be proved by substituting the given set into the equation and verifying that it is indeed a solution. The proof is omitted in this extended abstract due to its pure technicality.

The equation in Lemma 2 has a simple form corresponding to a conjunctive grammar. The result can thus be restated in the following form.

Example 2. The following one-nonterminal conjunctive grammar generates the language $\{ a^{4^n-8} \mid n \geqslant 3 \} \cup \{ a^{2 \cdot 4^n - 15} \mid n \geqslant 3 \} \cup \{ a^{3 \cdot 4^n - 11} \mid n \geqslant 3 \} \cup \{ a^{6 \cdot 4^n - 9} \mid n \geqslant 3 \}$:

$$S \to a^{22}SS\&a^{11}SS \mid a^9SS\&aSS \mid a^7SS\&a^{12}SS \mid a^{13}SS\&a^{14}SS \mid a^{56} \mid a^{113} \mid a^{181}$$

This example answers the question raised by Jeż [3] about the least number of nonterminals in a conjunctive grammar necessary to generate non-regular languages over $\{a\}$: *one is enough.*

4 Non-representability of fast growing sets

The set represented in Lemma 2 has exponential growth. It will now be shown that sets with asymptotically super-exponential growth cannot be represented by univariate equations. The following statement is also applicable to some sets that do not formally fit this description.

Theorem 3. *Let* $L = \{n_1, n_2, \ldots, n_i, \ldots\}$ *with* $0 \leqslant n_1 < n_2 < \ldots < n_i < \ldots$ *be an infinite set of natural numbers, for which* $\liminf_{i \to \infty} \frac{n_i}{n_{i+1}} = 0$*. Then* L *is not the least solution of any univariate equation* $X = \varphi(X)$*.*

In particular, the theorem asserts non-representability of sets like $\{ 2^{2^n} \mid n \geqslant 0 \}$ and $\{ n! \mid n \geqslant 1 \}$, as well as sets like $\{ n!, n!+1 \mid n \geqslant 1 \}$.

The assumption that limit inferior of $\frac{n_i}{n_{i+1}}$ as n approaches infinity is zero means that the size of gaps between consecutive numbers (measured relatively to the smaller number) is not bounded. That is, for every k there is $n \in L$ so that L does not contain any numbers between $n+1$ and kn.

If such a set is a least solution of an equation, then L can be expressed from itself and from ultimately periodic constants using union, intersection and addition. Then the gaps between elements of the set have to be bridged either by summing up several smaller elements of this set in an expression $X + \ldots + X$, or by adding an ultimately periodic constant to X. The expression φ contains only finitely many additions, and hence only a bounded number of smaller elements can be added up. Larger gaps can only be bridged by adding an ultimately periodic constant. However, this addition would make the sum ultimately periodic as well.

This reasoning is formalized in the following statement:

Lemma 3. *Let* $\varphi(X)$ *be an expression that contains instances of a unique variable* X *ultimately periodic constants with a common period* p *starting from* d*,* *and the operations of union, intersection and addition. Let h be the greatest number of nested additions in φ. Let a number n and a set of numbers L be such that $n \in \varphi(L)$, $L \cap \{\lceil \frac{n}{2^h} \rceil, \lceil \frac{n}{2^h} \rceil + 1, \ldots, n-1\} = \varnothing$ and $\frac{n}{2^h} \geqslant d+p$. Then $n \in L$ or $n - p \in \varphi(L)$.*

Proof. Induction on the structure of φ.

Basis I: $\varphi(X) = X$. Then $n \in \varphi(L)$ means $n \in L$.

Basis II: $\varphi(X) = C$, where C is an ultimately periodic set of natural numbers. Then $h = 0$ and hence $n \geqslant d + p$ by assumption. Since C has period p starting from d, $n \in \varphi(L) = C$ is equivalent to $n - p \in C = \varphi(L)$.

Induction step I: $\varphi(X) = \varphi_1(X) \cup \varphi_2(X)$. Then $n \in \varphi(L)$ implies that $n \in \varphi_i(L)$ for some $i \in \{1,2\}$. Assume without loss of generality that $n \in \varphi_1(L)$. Let h_1 be the greatest number of nested additions in φ_1; obviously, $h_1 \leqslant h$. Then $\frac{n}{2^{h_1}} \geqslant \frac{n}{2^h}$ and therefore $L \cap \{\lceil \frac{n}{2^{h_1}} \rceil, \lceil \frac{n}{2^{h_1}} \rceil + 1, \ldots, n-1\} = \varnothing$ and $\frac{n}{2^{h_1}} \geqslant d+p$. Thus the induction hypothesis is applicable to φ_1 and n, giving that $n \in L$ or $n - p \in \varphi_1(L) \subseteq \varphi(L)$.

Induction step II: $\varphi(X) = \varphi_1(X) \cap \varphi_2(X)$. In this case $n \in \varphi(L)$ implies both $n \in \varphi_1(L)$ and $n \in \varphi_2(L)$. Let h_1 and h_2 be the greatest numbers of nested additions in φ_1 and φ_2, respectively, for which it is known that $h_1 \leqslant h$ and $h_2 \leqslant h$. As in the case of union, the induction hypothesis is applicable to φ_1 and n, as well as to φ_2 and n, which gives $n \in L$ or $n - p \in \varphi_1(L)$, and at the same time $n \in L$ or $n - p \in \varphi_2(L)$. If either subexpression yields $n \in L$, this immediately proves the claim for φ and n. Otherwise the number $n - p$ is known to be both in $\varphi_1(L)$ and in $\varphi_2(L)$, which means $n - p \in \varphi(L)$.

Induction step III: $\varphi(X) = \varphi_1(X) + \varphi_2(X)$. Then it follows from $n \in \varphi(L)$ that there are two numbers $n_1, n_2 \geqslant 0$ with $n_1 + n_2 = n$ and $n_i \in \varphi_i(L)$ for $i \in \{1,2\}$. Assume without loss of generality that $n_1 \geqslant n_2$. Let h_1 be the greatest number of nested additions in φ_1, which is known to be at most $h-1$. Then $\frac{n_1}{2^{h_1}} \geqslant \frac{n_1}{2^{h-1}} \geqslant \frac{n}{2} \cdot \frac{1}{2^{h-1}} = \frac{n}{2^h}$, and therefore $L \cap \{\lceil \frac{n_1}{2^{h_1}} \rceil, \lceil \frac{n_1}{2^{h_1}} \rceil + 1, \ldots, n-1\} = \varnothing$ and $\frac{n_1}{2^{h_1}} \geqslant d+p$. By the induction hypothesis for φ_1 and n_1 it follows that $n_1 \in L$ or $n_1 - p \in \varphi_1(L)$. Consider each of these cases:

- In the former case, note that $\frac{n}{2} \leqslant n_1 \leqslant n$. Since $h \geqslant 1$ and $L \cap \{\lceil \frac{n}{2^h} \rceil, \lceil \frac{n}{2^h} \rceil + 1, \ldots, n-1\} = \varnothing$ by assumption, $n_1 \in L$ implies that n_1 must be equal to n, while n_2 must be zero. This proves that $n \in L$.
- If $n_1 - p \in \varphi_1(L)$, then $n - p = (n_1 - p) + n_2 \in \varphi(L)$.

This last case completes the proof of the lemma. □

Proof (Theorem 3). Suppose there exists an equation $X = \varphi(X)$ with the least solution L_0. Let $C_1, \ldots, C_m$ be all constants used in φ, and let each C_i have period p_i starting from d_i. Let $p = \mathrm{lcm}\{p_1, \ldots, p_m\}$ and $d = \max\{d_1, \ldots, d_m\}$; then all constants have period p starting from d. Denote the greatest number of nested additions in φ by h.

By the definition of limit inferior, there exist infinitely many numbers i with $\frac{n_i}{n_{i+1}} < \frac{1}{2^h}$. Then it is possible to choose a sufficiently large i so that $\frac{n_{i+1}}{2^h} \geqslant d+p$.

Now $n_i < \frac{n_{i+1}}{2^h} \leqslant \lceil \frac{n_{i+1}}{2^h} \rceil$, and since L_0 contains no elements between $n_i + 1$ and $n_{i+1} - 1$, it follows that $L_0 \cap \{\lceil \frac{n_{i+1}}{2^h} \rceil, \ldots, n_{i+1} - 1\} = \varnothing$.

Since L_0 is the least fixed point of φ, there exists a number of iterations ℓ, for which $n_{i+1} \notin \varphi^\ell(\varnothing)$ and $n_{i+1} \in \varphi^{\ell+1}(\varnothing)$. Denote $L = \varphi^\ell(\varnothing)$, that is, $n_{i+1} \notin L$ and $n_{i+1} \in \varphi(L)$. Since $L \subseteq L_0$, it is known that $L \cap \{\lceil \frac{n_{i+1}}{2^h} \rceil, \ldots, n_{i+1} - 1\} = \varnothing$. Therefore, Lemma 3 is applicable to φ, n and L, and it asserts that $n \in L$ or $n - p \in \varphi(L)$. The former contradicts the assumption, while the latter is not possible since $\lceil \frac{n}{2^h} \rceil \leqslant n - p \leqslant n - 1$. The contradiction obtained proves the theorem. □

Theorem 3 implies a separation of one-nonterminal conjunctive languages from conjunctive languages of the general form.

Theorem 4. *The following proper containments hold:*

$$\text{REG}_{\{a\}} \subset \text{CONJ}^1_{\{a\}} \subset \text{CONJ}_{\{a\}}$$

Proof. In particular, $\text{CONJ}^1_{\{a\}} \setminus Reg$ contains the language from Example 2, while $\text{CONJ}_{\{a\}} \setminus \text{CONJ}^1_{\{a\}}$ contains some languages growing faster than exponential (and as it will be demonstrated in the next section, also some languages with super-linearly growing complements). □

5 Non-representability of dense sets

In this section we derive non-representability results concerning a class of sets that are known as *additive bases*:

Definition 3. Let $S \subseteq \mathbb{N}$ be an infinite set of natural numbers, and let $k > 0$. For any $n \in \mathbb{N}$, define the number of its representations as a sum of k elements of S by

$$r_{k,S}(n) = |\{(a_1, \ldots, a_k) \in S^k : a_1 + \cdots + a_k = n\}|.$$

The set S is said to be a basis of order k if every sufficiently large natural number n can be represented as sum of k (not necessarily distinct) elements of S, or equivalently if $r_{k,S}(n) \geqslant 1$. In other words, S is a basis of order k if and only if $\underbrace{(S + \cdots + S)}_{k}$ is co-finite.

As an example, there is a well-known result, *Legendre's theorem*, that the set of squares of the natural numbers is a basis of order four.

Clearly, if a set S is a basis of order k then it is also a basis of every order $n > k$. The non-representability results we will obtain in this section, are for sets that are bases of order 2.

We start with a class of sets that are dense additive bases of order 2:

Definition 4. Given any $m, n \in \mathbb{N}$, let $[m, n]$ denote the discrete closed interval $[m, n] = \{i \in \mathbb{N} : m \leqslant i \leqslant n\}$.

A set $L \subseteq \mathbb{N}$ is said to be *dense* if $\lim_{n\to\infty} \frac{|L\cap[0,n]|}{n} = 1$.

For example, the set $\mathbb{N} \setminus \{ 2^n \mid n \geqslant 0 \}$ is obviously dense, and so is the set of composite numbers.

The following lemma is easy to establish using basic properties of limits:

Lemma 4. *Let L be a dense set. Then*

$$\lim_{n\to\infty} \frac{|(\mathbb{N} \setminus L) \cap [0, n]|}{|L \cap [0, n]|} = \lim_{n\to\infty} \frac{|(\mathbb{N} \setminus L) \cap [0, n]|}{n} = 0.$$

Similarly to Theorem 3, the following theorem states that sets of the above form cannot be represented using univariate equations that use finite or co-finite constants:

Theorem 5. *Let L be a dense non-ultimately periodic set. Then there is no univariate equation $X = \varphi(X)$ using finite and co-finite constants, which would have the least solution L.*

The proof of the theorem is based upon the following three lemmas.

Lemma 5. *Let $L_1 \subseteq \mathbb{N}$ and $L_2 \subseteq \mathbb{N}$ be dense sets. Then the set $L_1 + L_2$ is co-finite.*

Notice that this lemma implies that dense sets are additive bases of order 2 (just take $L_1 = L_2$).

Proof. The main idea of the proof is that every sufficiently large element of $\mathbb{N}$ can be written as the sum of two elements of $\mathbb{N}$ in *too many ways*. Now, since the sets $\mathbb{N} \setminus L_1$ and $\mathbb{N} \setminus L_2$ are "sparse", every sufficiently large element of $\mathbb{N}$ can also be written as the sum of at least two elements of L_1 and L_2. In other words, $\mathbb{N} \setminus (L_1 + L_2)$ is finite.

More formally now, it suffices to show that for every sufficiently large $n \in \mathbb{N}$ there exist $\ell_1 \in L_1$ and $\ell_2 \in L_2$ such that $n = \ell_1 + \ell_2$. Consider the number of ways in which a number n can be written as a sum of two numbers $n_1 \in L_1$ and $n_2 \in L_2$. More specifically, given $n \in \mathbb{N}$, define the functions:

$$\begin{aligned}
p(n) &= |\{(n_1, n_2) : (n_1 \in \mathbb{N}) \text{ and } (n_2 \in \mathbb{N}) \text{ and } (n_1 + n_2 = n)\}| \\
r_1(n) &= |\{k : (k \in \mathbb{N} \setminus L_1) \text{ and } (k \leqslant n)\}| \\
r_2(n) &= |\{k : (k \in \mathbb{N} \setminus L_2) \text{ and } (k \leqslant n)\}|
\end{aligned}$$

Now it is easy to see that every sufficiently large number n in $\mathbb{N}$ can be written as $n = \ell_1 + \ell_2$, with $\ell_1 \in L_1$ and $\ell_2 \in L_2$, in at least $p(n) - r_1(n) - r_2(n)$ ways. To prove that $p(n) - r_1(n) - r_2(n) > 0$ for large values of n, it suffices to show that $\lim_{n\to\infty} \frac{p(n)}{n} > 0$, while $\lim_{n\to\infty} \frac{r_1(n)}{n} = 0$ and $\lim_{n\to\infty} \frac{r_2(n)}{n} = 0$.

Notice now that $p(n) = n + 1$ since n can be written as the sum of two elements of $\mathbb{N}$ in the following ways: $(0, n), (1, n-1), \ldots, (n, 0)$. Therefore, $\lim_{n\to\infty} \frac{p(n)}{n} = 1$. Consider now the case of $r_1(n)$ (the case of $r_2(n)$ is identical). Since L_1 is a dense set, Lemma 4 asserts that $\lim_{n\to\infty} \frac{|(\mathbb{N}\setminus L_1)\cap[0,n]|}{n} = 0$, and therefore $\lim_{n\to\infty} \frac{r_1(n)}{n} = 0$. It follows that $p(n) - r_1(n) - r_2(n) > 0$ (that is, $n \in L_1 + L_2$) for all sufficiently large $n \in \mathbb{N}$. Therefore, $\mathbb{N} \setminus (L_1 + L_2)$ is a finite set. □

Lemma 6. *Let $S_1, S_2 \subseteq \mathbb{N}$ be dense sets, let $T \subseteq \mathbb{N}$ be any non-empty set. Then the sets $S_1 \cap S_2$, $S_1 \cup T$ and $S_1 + T$ are dense.*

The proof, which is omitted, proceeds by using simple set-theoretic arguments, and the basic properties of limits.

Lemma 7. *Let $\varphi(X)$ be an expression using the variable X, finite or co-finite constants, together with the operations of union, intersection and addition. Let L be a dense set and assume that $\varphi(L)$ is infinite. Then, $\varphi(L)$ is a dense set.*

Proof. Follows from Lemma 6 by a straightforward induction.

Proof (Proof of Theorem 5). Let $X = \varphi(X)$ be an equation. Let us prove that L cannot be its least solution. The proof is by an induction on the number of subexpressions of the form $\psi(X) + \xi(X)$ in φ, in which both ψ and ξ contain some instances of X.

Basis. If there are no such additions, then the least solution must be ultimately periodic by the known results on language equations with one-sided concatenation [2]. Since L is non-periodic, a contradiction is obtained.

Induction step. Consider any of the smallest such subexpressions of φ, that is, let $\varphi(X) = \widehat{\varphi}(X, \widetilde{\varphi}(X))$, where $\widetilde{\varphi} = \psi + \xi$.

Consider first the case where both $\psi(L)$ and $\xi(L)$ are infinite. Let us show that $\widetilde{\varphi}(L)$ is co-finite. Indeed, by Lemma 7, $\psi(L)$ is a dense set and $\xi(L)$ is also a dense set. Then Lemma 5 states that $\mathbb{N} \setminus (\psi(L) + \xi(L))$ is a finite set. In other words, $\psi(L) + \xi(L) = \mathbb{N} \setminus F$ for some finite $F \subset \mathbb{N}$. Denote $\mathbb{N} \setminus F$ by R'.

Then $\varphi(L) = \widehat{\varphi}(L, \widetilde{\varphi}(L)) = \widehat{\varphi}(L, R')$. Let $\varphi'(X)$ be a new expression defined as $\widehat{\varphi}(X, R')$. Then L should be the least solution of the equation $X = \varphi'(X)$. Since $\varphi'(X)$ contains fewer subexpressions of the form $\psi(X) + \xi(X)$, by the induction hypothesis, L cannot be the least solution of this equation. A contradiction.

Now consider the remaining case of $\psi(L)$ being a finite set, say F. Then $\varphi(L) = \widehat{\varphi}(L, \widetilde{\varphi}(L)) = \widehat{\varphi}(L, F + \xi(L))$. Define a new expression $\varphi'(X)$ as $\widehat{\varphi}(X, F + \xi(X))$; the set L should be the least solution of the equation $X = \varphi'(X)$. However, φ' contains fewer subexpressions of the form $\psi(X) + \xi(X)$, and hence L is not its least solution. This last contradiction establishes the induction step and concludes the proof. □

An immediate consequence of this result is that the class of sets of natural numbers that can be defined using univariate equations containing only finite or co-finite constants is not closed under complementation. Indeed, the complement of the language in Lemma 2 is dense and falls under Theorem 5. In particular, the class of unary languages generated by conjunctive grammars with one nonterminal is not closed under complementation.

Note that the equations corresponding to conjunctive grammars have a particular form, in which union and intersection may not be nested within addition. Further non-representability results for one-nonterminal conjunctive grammars can be obtained by using this form:

Theorem 6. *Let L be an additive basis of order 2 that is not ultimately periodic. Then L is not the least solution of any univariate equation $X = \varphi(X)$ that uses ultimately periodic constants, together with the operations of union, intersection and addition and in which union and intersection can not be nested within addition.*

Proof (a sketch). Let $X = \varphi(X)$ be an equation. Let us prove that L cannot be its least solution. Consider any subexpression of φ of the form $X + \cdots + X$. Since L is a basis, the corresponding sum $L + \cdots + L$ is co-finite and therefore ultimately periodic. Replace every such expression in φ by a corresponding constant. If there are no such additions left, then the least solution of the resulting equation must be ultimately periodic by the known results on language equations with one-sided concatenation [2], which is a contradiction. □

It follows that the family of unary languages generated by one-nonterminal conjunctive grammars does not contain any non-periodic additive bases of order 2.

6 Conclusions

It was shown that univariate equations $X = \varphi(X)$ with union, intersection and addition are, on one hand, nontrivial in the sense that they can represent some non-periodic sets. On the other hand, counting arguments were used to show that they cannot represent some sets that are known to be representable using systems of equations.

These non-representability results become the first of their kind, since no methods of proving sets non-representable by systems of equations with union, intersection and addition are currently known. This task appears challenging, though it is the authors' hope that the results obtained in this paper may also shed some light on this more general case.

Acknowledgement

We are grateful to an anonymous referee for his thorough comments.

References

1. A. Alhazov, personal communication, September 2007.
2. F. Baader, A. Okhotin, "Complexity of language equations with one-sided concatenation and all Boolean operations", *20th International Workshop on Unification* (UNIF 2006, Seattle, USA, August 11, 2006), 59–73.
3. A. Jeż, "Conjunctive grammars can generate non-regular unary languages", *Developments in Language Theory* (DLT 2007, Turku, Finland, July 3–6, 2007), LNCS 4588, 242–253.
4. A. Jeż, A. Okhotin, "Conjunctive grammars over a unary alphabet: undecidability and unbounded growth", *Computer Science in Russia* (CSR 2007, Ekaterinburg, Russia, September 3–7, 2007), LNCS 4649, 168–181.
5. M. Kunc, "What do we know about language equations?", *Developments in Language Theory* (DLT 2007, Turku, Finland, July 3–6, 2007), LNCS 4588, 23–27.
6. E. L. Leiss, "Unrestricted complementation in language equations over a one-letter alphabet", *Theoretical Computer Science*, 132 (1994), 71–93.
7. A. Okhotin, "Conjunctive grammars", *Journal of Automata, Languages and Combinatorics*, 6:4 (2001), 519–535.
8. A. Okhotin, O. Yakimova, "On language equations with complementation", *Developments in Language Theory* (DLT 2006, Santa Barbara, USA, June 26–29, 2006), LNCS 4036, 420–432.

References

[illegible]

Collisions and their Catenations: Ultimately Periodic Tilings of the Plane

Nicolas Ollinger and Gaétan Richard

Laboratoire d'informatique fondamentale de Marseille (LIF),
Aix-Marseille Université, CNRS,
39 rue Joliot-Curie, 13 013 Marseille, France
{nicolas.ollinger,gaetan.richard}@lif.univ-mrs.fr

Abstract. Motivated by the study of cellular automata algorithmic and dynamics, we investigate an extension of ultimately periodic words to two-dimensional infinite words: collisions. A natural composition operation on tilings leads to a catenation operation on collisions. By existence of aperiodic tile sets, ultimately periodic tilings of the plane cannot generate all possible tilings but exhibit some useful properties of their one-dimensional counterparts: ultimately periodic tilings are recursive, very regular, and tiling constraints are easy to preserve by catenation. We show that, for a given catenation scheme of finitely many collisions, the generated set of collisions is semi-linear.

1 Introduction

The theory of regular languages, sets of one-dimensional sequences of letters sharing some regularities, has been well studied since the fifties. Finite state machines [18], regular languages [14, 5], computing devices with bounded memory, monadic second-order logic [4]: various point of views lead to a same robust notion of regular languages. The concept extends to infinite words and various other one-dimensional structures. Unfortunately, when considering two-dimensional words – partial mappings from the plane $\mathbb{Z}^2$ to a finite alphabet – such a robust common object fails to emerge: automata on the plane, picture languages, second-order logic, all lead to different notions of regular languages [9]. A first difficulty arises from the definition of a finite word: should it be any partial mapping with a finite support? Should it be rectangles filled with letters? Should it be any mapping with a connected support for some particular connexity notion? A second difficulty arises from the complexity of two-dimensional patterns: in the simplest case of uniform local constraints, *i.e. tilings*, knowing whether a given finite pattern is a factor of a valid tiling (of the whole plane) is already undecidable [1].

In the present paper, we investigate a particular family of recursive tilings of the plane endowed with a catenation operation. Our definition of an ultimately periodic tiling, a *collision*, is inspired by geometrical considerations on one-dimensional cellular automata space-time diagrams and tilings. It can be

---

Please use the following format when citing this chapter:

Ollinger, N. and Richard, G., 2008, in IFIP International Federation for Information Processing, Volume 273; *Fifth IFIP International Conference on Theoretical Computer Science*; Giorgio Ausiello, Juhani Karhumäki, Giancarlo Mauri, Luke Ong; (Boston: Springer), pp. 229–240.

thought of as an extension of the notion of ultimately periodic bi-infinite words to two-dimensional words. These objects provide a convenient tool to describe synchronization problems in cellular automata algorithmic.

One-dimensional cellular automata [13] are dynamical systems whose configurations consist of bi-infinite words on a given finite alphabet. The system evolves by applying uniformly and synchronously a locally defined transition rule. The value at each position, or *cell*, of a configuration only depends on the values of the cells on its neighborhood at the previous time step. To discuss the dynamics or to describe algorithmic constructions, it is often convenient to consider space-time diagrams rather than configurations. A space-time diagram is a drawing of a particular orbit of the system: configurations are depicted one on top of the other, from bottom to top, by successively applying the transition rule, as depicted on Fig. 1. This representation permits to draw away the time-line and discuss the structure of emerging two-dimensional patterns. Formally, this is equivalent to consider tilings of half the plane with a special kind of local constraint, oriented by the time-line.

Time goes from bottom to top. Each letter is represented by a different color.

Fig. 1 Space-time diagram of a one-dimensional cellular automaton

Let us give first an informal overview of what collisions are and where they come from. An ultimately periodic configuration consists of two infinite periodic words separated by a finite non-periodic word. As transitions of cellular automata are locally defined, the image of an ultimately periodic configuration is an ultimately periodic configuration such that: for each periodic part, the period in the image divides the period in the preimage; for the non-periodic part, it can only grow by a finite size depending on the local rule. If, by iterating the transition rule of the cellular automaton, the size of the non-periodic part of the configurations remains bounded, then the orbit of the ultimately periodic configuration is, up to a translation, ultimately periodic. When considering this ultimately periodic behavior from the space-time diagram point of view, one can see some kind of *particle*: a localized structure moving with a rational slope in a periodic *background* environment, as depicted on Fig. 2a.

As particles are ultimately periodic configurations, one can construct more complicated configurations by putting particles side by side, ensuring that the non-periodic parts are far enough from each other, and that the periodic parts of two particles put side by side are the same and well aligned. If the non-periodic part of several particles (two or more) becomes near enough in the orbit, complex interactions might occur. If the interaction is localized in both space and time, as depicted on Fig. 2b, this interaction is called a *collision*.

Particles and collisions provide a convenient tool in the study of cellular automata. When constructing two-dimensional cellular automata, like in historical

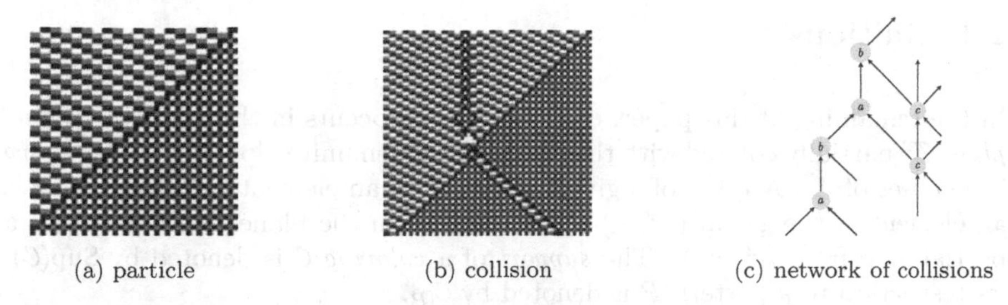
(a) particle (b) collision (c) network of collisions

Fig. 2 Particles and collisions generated by ultimately periodic configurations

constructions of von Neumann [20] and Codd [6], particles are a convenient way to convey quanta of information from place to place. The most well known example of particle is certainly the glider of the Game of Life used by Conway et al. to embed computation inside the Game of Life [2] by using particular behavior of glider collisions. When using one-dimensional cellular automata to recognize languages or to compute functions, a classical tool is the notion of signal introduced by Fischer [8] and later developed by Mazoyer and Terrier [16, 17]: signals and their interactions are simple kinds of particles and collisions. Particles appear even in the classification of cellular automata dynamics: in its classification [21], Wolfram identifies what he calls class 4 cellular automata where *"(...) localized structures are produced which on their own are fairly simple, but these structures move around and interact with each other in very complicated ways. (...)"* A first study of particles interaction was proposed by Boccara et al. [3], latter followed by Crutchfield et al. [12]: these works focus on particles and bounding the number of possible collisions they can produce. Finally, the proof by Cook of the universality of rule 110 [7] is a typical construction involving a huge number of particles and collisions: once the gadgets and the simulation are described, the main part of the proof consists of proving that particles are well synchronized and that collisions occur exactly as described in the simulation.

When dealing with space-time diagrams consisting of only particles and collisions, a second object is often used: a planar map describing the collisions and their interactions. When identifying particles and collisions in space-time diagrams, in the style of Boccara et al. [3], one builds the planar map to give a compact description of the diagram, as depicted on Fig. 2(c). When describing algorithmic computation, in the style of Fischer [8], one describes a family of planar maps as a scheme of the produced space-time diagrams.

The aim of the present paper is to define particles and collisions, describe how collisions can be catenated, introduce collisions schemes as planar maps and discuss the construction of finite catenations from collisions schemes. All the necessary material is defined in section 2 followed by basic catenation of tilings in section 3. Collisions and their catenations are formally introduced in section 4. The main result on catenation is presented in section 5.

2 Definitions

In the remaining of this paper, every discussion occurs in the *two-dimensional plane* $\mathbb{Z}^2$ partially colored with the letters of a given finite alphabet Σ. A *pattern* is a subset of $\mathbb{Z}^2$. A *cell* c of a given pattern P is an element $c \in P$. A *vector* is an element of the group $(\mathbb{Z}^2, +)$ of translations in the plane. A *coloring* $\mathcal{C}$ is a partial map from $\mathbb{Z}^2$ to Σ. The *support* of a *coloring* $\mathcal{C}$ is denoted by $\text{Sup}(\mathcal{C})$, its restriction to a pattern P is denoted by $\mathcal{C}_{|P}$.

The *translation* $u \cdot \mathcal{C}$ of a coloring $\mathcal{C}$ by a vector u is the coloring with support $\text{Sup}(\mathcal{C}) + u$ such that, for all $z \in \text{Sup}(\mathcal{C})$, it holds $(u \cdot \mathcal{C})(z + u) = \mathcal{C}(z)$. The *disjoint union* $\mathcal{C} \oplus \mathcal{C}'$ of two colorings $\mathcal{C}$ and $\mathcal{C}'$ is the coloring with support $\text{Sup}(\mathcal{C}) \cup \text{Sup}(\mathcal{C}')$ such that, for all $z \in \text{Sup}(\mathcal{C})$, it holds $\mathcal{C} \oplus \mathcal{C}'(z) = \mathcal{C}(z)$ and for all $z \in \text{Sup}(\mathcal{C}')$, it holds $\mathcal{C} \oplus \mathcal{C}'(z) = \mathcal{C}'(z)$. Colorings and their operations are depicted on Fig. 3.

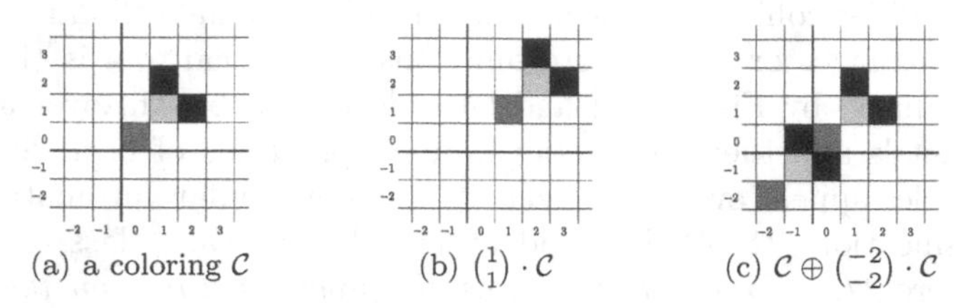

(a) a coloring $\mathcal{C}$ (b) $\binom{1}{1} \cdot \mathcal{C}$ (c) $\mathcal{C} \oplus \binom{-2}{-2} \cdot \mathcal{C}$

Fig. 3 Colorings, translations and disjoint unions

A *tiling constraint* is a pair (V, Υ) where V is a finite pattern and Υ is a subset of Σ^V. A coloring $\mathcal{C}$ *satisfies* a tiling constraint (V, Υ) if for each vector $u \in \mathbb{Z}^2$ such that V is a subset of $\text{Sup}(u \cdot \mathcal{C})$, it holds $(u \cdot \mathcal{C})_{|V} \in \Upsilon$. For now on we fix a tiling constraint (V, Υ). A *tiling* is a coloring with support $\mathbb{Z}^2$ that satisfies the tiling constraint. For any pattern P, the *neighborhood* along the constraint (V, Υ) is defined as $\partial P = P \cup \{p + v | p \in P \text{ and } v \in V\}$.

In the following, for geometrical considerations, we will implicitly use variations of discrete forms of the Jordan curve theorem [15]. Two points $\binom{x}{y}, \binom{x'}{y'} \in \mathbb{Z}^2$ are *4-connected* if $\binom{|x-x'|}{|y-y'|} \in \{\binom{1}{0}, \binom{0}{1}\}$, *8-connected* if $\binom{|x-x'|}{|y-y'|} \in \{\binom{1}{0}, \binom{0}{1}, \binom{1}{1}\}$. A pattern P is *4-connected, resp. 8-connected*, if for each pair of points $z, z' \in P$, there exists a 4-connected, *resp.* 8-connected, path of points of P from z to z'. The discrete Jordan curve theorem states that any non empty 4-connected closed path separates the plane into two 8-connected patterns, the interior and exterior of the path. More generally, a *frontier* is a 4-connected pattern separating the plane into n 8-connected patterns, its *borders*.

3 Catenation of tilings

Let (V, Υ) be a tiling constraint and C a set of colorings satisfying this constraint. To generate tilings by catenating colorings in C, the idea is to construct a patchwork of colorings by cutting portions of coloring and glue them together so that tiling constraints are preserved. A simple patchwork of 2 tilings is depicted on Fig. 4.

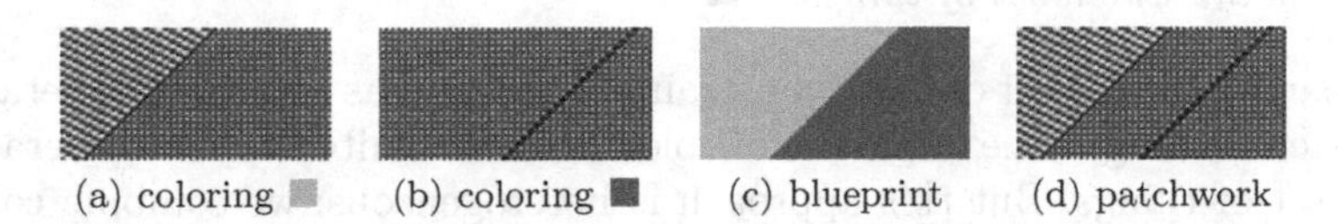

(a) coloring ■ (b) coloring ■ (c) blueprint (d) patchwork

Fig. 4 A patchwork

Definition 1. A *patchwork* is a tiling $\mathcal{T}_\phi$ defined for each $z \in \mathbb{Z}^2$ by $\mathcal{T}_\phi(z) = \phi(z)(z)$ where $\phi : \mathbb{Z}^2 \to C$ is the *blueprint* of the patchwork such that:

1. $\forall \mathcal{C} \in C, \quad \partial\phi^{-1}(\mathcal{C}) \subseteq \mathrm{Sup}(\mathcal{C})$;
2. $\forall z \in \mathbb{Z}^2, \forall v \in V, \quad \phi(z)(z+v) = \phi(z+v)(z+v)$.

Patchworks provide a convenient way to combinatorially generate tilings from a set of valid colorings without knowing explicitly the tiling constraint: it is sufficient to know a super-set of the tiling neighborhood V and to cut colorings on a big enough boundary containing the same letters.

Topology is a classical tool of symbolic dynamics [11], tilings being exactly the shifts of finite type for two-dimensional words. The set of colorings is endowed with the so called Cantor topology: the product of the discrete topology on $\Sigma \cup \{\perp\}$ where $\perp$ denotes undefined color. This topology is compatible with the following distance on colorings: $d(\mathcal{C}, \mathcal{C}') = 2^{-\min\{|z|, \mathcal{C}(z) \neq \mathcal{C}'(z)\}}$. Let $\mathcal{O}_\mathcal{C}$ be the set of colorings $\mathcal{C}'$ such that $\mathcal{C}'_{|\mathrm{Sup}(\mathcal{C})} = \mathcal{C}_{|\mathrm{Sup}(\mathcal{C})}$. The set of $\mathcal{O}_\mathcal{C}$ for colorings $\mathcal{C}$ with a finite support is a base of clopen sets for the given compact perfect topology.

Proposition 1. *The set of patchworks over C is a compact set. Furthermore, it contains the tilings of the closure of C.*

Proof. Let $\mathcal{T}_i$ be a sequence of patchworks over C converging to a limit tiling $\mathcal{T}$. Consider the blueprints ϕ_i of these patchworks. For each cell $z \in \mathbb{Z}^2$, let v_z be the element $(-z \cdot \mathcal{T})_{|V}$ of Υ. Let $\phi(z)$ be any $\phi_i(z)$ such that $(-z \cdot \phi_i(z))_{|V} = v_z$ – such a $\phi_i(z)$ always exists by definition of patchworks as $\mathcal{T}_i$ converges to $\mathcal{T}$. The map ϕ is a blueprint for $\mathcal{T}$.

Let $\mathcal{C}_i$ be a sequence of colorings in C converging to a limit tiling $\mathcal{T}$. For each $\mathcal{C}_i$, let P_i be the largest pattern, for inclusion, such that $\mathcal{C}_{i|P_i} = \mathcal{T}_{|P_i}$. As the sequence $\mathcal{C}_i$ converges to $\mathcal{T}$, the sequence P_i converges to $\mathbb{Z}^2$. Without loss of generality, consider that P_i is an increasing sequence of patterns. For each i let $\delta(i)$ be the smallest j such that $\partial P_i \subsetneq P_j$. Consider $P'_n = P_{\delta^n(1)}$, an increasing sub-sequence of P_i. Construct a blueprint ϕ as follows: for all $z \in \mathbb{Z}^2$, let $\phi(z) = P'_{\min\{n|z \in P'_n\}}$. By construction, this blueprint is valid and its patchwork is $\mathcal{T}$. ■

Corollary 1. *Let $\mathcal{O}_i$ be a base of open sets of colorings and C be a set of colorings containing at least one element of each $\mathcal{O}_i$. The set of patchworks over C is the whole set of tilings.* ■

In particular, the set of tiling constraints Υ, viewed as colorings, generates the whole set of tilings. The larger set of colorings with finite support generates the whole set of tilings. But this approach is heterogeneous: we combine colorings to obtain tilings. Can we restrict ourselves to combinations of tilings? More precisely, given a tiling constraint, can we recursively construct a recursive family of tilings T such that the set of patchworks over T is the whole family of tilings?

In the case of one-dimensional tilings, replacing $\mathbb{Z}^2$ by $\mathbb{Z}$, it is straightforward that the set of ultimately periodic tilings generates the whole set of tilings: the set of ultimately periodic tilings is a dense set – from any tiling $\mathcal{T}$ and any finite pattern P, one can construct an ultimately periodic tiling $\mathcal{T}'$ such that $\mathcal{T}_{|P} = \mathcal{T}'_{|P}$. In the case of two-dimensional tilings, due to the undecidability of the tiling problem [1, 19], there exists no such family. This result prohibits us to obtain a recursive set of tilings whose closure under catenation give us the whole set of tilings. Therefore, in the rest of the paper, we search for simplicity rather than being exhaustive.

4 Ultimately periodic tilings

Bi-periodic tilings are among the most regular ones and correspond to the idea of a *background* for cellular automata: a tiling $\mathfrak{B}$ with two non-co-linear periodicity vectors u and v such that $\mathfrak{B} = u \cdot \mathfrak{B} = v \cdot \mathfrak{B}$. As backgrounds are objects of dimension 2, if one wants to mix several backgrounds in a same tiling, the interface between two background is of dimension 1. The most regular kind of interface corresponds to the idea of a *particle*: a tiling $\mathfrak{P}$ with two non-co-linear vectors, the period u of the particle such that $\mathfrak{P} = u \cdot \mathfrak{P}$ and the period v of its backgrounds such that for all position $z \in \mathbb{Z}^2$, the extracted one-dimensional word $(\mathfrak{P}(z + vi))_{i \in \mathbb{Z}}$ is ultimately periodic. Of course, several particles might meet on the plane, leading to objects of dimension 0 that correspond to the idea of a *collision*. In this paper, an ultimately periodic tiling of the plane is such a collision.

Let $\sphericalangle_v(u, u')$ denote the angular portion of the plane, on the right hand side of u, starting in position $v \in \mathbb{Z}^2$ and delimited by the vectors $u, u' \in \mathbb{Z}^2$.

Formally, one might geometrically define a collision as follows (and depicted on Fig. 5):

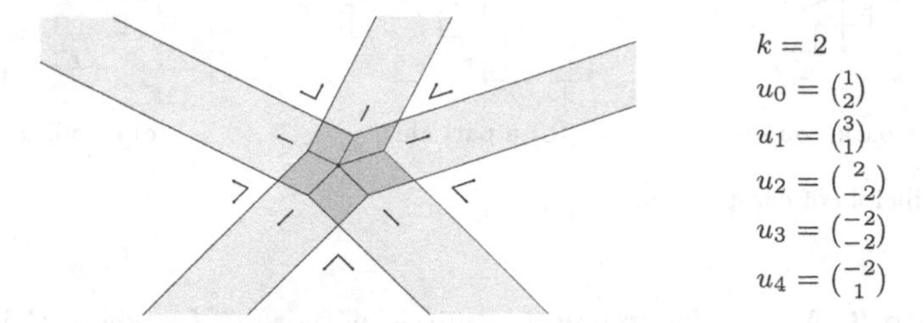

Fig. 5 Defining collisions through vectors

Definition 2. A *collision* is a tiling $\mathfrak{C}$ for which there exists an integer k and a finite cyclic sequence of n vectors $(u_i) \in \left(\mathbb{Z}^2\right)^{\mathbb{Z}_n}$ such that, for all $i \in \mathbb{Z}_n$, $\mathfrak{C}$ is u_i-periodic in z, *i.e.* $\mathfrak{C}(z) = \mathfrak{C}(z + u_i)$, for all positions z inside $\sphericalangle_{ku_i}(u_{i-1}, u_{i+1})$.

Although it corresponds to intuition, this definition made it difficult to effectively use collisions in constructions since it does not identify components of the collision. To overcome this problem, we introduce *constructive* versions of collisions. Ideas behind such definitions is that all elements can be represented with a finite description. A background is entirely determined by two non-collinear vectors of periodicity u and v and by a coloring of finite support $\mathcal{C}$ that tiles the plane along u and v (i.e.$\bigoplus_{i,j \in \mathbb{Z}^2} (iu + jv) \cdot \mathcal{C}$ is a tiling) (see Fig. 6). Such a triple $(\mathcal{C}, u, v)$ is called *background representation.*

The same way, in a particle, the uni-periodic part can be characterised by a vector u and a coloring with finite support $\mathcal{C}$ which repeats along u ($\mathcal{I} = \bigoplus_{k \in \mathbb{Z}} ku \cdot \mathcal{C}$) is a frontier with two borders (L and R). The rest of particle can be described using two backgrounds $\mathfrak{B}$ and $\mathfrak{B}'$. The resulting coloring $\mathfrak{P} = \mathfrak{B}_{|L} \oplus \mathcal{I} \oplus \mathfrak{B}'_{|R}$ is require to be a tiling. Furthermore, we require to have a condition ensuring that the different portion have some common "safety zone". This is done by adding the constraint that the function: $\phi : z \to \begin{cases} \mathfrak{P} & \text{if } z \in \operatorname{Sup}(I) \\ \mathfrak{B} & \text{if } z \in L \\ \mathfrak{B}' & \text{if } z \in R \end{cases}$ is the blueprint of a patchwork. Such a tuple $(\mathfrak{B}, \mathcal{C}, u, \mathfrak{B}')$ is called *particle representation.*

For collisions, the idea is basically the same (see Fig. 6) , the characterisation is based on a coloring with finite support $\mathcal{C}$ for the non-periodic part and a finite list of particles. Each particle defines a half-line starting form the center of the collision. The support of all the particles and the center must form a star and each consecutive pair of particles must have a common background to fill the space between them. Some safety zone is also required as in particle. This is formalised in the following definition:

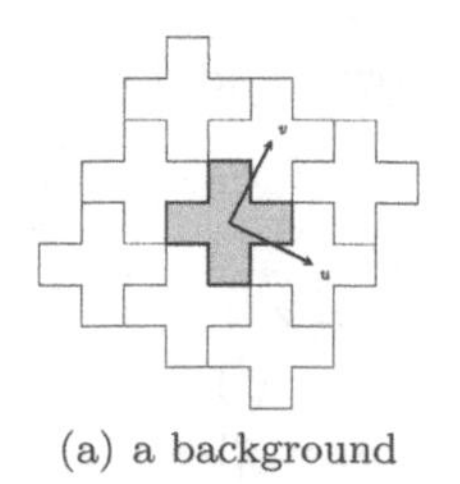

(a) a background

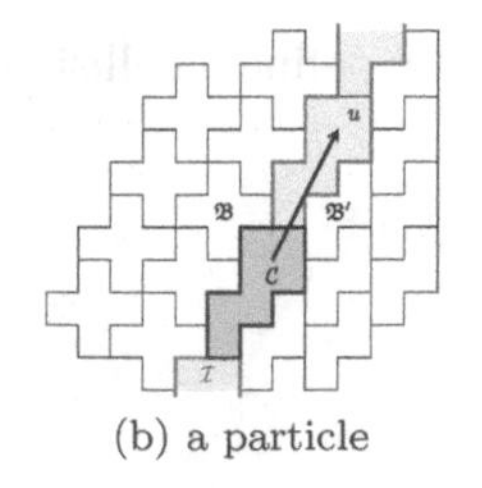

(b) a particle

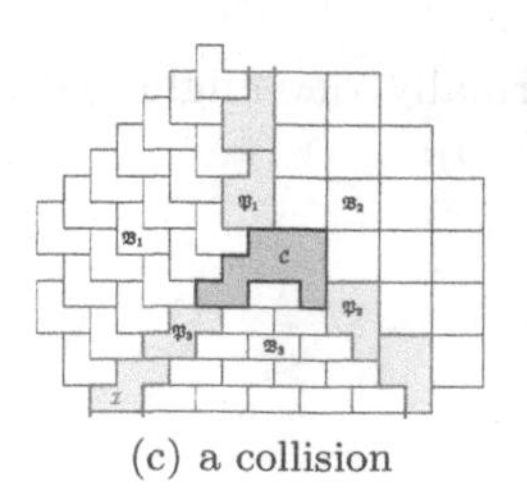
(c) a collision

Fig. 6 Principle of construction

Definition 3. A *collision representation* is a pair $(\mathcal{C}, L)$ where $\mathcal{C}$ is a finite pattern, L is a finite sequence of n particles $\mathfrak{P}_i = (\mathfrak{B}_i, \mathcal{C}_i, u_i, \mathfrak{B}'_i)$, satisfying:

1. $\forall i \in \mathbb{Z}_n, \quad \mathfrak{B}'_i = \mathfrak{B}_{i+1}$;
2. the support of $\mathcal{I} = \mathcal{C} \oplus \bigoplus_{i \in \mathbb{Z}_n, k \in \mathbb{N}} ku_i \cdot \mathcal{C}_i$ is a frontier with n borders;
3. For all $i \in \mathbb{Z}_n$, the support of $\mathcal{C} \oplus \bigoplus_{k \in \mathbb{N}} (ku_i \cdot \mathcal{C}_i \oplus ku_{i+1} \cdot \mathcal{C}_{i+1})$ is a frontier with two borders: let P_i be the border on the right of $\mathfrak{P}_i$;
4. $\mathfrak{C} = \mathcal{I} \oplus \bigoplus_i \mathfrak{B}_{i|P_i}$ is a tiling;
5. the function $\phi : z \to \begin{cases} \mathfrak{C} & \text{if } z \in \mathrm{Sup}(\mathcal{C}) \\ \mathfrak{P}_i & \text{if } z \in \mathrm{Sup}(\bigoplus_{k \in \mathbb{N}} ku_i \cdot \mathcal{C}_i) \\ \mathfrak{B}_i & \text{if } z \in P_i \end{cases}$ is the blueprint of a patchwork.

The set $\mathrm{Sup}(\mathcal{C})$ is called *perturbation* of the collision and $\mathrm{Sup}(\bigoplus_{k \in \mathbb{N}} ku_i \cdot \mathcal{C}_i)$ are called *perturbation* of the particle P_i.

The constructive definitions of particles, backgrounds and collisions provide us with a finite representation that allows us to recursively manipulate them. Contrary to intuition, representations are not invariant by translation. This seems unavoidable since we want to have means of expressing the relative position between two such representations. In the rest of the paper, we will always assume that background, particles and collisions are given by a representation.

5 Finite catenations

A blueprint of finitely many collisions might produce a tiling which is not a collision, however if the blueprint of the patchwork consists of finitely many 8-connected components, the patchwork is a collision. Using representations of collisions, a more regular family of patchworks can be defined: a catenation induces a patchwork combining collisions by binding pairs of similar particles as depicted on Fig. 2.

To "bind" collisions using particles, we need two identical particles facing each other such that the gap between them correspond to a integer number of particles n. Two particles $\mathfrak{P} = (\mathfrak{B}, \mathcal{C}, u, \mathfrak{B}')$ and $\tilde{\mathfrak{P}} = (\tilde{\mathfrak{B}}, \tilde{\mathcal{C}}, \tilde{u}, \tilde{\mathfrak{B}}')$ form a *n-binding* if $\tilde{u} = -u$ (particles are facing each other), $\tilde{\mathcal{C}} = (n-1)u \cdot \mathcal{C}$ (they have the same finite pattern and gap is n repetitions) , $\tilde{\mathfrak{B}} = (n-1)u \cdot \mathfrak{B}'$, $\tilde{\mathfrak{B}}' = (n-1)u \cdot \mathfrak{B}$ (backgrounds are the same). The set $\bigoplus_{0<i<n-1} iu \cdot \mathcal{C}$ is called the *perturbation* of the n-binding.

Since we want to get rid of positions, we introduce the *potential n-binding.* The idea is that given two collisions and one particle for each collision, the particles $\mathfrak{P}_1$ and $\mathfrak{P}_2$ form a *potential n-binding* if up to a translation z, the two particles form an n-binding (i.e. $\mathfrak{P}_1$ and $z \cdot \mathfrak{P}_2$ form a n-binding). One can remark in case of potential n-binding, the translation vector z is unique.

Now the idea is that we can use potential n-binding to construct patchworks since background is bi-periodic and does not cause heavy harm for checking properties on it. The description needs to have collisions as points and particles as lines. Particles can be half-infinite (if they are not part of potential n-binding) or link two collisions. Since we work in the plane, it is sound to require that the constructed element is planar and that the order of particles is compatible with the collisions. At last, we add a connected condition to avoid problem with free parts of the map. This leads to the following definition:

Definition 4. A *catenation* is a connected planar map where:

- vertices are labeled by collisions;
- edges are potentially semi infinite;
- edges extremities are labeled by particles;
- edges order in a vertex is compatible with the order on particles in the corresponding collision.
- finite edges (of extremities $\mathfrak{P}_1$ and $\mathfrak{P}_2$) are labeled with an integer n such that $\mathfrak{P}_1$ and $\mathfrak{P}_2$ form a potential n-binding.

At this point, we want to transform the catenation into a patchwork. For this, let us first study some necessary conditions. Since we deal with a planar map, it is possible to define faces as elements of the dual of catenation. To transform a catenation into a patchwork, it is necessary that every face can be transformed into a patchwork. Since we have potential n_i-bindings, the translation induced between two consecutive collisions is fixed. Since the sequence of collisions in a face is cyclic, it is sound to require that the sequence of corresponding translation sum up to zero when cycling. This will be the first condition. Now, with this condition, it is possible to assign (up to a global constant) a translation to every collision such that all edges are n_i-bindings. With those objects, the basic idea is to construct a patchwork that corresponds to each collision, particle or n_i-binding on its perturbation. This implies that all perturbations does not enforce contradictions. One easy way to get rid of this risk is just to require that all perturbations are distinct (this will be our second condition). If these conditions are met then we speak of *valid* catenation.

Proposition 2. *It is possible to associate a patchwork (and therefore a space-time diagram) to every valid catenation.*

Proof. To prove this result, we shall give a potential blueprint and show that it satisfies the conditions. First of all, the condition on null translation after a round on every faces induce a unique set of translation (up to a constant) for every collision in the map since the map is connected. At this point, let us consider the collisions with those translations.

The second condition ensure that perturbations of collisions, bindings and particles are disjoint. Thus it is possible to define a blueprint linking any point of such a perturbation to the corresponding collision, particle or binding. Let us now study the points that are not mapped. Since the map is planar and particle (and also bindings) are isolating, every left point belongs to one unique face. On this face, the associated background with particles or collision or bindings present is unique (bindings ensure that two consecutive collisions are the same and collisions ensure this for consecutive particles and bindings). So we map those points to the corresponding background.

The last point is to show that the constructed blueprint does really satisfy the properties for patchwork. The first condition on definition is trivial since the used valid coloration are tilings. Let us go now to the second and main point.

For this last part, let us study the different cases. For example, if we are in a collision $\mathfrak{C}$ perturbation. If the neighborhood is also in $\mathfrak{C}$ perturbation or in perturbation of binding, particle belonging to $\mathfrak{C}$ or even of background with this property, then the neighborhood is by definition equal to the original one of a collision. the only difficult case is when in the neighborhood, there is a perturbation originated from another element. For example let us suppose this elements is in the perturbation from $\mathfrak{C}'$. In this case, in $\mathfrak{C}$ we have in these points some backgrounds or particles. But since perturbations do not overlap, we are in the border of $\mathfrak{C}'$. As we have requested in our constructive version representation to be patchworks, the border of $\mathfrak{C}'$ does correspond to the value of backgrounds or particles present in $\mathfrak{C}'$. By definition of our catenations, the backgrounds and particles are the same so elements of $\mathfrak{C}'$ are the same of those in $\mathfrak{C}$.
The same arguments do also apply for other cases thus ending the proof. ∎

At this point, we have both a set of "simple" tilings (the collisions) and an operation generating new tilings from this set (the valid catenation). Despite being intuitive, catenations require to give explicitly the relative positions of collision via the number of repetitions of particles. Intuitively, we would like to give only the collisions involved and their organisation (as in Fig. 2c). With this approach, it is possible to define an alternative to catenation that does not require the number of repetitions to be given. The resulting element is called catenation scheme. Formally, a *catenation scheme* is a catenation whose label on finite edge where erased. Conversely, to go back from a catenation scheme to a catenation, one need to give every finite edge a label. Such elements of $\mathbb{N}^F$ where F is the set finite edges of the catenation scheme is called *affectation*. Moreover, it is called *valid* affectation if the resulting catenation is valid.

For a given catenation scheme, one natural question is whether it correspond to a tiling. To bring an answer one idea is to search for valid affectation of the scheme. In case of finite catenation scheme, we can achieve a very strong characterisation of this set and even compute it.

Theorem 1. *The set of valid affectation of a finite catenation is a recursive semi-linear set (i.e. a finite union of linear sets).*

Proof. To prove the main theorem, we will show that being a valid affectation of finite catenation scheme can be expressed with a formula in Presburger arithmetic (i.e first order logic on integer with addition and comparison). Since the set of solutions of formula in such arithmetic is a recursive semi-linear set [10] this will conclude the proof. One can note that the construction of the solution is explicit even if the complexity is non-elementary.

In our formula, the number of repetitions of each finite edge will correspond to free variables. let us call them $r_1, \ldots, r_n$. Since the conditions for valid catenation are for each face, the global formula F will consists on the conjunction of an elementary formula for each face: $F = \wedge_{f\text{face}} F_f$. For each face, let us look at the two conditions. First one (going back to the same point after a turn around the face) can be easily expressed: the translation induced by a particle i is just r_i times the vector of repetition of the particle u_i (just note that the direction of the particles is chosen in the face) which is a known constant. For the translation induced by collision ϵ_c they are know constant. So the formula is on the form $F_{f,1} = \Sigma_{i\text{particles in the face}} u_i r_i + \Sigma_{c\text{collisions}} \epsilon_c = \binom{0}{0}$. For the second condition (non overlap of perturbation) it can be expressed with the conjunction that any pair of points of different perturbations are distinct. In the case of collision perturbation, it is trivial since there is only a finite (and known) number of perturbation points. For bindings, it is more difficult since the set of points can be expressed with a universal quantifier with the following remark, the set of points in the binding's perturbation correspond to the set of points of the particle perturbation $\mathrm{Sup}(\mathcal{C}_i)$ (a finite number) for every integer n multiple of the vector of repetition u_i which is between 0 and the number of repetition r_i. thus the formula is on the form: $\forall x, 0 < x < r_i \Rightarrow \wedge_{p \in \mathrm{Sup}(\mathcal{C}_i)} p + u_i r_i \neq z$ where z are points for the other considered perturbation. The same applies for free particles (just omit the upper bound in the comparison).

With this, we have show how to construct the Presburger formula which conclude the proof. ■

With this theorem we achieve a very strong framework for cellular automata. After have extracted a set of collisions, one can give the desired finite catenation scheme and automatically check the necessary and sufficient conditions for that scheme to exists. This method would make proves far more understandable and could avoid the need to rely on combinatorial proves to ensure validity of intuition. For now, the main limitation of those results are that only the field of finite catenations are treated. One main goal of future work is to achieve such kind of result for infinite catenation schemes. Due to the infinite nature of such elements, such strong a characterisation is excluded but we hope to have sufficient computable conditions for affectation of a wide range of "regular" infinite catenations.

References

1. R. Berger. The undecidability of the domino problem. *Memoirs American Mathematical Society*, 66, 1966.
2. E. R. Berlekamp, J. H. Conway, and R. K. Guy. *Winning ways for your mathematical plays. Vol. 2.* Academic Press Inc. [Harcourt Brace Jovanovich Publishers], London, 1982. Games in particular.
3. N. Boccara, J. Nasser, and M. Roger. Particlelike structures and their interactions in spatiotemporal patterns generated by one-dimensional deterministic cellular-automaton rules. *Phys. Rev. A*, 44(2):866–875, 1991.
4. J. R. Büchi. On a decision method in restricted second order arithmetic. In *Proceedings of the International Congress on Logic, Methodology, and Philosophy of Science, Berkley, 1960*, pages 1–11. Standford University Press, 1962.
5. N. Chomsky. Three models for the description of language. *Information Theory, IEEE Transactions on*, 2(3):113–124, 1956.
6. E. F. Codd. *Cellular Automata.* Academic Press, New York, 1968.
7. M. Cook. Universality in elementary cellular automata. *Complex Systems*, 15:1–40, 2004.
8. P. C. Fischer. Generation of primes by a one-dimensional real-time iterative array. *Journal of the ACM*, 12(3):388–394, 1965.
9. D. Giammarresi and A. Restivo. Two-dimensional languages. In A. Salomaa and G. Rozenberg, editors, *Handbook of Formal Languages*, volume 3, Beyond Words, pages 215–267. Springer-Verlag, Berlin, 1997.
10. S. Ginsburg and E. H. Spanier. Semigroups, presburger formulas, and languages. *Pacific Journal of Mathematics*, 16:285–296, 1966.
11. G. A. Hedlund. Endormorphisms and automorphisms of the shift dynamical system. *Mathematical Systems Theory*, 3:320–375, 1969.
12. W. Hordijk, C. R. Shalizi, and J. P. Crutchfield. Upper bound on the products of particle interactions in cellular automata. *Phys. D*, 154(3-4):240–258, 2001.
13. J. Kari. Theory of cellular automata: a survey. *Theoretical Computer Science*, 334:3–33, 2005.
14. S. C. Kleene. Representation of events in nerve nets and finite automata. In C. Shannon and J. McCarthy, editors, *Automata Studies*, pages 3–41. Princeton University Press, 1956.
15. T. Y. Kong and A. Rosenfeld. Digital topology: introduction and survey. *Comput. Vision Graph. Image Process.*, 48(3):357–393, 1989.
16. J. Mazoyer. Computations on cellular automata: some examples. In *Cellular automata (Saissac, 1996)*, pages 77–118. Kluwer Acad. Publ., Dordrecht, 1999.
17. J. Mazoyer and V. Terrier. Signals in one-dimensional cellular automata. *Theoretical Computer Science*, 217(1):53–80, 1999. Cellular automata (Milan, 1996).
18. M. Minsky. *Computation: Finite and Infinite Machines.* Prentice Hall, Englewoods Cliffs, 1967.
19. R. M. Robinson. Undecidability and nonperiodicity for tilings of the plane. *Inventiones Mathematicae*, 12:177–209, 1971.
20. J. von Neumann. *Theory of Self-Reproducing Automata.* University of Illinois Press, Urbana, Ill., 1966.
21. S. Wolfram. Universality and complexity in cellular automata. *Physica D. Nonlinear Phenomena*, 10(1-2):1–35, 1984. Cellular automata (Los Alamos, N.M., 1983).

Cache-sensitive Memory Layout for Binary Trees*

Riku Saikkonen and Eljas Soisalon-Soininen

Helsinki University of Technology, Finland, {rjs,ess}@cs.hut.fi

Abstract We improve the performance of main-memory binary search trees (including AVL and red-black trees) by applying cache-sensitive and cache-oblivious memory layouts. We relocate tree nodes in memory according to a multi-level cache hierarchy, also considering the conflict misses produced by set-associative caches. Moreover, we present a method to improve one-level cache-sensitivity without increasing the time complexity of rebalancing. The empirical performance of our cache-sensitive binary trees is comparable to cache-sensitive B-trees. We also use the multi-level layout to improve the performance of cache-sensitive B-trees.

1 Introduction

Most of today's processor architectures use a hierarchical memory system: a number of caches are placed between the processor and the main memory. Caching has become an increasingly important factor in the practical performance of main-memory data structures. The relative importance of caching will likely increase in the future [1, 2]: processor speeds have increased faster than memory speeds, and many applications that previously needed to read data from disk can now fit all of the necessary data in main memory. In data-intensive main memory applications, reading from main memory is often a bottleneck similar to disk I/O for external-memory algorithms.

There are two types of cache-conscious algorithms. We will focus on the *cache-sensitive* (or cache-aware) model, where the parameters of the caches are assumed to be known to the implementation. In contrast, *cache-oblivious* algorithms attempt to optimize themselves to an unknown memory hierarchy.

The simplest cache-sensitive variant of the B-tree is an ordinary B^+-tree where the node size is chosen to match the size of a cache block (e.g., 64 or 128 bytes) [3]. A more advanced version called the Cache-Sensitive B^+-tree or CSB^+-tree [1] additionally removes pointers from internal nodes by storing the children of a node consecutively in memory. The CSB^+-tree has been further optimized using a variety of techniques, such as prefetching [4], storing only partial keys in nodes [5], and choosing the node size more carefully [2]. The above structures used a one-level cache model; B-trees in two-level cache models (one level of cache plus the TLB) are examined in [6, 7].

* This research was partially supported by the Academy of Finland.

Please use the following format when citing this chapter:

Saikkonen, R. and Soisalon-Soininen, E., 2008, in IFIP International Federation for Information Processing, Volume 273; *Fifth IFIP International Conference on Theoretical Computer Science*; Giorgio Ausiello, Juhani Karhumäki, Giancarlo Mauri, Luke Ong; (Boston: Springer), pp. 241–255.

A weight-balanced B-tree based on the cache-oblivious model has been proposed in [8]. Its simpler variants [9, 10] use an implicit binary tree (a complete binary tree stored in a large array without explicit pointers) whose structure and rebalancing operations are dictated by the cache-oblivious memory layout. In all three, update operations may rebuild parts of the tree, so most of the complexity bounds are amortized.

When using binary search trees, the node size cannot be chosen as freely as in B-trees. Instead, we will place the nodes in memory so that each cache block contains nodes that are close to each other in the tree. Binary search tree nodes are relatively small; for example, AVL and red-black tree nodes can fit in about 16 or 20 bytes using 4-byte keys and 4-byte pointers, so 3–8 nodes fit in one 64-byte or 128-byte cache block. (We assume that the nodes contain only small keys. Larger keys could be stored externally with the node storing a pointer to the key.)

Caching and explicit-pointer binary search trees have been previously considered in [11], which presents a cache-oblivious splay tree based on periodically rearranging all nodes in memory. In addition, [12] presents a one-level cache-sensitive periodic rearrangement algorithm for explicit-pointer binary trees. A similar one-level layout (extended to unbalanced trees) is analyzed in [13], which also discusses the multi-level cache-oblivious layout known as the van Emde Boas layout. The latter is analyzed in detail in [14].

We give an algorithm that preserves cache-sensitivity in binary trees in the dynamic case, i.e., during insertions and deletions. Our algorithm retains single-level cache-sensitivity using small worst-case constant-time operations executed when the tree changes. In addition, we give an explicit algorithm for multi-level cache-sensitive global rearrangement, including a variation that obtains a cache-oblivious layout. We also investigate a form of conflict miss caused by cache-sensitive memory layouts that interact poorly with set-associative caches.

Our approach does not change the internal structure of the nodes nor the rebalancing strategy of the binary search tree. The approach is easy to implement on top of an existing implementation of any tree that uses rotations for balancing, e.g., red-black trees and AVL trees. Our global rearrangement algorithm can also be applied to cache-sensitive B-trees, and our empirical results indicate that the multi-level memory layout improves the performance of both B^+-trees with cache-block-sized nodes and CSB^+-trees.

2 Cache model

We define a multi-level cache model as follows. We have a k-level cache hierarchy with block sizes $B_1, \ldots, B_k$ at each level. We also define B_0 = node size in bytes, $B_{k+1} = \infty$. We assume that our algorithms know these cache parameters. (In practice, they can be easily inferred from the CPU model or from metadata stored in the CPU.) To keep the model simple, we do not model any other features of the cache, such as the capacity.

Our algorithms shorten the B_i*-block search path length*, denoted P_i and defined as the length of a root-to-leaf path measured in the number of separate cache blocks of size B_i encountered on the path. Using this terminology, P_0 is the traditional search path length in nodes (assuming that the search does not end before the leaf level), P_1 is the length counted in separate B_1-sized cache blocks encountered on the path, and so on.

We assume that for $i > 1$, each block size B_i is an integer multiple of B_{i-1}. Additionally, if B_1 is not an integer multiple of the node size B_0, a node should not cross a B_1-block boundary (so that it is never necessary to fetch two cache blocks from memory in order to access a single node). In practice, this is achieved by not using the last $B_1 \mathbf{~mod~} B_0$ bytes of each B_1-block. (In practice, B_i, $i > 0$, is almost always a power of 2.)

A typical modern computer employs two levels of caches: a relatively small and fast level 1 ("L1") cache, and a larger and slower level 2 ("L2") cache. In addition, the mapping of virtual addresses to physical addresses used by multitasking operating systems employs a third hardware cache: the Translation Lookaside Buffer or TLB cache.

Currently the cache block size is often the same in the L1 and L2 caches. They then use only one level of our hierarchy. For example, the cache model used in the experiments in Section 5 is $k = 2$, $B_0 = 16$ (16-byte nodes), $B_1 = 64$ (the block size of the L1 and L2 caches in an AMD Athlon XP processor), $B_2 = 4096$ (the page size of the TLB cache), $B_3 = \infty$. However, our algorithms can be applied to an arbitrary hierarchy of cache block sizes.

3 Global relocation

Figure 1 gives an algorithm that rearranges the nodes of a tree in memory into a multi-level cache-sensitive memory layout. The algorithm can be used for any kind of balanced tree with fixed-size nodes.

The produced layout can be considered to be a generalization of the one-level cache-sensitive layouts of [12, 13] and the two-level layouts of [6, 7] to an arbitrary hierarchy of block sizes. It is different from the multi-level "van Emde Boas" layout (see [13]) in that the recursive placement of smaller blocks in larger ones is more complex, because, in the cache-sensitive model, we cannot choose the block sizes according to the structure of the tree, as is done in the cache-oblivious van Emde Boas layout.

In the produced layout, the first lowest-level ($l = 1$) block is filled by a breadth-first traversal of the tree starting from the root r. When this "root block" is full, each of its children (i.e., the "grey" or border nodes in the breadth-first search) will become the root node of its own level 1 block, and so on. On levels $l > 1$, level $l - 1$ blocks are allocated to level l blocks in the same manner.

```
RELOC-BLOCK(l, r):
 1: if l = 0 then
 2:    Copy node r to address A, and update the link in its parent.
 3:    A ← A + B_0
 4:    return children of r
 5: else
 6:    S ← A
 7:    E ← A + F(A, l) − B_{l−1}
 8:    Q ← empty queue
 9:    put(Q, r)
10:    while Q is not empty and A ≤ E do
11:       n ← get(Q)
12:       c ← RELOC-BLOCK(l − 1, n)
13:       put(Q, all nodes in c)
14:    end while
15:    if Q is not empty then
16:       A ← start of next level l block (= E + B_{l−1})
17:       if F(S, l) < B_l/2 then       {less than half of the block was free}
18:          Free the copies made above, i.e., all nodes at addresses S to A − 1.
19:          return r      {our caller will try to relocate r again later}
20:       end if
21:    end if
22:    return remaining nodes in Q
23: end if

RELOCATE(r):
 1: A ← beginning of a new memory area, aligned at a level k block boundary
 2: RELOC-BLOCK(k + 1, r)       {B_{k+1} = ∞, so this relocates everything}
```

Fig. 1 The global relocation algorithm. The address A of the next available position for a node is a global variable. $F(A,l) = B_l - A \textbf{ mod } B_l$ is the number of bytes between A and the end of the level l block containing A. (To be able to update the link in a parent when a node is copied, the algorithm actually needs to store (node, parent) pairs in the queue Q, unless the tree structure contains parent links. This was left out of the pseudocode for clarity.)

The algorithm of Figure 1 produces this layout using a single traversal of the tree using auxiliary queues that store border nodes for each level of the breadth-first search. Lines 17–20 are an optional space optimization: at the leaf level, there may not be enough nodes to fill a block. Lines 17–20 ensure that each level l block will be at least half full by trying to allocate the next available subtree in the remaining space in a non-full block.

Theorem 1. *Assume that the global relocation algorithm of Figure 1 is executed on a complete binary tree of height h. Then the worst-case B_i-block path length will be $P_i = \lceil h/h_i \rceil$, where $h_i = h_{i-1} \cdot \lfloor \log_{d_{i-1}}(B_i/B_{i-1} + 1) \rfloor$, $h_0 = 1$. If B_1 is an integer multiple of B_0, then $d_i = B_i/B_0 + 1$; otherwise, $d_0 = 2$ and $d_i = (d_{i-1} - 1) \cdot \lfloor B_i/B_{i-1} \rfloor + 1$.*

Proof. Consider a cache block level $i \in \{1, \ldots, k\}$. Each level $i-1$ block produced by the layout (except possibly for blocks that contain leaves of the tree) contains a connected part of the tree with $d_{i-1} - 1$ binary tree nodes. These blocks can be thought of as "super-nodes" with fanout d_{i-1}. The algorithm of Figure 1 produces a level i block by allocating B_i/B_{i-1} of these super-nodes in breadth-first order (i.e., highest level $i-1$ block first). The shortest root-to-leaf path of the produced level i block has h_i binary tree nodes. □

The produced layout is optimal on the level of B_1-blocks: it is not possible to produce a larger h_1. It is not possible to be optimal on all levels [14], and we resolved this tradeoff by preferring the lowest level. Knowledge of the relative costs of cache misses at each level could in theory be used to produce a more optimal layout, but we did not want our cache-sensitive algorithms to depend on these kinds of additional parameters.

Theorem 2. *The algorithm of Figure 1 rearranges the nodes of a tree into a multi-level cache-sensitive memory layout in time $O(nk)$, where n is the number of nodes in the tree and k is the number of memory-block levels.*

Proof. Each node in the tree is normally copied to a new location only once. However, the memory-usage optimization in line 18 may "undo" (free) some of these copies. The undo only happens when filling a level l cache block that was more than half full, and the layout is then restarted from an empty level l cache block. Thus, an undo concerning the same nodes cannot happen again on the same level l. However, these nodes may already have taken part in an undo on a smaller level $l' < l$. In the worst case, a node may have taken part in an undo on all k memory-block levels. Each of the n nodes can then be copied at most k times.

Consider then the queues Q at various levels of recursion. Each node enters a queue at level $l = 1$ (line 13, using c from line 4), and travels up to a level $l' \leq k+1$, where it becomes the root of a level $l'-1$ subtree and descends to level 0 in the recursion. Thus, each node is stored in $O(k)$ queues. □

Cache-oblivious layout. Though cache-sensitive, the produced layout is similar to the "van Emde Boas" layout used as the basis of many cache-oblivious algorithms. In fact, our algorithm can produce the van Emde Boas layout: simply use the block sizes $B_i = (2^{2^i} - 1) \cdot B_0$ ($i = 1, \ldots, k$ where $k = 4$ or $k = 5$ is enough for trees that fit in main memory). The only difference between the layout thus produced and the van Emde Boas layout (as described in, e.g., [13]) is that the recursive subdivision is done top-down instead of bottom-up, and some leaf-level blocks may not be full. (These differences are unavoidable because the van Emde Boas layout is defined only for complete trees.)

Aliasing correction. While experimenting with the global relocation algorithm, we found that multi-level cache-sensitive layouts can suffer from a problem called aliasing, a kind of repeated conflict miss. Many hardware caches are d-way set associative ($d \in \{2, 4, 8\}$ are common), i.e., there are only d possible places in the cache for a block with a given address A. The problem is that, for instance, the ith cache block in each TLB page is often mapped to the same set of d places. Therefore, if the ith cache blocks of several TLB pages are accessed, the cache can only hold d of these blocks.

A straightforward multi-level cache-sensitive layout (including the one produced by the above algorithm) fills a TLB page (of size B_l for some l) with a subtree so that the root of the subtree is placed at the beginning of the TLB page (i.e., in the first B_{l-1}-sized cache block). Then, for example, when a particular root-to-leaf path is traversed in a search, only d root nodes of these TLB-sized subtrees can be kept in the (set associative) B_{l-1}-block cache. (The root of the TLB-sized subtree is not of course the only problematic node, but the problem is most pronounced at the root.)

The problem can be fixed by noting that we can freely reorder the cache blocks inside a TLB page. The B_l-sized TLB page consists of B_l/B_{l-1} cache blocks, and the subtree located in the TLB page can use these cache blocks in any order. We simply use a different ordering for separate TLB pages, so the root node of the subtree will not always be located in the first cache block.

We implement the reordering by a simple cache-sensitive translation of the addresses of each node allocated by the global relocation algorithm, as follows.[2] Every address A can be partitioned into components according to the cache block hierarchy: $A = A_k \ldots A_2 A_1 A_0$, where each A_i, $i \in \{1, \ldots, k-1\}$, has $\log_2 B_i/B_{i-1}$ bits of A, and A_0 and A_k have the rest. For each level $i = \{1, \ldots, k\}$, we simply add the upper portion $A_k \ldots A_{i+1}$ to A_i, modulo B_i/B_{i-1} (so that only the A_i part is changed).

For example, if B_l is the size of the TLB page, the root of the first allocated TLB page ($A_k \ldots A_{l+1} = 0$) will be on the first cache block (the translated portion $A'_l = 0$), but the root of the second TLB page (which is a child of the first page) will be on the second cache block ($A_k \ldots A_{l+1} = 1$, so $A'_l = 1$) of its page.

It would be enough to apply this correction to those memory-block levels with set associative caches on the previous level (i.e., only level l in the above example, since level $l-1$ has the set associative cache). However, we do it on all levels, because then our cache-sensitive algorithms only need knowledge of the block sizes and not any other parameters of the cache hierarchy. Applying the translation on every level increases the time complexity of the relocation algorithm to $O(nk^2)$, but this is not a problem in practice, since k is very small (e.g., $k = 2$ was discussed above).

4 Local relocation

When updates (insertions and deletions) are performed on a tree which has been relocated using the global algorithm of the previous section, each update may disrupt the cache-sensitive memory layout at the nodes that are modified in the update. In this section, we present modifications to the insert and delete algo-

[2] The translation is applied to every address used in lines 2 and 18 of the algorithm of Figure 1. The other addresses S, A and E in the algorithm do not need to be translated, because they are only used to detect block boundaries.

rithms that try to preserve a good memory layout without increasing the time complexity of insertion and deletion in a binary search tree that uses rotations for balancing. These algorithms can be used either together with the global relocation algorithm of the previous section (which could be run periodically) or completely independently.

Our approach preserves the following memory-layout property:

Invariant 1 *For all non-leaf nodes x, either the parent or one of the children of x is located on the same B_1-sized cache block as x.*

This property reduces the average B_1-block path length even in a worst-case memory layout. For simplicity, the proof only considers a complete binary tree of height h. (To see that Invariant 1 improves the memory layout of, e.g., a red-black tree, note that the top part of a red-black tree of height h is a complete tree of height at least $h/2$.)

Theorem 3. *Assume that Invariant 1 holds in a complete binary tree of height h. Then the average B_1-block path length $\overline{P_1} \leq 2h/3 + 1/3$.*

Proof. In the worst-case memory layout, each B_1-sized cache block contains only nodes prescribed by Invariant 1, i.e., a single leaf or a parent and child.

By Invariant 1, the root r of the tree (with height h) is on the same cache block as one of its children. Considering all possible paths down from r leads to the following recurrence for the expected value of the B_1-block path length: $P(h) = 1/2 \cdot (1 + P(h-2)) + 1/2 \cdot (1 + P(h-1))$ (with $P(1) = 1$ and $P(0) = 0$). Solving gives $E[P_1 | \text{worst-case memory layout}] = P(h) = 2h/3 + 2/9 - 2(-1)^h/(9 \cdot 2^h) \leq 2h/3 + 1/3$. In any memory layout, the average $\overline{P_1} \leq E[P_i | \text{worst-case memory layout}]$. □

We say that a node x is *broken* if Invariant 1 does not hold for x. To analyze how this can happen, denote $N(x) =$ the set of "neighbors" of node x, i.e., the parent of x and both of its children (if they exist). Furthermore, say that x *depends* on y if y is the only neighbor of x that keeps x non-broken (i.e., the only neighbor on the same cache block).

Our local relocation approach works as follows. We do the standard binary search tree structure modification operations (insertion, deletion, rotations) as usual, but after each such operation, we collect a list of nodes that can potentially be broken (Figure 2), and use the algorithm given below to re-establish Invariant 1 before executing the next operation.

The nodes that can break are exactly those whose parent or either child changes in the structure modification, since a node will break if it depended on a node that was moved away or deleted. As seen from Figure 2, 1 to 6 nodes can be broken by one structure modification. We explain the various cases in Figure 2 below.

In internal trees, actual insertion is performed by adding a new leaf node to an empty location in the tree. If the parent of the new node was previously a leaf, it may now be broken; thus, the parent is marked as potentially broken in Figure 2(c).

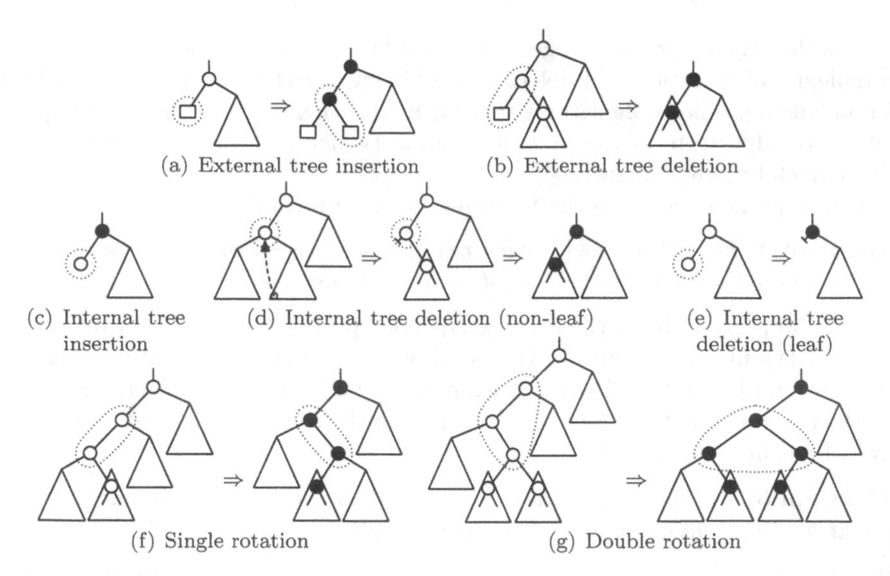

(a) External tree insertion
(b) External tree deletion
(c) Internal tree insertion
(d) Internal tree deletion (non-leaf)
(e) Internal tree deletion (leaf)
(f) Single rotation
(g) Double rotation

Fig. 2 Broken nodes in actual insertion, actual deletion and rotations. Potentially broken nodes are filled black; the dotted lines indicate the nodes that the operation works on.

In external (leaf-oriented) trees, actual insertion replaces a leaf node by a new internal node with two children: the old leaf and a new one (see Figure 2(a)). The new internal node is potentially broken (if it was not allocated on the same cache block as one of the other nodes), and its parent may become broken, if the parent depended on the old leaf node.

Actual deletion in external trees deletes a leaf and its parent and replaces the parent with its other child (Figure 2(b)). The parent of the deleted internal node and the other child can become broken, since they could have depended on the deleted internal node.

Actual deletion in internal trees is slightly more complicated, with two cases. In the simple case (Figure 2(e)), a leaf is deleted, and its parent becomes broken, if it depended on the deleted leaf. The more complicated case arises when a non-leaf node x needs to be deleted (Figure 2(d)). The standard way of doing the deletion is to locate the node y with the next-larger key from the right child of x, copy the key and possible associated data fields to x, and then delete y by replacing it with its right child (if any). In this process, the parent of y and the right child can become broken (if they depended on y). The node x cannot become broken, since it or its neighbors were not moved in memory. (The equivalent implementation that looks for the next-smaller key of x in its left child is completely symmetric with regard to broken nodes.)

When a single or double rotation is performed, the nodes that can break are those whose parent or either child changes in the rotation, since a node will break if it depended on a node that was moved away by the rotation.

FIX-BROKEN(B):

1: **while** $B \neq \emptyset$ **do**
2: Remove any non-broken nodes from B (and exit if B is emptied).
3: **if** a node in $N(B)$ has free space in its cache block **then**
4: Select such a node x and a broken neighbor $b \in B$. (Prefer the x with the most free space and a b with no broken neighbors.)
5: Move b to the cache block containing x.
6: **else if** a node $b \in B$ has enough free space in its cache block **then**
7: Select the neighbor $x \in N(b)$ with the smallest $|D(x)|$.
8: Move x and all nodes in $D(x)$ to the cache block containing b.
9: **else**
10: Select a node $x \in N(B)$ and its broken neighbor $b \in B$. (Prefer a broken x, and after that an x with small $|D(x)|$. If there are multiple choices for b, prefer a b with $N(b) \setminus x$ non-broken.)
11: Move b, x and all nodes in $D(x)$ to a newly-allocated cache block.
12: **end if**
13: **end while**

Fig. 3 The local relocation algorithm. B is a set of potentially broken nodes which the algorithm will make non-broken; $N(B) = \bigcup_{b \in B} N(b)$. An implementation detail is that the algorithm needs access to the parent, grandparent and great grandparent of each node in B, since the grandparent may have to be moved in lines 8 and 11.

We can optimize the memory layout somewhat further with a simple heuristic (not required for Invariant 1): In insertion, a new node should be allocated in the cache block of its parent, if it happens to have enough free space.

We need an additional definition for the algorithm of Figure 3: $D(x)$ is the set of neighbors of node x that depend on node x (i.e., will be broken if x is moved to another cache block). Thus, $D(x) \subset N(x)$ and $0 \leq |D(x)| \leq |N(x)| \leq 3$. A crucial property is that nothing depends on a broken node (because no neighbor is on the same cache block), and thus broken nodes can be moved freely.

The algorithm of Figure 3 repeats three steps until the set of broken nodes B is empty. First, all neighbors of the broken nodes are examined to find a neighbor x with free space in its cache block. If such a neighbor is found, a broken node $b \in N(x)$ is fixed by moving it to this cache block. If no such neighbor was found, then the cache blocks of the nodes in B are examined; if one of them has enough space for a neighbor x and its dependants $D(x)$, they are moved to this cache block. Otherwise, if nothing was moved in the previous steps, then a broken node b is forcibly fixed by moving it and some neighboring nodes to a newly allocated cache block. At least one neighbor x of b needs to be moved along with b to make b non-broken; but if x was not broken, some of its other neighbors may depend on x staying where it is – these are exactly the nodes in $D(x)$, and we move all of them to the new cache block. (It is safe to move the nodes in $D(x)$ together with x, because their other neighbors are not on the same cache block.)

Theorem 4. *Assume that Invariant 1 holds in all nodes in a tree, except for a set B of broken nodes. Then giving B to the algorithm of Figure 3 will establish Invariant 1 everywhere.*

Theorem 5. *The algorithm of Figure 3 moves at most $4|B| = O(|B|)$ nodes in memory. The total time complexity of the algorithm is $O(|B|^2)$.*

Proof. Each iteration of the loop in the algorithm of Figure 3 fixes at least one broken node. Line 5 does this by moving one node; line 11 moves at most 4 nodes (b, x, and the two other neighbors of x), and line 8 moves at most 3 nodes (x and two neighbors). Thus, at most $4|B|$ nodes are moved in the at most $|B|$ iterations that the algorithm executes.

Each iteration looks at $O(|B|)$ nodes; thus, the total time complexity is $O(|B|^2)$. Additionally, looking for free nodes in a B_1 cache block can require more time. A naïve implementation looks at every node in the B_1-block to locate the free nodes, thus increasing the time complexity to $O(|B|^2 \cdot B_1/B_0)$. This may actually be preferable with the small B_1 of current processors. (The implementation we describe in Section 5 did this, with $B_1/B_0 = 4$.)

With larger B_1/B_0, the bound of the theorem is reached simply by keeping track of the number of free nodes in an integer stored somewhere in the B_1-sized block. To find a free node in constant time, a doubly-linked list of free nodes can be stored in the (otherwise unused) free nodes themselves, and a pointer to the head of this list is stored in a fixed location of the B_1-block. □

Remember that $|B| \leq 6$ always when we execute the algorithm.

A space-time tradeoff is involved in the algorithm of Figure 3: we sometimes allocate a new cache block to get two nodes on the same cache block (thus improving cache locality), even though two existing cache blocks have space for the nodes. Since our relocation algorithm always prefers an unused location in a previously allocated cache block, it is to be hoped that the cache blocks do not become very empty on average. (Moving unrelated nodes on the existing cache blocks "out of the way" is not practical: to move a node x in memory, we need access to the parent of x to update the link that points to the node, and our small-node trees do not store parent links.)

We get a lower limit for the cache block fill ratio from the property that our algorithm preserves: each non-leaf node has at least the parent or one child accompanying it on the same cache block. (Empty cache blocks should of course be reused by new allocations.)

5 Experiments

We implemented the algorithms of Sections 3 and 4 on internal AVL and red-black trees, and compared them to the cache-sensitive B^+-tree (the "full CSB^+-tree" of [1]) and to a standard B^+-tree with cache block-sized nodes (called a

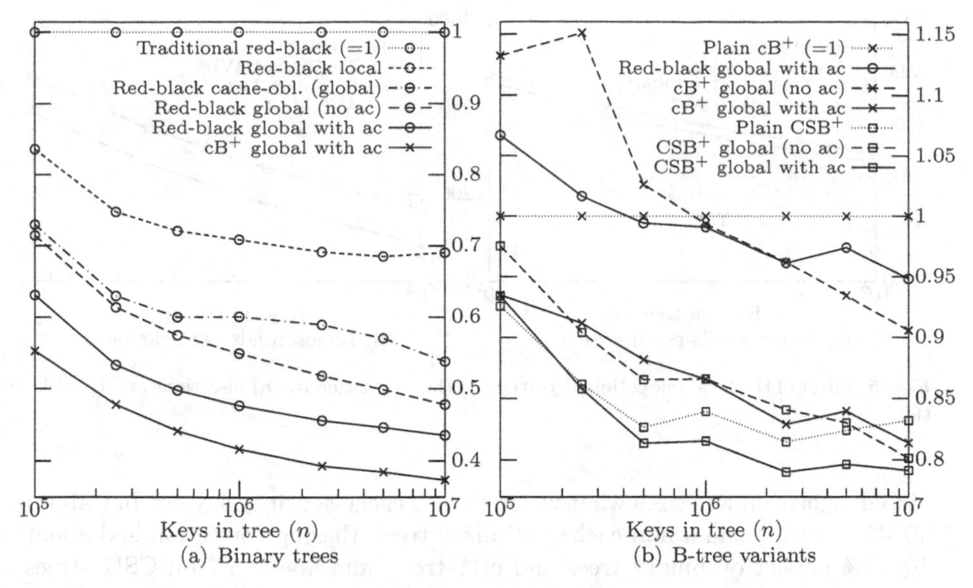

Fig. 4 Effect of global and local relocation and aliasing correction (="ac"). The figures give the search time relative to (a) the traditional red-black tree, (b) the cB+-tree. The trees marked "global" have been relocated using the global algorithm. "Red-black local" uses local relocation; the others use neither global nor local relocation. AVL trees (not shown) performed almost identically to red-black trees.

"cB+-tree" below for brevity).[3] As noted in Section 2, we used the following cache parameters: $k = 2$, $B_0 = 16$, $B_1 = 64$, $B_2 = 4096$, $B_3 = \infty$. The tree implementations did not have parent links: rebalancing was done using an auxiliary stack.[4]

Figure 4 examines the time taken to search for 10^5 uniformly distributed random keys in a tree initialized by n insertions of random keys. (Before the 10^5 searches whose time was measured, the cache was "warmed up" with 10^4 random searches.) The search performance of red-black and AVL trees relocated using the global algorithm was close to the cB+-tree. The local algorithm was not quite as good, but still a large (about 30%) improvement over a traditional non-cache-optimized binary tree. The cache-oblivious layout produced by the

[3] We executed our experiments on an AMD Athlon XP processor running at 2167 MHz, with 64 Kb L1 data cache (2-way associative) and 512 Kb L2 cache (8-way associative). Our implementation was written in C, compiled using the GNU C compiler version 4.1.1, and ran under the Linux kernel version 2.6.18. Each experiment was repeated 15 times; we report averages.

[4] The binary tree node size $B_0 = 16$ bytes was reached by using 4-byte integer keys, 4-byte data fields and 4-byte left and right children. The balance and color information for the AVL and red-black tree was encoded in the otherwise unused low-order bits of the child pointers. The nodes of the B-trees were structured as simple sorted arrays of keys and pointers. The branching factor of a non-leaf node was 7 in the cB+-tree and 14 in the CSB+-tree.

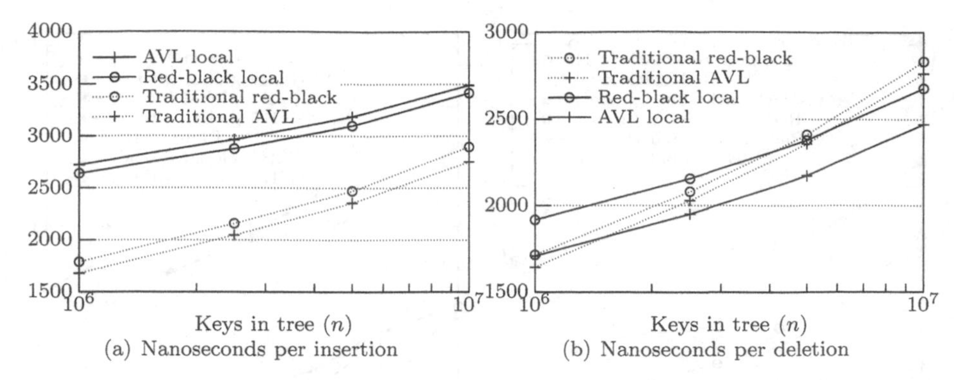

Fig. 5 Effect of the local relocation algorithm on the time taken by (a) insertions and (b) deletions.

global algorithm was somewhat worse than a cache-sensitive layout, but about 40–45% better than a non-cache-optimized tree. Aliasing correction had about 10–15% impact on binary trees and cB^+-trees, and about 5% on CSB^+-trees (which don't always access the first B_1-sized node of a TLB page). Especially in the B-trees, global relocation was not very useful without aliasing correction. In summary, the multi-level cache-sensitive layout improved binary search trees by 50–55%, cB^+-trees by 10–20% and CSB^+-trees by 3–5% in these experiments.

Figure 5 examines the running time of updates when using the local algorithm. Here the tree was initialized with n random insertions, and then 10^4+10^5 uniformly distributed random insertions or deletions were performed. The times given are averaged from the 10^5 updates (the 10^4 were used to "warm up" the cache). The local algorithm increased the insertion time by about 20–70% (more with smaller n). The deletion time was affected less: random deletions in binary search trees produce less rotations than random insertions, and the better memory layout produced by the local algorithm decreases the time needed to search for the key to be inserted or deleted.

In addition, we combined the global and local algorithms and investigated how quickly updates degrade the cache-sensitive memory layout created by the global algorithm. In Figure 6, we initialized the tree using $n = 10^6$ random insertions, executed the global algorithm once, and performed a number of random updates (half insertions and half deletions). Finally, we measured the average search time from 10^5 random searches (after a warmup period of 10^4 random searches), and the average B_1-block path length. The results indicate that the cache-sensitivity of the tree decreases significantly only after about n updates have been performed. The local algorithm keeps a clearly better memory layout, though it does not quite match the efficiency of the global algorithm.

Our experiments, as well as those in [6, 9], support the intuition that multi-level cache-sensitive structures are more efficient than cache-oblivious ones. It has been shown in [14] that a cache-oblivious layout is never more than 44%

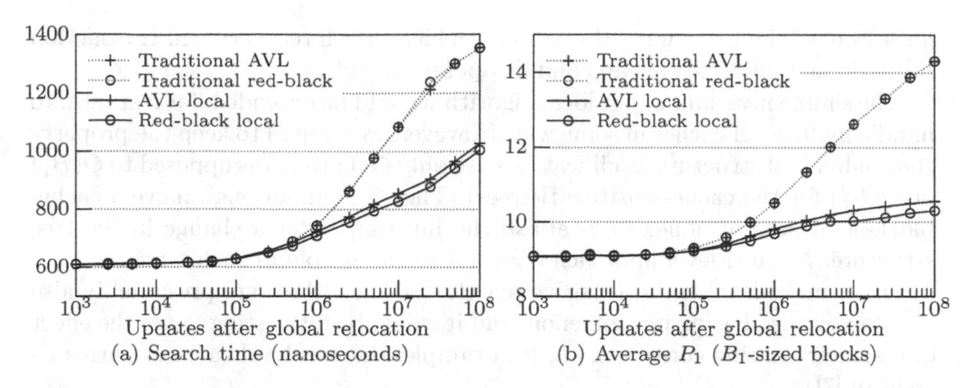

Fig. 6 Degradation of locality when random insertions and deletions are performed after global relocation of a tree with $n = 10^6$ initial keys: (a) average search time from 10^5 searches, (b) B_1-block path length. For 0 to 10^4 updates after relocation, there was practically no change in the values; here, the x axis begins from 10^3 for clarity.

worse in the number of block transfers than an optimal cache-sensitive layout, and that the two converge when the number of levels of caches increases. However, the cache-sensitive model is still important, because the number of levels of caches with different block sizes is relatively small in current computers (e.g., only two in the one we used for our experiments).

6 Conclusions

We have examined binary search trees in a k-level cache memory hierarchy with block sizes $B_1, \ldots, B_k$. We presented an algorithm that relocates tree nodes into a multi-level cache-sensitive memory layout in time $O(nk)$, where n is the number of nodes in the tree. Moreover, our one-level local algorithm preserves an improved memory layout for binary search trees by executing a constant-time operation after each structure modification (i.e., actual insertion, actual deletion or individual rotation).

Although cache-sensitive binary trees did not quite match the speed of the B^+-tree variants in our experiments, in practice there may be other reasons than average-case efficiency to use binary search trees. For instance, the worst-case (as opposed to amortized or average) time complexity of updates in red-black trees is smaller than in B-trees: $O(\log_2 n)$ vs. $O(d \log_d n)$ time for a full sequence of page splits or merges in a d-way B-tree, $d \geq 5$. Red-black tree rotations are constant-time operations, unlike B-tree node splits or merges (which take $O(B_1)$ time in B-trees with B_1-sized nodes, or $O(B_1^2)$ in the full CSB^+-tree). This may improve concurrency: nodes are locked for a shorter duration. In addition, it has been argued in [15] that, in main-memory databases, binary trees are optimal for a form of shadow paging that allows efficient crash recovery and transaction rollback, as well as the group commit operation [16].

The simple invariant of our local algorithm could be extended, for instance, to handle multi-level caches in some way. However, we wanted to keep the property that individual structure modifications use only $O(1)$ time (as opposed to $O(B_1)$ or $O(B_1^2)$ for the cache-sensitive B-trees). Then we cannot, e.g., move a cache-block-sized area of nodes to establish the invariant after a change in the tree structure. A multi-level approach does not seem feasible in such a model.

Other multi-level cache-sensitive search tree algorithms are presumably also affected by the aliasing phenomenon, and it would be interesting to see the effect of a similar aliasing correction on, for example, the two-level cache-sensitive B-trees of [7].

References

1. Rao, J., Ross, K.A.: Making B+-trees cache conscious in main memory. In: 2000 ACM SIGMOD International Conference on Management of Data, ACM Press (2000) 475–486
2. Hankins, R.A., Patel, J.M.: Effect of node size on the performance of cache-conscious B+-trees. In: 2003 ACM SIGMETRICS International Conference on Measurement and Modeling of Computer Systems, ACM Press (2003) 283–294
3. Rao, J., Ross, K.A.: Cache conscious indexing for decision-support in main memory. In: 25th International Conference on Very Large Data Bases (VLDB 1999), Morgan Kaufmann (1999) 78–89
4. Chen, S., Gibbons, P.B., Mowry, T.C.: Improving index performance through prefetching. In: 2001 ACM SIGMOD International Conference on Management of Data, ACM Press (2001) 235–246
5. Bohannon, P., McIlroy, P., Rastogi, R.: Main-memory index structures with fixed-size partial keys. In: 2001 ACM SIGMOD International Conference on Management of Data, ACM Press (2001) 163–174
6. Rahman, N., Cole, R., Raman, R.: Optimised predecessor data structures for internal memory. In: 5th Workshop on Algorithm Engineering (WAE 2001). Volume 2141 of Lecture Notes in Computer Science., Springer-Verlag (2001) 67–78
7. Chen, S., Gibbons, P.B., Mowry, T.C., Valentin, G.: Fractal prefetching B+-trees: Optimizing both cache and disk performance. In: 2002 ACM SIGMOD International Conference on Management of Data, ACM Press (2002) 157–168
8. Bender, M.A., Demaine, E.D., Farach-Colton, M.: Cache-oblivious B-trees. SIAM Journal on Computing **35**(2) (2005) 341–358
9. Brodal, G.S., Fagerberg, R., Jacob, R.: Cache oblivious search trees via binary trees of small height. In: 13th Annual ACM-SIAM Symposium on Discrete Algorithms (SODA 2002), Society for Industrial and Applied Mathematics (2002) 39–48
10. Bender, M.A., Duan, Z., Iacono, J., Wu, J.: A locality-preserving cache-oblivious dynamic dictionary. In: 13th Annual ACM-SIAM Symposium on Discrete Algorithms (SODA 2002), Society for Industrial and Applied Mathematics (2002) 29–38
11. Jiang, W., Ding, C., Cheng, R.: Memory access analysis and optimization approaches on splay trees. In: 7th Workshop on Languages, Compilers and Run-time Support for Scalable Systems, ACM Press (2004) 1–6
12. Oksanen, K., Malmi, L.: Memory reference locality and periodic relocation in main memory search trees. In: 5th Hellenic Conference of Informatics, Greek Computer Society (1995)

13. Bender, M.A., Demaine, E.D., Farach-Colton, M.: Efficient tree layout in a multilevel memory hierarchy. In: 10th Annual European Symposium on Algorithms (ESA 2002). Volume 2461 of Lecture Notes in Computer Science., Springer-Verlag (2002) 165–173
14. Bender, M.A., Brodal, G.S., Fagerberg, R., Ge, D., He, S., Hu, H., Iacono, J., López-Ortiz, A.: The cost of cache-oblivious searching. In: 44th Annual IEEE Symposium on Foundations of Computer Science (FOCS 2003), IEEE Computer Society (2003) 271–282
15. Soisalon-Soininen, E., Widmayer, P.: Concurrency and recovery in full-text indexing. In: String Processing and Information Retrieval Symposium (SPIRE 1999), IEEE Computer Society (1999) 192–198
16. Gray, J., Reuter, A.: Transaction Processing: Concepts and Techniques. Morgan Kaufmann (1993)

Part II
Track B

Invited talks

From Processes to ODEs by Chemistry

Luca Cardelli

Microsoft Research luca@microsoft.com

Abstract. We investigate the collective behavior of processes in terms of differential equations, using chemistry as a stepping stone. Chemical reactions can be converted to ordinary differential equations, and also to processes in a stochastic process algebra. Conversely, certain stochastic processes (in *Chemical Parametric Form*, or CPF) can be converted to chemical reactions. CPF is a subset of π-calculus, but is already more powerful that what is strictly needed to represent chemistry: it supports also parameterization and compositional reuse of models. The mapping of CPF to chemistry thus induces a parametric and compositional mapping of CPF to differential equations; the indirect mapping through chemistry is easier to define and understand than a direct mapping. As an example, we derive a quantitative interleaving law from the differential equations.

1 Introduction

In Systems Biology, biochemical systems are routinely described as large state transition diagrams with rates on transition [20], with emphasis on the graphical and database-oriented representation of the models. Graphical representation has advantages in terms of readability and sharing of information, but has obvious disadvantages in terms of precision, scalability (compositionality), and analyzability. Given that such models are presented already as state transition systems, it is natural to interpret them as term-rewriting systems or process algebras. These interpretations have a better chance than diagrams of satisfying scalability requirements, and can be mapped to increasingly promising analysis tools and techniques [23,17], including ones such as Petri Nets that already have a long tradition in other areas. This is not to say that current modeling approaches lack formality: biochemical systems, including the diagrammatic ones just discussed, are usually interpreted as systems of chemical reactions, and ultimately as systems of ordinary differential equations. A question then arises: what is the relationship between transition-based models and differential equation models, and even more fundamentally with chemistry?

Our starting point is stochastic process algebra, which provides us with a quantitative compositional semantics, and with simulation and analysis techniques [15,25]. Our goal is to relate process algebra to ordinary differential equations (ODEs), so that one can hopefully use techniques and tools from both camps [16]. We cannot carry out this program yet for general process algebras (e.g. π-calculus [21]) where some features go "beyond ordinary chemistry", but we can do it for interesting fragments that include unbounded-state systems, and that provide rich model parameterization. In this paper we establish a mapping from processes to ODEs via a detour though

Please use the following format when citing this chapter:

Cardelli, L., 2008, in IFIP International Federation for Information Processing, Volume 273; *Fifth IFIP International Conference on Theoretical Computer Science*; Giorgio Ausiello, Juhani Karhumäki, Giancarlo Mauri, Luke Ong; (Boston: Springer), pp. 261–281.

chemical reactions, primarily to obtain an easy two-step translation, but also to build a connection with chemistry. The foundations of this work are covered in more detail elsewhere [6]; here we emphasize the intuitive connection to chemical reactions by a number of examples, and we additionally handle parametric models.

The basic chemistry of well-mixed solutions can be described, at the molecular level, in terms of random molecular collisions and subsequent reactions. From this *microscopic* point of view, it can be modeled by Markov chains in continuous time (with real-valued reaction rates), with a discrete unbounded state space (an unbounded number of molecules that may flow in or be produced), and with a bounded number of chemical species (kinds of molecules). Alternatively, *macroscopically*, the fundamental law of interaction in chemistry is the *law of mass action*, which quantitatively determines the speed of reactions based on the continuous state space of concentrations of chemical species. The relationships between the discrete-space and the continuous-space views are subtle [10, 30, 6], and must be considered when relating discrete-state stochastic process algebra to continuous-state differential equations.

This paper is organized as follows. In Section 2 we introduce the notation of chemical reactions and its standard interpretation in terms of ordinary differential equations, relating changes of concentration of chemical species over time [18]. We also provide an interpretation of chemical reactions as a stochastic process algebra (CGF) that is a fragment of well-known ones. Stochastic processes can in turn be mapped to continuous-time Markov chains [12], which provide another standard interpretation of chemical reactions [11]. In Section 3 we translate stochastic processes (CGF) back to chemical reactions. We also consider a more general process algebra (CPF) that supports various kinds of parameterization. We show how to translate CPF down to CGF, and we provide an incremental algorithm for doing so. We thus obtain a systematic way of translating CPF parametric stochastic processes, through chemical systems, to differential equations. In section 4 we give various examples of the translations, including non-chemical ODE systems such as Kermack-McKendrick epidemics and Lotka-Volterra predation. The mapping to ODEs can be used also as a semantics of processes, and we show how to derive a quantitative interleaving law from it.

2 From Chemistry to ODEs and Processes

2.1 Chemical Reactions

We consider three basic kinds of chemical reactions. First, in *unary reactions*, a chemical species A may spontaneously degrade into other species; the rate of such a reaction is given by the *exponential decay law*: the rate is proportional to the concentration of the species A. Second, in *hetero binary reactions*, two chemical species A_1 and A_2 may react and produce other species; the rate of such a reaction is given by the *law of mass action*: the rate is proportional to the product of the concentrations of A_1 and A_2. Third, in *homeo binary reactions*, A_1 and A_2 are the same species A, and the rate is then proportional to the square of the concentration of A. We write $[A]$ for the concentration of A in moles per liter as a function of continuous time, and $[A]^\bullet$ for its time derivative.

A	$\rightarrow^k B_1 + \ldots + B_n$	Unary	$k : s^{-1}$	$[A]^\bullet = -k[A]$
$A_1 + A_2$	$\rightarrow^k B_1 + \ldots + B_n$	Hetero Binary	$k : M^{-1}s^{-1}$	$[A_i]^\bullet = -k[A_1][A_2]$
$A + A$	$\rightarrow^k B_1 + \ldots + B_n$	Homeo Binary	$k : M^{-1}s^{-1}$	$[A]^\bullet = -2k[A]^2$
				(assuming $A \neq B_i \neq A_j$ for all i, j)

Table 1 The Three Kinds of Chemical Reactions

Chemical reactions and the law of mass action can be presented in a more general form, with any number of molecules on the left-hand side. Still, the only chemical reactions of interest to us are unary and binary, in view of the molecular interpretation of interactions between chemical species. For example, we can ignore unlikely reactions that require three molecules to collide at the same time: "Genuinely *trimolecular* reactions do not physically occur in dilute fluids with any appreciable frequency. *Apparently* trimolecular reactions in a fluid are usually the combined result of two bimolecular reactions and one monomolecular reaction, and involve an additional short-lived species."[11]

A *system of chemical reactions* is a finite set of reactions between a finite set of chemical species[1]. We assume, as is common, that our reactions happen in a *well-stirred solution*, that is, that the dynamics of chemical reactions depends only on concentrations of the species (and on other factors, such as temperature, that are assumed fixed), and not on the positions of the molecules. Each reaction, $\rightarrow^k$, has a *(base) rate*, k, which is a proportionality constant used in the corresponding rate law, with bigger base rates meaning faster reactions. The *initial conditions* of the system, that is, the initial concentrations of the chemical species, are specified separately from the reactions.

2.2 From Chemistry to ODEs

A system of ordinary differential equations can be extracted from any system of chemical reactions, to describe the rate of change in concentration of chemical species over time. The ODEs provides the *kinetics* of the chemical system, that is, they completely describe the dynamic time evolution of the various quantities.

The procedure for extracting ODEs is standard [18]. Consider, as an example, the following system of 4 chemical reactions v_1, v_2, v_3, v_4 with corresponding reaction rates k_1, k_2, k_3, k_4, between 6 chemical species A, B, C, D, E, F.

[1] More generally, it could be a collection of reactions and chemical species indexed by an infinite set; this is necessary, e.g., to describe polymerization. A corresponding effect can be obtained within -calculus [26], but here we consider only finite systems of reactions.

Chemical reactions	
$v_1 : A + B \rightarrow^{k_1} C + C$	Hetero
$v_2 : A + C \rightarrow^{k_2}$ D	Hetero
$v_3 : C \rightarrow^{k_3}$ E+F	Unary
$v_4 : F + F \rightarrow^{k_4} B$	Homeo

N	v_1	v_2	v_3	v_4
A	-1	-1		
B	-1			1
C	2	-1	-1	
D		1		
E			1	
F			1	-2

Stoichiometry, **N**

	l
l_1	$k_1[A][B]$
l_2	$k_2[A][C]$
l_3	$k_3[C]$
l_4	$k_4[F]^2$

Rate laws, **l**

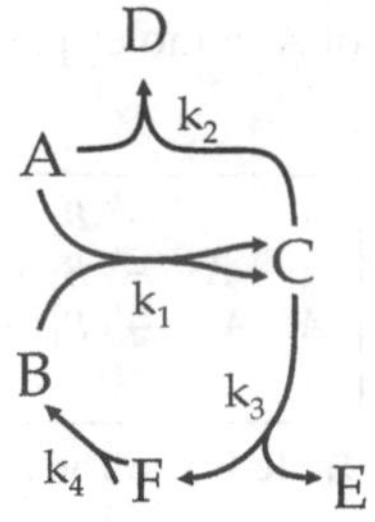

Flux

We first build the *stoichiometric matrix*, **N**, which has one row for each species and one column for each reaction. Each cell $\langle S, v\rangle$ in the matrix contains a positive number n if n molecules of species S are produced (overall) in reaction v; it contains a negative number $-n$ if n molecules of species S are removed (overall) in reaction v, and otherwise it contains 0.

Then, we build the *vector of rate laws*, **l**: for each reaction it specifies the rate law for that reaction. In our case, v_1,v_2 have the hetero rate law, v_3 has the unary rate law, and v_4 has the homeo rate law. (In general, other rate laws may include steady-state approximations, such as the Michaelis-Menten law for enzymatic reactions, or empirical rate laws).

Finally let **X** be the vector of chemical species (A,B,C,D,E,F). The system of ODEs is then, in general, given by the following *rate equation*:

$$[\mathbf{X}]^\bullet = \mathbf{N} \cdot \mathbf{l}$$

Table 2 From Chemical Reactions to Ordinary Differential Equations

Expanding for our set of reactions we obtain:

$$[A]^\bullet = -l_1 - l_2 = -k_1[A][B] - k_2[A][C] \qquad [D]^\bullet = l_2 = k_2[A][C]$$

$$[B]^\bullet = -l_1 + l_4 = -k_1[A][B] + k_4[F]^2 \qquad [E]^\bullet = l_3 = k_3[C]$$

$$[C]^\bullet = 2l_1 - l_2 - l_3 = 2k_1[A][B] - k_2[A][C] - k_3[C] \qquad [F]^\bullet = l_3 - 2l_4 = k_3[C] - 2k_4[F]^2$$

The rate law l_4 for the homeo reactions is $k_4[F]^2$, but the contribution of v_4 to $[F]^\bullet$ is $-2k_4[F]^2$ because two F are consumed in that reaction (hence the rate law shown in table 2). Compare this with the contribution of v_4 to $[B]^\bullet$, which is $k_4[F]^2$.

2.3 Processes in Chemical Ground Form

We introduce a subset of π-calculus (and of CCS) [21]: the *Chemical Ground Form* (CGF) [6], which is sufficient, in stochastic version, for translating chemical reactions to processes. See [25, 23, 3, 4, 12, 13, 14, 15] for the semantics of stochastic π-calculus and other stochastic process algebras.

$E ::= 0 \vdots X = M, E$	Reagents	(empty, or a reagent X=M and Reagents)
$M ::= 0 \vdots \pi; P \oplus M$	Molecule	(empty, or an interaction $\pi; P$ and Molecule)
$P ::= 0 \vdots X \vert P$	Solution	(empty, or a variable X and Solution)
$\pi ::= \tau_{(r)} \vdots ?n_{(r)} \vdots !n_{(r)}$	Interaction prefix	(delay, input, output)
$CGF ::= E, P$	Chemical Ground Form	(Reagents with initial Solution)

Table 3 Chemical Ground Form (CGF)

A chemical ground form CGF has a finite set E of *reagents* $X_i = M_i$ (named molecules) for distinct *variables* X_i naming chemical species, and *molecules* M_i describing the interaction capabilities of the corresponding species. The possible process *interactions* π are: *delay* $\tau_{(r)}$ at *rate r* (where r is a positive real), *input* $?n_{(r)}$ on *channel* n at rate r, and *output* $!n_{(r)}$ on channel n at rate r (each channel always has the same rate). In the syntax of molecules, each interaction π leads to releasing a *solution P* (a multiset of variables). We use $\oplus$ for *choice*, $\vert$ for *parallel composition*, and 0 for the empty reagent, the empty molecule, and the empty solution. Trailing 0's are usually left implicit, and we use $\vert$ also as an operator over the syntax: if P and P' are 0-terminated lists of variables, according to the syntax above, then $P \vert P'$ means appending the two lists into a single 0-terminated list. Therefore, if P is a solution, then $0 \vert P$, $P \vert 0$, and P are syntactically equal.

A CGF (E, P) is a set of reagents E together with *initial conditions*, which are a solution P. If a variable X occurs in some M_i or initial conditions P, but X is not defined in E, we can assume the existence of an additional reagent $X = 0$. The meaning of a CGF can be given by directly extracting a continuous time Markov chain from it [6].

Some CGFs can be drawn conveniently as stochastic interacting automata; for example here is a two-state automaton, with states X, Y, which interacts with copies of itself over the channels $a_{(r)}, b_{(s)}$:

$X = !a_{(r)}; X \oplus ?b_{(s)}; Y$
$Y = !b_{(s)}; Y \oplus ?a_{(r)}; X$
$X \mid X \mid X \mid Y \mid Y$ (initial conditions)

?b
!a X Y !b
?a

In general, however, a CGF can "split" by parallel composition after an interaction, and then some less standard graphical notation (similar to Petri Net transitions) must be used to represent such splitting.

2.4 From Chemistry to Processes

We now discuss how to produce CGF processes from systems of chemical reactions. We need, in particular, to convert concentrations of chemical species, of dimension M (*molarity*), to discrete numbers of molecules/processes, for which we need a conversion factor γ of dimension M^{-1}. In chemistry, $\gamma = N_A V$, where N_A is Avogadro's number, and V is the volume of the solution; if we take $\gamma = 1.0$, for example, it means that we are considering a chemical solution of volume $1/N_A$.

The factor γ has other uses too. The mass action rates "k" have dimension s^{-1} for unary reactions and $M^1 s^{-1}$ for binary reactions. Stochastic processes, instead, operate on molecule counts, and the stochastic rates "r" always have dimension s^{-1}. Therefore, an appropriate M^{-1} conversion factor is needed for the rates of binary reactions. In particular, the conversions between stochastic "r" and mass action "k" rates are: $r = k$ for unary reactions, $r = k/\gamma$ for hetero reactions, and $r = 2k/\gamma$ for homeo reactions [10, 30]. There is, however, an additional twist for homeo reactions. The natural encoding of homeo reactions, as processes that offer both an input and an output on the same channel, produces an artificial doubling of interactions; see for example F below, where 2 copies of F have 2 interactions on channel v_4, instead of 1 "collision" [25]. We can compensate for this doubling by halving the stochastic rate of the interaction channel, with the net effect that homeo *channels* too end up with $r = k/\gamma$. We keep the two contributions to the rate of homeo reactions separate in Table 4 for emphasis.

To convert chemical reactions to process reagents, we first prepare a separate channel $v_{(r)}$ of rate r for each binary reaction v of rate k, setting $r = k/\gamma$ as discussed. The unary reactions do not need channels, and use a τ delay with $r = k$. Setting up such channels is similar to setting up the vector of rate laws in Section 2, but fixing the base rates is sufficient here because the semantics of the intended process algebra already incorporates the decay law and the mass action law [23].

With these channels, we can produce the CGF reagents for the reactions from Section 2:

$$
\begin{aligned}
A &= ?v_{1(k_1/\gamma)};(C|C) \oplus ?v_{2(k_2/\gamma)};D && \oplus ?cA_{(0)};0 \\
B &= !v_{1(k_1/\gamma)};0 && \oplus ?cB_{(0)};0 \\
C &= !v_{2(k_2/\gamma)};0 \oplus \tau_{(k_3)};(E|F) && \oplus ?cC_{(0)};0 \\
D &= 0 && \oplus ?cD_{(0)};0 \\
E &= 0 && \oplus ?cE_{(0)};0 \\
F &= ?v_{4(k_4/\gamma)};B \oplus !v_{4(k_4/\gamma)};0 && \oplus ?cF_{(0)};0
\end{aligned}
$$

That is done as follows. For each species X we produce an initially empty reagent, $X = 0$. Then we scan each chemical reaction to gradually populate the reagents with summands. For a degradation reaction $v : X \rightarrow^k P$ we add a summand $\tau_{(r)};P$ with $r = k$ to the reagent X. For a hetero reaction $v : X + Y \rightarrow^k P$ we add a summand $?v_{(r)};P$ with $r = k/\gamma$ to the reagent X and a summand $!v_{(r)};0$ to the reagent Y, using the reaction names as the channel names. For a homeo reaction $v : X + X \rightarrow^k P$ we add two summands $?v_{(r/2)};P$ and $!v_{(r/2)};0$ to the reagent X, with $r = 2k/\gamma$. (We also change all chemical "+" to process "|".)

We may optionally add an extra summand $?cX_{(0)};0$ to the definition of each X, where cX is a channel where no interaction ever happens. This is useful if we want to *observe* the system (e.g., counting the number of X for plotting), by observing how many $?cX$ are being generated [23].

The formal procedure for obtaining processes $Pi_\gamma(C)$ from a chemical system C is finally given in Table 4, assuming that the reactions in C are uniquely named. The initial conditions of a chemical system consist of a vector V of concentrations $V_{X_i} : M = [X_i]$ for the various species X_i; these can be converted to CGF initial conditions P with $\#X_i(P) = \lceil \gamma [X_i] \rceil$, with a rounding error. $Pi_\gamma(C)$ has the same dynamics as C [6].

$$\begin{array}{ll}
Pi_\gamma(C) = \{(X = \oplus((v : X \rightarrow^k P) \in C) \text{ of } (\tau_{(r)};P) & \oplus \text{ with } r = k \\
\quad \oplus((v : X + Y \rightarrow^k P) \in C \text{ and } Y \neq X) \text{ of } (?v_{(r)};P) & \oplus \text{ with } r = k/\gamma \\
\quad \oplus((v : Y + X \rightarrow^k P) \in C \text{ and } Y \neq X) \text{ of } (!v_{(r)};0) & \oplus \text{ with } r = k/\gamma \\
\quad \oplus((v : X + X \rightarrow^k P) \in C) \text{ of } (?v_{(r/2)};P \oplus !v_{(r/2)};0)\ ) & \text{with } r = 2k/\gamma \\
\quad \text{s.t. } X \text{ is a species in } C\} & \\
Pi_\gamma(C,V) = E,P \text{ where } E = Pi^\gamma(C) \text{ and } \#X(P) = \lceil \gamma V^X \rceil \text{ for all } X \in E &
\end{array}$$

Table 4 From a Chemical Reaction System C to a Chemical Ground Form $Pi_\gamma(C)$

(C,V)	$Pi\gamma(C,V)$	where	initially	ODEs
$n : X \rightarrow^k 0,\quad V$	$X = \tau_{(r)};0,\quad P$	$r = k$	$\#X(P) = \lceil \gamma V_X \rceil$	$[X]^\bullet = -k[X] = -r[X]$
$n : X + Y \rightarrow^k 0,\quad V$	$X = ?n_{(r)};0,$ $Y = !n_{(r)};0,\quad P$	$r = k/\gamma$	$\#X(P) = \lceil \gamma V_X \rceil$ $\#Y(P) = \lceil \gamma V_Y \rceil$	$[X]^\bullet = -k[X][Y] = -r\gamma[X][Y]$ $[Y]^\bullet = -k[X][Y] = -r\gamma[X][Y]$
$n : X + X \rightarrow^k 0,\quad V$	$X = ?n_{(r/2)};0 \oplus$ $!n_{(r/2)};0,\quad P$	$r = 2k/\gamma$	$\#X(P) = \lceil \gamma V_X \rceil$	$[X]^\bullet = -2k[X]^2 = -r\gamma[X]^2$

Table 5 Examples: from (C,V) to $Pi_\gamma(C,V)$

3 From Processes to Chemistry and ODEs

We have seen that we can convert chemical reactions to ODEs. Therefore, a mapping from a stochastic process algebra to chemical reactions, which we study in this section, induces in two steps a mapping from that process algebra to ODEs. A two-step approach is desirable, because a direct mapping from a process algebra to ODEs, although possible and intuitively understandable, is more challenging [16, 6]. The first step, from process algebra to chemical reactions, has the effect of identifying the transitions that the system performs, and the second step, from chemical reactions to ODEs, identifies the rate of change of populations of processes.

3.1 Processes in Chemical Parametric Form

We begin by defining a more general subset of π-calculus, the *Chemical Parametric Form* (CPF), which extends the CGF with parameterization and communication. CPF is not technically a subset of CCS, since it allows channel passing, but the subsequent translation of CPF to CGF essentially amounts to a translation of CPF to CCS. The reason for these subsets is that, in general, it is not possible to translate an arbitrary π-calculus process to a system of chemical reactions with a finite set of chemical species, because in full π-calculus, via name generation, we can generate unboundedly many species. Therefore, the CPF incorporates limitations that, as we shall see, are sufficient to enable the translation to chemistry. The limitations are not that our systems be *finite state* (since it is convenient to abstract from detailed accounting of energy and to express chemical systems that produce unbounded quantities of product), nor that they be *finite control* [8] (since parallel composition within recursive definitions models chemical reactions that generate multiple products). However, it is essential that our systems have a *finite number of species.*

$E ::= 0 \vdots X(\mathbf{p}) = M, E$	Reagents (empty, or a parametric reagent $X(\mathbf{p}) = M$ and Reagents)
$M ::= 0 \vdots \pi; P \oplus M$	Molecule (empty, or an interaction $\pi; P$ and Molecule)
$P ::= 0 \vdots X(\mathbf{p})\|P$	Solution (empty, or an instanced variable $X(\mathbf{p})$ and Solution)
$\pi ::= \tau_{(r)} \vdots ?n(\mathbf{p}) \vdots !n(\mathbf{p})$	Interaction prefix (delay, parametric input, instanced output)
$CPF ::= E, P$	Chemical Parametric Form (Reagents with initial Solution)

Table 6 Chemical Parametric Form (CPF)

The syntax of the CPF is the same as that of the CGF, but with additional parameter lists $\mathbf{p}$. There is a finite set E of parametric reagents $X_i(\mathbf{p}_i) = M_i$ for distinct variables X_i. Each $\mathbf{p}$ is a vector of distinct channel names, and $\#\mathbf{p}$, the length of $\mathbf{p}$, is the arity of the corresponding X, which must be used consistently through E, P. If $X(\mathbf{p})$ occurs in some M_i or initial conditions P, but X is not defined in E, we can assume the existence of an additional reagent of the form $X(\mathbf{q}) = 0$.

A name n in E is *free* if it occurs in some M_i but is not bound by the corresponding $\mathbf{p}_i$ or any enclosing $?n(\mathbf{p})$; we then say that $n \in \mathrm{fn}(E)$. Moreover, any name n occurring in the initial conditions P is free ($n \in \mathrm{fn}(P)$). Each free name n is uniformly associated with a fixed rate $n_{(r)}$; we may also keep track of this information separately by saying that $\rho_{(E,P)}(n) = r$ for $n \in \mathrm{fn}(E,P)$. The non-free (*parametric*) names are not annotated with a rate, and simply acquire the rate of the free names that they must be replaced with before any interaction can happen. Therefore, the possible process interactions are: stochastic delay $\tau_{(r)}$ at rate r, input $?n_{(r)}(\mathbf{p})$ of names $\mathbf{p}$ (parametric and distinct) on channel n at rate r, and output $!n_{(r)}(\mathbf{p})$ of names $\mathbf{p}$ (free at the time of interaction) on channel n at rate r.

An example of a CPF system (with no initial conditions, no free names, and no input and output parameters), is the following gene gate $\mathrm{Neg}(a,b)$ [1]. This is a process that at stochastic intervals produces copies of $\mathrm{Tr}(b)$, unless it is inhibited (for some time) on channel a. $\mathrm{Tr}(b)$ can in turn inhibit other gates that accept input on channel b, or decay.

$$\begin{aligned}
\mathrm{Neg}(a,b) &= ?a();\mathrm{Inh}(a,b) \oplus \tau_{(\varepsilon)};(\mathrm{Tr}(b)|\mathrm{Neg}(a,b)) \\
\mathrm{Inh}(a,b) &= \tau_{(\eta)};\mathrm{Neg}(a,b) \\
\mathrm{Tr}(b) &= !b();\mathrm{Tr}(b) \oplus \tau_{(\delta)};0
\end{aligned}$$

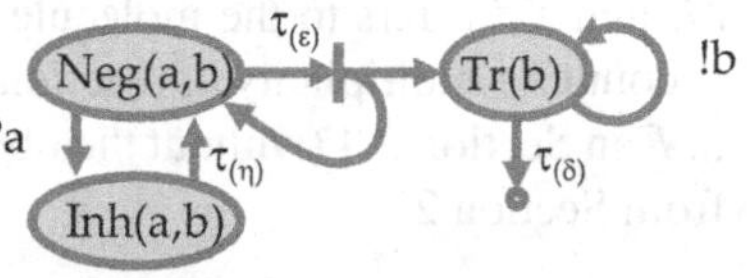

This description is parametric in that it defines the behavior of a gate in a network without specifying its connectivity; the connectivity of the network is then given in the initial conditions. Initial conditions for this CPF system could be given by $\mathrm{Neg}(x_{(r)},x_{(r)})$: a single gate with a self loop (with free name $x_{(r)}$), or by $\mathrm{Neg}(x_{(r)},y_{(s)})|$ $\mathrm{Neg}(y_{(s)},z_{(t)})|\mathrm{Neg}(z_{(t)},x_{(r)})$: a network of three gates (with free names $\{x_{(r)},y_{(s)},z_{(t)}\}$) which can function as a stochastic oscillator.

A more general normal form, that can represent any π-calculus process, can be obtained by allowing π-calculus restriction in reagents: $X(\mathbf{p}) = (\nu\mathbf{q})M$. This way we can express complexation and polymerization by channel passing [25], but ODE translations are not currently known.

3.2 From CGF to Chemistry

The chemical ground form, CGF, from Section 2.3 is a restricted version of the CPF, where there are zero parameters in definitions, inputs and outputs. Empty parameters, (), are omitted.

We first consider the problem of converting a CGF (E,P) to a system of chemical reactions $Ch_Y(E,P)$ (the resulting $Ch_Y(E)$ has the same dynamics as E [6]). This is achieved by producing a degradation reaction for each $\tau_{(r)}$ delay in E, a hetero reaction for each pair $?a$, $!a$ of interactions in different molecules of E, and a rate-doubled homeo reaction for each pair of interactions $?a$, $!a$ in the same molecule of E. Several examples are shown in section 4. The mass action rate for homeo reactions is $r\gamma$, but we keep the factors contributing to it ($r\gamma = 2(r\gamma/2)$) separate in Table 7 for emphasis.

$$
\begin{array}{ll}
Ch_\gamma(E) = & \\
\quad \{(\{X.i\}X \to^k P) \text{ s.t. } E(X).i = \tau_{(r)};P\} \cup & \text{with } k = r \\
\quad \{(\{X.i,Y.j\}X+Y \to^k P+Q) \text{ s.t. } X \neq Y \text{ and } E(X).i = ?n_{(r)};P \text{ and } E(Y).j = !n_{(r)};Q\} \cup & \text{with } k = r\gamma \\
\quad \{(\{X.i,X.j\}X+X \to^{2k} P+Q) \text{ s.t. } E(X).i = ?n_{(r)};P \text{ and } E(X).j = !n_{(r)};Q\} & \text{with } k = r\gamma/2 \\
Ch_\gamma(E,P) = C,V \text{ where } C = Ch_\gamma(E) \text{ and } V_X = \#X(P)/\gamma \text{ for all } X \in E &
\end{array}
$$

Table 7 From a Chemical Ground Form E to a Chemical Reaction System $Ch_\gamma(E)$

When inserting a P into a chemical reaction, we change all process "|" to chemical "+". The initial conditions of the chemical system can be obtained from the initial conditions P of the CGF by setting $[X_i] = \#X_i(P)/\gamma$ for each species X_i. Note how we have tagged the resulting reactions (by $\{...\}$): here $M.i$ is the i-th summand in molecule M, and $X.i$ refers to the molecule summand $E(X).i$. This tagging allows us to easily account for multiplicity of reactions. Applying this procedure to the process reagents A ... F in Section 2.4 (without the observer channels), reproduces the system of reactions from Section 2.

(E,P)	$Ch_\gamma(E,P)$	where	initially	ODEs
$X = \tau_{(r)};0, \ P$	$\{X.1\} \ X \to^k 0, \ V$	$k = r$	$V_X = \#X(P)/\gamma$	$[X]^\bullet = -k[X] = -r[X]$
$X = ?n_{(r)};0,$ $Y = !n_{(r)};0, \ P$	$\{X.1,Y.1\} \ X+Y \to^k 0, \ V$	$k = r\gamma$	$V_X = \#X(P)/\gamma$ $V_Y = \#Y(P)/\gamma$	$[X]^\bullet = -k[X][Y] = -r\gamma[X][Y]$ $[Y]^\bullet = -k[X][Y] = -r\gamma[X][Y]$
$X = ?n_{(r)};0 \oplus$ $!n_{(r)};0, \ P$	$\{X.1,X.2\} \ X+X \to^{2k} 0, \ V$	$k = r\gamma/2$	$V_X = \#X(P)/\gamma$	$[X]^\bullet = -4k[X]^2 = -2r\gamma[X]^2$

Table 8 Examples: from (E,P) to $Ch_\gamma(E,P)$

3.3 From CGF to ODEs directly

We have seen how to convert CGF to chemistry (Section 3.2) and how to convert chemistry to ODEs (Section 2). We can obviously compose the two conversions to go from CGF to ODEs, but we can also do it more directly via a stoichiometric matrix technique. In comparison to the chemical technique, the role of the set of chemical reactions is replaced by the following set:

$$
\begin{aligned}
\mathfrak{I} = {} & \{\{X.i\} \text{ s.t. } E.X.i = \tau_{(r)}.Q\} \\
& \cup \{\{X.i, Y.j\} \text{ s.t. } E.X.i = ?n_{(r)}.Q \text{ and } E.Y.j = !n_{(r)}.R\} \quad \text{(for any } r,n,Q,R)
\end{aligned}
$$

$\mathfrak{I}$ is the finite set of *possible interactions* arising from a set of reagents E, where $X.i$ is an ordered pair identifying a molecule summand in E, and $E.X.i$ is a molecule summand as previously defined.

The stoichiometric matrix used in the conversion has as many rows as species, and as many columns as interactions $\mathfrak{I}$: each column contains coefficients for the reagents that are gained or lost in that interaction. The corresponding vector of rate laws has as many rows as interactions, and contains the rate laws for the interactions. For example, in volume $\gamma : M^{-1} = N_A V$ where N_A:mol^{-1} and V:L, and $r,t,u : s^{-1}$, we have:

$$X = \tau_{(t)}.Y \oplus ?a_{(r)}.X$$
$$Y = !a_{(r)}.(Y|Y) \oplus \tau_{(u)}.Y$$

Processes

$$\{\{X.1\},\{X.2,Y.1\},\{Y.2\}\}$$

Interactions $\mathfrak{I}$

N	$\{X.1\}$	$\{X.2,Y.1\}$	$\{Y.2\}$
X	-1	0	0
Y	$+1$	$+1$	-1

Stoichiometry, **N**

	l
$\{X.1\}$	$t[X] : M \cdot s^{-1}$
$\{X.2,Y.1\}$	$r\gamma[X][Y] : M \cdot s^{-1}$
$\{Y.2\}$	$u[Y] : M \cdot \cdot \cdot^{-1}$

Rate laws, **l**

The ODEs are then obtained, as in the chemical technique, as $\mathbf{N} \cdot \mathbf{l}$:

$$[X]^{\bullet} : M \cdot s^{-1} = -t[X]$$
$$[Y]^{\bullet} : M \cdot s^{-1} = t[X] + r\gamma[X][Y] - u[Y]$$

3.4 From CPF to CGF

The procedure in Section 3.2 allows us to obtain chemical systems from ground forms. But we can use it also for the more general parametric forms, if we can first convert a CPF to a CGF. To that end, *grounding* ($/_N$) is a process that converts molecules and solutions of a CPF to those of a CGF. It eliminates parameters on the basis of a set of free names N (covering all free names), which is initially chosen to be that of the CPF. Here $n/\mathbf{p}$ denote (single) channel names in bijection with pairs $\langle n, \mathbf{p}\rangle$, and $X/\mathbf{p}$ denote species names in bijection with pairs $\langle X, \mathbf{p}\rangle$ (any rate annotations in $\mathbf{p}$ are ignored). Each $X/\mathbf{p}$ has the role of a separate chemical species for the parameter instantiation given by $\mathbf{p}$.

In Table 9, $n/\mathbf{p}_{(r)}$ means that the (single) name $n/\mathbf{p}$ is annotated with r. The notation $\{\mathbf{p} \leftarrow \mathbf{q}\}$ is the simultaneous substitution of name vectors, where $\mathbf{q} \in N^{\#\mathbf{p}}$ are vectors of free names (of the same size as $\mathbf{p}$) and hence annotated with rates. As an invariant of the definition, the names in channel position and in output must be annotated with rates; this is maintained by $\{\mathbf{p} \leftarrow \mathbf{q}\}$.

Then, a process of *parametric explosion* converts a parametric form E, to a ground form E_G, by instantiating all possible parameter lists with respect to the set N of free names of E. Grounding is used in such a process. The initial conditions are simply grounded once.

$$
\begin{aligned}
/_N(\tau_{(r)};P) &= \tau_{(r)};/_N(P)\\
/_N(!n_{(r)}(\mathbf{p});P) &= !n/\mathbf{p}_{(r)};/_N(P)\\
/_N(?n_{(r)}(\mathbf{p});P) &= \oplus(\mathbf{q}\in N^{\#\mathbf{p}})\text{ of }?n/\mathbf{q}_{(r)};/_N(P\{\mathbf{p}\leftarrow\mathbf{q}\})\\
/_N(\pi_1;P_1\oplus\ldots\oplus\pi_n;P_n) &= /_N(\pi_1;P_1)\oplus\ldots\oplus/_N(\pi_n;P_n)\\
/_N(X_1(\mathbf{p}_1)|\ldots|X_n(\mathbf{p}_n)) &= X_1/\mathbf{p}_1|\ldots|X_n/\mathbf{p}_n
\end{aligned}
$$

Table 9 Grounding

$$E_G = \{(X/\mathbf{q} = /_N(M\{\mathbf{p}\leftarrow\mathbf{q}\}))\text{ s.t. }(X(\mathbf{p}) = M)\in E\text{ and }N = \mathrm{fn}(E,P)\text{ and }\mathbf{q}\in N^{\#\mathbf{p}}\}$$

$$P_G = /_N(P) \qquad \text{where } N = \mathrm{fn}(E,P)$$

Table 10 Parametric Explosion: From a CPF (E,P) to a CGF (E_G,P_G)

Finally, we can convert a CPF to chemical reactions simply by first exploding it into a CGF, and then applying the previous Ch_γ procedure. See section 4.5 for an example.

$$Cp_\gamma(E,P) = Ch_\gamma(E_G,P_G)$$

Table 11 From a Chemical Parametric Form (E,P) to a Chemical Reaction System $Cp_\gamma(E,P)$

3.5 Iterative CPF to CGF Algorithm

A system $Cp_\gamma(E,P)$ computed from E_G,P_G is highly redundant because it includes all the parameter permutation symmetries, many of which are not needed for any given set of initial conditions. The following iterative algorithm for the CPF case, combining definitions 7 and 10, computes a subset of E_G from the initial conditions of P_G. It produces a (usually) much smaller although not necessarily minimal set C. Again, see section 4.5 for an example.

The algorithm terminates: E_C never shrinks and is always a subset of E_G, which is finite.

initialization for a CPF (E,P)
$N = \mathrm{fn}(E,P)$
$E_C := \{X/\mathbf{q} = /_N(M\{\mathbf{p} \leftarrow \mathbf{q}\}))$ *s.t.* $X(\mathbf{q})$ occurs in P and $(X(\mathbf{p}) = M) \in E\}$ (initial conditions)
iteration
$C := Ch_\gamma(E_C)$
$E'_C := E_C \cup X/\mathbf{q} = /_N(M\{\mathbf{p} \leftarrow \mathbf{q}\}))$ *s.t.* $X/\mathbf{q}$ occurs in C and $(X(\mathbf{p}) = M) \in E\}$
termination
if $E'_C = E_C$ then stop and return $(C, /_N(P))$, else $E_C := E'_C$ and iterate.

Table 12 Algorithm: Chemical Reaction System from CPF Initial Conditions

4 Examples

We illustrate the translations $Pi_\gamma(-)$ from 4, $Ch_\gamma(-)$ from 7 and $Cp_\gamma(-)$ from 11, 12. There are natural issues about correctness of these translations, which are investigated in detail in [6]; the examples are provided to give some appreciation of the expected properties of the translations.

4.1 Unary Reactions

We begin with a degradation reaction that is not finite-control (because parallel splitting occurs) and is not finite-state (because the cardinality of X grows over time). However, the set of species is fixed ($\{X\}$), so we can still carry out translations between processes and reactions.

Chemistry (C)	to Processes $(Pi_\gamma(C))$	to Chemistry $(Ch_\gamma(Pi_\gamma(C)))$
$v : X \rightarrow^r X + X$	$X = \tau_{(r)}; (X\|X)$	$v : X \rightarrow^r X + X$

Next is a similar unbounded-state system, but its size may grow or shrink depending on the rates r, s.

Chemistry (C)	to Processes $(Pi_\gamma(C))$	to Chemistry $(Ch_\gamma(Pi_\gamma(C)))$
$v : X \rightarrow^r X + X$ $d : X \rightarrow^s 0$	$X = \tau_{(r)}; (X\|X) \oplus \tau_{(s)}; 0$	$v : X \rightarrow^r X + X$ $d : X \rightarrow^s 0$

4.2 Hetero Binary Reactions

The translation of reversible ionization reactions between Na and Cl is shown below. Note that a more natural version of $Pi_\gamma(C)$ would map Na to Na^+ and Cl to Cl^-, but that is not what the default translation produces. The back translation $(Ch_\gamma(Pi_\gamma(C)))$ yields the initial reactions (once retagged).

C	$Pi_\gamma(C)$	$Ch_\gamma(Pi_\gamma(C))$
$i: Na+Cl \to^k Na^+ + Cl^-$ $d: Na^+ + Cl^- \to^v Na+Cl$	$Na = ?i_{(k/\gamma)};(Na^+ \mid Cl^-)$ $Cl = !i_{(k/\gamma)};0$ $Na^+ = ?d_{(v/\gamma)};(Na \mid Cl)$ $Cl^- = !d_{(v/\gamma)};0$	$i: Na+Cl \to^k Na^+ + Cl^-$ $d: Na^+ + Cl^- \to^v Na+Cl$

The next example starts from a two state process (from Section 2.3), and translates it to chemistry and back; the result is an equivalent but not identical process.

E	$Ch_\gamma(E)$ $(= Ch_\gamma(Pi_\gamma(Ch_\gamma(E))))$	$Pi_\gamma(Ch_\gamma(E))$
$X = ?a_{(r)};Y \oplus !b_{(s)};X$ $Y = ?b_{(s)};X \oplus !a_{(r)};Y$	$a: X+Y \to^{r\gamma} Y+Y$ $b: Y+X \to^{s\gamma} X+X$	$X = ?a_{(r)};(Y \mid Y) \oplus !b_{(s)};0$ $Y = !a_{(r)};0 \oplus ?b_{(s)};(X \mid X)$

4.3 Homeo Binary Reactions

The inverse translation of a homeo chemical reaction gives back in the original reaction, and in particular it reproduces the original rate $k = 2((k/\gamma)\gamma/2)$.

C	$Pi_\gamma(C)$	$Ch_\gamma(Pi_\gamma(C))$
$v: X+X \to^k Y$	$X = ?v_(k/\gamma);Y \oplus !v_{(k/\gamma)};0$ $Y = 0$	$v: X+X \to^k Y$

Conversely, starting from processes that self-interact, we produce homeo reactions, and then we go back again to equivalent but not identical processes.

E	$Ch_\gamma(E)$ (and $Ch_\gamma(Pi_\gamma(Ch_\gamma(E)))$)	$Pi_\gamma(Ch_\gamma(E))$
$X = ?a_{(r)};Y \oplus !a_{(r)};X$ $Y = ?b_{(s)};X \oplus !b_{(s)};Y$	$a: X+X \to^{r\gamma} Y+X$ $b: Y+Y \to^{s\gamma} X+Y$	$X = ?a_{(r)};(Y \mid X) \oplus !a_{(r)};0$ $Y = ?b_{(s)};(X \mid Y) \oplus !b_{(s)};0$

4.4 Hetero and Homeo Reactions on a Shared Channel

This example involves both homeo and hetero reactions. We start with processes E and we obtain $Ch_\gamma(E)$, assigning unique reaction names b,c (this is a precondition for applying $Pi_\gamma(-)$). The translation back, $Pi_\gamma(Ch_\gamma(E))$, produces different-looking processes, but both in E and in $Pi_\gamma(Ch_\gamma(E))$, the interaction of X with Y produces $X|X$, and the interaction of X with X produces $Y|X$. We show stochastic simulations of E and $Pi_\gamma(Ch_\gamma(E))$, and Matlab simulations of ODE($Ch_\gamma(E)$) for two values of γ.

E	$Ch_\gamma(E)$ (and $Ch_\gamma(Pi_\gamma(Ch_\gamma(E)))$)	$Pi_\gamma(Ch_\gamma(E))$
$X = !a_{(r)};X \oplus ?a_{(r)};Y$ $Y = ?a_{(r)};X$	$b: Y+X \to^{r\gamma} X+X$ $c: X+X \to^{r\gamma} Y+X$	$X = !b_{(r)};0 \oplus ?c_{(r)};(Y \mid X) \oplus !c_{(r)};0$ $Y = ?b_{(r)};(X \mid X)$

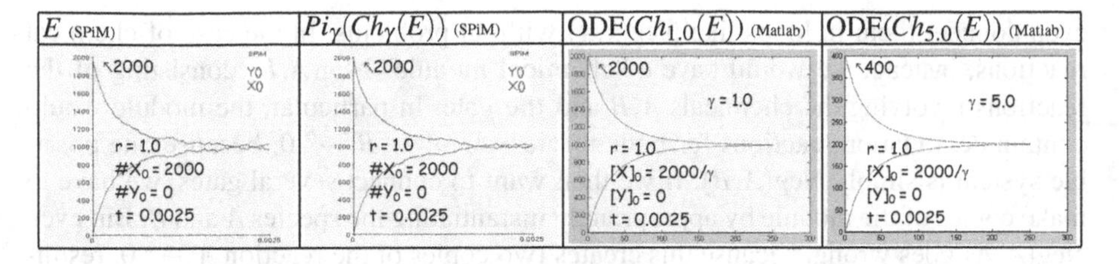

4.5 A Parametric Example: Gene Networks

We compute the reactions for the parametric gate of Section 3.1, with initial conditions Neg(x,x). Since this system E,P has a single free name, x, the parametric explosion does not actually increase its size, and we can easily show its expansion E_G, P_G. We could then compute $Ch_\gamma(E_G)$ directly. For illustration, though, we convert E to chemical reactions E_C using the iterative algorithm of Section 3.5. With initial conditions $\text{Tr}(x)|\text{Neg}(x,y)$ (not shown), with two free variables, the parametric explosion is larger, but the algorithm terminates in just two iterations with an output that is much smaller than E_G.

E,P (Input)	E_G,P_G (directly obtained, for comparison)
$\text{Neg}(a,b) = ?a();\text{Inh}(a,b) \oplus \tau_{(\varepsilon)};(\text{Tr}(b)\|\text{Neg}(a,b))$	$\text{Neg}/x,x = ?x/_{(r)};\text{Inh}/x,x \oplus \tau_{(\varepsilon)};(\text{Tr}/x\|\text{Neg}/x,x)$
$\text{Inh}(a,b) = \tau_{(\eta)};\text{Neg}(a,b)$	$\text{Inh}/x,x = \tau_{(\eta)};\text{Neg}/x,x$
$\text{Tr}(b) = !b();\text{Tr}(b) \oplus \tau_{(\delta)};0$	$\text{Tr}/x = !x/_{(r)};\text{Tr}/x \oplus \tau_{(\delta)};0$
$\text{Neg}(x_{(r)},x_{(r)}) \qquad N = \text{fn}(E,P) = \{x_{(r)}\}$	$\text{Neg}/x,x$

Iterative Algorithm for E,P	Initialization: E_C
	$\text{Neg}/x,x = ?x/_{(r)};\text{Inh}/x,x \oplus \tau_{(\varepsilon)};(\text{Tr}/x\|\text{Neg}/x,x)$

Iteration 1: C	Iteration 1: E_C
$\text{Neg}/x,x \to^{\varepsilon} \text{Tr}/x + \text{Neg}/x,x$	$\text{Neg}/x,x = ?x/_{(r)};\text{Inh}/x,x \oplus \tau_{(\varepsilon)};(\text{Tr}/x\|\text{Neg}/x,x)$
	$\text{Tr}/x = !x/_{(r)};\text{Tr}/x \oplus \tau_{(\delta)};0$

Iteration 2: C	Iteration 2: E_C
$\text{Neg}/x,x \to^{\varepsilon} \text{Tr}/x + \text{Neg}/x,x$	$\text{Neg}/x,x = ?x/_{(r)};\text{Inh}/x,x \oplus \tau_{(\varepsilon)};(\text{Tr}/x\|\text{Neg}/x,x)$
$\text{Tr}/x \to^{\delta} 0$	$\text{Tr}/x = !x/_{(r)};\text{Tr}/x \oplus \tau_{(\delta)};0$
$\text{Tr}/x + \text{Neg}/x,x \to^{r\gamma} \text{Tr}/x + \text{Inh}/x,x$	$\text{Inh}/x,x = \tau_{(\eta)};\text{Neg}/x,x$

Iteration 3: C (Result)	Iteration 3: E_C (Termination)
$\text{Neg}/x,x \to^{\varepsilon} \text{Tr}/x + \text{Neg}/x,x$	no change
$\text{Tr}/x \to^{\delta} 0$	
$\text{Tr}/x + \text{Neg}/x,x \to^{r\gamma} \text{Tr}/x + \text{Inh}/x,x$	Resulting initial conditions: $\text{Neg}/x,x$
$\text{Inh}/x,x \to^{\eta} \text{Neg}/x,x$	

In the context of this example we can discuss a basic question: if we can translate freely between processes and chemistry, why are processes better than chemistry? One answer, which is at the core of managing large models, is: consistent parameterization and modularization. With processes, we can describe a gate $\text{Neg}(a,b)$ as a module with input a and output b. We can later connect several of these modules, and even connect them in loops like Neg(x,x) without problems. In the case of chemical reactions, instead, we would have a "chemical module" Neg(A,B) consisting of the reactions involving the chemicals A,B and the gate. In particular, the module would contain degradation reactions for both species: $A \to^{\delta} 0$, $B \to^{\delta} 0$, because one possible system is simply Neg(A,B). If we then want to connect several gates, we have to make copies of the module by appropriately instantiating the species A and B. But even Neg(A,A) goes wrong, because this creates two copies of the reaction $A \to^{\delta} 0$, resulting in a doubled degradation rate for A. This problem does not occur with the processes as defined above (and does not even require any planning), because each molecule is a process that knows how to degrade. Parameterized chemical reactions are not as good as parameterized processes.

4.6 Processes to ODEs: The Kermack-McKendrick Model of Epidemics

This example, and the next one, are not about chemistry, but they describe interactions governed by laws equivalent to the law of mass action ("chance of collision"), and use reactions to derive the ODEs.

E	$Ch_\gamma(E)$	ODE$(Ch_\gamma(E))$
$S = ?i_{(t)};I$ $I = !i_{(t)};I \oplus ?i_{(t)};I \oplus \tau_{(r)};R$ $R = ?i_{(t)};R$	$v_1 : S+I \to^{t\gamma} I+I$ $v_2 : I+I \to^{t\gamma} I+I$ $v_3 : I \to^{r} R$ $v_4 : R+I \to^{t\gamma} R+I$	$[S]^\bullet = -t\gamma[S][I]$ (v_1) $[I]^\bullet = t\gamma[S][I] - r[I]$ (v_1, v_3) $[R]^\bullet = r[I]$ (v_3)

In this *SIR* model, inspired by [22], we map out the behavior of individuals during an epidemic. A *Susceptible* individual may become *Infected* by interaction with an Infected at rate t. A *Recovered* may be infected (with no effect). An Infected may spontaneously become Recovered at rate r, or may infect a Susceptible, or a Recovered, or another Infected (with no effect).

Although we start with an intuitive process-oriented description of individual behavior, the resulting ODE system is exactly the Kermack-McKendrick population model [19]. Moreover, we may notice that reactions v_2 and v_4 do not contribute to the ODE translation. This suggests that the process model can in fact be simplified to $S = ?i_{(t)};I, I = !i_{(t)};I \oplus \tau_{(r)};R, R = 0$, which produces only (v_1, v_3) and hence the same ODE system. Below we run the processes with SPiM [23], the chemical reactions with CellDesigner (which converts them to ODEs) [9], and the ODEs with Matlab [28].

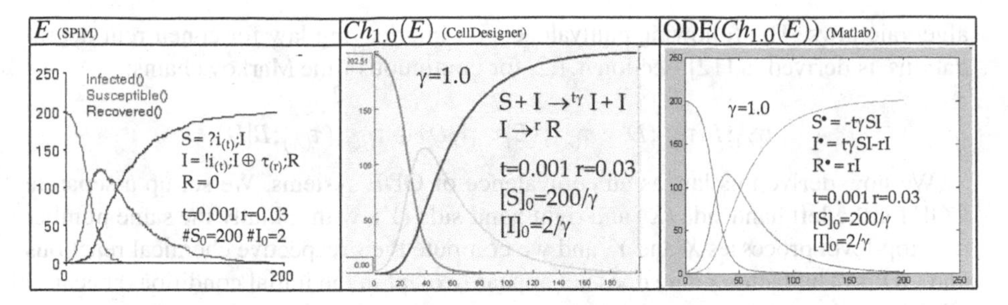

4.7 Unbounded Processes to ODEs: The Lotka-Volterra Model of Predation

In the Lotka-Volterra model of predation, prey (H) can breed without bounds, but is culled by predators (C), which can reproduce by predation, but have a regular mortality rate.

E	$Ch_\gamma(E)$	$ODE(Ch_\gamma(E))$
$H = \tau_{(b)};(H\|H) \oplus ?c_{(p)};0$ $C = \tau_{(m)};0 \oplus !c_{(p)};(C\|C)$	$v_1 : C \rightarrow^m 0$ $v_2 : H \rightarrow^b H+H$ $v_3 : H+C \rightarrow^{p\gamma} C+C$	$[H]^\bullet = b[H] - p\gamma[H][C] \quad (v_2,v_3)$ $[C]^\bullet = -m[C] + p\gamma[H][C] \quad (v_1,v_3)$

The resulting ODEs are the Lotka-Volterra equations [2], for the case where the rates at which prey decrease and predators increase are equal (p). Different ratios can be modeled (e.g., a normal form of: $C = \tau_{(m)};0\oplus!c_{(p)};!c_{(p)};!c_{(p)};(C|C)$), but such a model of sequential predation no longer corresponds exactly to the original Lotka-Volterra. A SPiM stochastic simulation of E is shown: it quickly leads to extinction, unlike the ODE model that with the same parameters oscillates indefinitely.

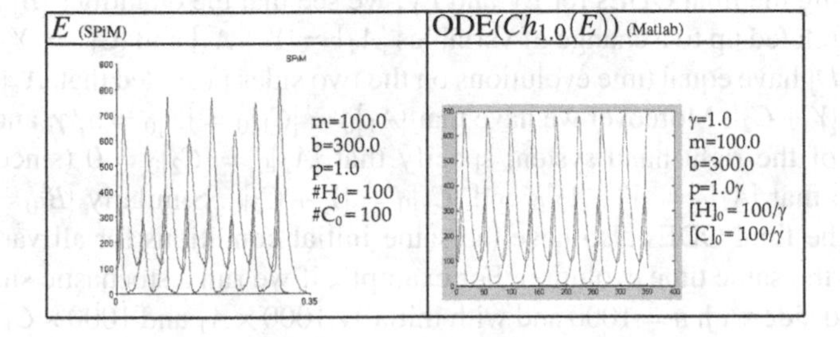

4.8 Process Equivalences from ODEs: The Markovian Interleaving Law

The translation of processes to ODEs can be regarded as a semantics of processes, and in particular it induces an equivalence over processes that can be used to derive algebraic laws. The following equivalence, an interleaving law for concurrent degradations, is derived in [12], section 4.1.2, for continuous-time Markov chains:

$$\tau_{(\lambda)};B\mid\tau_{(\mu)};D = \tau_{(\lambda)};(B\mid\tau_{(\mu)};D) \oplus \tau_{(\mu)};(\tau_{(\lambda)};B\mid D)$$

We now derive this law as an equivalence of ODE systems. We set up a separate CGF for the left hand side (X) and right hand side (Y), with initially the same number n of top-level processes X and Y, and we compute their respective chemical reactions and ODEs. The factor γ here does not appear (except in the initial conditions) because all the reactions are exponential decays:

Left hand side: $X = \tau_{(\lambda)};B\mid\tau_{(\mu)};D$		Right hand side: $Y = \tau_{(\lambda)};(B\mid\tau_{(\mu)};D)\oplus\tau_{(\mu)};(\tau_{(\lambda)};B\mid D)$	
$A_1 = \tau_{(\lambda)};B$ $C_1 = \tau_{(\mu)};D$ $n\times A_1\mid n\times C_1$	The CGF E_X,P_X (Sec 2.3) ←initial conditions	$Y = \tau_{(\lambda)};(B\mid C_2)\oplus\tau_{(\mu)};(A_2\mid D)$ $C_2 = \tau_{(\mu)};D$ $A_2 = \tau_{(\lambda)};B$ $n\times Y$	The CGF E_Y,P_Y ←initial conditions
$A_1 \rightarrow^{\lambda} B$ $C_1 \rightarrow^{\mu} D$ $[A_1]_0 = n/\gamma$ $[C_1]_0 = n/\gamma$	$Ch_\gamma(E_X,P_X)$: (Sec 2)	$Y \rightarrow^{\lambda} B + C_2$ $Y \rightarrow^{\mu} A_2 + D$ $C_2 \rightarrow^{\mu} D$ $A_2 \rightarrow^{\lambda} B$ $[Y]_0 = n/\gamma$	$Ch_\gamma(E_Y,P_Y)$
$[A_1]^\bullet = -\lambda[A_1]$ $[B]^\bullet = \lambda[A_1]$ $[C_1]^\bullet = -\mu[C_1]$ $[D]^\bullet = \mu[C_1]$	ODE for $Ch_\gamma(E_X)$ (Sec 3.2)	$[Y]^\bullet = -\lambda[Y] - \mu[Y]$ $[A_2]^\bullet = \mu[Y] - \lambda[A_2]$ $[B]^\bullet = \lambda[Y] + \lambda[A_2]$ $[C_2]^\bullet = \lambda[Y] - \mu[C_2]$ $[D]^\bullet = \mu[Y] + \mu[C_2]$	ODE for $Ch_\gamma(E_Y)$
		$[Y+A_2]^\bullet = -\lambda[Y+A_2]$ $[B]^\bullet = \lambda[Y+A_2]$ $[Y+C_2]^\bullet = -\mu[Y+C_2]$ $[D]^\bullet = \mu[Y+C_2]$	Derived ODE

The final ODE on the right is derived from the one above it, because $[Y+A_2]^\bullet = [Y]^\bullet + [A_2]^\bullet = (-\lambda[Y] - \mu[Y]) + (\mu[Y] - \lambda[A_2]) = -\lambda[Y] - \lambda[A_2] = -\lambda[Y+A_2]$, and $[B]^\bullet = \lambda[Y] + \lambda[A_2] = \lambda[Y+A_2]$.

Comparing the final ODEs for E_X and E_Y, we see that the quantities $[B]$ and $[D]$ are identically related up to a change of variables $[A_1] = [Y+A_2]$ and $[C_1] = [Y+C_2]$. That is, $[B]$ and $[D]$ have equal time evolutions on the two sides provided that $[A_1] = [Y+A_2]$ and $[C_1] = [Y+C_2]$. Moreover we have that $[A_1]_0 = [C_1]_0 = [Y]_0 = n/\gamma$, and the initial conditions of the right hand system specify that $[A_2]_0 = [C_2]_0 = 0$ (since only Y is present), so that $[A_1]_0 = [Y+A_2]_0$ and $[C_1]_0 = [Y+C_2]_0$. Similarly $[B]_0 = [D]_0 = 0$. Therefore the final ODEs also have the same initial conditions for all variables, and hence have the same time evolution. For example, if we run a stochastic simulation of the left hand side with $n = 1000$ and with initially $1000\times A_1$ and $1000\times C_1$, we obtain the same curves for B and D than a simulation of the right hand side with initially $1000\times Y$. The figure below illustrates the case with rates $\lambda = 1.0$, $\mu = 2.0$.

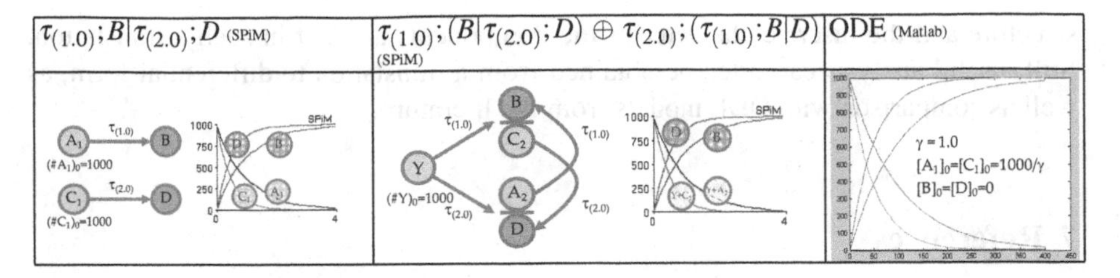

5 Related Work

A direct translation from a stochastic process algebra (PEPA) to ODEs is presented in [5,16]. The rate equation and the *activity matrix* techniques presented there, are similar to the stoichiometric techniques from chemistry (section 2). Differences in apparent rate computations between PEPA and the law of mass action can be easily adapted either way.

A main difference between our CGF and PEPA, however, is our ability to represent chemical reactions of the form $A \rightarrow^r B+C$: these reactions require a process to "split" in two. PEPA, instead, is intentionally restricted to the composition of purely sequential processes, to enable Markov chain analysis by linear algebra. At the level of ODEs, this ability allows us, for example, to give a process model corresponding to the Lotka-Volterra equations, which are based on an unbounded growth of prey ($H \rightarrow^b H + H$, example 4.7). In addition, the translation of parametric processes (CPF) does not appear to have been considered before, and this is very useful for parameterizing and modularizing models of even modest size [1].

6 Conclusions

We have shown how to go from a subset of π-calculus to ODEs, and how to use chemistry as a stepping stone to simplify the translation, avoiding a direct expression of a rate equation for processes [6]. A mapping to ODEs is necessary to validate and compare models of biochemistry written in process calculi, with respect to the wider and deeper literature of chemical and ODE models.

An advantage of process algebra modeling shines through from this analysis: compositionality. Parametric models can be written in process calculi without reference to initial conditions (e.g., a library of gene gates, as in example 4.5), and then reused without change by adding specific initial conditions (e.g., by wiring gene gates to produce specific gene networks). In contrast, under translations such as the one shown here, any given network expands to a different set of chemical or differential equations. The resulting model consists of a large "flat" set of equations that unrolls the state space (as shown in [6]), and that may oddly relate in differential form entities that actually only exist in discrete quantity (the genes of example 4.5). Both the network structure and the discrete character of the components are lost in such a translation. Still, useful analysis can often be obtained from a translation to differential form, as well as comparison with ODE models from the literature.

7 References

1. R. Blossey, L. Cardelli, A. Phillips: A Compositional Approach to the Stochastic Dynamics of Gene Networks. Transactions on Computational Systems Biology IV, LNCS 3939, 99-122. Springer 2006.
2. W.E. Boyce, R.C. DiPrima. Elementary Differential Equations and Boundary Value Problems, 5th ed. New York: Wiley, p. 494, 1992.
3. M. Bernardo, L. Donatiello, R. Gorrieri. MPA: a Stochastic Process Algebra. Technical Report UBLCS-94-10, University of Bologna, Laboratory of Computer Science. 1994
4. P. Buchholz. Markovian Process Algebra: Composition and Equivalence. In Proc. PAPM '94, Erlangen (Germany), 11-30, 1994.
5. M. Calder, S. Gilmore, J. Hillston: Automatically Deriving ODEs from Process Algebra Models of Signalling Pathways. Proc. Computational Methods in Systems Biology 2005, pp 204-215.
6. L. Cardelli: On Process Rate Semantics. Theoretical Computer Science 391(3) 190-215, Elsevier, 2008. DOI: <http://dx.doi.org/10.1016/j.tcs.2007.11.012>.
7. N. Chabrier, M. Chiaverini, V. Danos, F. Fages and V. Schchter. Modeling and querying biomolecular interaction networks. Theoretical Computer Science, 2004.
8. M. Dam: On the Decidability of Process Equivalences for the pi-Calculus. Theoretical Computer Science 183, 215-228, 1997.
9. A. Funahashi, N. Tanimura, M. Morohashi, H. Kitano. CellDesigner: a process diagram editor for gene-regulatory and biochemical networks, BIOSILICO, 1:159-162, 2003.
10. D.T. Gillespie. Exact Stochastic Simulation of Coupled Chemical Reactions. Journal of Physical Chemistry 81, 2340–2361. 1977.
11. D.T. Gillespie: The chemical Langevin equation. Journal of Chemical Physics 113(1), 297-306, 2000.
12. H. Hermanns: Interactive Markov Chains. Springer LNCS, vol 2428, 2002.
13. H. Hermanns, M. Rettelbach. Syntax, Semantics, Equivalences, and Axioms for MTIPP. Proc. of PAPM '94, Erlangen (Germany), pp 71-87. 1994.
14. N. Gtz, H. Hermanns, U. Herzog, V. Mertsiotakis, M. Rettelbach. Stochastic Process Algebras: Constructive Specification Techniques Integrating Functional, Performance and Dependability. In Baccelli and Mitrani (eds): Quantitative Modelling in Parallel Systems. Chapter 1, Springer 1995.
15. J. Hillston. A compositional approach to performance modelling. Cambridge University Press, 1996.

16. J. Hillston: Fluid flow approximation of PEPA models. In Proceedings of the Second International Conference on the Quantitative Evaluation of Systems, 33-43. IEEE Press, 2005.
17. A. Hinton, M. Kwiatkowska, G. Norman, D. Parker. PRISM: A Tool for Automatic Verification of Probabilistic Systems. In H. Hermanns, J. Palsberg (Eds.): Proc. TACAS'06. Springer LNCS 3920, 441-444, 2006.
18. F. Horn, R. Jackson. General mass action kinetics. Arch. Rational Mech. Anal. 47, 81–116, 1972.
19. W.O. Kermack, A.G. McKendrick: A Contribution to the Mathematical Theory of Epidemics. Proc. Roy. Soc. Lond. A 115, 700-721, 1927.
20. H. Kitano: A graphical notation for biochemical networks. BioSilico 1(5): 169-76. 2003.
21. R. Milner: Communicating and Mobile Systems: The p-Calculus. Cambridge University Press, 1999.
22. R. Norman, C. Shankland. Developing the Use of Process Algebra in the Derivation and Analysis of Mathematical Models of Infectious Disease. Proc. Computer Aided Systems Theory - EUROCAST 2003. Springer LNCS 2809, 404-414, 2003.
23. A. Phillips, L. Cardelli: A Correct Abstract Machine for the Stochastic Pi-calculus. Proc. BioConcur'04.
24. C. Priami. Stochastic p-calculus. The Computer Journal, 38, 578–589, 1995.
25. C. Priami, A. Regev, E. Shapiro, W. Silverman: Application of a stochastic name-passing calculus to representation and simulation of molecular processes. Information Processing Letters 80, 25-31. 2001.
26. A. Regev. Computational systems biology: a calculus for biomolecular knowledge. Ph.D. Thesis, Tel Aviv University, 2002.
27. A. Regev, E. Shapiro. Cellular abstractions: Cells as computation. Nature 419 343. 2002.
28. The Mathworks: Matlab. http://www.mathworks.com.
29. J.M.G. Vilar, H.Y. Kueh, N. Barkai, S. Leibler: Mechanisms of noise-resistance in genetic oscillators. PNAS 99(9) 5988-5992. 2002.
30. O. Wolkenhauer, M. Ullah, W. Kolch, K. Cho. Modelling and simulation of intracellular dynamics: Choosing an appropriate framework. IEEE Transactions on NanoBioscience 3, 200-207. 2004.

Differential Linear Logic and Processes

Thomas Ehrhard

Preuves, Programmes & Systèmes
Université Paris Diderot and CNRS

In Linear Logic, the tensor/par and plus/with dualities are lost when exponentials come in.

- The "?" modality is introduced by the *weakening* and *dereliction* rules, and *contraction* allow to contract two occurrences of a formula $?A$ (from an unique premise sequent) into a single one
- whereas the "!" modality can be introduced only by mean of a *promotion* rule.

By adding new rules for the "!" modality, one retrieves, in the exponential fragment, a duality and a symmetry similar to that of the multiplicative fragment. These new rules are

- *coweakening* and *cocontraction* which are new ways of introducing "!" formulas
- and *cocontaction*, which allow to contract two occurrences of $!A$ (from two different premise sequents) into a single one.

Corresponding reduction (cut-elimination) rules are added, which express operationally this new $!/?$ symmetry. These reduction rules are semantically justified, when interpreting the new logical rules for "!" as standard operations on functions (in particular, codereliction corresponds to differentiation of a function at point 0 of a vector space). This extended linear logic is called Differential Linear Logic (DLL).

This new symmetry adds expressive power to linear logic. In particular, we show how a fragment of the π-calculus can be translated into *differential interaction nets* (a system if interaction nets where cells correspond to rules of DLL) and how the dereliction/codereliction reductions of this differential interaction net simulate the reductions of the process. Last, we present a simple denotational model of differential interaction nets, in a category of sets and relations. This model, which is also a natural model of the pure lambda-calculus (β and η), becomes therefore a model of the considered fragment of the π-calculus, and we explore some of its properties.

Please use the following format when citing this chapter:

Ehrhard, T., 2008, in IFIP International Federation for Information Processing, Volume 273; *Fifth IFIP International Conference on Theoretical Computer Science*; Giorgio Ausiello, Juhani Karhumäki, Giancarlo Mauri, Luke Ong; (Boston: Springer), pp. 283.

Solving Monotone Polynomial Equations

Javier Esparza, Stefan Kiefer, and Michael Luttenberger

Institut für Informatik, Technische Universität München, 85748 Garching, Germany
{esparza,kiefer,luttenbe}@in.tum.de

Abstract. We survey some recent results on iterative methods for approximating the least solution of a system of monotone fixed-point polynomial equations.

1 Introduction

Consider the following problem formulated by Francis Galton in the (politically incorrect) 19th century [26], and quoted by Thomas Harris in his classical text on branching stochastic processes [20]:

> Let $p_0, p_1, p_2 \ldots$ be the respective probabilities that a man has $0, 1, 2, \ldots$ sons, let each son have the same probability for sons of his own, and so on. What is the probability that the male line is extinct after r generations, and more generally what is the probability for any given number of descendants in the male line in any given generation?

We are interested here in the probability that the male line *eventually* becomes extinct. A little thought shows that this probability is a solution of the fixed-point equation

$$X = \sum_{n \geq 0} p_n X^n \tag{1}$$

and after some more thought one concludes that it is in fact the smallest solution.

Consider now the following stochastic context-free grammar (i.e., a grammar whose productions are annotated with probabilities) with axiom X:

$$\begin{array}{lll} X \xrightarrow{0.4} XY, & X \xrightarrow{0.6} a & \\ Y \xrightarrow{0.3} XY, & Y \xrightarrow{0.4} YZ, & Y \xrightarrow{0.3} b \\ Z \xrightarrow{0.3} XZ, & Z \xrightarrow{0.7} b & \end{array}$$

What is the probability that the grammar eventually generates a word, i.e., a string of non-terminals? Again, it is not to difficult to show that it is equal to the X-component of the least solution of the following system of equations.

$$\begin{aligned} X &= 0.4XY + 0.6 \\ Y &= 0.3XY + 0.4YZ + 0.3 \\ Z &= 0.3XZ + 0.7 \end{aligned} \tag{2}$$

Please use the following format when citing this chapter:

Esparza, J., Kiefer, S. and Luttenberger, M., 2008, in IFIP International Federation for Information Processing, Volume 273; *Fifth IFIP International Conference on Theoretical Computer Science*; Giorgio Ausiello, Juhani Karhumäki, Giancarlo Mauri, Luke Ong; (Boston: Springer), pp. 285–289.

Notice that the vector $(1,1,1)$ is a solution of the system. We will later investigate whether it is the least solution or not.

Equations (1) and (2) are two examples of *monotone systems of polynomial equations* (MSPEs for short). MSPEs are systems of the form

$$X_1 = f_1(X_1,\ldots,X_n)$$
$$\vdots$$
$$X_n = f_n(X_1,\ldots,X_n)$$

where $f_1,\ldots,f_n$ are polynomials with *positive* real coefficients. In vector form we denote an MSPE by $X = f(X)$. We call the vector $f(X)$ of polynomials a *monotone system of polynomials*, or MSP. Obviously, a solution of $X = f(X)$ is a fixed-point of $f(X)$, and vice versa. Further, any solution of $X = f(X)$ can be visualized as a point of intersection of the submanifolds defined by the n implicit functions $f_i(X) - X_i = 0$. In particular, when the polynomials of $f(X)$ are quadratic the solutions of $X = f(X)$ correspond to the intersection of n quadrics. Figure 1 shows the graph of such a quadratic MSPE with $n = 2$.

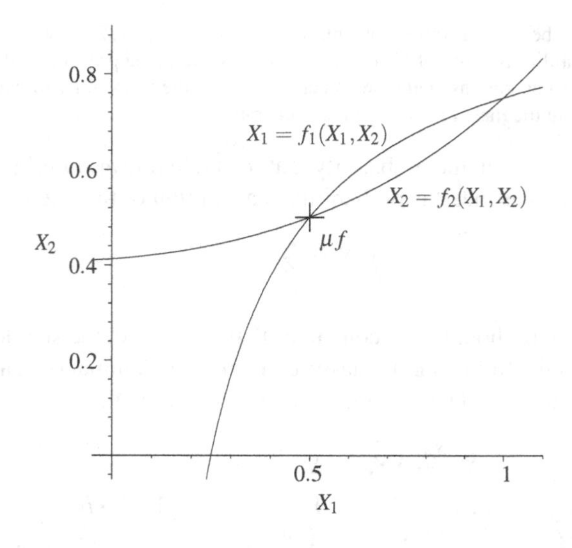

Fig. 1 Graphs of the equations $X_1 = f_1(X_1,X_2)$ and $X_2 = f_2(X_1,X_2)$ with $f_1(X_1,X_2) = X_1X_2 + \frac{1}{4}$ and $f_2(X_1,X_2) = \frac{1}{6}X_1^2 + \frac{1}{9}X_1X_2 + \frac{2}{9}X_2^2 + \frac{3}{8}$. There are two real solutions in $\mathbb{R}^2_{[0,\infty]}$, the least one is labelled with μf.

We call MSPEs and MSPs "monotone" because $x \leq x'$ implies $f(x) \leq f(x')$ for every $x, x' \in \mathbb{R}^n_{\geq 0}$. This is a bit imprecise, because not every monotone polynomial has positive coefficients. Perhaps "positive systems" would be a better name, but since we have used the term "monotone" in several papers we stick to it.

MSPEs appear naturally in the analysis of many stochastic models, such as stochastic context-free grammars (with numerous applications to natural language processing

[23, 19], and computational biology [24, 5, 4, 22]), probabilistic programs with procedures [9, 2, 13, 11, 10, 12, 14], web-surfing models with back buttons [16, 17], and branching processes [20], a topic in stochastic theory that can be traced back to Galton's problem.

In the last years Etessami and Yannakakis [13] and ourselves [21, 8] have studied the problem of solving MSPEs. This paper gives a succinct—and informal—overview of our results.

2 Some Definitions and Facts

Let $\mathbb{R}_{[0,\infty]}$ denote the set of non-negative reals extended with ∞. We extend the definitions of sum and product as usual: $\infty + k = \infty$ for every $k \in \mathbb{R}_{[0,\infty]}$, $\infty \cdot 0 = 0$, and $\infty \cdot k = \infty$ for every $k \in \mathbb{R}_{[0,\infty]} \setminus \{0\}$. The resulting algebraic structure is the *real semiring*. MSPEs are systems of fixed-point equations over the real semiring.

Given two vectors $u, v \in \mathbb{R}^n_{[0,\infty]}$, we say that $u \leq v$ holds if $u_i \leq v_i$ holds for every $1 \leq i \leq n$, where u_i, v_i are the i-th components of u and v, respectively. This is the pointwise order on vectors of reals. The first positive result on MSPEs is a direct consequence of Kleene's theorem:

Theorem 1 (Kleene's fixed-point theorem). *Every MSP $f(X)$ has a least fixed-point μf in $\mathbb{R}^n_{[0,\infty]}$ with respect to the pointwise order. Moreover, the sequence $(\kappa_f^{(k)})_{k\in\mathbb{N}}$ given by*

$$\begin{aligned} \kappa_f^{(0)} &:= 0 \\ \kappa_f^{(k+1)} &:= f(\kappa_f^{(k)}) = f^{k+1}(0) \end{aligned}$$

is non-decreasing with respect to $\leq$ (i.e., $\kappa_f^{(k)} \leq \kappa_f^{(k+1)}$) and converges to μf.

We call $(\kappa_f^{(k)})_{k\in\mathbb{N}}$ the *Kleene sequence*, and its elements the *Kleene approximants of μf*.

Example 1. For the system (2) we obtain:

$$\begin{aligned} \kappa_f^{(0)} &= (0,0,0), & \kappa_f^{(1)} &= (0.6,0.3,0.7), \\ \kappa_f^{(2)} &= (0.672,0.438,0.826), & \kappa_f^{(3)} &= (0.718,0.533,0.867), \\ \kappa_f^{(4)} &= (0.753,0.600,0.887), & \kappa_f^{(5)} &= (0.781,0.648,0.900), & \ldots \end{aligned}$$

The least solution of a system of *linear* equations (monotone or not) satisfies some good properties that no longer hold for MSPEs. It is easy to show (using for instance Cramer's rule) that if the coefficients are rationals, then the least solution is also rational. However, using Galois theory one can prove that the least solution of a polynomial system may not be expressible by radicals. For instance:

Fact 1. *The least fixed-point of*

$$X = \frac{1}{6}X^6 + \frac{1}{2}X^5 + \frac{1}{3}\,. \tag{3}$$

is not expressible by radicals.

This fact also holds for quadratic systems, i.e., systems in which all polynomials have at most degree 2. Given an MSP f over a set $\mathscr{X}$ of variables, it is easy to construct a quadratic MSP g over a larger set $\mathscr{X} \cup \mathscr{Y}$ such that the projection of μg onto $\mathscr{X}$ is equal to μf. The construction is very similar to the one that brings a context-free grammar in Chomsky normal form. For instance, it "expands" Equation (3) into the system

$$\begin{aligned} X &= \frac{1}{6}XX_5 + \frac{1}{2}XX_4 + \frac{1}{3} \\ X_n &= XX_{n-1} \qquad (\text{for } n = 5,4,3) \\ X_2 &= X^2 \end{aligned}$$

Since this expansion involves only a linear blowup, we can take quadratic MSPEs as a normal form of MSPEs.

The least solution of linear MSPEs is not only rational, but a succinct rational. Consider a system of dimension n (i.e., with n equations) whose coefficients are given as ratios of m-bit integers. It is easy to show using Cramer's rule that the least solution can be written as the quotient of two natural numbers with at most $O(n^2m + n\log n)$ bits. As a consequence, we get

Fact 2. *Let $X = f(X)$ be a linear MSPE of dimension n whose coefficients are given as ratios of m-bit integers. For every component μf_i of the least fixed-point of f: if $0 < \mu f_i < \infty$ then*

$$\frac{1}{2^{O(n^2m+n\log n)}} \leq \mu f_i \leq 2^{O(n^2m+n\log n)}$$

(where the constant of the Big-Oh notation is independent of f).

Since the least fixed-point of a MSP can be irrational, there is no bound on the number of digits needed to write it down. However, using results of [8] we can still give a lower and an upper bound:

Fact 3. *Let f be a quadratic MSP of of dimension n whose coefficients are given as ratios of m-bit integers. For every component μf_i of the least fixed-point of f: if $0 < \mu f_i < \infty$ then*

$$\frac{1}{2^{m\cdot 2^{O(n)}}} \leq \mu f_i \leq 2^{m\cdot 2^{O(n)}}$$

(where the constant of the Big-Oh notation is independent of f).

So, loosely speaking, while the least fixed-point of a linear system is at most exponential in the dimension of the system, the least solution of a quadratic system is at most double exponential.

It is easy to find examples of quadratic MSPs in which the least fixed-point is rational and double exponential. The n-th component of the least solution of system

$$\begin{aligned} X_1 &= k \\ X_2 &= X_1^2 \\ &\vdots \\ X_n &= X_{n-1}^2 \end{aligned}$$

is equal to $k^{2^{(n-1)}}$.

3 Computational complexity

The fundamental decision problem for MSPs is whether $(\mu f)_i \sim a$ holds for a given MSP f and a component i, where a is some positive rational number and $\sim \in \{\leq, =, \geq\}$. Let us call this problem *MSP-DECISION*. Little is known about its computational complexity. The problem lies in PSPACE:

Consider e.g. a two-dimensional MSPE $X_1 = f_1(X_1, X_2), X_2 = f_2(X_1, X_2)$. To decide whether $(\mu f)_1 \leq a$ holds one can equivalently decide if the following formula is true:

$$\exists x_1 \in \mathbb{R}, x_2 \in \mathbb{R} : x_1 = f_1(x_1, x_2) \wedge x_2 = f_2(x_1, x_2) \wedge x_1, x_2 \geq 0 \wedge x_1 \leq a$$

Such formulas can be decided in PSPACE, because the first-order theory of the reals is decidable, and its existential fragment is even in PSPACE [3].

For a lower bound, we introduce the problem *SQUARE-ROOT-SUM*:

Given $k+1$ natural numbers $n_1, \ldots, n_k$ and b, determine whether $\sum_{i=1}^{k} \sqrt{n_i} \leq b$ holds.

The SQUARE-ROOT-PROBLEM is a natural subproblem of many questions in computational geometry. For instance, the length of the boundary of a polygon whose vertices lie in $\mathbb{Z}^2$ is a sum of square roots of integers. It has been a major open problem since the 70s whether SQUARE-ROOT-SUM belongs to NP. The problem can easily be reduced to MSP-DECISION:

Suppose we are given $n_1 = 2$, $n_2 = 3$, and $b = 3$, and we want to decide if $\sqrt{2} + \sqrt{3} \leq 3$. One would like to come up with an MSP $f(X)$ such that $(\mu f)_1 = \sqrt{2}, (\mu f)_2 = \sqrt{3}, (\mu f)_3 = \sqrt{2} + \sqrt{3}$, so that deciding $\sqrt{2} + \sqrt{3} \leq 3$ is equivalent to deciding $(\mu f)_3 \leq 3$. One has to be careful though, because for instance the equation $X_1 = X_1^2 + X_1 - 2$ is not an MSPE. It was shown in [13] how to overcome this problem: Instead of encoding e.g. $\sqrt{2}$ directly, it suffices to encode $a + b \cdot \sqrt{2}$ for some rationals a, b.

The least solution of the equation $X = X^2 + (1 - \lambda^2 \cdot n)/4$ equals $(1 - \lambda\sqrt{n})/2$. So, by choosing for λ a small enough rational number we get a 1-dimensional MSP whose least solution is $a + b \cdot \sqrt{n}$ for some rationals a, b. In our example we can set $\lambda = \frac{1}{\max(2,3)} = \frac{1}{3}$ which leads to the following MSPE.

$$\begin{aligned} X_1 &= X_1^2 + \tfrac{1-\frac{2}{9}}{4} \\ X_2 &= X_2^2 + \tfrac{1-\frac{3}{9}}{4} \\ X_3 &= X_1 + X_2 \end{aligned}$$

Its least solution is

$$\mu f = \left(\frac{1}{2} - \frac{1}{6}\sqrt{2}, \frac{1}{2} - \frac{1}{6}\sqrt{3}, 1 - \frac{1}{6}(\sqrt{2}+\sqrt{3})\right) .$$

So, the question whether $\sqrt{2}+\sqrt{3} \leq 3$ holds can be translated into the question whether $(\mu f)_3 \geq 1 - \frac{1}{6} \cdot 3 = \frac{1}{2}$ holds.

It follows from this reduction that proving membership of MSP-DECISION in NP would be a major breakthrough.

An interesting issue is the complexity of MSP-DECISION in the Blum-Shub-Smale computational model, in which all operations on rationals take unit time independently of their size. SQUARE-ROOT-SUM can be decided in polynomial time in this model [25], but it is open whether the result extends to MSP-DECISION.

4 Approximating the Least Fixed-Point: Newton's Method

For most practical purposes, the main computational problem concerning MSPs is the approximation of the least fixed-point up to a given accuracy. Kleene's method can be applied, and it is very robust: it always converges when started at 0, for any MSP. On the other hand, the convergence speed of the Kleene sequence can be very poor. Before presenting an example, we define a notion of convergence order that differs from the one commonly used in numerical mathematics, but is particularly natural for computer science.

Let $(a_k)_{k\geq 0}$ be a non-decreasing sequence of vectors over the real semiring such that $\lim_{k\to\infty} a_k = a < \infty$. The *convergence order* of the sequence is the function $\beta : \mathbb{N} \to \mathbb{N}$ defined as follows: $\beta(k)$ is the greatest natural number i such that

$$\frac{\|a - a_k\|}{\|a\|} \leq 2^{-i}$$

where $\|\cdot\|$ is some norm. We say that a sequence has linear, exponential, logarithmic, etc. convergence order if the function $\beta(k)$ grows linearly, exponentially, or logarithmically in k, respectively. Notice that the asymptotic behaviour of $\beta(k)$ is independent of the norm, because all norms are equivalent up to a constant. In the univariate case, $\beta(k)$ is the number of bits of a_k that coincide with the corresponding bits of a (the formalization of this intuition requires some care, like identifying 1 and 0.999...). For instance, for the sequence $(1-2^{-k})_{k\geq 0}$ we have $\beta(k) = k$, i.e., the first k bits of the k-th element of the sequence coincide with the first k bits of the limit.

Consider now this very simple but at the same time very illustrative quadratic MSPE in one variable:

$$X = 1/2 + 1/2X^2 \tag{4}$$

In Galton's problem, the least solution of this equation gives the extinction probability of an individual's descent line when every individual has 0 or 2 children with probability $1/2$. The least solution is 1. We have:

Fact 4. *The i-th Kleene approximant of $X = 1/2 + 1/2X^2$ satisfies $\kappa^{(i)} \leq 1 - \frac{1}{i+1}$ for every $i \geq 0$. So the Kleene sequence only has logarithmic convergence order.*

Example 2. Here are some of the Kleene iterates.

$$\begin{array}{lll} \kappa^{(0)} = 0, & \kappa^{(1)} = 0.5, & \kappa^{(2)} = 0.625 \\ \kappa^{(3)} = 0.695, & \kappa^{(4)} = 0.742, & \kappa^{(5)} = 0.775 \\ \ldots & & \end{array}$$

$$\kappa^{(20)} = 0.920,\ \ldots,\ \kappa^{(200)} = 0.990,\ \ldots,\ \kappa^{(2000)} = 0.9990,\ \ldots$$

Faster approximation techniques have been known for a long time. In particular, Newton's method, suggested by Isaac Newton more than 300 years ago, is a standard efficient technique for approximating a zero of a differentiable function. Since the least solution of a fixed-point equation $X = f(X)$ is a zero of $g(X) = f(X) - X$, the method can be applied to search for fixed-points of $f(X)$.

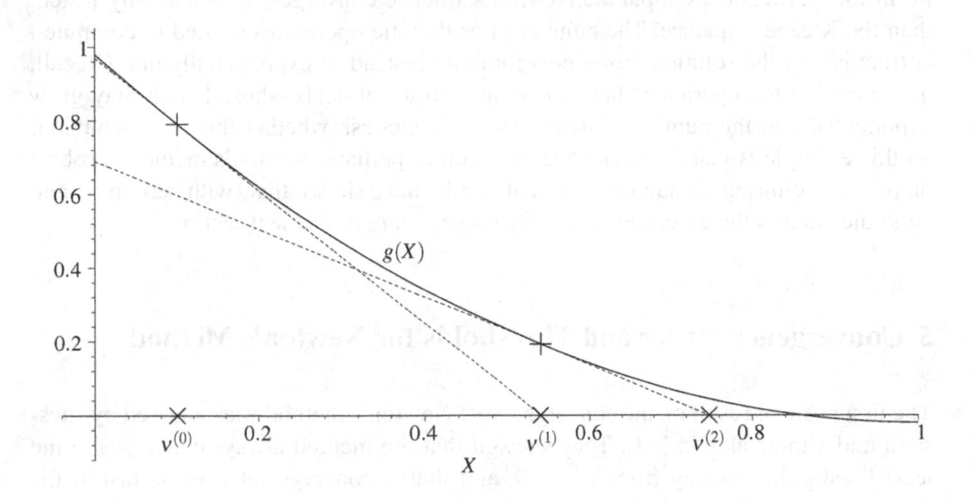

Fig. 2 Newton's method to find a zero of a one-dimensional function $g(X)$

We briefly recall the method for the case of one variable, see Fig. 2 for an illustration. Starting at some value $\nu^{(0)}$ "close enough" to the zero of $g(X)$, we proceed iteratively: given $\nu^{(i)}$, we compute a value $\nu^{(i+1)}$ closer to the zero than $\nu^{(i)}$. For that, we compute the tangent to $g(X)$ passing through the point $(\nu^{(i)}, g(\nu^{(i)}))$, and take $\nu^{(i+1)}$

as the zero of the tangent (i.e., the X-coordinate of the point at which the tangent cuts the X-axis). A little arithmetic leads to:

$$v^{(i+1)} = v^{(i)} + \frac{f(v^{(i)}) - v^{(i)}}{1 - f'(v^{(i)})}$$

Newton's method can be easily generalized to the multivariate case:

$$v^{(i+1)} = v^{(i)} + (\mathrm{Id} - f'(v^{(i)}))^{-1}(f(v^{(i)}) - v^{(i)})$$

where $f'(X)$ is the Jacobian of f, i.e., the matrix of partial derivatives of f, and Id is the identity matrix.

Notice that Newton's method is not restricted to the real semiring, it can be applied to any differentiable functions over the real field. However, when applied with this generality it is far less robust than Kleene's method: it may converge very slowly, converge only when started at a point very close to the zero—which must be guessed—or even not converge at all.

However, if we apply Newton's method to $f(X) = 1/2 + 1/2X^2$, starting at $v^{(0)} = 0$, we obtain:

Fact 5. *The i-th Newton approximant of $X = 1/2 + 1/2X^2$ satisfies $v^{(i)} = 1 - \frac{1}{2^i}$ for every $i \geq 0$. The i-th approximant has i correct bits, i.e., the Newton sequence has linear convergence.*

So in this particular example the Newton sequence converges "exponentially faster" than the Kleene sequence. The number of arithmetic operations needed to compute i correct bits of the solution grows polynomially instead of exponentially in i. (Recall, however, that the operations have to be applied to rationals whose length may grow exponentially in the number of iterations.) One can ask whether the good behaviour on this example is just a coincidence, or whether perhaps Newton's method is robust on the real semiring. A number of recent results have shown that (with certain ifs and buts) the latter is the case, and we briefly survey them in the next section.

5 Convergence Order and Thresholds for Newton's Method

The first positive result on the convergence of Newton's method was obtained by Etessami and Yannakakis in [13]. They showed that the method always converges to the least fixed-point starting from $v^{(0)} = 0$, and that it converges at least as fast at the Kleene sequence.[1]

Inspired by this positive result, we started to study the convergence order. Given an MSPE $X = f(X)$ whose least solution μf is finite, it is well-known that the convergence order depends critically on the Jacobian matrix at the least fixed-point, i.e., on

[1] More precisely, Etessami and Yannakakis proved the result for a structured version of the method, and we showed in [8] that this additional structure is not required for convergence (although it is convenient for efficiency).

$f'(\mu f)$. Every textbook proves that the method performs brilliantly when the matrix $(\mathrm{Id} - f'(\mu f))$ is non-singular: it exhibits *exponential* convergence order. So we focused our attention on the singular case, of which $f(X) = 1/2 + 1/2X^2$ is an example. By Fact 5 we can expect at most linear convergence. But perhaps the method converges more slowly on other examples?

It is convenient to start with the special case of *strongly connected* MSPEs. Loosely speaking, an MSPE is strongly connected if every variable depends on any other variable, where dependence is defined as follows. Given two variables X and Y, X *depends on* Y if either Y appears on the right-hand-side of the equation for X, or if there is a variable Z such that X depends on Z and Z depends on Y.

5.1 Strongly Connected MSPEs

We proved the following theorem in [21].

Theorem 2. *Let $f(X)$ be a strongly connected MSP such that μf is finite. There is a number t_f such that for every $i \geq 0$:*

$$\beta(t_f + i) \geq i\,.$$

In particular, the Newton sequence has linear convergence order.

We call t_f the *threshold* of $f(X)$. Loosely speaking, the theorem states that after crossing the threshold (i.e., from the t_f-th approximant onwards) the Newton sequence gains at least one bit of accuracy per iteration. The threshold itself is an upper bound on the number of iterations needed to obtain the first bit of the least fixed-point.

The proof of [21] was based on the following topological property of $\mathbb{R}^n$: if the infimum of the distances between points of two sets is 0, then the two sets have at least one common point. As a consequence, it was a purely existential proof, and provided no information on the size of the threshold. In [7] we obtained the following relation between the threshold and the minimal component of μf.

Theorem 3. *Let $f(X)$ be a quadratic strongly connected MSP of dimension n whose coefficients are given as ratios of m-bit integers. Let μ_{min} be the minimal component of μf. The threshold t_f of Theorem 2 satisfies*

$$t_f \leq 3n^2 m + 2n^2 \left|\log \mu_{min}\right|\,.$$

Moreover, if $f_i(0) > 0$ holds for every $1 \leq i \leq n$, then $t_f \leq 3mn$.

Example 3. Consider again the following MSPE, which was given as Equation (2) on page 285.

$$\begin{aligned} X &= 0.4XY + 0.6 \\ Y &= 0.3XY + 0.4YZ + 0.3 \\ Z &= 0.3XZ + 0.7 \end{aligned}$$

Using a result from [8], slightly stronger than Theorem 3 but technically more difficult to state, one can prove that the threshold of this system satisfies $t_f \leq 6$ for the maximum-norm (i.e., the norm of a vector is the absolute value of its maximal component). So $\beta(14) \geq 8$. After computing 14 Newton iterates we get $\nu^{(14)} = (0.983, 0.974, 0.993)$. As we have computed at least $\beta(14) \geq 8$ bits, we know that μf is at most $\nu^{(14)} + (2^{-8}, 2^{-8}, 2^{-8})$ which is strictly less than 1 in every component. Therefore, the stochastic context-free grammar from the introduction produces a terminal string with probability less than 1.

Combining Theorem 3 with Fact 3 we obtain:

Corollary 1. *Let $X = f(X)$ be a quadratic strongly connected MSPE of dimension n whose coefficients are given as ratios of m-bit integers. The threshold t_f of Theorem 2 satisfies $t_f \in m2^{O(n)}$.*

This corollary gives an exponential bound on the number of iterations needed to compute the first bit of the least fixed-point. It is open whether this bound is tight.

5.2 General MSPEs

The following example shows that an exponential number of iterations is sometimes needed for the first bit, if the MSPE is *not strongly connected*. We give a family of MSPEs in which the number of iterations needed to compute the first bit grows exponentially in the dimension of the system.

Example 4. Consider the following family of MSPEs.

$$\begin{aligned} X_1 &= 1/2 + 1/2 \cdot X_1^2 \\ X_2 &= 1/4 \cdot X_1^2 + 1/2 \cdot X_1 X_2 + 1/4 \cdot X_2^2 \\ &\vdots \\ X_n &= 1/4 \cdot X_{n-1}^2 + 1/2 \cdot X_{n-1} X_n + 1/4 \cdot X_n^2 \end{aligned} \tag{5}$$

The variable X_i depends on X_j if and only if $j \leq i$. So the dependence graph contains n strongly connected components, one for each variable. The least fixed-point of the system is the vector $(1, 1, \ldots, 1)$. We show in [21] that $\nu_n^{(2^{n-1})} \leq 1/2$ holds, and so that at least 2^{n-1} iterations of Newton's method are needed to obtain the first bit of X_n. The proof goes as follows. We consider a decomposed version of Newton's method, in which for a given k we perform k iterations of the normal method on the first equation, yielding a lower bound $a_1^{(k)}$ of the first component of the least fixed-point. Then we perform k iterations on the second equation *after setting* $X_1 := a_1^{(k)}$; by monotonicity, this yields a lower bound $a_2^{(k)}$ of the second component. Repeating this procedure we finally obtain a lower bound $a_n^{(k)}$ of the n-th component. It is easy to see that $\nu_i^{(k)} \leq$

$a_i^{(k)}$ holds, i.e, the decomposed method converges at least as fast as the method that performs k iterations on the whole system. Now, let $\delta_i^{(k)} = 1 - a_i^{(k)}$ be the error of the decomposed method. A simple analysis reveals that $\delta_{i+1}^{(k)} \geq \sqrt{\delta_i^{(k)}}$ holds for every $1 \leq i < n$. By Fact 5 we have $\delta_1^{(2^{n-1})} = (1/2)^{2^{n-1}}$, and so we get $\delta_n^{(2^{n-1})} \geq 1/2$, i.e., $\nu_n^{(2^{n-1})} \leq 1/2$.

So, intuitively, the problem of non-strongly connected systems is that the error gets "amplified" when we move up the graph of strongly connected components.

For MSPs that are not strongly connected, Newton's method still has linear convergence order, but a worse rate [8]:

Theorem 4. *Let $f(X)$ be a* clean *(see below) MSP such that μf is finite. There is a number t_f such that for every $i \geq 0$:*

$$\beta(t_f + i \cdot (n+1) \cdot 2^n) \geq i\,.$$

In particular, the Newton sequence has linear convergence order.

In order to make sure that Newton's method stays well-defined (i.e. that the matrix inverses exist) Theorem 4 assumes that the MSP is clean, i.e., $(\mu f)_i > 0$ for all i. An MSP can easily be made clean in linear time by identifying and removing the components with $(\mu f)_i = 0$: $(\mu f)_i = 0$ holds iff $(\kappa^{(n)})_i = 0$.

The rate in Theorem 4 is worse than in the strongly connected case: Newton's method needs (in the worst case) about 2^n iterations per bit, instead of only 1 as in the strongly connected case. This worst case is attained by the MSPE in Equation (5) above, so the exponential rate in Theorem 4 cannot be avoided. Unfortunately, we do not have an upper bound on the threshold t_f in this general case.

5.3 min-max-MSPEs

Theorem 4 forms the basis for the convergence analysis of a recent extension [6] of Newton's method to min-max-MSPEs, i.e., MSPEs where minimum and maximum are allowed as additional operators. Here is an example of a min-max-MSPE:

$$\begin{aligned} X &= \max\{0.7Y + 0.3\,, \quad 0.6XY + 0.4\} \\ Y &= \min\{X\,, \quad 0.8Y^2 + 0.2\} \end{aligned}$$

Such systems arise, for instance, in *extinction games*. Those games add two adversarial players to Galton's setting from the beginning: There are n species, each of which is controlled by one of two players, the *terminator* and the *rescuer*. Each player can apply actions to the individuals controlled by her; an action transforms an individual (probabilistically) into zero or more individuals. The terminator tries to extinguish all individuals, whereas the rescuer tries to save them. Natural questions are: What are optimal strategies[2] for the terminator and the rescuer, and what is the probability of extinction of all individuals, assuming that there is a single initial individual and the players follow optimal strategies?

The MSPE above can be thought of as an equation system for the extinction probabilities of two species X and Y. Species X is controlled by the terminator, whereas Y is controlled by the rescuer. The terminator can apply one of two possible actions to an X-individual: the first one kills the X-individual with probability 0.3, but with probability 0.7 transforms it to a Y-individual; the second action kills the X-individual with probability 0.4, but, with probability 0.6, keeps the X-individual and creates a Y-individual. What can the rescuer do with a Y-individual? She can choose between transforming it to an X-individual and a second action which kills the Y-individual with probability 0.3 and adds another Y-individual with probability 0.7.

It turns out that the X-component (resp. Y-component) of the least solution of the MSPE above equals the extinction probability assuming a single initial X-individual (resp. Y-individual) if both the terminator and the rescuer follow optimal strategies. Such systems also arise in the analysis of recursive simple stochastic games [14, 15].

In order to approximate the least solution of a min-max-MSPE, one could use Kleene iteration. But, as we have seen before (Fact 4), Kleene iteration may converge very slowly even without minimum and maximum. Therefore, in [6] we propose two methods for approximating the least solution of a min-max-MSPE. Both are iterative procedures based on Newton's method.

- The first method linearizes each polynomial appearing in the system (possibly inside a minimum or a maximum expression) by computing the "tangent" at the current iterate. One obtains a min-max-MSP whose polynomials have degree at most 1. Its least fixed-point can be computed exactly by a method from [18] that uses strategy iteration and linear programming. The result is the next iterate.
- The second method linearizes each max-polynomial appearing in the system (possibly inside a minimum expression) by computing the "tangent" at the current iterate. (A special "tie breaking" policy must be adhered to if the current iterate is at the "edge" between two polynomials inside a maximum expression.) One obtains a min-MSP whose polynomials have degree at most 1. Its least fixed-point can be computed exactly by solving a single linear program. The result is the next iterate.

Both methods have at least linear convergence order [6]:

Theorem 5. *Let $f(X)$ be a min-max-MSP such that μf is finite. There is a number t_f such that for every $i \geq 0$:*

$$\beta(t_f + i \cdot m \cdot (n+1) \cdot 2^n) \geq i\,,$$

where m is the number of possible strategies of the players. In particular, the two extensions of Newton's method have linear convergence order.

The first method converges somewhat faster whereas a single step of the second method is cheaper. The second method also computes ε-optimal strategies for the

[2] A strategy tells a player which action to apply to the individuals controlled by her.

terminator, i.e., strategies that achieve as extinction probabilities at least the current iterate.

We have used the second method to approximate the extinction probabilities assuming perfect strategies: A population that starts with a single X-individual (resp. Y-individual) becomes extinct with probability 0.475 (resp. 0.250). We have obtained those numbers after performing 3 iterations and then rounding, but in this case those numbers are already the exact solution. The optimal strategy for the terminator is to apply the first action to the X-individuals. The rescuer should choose her second action for her Y-individuals.

6 Conclusions

We have shown that Newton's method is not only efficient but also remarkably robust when applied to monotone systems of fixed-point equations (MSPEs). Unlike for arbitrary systems, the method always converges when started at 0. For strongly connected systems the method always reaches a point, the threshold, after which it is guaranteed to gain at least one bit of accuracy per iteration (in favourable cases it *doubles* the number per iteration). In fewer words, after crossing the threshold the method has linear convergence order with rate 1. If the system is not strongly connected the method still has linear convergence, but the rate deteriorates.

The threshold of the strongly connected case is inversely proportional to the logarithm of the minimal component of the least fixed-point. Therefore, if some kind of analysis can establish that the least fixed-point is not very small, then the method quickly enters the one-bit-per-iteration zone. We still do not have any threshold for the general, non-strongly-connected case.

Newton's method still works for MSPEs that are not strongly connected. We have shown that the convergence order is still linear, albeit the rate may deteriorate exponentially with the dimension.

Newton's method can be extended to min-max-MSPEs, preserving its linear convergence order.

MSPEs appear in a large number of stochastic systems. In [1] we have designed a formal system for establishing the reputation of the individuals of a social network. The reputation of the individuals (defined as the stationary distribution of a Markov chain) is the least solution of a MSPE. These case studies lead to very large MSPEs, and computing their least solutions is an exciting challenge for future research.

References

1. Bouajjani, A., Esparza, J., Schwoon, S., Suwimonteerabuth, D.: SDSIrep: A reputation system based on SDSI. In: Proceedings of TACAS (Tools and Algorithms for the Construction and Analysis of Systems), *LNCS*, vol. 4963, pp. 501–516. Springer (2008)

2. Brázdil, T., Kučera, A., Stražovský, O.: On the decidability of temporal properties of probabilistic pushdown automata. In: Proceedings of STACS'2005, *LNCS*, vol. 3404, pp. 145–157. Springer (2005)
3. Canny, J.: Some algebraic and geometric computations in PSPACE. In: Proceedings of STOC, pp. 460–467 (1988)
4. Dowell, R., Eddy, S.: Evaluation of several lightweight stochastic context-free grammars for RNA secondary structure prediction. BMC Bioinformatics **5**(71) (2004)
5. Durbin, R., Eddy, S., Krogh, A., Michison, G.: Biological Sequence Analysis: Probabilistic Models of Proteins and Nucleic Acids. Cambridge University Press (1998)
6. Esparza, J., Gawlitza, T., Kiefer, S., Seidl, H.: Approximative methods for monotone systems of min-max-polynomial equations. In: Proceedings of ICALP 2008 (to appear)
7. Esparza, J., Kiefer, S., Luttenberger, M.: On fixed point equations over commutative semirings. In: Proceedings of STACS, LNCS 4397, pp. 296–307. Springer (2007)
8. Esparza, J., Kiefer, S., Luttenberger, M.: Convergence thresholds of Newton's method for monotone polynomial equations. In: Proceedings of STACS, pp. 289–300 (2008)
9. Esparza, J., Kučera, A., Mayr, R.: Model-checking probabilistic pushdown automata. In: Proceedings of LICS 2004, pp. 12–21 (2004)
10. Esparza, J., Kučera, A., Mayr, R.: Quantitative analysis of probabilistic pushdown automata: Expectations and variances. In: Proceedings of LICS 2005, pp. 117–126. IEEE Computer Society Press (2005)
11. Etessami, K., Yannakakis, M.: Algorithmic verification of recursive probabilistic systems. In: Proceedings of TACAS 2005, LNCS 3440, pp. 253–270. Springer (2005)
12. Etessami, K., Yannakakis, M.: Checking LTL properties of recursive Markov chains. In: Proceedings of 2nd Int. Conf. on Quantitative Evaluation of Systems (QEST'05) (2005)
13. Etessami, K., Yannakakis, M.: Recursive Markov chains, stochastic grammars, and monotone systems of nonlinear equations. In: Proceedings of STACS, pp. 340–352. Springer (2005)
14. Etessami, K., Yannakakis, M.: Recursive Markov decision processes and recursive stochastic games. In: Proceedings of ICALP 2005, *LNCS*, vol. 3580, pp. 891–903. Springer (2005)
15. Etessami, K., Yannakakis, M.: Efficient qualitative analysis of classes of recursive Markov decision processes and simple stochastic games. In: STACS, pp. 634–645 (2006)
16. Fagin, R., Karlin, A., Kleinberg, J., Raghavan, P., Rajagopalan, S., Rubinfeld, R., Sudan, M., Tomkins, A.: Random walks with "back buttons" (extended abstract). In: STOC, pp. 484–493 (2000)
17. Fagin, R., Karlin, A., Kleinberg, J., Raghavan, P., Rajagopalan, S., Rubinfeld, R., Sudan, M., Tomkins, A.: Random walks with "back buttons". Annals of Applied Probability **11**(3), 810–862 (2001)
18. Gawlitza, T., Seidl, H.: Precise relational invariants through strategy iteration. In: CSL, pp. 23–40 (2007)
19. Geman, S., Johnson, M.: Probabilistic grammars and their applications (2002)
20. Harris, T.: The Theory of Branching Processes. Springer (1963)
21. Kiefer, S., Luttenberger, M., Esparza, J.: On the convergence of Newton's method for monotone systems of polynomial equations. In: Proceedings of STOC, pp. 217–226. ACM (2007)
22. Knudsen, B., Hein, J.: Pfold: RNA secondary structure prediction using stochastic context-free grammars. Nucleic Acids Research **31**(13), 3423–3428 (2003)
23. Manning, C., Schütze, H.: Foundations of Statistical Natural Language Processing. MIT Press (1999)
24. Sakabikara, Y., Brown, M., Hughey, R., Mian, I., Sjolander, K., Underwood, R., Haussler, D.: Stochastic context-free grammars for tRNA. Nucleic Acids Research **22**, 5112–5120 (1994)
25. Tiwari, P.: A problem that is easier to solve on the unit-cost algebraic RAM. J. Complexity **8**(4), 393–397 (1992)
26. Watson, H., Galton, F.: On the probability of the extinction of families. J. Anthropol. Inst. Great Britain and Ireland **4**, 138–144 (1874)

Contributed talks

Lifting Non-Finite Axiomatizability Results to Extensions of Process Algebras

Luca Aceto[1], Wan Fokkink[2], Anna Ingolfsdottir[1], and MohammadReza Mousavi[3]

[1] Reykjavík University, Kringlan 1, IS-103, Reykjavík, Iceland
[2] Vrije Universiteit Amsterdam, NL-1081HV, The Netherlands
[3] Eindhoven University of Technology, NL-5600MB Eindhoven, The Netherlands

Abstract. This paper presents a general technique for obtaining new results pertaining to the non-finite axiomatizability of behavioral semantics over process algebras from old ones. The proposed technique is based on a variation on the classic idea of reduction mappings. In this setting, such reductions are translations between languages that preserve sound (in)equations and (in)equational proofs over the source language, and reflect families of (in)equations responsible for the non-finite axiomatizability of the target language. The proposed technique is applied to obtain a number of new non-finite axiomatizability theorems in process algebra via reduction to Moller's celebrated non-finite axiomatizability result for CCS. The limitations of the reduction technique are also studied.
This paper presents a general technique for obtaining new results pertaining to the non-finite axiomatizability of behavioral semantics over process algebras from old ones. The proposed technique is based on a variation on the classic idea of reduction mappings. In this setting, such reductions are translations between languages that preserve sound (in)equations and (in)equational proofs over the source language, and reflect families of (in)equations responsible for the non-finite axiomatizability of the target language. The proposed technique is applied to obtain a number of new non-finite axiomatizability theorems in process algebra via reduction to Moller's celebrated non-finite axiomatizability result for CCS. The limitations of the reduction technique are also studied.

1 Introduction

A classic and fundamental theoretical question in the study of algebras of processes is whether they afford a finite (in)equational axiomatization. Apart from being of foundational importance, (finite) axiomatizations of process semantics may form the basis for implementation verification using tools based on theorem-proving technology [10]. The first negative results concerning finite axiomatizability of process algebras go back to the Ph.D. thesis of Faron Moller [20], who showed that strong bisimilarity is not finitely based over CCS and over ACP without the left-merge operator. Since then, several other non-finite axiomatizability results have been obtained for a wide collection of very basic process algebras—see, e.g., [4] for a survey of such results.

In general, results concerning (non-)finite axiomatizability are very vulnerable to small changes in, and extensions of, the formalism under study. The addition of a single operator to a non-finitely axiomatizable formalism may make it finitely axiomatizable (e.g., adding the left-merge operator to the synchronization-free subset

---

Please use the following format when citing this chapter:

Aceto, L., et al., 2008, in IFIP International Federation for Information Processing, Volume 273; *Fifth IFIP International Conference on Theoretical Computer Science*; Giorgio Ausiello, Juhani Karhumäki, Giancarlo Mauri, Luke Ong; (Boston: Springer), pp. 301–316.

of CCS [9]). Conversely, the addition of a single operator may ruin the finite axiomatizability of a calculus (e.g., adding parallel composition to the sequential subset of CCS [19, 21]). Also, apparently simple changes to the semantics of process calculi, e.g., adding aspects such as timing, may ruin the (non-)finite axiomatizability results and make their proofs obsolete (e.g., adding timing to synchronization-free CCS with left merge makes it non-finitely axiomatizable, as shown in [8]). Furthermore, proofs of non-finite axiomatizability results in the concurrency-theory literature are extremely delicate and error-prone; they are often rather long, and involve intricate syntactic arguments. Hence, we believe that it would be useful to find some general theorems that can be used to prove non-finite axiomatizability results. Such a general theory would allow one to relate non-finite axiomatizability theorems for different formalisms, and spare researchers (some of) the delicate technical analysis needed to adapt the proofs of such results. Despite some initial proposals, like the one in [2], it is fair to say that such a general theory is missing to date.

In this paper, we present a meta-theorem offering a general technique that can be used to prove non-finite axiomatizability results, and present some of its applications within concurrency theory. In this meta-theorem, we give sufficient criteria to obtain new non-finite axiomatizability results from known ones. The proposed technique is based on a variation on the classic idea of reduction mappings, which underlies the proofs of many classic undecidability results in computability theory and of lower bounds in complexity theory—see, e.g., [26] for a textbook presentation. In this setting, reductions are translations between languages that preserve sound (in)equations and (in)equational proofs over the source language, and reflect families of (in)equations responsible for the non-finite axiomatizability of the target language. We show the applicability of our reduction-based technique by obtaining several, to our knowledge novel, non-finite axiomatizability results for timed and stochastic process algebras. All these results are proved by showing that the existence of a finite axiomatization for the seven calculi we consider in this extended abstract would contradict a result of Moller's that entails the non-finite axiomatizability of strong bisimilarity over CCS. We also investigate some of the limitations of our reduction-based technique. In particular we exhibit a classic variation on CCS that is not finitely based, but whose non-finite axiomatizability cannot be shown by reduction to CCS modulo bisimilarity.

The paper is organized as follows. In Section 2, we review some preliminary definitions from universal algebra. Section 3 presents our reduction-based technique for proving non-finite axiomatizability results. In Section 4 we apply our approach to obtain seven new non-finite axiomatizability results. In Section 5, we illustrate the limitations of our proof methodology by presenting a non-finite axiomatizability result that cannot be proved using the strategy we employed to obtain the results in Section 4. These limitations can provide sources of inspiration for future improvements on our techniques. Finally, Section 6 concludes the paper and presents some directions for future and ongoing research.

Due to space restrictions, we have omitted most of the proofs of our results in this extended abstract. The reader is referred to [5] for full details and an in-depth coverage of the issues discussed in this paper.

2 Preliminaries

We begin by recalling some basic notions from universal algebra that will be used throughout the paper. We refer the interested reader to, e.g., [14] for more information.

A signature Σ is a set of function symbols $f, g, \ldots$ with fixed arities. A function symbol of arity zero is often called a *constant (symbol)*. Given a signature Σ and a set of variables V, terms $t, u, \ldots \in \mathscr{T}(\Sigma)$ are constructed inductively (from function symbols and variables) while respecting the arities of the function symbols. (In what follows, whenever we write a term $f(t_1, \ldots, t_n)$ we tacitly assume that the arity of f is n.) Closed terms $p, q, \ldots \in \mathscr{C}(\Sigma)$ are terms that do not contain variables. We write $\equiv$ for syntactic equality over terms.

A *precongruence* $\precsim$ over $\mathscr{C}(\Sigma)$ is a substitutive preorder over $\mathscr{C}(\Sigma)$—that is, a preorder over $\mathscr{C}(\Sigma)$ that is preserved by all the function symbols in Σ. A *congruence* $\sim$ over $\mathscr{C}(\Sigma)$ is a substitutive equivalence relation. Each precongruence $\precsim$ over $\mathscr{C}(\Sigma)$ induces a congruence $\sim$ thus: $p \sim q$ iff $p \precsim q \precsim p$.

A (closed) substitution maps variables in V to (closed) terms. For every term t and substitution σ, the term $\sigma(t)$ is obtained by replacing every occurrence of a variable x in t by $\sigma(x)$. Note that $\sigma(t)$ is closed if σ is a closed substitution. We write $[t_1/x_1, \ldots, t_n/x_n]$, where the x_i $(1 \leq i \leq n)$ are distinct variables, for the substitution mapping each variable x_i to t_i, and acting like the identity function on all the other variables.

Given a relation R over closed terms, for open terms t and u, we define $t\ R\ u$ if $\sigma(t)\ R\ \sigma(u)$ for each closed substitution σ.

Consider a signature Σ. A set E of equations $t = t'$, where $t, t' \in \mathscr{T}(\Sigma)$, is called an *axiomatization (on $\mathscr{T}(\Sigma)$)*. We write $E \vdash t = t'$ when $t = t'$ is derivable from E by the following set of inference rules.

$$(\textbf{refl})\frac{}{E \vdash t = t} \qquad (\textbf{trans})\frac{E \vdash t_0 = t_1 \quad E \vdash t_1 = t_2}{E \vdash t_0 = t_2}$$

$$(\textbf{cong})\frac{E \vdash t_1 = t_1' \quad \ldots \quad E \vdash t_n = t_n'}{E \vdash f(t_1, \ldots, t_n) = f(t_1', \ldots, t_n')} \qquad (\textbf{E})\frac{}{E \vdash \sigma(t) = \sigma(t')}\ t = t' \in E$$

(Deduction rule (**cong**) is a rule schema with one instance for each function symbol f in the signature Σ.) For axiomatizations E and E', we write $E' \vdash E$ when $E' \vdash t = u$ for each $t = u \in E$. Above, we intentionally did not include the inference rule for symmetry, i.e., $(\textbf{symm})\frac{E \vdash t = t'}{E \vdash t' = t}$. Excluding (**symm**) does not restrict the applicability of our results by any measure. Any set of equations can be closed under symmetry by simply adding to it a symmetric copy of each equation, and this transformation preserves finiteness. (In what follows, we shall tacitly assume that each equational axiomatization is closed with respect to symmetry.) Furthermore, the omission of the rule for symmetry allows us to deal with axiomatizations for precongruences, which are not necessarily symmetric relations. When working with precongruences, our axiomatizations consist of inequations $t \leq t'$ between terms.

Given a congruence $\sim \subseteq \mathscr{T}(\Sigma) \times \mathscr{T}(\Sigma)$, an equation $t = t'$ is *sound* modulo $\sim$ when $t \sim t'$. An axiomatization is *sound* modulo $\sim$ if each of its equations is sound modulo $\sim$. An axiomatization E is *complete* modulo $\sim$ if for each sound equation $t = t'$, it holds that $E \vdash t = t'$. E is *ground-complete* modulo $\sim$ if for each *closed* sound equation $p = q$, it holds that $E \vdash p = q$. We say that $\sim$ is *finitely based* over $\mathscr{T}(\Sigma)$ if there is a finite, sound and complete axiomatization for $\mathscr{T}(\Sigma)$ modulo $\sim$. Similar definitions apply to precongruences and inequational axiomatizations.

3 The Reduction Theorem

Our aim in this section will be to present a general result that will allow us to lift non-finite axiomatizability results from one process algebra to another. Throughout this section, we fix two signatures Σ_o and Σ_e, a common set of variables V and two precongruences $\precsim_o$ and $\precsim_e$ over $\mathscr{T}(\Sigma_o)$ and $\mathscr{T}(\Sigma_e)$, respectively. Intuitively, the signature Σ_o stands for the collection of operations in an *o*riginal process language for which we already have a non-finite axiomatizability result modulo the precongruence $\precsim_o$. On the other hand, the signature Σ_e stands for the collection of operations in an *e*xtended process language for which we intend to prove a non-finite axiomatizability result modulo the precongruence $\precsim_e$. Since a congruence is a symmetric precongruence, all the results we present in the remainder of this section apply equally well when any of $\precsim_o$ and $\precsim_e$ is a congruence relation.

Consider a mapping $\widehat{\cdot}: \mathscr{T}(\Sigma_e) \to \mathscr{T}(\Sigma_o)$. For an axiomatization E over $\mathscr{T}(\Sigma_e)$, we define the axiomatization $\widehat{E}$ over $\mathscr{T}(\Sigma_o)$ to be $\{\widehat{t} \leq \widehat{u} \mid t \leq u \in E\}$.

Definition 1. A function $\widehat{\cdot}: \mathscr{T}(\Sigma_e) \to \mathscr{T}(\Sigma_o)$ is a *reduction* from $\mathscr{T}(\Sigma_e)$ to $\mathscr{T}(\Sigma_o)$, when for all $t, u \in \mathscr{T}(\Sigma_e)$,

1. $t \precsim_e u \Rightarrow \widehat{t} \precsim_o \widehat{u}$ (that is, $\widehat{\cdot}$ preserves sound inequations), and
2. $E \vdash t \leq u \Rightarrow \widehat{E} \vdash \widehat{t} \leq \widehat{u}$, for each axiomatization E on $\mathscr{T}(\Sigma_e)$ (that is, $\widehat{\cdot}$ preserves provability).

Definition 2. Let E be an axiomatization over $\mathscr{T}(\Sigma_o)$. A reduction $\widehat{\cdot}$ is *E-reflecting*, when for each $t \leq u \in E$, there exists an inequation $t' \leq u'$ over $\mathscr{T}(\Sigma_e)$ that is sound modulo $\precsim_e$ such that $\widehat{t'} \equiv t$ and $\widehat{u'} \equiv u$. The reduction $\widehat{\cdot}$ is called *ground E-reflecting* if for each *closed* inequation $p \leq q \in E$, there exists a *closed* inequation $p' \leq q'$ on $\mathscr{T}(\Sigma_e)$ that is sound modulo $\precsim_e$ such that $\widehat{p'} \equiv p$ and $\widehat{q'} \equiv q$.

We are now ready to state the general tool that we shall use in this paper to lift non-finite axiomatizability results from $\mathscr{T}(\Sigma_o)$ modulo $\precsim_o$ to $\mathscr{T}(\Sigma_e)$ modulo $\precsim_e$.

Theorem 1. *Assume that there is a set of inequations E on $\mathscr{T}(\Sigma_o)$ that is sound modulo $\precsim_o$ and that is not provable from any finite sound axiomatization on $\mathscr{T}(\Sigma_o)$. If there exists an E-reflecting reduction from $\mathscr{T}(\Sigma_e)$ to $\mathscr{T}(\Sigma_o)$, then $\precsim_e$ is not finitely based over $\mathscr{T}(\Sigma_e)$.*

The above theorem gives us a general technique to lift non-finite axiomatizability results from a language $\mathscr{T}(\Sigma_o)$ modulo $\precsim_o$ to a language $\mathscr{T}(\Sigma_e)$ modulo $\precsim_e$. Indeed, suppose that we know that a precongruence $\precsim_o$ is not finitely based over $\mathscr{T}(\Sigma_o)$. Typically, such a negative result is shown by exhibiting an infinite collection E of sound inequations that cannot be proved from any finite sound axiomatization over Σ_o. (See, e.g., [1, 3, 4, 6, 8, 11, 12, 20, 22] and the references therein.) In the light of the above theorem, to show that $\precsim_e$ is not finitely based over $\mathscr{T}(\Sigma_e)$ it suffices only to exhibit an E-reflecting reduction from $\mathscr{T}(\Sigma_e)$ to $\mathscr{T}(\Sigma_o)$.

As the examples we present in Section 4 will show, Theorem 1, albeit not technically complex, is widely applicable. In all our applications of Theorem 1, the reduction from Σ_e to Σ_o is defined inductively on the structure of terms. Since such "structural" reductions play an important role in the remainder of the paper, we now proceed to define them precisely and to state a very useful property such reductions afford.

Definition 3. A mapping $\widehat{\cdot}: \mathscr{T}(\Sigma_e) \to \mathscr{T}(\Sigma_o)$ is *structural* if

1. it is the identity on variables, i.e., $\widehat{x} \equiv x$ for each $x \in V$,
2. it does not introduce new variables, i.e., $vars(\widehat{f(x_1,\ldots,x_n)}) \subseteq \{x_1,\ldots,x_n\}$, for each $f \in \Sigma_e$ and sequence of distinct $x_1,\ldots,x_n \in V$, and
3. it is defined compositionally, i.e., $\widehat{f(t_1,\ldots,t_n)} \equiv \widehat{f(x_1,\ldots,x_n)}\ [\widehat{t_1}/x_1,\ldots,\widehat{t_n}/x_n]$, for each $f \in \Sigma_e$, and sequences of distinct $x_1,\ldots,x_n \in V$ and of $t_1,\ldots,t_n \in \mathscr{T}(\Sigma_e)$.

Lemma 1. *Let $\widehat{\cdot}: \mathscr{T}(\Sigma_e) \to \mathscr{T}(\Sigma_o)$ be a structural mapping. Then $\widehat{\sigma(t)} \equiv \widehat{\sigma}(\widehat{t})$, for each term $t \in \mathscr{T}(\Sigma_e)$ and each substitution σ over Σ_e.*

The following theorem shows that, if the reduction is structural, one can dispense with proving item 2 of Definition 1. Since each reduction we consider in this paper is structural, this result eases our applications of Theorem 1 considerably.

Theorem 2. *A structural mapping satisfies item 2 of Definition 1.*

If the collection of equations E mentioned in the statement of Theorem 1 is closed, then one can prove impossibility of a finite ground-complete axiomatization of $\precsim_e$ over $\mathscr{T}(\Sigma_e)$, which is a stronger result than Theorem 1.

Theorem 3. *Assume that there is a set of closed equations E that is sound modulo $\precsim_o$, and that is not provable from any finite axiomatization over $\mathscr{T}(\Sigma_o)$ that is sound modulo $\precsim_o$. If there exists a ground E-reflecting reduction from Σ_e to Σ_o, then there exists no sound and ground-complete finite axiomatization for $\precsim_e$ over $\mathscr{T}(\Sigma_e)$.*

For structural reductions whose source is a language over a signature that contains at least one constant, in order to apply Theorem 3 it suffices to show that the reduction is E-reflecting by the following theorem. Thus, if the collection of equations E is closed and the reduction is structural, one can readily obtain impossibility of a finite ground-complete axiomatization without any further work (by showing that the premises of Theorem 1 hold).

Theorem 4. *An E-reflecting structural reduction $\widehat{\cdot}$ is also ground E-reflecting, provided that the signature Σ_e contains at least one constant symbol.*

The set of basic equations that we shall use throughout the rest of this paper in our applications of Theorem 1 is closed and, furthermore, all our reductions are structural; thus, all the impossibility results we present in the subsequent section hold for ground-complete as well as complete axiomatizations.

4 Applications

In this section, we take a well-known non-finite axiomatizability result in the setting of process algebra due to Moller [20, 21], and use Theorem 1 to establish other, to the best of our knowledge novel, non-finite axiomatizability results for several notions of behavioral (pre)congruences over other process algebras. A brief comparison between the full proof of the original result in [20, 21] and those based on Theorem 1 presented here (and in the full version of this paper [5]) reveals that our proofs are substantially more concise and simpler than direct proofs. This is despite the fact that the calculi and notions of (pre)congruence treated henceforth are more sophisticated than the ones treated in [20, 21].

Consider the subset of CCS [19] with the following syntax.

$$P ::= \mathbf{0} \mid a.P \mid P+P \mid P||P$$

Note that here $a.P$ stands for *one unary operator* (action-prefixing with one particular action a) and not, as it is customary, for a collection of unary operators. Henceforth, we denote the signature of the above-mentioned calculus by Σ_o since that fragment of CCS will be the target language in all the applications of Theorem 1 to follow.

The operational semantics of the calculus above is given by the following SOS rules.

$$(\mathbf{a})\frac{}{a.x \xrightarrow{a} x} \quad (\mathbf{c0})\frac{x_0 \xrightarrow{a} y}{x_0 + x_1 \xrightarrow{a} y} \quad (\mathbf{p0})\frac{x_0 \xrightarrow{a} y_0}{x_0 \,||\, x_1 \xrightarrow{a} y_0 \,||\, x_1}$$

Note that we have omitted the symmetric versions of $(\mathbf{c0})$ and $(\mathbf{p0})$, for brevity; furthermore, since there is only one action (and no co-action) in our signature, the standard SOS rule for communication in CCS can be safely omitted.

Definition 4. A symmetric relation $R \subseteq \mathscr{C}(\Sigma_o) \times \mathscr{C}(\Sigma_o)$ is a *strong bisimulation* when for all $(p,q) \in R$ and $p' \in \mathscr{C}(\Sigma_o)$, if $p \xrightarrow{a} p'$ then there exists a q' such that $q \xrightarrow{a} q'$ and $(p',q') \in R$. Two closed terms p and q are *strongly bisimilar* (or just bisimilar), denoted by $p \leftrightarrow_b q$, when there exists a strong bisimulation R such that $(p,q) \in R$.

Moller showed in [20, 21] that strong bisimilarity affords no finite ground-complete axiomatization over the above calculus. His negative result was a corollary of the following stronger theorem.

Theorem 5 (Moller [20, 21]). *There is no finite axiomatization over the signature Σ_o that is sound modulo strong bisimilarity and proves all the equations in the set $\mathscr{M}$ defined below:*

$$\{a^1 \,||\, (a^1 + a^2 + \cdots + a^n) = a.(a^1 + a^2 + \cdots + a^n) + a^2 + a^3 + \cdots + a^{n+1} \mid n \geq 1\} \ ,$$

where $a^i = \underbrace{a.\ldots.a}_{i\ times}.\mathbf{0}$, *for each* $i \geq 1$.

In the remainder of this section, we use Theorems 1 and 5 to obtain other non-finite axiomatizability results, with the aforementioned fragment of CCS as the target language for our reductions.

4.1 Discrete-time CCS and Timed Bisimilarity

Timed CCS is a timed extension of CCS proposed by Wang Yi [27]. In [8], we proved some non-finite axiomatizability results for Timed CCS modulo timed bisimilarity under the assumption that the underlying time domain satisfy a density property, and left open whether those results carry over to the discrete-time fragment of Timed CCS (referred to as DiTCCS in what follows). In this section, we instantiate our reduction theorem to show that a finite sound and ground-complete axiomatization for DiTCCS modulo timed bisimilarity does not exist.

Let A be a set of actions that contains the action a. Following Milner, we write $\overline{A}$ for the set of complementary actions $\{\overline{b} \mid b \in A\}$, and assume that $\overline{\overline{\alpha}} = \alpha$ for each $\alpha \in A \cup \overline{A}$. The internal action is denoted by $\tau \notin A \cup \overline{A}$. The syntax of DiTCCS is given by the grammar:

$$P ::= \mathbf{0} \mid \mu.P \mid \varepsilon(d).P \mid P + P \mid P || P \ ,$$

where $\mu.P$ is a set of unary operators, one for each $\mu \in A \cup \overline{A} \cup \{\tau\}$, and $\varepsilon(d).P$ is a set of unary operators, one for each $d \in \mathbb{N} = \{1, 2, \ldots\}$. In this subsection, we refer to the signature of DiTCCS as Σ_e since we use this language as our source language in applying Theorem 1. The operational semantics of DiTCCS is given by the following set of SOS rules, where $\alpha \in A \cup \overline{A}$, $\mu \in A \cup \overline{A} \cup \{\tau\}$ and $d, e \in \mathbb{N}$.

$$(\mathbf{tn})\frac{}{\mathbf{0} \xrightarrow{\varepsilon(d)} \mathbf{0}} \qquad (\mathbf{a})\frac{}{\mu.x \xrightarrow{\mu} x} \qquad (\mathbf{ta})\frac{}{\alpha.x \xrightarrow{\varepsilon(d)} \alpha.x}$$

$$(\mathbf{td0})\frac{}{\varepsilon(d).x \xrightarrow{\varepsilon(d)} x} \qquad (\mathbf{td1})\frac{}{\varepsilon(d+e).x \xrightarrow{\varepsilon(d)} \varepsilon(e).x} \qquad (\mathbf{td2})\frac{x \xrightarrow{\varepsilon(e)} y}{\varepsilon(d).x \xrightarrow{\varepsilon(d+e)} y}$$

$$(\mathbf{c0})\frac{x_0 \xrightarrow{\mu} y}{x_0 + x_1 \xrightarrow{\mu} y} \qquad (\mathbf{tc})\frac{x_0 \xrightarrow{\varepsilon(d)} y_0 \quad x_1 \xrightarrow{\varepsilon(d)} y_1}{x_0 + x_1 \xrightarrow{\varepsilon(d)} y_0 + y_1} \qquad (\mathbf{p0})\frac{x_0 \xrightarrow{\mu} y_0}{x_0 || x_1 \xrightarrow{\mu} y_0 || x_1}$$

$$(\mathbf{p2})\frac{x_0 \xrightarrow{\alpha} y_0 \quad x_1 \xrightarrow{\overline{\alpha}} y_1}{x_0 || x_1 \xrightarrow{\tau} y_0 || y_1} \qquad (\mathbf{tp})\frac{x_0 \xrightarrow{\varepsilon(d)} y_0 \quad x_1 \xrightarrow{\varepsilon(d)} y_1}{x_0 || x_1 \xrightarrow{\varepsilon(d)} y_0 || y_1} \ \mathrm{Sort}_d(x_0) \cap \overline{\mathrm{Sort}_d(x_1)} = \emptyset$$

These rules define transitions between closed DiTCCS terms. (Again, we have omitted the symmetric versions of (**c0**) and (**p0**).) The side condition in rule (**tp**) uses the *timed sort* $\mathrm{Sort}_d(p)$, where p is a closed DiTCCS term and $d \in \mathbb{N}$, which is defined thus: $\mathrm{Sort}_d(p) = \{\alpha \in A \cup \overline{A} \mid p \stackrel{\varepsilon(e)}{\rightarrow} p' \stackrel{\alpha}{\rightarrow}$ for some p' and $e < d\}$.

The notion of equivalence over DiTCCS we shall consider in what follows is *timed bisimilarity*, denoted by $\leftrightarrow_t$. Timed bisimilarity is just bisimilarity over the labelled transition system whose states are terms in $\mathscr{C}(\Sigma_e)$ and whose transitions are of the form $p \stackrel{\chi}{\rightarrow} p'$, where $\chi \in A \cup \overline{A} \cup \{\tau\} \cup \{\varepsilon(d) \mid d \in \mathbb{N}\}$. It is well known that $\leftrightarrow_t$ is a congruence over DiTCCS; see, e.g., [27, Theorem 5.1], where the congruence result is stated for dense-time Timed CCS.

Theorem 6. *DiTCCS affords no finite ground-complete axiomatization modulo* $\leftrightarrow_t$.

We prove the above result using Theorem 1. To this end, we begin by defining the following translation $\widehat{\_}: \mathscr{T}(\Sigma_e) \rightarrow \mathscr{T}(\Sigma_o)$.

$$\widehat{\mathbf{0}} = \mathbf{0} \qquad \widehat{x} = x \qquad \widehat{\mu.p} = \begin{cases} a.\widehat{p} & \text{if } \mu = a, \\ \mathbf{0} & \text{if } \mu \neq a. \end{cases}$$

$$\widehat{\varepsilon(d).p} = \mathbf{0} \qquad \widehat{p+q} = \widehat{p} + \widehat{q} \qquad \widehat{p \| q} = \widehat{p} \| \widehat{q}$$

Lemma 2. *The mapping* $\widehat{\_}$ *defined above is structural.*

Consider now the set of Moller's equations $\mathscr{M}$, which are sound over CCS modulo bisimilarity. In order to prove that timed bisimilarity is not finitely based over DiTCCS, by Theorem 1 it suffices only to show the following statements:

1. $t \leftrightarrow_t u \Rightarrow \widehat{t} \leftrightarrow_b \widehat{u}$, for each $t, u \in \mathscr{T}(\Sigma_e)$, and
2. $\widehat{\_}$ is $\mathscr{M}$-reflecting.

Note that, for each axiomatization E over the signature of DiTCCS,

$$E \vdash t = u \Rightarrow \widehat{E} \vdash \widehat{t} = \widehat{u}$$

holds by Theorem 2 since $\widehat{\_}$ is structural (Lemma 2). Therefore, once we prove the two statements above, Theorem 6 indeed follows as a corollary of Theorem 1.

Next, we give the proofs of the above two statements.

1. Proof of $t \leftrightarrow_t u \Rightarrow \widehat{t} \leftrightarrow_b \widehat{u}$.
   In order to prove this statement, it suffices to show that the relation

   $$R = \{(\sigma(\widehat{t}), \sigma(\widehat{u})) \mid t \leftrightarrow_t u \wedge \sigma : V \rightarrow \mathscr{C}(\Sigma_o)\}$$

   is a bisimulation. To this end, observe, first of all, that R is symmetric. In order to prove that R satisfies the transfer property in Definition 4, we shall make use of the following two claims, whose proof will be given later.

   a. For all $p \in \mathscr{C}(\Sigma_e)$ and $p' \in \mathscr{C}(\Sigma_o)$, if $\widehat{p} \stackrel{a}{\rightarrow} p'$ with respect to the operational semantics of CCS, then there exists some $p'' \in \mathscr{C}(\Sigma_e)$ such that $p \stackrel{a}{\rightarrow} p''$, with respect to the operational semantics of DiTCCS, and $\widehat{p''} \equiv p'$.

b. For all $p, p' \in \mathscr{C}(\Sigma_e)$, if $p \xrightarrow{a} p'$ with respect to the operational semantics of DiTCCS, then $\widehat{p} \xrightarrow{a} \widehat{p'}$ with respect to the operational semantics of CCS.

Assume now that $\sigma(\widehat{t}) \; R \; \sigma(\widehat{u})$ and $\sigma(\widehat{t}) \xrightarrow{a} p'_0$. By Lemmas 1 and 2, $\sigma(\widehat{t}) \equiv \widehat{\sigma(t)}$. It follows from item 1a above that $\sigma(t) \xrightarrow{a} p_0$, for some p_0 such that $\widehat{p_0} \equiv p'_0$. Furthermore, as t and u are timed bisimilar, $\sigma(u) \xrightarrow{a} p_1$, for some p_1 such that $p_0 \leftrightarrow_t p_1$. From item 1b and Lemmas 1–2, we have that $\sigma(\widehat{u}) \equiv \widehat{\sigma(u)} \xrightarrow{a} \widehat{p_1}$ and, by the definition of R, we may conclude that $p'_0 = \widehat{p_0} \; R \; \widehat{p_1}$, which was to be shown.
In order to complete the proof of this statement, we are therefore left to show items 1a and 1b. This we now proceed to do.

a. Proof of item 1a.
   We prove this claim by an induction on the structure of p, and only detail the argument for two representative cases.
   – Assume that $p \equiv \mu.p_0$. Then p must be of the form $a.p_0$ (in order for $\widehat{p}$ to make an a-transition) and thus, $\widehat{p} = a.\widehat{p_0} \xrightarrow{a} \widehat{p_0} = p'$. The claim then follows since $a.p_0 \xrightarrow{a} p_0$.
   – Assume that $p \equiv p_0 + p_1$. Then $\widehat{p} \equiv \widehat{p_0} + \widehat{p_1}$. Suppose, without loss of generality, that the transition $\widehat{p_0} + \widehat{p_1} \xrightarrow{a} p'$ is due to an application of rule (**c0**); thus, $\widehat{p_0} \xrightarrow{a} p'$. It then follows from the induction hypothesis that $p_0 \xrightarrow{a} p''$ for some p'' such that $\widehat{p''} \equiv p'$. By applying deduction rule (**c0**), we obtain $p \equiv p_0 + p_1 \xrightarrow{a} p''$.
b. Proof of item 1b.
   By an induction on the depth of the proof for $p \xrightarrow{a} p'$. We distinguish the following cases based on the last deduction rule applied to obtain $p \xrightarrow{a} p'$.
   (**a**) In this case, p is of the form $a.p_0$ and $p' \equiv p_0$ Thus, using to the same deduction rule in the semantics of CCS, we have $\widehat{p} \equiv a.\widehat{p_0} \xrightarrow{a} \widehat{p_0}$.
   (**c0**) Then $p \equiv p_0 + p_1$ and $p_0 \xrightarrow{a} p'$ by a shorter inference. It follows from the induction hypothesis that $\widehat{p_0} \xrightarrow{a} \widehat{p'}$ and, using rule (**c0**) in the semantics of CCS, we infer that $\widehat{p_0} + \widehat{p_1} \xrightarrow{a} \widehat{p'}$. Furthermore, by the definition of $\widehat{\_}$, we have that $\widehat{p} \equiv \widehat{p_0} + \widehat{p_1}$.
   The cases for deduction rules (**c1**), (**p0**) and (**p1**) are similar to the case of (**c0**).

The proof of the first statement is now complete.

2. Proof of the fact that $\widehat{\_}$ is $\mathscr{M}$-reflecting.
   We show that all axioms in $\mathscr{M}$ are sound modulo $\leftrightarrow_t$. Since $\widehat{\_}$ is the identity over CCS terms, the statement then follows immediately. To this end, we prove the following two claims.

   a. For each $p \in \mathscr{C}(\Sigma_o)$ and positive integer d, $p \xrightarrow{\varepsilon(d)} p'$ iff $p \equiv p'$. We prove this claim by an induction on the structure of p. The cases for **0** and $a.p_0$ follow from deduction rules (**tn**) and (**ta**), respectively. The cases for $p_0 + p_1$ and $p_0 \,||\, p_1$ follow from the induction hypothesis, and (**tc**) and (**tp**), respectively.
   b. For each $p, q \in \mathscr{C}(\Sigma_o)$, if $p \leftrightarrow_b q$ then $p \leftrightarrow_t q$.

We show that $\leftrightarrow_b$ is a timed bisimulation. To this end, note, first of all, that the relation $\leftrightarrow_b$ is symmetric. Assume now that $p \xrightarrow{a} p'$ and $p \leftrightarrow_b q$. Since $\leftrightarrow_b$ is a bisimulation, it follows that $q \xrightarrow{a} q'$ (with respect to the semantics of CCS, and thus of DiTCCS using the same deduction rules) for some q' such that $p' \leftrightarrow_b q'$, and we are done. That delay transitions of p may be matched by q follows trivially from the previous item.

Since all the provisos of Theorem 1 are met, Theorem 6 follows.

4.2 Temporal CCS

In the paper [23], Moller and Tofts proposed another timed extension of Milner's CCS, which they called *Temporal Calculus of Communicating Systems* (referred to as TCCS_{MT} in what follows to avoid any confusion with Wang Yi's Timed CCS), and studied its semantics theory modulo timed bisimilarity. Our order of business in this section is to use our reduction-based method to show that timed bisimilarity affords no finite ground-complete axiomatization over TCCS_{MT}.

For our purposes in this section, TCCS_{MT} is the language generated by the following grammar:

$$P ::= \mathbf{0} \mid \mu.P \mid (d).P \mid \delta.P \mid P+P \mid P \oplus P \mid P || P \ ,$$

where $\mu.P$ is a set of unary operators, one for each $\mu \in A \cup \overline{A} \cup \{\tau\}$, and $(d).P$ is a set of unary operators, one for each positive integer d. The intuition underlying each of the operators in the signature of TCCS_{MT} is carefully described in [23, Pages 402–403]. For the sake of clarity, however, we find it useful to mention that:

- process terms of the form $\mathbf{0}$ or $\alpha.p$ *cannot* delay, unlike in DiTCCS;
- $(d).p$ behaves exactly like $\varepsilon(d).p$ in DiTCCS;
- $\delta.p$ describes a process which behaves like p, but is willing to wait any amount to time before doing so; and
- $p \oplus q$ is a "weak choice" between p and q. The choice between p and q is made upon performance of an action from either of the two processes, or at the occurrence of a time delay which can only be performed by one of the processes. By way of example, as $a.p$ cannot delay, a process of the form $a.p \oplus (1).\mathbf{0}$ will be transformed into $\mathbf{0}$ after a delay of one time unit.

In order to define the operational semantics of the weak choice operator, the Plotkin-style rules for that operator from [23] make use of the function maxdelay(), which associates a non-negative integer or ω with each closed TCCS_{MT} term. The function maxdelay() is defined by structural induction on terms as follows:

$$\begin{array}{l}\text{maxdelay}(\mathbf{0}) = \text{maxdelay}(\mu.p) = 0 \quad \text{maxdelay}(\delta.p) = \omega \\ \text{maxdelay}(p+q) = \text{maxdelay}(p || q) = \min(\text{maxdelay}(p), \text{maxdelay}(q)) \\ \text{maxdelay}(p \oplus q) = \max(\text{maxdelay}(p), \text{maxdelay}(q)) \ .\end{array}$$

$$\frac{}{\delta.x \xrightarrow{\varepsilon(d)} \delta.x} \qquad \frac{}{(d).x \xrightarrow{\varepsilon(d)} x} \qquad \frac{}{(d+e).x \xrightarrow{\varepsilon(d)} (e).x}$$

$$\frac{x \xrightarrow{\varepsilon(e)} y}{(d).x \xrightarrow{\varepsilon(d+e)} y} \qquad \frac{x_0 \xrightarrow{\varepsilon(d)} y_0 \quad x_1 \xrightarrow{\varepsilon(d)} y_1}{x_0 \oplus x_1 \xrightarrow{\varepsilon(d)} y_0 \oplus y_1} \qquad \frac{x_0 \xrightarrow{\varepsilon(d)} y_0 \quad \text{maxdelay}(x_1) < d}{x_0 \oplus x_1 \xrightarrow{\varepsilon(d)} y_0}$$

$$\frac{x_1 \xrightarrow{\varepsilon(d)} y_1 \quad \text{maxdelay}(x_0) < d}{x_0 \oplus x_1 \xrightarrow{\varepsilon(d)} y_1} \qquad \frac{x_0 \xrightarrow{\varepsilon(d)} y_0 \quad x_1 \xrightarrow{\varepsilon(d)} y_1}{x_0 + x_1 \xrightarrow{\varepsilon(d)} y_0 + y_1} \qquad \frac{x_0 \xrightarrow{\varepsilon(d)} y_0 \quad x_1 \xrightarrow{\varepsilon(d)} y_1}{x_0 || x_1 \xrightarrow{\varepsilon(d)} y_0 || y_1}$$

Table 1 Rules defining the delay transitions $\xrightarrow{\varepsilon(d)}$ over TCCS$_{MT}$ ($d \in \mathbb{N}$)

The operational semantics of closed TCCS$_{MT}$ terms is given by means of two types of transitions, namely actions transitions $\xrightarrow{\mu}$ with $\mu \in A \cup \overline{A} \cup \{\tau\}$ and delay transitions $\xrightarrow{\varepsilon(d)}$, with $d \in \mathbb{N}$. The transition relations $\xrightarrow{\mu}$ are defined as for DiTCCS, with the proviso that

- $(d).p$ has no outgoing action transitions,
- $p \oplus q$ has the same outgoing action transitions as $p+q$, and
- the action transitions of $\delta.p$ are exactly those of p—i.e., they are those provable using the rules

$$\frac{x \xrightarrow{\mu} y}{\delta.x \xrightarrow{\mu} y} \quad (\mu \in A \cup \overline{A} \cup \{\tau\}) .$$

On the other hand, the transition relations $\xrightarrow{\varepsilon(d)}$ are the least relations satisfying the rules on Table 1. Closed TCCS$_{MT}$ terms are considered modulo timed bisimilarity $\leftrightarrow_t$ (as defined in Section 4.1). Timed bisimilarity is a congruence over TCCS$_{MT}$ as shown in [23, Proposition 3.4].

Theorem 7. *TCCS$_{MT}$ affords no finite ground-complete axiomatization modulo $\leftrightarrow_t$.*

In the remainder of this subsection, we prove the above result using Theorem 1. To this end, we begin by defining the following translation $\widehat{\ }$ from open TCCS$_{MT}$ terms to open CCS terms.

$$\widehat{\mathbf{0}} = \mathbf{0} \qquad \widehat{x} = x \qquad \widehat{\delta.p} = \widehat{p}$$
$$\widehat{a.p} = a.\widehat{p} \qquad \widehat{\mu.p} = \mathbf{0} \text{ for } \mu \neq a \qquad \widehat{(d).p} = \mathbf{0}$$
$$\widehat{p+q} = \widehat{p}+\widehat{q} \qquad \widehat{p \oplus q} = \widehat{p}+\widehat{q} \qquad \widehat{p||q} = \widehat{p}||\widehat{q}$$

Remark 1. Note that the mapping obtained from the one defined above by associating $\mathbf{0}$ to $\widehat{p \oplus q}$ would *not* be a reduction, since it does not preserve valid equations. For example, the valid equation $x \oplus x = x$ would not be preserved by such a mapping.

Lemma 3. *The mapping $\widehat{\ }$ defined above is structural.*

Consider now the set of Moller's equations $\mathscr{M}$, which are sound over CCS modulo bisimilarity. In order to prove that timed bisimilarity is not finitely based over TCCS_{MT}, by Theorem 1 it suffices only to show the following statements:

1. $t \leftrightarrow_t u$ implies $\widehat{t} \leftrightarrow_b \widehat{u}$, for all TCCS_{MT} terms t, u, and
2. $\widehat{\_}$ is $\mathscr{M}$-reflecting.

Note that, for all TCCS_{MT} terms t, u and axiomatization E,

$$E \vdash t = u \Rightarrow \widehat{E} \vdash \widehat{t} = \widehat{u}$$

holds by Theorem 2 since $\widehat{\_}$ is structural (Lemma 3). Therefore, once we prove the two statements above, Theorem 7 indeed follows as a corollary of Theorem 1.

We establish the two statements above in turn. The following lemma will be useful.

Lemma 4.

1. *Assume that $\widehat{p} \xrightarrow{a} r$ holds with respect to the operational semantics of CCS for some closed $TCCS_{MT}$ term p and CCS term r. Then $p \xrightarrow{a} p'$ holds with respect to the operational semantics of $TCCS_{MT}$ for some closed $TCCS_{MT}$ term p' such that $\widehat{p'} = r$.*
2. *If $p \xrightarrow{a} p'$ holds with respect to the operational semantics of $TCCS_{MT}$ for some closed $TCCS_{MT}$ terms p, p' then $\widehat{p} \xrightarrow{a} \widehat{p'}$ holds with respect to the operational semantics of CCS.*

We are now ready to show that $\widehat{\_}$ preserves sound equations.

Proposition 1. *$t \leftrightarrow_t u$ implies $\widehat{t} \leftrightarrow_b \widehat{u}$, for all $TCCS_{MT}$ terms t, u.*

Proof. It suffices to show that the relation

$$R = \{(\widehat{p}, \widehat{q}) \mid p \leftrightarrow_t q, \text{ with } p, q \text{ closed } \text{TCCS}_{\text{MT}} \text{ terms}\}$$

is a strong bisimulation. Indeed, assuming that R is a strong bisimulation, we can show the proposition as follows.

Suppose that $t \leftrightarrow_t u$ holds for some TCCS_{MT} terms t, u. Let σ be a closed CCS substitution. We shall argue that $\sigma(\widehat{t}) \leftrightarrow_b \sigma(\widehat{u})$ holds. This follows because

- $\sigma(\widehat{t}) = \widehat{\sigma(t)}$ and $\sigma(\widehat{u}) = \widehat{\sigma(u)}$ (by Lemma 1, as $\widehat{\_}$ is structural and $\sigma = \widehat{\sigma}$), and
- $\widehat{\sigma(t)} \leftrightarrow_b \widehat{\sigma(u)}$ (since $\widehat{\sigma(t)}\ R\ \widehat{\sigma(u)}$ and R is a strong bisimulation).

So we are left to show that R is indeed a strong bisimulation. This can be easily checked using Lemma 4. □

To complete the proof of Theorem 7, we now show that $\widehat{\_}$ is $\mathscr{M}$-reflecting. Since $\widehat{\_}$ is the identity function over CCS terms, it suffices to prove the following result. (Note that, since CCS is a reduct of the language TCCS_{MT}, it makes sense to consider CCS terms modulo $\leftrightarrow_t$.)

Proposition 2. *The relations $\leftrightarrow_t$ and $\leftrightarrow_b$ coincide over CCS terms.*

Proof. The relation $\underline{\leftrightarrow}_t$ is included in $\underline{\leftrightarrow}_b$ over the collection of CCS terms by Proposition 1. The converse inclusion follows because $\underline{\leftrightarrow}_b$ is a timed bisimulation. This can be shown using Lemma 4 and observing that $p \stackrel{\varepsilon(d)}{\nrightarrow}$ holds for each closed CCS term p and positive integer d. □

Since all the provisos of Theorem 1 are met by our reduction, Theorem 7 follows.

4.3 Other Calculi, Equivalences and Preorders

There are many other extensions of process algebras in the literature, and each of these languages comes equipped with notions of behavioral equivalence and/or preorder. In this section, we briefly review the results we obtained using our reduction technique for a few such extensions and refer the reader to the extended version of this paper [5] for the full treatment of these cases. Here we limit ourselves to remarking that all the non-finite axiomatizability results covered by the following theorem are proved using $\mathscr{M}$-reflecting reductions to CCS.

Theorem 8. *The following process algebras afford no finite (ground-)complete axiomatization: ATP modulo timed bisimilarity [25]; TACS^{UT} modulo the faster-than preorder [16]; TACS^{LT} modulo the MT-preorder [17]; TACS modulo urgent timed bisimilarity [18]; and IMC modulo strong Markovian bisimilarity [15].*

5 Limitations of Our Approach

As witnessed by the applications described in the previous section, our reduction-based method for proving non-finite axiomatizability results, based on Theorem 1, is widely applicable. Moreover, in all of the applications of Theorem 1 we presented in Section 4, we used CCS modulo bisimilarity as our target language for an $\mathscr{M}$-reflecting reduction. In this section, we give an example of an equational theory within the realm of classic process algebra, whose non-finite axiomatizability cannot be shown in that fashion.

The language CCS_Ω (a variant of the calculus presented in [7]) is obtained by adding the constant Ω to the fragment of CCS introduced in Section 4. Intuitively, Ω stands for a process whose behavior is completely unspecified. The operational semantics of CCS_Ω is given by two ingredients: $\stackrel{a}{\rightarrow}$ transitions, which are defined by the same deduction rules used for CCS (thus, Ω has no outgoing transitions), and a *convergence predicate* $\downarrow$, which is the least predicate over closed CCS_Ω satisfying the rules given below.

$$\frac{}{\mathbf{0}\downarrow} \qquad \frac{}{a.p\downarrow} \qquad \frac{p\downarrow \quad q\downarrow}{p+q\downarrow} \qquad \frac{p\downarrow \quad q\downarrow}{p||q\downarrow}$$

So, for instance $a.\Omega\downarrow$, but neither $\Omega\downarrow$ nor $a||\Omega\downarrow$ hold.

The following notion of prebisimilarity is a relevant notion of behavioral preorder in the presence of divergence as adopted in, e.g., [13]. We refer the interested reader to that paper and the references therein for a wealth of results on the semantic theory of CCS_Ω modulo prebisimilarity.

Definition 5. The relation $\precsim_{pre}$ is the largest relation over the closed terms of CCS_Ω satisfying the following clauses, whenever $p \precsim_{pre} q$,

1. for each p', if $p \xrightarrow{a} p'$ then there exists a q' such that $q \xrightarrow{a} q'$ and $p' \precsim_{pre} q'$;
2. if $p \downarrow$, then
   a. $q \downarrow$ and
   b. for each q', if $q \xrightarrow{a} q'$, then there exists a p' such that $p \xrightarrow{a} p'$ and $p' \precsim_{pre} q'$.

The relation $\precsim_{pre}$ is a preorder and a precongruence over closed CCS_Ω terms. Moreover, it coincides with bisimilarity over CCS terms.

Using an argument based on the soundness of the equations in the set $\mathcal{M}$ over CCS_Ω modulo $\precsim_{pre}$, we can show the following theorem.

Theorem 9. *CCS_Ω affords no finite sound and ground-complete axiomatization modulo $\precsim_{pre}$.*

It is natural to wonder whether the above result can be established, like all those we presented in Section 4, by using CCS modulo bisimilarity as our target language for an $\mathcal{M}$-reflecting reduction. The following theorem shows that this is not possible, and highlights a limitation of our present proof strategy based on reductions to CCS modulo bisimilarity.

Theorem 10. *There is no $\mathcal{M}$-reflecting reduction from CCS_Ω modulo $\precsim_{pre}$ to CCS modulo strong bisimilarity.*

6 Conclusions

In this paper, we have proposed a meta-theorem for proving non-finite axiomatizability results. This theorem can be used to show such results when there exists a reduction from the calculus under consideration to a calculus for which non-finite axiomatizability is known. If the reduction is defined structurally (in the sense of Definition 3), then one only needs to prove that the reduction preserves sound (in)equalities and that it reflects a set of "difficult" (in)equations that form the core of the non-finite axiomatizability result over the target calculus. We have shown seven new non-finite axiomatizability results in process algebra by applying our meta-theorem and reducing different calculi (modulo their respective notion of equivalence or preorder) to a subset of CCS. We intend to apply our reduction technique to obtain several other new non-finite axiomatizability results in process algebra.

The above-mentioned conditions on the reductions can be established following similar lines for the different calculi and different notions of (pre)congruence studied in this paper. The resulting proofs are substantially more concise and simpler than typical proofs of non-finite axiomatizability. We believe that the proofs of the aforementioned two conditions can be further simplified if one commits to particular models such as those given by Plotkin-style SOS rules. A promising future research direction is to study whether one can apply our meta-theorem in conservative and orthogonal language extensions. Using the SOS meta-theory, one can seek sufficient syntactic conditions on the reduction function that would automatically provide us with the properties required by our meta-theorem. Furthermore, in this paper, we pointed out a limitation of our meta-theorem by presenting a non-finite axiomatizability result that cannot be proved using our general strategy of reducing calculi to CCS. Studying the roots of such limitations may lead to improvements upon the meta-theorem presented in this paper.

Acknowledgements The work of Aceto, Ingolfsdottir and Mousavi has been partially supported by the projects "The Equational Logic of Parallel Processes" (nr. 060013021), "A Unifying Framework for Operational Semantics" (nr. 070030041) and "New Developments in Operational Semantics" (nr. 080039021) of the Icelandic Research Fund.

References

1. L. Aceto, T. Chen, W. Fokkink, and A. Ingolfsdottir. On the axiomatizability of priority. Proceedings of Automata, Languages and Programming, 33rd International Colloquium, ICALP 2006, Venice, Italy, July 10-14, 2006, Part II, volume 4052 of *LNCS*, pages 480–491, Springer, 2006.
2. L. Aceto, W. Fokkink and A. Ingolfsdottir. Ready to preorder: Get your BCCSP axiomatization for free! Proceedings of CALCO'07, volume 4624 of *LNCS*, pages 338–367. Springer, 2007.
3. L. Aceto, W. Fokkink, A. Ingolfsdottir, and B. Luttik. CCS with Hennessy's merge has no finite equational axiomatization. *Theoretical Computer Science*, 330(3):377–405, 2005.
4. L. Aceto, W. Fokkink, A. Ingolfsdottir, and B. Luttik. Finite equational bases in process algebra: Results and open questions. In A. Middeldorp, V. van Oostrom, F. van Raamsdonk, and R. C. de Vrijer, editors, *Processes, Terms and Cycles: Steps on the Road to Infinity, Essays Dedicated to Jan Willem Klop, on the Occasion of His 60th Birthday*, volume 3838 of *LNCS*, pages 338–367. Springer, 2005.
5. L. Aceto, W. Fokkink, A. Ingolfsdottir, and M.R. Mousavi. Lifting non-finite axiomatizability results to extensions of process algebras. Technical Report CSR-08-05, TU/Eindhoven, 2008.
6. L. Aceto, W. Fokkink, A. Ingolfsdottir, and S. Nain. Bisimilarity is not finitely based over BPA with interrupt. *Theoretical Computer Science*, 366(1–2):60–81, 2006.
7. L. Aceto and M. Hennessy. Termination, deadlock, and divergence. Journal of the ACM, 39(1):147–187, 1992.
8. L. Aceto, A. Ingolfsdottir and M. R. Mousavi. Impossibility results for the equational theory of Timed CCS. Proceedings of CALCO'07, volume 4624 of *LNCS*, pages 80–95, Springer, 2007.
9. J.A. Bergstra and J.W. Klop. Process algebra for synchronous communication. *Information and Control*, 60(1–3):109–137, 1984.
10. S.C.C. Blom, W.J. Fokkink, J.F. Groote, I. van Langevelde, B. Lisser, and J.C. van de Pol. mCRL: A toolset for analysing algebraic specifications. Proceedings of CAV'01, volume 2102 of *LNCS*, pages 250-254, Springer, 2001.

11. S. L. Bloom and Z. Ésik. Nonfinite axiomatizability of shuffle inequalities. Proceedings of TAPSOFT'95, volume 915 of *LNCS*, pages 318–333. Springer, 1995.
12. Z. Ésik and M. Bertol. Nonfinite axiomatizability of the equational theory of shuffle. *Acta Informatica*, 35(6):505–539, 1998.
13. M. Hennessy. A term model for synchronous processes. *Information and Control*, 51(1):58–75, 1981.
14. M. Hennessy. *Algebraic Theory of Processes.* Foundations of Computing Series. MIT Press, 1988.
15. H. Hermanns. *Interactive Markov Chains and the Quest for Quantified Quality.* Volume 2428 of LNCS, Springer, 2002.
16. G. Lüttgen and W. Vogler. Bisimulation on speed: worst-case efficiency. *Information and Computation*, 191(2): 105–144, 2004.
17. G. Lüttgen and W. Vogler. Bisimulation on speed: Lower time bounds. *Proceedings of FOSSACS'04*, volume 2987 of *LNCS*, Springer, 2004.
18. G. Lüttgen and W. Vogler. Bisimulation on speed: A unified approach. *Theoretical Computer Science*, 360(1–3):209-227, 2006.
19. R. Milner. *Communication and Concurrency.* Prentice Hall, 1989.
20. F. Moller. *Axioms for Concurrency.* Ph.D. Thesis, University of Edinburgh, 1989.
21. F. Moller. The nonexistence of finite axiomatisations for CCS congruences. In *Proceedings, Fifth Annual IEEE Symposium on Logic in Computer Science*, pages 142–153. IEEE Computer Society, 1990.
22. F. Moller. The importance of the left merge operator in process algebras. In *Proceedings of ICALP'90*, volume 443 of LNCS, pages 752–764. Springer, 1990.
23. F. Moller and C.M.N. Tofts. A temporal calculus of communicating systems. In *Proceedings of CONCUR'90*, volume 458 of *LNCS*, pages 401–415. Springer, 1990.
24. F. Moller and C. Tofts. Relating processes with respect to speed. In *Proceedings of CONCUR '91*, volume 527 of *LNCS*, pages 424–438. Springer, 1991.
25. X. Nicollin, J. Sifakis. The algebra of timed processes, ATP: Theory and application. *Information and Computation*, 114(1):131–178, 1994.
26. M. Sipser. *Introduction to the Theory of Computation.* 2nd Ed., PWS Publishing, 2005.
27. W. Yi. Real-time behaviour of asynchronous agents. In *Proceedings of CONCUR'90*, volume 458 of *LNCS*, pages 502–520. Springer, 1990.

Finite Equational Bases for Fragments of CCS with Restriction and Relabelling*

Luca Aceto[1], Anna Ingólfsdóttir[1], Bas Luttik[2], and Paul van Tilburg[2]

[1] School of Computer Science, Reykjavík University, Kringlan 1, 103 Reykjavík, Iceland, {luca, annai}@ru.is

[2] Department of Mathematics and Computer Science, Eindhoven University of Technology, P.O. Box 513, 5600 MB Eindhoven, The Netherlands, {s.p.luttik, p.j.a.v.tilburg}@tue.nl

Abstract We investigate the equational theory of several fragments of CCS modulo (strong) bisimilarity with special attention to restriction and relabelling. The largest fragment we consider includes action prefixing, choice, parallel composition without communication, restriction and relabelling. We present a finite equational base (i.e., a finite ground-complete and omega-complete axiomatisation) for it, including the left merge from ACP as auxiliary operation to facilitate the axiomatisation of parallel composition.

1 Introduction

The Calculus of Communicating Systems (CCS) was developed by Robin Milner in the late 1970s [8]. This calculus introduced a formal language for describing processes, using a transition system to give an operational meaning to the expressions in the language. In this paper we pay special attention to the restriction and relabelling operators of CCS.

The restriction operator takes a process and a set of actions as arguments. It delimits the scope of actions by preventing the execution by the process of the actions in the set. Restriction is often used to specify the communication topology of a system by blocking the execution of interleaving actions of parallel processes so that only the result of (synchronous) communication remains. Restriction is also present in ACP [3], where it is called encapsulation.

The relabelling operator takes a process and a function from actions to actions. It renames the actions in the process according to the function, and can be used to instantiate a generic specification for specific needs. In CCS, relabelling is, e.g., used in defining the so-called linking operation, which is at the core of many of the specifications offered in [9]. Relabelling is not present in ACP, but it can be added and then it increases the expressiveness of the language. Namely, Baeten and Bergstra prove in [2] that the process Queue cannot be specified by means of a finite guarded

* A full version of this paper, including omitted proofs and auxilary lemmas is available as CS-Report 08-08, Eindhoven University of Technology.

Please use the following format when citing this chapter:

Aceto, L., et al., 2008, in IFIP International Federation for Information Processing, Volume 273; *Fifth IFIP International Conference on Theoretical Computer Science*; Giorgio Ausiello, Juhani Karhumäki, Giancarlo Mauri, Luke Ong; (Boston: Springer), pp. 317–332.

recursive specification over ACP, whereas it can be specified by means of a finite guarded recursive specification over ACP with renaming.

In [6] (see also [9]), Hennessy and Milner propose an axiomatisation for CCS modulo bisimilarity that they prove ground-complete (i.e., all valid equations involving terms without variables are equationally derivable from it). Their axiomatisation is infinite, which is unavoidable as proved by Moller [11]. For a finite axiomatisation it is necessary to add auxiliary operators, e.g., the left merge and communication merge of ACP [3].

We want to give an equational base (i.e., an axiomatisation that is not just ground-complete but complete also for equations involving terms with variables) for CCS modulo bisimilarity. Perhaps surprisingly, no complete axiomatisations of bisimilarity over languages including restriction and relabelling have been given to date. In [7], Milner studied an algebra of flowgraphs with operations of (parallel) composition, restriction and relabelling, and provided a complete axiomatisation for it. In that reference, however, the notion of equivalence between expressions is purely "structural", since two expressions are equated when they denote the same flowgraph up to isomorphism.

In this paper we present finite equational bases for fragments of CCS modulo bisimilarity that include restriction and relabelling operators. The largest fragment we consider here includes all the operators from recursion-free CCS, but the parallel composition operator is limited to pure interleaving and does not allow for synchronisation between parallel components. Our completeness proofs build on results and techniques developed in [1], where a finite axiomatisation for the fragment of CCS without restriction and relabelling operators is proved complete.

For our completeness proofs we adopt the classic normal form strategy. This entails showing that all process terms can be proved equal to some normal form using the axioms, followed by the construction of a distinguishing valuation that ensures that two normal forms are equal under this valuation only if they can be proved equal. Both the above-mentioned steps involve non-trivial extensions of the techniques from [1] for the languages we consider because, unlike for ground-complete axiomatisations, restriction and relabelling cannot be eliminated from terms. This means that normal forms may contain occurrences of these operations, and their form will be more complicated than that considered thus far in the literature. Moreover, in order to implement the latter step in the above-mentioned proof technique, distinguishing valuations will need to be defined in such a way that they allow us to detect the restrictions and relabellings that occur in the normal forms.

For the shape of the normal forms in the present paper it is crucial that restriction and relabelling distribute over parallel composition. This is the reason that we now only consider an operator for parallel composition that is limited to pure interleaving; neither restriction nor relabelling distribute over parallel composition in the presence of synchronisation. So an obvious avenue for future work is the technically challenging problem of giving a complete axiomatisation of full CCS modulo bisimilarity, with restriction, relabelling and parallel composition that allows for synchronisation.

The paper is organised as follows. In Sect. 2 we introduce the fragments of CCS that will be discussed in this paper. In Sects. 3–5 we propose equational bases for three fragments of CCS: first only with the restriction operator, then only with the relabelling operator, and finally with both operators.

2 Preliminaries

In this section we introduce a process calculus that is obtained from Milner's pure CCS [9] by omitting recursion, replacing parallel composition by an operation for pure interleaving (i.e., which does not include synchronisation between components), and adding the left merge of Bergstra and Klop [3]. The calculus gives rise to a process algebra **P** for which we will present a (finite) axiomatisation. The main result of this paper states that this axiomatisation is complete.

We fix a set of *action labels* $\mathscr{L}$, a set of *co-action labels* $\overline{\mathscr{L}}$ disjoint from $\mathscr{L}$ and a bijection $\bar{\cdot} : \mathscr{L} \to \overline{\mathscr{L}}$. We define the set of *actions* $\mathscr{A}$ as $\mathscr{L} \cup \overline{\mathscr{L}}$. The inverse of $\bar{\cdot}$ we shall also denote by $\bar{\cdot}$, and thus $\overline{\overline{a}} = a$ for each $a \in \mathscr{A}$. In [9], Milner assumes that $\mathscr{L}$ and $\overline{\mathscr{L}}$ are infinite. However, to obtain a finite axiomatisation, we need to require that the sets $\mathscr{L}$ and $\overline{\mathscr{L}}$ are finite. We also fix a countably infinite set of *variables* $\mathscr{V}$. The meta-variables a, b, and c generally range over $\mathscr{A}$; x, y, and z range over $\mathscr{V}$.

A *relabelling function* is a function $f : \mathscr{A} \to \mathscr{A}$ such that $f(\overline{a}) = \overline{f(a)}$ for each $a \in \mathscr{A}$. With every relabelling function $f : \mathscr{A} \to \mathscr{A}$ we associate a function $f^{-1} : \mathscr{P}(\mathscr{A}) \to \mathscr{P}(\mathscr{A})$ such that $f^{-1}(\mathscr{A}') = \{a \mid f(a) \in \mathscr{A}'\}$ for each $\mathscr{A}' \subseteq \mathscr{A}$. The identity relabelling function *Id* is defined by $Id(a) = a$ for each $a \in \mathscr{A}$. For each relabelling function f and $L \subseteq \mathscr{L}$, we write $f \upharpoonright L$ for the relabelling function defined by

$$(f \upharpoonright L)(a) = \begin{cases} f(a) & \text{if } a \in L \text{ or } \overline{a} \in L, \\ a & \text{otherwise.} \end{cases}$$

The meta-variables f and g generally refer to relabelling functions, and K and L refer to subsets of $\mathscr{L}$.

The set of *process terms* $\mathscr{T}_{\backslash,[]}$ is generated by the following grammar:

$$\mathrm{T} ::= \mathbf{0} \mid x \mid a.\mathrm{T} \mid \mathrm{T}+\mathrm{T} \mid \mathrm{T} \parallel \mathrm{T} \mid \mathrm{T} \mathbin{\|\!\!\lfloor} \mathrm{T} \mid \mathrm{T} \backslash L \mid \mathrm{T}[f]$$

where $a \in \mathscr{A}$, $x \in \mathscr{V}$, $L \subseteq \mathscr{L}$, and $f : \mathscr{A} \to \mathscr{A}$ is a relabelling function. The meta-variables p, q and r generally range over $\mathscr{T}_{\backslash,[]}$. We use the following convention for the binding power of the operators in decreasing order: relabelling $\_[f]$ and restriction $\_\backslash L$ (tightest binding), prefixing $a.\_$, parallel composition $\_ \parallel \_$ and left merge $\_ \mathbin{\|\!\!\lfloor} \_$, alternative composition $\_+\_$. In the remainder of the paper we also need notation for the following subsets of $\mathscr{T}_{\backslash,[]}$: we use $\mathscr{T}_{\backslash}$ to denote the set of all process terms without occurrences of relabelling operators, and $\mathscr{T}_{[]}$ to denote the set of all process terms without occurrences of restriction operators.

Process terms that do not contain any variables are called *closed*. The set of closed process terms is denoted by $\mathscr{T}^{C}_{\backslash,[]}$. We give an operational semantics to closed terms using the binary relations $\xrightarrow{a}$ $(a \in \mathscr{A})$ on $\mathscr{T}^C_{\backslash,\square}$ defined by means of the specification in Table 2.1.

1 $\dfrac{}{a.p \xrightarrow{a} p}$	2 $\dfrac{p \xrightarrow{a} p'}{p+q \xrightarrow{a} p'}$	3 $\dfrac{q \xrightarrow{a} q'}{p+q \xrightarrow{a} q'}$
4 $\dfrac{p \xrightarrow{a} p'}{p \parallel q \xrightarrow{a} p' \parallel q}$	5 $\dfrac{q \xrightarrow{a} q'}{p \parallel q \xrightarrow{a} p \parallel q'}$	6 $\dfrac{p \xrightarrow{a} p'}{p \lfloor\!\lfloor q \xrightarrow{a} p' \parallel q}$
7 $\dfrac{p \xrightarrow{a} p' \quad a,\overline{a} \notin L}{p \backslash L \xrightarrow{a} p' \backslash L}$	8 $\dfrac{p \xrightarrow{a} p'}{p[f] \xrightarrow{f(a)} p'[f]}$	

Table 2.1: Operational semantics

If $p \xrightarrow{a} p'$ for some $a \in \mathscr{A}$, then we call p' a *residual* of p. If for a term p and an action a there does not exist a term p' such that $p \xrightarrow{a} p'$, then we write $p \not\xrightarrow{a}$.

It is technically convenient to extend the usage of the rules in Table 2.1 by letting them define binary relations $\xrightarrow{a}$ $(a \in \mathscr{A})$ on the full set of terms $\mathscr{T}_{\backslash,\square}$. (Since there are no operational rules for variables, this effectively means that variables are assigned the "same behaviour" as $\mathbf{0}$.)

The *depth* $\mathrm{d}(p)$ can then be defined for all process terms $p \in \mathscr{T}_{\backslash,\square}$ as the maximum number of consecutive transitions that can be performed starting from p, i.e.,

$$\mathrm{d}(p) = \max\{n \mid \exists_{p_1,\dots,p_n \in \mathscr{T}_{\backslash,\square}} \text{ s.t. } p \xrightarrow{a_1} p_1 \xrightarrow{a_2} \dots \xrightarrow{a_n} p_n\}.$$

The operational semantics assigns behaviour to closed terms. The notion of bisimilarity [12] relates closed process terms that exhibit equal behaviour.

Definition 1. A *bisimulation* is a symmetric binary relation $\mathscr{R}$ on $\mathscr{T}^C_{\backslash,\square}$ such that $p \mathrel{\mathscr{R}} q$ implies

$$\text{if } p \xrightarrow{a} p', \text{ then there exists some } q' \in \mathscr{T}^C_{\backslash,\square} \text{ such that } q \xrightarrow{a} q' \text{ and } p' \mathrel{\mathscr{R}} q'.$$

Closed process terms $p,q \in \mathscr{T}^C_{\backslash,\square}$ are said to be *bisimilar* (notation: $p \leftrightarroweq q$) if a bisimulation relation $\mathscr{R}$ exists such that $p \mathrel{\mathscr{R}} q$.

It is well-known that $\leftrightarroweq$ is an equivalence relation. We denote by $[p]$ the *equivalence class* of a closed process term $p \in \mathscr{T}^C_{\backslash,\square}$ modulo bisimilarity, and by $\mathscr{T}^C_{\backslash,\square}/\!\leftrightarroweq$ the set of all such equivalence classes. The rules in Table 2.1 are all in de Simone's format [13], and from this it follows that bisimilarity is compatible with the syntactic constructs of our process calculus. So $\mathscr{T}^C_{\backslash,\square}/\!\leftrightarroweq$ is the universe of a process algebra with a distinguished element $\mathbf{0}$, unary operators $a.\_$ (for all $a \in \mathscr{A}$), $\_[f]$ (for all relabelling functions $f : \mathscr{A} \to \mathscr{A}$), and $\_ \backslash L$ (for all $L \subseteq \mathscr{L}$), and binary operators $\_ + \_$, $\_ \parallel \_$ and $\_ \lfloor\!\lfloor \_$ defined as follows:

$$\begin{array}{lll}
\mathbf{0} = [\mathbf{0}] \ , & [p] \parallel [q] = [p \parallel q] \ , & [p] \backslash L = [p \backslash L] \ , \\
a.[p] = [a.p] \ , & [p] \lfloor\!\lfloor [q] = [p \lfloor\!\lfloor q] \ , & [p][f] = [p[f]] \ , \\
[p] + [q] = [p+q] \ . & &
\end{array}$$

Henceforth we shall denote this process algebra by **P**. Members of **P** are called *processes* and will be ranged over by p, q and r like process terms. This convention will not lead to confusion because it will be clear from the context which is meant.

To be able to reason syntactically about **P**, we define how process terms can be used to denote elements of **P** and present an inference system for the derivation of equations between process terms that are valid in **P**.

Definition 2. A *valuation* is a mapping $\nu : \mathscr{V} \to \mathbf{P}$. Such a mapping may be applied to process terms in $\mathscr{T}_{\backslash,[]}$ using the *evaluation mapping* $[\![\cdot]\!]_\nu : \mathscr{T}_{\backslash,[]} \to \mathbf{P}$ defined inductively by:

$$
\begin{array}{lll}
[\![\mathbf{0}]\!]_\nu = \mathbf{0} \ , & [\![q+r]\!]_\nu = [\![q]\!]_\nu + [\![r]\!]_\nu \ , & [\![q \backslash L]\!]_\nu = [\![q]\!]_\nu \backslash L \ , \\
[\![x]\!]_\nu = \nu(x) \ , & [\![q \parallel r]\!]_\nu = [\![q]\!]_\nu \parallel [\![r]\!]_\nu \ , & [\![q[f]]\!]_\nu = [\![q]\!]_\nu[f] \ , \\
[\![a.q]\!]_\nu = a.[\![q]\!]_\nu \ , & [\![q \mathbin{\|\!\lfloor} r]\!]_\nu = [\![q]\!]_\nu \mathbin{\|\!\lfloor} [\![r]\!]_\nu \ . &
\end{array}
$$

Note that the evaluation mapping maps process terms to members of the algebra **P**, given an assignment of processes to variables. When an evaluation mapping is applied to a closed process term, the assignment is irrelevant and the evaluation mapping amounts to interpreting the syntactic constructs as the corresponding operations of the algebra. Thus, without fixing a specific evaluation mapping, we can use a closed term to denote an element of **P**; this element of **P** is then, of course, the equivalence class that contains the particular closed term. For example, the closed term $a.\mathbf{0}+b.\mathbf{0}$ denotes the element $[\![a.\mathbf{0}+b.\mathbf{0}]\!]_\nu$ of **P**.

A *process equation* is a pair of process terms (p, q) written as $p \approx q$. The equation $p \approx q$ is *valid* in **P** if $[\![p]\!]_\nu = [\![q]\!]_\nu$ for all valuations $\nu : \mathscr{V} \to \mathbf{P}$. Henceforth, we write $p \leftrightarrows q$ if $p \approx q$ is valid in **P**.

(A1) $x+y \approx y+x$	(LM1) $x \mathbin{\|\!\lfloor} \mathbf{0} \approx x$
(A2) $(x+y)+z \approx x+(y+z)$	(LM2) $\mathbf{0} \mathbin{\|\!\lfloor} x \approx \mathbf{0}$
(A3) $x+x \approx x$	(LM3) $a.x \mathbin{\|\!\lfloor} y \approx a.(x \parallel y)$
(A4) $x+\mathbf{0} \approx x$	(LM4) $(x+y) \mathbin{\|\!\lfloor} z \approx x \mathbin{\|\!\lfloor} z + y \mathbin{\|\!\lfloor} z$
	(LM5) $(x \mathbin{\|\!\lfloor} y) \mathbin{\|\!\lfloor} z \approx x \mathbin{\|\!\lfloor} (y \parallel z)$
	(M) $x \parallel y \approx x \mathbin{\|\!\lfloor} y + y \mathbin{\|\!\lfloor} x$
(RS1a) $x \backslash \emptyset \approx x$	(RL1) $x[Id] \approx x$
(RS1b) $x \backslash \mathscr{L} \approx \mathbf{0}$	(RL2) $\mathbf{0}[f] \approx \mathbf{0}$
(RS2) $\mathbf{0} \backslash L \approx \mathbf{0}$	(RL3) $(a.x)[f] \approx f(a).(x[f])$
(RS3) $a.x \backslash L \approx \begin{cases} \mathbf{0} & \text{if } a, \overline{a} \in L \\ a.(x \backslash L) & \text{if } a, \overline{a} \notin L \end{cases}$	(RL4) $(x+y)[f] \approx x[f]+y[f]$
	(RL5) $(x \mathbin{\|\!\lfloor} y)[f] \approx x[f] \mathbin{\|\!\lfloor} y[f]$
(RS4) $(x+y) \backslash L \approx x \backslash L + y \backslash L$	(RL6) $(x[f])[g] \approx x[g \circ f]$
(RS5) $(x \mathbin{\|\!\lfloor} y) \backslash L \approx x \backslash L \mathbin{\|\!\lfloor} y \backslash L$	
(RS6) $(x \backslash L) \backslash K \approx x \backslash (L \cup K)$	
(RR1) $x[f] \backslash L \approx (x \backslash f^{-1}(L))[f]$	
(RR2) $(x \backslash L)[f] \approx (x \backslash L)[g]$ if $f \restriction (\mathscr{L} - L) = g \restriction (\mathscr{L} - L)$	

Table 2.2: The set of axioms $\mathscr{E}$

Table 2.2 presents a set of process equations $\mathscr{E}$ that are all well-known to be valid in $\mathbf{P}$ (see, e.g., [6, 9, 5, 3]). We shall use the process equations in $\mathscr{E}$ as the axioms of an inference system with as rules the familiar rules of equational logic [4]. Henceforth, whenever we write $p \approx q$ we mean that the process equation $p \approx q$ is derivable within this inference system. (In the cases in which we intend to highlight that only a proper subset of the axioms in $\mathscr{E}$ is needed to derive $p \approx q$, we shall explicitly mention the needed axioms.)

Since the axioms are valid in $\mathbf{P}$ and the rules of equational logic preserve validity, we have the following soundness result.

Proposition 1 (Soundness). *For all process terms $p,q \in \mathscr{T}_{\backslash,[]}$, if $p \approx q$, then $p \leftrightarrows q$.*

The main goal of this paper is to prove that the inference system is also complete, i.e., that, for all process terms $p,q \in \mathscr{T}_{\backslash,[]}$, if $p \leftrightarrows q$ then $p \approx q$; if this is the case, then it follows that $\mathscr{E}$ is an *equational base* for the algebra $\mathbf{P}$. Our completeness proof proceeds according to the following strategy:

1. Identify an appropriate notion of normal form and prove that every term in $\mathscr{T}_{\backslash,[]}$ is rewritable according to the axioms in $\mathscr{E}$ to a normal form. To establish completeness, it is then enough to prove that $s \leftrightarrows t$ implies $s \approx t$ for all normal forms s and t.
2. Associate with every two normal forms s and t a distinguishing valuation, i.e., a valuation $* : \mathscr{V} \to \mathbf{P}$ such that if $s \not\approx t$, then $[\![s]\!]_* \neq [\![t]\!]_*$. From this it follows that $s \leftrightarrows t$ implies $s \approx t$ for all normal forms s and t.

The first step is fairly straightforward, even though the normal forms we need to consider involve all the operations in the calculus; the crux of our completeness proof is to find a suitable distinguishing valuation and prove the property described in the second step. Our distinguishing valuation combines several ideas that are best explained separately. To this end, we shall, as stepping stones towards our main result, first apply the aforementioned strategy to obtain completeness results for the fragments $\mathscr{T}_{\backslash}$ and $\mathscr{T}_{[]}$ of our calculus. In Sect. 3 we consider the fragment without relabelling. In Sect. 4 we study the fragment without restriction. Finally, in Sect. 5 we consider the full calculus.

We use the summation $\sum_{i \in I} p_i$ (modulo A1, A2 and A4) to denote an alternative composition of the form $p_1 + p_2 + \ldots$ for a finite set I and processes p_i $(i \in I)$. We also define $\mathbf{0} = \sum_{i \in \emptyset} p_i$ for the empty index set. Furthermore, we shall use an abbreviation for iterated prefixing, defining $a^0.\mathbf{0} = \mathbf{0}$ and $a^{i+1}.\mathbf{0} = a.(a^i.\mathbf{0})$.

We conclude this section with a few properties pertaining to the algebra $\mathbf{P}$ that we shall need in the rest of the paper.

The binary relations $\xrightarrow{a}$ $(a \in \mathscr{A})$ defined earlier for $\mathscr{T}^C_{\backslash,[]}$ induce binary relations $\xrightarrow{a}$ $(a \in \mathscr{A})$ on $\mathbf{P}$ as follows: for all $p, p' \in \mathscr{T}^C_{\backslash,[]}$ we define that $[p] \xrightarrow{a} [p']$ iff for all $q \in [p]$ there exists a $q' \in [p']$ such that $q \xrightarrow{a} q'$.

Proposition 2. *For all $p,q,r \in \mathbf{P}$*

1. $p = \mathbf{0}$ *iff* $p \not\xrightarrow{a}$ *for all* $a \in \mathscr{A}$*;*
2. $a.p \xrightarrow{b} r$ *iff* $a = b$ *and* $p = r$*;*
3. $p + q \xrightarrow{a} r$ *iff* $p \xrightarrow{a} r$ *or* $q \xrightarrow{a} r$*;*

4. $p \lfloor\!\lfloor q \xrightarrow{a} r$ *iff there exists some* $p' \in \mathbf{P}$ *such that* $p \xrightarrow{a} p'$ *and* $r = p' \parallel q$;
5. $p \parallel q \xrightarrow{a} r$ *iff* $p \lfloor\!\lfloor q \xrightarrow{a} r$ *or* $q \lfloor\!\lfloor p \xrightarrow{a} r$;
6. $p \setminus L \xrightarrow{a} r$ *iff* $a, \overline{a} \notin L$ *and there exists some* $q \in \mathbf{P}$ *such that* $p \xrightarrow{a} q$ *and* $r = q \setminus L$;
7. $p[f] \xrightarrow{b} r$ *iff there exist some* $a \in \mathscr{A}$ *and* $q \in \mathbf{P}$ *such that* $f(a) = b$, $p \xrightarrow{a} q$ *and* $r = q[f]$.

Bisimulation equivalence preserves the notion of depth (i.e., the closed process terms in an equivalence class have the same depth). Therefore we can define the depth $\mathrm{d}(p)$ of a process $p \in \mathbf{P}$ as the depth of any of its members. As a technical tool we shall also need the notion of *branching degree* $\mathrm{b}(p)$ of a process $p \in \mathbf{P}$ defined by

$$\mathrm{b}(p) = |\{(a, p') \mid p \xrightarrow{a} p'\}|.$$

Lemma 1. *For all* $p, q \in \mathbf{P}$, *it holds that*
1. $\mathrm{b}(\mathbf{0}) = 0$;
2. $\mathrm{b}(a.p) = 1$;
3. $\mathrm{b}(p+q) \leq \mathrm{b}(p) + \mathrm{b}(q)$;
4. $\mathrm{b}(p \lfloor\!\lfloor q) = \mathrm{b}(p)$;
5. $\mathrm{b}(p \parallel q) \geq \mathrm{b}(p)$ *and* $\mathrm{b}(p \parallel q) \geq \mathrm{b}(q)$.

An element $p \in \mathbf{P}$ is *parallel prime* if $p \neq \mathbf{0}$, and $p = q \parallel r$ implies $q = \mathbf{0}$ or $r = \mathbf{0}$. A *parallel decomposition* of p is a finite multiset $[p_1, \ldots, p_n]$ of parallel primes such that $p = p_1 \parallel \cdots \parallel p_n$. The following theorem and corollary are proved in [10].

Theorem 1. *Every element of* $\mathbf{P}$ *has a unique parallel decomposition.*

Corollary 1. *Let* $p, q, r \in \mathbf{P}$. *If* $p \parallel q = p \parallel r$, *then* $q = r$.

3 Restriction

In this section we establish a completeness result for the fragment of our process calculus that includes the restriction operators, but excludes relabelling operators.

The set of normal forms $\mathscr{N}_{\setminus}$ is generated by the following grammar:

$$\mathrm{N} ::= \mathbf{0} \mid a.\mathrm{N} \mid (x \setminus L) \lfloor\!\lfloor \mathrm{N} \mid \mathrm{N} + \mathrm{N}$$

where $a \in \mathscr{A}$, $x \in \mathscr{V}$, and $L \subset \mathscr{L}$. We refer to $a.s$ and $(x \setminus L) \lfloor\!\lfloor s$ as simple normal forms.

Lemma 2. *Every process term* $p \in \mathscr{T}_{\setminus}$ *has a normal form* $s \in \mathscr{N}_{\setminus}$ *such that* $p \approx s$ *is provable using RS1a–RS6, LM1–LM5, and M.*

Because of Lemma 2, each term can be written using the following general form:

$$\sum_{i \in I} a_i.s_i + \sum_{j \in J} (x_j \setminus L_j) \lfloor\!\lfloor s_j \quad \text{(modulo A1, A2 and A4)}$$

for finite index sets I, J and with $a_i \in \mathscr{A}$, $s_i, s_j \in \mathscr{N}_\setminus$, $x_j \in \mathscr{V}$, and $L_j \subset \mathscr{L}$.

For our completeness proof, we define a valuation that allows us to distinguish non-bisimilar normal forms. The definitions of the distinguishing valuations we use in this paper are geared towards achieving the properties stated in Lemmas 5 and 6 to follow (or similar lemmas in the subsequent sections). In particular, distinguishing valuations will allow us to tell apart the different types of simple normal forms (Lemma 5).

Definition 3. Let $w \geq 1$ and let $\lceil\cdot\rceil : \mathscr{V} \to (\mathbb{N} - \{0,1\})$ be an injective function. We define the valuation $\diamond_w$ for each variable $x \in \mathscr{V}$ by:

$$\diamond_w(x) = \sum_{a \in \mathscr{L}} a.\xi_{\lceil x \rceil \cdot w} \text{ with } \xi_i = \sum_{a \in \mathscr{L}} \sum_{j=1}^{i} a^i.\mathbf{0}.$$

Note that if s is a simple normal form, then $[\![s]\!]_{\diamond_w}$ has a unique residual. In the following lemmas we establish some special properties pertaining to the valuation $\diamond_w$ when applied to normal forms. These properties will be used to show that $\diamond_w$ is indeed a distinguishing valuation.

First we state two properties of the process $\xi_{\lceil x \rceil \cdot w} \setminus L$, which is a parallel component of the unique residual of $[\![(x \setminus L) \mathbin{\lfloor\!\lfloor} s]\!]_{\diamond_w}$.

Lemma 3. *For all $i \geq 1$ and $L \subset \mathscr{L}$, the process $\xi_i \setminus L$ is parallel prime, and its branching degree $\mathrm{b}(\xi_i \setminus L)$ is $i \cdot |\mathscr{L} - L|$.*

The valuation $\diamond_w$ is such that if the parameter w is greater than an estimated highest branching degree occurring already in s, then it is possible to determine from the process $[\![s]\!]_{\diamond_w}$ whether s has action prefixing or a left merge as head operator. This will be explained in Lemma 5 below; first we formalise an appropriate estimation of the highest branching degree occurring in a normal form s.

Definition 4. For all $s \in \mathscr{N}_\setminus$, the estimated highest branching degree $\mathrm{esb}(s)$ occurring in s is defined inductively as follows:

$$\mathrm{esb}(\mathbf{0}) = 0, \qquad \mathrm{esb}(s+t) = \mathrm{esb}(s) + \mathrm{esb}(t),$$
$$\mathrm{esb}(a.t) = \max(1, \mathrm{esb}(t)), \qquad \mathrm{esb}((x \setminus L) \mathbin{\lfloor\!\lfloor} t) = \max(|\mathscr{L} - L|, \mathrm{esb}(t)),$$

with $a \in \mathscr{A}$, $x \in \mathscr{V}$, $L \subset \mathscr{L}$ and $t \in \mathscr{N}_\setminus$.

Note 1. The lower bound $|\mathscr{L} - L|$ in the definition of $\mathrm{esb}((x \setminus L) \mathbin{\lfloor\!\lfloor} t)$ follows from the definition of $\diamond_w$ (see Definition 3), since $[\![x \setminus L]\!]_{\diamond_w} \xrightarrow{a} \xi_{\lceil x \rceil \cdot w} \setminus L$ for all $a \in \mathscr{L} - L$.

The following lemma shows that the estimated branching degree of s is an upper bound on the branching degree of $[\![s]\!]_{\diamond_w}$.

Lemma 4. *For every normal form $s \in \mathscr{N}_\setminus$, $\mathrm{b}([\![s]\!]_{\diamond_w}) \leq \mathrm{esb}(s)$.*

Lemma 5. *Let $s, s' \in \mathscr{N}_\setminus$ with s simple, and let $x \in \mathscr{V}$, $L \subset \mathscr{L}$ and $w > \mathrm{esb}(s)$.*

1. If $s = a.s'$, then the branching degree of the unique residual of $[\![s]\!]_{\diamond_w}$ is smaller than w.

2. *If $s = (x \setminus L) \lfloor\!\lfloor s'$, then the branching degree of the unique residual of $[\![s]\!]_{\diamond_w}$ is larger than w.*

Proof. *Assume that p is the residual of s: $[\![s]\!]_{\diamond_w} \xrightarrow{a} p$ for some $a \in \mathscr{L}$. We have the following two cases:*

1. *If $s = a.s'$, then $p = [\![s']\!]_{\diamond_w}$. Because $\mathrm{esb}(s) < w$, by Definition 4 $\mathrm{esb}(s') \leq \mathrm{esb}(s) < w$. Hence, by Lemma 4, the branching degree of $[\![s']\!]_{\diamond_w}$ is smaller than w.*
2. *If $s = (x \setminus L) \lfloor\!\lfloor s'$, then $p = (\xi_{\lceil x \rceil \cdot w} \setminus L) \parallel [\![s']\!]_{\diamond_w}$. We have by Lemma 3 that $\mathrm{b}(\xi_{\lceil x \rceil \cdot w} \setminus L) = \lceil x \rceil \cdot w \cdot |\mathscr{L} - L| > w$ (given that $L \subset \mathscr{L}$ and $\lceil x \rceil > 1$). Because $[\![s']\!]_{\diamond_w}$ does not decrease the branching degree of the residual p (by Lemma 1), we may conclude that the residual p has a branching degree that exceeds w.* □

When it has been determined from the unique residual of $[\![s]\!]_{\diamond_w}$ that s has a left merge as head operator, then the following key lemma allows us to determine which variable occurs in its left argument, and by which proper subset of $\mathscr{L}$ this variable is restricted.

Lemma 6. *For $i, j \geq 1$ and $K, L \subset \mathscr{L}$, if $\xi_i \setminus K = \xi_j \setminus L$, then $K = L$ and $i = j$.*

Proof. *We first show that $K = L$. Assume that $a \in \mathscr{L} - K$. By Definition 3 and Proposition 2(6) there exists some $r \in \mathbf{P}$ such that $\xi_i \setminus K \xrightarrow{a} r$. Therefore $\xi_j \setminus L \xrightarrow{a} r$ also holds. However, by Proposition 2(6) this also means that $a \in \mathscr{L} - L$. The case that $a \in \mathscr{L} - L$ is symmetrical. Hence, since $a \in \mathscr{L} - K$ iff $a \in \mathscr{L} - L$, it follows that $K = L$.*

Because $K = L$ and $\xi_i \setminus K = \xi_j \setminus L$, we know that $\mathrm{b}(\xi_i \setminus K) = \mathrm{b}(\xi_j \setminus K)$ and therefore $i \cdot |\mathscr{L} - K| = j \cdot |\mathscr{L} - K|$ by Lemma 3. Since $K \subset \mathscr{L}$, it follows that $i = j$. □

The following result states that the valuation $\diamond_w$ is indeed distinguishing.

Theorem 2. *For every two normal forms $s, t \in \mathscr{N}$ with $w > \mathrm{esb}(s), \mathrm{esb}(t)$, it holds that if $[\![s]\!]_{\diamond_w} = [\![t]\!]_{\diamond_w}$, then $s \approx t$ modulo A1–A4.*

Proof. *Assume that $[\![s]\!]_{\diamond_w} = [\![t]\!]_{\diamond_w}$ holds; we prove that $s \approx t$ is derivable using A1–A4 by induction on the sum of the depths of s and t. We do this by showing that for every summand s_i of s there exists a summand t_j of t such that $s_i \approx t_j$ modulo A1–A4. Consider the following case analysis based on the syntax of an arbitrary summand s_i of s:*

1. *If $s_i = a.s_i'$, then $[\![s_i]\!]_{\diamond_w} \xrightarrow{a} [\![s_i']\!]_{\diamond_w}$. Because $[\![s]\!]_{\diamond_w} = [\![t]\!]_{\diamond_w}$, there must also be a summand t_j of t such that $[\![t_j]\!]_{\diamond_w} \xrightarrow{a} [\![s_i']\!]_{\diamond_w}$. By Lemma 5 we know that t_j must have the form $b.t_j'$, because the branching degree of the unique residual of $[\![t_j]\!]_{\diamond_w}$ does not exceed w.*
   Given that t_j has this form, it can only perform one transition: $[\![t_j]\!]_{\diamond_w} \xrightarrow{b} [\![t_j']\!]_{\diamond_w}$. Since also $[\![t_j]\!]_{\diamond_w} \xrightarrow{a} [\![s_i']\!]_{\diamond_w}$ it follows that $a = b$ and $[\![s_i']\!]_{\diamond_w} = [\![t_j']\!]_{\diamond_w}$. By induction hypothesis we have that $s_i' \approx t_j'$ modulo A1–A4. Hence, we may conclude that $s_i = a.s_i' \approx b.t_j' = t_j$.

2. *If $s_i = (x \setminus K) \mathbin{⌊⌊} s'_i$, then, since $K \subset \mathscr{L}$, $[\![s_i]\!]_{\diamond_w} \xrightarrow{a} p$ for some $a \in \mathscr{L} - K$. We know that also a summand t_j of t exists such that $[\![t_j]\!]_{\diamond_w} \xrightarrow{a} p$. Definition 3 gives us that $p = (\xi_{\lceil x \rceil \cdot w} \setminus K) \parallel [\![s'_i]\!]_{\diamond_w}$. Similarly to the previous case, by Lemma 5 we also know that t_j must have the form $(y \setminus L) \mathbin{⌊⌊} t'_j$ for some $y \in \mathscr{V}$ and $L \subset \mathscr{L}$. The residual of t_j after performing an action $a \in \mathscr{L} - L$ is $(\xi_{\lceil y \rceil \cdot w} \setminus L) \parallel [\![t'_j]\!]_{\diamond_w}$ (also by Definition 3). This residual is equal to p, so we know that $(\xi_{\lceil x \rceil \cdot w} \setminus K) \parallel [\![s'_i]\!]_{\diamond_w} = (\xi_{\lceil y \rceil \cdot w} \setminus L) \parallel [\![t'_j]\!]_{\diamond_w}$. By Lemma 3 we have that the process $\xi_{\lceil x \rceil \cdot w} \setminus K$ is parallel prime and has a branching degree that exceeds w. This process cannot occur in the unique parallel decomposition of $[\![t'_j]\!]_{\diamond_w}$ because, by Lemmas 1 and 4, and the assumption of the theorem that $w > \mathrm{esb}(t)$, the branching degrees of all processes in the decomposition of $[\![t'_j]\!]_{\diamond_w}$ do not exceed w. Conversely, this also holds in a symmetric way for the process $\xi_{\lceil y \rceil \cdot w} \setminus L$ with respect to the unique parallel decomposition of $[\![s'_i]\!]_{\diamond_w}$. Hence, $\xi_{\lceil x \rceil \cdot w} \setminus K = \xi_{\lceil y \rceil \cdot w} \setminus L$.*
From $\xi_{\lceil x \rceil \cdot w} \setminus K = \xi_{\lceil y \rceil \cdot w} \setminus L$ it follows by Lemma 6 that $K = L$ and $\lceil x \rceil \cdot w = \lceil y \rceil \cdot w$. Therefore, $x = y$ by injectivity of $\lceil \cdot \rceil$.
We have established that $K = L$ and $x = y$, so $(\xi_{\lceil x \rceil \cdot w} \setminus K) \parallel [\![s'_i]\!]_{\diamond_w} = (\xi_{\lceil y \rceil \cdot w} \setminus L) \parallel [\![t'_j]\!]_{\diamond_w} = (\xi_{\lceil x \rceil \cdot w} \setminus K) \parallel [\![t'_j]\!]_{\diamond_w}$, and hence, by Corollary 1, $[\![s'_i]\!]_{\diamond_w} = [\![t'_j]\!]_{\diamond_w}$. By induction hypothesis it follows that $s'_i \approx t'_j$ modulo A1–A4, so we may conclude that $s_i = (x \setminus K) \mathbin{⌊⌊} s'_i \approx (y \setminus L) \mathbin{⌊⌊} t'_j = t_j$ modulo A1–A4.

It follows by a symmetric argument that every summand of t is also provably equal to a summand of s using the above mentioned equations. Hence, $s \approx s + t \approx t$ modulo A1–A4. □

Corollary 2. *For all $p, q \in \mathscr{T}_{\setminus}$ it holds that $p \approx q$ is provable using A1–A4, RS1a–RS6, LM1–LM5, and M if, and only if, $p \leftrightarrows q$.*

Proof. The implication from left to right follows from Proposition 1.

For the proof of the implication from the right to the left, we assume that $p \leftrightarrows q$. By Lemma 2, there are two normal forms s and t such that the equations $p \approx s$ and $q \approx t$ are provable using RS1a–RS6, LM1–LM5, and M. If $p \leftrightarrows q$, then by Proposition 1 and transitivity of $\leftrightarrows$ we also know that $s \leftrightarrows t$ and thus $[\![s]\!]_{\diamond_w} = [\![t]\!]_{\diamond_w}$. Hence, by Theorem 2 we know that $s \approx t$ is provable using A1–A4 and we can conclude that $p \approx s \approx t \approx q$. □

4 Relabelling

In this section we establish a completeness result for the fragment of our process calculus that includes relabelling operators, but excludes restriction operators.

The set of normal forms $\mathscr{N}_{[]}$ is generated by the following grammar:

$$\mathrm{N} ::= \mathbf{0} \mid a.\mathrm{N} \mid x[f] \mathbin{⌊⌊} \mathrm{N} \mid \mathrm{N} + \mathrm{N},$$

where $a \in \mathscr{A}$, $x \in \mathscr{V}$, and $f : \mathscr{L} \to \mathscr{L}$ is a relabelling function. We refer to $a.\mathrm{N}$ and $x[f] \mathbin{⌊⌊} \mathrm{N}$ as simple normal forms.

Lemma 7. *Every process term $p \in \mathscr{T}_{[]}$ has a normal form $s \in \mathscr{N}_{[]}$ such that $p \approx s$ is provable using RL1–RL6, LM1–LM5, and M.*

Because of Lemma 7, each term can be written using the following general form:

$$\sum_{i \in I} a_i.s_i + \sum_{j \in J} (x_j[f_j]) \,\lfloor\!\lfloor\, s_j \quad \text{(modulo A1, A2 and A4)}$$

for finite index sets I, J and with $a_i \in \mathscr{A}$, $s_i, s_j \in \mathscr{N}_{[]}$, $x_j \in \mathscr{V}$, and relabelling functions $f_j : \mathscr{L} \to \mathscr{L}$.

Our goal now is to find a distinguishing valuation for each pair of non-bisimilar normal forms. In the following definitions and lemmas $\mathbb{P}$ denotes the set of prime numbers.

Definition 5. Let $\lfloor \cdot \rfloor : \mathscr{L} \to \mathbb{P}$ be an injective function, w a prime number larger than any prime number in the range of $\lfloor \cdot \rfloor$, and let $\lceil \cdot \rceil : \mathscr{V} \to \{m \in \mathbb{P} \mid m > w\}$ be another injective function. We define the valuation $\diamond_w$ for each variable $x \in \mathscr{V}$ by:

$$\diamond_w(x) = a.\zeta_{\lceil x \rceil, w} \text{ with } \zeta_{i,w} = a.\mathbf{0} + \sum_{b \in \mathscr{L}} \sum_{j=1}^{w} b^{i \cdot \lfloor b \rfloor^j}.\mathbf{0},$$

where a is an arbitrary action in $\mathscr{A}$.

Our aim in defining the valuation $\diamond_w$ is again to be able to distinguish the different types of simple normal forms that may occur as summands of a normal form. As in Sect. 3, we will be able to distinguish summands of the form $a.s$ from those of the form $x[f] \,\lfloor\!\lfloor\, s'$ since the unique residual of terms with the latter form will have a larger branching degree than the unique residual of action-prefixed terms—see Lemma 10 to follow. However, in the definition of $\diamond_w$ we also want to ensure that terms of the form $\zeta_i, w[f]$ are prime, and that the sequences of actions those terms afford "encode" the relabelling function f. We obtain the primality of $\zeta_{i,w}[f]$ by means of the summand $a.\mathbf{0}$ of $\zeta_{i,w}$, whereas we encode relabelling functions by taking sequences of actions whose lengths are powers of distinct prime numbers. This is enough to ensure that if $\zeta_{i,w}[f]$ and $\zeta_{i,w}[g]$ are bisimilar, then $f = g$—see Lemma 11 to follow.

Lemma 8. *For all $i \geq 1$ and relabelling functions $f : \mathscr{L} \to \mathscr{L}$, the process $\zeta_{i,w}[f]$ is parallel prime, and its branching degree is* $\mathrm{b}(\zeta_{i,w}[f]) = 1 + |\mathscr{L}| \cdot w$.

Again, the distinguishing ability of the valuation $\diamond_w$ depends on the value of the parameter w being greater than an estimated highest branching degree occurring already in s. This is explained in Lemma 10 below; first we formalise an appropriate estimation of the highest branching degree occurring in a normal form s.

Definition 6. For all $s \in \mathscr{N}_{[]}$, the estimated highest branching degree $\mathrm{esb}(s)$ is defined inductively as follows:

$$\mathrm{esb}(\mathbf{0}) = 0, \qquad \mathrm{esb}(s+t) = \mathrm{esb}(s) + \mathrm{esb}(t),$$
$$\mathrm{esb}(a.t) = \max(1, \mathrm{esb}(t)), \qquad \mathrm{esb}(x[f] \,\lfloor\!\lfloor\, t) = \max(1, \mathrm{esb}(t)),$$

with $a \in \mathscr{A}$, $x \in \mathscr{V}$, relabelling function $f : \mathscr{L} \to \mathscr{L}$ and $t \in \mathscr{N}_{[]}$.

The following lemma shows that the estimated branching degree of s is an upper bound on the branching degree of $[\![s]\!]_{\diamond_w}$.

Lemma 9. *For every normal form $s \in \mathscr{N}_{[]}$, $\mathrm{b}([\![s]\!]_{\diamond_w}) \leq \mathrm{esb}(s)$.*

Lemma 10. *Let $s, s' \in \mathscr{N}_{[]}$ be simple normal forms, $x \in \mathscr{V}$, $f : \mathscr{L} \to \mathscr{L}$ a relabelling function and let $w > \mathrm{esb}(s)$.*

1. *If $s = a.s'$, then the unique residual of $[\![s]\!]_{\diamond_w}$ has a branching degree smaller than w.*
2. *If $s = (x[f]) \mathbin{⌊⌊} s'$, then the unique residual of $[\![s]\!]_{\diamond_w}$ has a branching degree larger than w.*

When it has been determined from the unique residual of $[\![s]\!]_{\diamond_w}$ that s has a left merge as head operator, then the following key lemma allows us to determine which variable and which relabelling function occur in its left argument.

Lemma 11. *For $i, j \geq 1$ and relabelling functions $f, g : \mathscr{L} \to \mathscr{L}$, if $\zeta_{i,w}[f] = \zeta_{j,w}[g]$, then $i = j$ and $f = g$.*

Proof. From $\zeta_{i,w}[f] = \zeta_{j,w}[g]$ it follows that $\mathrm{d}(\zeta_{i,w}[f]) = \mathrm{d}(\zeta_{j,w}[g])$ and therefore $i \cdot \lfloor b \rfloor^w = j \cdot \lfloor b \rfloor^w$ for that $b \in \mathscr{L}$ for which $\lfloor b \rfloor$ is largest. Since $\lfloor b \rfloor$ is positive, $i = j$. It remains to prove that $f = g$. Let $b \in \mathscr{L}$. Then $\zeta_{i,w}[f] \xrightarrow{f(b)} \left(b^{(i \cdot \lfloor b \rfloor^w) - 1}\right)[f]$. By the assumption that $\zeta_{i,w}[f] = \zeta_{j,w}[g]$ and since $i = j$, it follows that there also exists some $c \in \mathscr{L}$ such that $f(b) = g(c)$, $\zeta_{i,w}[g] \xrightarrow{f(c)} \left(c^{(i \cdot \lfloor c \rfloor^v) - 1}\right)[f]$ and $\left(b^{(i \cdot \lfloor b \rfloor^w) - 1}\right)[f] = \left(c^{(i \cdot \lfloor c \rfloor^v) - 1}\right)[g]$. Hence $i \cdot \lfloor b \rfloor^w = i \cdot \lfloor c \rfloor^v$ and since $\lfloor b \rfloor$ and $\lfloor c \rfloor$ are prime, it follows that $b = c$ and $w = v$. Therefore, $f(b) = g(b)$. □

Using the previous lemmas, and reasoning as in the proof of Theorem 2, we can now prove that the valuation defined in Definition 5 is indeed distinguishing.

Theorem 3. *For every two normal forms $s, t \in \mathscr{N}_{[]}$ with $w > \mathrm{esb}(s), \mathrm{esb}(t)$, it holds that if $[\![s]\!]_{\diamond_w} = [\![t]\!]_{\diamond_w}$, then $s \approx t$ modulo A1–A4.*

Proof. Assume that $[\![s]\!]_{\diamond_w} = [\![t]\!]_{\diamond_w}$ holds; we prove that $s \approx t$ is derivable using A1–A4 by induction on the sum of the depths of s and t. We do this by showing that for each summand s_i of s there exists a summand t_j of t such that $s_i \approx t_j$ modulo A1–A4. By symmetry this suffices to prove the claim.*

1. *If $s_i = a.s'_i$, then $[\![s_i]\!]_{\diamond_w} \xrightarrow{a} [\![s'_i]\!]_{\diamond_w}$. Because $[\![s]\!]_{\diamond_w} = [\![t]\!]_{\diamond_w}$ there must also be a summand t_j of t such that $[\![t_j]\!]_{\diamond_w} \xrightarrow{a} [\![s'_j]\!]_{\diamond_w}$. By Lemma 10 we know that t_j must have the form $b.t'_j$, because the branching degree of the unique residual of $[\![t'_j]\!]_{\diamond_w}$ does not exceed w.*
   Given that t_j has this form, it can only perform one transition: $[\![t_j]\!]_{\diamond_w} \xrightarrow{b} [\![t'_j]\!]_{\diamond_w}$. Since also $[\![t_j]\!]_{\diamond_w} \xrightarrow{a} [\![s'_i]\!]_{\diamond_w}$ it follows that $a = b$ and $[\![s'_i]\!]_{\diamond_w} = [\![t'_j]\!]_{\diamond_w}$. By induction hypothesis we have that $s'_i \approx t'_j$ modulo A1–A4. Hence, we may conclude that $s_i = a.s'_i \approx b.t'_j = t_j$.

2. *If $s_i = x[f] \,\|\!\_\, s'_i$, then $[\![s_i]\!]_{\diamond_w} \xrightarrow{f(a)} \zeta_{\lceil x \rceil, w}[f] \parallel [\![s'_i]\!]_{\diamond_w} = p$. Since $[\![s]\!]_{\diamond_w} = [\![t]\!]_{\diamond_w}$, there must be a summand $t_j = y[g] \,\|\!\_\, t'_j$ of t such that $[\![t_j]\!]_{\diamond_w} \xrightarrow{g(b)} \zeta_{\lceil y \rceil, w}[g] \parallel [\![t'_j]\!]_{\diamond_w} = q$ and $p = q$. By Lemma 9, the right-hand side parallel components of p and q have branching degrees not exceeding w whereas, by Lemma 8, the left-hand side parallel components are parallel prime and have branching degree $1 + |\mathscr{L}| \cdot w$. Using Theorem 1 it follows that $\zeta_{\lceil x \rceil, w}[f] = \zeta_{\lceil y \rceil, w}[g]$ and $[\![s'_j]\!]_{\diamond_w} = [\![t'_j]\!]_{\diamond_w}$. By Lemma 11 we have that $\lceil x \rceil = \lceil y \rceil$ and $f = g$. Hence, $x = y$ by injectivity of $\lceil \cdot \rceil$. By induction, we have that $s'_i \approx t'_j$ modulo A1–A4. Therefore $x[f] \,\|\!\_\, s'_j$ is provably equal to a summand of t.*

It follows by a symmetric argument that every summand of t is also provably equal to a summand of s using the above mentioned equations. Hence, $s \approx s + t \approx t$ modulo A1–A4. □

Corollary 3. *For all process terms $p, q \in \mathscr{T}_{[]}$ it holds that $p \approx q$ is provable using A1–A4, RL1–RL6, LM1–LM5, and M if, and only if, $p \leftrightarrows q$.*

5 Restriction and Relabelling

In this section, we consider the language that includes both restriction and relabelling operators.

The set of normal forms $\mathscr{N}_{\backslash,[]}$ is generated by the following grammar:

$$\mathrm{N} ::= \mathbf{0} \mid a.\mathrm{N} \mid (x \backslash L)[f] \,\|\!\_\, \mathrm{N} \mid \mathrm{N} + \mathrm{N}$$

where $a \in \mathscr{A}$, $x \in \mathscr{V}$, $L \subset \mathscr{L}$, and $f : \mathscr{L} \to \mathscr{L}$ is a relabelling function satisfying $f = f \restriction (\mathscr{L} - L)$ (i.e., f is the identity on all $a \in L$). We refer to the normal forms $a.\mathrm{N}$ and $(x \backslash L)[f] \,\|\!\_\, \mathrm{N}$ as simple normal forms.

Lemma 12. *Every process term $p \in \mathscr{T}_{\backslash,[]}$ has a normal form $s \in \mathscr{N}_{\backslash,[]}$ such that $p \approx s$ is provable using RS1a–RS6, RL1–RL6, RR1, RR2, LM1–LM5, and M.*

Now, using the previous lemma, each term can be written using the following general form:

$$\sum_{i \in I} a_i.s_i + \sum_{j \in K} (x_j \backslash L_j)[f_j] \,\|\!\_\, s_j \quad \text{(modulo A1, A2, A4, and RR2)}$$

for finite index sets I, J and with $a_i \in \mathscr{A}$, $s_i, s_j \in \mathscr{N}_{\backslash,[]}$, $x_j \in \mathscr{V}$, $L_j \subset \mathscr{L}$, and relabelling functions $f_j : \mathscr{L} \to \mathscr{L}$ with $f_j = f_j \restriction (\mathscr{L} - L_j)$.

A valuation that distinguishes an action prefix from a variable under restriction and relabelling can be constructed by combining the ideas underlying the valuations presented in Definitions 3 and 5. The result shown below uses powers of distinct prime numbers to "encode" the relabelling function and employs a summation over all actions to allow for the detection of the restricting set.

Definition 7. Let $\lfloor\cdot\rfloor : \mathscr{L} \to \mathbb{P}$ be an injective function, w a prime number larger than any prime number in the range of $\lfloor\cdot\rfloor$, and let $\lceil\cdot\rceil : \mathscr{V} \to \{m \in \mathbb{P} \mid m > w\}$ be another injective function. We define the valuation $\diamond_w$ for each variable $x \in \mathscr{V}$ by:

$$\diamond_w(x) = \sum_{a \in \mathscr{L}} a.\chi_{\lceil x \rceil, w} \text{ with } \chi_{i,w} = \sum_{a \in \mathscr{L}} \left(a.\mathbf{0} + \sum_{j=1}^{w} a^{i \cdot \lfloor a \rfloor^j}.\mathbf{0} \right).$$

First, we establish two properties of the process $(\chi_{\lceil x \rceil, w} \setminus L)[f]$, which is a parallel component of the unique residual of $[\![(x \setminus L)[f] \mathbin{\|\!\!\lfloor} s]\!]_{\diamond_w}$.

Lemma 13. *For all $i > 1$, $L \subset \mathscr{L}$, and relabelling functions $f : \mathscr{L} \to \mathscr{L}$, the process $(\chi_{i,w} \setminus L)[f]$ is parallel prime, and its branching degree is $|f(\mathscr{L} - L)| + |\mathscr{L} - L| \cdot w$.*

To enable the valuation $\diamond_w$ to distinguish between an action prefix and a term with the left merge as head operator, as explained in Lemma 15 below, we need an appropriate estimation of the highest branching degree occurring in a normal form s.

Definition 8. For all $s \in \mathscr{N}_{\setminus, []}$, the lower bound estimate of the branching degree of s, denoted with $\text{esb}(s)$, is defined inductively as follows:

$$\begin{aligned} &\text{esb}(\mathbf{0}) = 0, && \text{esb}(s+t) = \text{esb}(s) + \text{esb}(t), \\ &\text{esb}(a.t) = \max(1, \text{esb}(t)), && \text{esb}((x \setminus L)[f] \mathbin{\|\!\!\lfloor} t) = \max(|\mathscr{L}|, \text{esb}(t)). \end{aligned}$$

with $a \in \mathscr{A}$, $x \in \mathscr{V}$, $L \subset \mathscr{L}$, relabelling function $f : \mathscr{L} \to \mathscr{L}$ and $t \in \mathscr{N}_{\setminus, []}$.

The following lemma shows that the estimated branching degree of s is an upper bound on the branching degree of $[\![s]\!]_{\diamond_w}$.

Lemma 14. *For every normal form $s \in \mathscr{N}_{\setminus, []}$, $\text{b}([\![s]\!]_{\diamond_w}) \leq \text{esb}(s)$.*

Lemma 15. *Let $s, s' \in \mathscr{N}_{\setminus, []}$ be simple normal forms, $x \in \mathscr{V}$, $L \subset \mathscr{L}$, $f : \mathscr{L} \to \mathscr{L}$ a relabelling function and let $w > \text{esb}(s)$.*

1. *If $s = a.s'$, then the unique residual of $[\![s]\!]_{\diamond_w}$ has a branching degree smaller than w.*
2. *If $s = (x \setminus L)[f] \mathbin{\|\!\!\lfloor} s'$, then the unique residual of $[\![s]\!]_{\diamond_w}$ has a branching degree larger than w.*

The following lemma allows us to determine the variable, the restriction set and relabelling function in a simple normal form of the shape $(x \setminus L)[f] \mathbin{\|\!\!\lfloor} s$.

Lemma 16. *For $w \in \mathbb{P}$, $i, j \in \{m \in \mathbb{P} \mid m > w\}$, $K, L \subset \mathscr{L}$, and relabelling functions $f, g : \mathscr{L} \to \mathscr{L}$, if $(\chi_{i,w} \setminus K)[f] = (\chi_{j,w} \setminus L)[g]$, then $K = L$, $f \restriction (\mathscr{L} - K) = g \restriction (\mathscr{L} - K)$ and $i = j$.*

By following the strategy we adopted in the proofs of Theorems 2 and 3, we can show that the valuation defined above is indeed distinguishing.

Theorem 4. *For every two normal forms $s, t \in \mathscr{N}_{\setminus, []}$ with $w > \text{esb}(s), \text{esb}(t)$, it holds that if $[\![s]\!]_{\diamond_w} = [\![t]\!]_{\diamond_w}$, then $s \approx t$ modulo A1–A4 and RR2.*

Proof. *We now prove that $s \approx t$ assuming that $[\![s]\!]_{\diamond w} = [\![t]\!]_{\diamond w}$ by induction on the sum of the depths of s and t. We do so by proving that for every summand s_i of s a summand t_j of t exists such that $s_i \approx t_j$ modulo A1–A4 and RR2. Consider the following case analysis based on the syntax of an arbitrary summand s_i of s.*

1. *If $s_i = a.s'_i$, then $[\![s_i]\!]_{\diamond w} \xrightarrow{a} [\![s'_i]\!]_{\diamond w}$. Because $[\![s]\!]_{\diamond w} = [\![t]\!]_{\diamond w}$, there also must be a summand t_j of t such that $[\![t_j]\!]_{\diamond w} \xrightarrow{a} [\![s'_i]\!]_{\diamond w}$. By Lemma 15 we know that t_j must have the form $b.t'_j$, because the branching degree of the residual $[\![t'_j]\!]_{\diamond w}$ does not exceed w.*

   Given that t_j has this form, it can only perform one transition: $[\![t_j]\!]_{\diamond w} \xrightarrow{b} [\![t'_j]\!]_{\diamond w}$. Since also $[\![t_j]\!]_{\diamond w} \xrightarrow{a} [\![s'_i]\!]_{\diamond w}$ if follows that $a = b$ and $[\![s'_i]\!]_{\diamond w} = [\![t'_j]\!]_{\diamond w}$. By induction hypothesis we have that $s'_i \approx t'_j$ modulo A1–A4 and RR2. Hence, we may conclude that $s_i = a.s'_i \approx b.t'_j = t_j$.

2. *If $s_i = (x \setminus K)[f] \,\|\!\!\lfloor\, s'_i$, then, since $K \subset \mathscr{L}$, $[\![s_i]\!]_{\diamond w} \xrightarrow{f(a)} (\chi_{\lceil x \rceil, w} \setminus K)[f] \parallel [\![s'_i]\!]_{\diamond w} = p$ (by Definition 7) for some $a \in \mathscr{L} - K$. We know that also a summand t_j of t exists such that $[\![t_j]\!]_{\diamond w} \xrightarrow{f(a)} p$. Similarly as in previous case, by Lemma 15, we also know that t_j must have the form $(y \setminus L)[g] \,\|\!\!\lfloor\, t'_j$ for some $y \in \mathscr{V}$, $L \subset \mathscr{L}$, $g : \mathscr{L} \to \mathscr{L}$ such that $g(b) = f(a)$ for some $b \in \mathscr{L} - L$, and t'_j. The residual of t_j after performing an action $g(b)$ with $b \in \mathscr{L} - L$ is $(\chi_{\lceil y \rceil, w} \setminus L) \parallel [\![t'_j]\!]_{\diamond w}$ (also by Definition 7). This residual is equal to p, so we know that $(\chi_{\lceil x \rceil, w} \setminus K)[f] \parallel [\![s'_i]\!]_{\diamond w} = (\chi_{\lceil y \rceil, w} \setminus L)[g] \parallel [\![t'_j]\!]_{\diamond w}$.*
   By Lemma 13 we have that the process $(\chi_{\lceil x \rceil, w} \setminus K)[f]$ is parallel prime and has a branching degree that exceeds w. This process cannot occur in the unique parallel decomposition of $[\![t'_j]\!]_{\diamond w}$ because, by Lemma 1 and the fact that $w > \mathrm{esb}(t) \geq \mathrm{esb}(t'_j)$, the branching degrees of all processes in the parallel decomposition of $[\![t'_j]\!]_{\diamond w}$ do not exceed w. Conversely, this also holds in a symmetric way for the process $(\chi_{\lceil y \rceil, w} \setminus L)[g]$ with respect to the unique parallel decomposition of $[\![s'_i]\!]_{\diamond w}$. Hence by Theorem 1, $(\chi_{\lceil x \rceil, w} \setminus K)[f] = (\chi_{\lceil y \rceil, w} \setminus L)[g]$ and $[\![s'_i]\!]_{\diamond w} = [\![t'_j]\!]_{\diamond w}$.
   From $(\chi_{\lceil x \rceil, w} \setminus K)[f] = (\chi_{\lceil y \rceil, w} \setminus L)[g]$ it follows by Lemma 16 that $\lceil x \rceil = \lceil y \rceil$, $K = L$ and $f \upharpoonright (\mathscr{L} - K) = g \upharpoonright (\mathscr{L} - K)$. By the injectivity of $\lceil \cdot \rceil$ we know also that $x = y$. Since $[\![s'_i]\!]_{\diamond w} = [\![t'_j]\!]_{\diamond w}$, by induction hypothesis it follows that $s'_i \approx t'_j$ modulo A1–A4 and RR2.
   Summing up, we have established that $K = L$, $f \upharpoonright (\mathscr{L} - K) = g \upharpoonright (\mathscr{L} - K)$, $x = y$, and $s'_i \approx t'_j$ modulo A1–A4 and RR2, We may therefore conclude that $s_i = (x \setminus K)[f] \,\|\!\!\lfloor\, s'_i \approx (y \setminus L)[g] \,\|\!\!\lfloor\, t'_j = t_j$.

The above analysis shows that for each summand s_i of s there exists a summand t_j of t such that $s_i \approx t_i$ modulo A1–A4 and RR2. It follows by a symmetric argument that every summand of t is also provably equal to a summand of s using the above mentioned equations. Hence, $s \approx s + t \approx t$ modulo A1–A4 and RR2. □

Corollary 4. *For all process terms $p, q \in \mathscr{T}_{\setminus, []}$ it holds that $p \approx q$ if, and only if, $p \leftrightarrows q$.*

Acknowledgements

The work of Aceto and Ingólfsdóttir has been partially supported by the projects "The Equational Logic of Parallel Processes" (nr. 060013021) and "New Developments in Operational Semantics" (nr. 080039021) of the Icelandic Research Fund.

The research of van Tilburg was supported by the project "Models of Computation: Automata and Processes" (nr. 612.000.630) of the Netherlands Organisation for Scientific Research (NWO).

References

1. Aceto, L., Fokkink, W., Ingólfsdóttir, A., Luttik, B.: A finite equational base for CCS with left merge and communication merge. ACM Trans. Comput. Log. (2008). To appear (available as `http://tocl.acm.org/accepted/310luttik.pdf`).
2. Baeten, J.C.M., Bergstra, J.A.: Global renaming operators in concrete process algebra. Information and Computation **78**(3), 205–245 (1988)
3. Bergstra, J., Klop, J.: Process algebra for synchronous communication. Information and Control **60**(1-3), 109–137 (1984)
4. Birkhoff, G.: On the structure of abstract algebras. Proceedings of the Cambridge Philosophical Society **31**, 433–454 (1935)
5. Christensen, S., Hirshfeld, Y., Moller, F.: Decidable subsets of CCS. The Computer Journal **37**(4), 233–242 (1994)
6. Hennessy, M., Milner, R.: Algebraic laws for nondeterminism and concurrency. Journal of the ACM (JACM) **32**(1), 137–161 (1985)
7. Milner, R.: Flowgraphs and flow algebras. Journal of the ACM (JACM) **26**(4), 794–818 (1979)
8. Milner, R.: A Calculus of Communicating Systems. Springer (1980)
9. Milner, R.: Communication and Concurrency. Prentice-Hall (1989)
10. Milner, R., Moller, F.: Unique decomposition of processes. Theoret. Comput. Sci. **107**, 357–363 (1993)
11. Moller, F.: Axioms for Concurrency. Ph.D. thesis (1989)
12. Park, D.: Concurrency and automata on infinite sequences. In: P. Deussen (ed.) 5th GI Conference, *Lecture Notes in Computer Science*, vol. 104, pp. 167–183. Springer (1981)
13. de Simone, R.: Higher level synchronization devices in SCCS-Meije. Theoretical Computer Science **37**, 245–267 (1985)

μ-calculus Pushdown Module Checking with Imperfect State Information

Benjamin Aminof[1], Axel Legay[2], Aniello Murano[3], and Olivier Serre[4]

[1] Hebrew University, Jerusalem 91904, Israel.
[2] University of Liège, Belgium.
[3] Università degli Studi di Napoli "Federico II", 80126 Napoli, Italy.
[4] LIAFA, CNRS & Université Paris VII, France.

Abstract. The model checking problem for open systems (*module checking*) has recently been the subject of extensive study. The problem was first studied by Kupferman, Vardi, and Wolper for finite-state systems and properties expressed in the branching time logics *CTL* and *CTL**. Further study continued mainly in two directions: considering systems equipped with a pushdown store, and considering environments with imperfect information about the system. A recent paper combined the two directions and considered the *CTL* pushdown module checking problem in the imperfect information setting, i.e., in the case where the environment has only a partial view of the system control states and pushdown store content. It has been shown that this problem is undecidable when the environment has imperfect information about the pushdown store content, while it is decidable and 2EXPTIME-complete when the imperfect information only concerns control states. It was left open whether the latter remains decidable also for more expressive logics. In this paper, we answer this question in the affirmative, showing that the pushdown module checking problem with imperfect information about the control states is decidable and 2EXPTIME-complete for the propositional and the graded μ-calculus, and 3EXPTIME-complete for *CTL**.

1 Introduction

A main distinction in system modeling is between closed systems, whose behavior is totally determined by the program, and open systems, which are systems where the program interacts with an external environment [HP85, Hoa85]. In order to check whether a closed system satisfies a required property we translate the system into a formal model (such as a transition system), specify the property with a temporal-logic formula (such as *CTL* [CE81], *CTL** [EH86], and μ-calculus [Koz83]), and check formally that the model satisfies the formula. This process is called *model checking* ([CE81, QS81]). Checking whether an open system satisfies a required temporal logic formula is much harder, as one has to consider the interaction of the system with all possible environments.

In this paper, we consider open systems which are modeled in the framework introduced by Kupferman, Vardi, and Wolper. Concretely, in [KV96, KVW01], an open finite-state system is described by an extended transition system called a *module*, whose set of states is partitioned into *system states* (where the

---

Please use the following format when citing this chapter:

Aminof, B., et al., 2008, in IFIP International Federation for Information Processing, Volume 273; *Fifth IFIP International Conference on Theoretical Computer Science*; Giorgio Ausiello, Juhani Karhumäki, Giancarlo Mauri, Luke Ong; (Boston: Springer), pp. 333–348.

system makes a transition) and *environment states* (where the environment makes a transition). Given a module $\mathcal{M}$, describing the system to be verified, and a branching time temporal logic formula φ, specifying the desired behavior of the system, the problem of model checking a module, called *module checking*, asks whether for all possible environments, $\mathcal{M}$ satisfies φ. In particular, it might be that the environment does not enable all the external choices. Module checking thus involves not only checking that the full computation tree obtained by unwinding $\mathcal{M}$ (which corresponds to the interaction of $\mathcal{M}$ with a maximal environment) satisfies the specification φ, but also that every tree obtained from it by pruning children of environment nodes (this corresponds to the different choices of different environments) satisfies φ. For example, consider an ATM machine that allows customers to deposit money, withdraw money, check balance, etc. The machine is an open system, and an environment for it is a subset of the set of all possible infinite lines of customers, each with their own plans. Accordingly, there are many different possible environments to consider.

The finite-state system module checking problem, for *CTL* and *CTL** formulas, has been investigated in [KV96,KVW01]; while for propositional μ-calculus formulas it has been investigated in [FM07]. In all these cases, it has been shown that module checking is exponentially harder than model checking. However, an interesting aspect of these results is that they bear on the corresponding automata-based results for closed systems [KVW00], which gives the hope for practical implementations and applications.

Recently, the module checking idea has been extended to pushdown systems [BMP05], and it has been shown that *CTL* and μ-calculus pushdown module checking is 2EXPTIME-complete, while *CTL** pushdown module checking is 3EXPTIME-complete [BMP05,FMP07]. Another extension of the module checking idea has been the investigation of environments with *imperfect information*. The first results on the subject were dedicated to finite-state systems [KV97]. In this framework, every state of the module is a composition of *visible* and *invisible* variables, where the latter are hidden from the environment. While a composition of a module $\mathcal{M}$ with an environment with perfect information corresponds to arbitrary disabling of transitions in $\mathcal{M}$, the composition of $\mathcal{M}$ with an environment with imperfect information is such that whenever two computations of the system differ only in the values of invisible variables along them, the disabling of transitions along them coincide. In [KV97] it has been shown that *CTL* and *CTL** module checking with imperfect information is harder than module checking with perfect information. The results in [KV97] were recently extended in [AMV07] to pushdown systems. In this framework, environments with imperfect information about the system's control state and pushdown store content are considered. Like in the finite-state case, the control states are assignments to Boolean *control variables*, some of which are visible and some of which are not. Similarly, symbols of the pushdown store are assignments to Boolean visible and invisible *pushdown store variables*. It has been shown in [AMV07] that in the presence of imperfect information, *CTL* pushdown module-checking becomes undecidable, and that the undecidability relies upon hiding information about the pushdown store. Indeed, it was shown that *CTL* pushdown module checking with imperfect state information but visible pushdown store is decidable and 2EXPTIME-complete.

[AMV07] left open the question whether the pushdown module checking problem with imperfect state information, but visible pushdown store, is still decidable when more expressive logics are considered. In this paper we answer this question in the affirmative. Our main contribution is showing that this problem is decidable and 2EXPTIME-complete for the propositional μ-calculus and the graded μ-calculus [KSV02][1], and 3EXPTIME-complete for CTL^*. The lower bound follows from the known perfect information case. For the upper bound we use an automata theoretic approach, and reduce the problem to the emptiness problem of a *semi-alternating pushdown parity tree automaton* (PD-SPT). These are alternating pushdown parity tree automata that behave deterministically on the pushdown store content. That is, two copies of the automaton that read the same input, from two configurations that have the same top of pushdown store, must push the same value into the pushdown store. In this paper, we show that unlike alternating pushdown parity tree automata, for which the emptiness problem is undecidable[2], the emptiness problem for PD-SPT is solvable in 2EXPTIME, which allows us to get the required upper bound for our problem.

2 Preliminaries

In this section, we first recall the concept of open system. Then, we introduce the logics that will be model checked.

2.1 Open Systems.

Let Υ be a finite set. An *Υ-tree* is a prefix closed subset $T \subseteq \Upsilon^*$. The elements of T are called *nodes* and the empty word ε is the *root* of T. For $v \in T$, the set of *children* of v (in T) is $child(T, v) = \{v \cdot x \in T \mid x \in \Upsilon\}$. Given a node $v = u \cdot x$, with $u \in \Upsilon^*$ and $x \in \Upsilon$, we define $last(v)$ to be x. The *complete* Υ-tree is the tree Υ^*. For $v \in T$, a (full) path π of T from v is a *minimal* set $\pi \subseteq T$, such that $v \in \pi$ and for each $v' \in \pi$, such that $child(T, v') \neq \emptyset$, there is exactly one node in $child(T, v')$ belonging to π. Note that every $w \in \Upsilon^\omega$ can be thought of as an infinite path in the tree Υ^*, namely the path containing all

[1] The graded μ-calculus extends the propositional μ-calculus by allowing graded modalities, which enable statements about the number of successors of a state.

[2] Since the emptiness problem of the intersection of two context free languages is undecidable [HU79], the emptiness problem of alternating pushdown automata is undecidable, already in the case of finite words.

the finite prefixes of w. For an alphabet Σ, a Σ-labeled Υ-tree is a pair $\langle T, V\rangle$ where T is an Υ-tree and $V : T \rightarrow \Sigma$ maps each node of T to a symbol in Σ.

An *open system* is a system that interacts with its environment and whose behavior depends on this interaction. We consider the case where the environment has imperfect information about the system, i.e., when the system has internal variables that are not visible to its environment. We describe such a system by a *module* $\mathcal{M} = \langle AP, W_s, W_e, w_0, R, L, \cong\rangle$, where AP is a finite set of *atomic propositions*, W_s is a set of *system states*, and W_e is a set of *environment states*. We assume that $W_s \cap W_e = \emptyset$, and call $W = W_s \cup W_e$ the set of $\mathcal{M}$'s *states*. The state $w_0 \in W$ is the *initial state*, $R \subseteq W \times W$ is a total *transition relation*, $L : W \rightarrow 2^{AP}$ is a labeling function that maps each state of $\mathcal{M}$ to the set of atomic propositions that hold in it, and $\cong$ is an equivalence relation on W. A module $\mathcal{M}$ is *closed* if $W_e = \emptyset$. States that are indistinguishable by the environment are equivalent according to $\cong$. We write $[W]$ for the set of equivalence classes of W under $\cong$. For the environment, the states of the system are actually the equivalence classes themselves. The equivalence class $[w]$ of a state $w \in W$ is called the *visible part* of w. We write $vis(w)$, instead of $[w]$, to emphasize this.

Given $\langle w, w'\rangle \in R$, w' is a *successor* of w. For each state $w \in W$, we denote by $succ(w)$ the set (possibly empty) of w's successors. A *computation* of $\mathcal{M}$ is a sequence $w_0 \cdot w_1 \cdots$ of states, such that for all $i \geq 0$ we have $\langle w_i, w_{i+1}\rangle \in R$. The set of all (maximal) computations of $\mathcal{M}$ starting from the initial state w_0 can be described by an AP-labeled W-tree $\langle T_{\mathcal{M}}, V_{\mathcal{M}}\rangle$ called a *computation tree*, which is obtained by unwinding $\mathcal{M}$ in the usual way. Each node $v = v_1 \cdots v_k$ of $\langle T_{\mathcal{M}}, V_{\mathcal{M}}\rangle$ describes the (partial) computation $w_0 \cdot v_1 \cdots v_k$ of $\mathcal{M}$, with the root ε corresponding to w_0. The children of v are exactly all nodes of the form $v_1 \cdots v_k \cdot w$, where w ranges over all the successors of v_k in $\mathcal{M}$. We extend the definition of vis to nodes in the natural way. Thus, the visible part of a node v is $vis(v) = vis(v_1) \cdots vis(v_k)$. The labeling $V_{\mathcal{M}}$ of a node v depends on the state it corresponds to (its last state), i.e., $V_{\mathcal{M}}(v) = L(last(v))$. Also, if v corresponds to an environment state we say that v is an *environment node*.

Whenever $\mathcal{M}$ interacts with an environment ξ, its possible moves from environment states (i.e., states in W_e) depend on the behavior of ξ. We can think of an environment to $\mathcal{M}$ as a strategy $\xi : [W]^* \rightarrow \{\top, \bot\}$ that maps a finite history s of a computation, as seen by the environment, to either $\top$ or $\bot$, meaning that the environment respectively allows or disallows $\mathcal{M}$ to trace s (obviously, if s is a successor of a system state, the decision whether to trace s or not is made by the system, and we ignore the environment's value of $\xi(s)$). Observe that if an environment disallows $\mathcal{M}$ to trace s, it effectively disallows $\mathcal{M}$ to trace any of the successors of s. Note that one can either require that for every $y \in [W]^*$, if $\xi(x) = \bot$ then $\xi(x \cdot y) = \bot$, or simply ignore the value $\xi(x \cdot y)$. We chose the latter. We say that the tree $\langle [W]^*, \xi\rangle$ maintains the strategy applied by ξ, and we call it a *strategy tree*. We denote by $\mathcal{M} \lhd \xi$ the AP-labeled W-tree induced by the composition of $\langle T_{\mathcal{M}}, V_{\mathcal{M}}\rangle$ with ξ; that is, the AP-labeled W-tree obtained by pruning from $\langle T_{\mathcal{M}}, V_{\mathcal{M}}\rangle$ subtrees according to ξ. Note that by the definition above, ξ may disable all the children of a node v. Since we usually do not want the environment to completely block the system, we require that at least one child of each node is enabled. In this case, we say that the composition $\mathcal{M} \lhd \xi$ is *deadlock-free*. Given a module $\mathcal{M}$, and a strategy tree $\langle [W]^*, \xi \rangle$ for an environment ξ, an AP-labeled W-tree $\langle T, V \rangle$ corresponds to $\mathcal{M} \lhd \xi$ iff:

- The root of T corresponds to w_0.
- For $v \in T$ with $last(v) \in W_s$, we have $child(T, v) = \{v \cdot w_1, \ldots, v \cdot w_n\}$, where $succ(last(v)) = \{w_1, \ldots, w_n\}$.
- For $v \in T$ with $last(v) \in W_e$, there exists a nonempty subset $\{w_1, \ldots, w_k\}$ of $succ(last(v))$ such that $child(T, v) = \{v \cdot w_1, \ldots, v \cdot w_k\}$. Furthermore, for all w in $\{w_1, \ldots, w_k\}$ we have that $\xi(vis(v \cdot w)) = \top$, while for all w in $succ(last(v)) \setminus \{w_1, \ldots, w_k\}$ we have that $\xi(vis(x \cdot w)) = \bot$.
- For every node $v \in T$, we have that $V(v) = L(last(v))$.

For a module $\mathcal{M}$ and a temporal logic formula φ defined over AP, we say that $\mathcal{M}$ *reactively satisfies* φ, denoted $\mathcal{M} \models_r \varphi$, if $\mathcal{M} \lhd \xi$ satisfies φ, for every environment ξ for which $\mathcal{M} \lhd \xi$ is deadlock-free. The problem of deciding whether $\mathcal{M} \models_r \varphi$ is called the *module checking problem with imperfect information.*

2.2 Logics.

In this paper, we consider φ to be either a *CTL** or a propositional/graded μ-calculus formula. The syntax and semantics of *CTL** and μ-calculus are well known, and we assume that the reader is familiar with them (for references, see [Koz83] and [KVW00]). In the rest of this section, we focus on graded μ-calculus, which is an extension of the propositional μ-calculus that allows *graded modalities.* These modalities are denoted by $\langle n \rangle$ ("*exist at least n-successors*") and $[n]$ ("*all but at most n successors*"), respectively.

Formally, we have the following. Let AP and *Var* be finite and pairwise disjoint sets of *atomic propositions* and *propositional variables.* The set of *graded μ–calculus* formulas is the smallest set such that (i) **true** and **false** are formulas; (ii) p and $\neg p$, for $p \in AP$, are formulas; (iii) $x \in$ *Var* is a formula; (iv) if φ_1 and φ_2 are formulas, n is a non negative integer, and $y \in Var$, then $\varphi_1 \vee \varphi_2$, $\varphi_1 \wedge \varphi_2$, $\langle n \rangle \varphi_1$, $[n]\varphi_1$, $\mu y.\varphi_1(y)$, and $\nu y.\varphi_1(y)$ are also formulas. Observe that we use positive normal form, i.e., negation is applied only to atomic propositions. We often refer to the *graded modalities* $\langle n \rangle \varphi_1$ and $[n]\varphi_1$ as, respectively, *atleast formulas* and *allbut formulas,* and assume that the integers in these operators are given in binary coding: the contribution of n to the length of each of the formulas $\langle n \rangle \varphi$ and $[n]\varphi$ is $\lceil \log n \rceil$, rather than n.

The definition of the semantics of graded μ-calculus w.r.t an AP-labeled W-tree $\langle T, V \rangle$ is similar to that of the standard μ-calculus, except for the graded modalities. Informally, an *atleast* formula $\langle n \rangle \varphi$ holds at a node w of the tree if φ holds in at least $n + 1$ children of the node. Dually, the *allbut* formula $[n]\varphi$

holds in a node of the tree $\mathcal{K}$ if φ holds in all but at most n of its successors. Due to space limitation, we refer the reader to [KSV02] (also [BLMV06]) for a formal description of the full semantics.

3 Imperfect Information Pushdown Module Checking

In this section, we consider infinite-state modules which are induced by *Open Pushdown Systems* (*OPD*) [AMV07]. In our framework, the environment has imperfect information about the internal control states of the system, but the pushdown store is visible.

Definition 1. An *OPD* is a tuple $\mathcal{S} = \langle AP, Q, q_0, \Gamma, \flat, \delta, \eta, Env \rangle$, where AP is a finite set of atomic propositions; Q is a finite set of (*control*) *states*; and $q_0 \in Q$ is an *initial state*. We assume that $Q \subseteq 2^{V \cup H}$, where V and H are disjoint finite sets of *visible* and *invisible control variables*, respectively. Γ is a finite pushdown store alphabet; $\flat \notin \Gamma$ is the *pushdown store bottom symbol*, and we use $\Gamma_\flat$ to denote $\Gamma \cup \{\flat\}$. The transition relation $\delta \subseteq (Q \times \Gamma_\flat) \times (Q \times \Gamma_\flat^*)$ is finite; $\eta : Q \times \Gamma_\flat \to 2^{AP}$ is a labeling function; and $Env \subseteq Q \times \Gamma_\flat$ is used to specify the set of *environment configurations*. The *size* $|\mathcal{S}|$ of $\mathcal{S}$ is $|Q| + |\Gamma| + |\delta|$, with $|\delta| = \sum_{((p,\gamma),(q,\beta)) \in \delta} |\beta|$.

A *configuration* of $\mathcal{S}$ is a pair (q, α), where q is a control state and $\alpha \in \Gamma^* \cdot \flat$ is a pushdown store content. We write $top(\alpha)$ for the leftmost symbol of α, and call it the *top of the pushdown store* α. The *OPD* moves according to the transition relation. Thus, $((p, \gamma), (q, \beta)) \in \delta$ implies that if the *OPD* is in state p, and the top of the pushdown store is γ, then it can move to state q, pop γ and push β. We assume that $\flat$ is always present at the bottom of the pushdown store, and nowhere else. Note that we make this assumption also about the various pushdown automata we use later. For a control state $q \in Q$, the visible part of q is $vis(q) = q \cap V$. The visible part of a configuration (q, α), is thus $vis((q, \alpha)) = (vis(q), \alpha)$. As for modules, the designation of a configuration of an *OPD* as an environment configuration is known to the environment. Thus, we require that for every two configurations (q, α) and (q', α'), such that $vis(q) = vis(q')$, it holds that $(q, top(\alpha)) \in Env$ iff $(q', top(\alpha')) \in Env$.

Definition 2. An *OPD* $\mathcal{S} = \langle AP, Q, q_0, \Gamma, \flat, \delta, \eta, Env \rangle$ induces an infinite-state module $\mathcal{M}_\mathcal{S} = \langle AP, W_s, W_e, w_0, R, L, \cong \rangle$, possibly with invisible information, where AP is a set of atomic propositions; $W_s \cup W_e = Q \times \Gamma^* \cdot \flat$ is the set of configurations; W_e is the set of configurations (q, α) such that $(q, top(\alpha)) \in Env$; $w_0 = (q_0, \flat)$ is the initial configuration; R is a transition relation, where $((q, \gamma \cdot \alpha), (q', \beta)) \in R$ *iff* there exist $((q, \gamma), (q', \beta')) \in \delta$ such that $\beta = \beta' \cdot \alpha$; $L((q, \alpha)) = \eta(q, top(\alpha))$ for all $(q, \alpha) \in W$; and for every two configurations $w, w' \in W$, we have that $w \cong w'$ iff $vis(w) = vis(w')$.

To describe the interaction of an OPD $\mathcal{S}$ with its environment we consider the interaction of the environment with the induced module $\mathcal{M}_\mathcal{S}$. Indeed, every environment ξ of $\mathcal{S}$ can be represented by a strategy tree $\langle [W]^*, \xi\rangle$, and the composition $\mathcal{M}_\mathcal{S} \lhd \xi$ of $\langle [W]^*, \xi\rangle$ with $\langle T_{\mathcal{M}_\mathcal{S}}, V_{\mathcal{M}_\mathcal{S}}\rangle$ describes all the computations of $\mathcal{S}$ allowed by the environment ξ.

We consider the *pushdown module checking problem with imperfect state information*, i.e., given an OPD $\mathcal{S}$ and a formula φ, decide whether $\mathcal{M}_\mathcal{S} \models_r \varphi$.

The pushdown module checking problem with imperfect state information is known to be 2EXPTIME-complete when φ is a CTL formula [AMV07]. In this paper, we answer an open question of [AMV07] and show that the problem remains 2EXPTIME-complete when considering φ to be a propositional or a graded μ-calculus formula, and that it becomes 3EXPTIME-complete when φ is a CTL^* formula. For the upper bound, we reduce our problem to the emptiness problem of a semi-alternating pushdown parity tree automata.

3.1 Semi-Alternating Pushdown Tree Automata.

We start with the definition of *semi-alternating pushdown parity tree automata* (*PD-SPT*), first introduced in [AMV07] w.r.t. a Büchi acceptance condition. A PD-SPT is a tuple $\mathcal{A} = \langle \Sigma, D, \Gamma, Q, q_0, \flat, \delta, F\rangle$, where Σ is a finite input alphabet, D is a finite set of *directions*, Γ is a finite pushdown store alphabet, Q is a finite set of states, $q_0 \in Q$ is the initial state, $\flat \notin \Gamma$ is the pushdown store bottom symbol, and F is a parity acceptance condition (to be defined later). Moreover, δ is a finite transition relation defined as a function $\delta : Q \times \Sigma \times \Gamma_\flat \rightarrow \mathcal{B}^+(D \times Q \times \Gamma_\flat^*)$, where, as usual, $\Gamma_\flat = \Gamma \cup \{\flat\}$, and $\mathcal{B}^+(D \times Q \times \Gamma_\flat^*)$ is the set of all finite positive Boolean combinations of triples (d, q, β), where d is a direction, q is a state, and β is a word made of pushdown store symbols. We also allow the formulas **true** and **false**. We write $S \in \delta(p, \sigma, \gamma)$ to denote that S is a set of tuples (d, q, β) that satisfy $\delta(p, \sigma, \gamma)$.

What makes the automaton semi-alternating is the requirement that for every $d \in D$, $\sigma \in \Sigma$, $p, p' \in Q$ (possibly the same state), and $\gamma \in \Gamma$, if (d, q, β) appears in $\delta(p, \sigma, \gamma)$, and (d, q', β') appears in $\delta(p', \sigma, \gamma)$, then $\beta = \beta'$. That is, two copies of the automaton that read the same input, from two configurations that have the same top symbol of the pushdown store, and proceed in the same direction, must push the same value into the pushdown store. In particular, it follows that in every run, two copies of the automaton that are reading the same node of an input tree have the same pushdown store content. Note that if we remove the semi-alternation requirement the resulting automaton is called *alternating pushdown parity tree automaton* (*PD-APT*).

As a special case of PD-APT, we consider *nondeterministic pushdown parity tree automata* (*PD-NPT*), where the concurrency feature (i.e., the $\wedge$ operator in δ) is not allowed. That is, whenever a PD-NPT visits a node x of the input tree, it sends to each successor (direction) of x at most one copy of itself. More formally, a PD-NPT is a PD-APT in which δ is in disjunctive normal form, and in each conjunct each direction appears at most once. Note that if $\mathcal{A}$ is a PD-APT with $\Gamma = \emptyset$, its pushdown store is neutralized, hence, $\mathcal{A}$ is simply called an *alternating parity tree automaton* (*APT*), and we can abbreviate and write $\mathcal{A} = \langle \Sigma, D, Q, q_0, \delta, F \rangle$, where $\delta : Q \times \Sigma \rightarrow \mathcal{B}^+(D \times Q)$. Similarly, a PD-NPT with an empty pushdown store alphabet is called a *nondeterministic parity tree automaton* (*NPT*).

A run of a PD-SPT $\mathcal{A}$, on a Σ-labeled tree $\langle T, V \rangle$, with $T = D^*$, is a $(D^* \times Q \times \Gamma^* \cdot \flat)$-labeled $\mathbb{N}$-tree $\langle T_r, r \rangle$, such that the root is labeled with $(\varepsilon, q_0, \flat)$ and the labels of each node and its successors satisfy the transition relation. Formally, a $(D^* \times Q \times \Gamma^* \cdot \flat)$-labeled tree $\langle T_r, r \rangle$ is a run of $\mathcal{A}$ on $\langle T, V \rangle$ iff

- $r(\varepsilon) = (\varepsilon, q_0, \flat)$, and
- for all $x \in T_r$ such that $r(x) = (y, p, \gamma \cdot \alpha)$, there is an $n \in \mathbb{N}$ such that the successors of x are exactly $x \cdot 1, \ldots x \cdot n$, and for all $1 \leq i \leq n$ we have $r(x \cdot i) = (y \cdot d_i, p_i, \beta_i \cdot \alpha)$ for some $\{(d_1, p_1, \beta_1), \ldots, (d_n, p_n, \beta_n)\} \in \delta(p, V(y), \gamma)$.

For a path $\pi \subseteq T_r$, let $inf_r(\pi) \subseteq Q$ be the set of states that appear in the labels of infinitely many nodes in π. For a parity condition $F = \{F_1, F_2, \ldots, F_k\}$, with $F_1 \subseteq F_2 \subseteq \cdots \subseteq F_k = Q$, we have that π is *accepting* iff the minimal index i, for which $inf_r(\pi) \cap F_i \neq \emptyset$, is even. The number k is called the *index* of the automaton. A run $\langle T_r, r \rangle$ is *accepting* iff all its paths are accepting. The automaton $\mathcal{A}$ accepts an input tree $\langle T, V \rangle$ iff there is an accepting run of $\mathcal{A}$ on $\langle T, V \rangle$. The language of $\mathcal{A}$, denoted $L(\mathcal{A})$, is the set of Σ-labeled trees with branching degree D accepted by $\mathcal{A}$. We say that an automaton $\mathcal{A}$ is nonempty iff $L(\mathcal{A}) \neq \emptyset$. Given a PD-SPT $\mathcal{A} = \langle \Sigma, D, \Gamma, Q, q_0, \flat, \delta, F \rangle$, we define the size of δ as the sum of the lengths of the satisfiable (i.e., not **false**) formulas that appear in $\delta(q, \sigma, \gamma)$, for some q, σ, and γ.

3.2 Simulating a PD-SPT by a PD-NPT.

As mentioned in [AMV07], alternating pushdown automata are not equivalent to nondeterministic ones. However, as we show here, the limitations imposed on semi-alternating automata allow us to translate a PD-SPT to an equivalent PD-NPT[3]. A key observation is that since a pushdown store operation performed by a semi-alternating automaton does not depend on the current (or next) control states, we can split the transition function of a PD-SPT into two functions: a *state transition function* δ_Q, and a *pushdown store update function* δ_Γ, as follows. Given a PD-SPT $\mathcal{A} = \langle \Sigma, D, \Gamma, Q, q_0, \flat, \delta, F \rangle$, let $\delta_Q : Q \times \Sigma \times \Gamma_\flat \rightarrow \mathcal{B}^+(D \times Q)$ be the projection of δ on $\mathcal{B}^+(D \times Q)$. That is, $\delta_Q(q, \sigma, \gamma)$ is obtained

[3] The translation used in [AMV07], for semi-alternating pushdown Büchi tree automata, made a crucial use of the Büchi acceptance condition, and can not be extended to the parity acceptance condition.

from $\delta(q, \sigma, \gamma)$ by replacing every element (d, q, β) that appears in $\delta(q, \sigma, \gamma)$ with (d, q). The pushdown store update function $\delta_\Gamma : \Sigma \times \Gamma_\flat \times D \to \Gamma_\flat^*$, is a partial function; for every $(p, \sigma, \gamma) \in Q \times \Sigma \times \Gamma_\flat$ and every $(d, q, \beta) \in D \times Q \times \Gamma_\flat^*$, such that (d, q, β) appears in $\delta(p, \sigma, \gamma)$, we let $\delta_\Gamma(\sigma, \gamma, d) = \beta$. Since $\mathcal{A}$ is semi-alternating, δ_Γ is well defined. Observe that for every $(p, \sigma, \gamma) \in Q \times \Sigma \times \Gamma_\flat$ we have that $\delta(p, \sigma, \gamma)$ can be obtained from $\delta_Q(p, \sigma, \gamma)$ by replacing every (d, q) that appears in $\delta_Q(p, \sigma, \gamma)$ with $(d, q, \delta_\Gamma(\sigma, \gamma, d))$.

Consider a Σ-labeled tree $\langle T, V \rangle$, with $T = D^*$. Note that for every node $x \in T$ and every run of $\mathcal{A}$ on $\langle T, V \rangle$, the pushdown store content of all the copies of $\mathcal{A}$ that visit x is the same, and only depends on x. We can thus define a function $\Delta_\Gamma : T \to \Gamma_\flat^*$, giving for every node x its associated pushdown store content, as follows: (1) $\Delta_\Gamma(\varepsilon) = \flat$, and (2) for all $x \cdot d \in T$ we have $\Delta_\Gamma(x \cdot d) = \delta_\Gamma(V(x), \gamma, d) \cdot \beta$, where $\Delta_\Gamma(x) = \gamma \cdot \beta$, and $\gamma \in \Gamma_\flat$.

Annotating input trees with pushdown store symbols enables us to simulate a PD-SPT by an APT running on the annotated version of an input tree. Given a Σ-labeled tree $\langle T, V \rangle$, we define its *$\Gamma_\mathcal{A}$-annotation* to be the $(\Sigma \times \Gamma_\flat)$-labeled tree $\langle T, U \rangle$, obtained by letting $U(x) = (V(x), top(\Delta_\Gamma(x)))$, for every $x \in T$.

Lemma 1. *Let $\mathcal{A} = \langle \Sigma, D, \Gamma, Q, q_0, \flat, \delta, F \rangle$ be a PD-SPT. There is an APT $\tilde{\mathcal{A}}$, such that $\mathcal{A}$ accepts $\langle T, V \rangle$ iff $\tilde{\mathcal{A}}$ accepts the $\Gamma_\mathcal{A}$-annotation of $\langle T, V \rangle$.*

Proof. Consider the APT $\tilde{\mathcal{A}} = \langle \Sigma \times \Gamma_\flat, D, Q, q_0, \tilde{\delta}, F \rangle$, where $\tilde{\delta}(q, (\sigma, \gamma)) = \delta_Q(q, \sigma, \gamma)$. It is not hard to see that every run $r = \langle T_r, r \rangle$ of $\mathcal{A}$ on $\langle T, V \rangle$ induces a corresponding run $r' = \langle T_r, r' \rangle$ of $\tilde{\mathcal{A}}$ on the $\Gamma_\mathcal{A}$-annotation of $\langle T, V \rangle$, and vice versa. The connection between r and r' being that for every $x \in T_r$, we have that $r(x) = (y, p, \alpha)$ iff $r'(x) = (y, p)$ and $\Delta_\Gamma(x) = \alpha$. $\square$

By [MS87], every APT can be translated to an equivalent NPT. Hence, Lemma 1 implies that if $\mathcal{A}$ is a PD-SPT, then there is an NPT $\mathcal{A}'$ such that $\mathcal{A}$ accepts $\langle T, V \rangle$ iff $\mathcal{A}'$ accepts the $\Gamma_\mathcal{A}$-annotation of $\langle T, V \rangle$. This allows us to translate $\mathcal{A}$ to an equivalent PD-NPT $\mathcal{A}''$ (running on the same input trees as $\mathcal{A}$). Given a Σ-labeled tree, $\mathcal{A}''$ generates on the fly its $\Gamma_\mathcal{A}$-annotation and runs $\mathcal{A}'$ on the annotated tree. Formally, we have the following:

Theorem 1. *Every PD-SPT can be translated to an equivalent PD-NPT.*

Proof. Let $\mathcal{A} = \langle \Sigma, D, \Gamma, Q, q_0, \flat, \delta, F \rangle$ be a PD-SPT and $\tilde{\mathcal{A}} = \langle \Sigma \times \Gamma_\flat, D, Q, q_0, \tilde{\delta}, F \rangle$ be an APT derived from $\mathcal{A}$ by Lemma 1. By [MS87], $\tilde{\mathcal{A}}$ has an equivalent NPT $\mathcal{A}' = \langle \Sigma \times \Gamma_\flat, D, Q', q_0', \delta', F' \rangle$. Consider the PD-NPT $\mathcal{A}'' = \langle \Sigma, D, \Gamma, Q', q_0', \flat, \delta'', F' \rangle$, where for every $(p, \sigma, \gamma) \in Q' \times \Sigma \times \Gamma_\flat$, we have that $\delta''(p, \sigma, \gamma)$ is obtained from $\delta'(p, (\sigma, \gamma))$ by replacing every (d, q) that appears in $\delta'(p, (\sigma, \gamma))$, with $(d, q, \delta_\Gamma(\sigma, \gamma, d))$. Since $\mathcal{A}'$ is nondeterministic, so is $\mathcal{A}''$. Given a Σ-labeled tree $\langle T, V \rangle$, it is not hard to see that for every $x \in T$, the pushdown store of every copy of $\mathcal{A}''$ that visits x contains exactly $\Delta_\Gamma(x)$. Hence, $\mathcal{A}''$ accepts $\langle T, V \rangle$ iff $\mathcal{A}'$ accepts the $\Gamma_\mathcal{A}$-annotation of $\langle T, V \rangle$, i.e., iff $\mathcal{A}$ accepts $\langle T, V \rangle$. $\square$

3.3 The Emptiness Problem of PD-SPT.

Looking at the automata transformations performed in Theorem 1 and Lemma 1 we see that the only transformation that incurs a blowup in the size of the automaton is the transformation of the APT $\tilde{\mathcal{A}}$ to the NPT $\mathcal{A}'$. By [MS87], given an APT with n states and index k, running over D^* trees, one can build an equivalent NPT with $(nk)^{O(nk)}$ states, an $O(nk)$ index, and a transition relation of size $(nk)^{O(|D|nk)}$. Hence, starting with a PD-SPT $\mathcal{A}$ with n states and index k, our algorithm yields an equivalent PD-NPT $\mathcal{A}''$ with $(nk)^{O(nk)}$ states, an $O(nk)$ index, and a transition relation of size $(nk)^{O(|D|nk)}$. It is worth noting that the blowup is independent of the size of the transition relation of $\mathcal{A}$. By [KPV02], the emptiness of $\mathcal{A}''$ can be decided in time exponential in the product of the number of states, the index, and the size of the transition relation of $\mathcal{A}''$. Overall, we get the following corollary:

Corollary 1. *The emptiness problem for a PD-SPT with n states and index k, running on D^* trees, can be solved in time double-exponential in $|D|nk$.*

4 Solving Pushdown Module Checking with Imperfect State Information

We first show that the pushdown module checking problem with imperfect state information, for μ-calculus, graded μ-calculus, and *CTL**, can be reduced to the emptiness problem of PD-SPT.

Basically, we extend the automata theoretic approach used in [AMV07] for *CTL* pushdown module checking with imperfect state information. Before presenting the formal reduction, let us briefly recap the approach taken by [AMV07], and discuss the main changes required to adapt it to the problem we address. Given an OPD $\mathcal{S}$, and a *CTL* formula φ, one builds an automaton $\mathcal{A}_{\mathcal{S},\varphi}$ that accepts $\{\top, \bot\}$-labeled trees corresponding to strategies ξ, whose composition with $\mathcal{M}_{\mathcal{S}}$ is deadlock-free and satisfies φ. Intuitively, a run of $\mathcal{A}_{\mathcal{S},\varphi}$ on an input strategy tree ξ proceeds by simulating an unwinding of the module $\mathcal{M}_{\mathcal{S}}$, pruned at each step accordingly to the strategy ξ; copies of the automaton which simulate nodes in the computation tree of $\mathcal{M}_{\mathcal{S}}$ that are indistinguishable by the environment are sent to the same direction in the input tree. The resulting run tree of $\mathcal{A}_{\mathcal{S},\varphi}$ on ξ is basically a replica of the composition $\mathcal{M}_{\mathcal{S}} \lhd \xi$, and the fact that it satisfies the formula φ is checked on the fly, by employing in $\mathcal{A}_{\mathcal{S},\varphi}$ the classical alternating-automata approach for model checking *CTL*.

When considering *CTL** or μ-calculus, adapting the construction used in [AMV07] basically amounts to replacing the embedded alternating automaton that does the on-the-fly model checking: instead of using an automaton that handles *CTL*, one uses an automaton that handles *CTL** (or μ-calculus). Since an alternating automaton that does μ-calculus model checking is linear in the size of the formula, while one that does CTL^* model checking is exponential in the size of the formula [KVW00], the automaton $\mathcal{A}_{\mathcal{S},\varphi}$ has $O(|\mathcal{S}| * |\varphi|)$ states in the case of μ-calculus, and $O(|\mathcal{S}| * 2^{|\varphi|})$ states in the case of CTL^*. It is important to note that the acceptance condition of $\mathcal{A}_{\mathcal{S},\varphi}$ is essentially that of the embedded model checking automaton. Hence, unlike in [AMV07], where a Büchi condition was enough, for the more expressive logics that we consider here, we need a stronger acceptance condition, namely, a parity condition, for which solving the emptiness problem requires stronger machinery.

The extension of the construction used in [AMV07] is slightly more delicate when considering graded μ-calculus. Given a graded μ-calculus formula φ, one possible approach is to translate φ into an equivalent μ-calculus formula (without graded modalities). Essentially, as pointed out in [KSV02], one introduces new atomic propositions $p_1, \ldots, p_b$, (where b is the largest number used in the graded modalities in φ) and replaces every atleast formula $\langle n \rangle \psi$ by $\bigvee_{\{i_1,\ldots,i_{n+1}\} \subseteq \{1,\ldots,b\}} \bigwedge_{1 \le j \le n+1} \langle 0 \rangle (\psi \wedge p_{i_j})$, and dually for allbut formulas. One also has to conjoin φ with a formula stating that exactly one of the $p_1, \ldots, p_b$ holds at each state, and that successors that are labeled with the same p_i agree on their label with respect to all the formulas in the closure of φ. Unfortunately, since the numbers in the graded modalities are coded in binary, such a translation may result in a μ-calculus formula which is exponentially larger than φ; resulting in an overall exponentially worse complexity for the graded μ-calculus, compared to the un-graded one. In order to avoid this extra exponent, in the context of satisfiability, [KSV02] introduced graded automata. However, graded automata do not transfer directly to the imperfect information setting. Fortunately, there is another solution. Instead of expanding the graded modalities at the formula stage, as suggested above, we can expand them as we build the transition relation of $\mathcal{A}_{\mathcal{S},\varphi}$. Thus, for example, the transition relation of $\mathcal{A}_{\mathcal{S},\varphi}$ will specify that a copy of the automaton, that is responsible for verifying that an atleast formula $\langle n \rangle \psi$ holds at a certain configuration of the OPD, should send $n+1$ copies of itself to one of the exponentially many possible subsets of $n+1$ successors of the current configuration. This expansion of the graded modalities allows $\mathcal{A}_{\mathcal{S},\varphi}$ to handle graded μ-calculus formulas using an embedded regular μ-calculus model checker (without graded modalities). This comes at the price of $\mathcal{A}_{\mathcal{S},\varphi}$ having an exponentially larger transition relation than if graded modalities were not present; but does not affect the number of states, or the index, of $\mathcal{A}_{\mathcal{S},\varphi}$. Since our algorithm for checking the emptiness of PD-SPT is such that its complexity does not depend on the size of the transition relation of the PD-SPT, we handle graded μ-calculus formulas with the same complexity as we do regular μ-calculus formulas.

Theorem 2. *Consider an OPD $\mathcal{S}$ and a propositional, or a graded, μ-calculus (resp.* CTL**) formula φ, over $\mathcal{S}$'s atomic propositions. There is a PD-SPT $\mathcal{A}_{\mathcal{S},\varphi}$ with $O(|\mathcal{S}| * |\varphi|)$ states (resp. $O(|\mathcal{S}| * 2^{|\varphi|})$), and an index $O(|\varphi|)$, such that $L(\mathcal{A}_{\mathcal{S},\varphi})$ is exactly the set of strategies ξ for which $\mathcal{M}_S \lhd \xi$ is deadlock-free and satisfies φ.*

Proof (Sketch). We give the construction of $\mathcal{A}_{\mathcal{S},\varphi}$ for the graded μ-calculus. The construction for the propositional μ-calculus is very similar, and the one for CTL^* is obtained by replacing the embedded classical alternating-automata model checker with a CTL^* one.

We first give some extra definitions regarding graded μ-calculus. From now on, we refer to μ and ν as *fixpoint operators.* A propositional variable y occurs *free* in a formula if it is not in the scope of a fixpoint operator, and *bounded* otherwise. We use λ to denote a fixpoint operator μ or ν. For a formula $\lambda y.\varphi(y)$, we write $\varphi(\lambda y.\varphi(y))$ to denote the formula that is obtained from $\lambda y.\varphi(y)$ by one-step unfolding; i.e., $\varphi(\lambda y.\varphi(y))$ is obtained by replacing each free occurrence of y in φ with $\lambda y.\varphi(y)$. For technical convenience, we restrict our attention to formulas without free variables (also called *sentences*). The closure $cl(\varphi)$ of a graded μ-calculus sentence φ is the smallest set of graded μ-calculus formulas that contains φ and is closed under sub-formulas (that is, if ψ is in the closure, then so do all its sub-formulas that are sentences) and fixpoint applications (that is, if $\lambda y.\varphi(y)$ is in the closure, then so is $\varphi(\lambda y.\varphi(y))$). As proved in [BLMV06], for every graded μ-calculus formula φ, the number of elements in $cl(\varphi)$ is linear in the length of φ. Accordingly, we define the size $|\varphi|$ of φ to be the number of elements in $cl(\varphi)$.

Let $\mathcal{S} = \langle AP, Q, q_0, \Gamma, \flat, \delta, \eta, Env\rangle$ be an OPD, let φ be a graded μ-calculus formula (guarded[4], without free variables, and in positive normal form), and let $\mathcal{M}_{\mathcal{S}} = \langle AP, W_s, W_e, w_0, R, L, \cong\rangle$ be the module induced by $\mathcal{S}$. We build an automaton $\mathcal{A}_{\mathcal{S},\varphi}$ that accepts $\{\top, \bot\}$-labeled trees corresponding to strategies ξ, whose composition with $\mathcal{M}_{\mathcal{S}}$ is deadlock-free and satisfy φ. Intuitively, a run of $\mathcal{A}_{\mathcal{S},\varphi}$ on an input strategy tree ξ, proceeds by simulating an unwinding of the module $\mathcal{M}_{\mathcal{S}}$, pruned at each step according to the strategy ξ; copies of the automaton simulating nodes in the computation tree of $\mathcal{M}_{\mathcal{S}}$ that are indistinguishable by the environment are sent to the same direction in the input tree. The resulting run tree of $\mathcal{A}_{\mathcal{S},\varphi}$ on ξ is essentially a replica of the composition $\mathcal{M}_{\mathcal{S}} \lhd \xi$, and the fact that it satisfies the formula φ is checked on the fly by employing in $\mathcal{A}_{\mathcal{S},\varphi}$ the usual alternating-automata approach for μ-calculus model checking. In the full computation tree of $\mathcal{M}_{\mathcal{S}}$, the set of directions is $G = \{(q, \beta) \mid ((p, \alpha), (q, \beta)) \in R$ for some p, α and $\beta\}$. Since in $\mathcal{S}$ the pushdown store is completely visible to the environment, the set of directions of the input strategy trees is $D = \{(vis(q), \beta) \mid ((p, \alpha), (q, \beta)) \in R$ for some p, q, α and $\beta\}$.

Finally, due to the fact that all copies of the automaton sent to direction $(vis(q), \beta)$ push β into the pushdown store, the resulting automaton $\mathcal{A}_{\mathcal{S},\varphi}$ is

[4] A graded μ-calculus formula is *guarded* if for every variable y, all the occurrences of y that are in a scope of a fixpoint modality λ are also in the scope of a graded modality that is itself in the scope of λ. For example, the formula $\nu y.(p \vee [0]y)$ is guarded, but the formula $\nu y.(p \vee y)$ is not. Given a graded μ-calculus formula, we can construct in linear time an equivalent guarded formula (see [KVW00] for a proof for μ-calculus, which is easily extendible to the graded setting). Accordingly, we assume that all formulas are guarded. This is essential for the correctness of our construction (it guarantees that transitions involving fixpoint formulas are well defined).

semi-alternating. As in [KVW00] we are going to use a function split to avoid the problem of having states with a component in $cl(\varphi)$ that is a disjunction or a conjunction. Without the use of split, a run of the automaton may have no states that correspond to a fixpoint sub-formula of φ that is part of a conjunction or a disjunction, which makes it impossible to correctly define the acceptance condition.

We formally define $\mathcal{A}_{S,\varphi} = \langle \{\top, \bot\}, D, \Gamma, Q', q'_0, \flat, \delta', F \rangle$, where

- $Q' = (Q \times (cl(\varphi) \cup \{p_\top\}) \times \{\forall, \exists\} \times \{p_e, p_s\}) \cup \{q'_0\}$. States with the component $p_\top$ are used to check that the composition of $\mathcal{M}_S$ with the strategy is deadlock-free, while states with a component in $cl(\varphi)$ check that this composition satisfies φ. The components p_e and p_s are used to flag that the currently simulated node, of the computation tree of $\mathcal{M}_S$, is a child of an environment or a system node, respectively. Clearly, the simulation should respect the strategy pruning specifications only if they correspond to children of environment nodes; that is, only if the current state q contains p_e. Every state is either in an existential or a universal mode, as specified by the $\forall$ and $\exists$ components. When the automaton is in a universal state $(q, \varphi, \forall, p_e)$ with a pushdown store content α, it accepts all strategies for which (q, α) in $\mathcal{M}_S$ is either pruned or satisfies φ (where $p_\top$ is satisfied iff the root of the strategy is labeled $\top$). When the automaton is in an existential state $(q, \varphi, \exists, p_e)$ with a pushdown store content α, it accepts all strategies for which (q, α) in $\mathcal{M}_S$ is not pruned and satisfies φ.
- δ' is a function $\delta' : Q' \times \Sigma \times \Gamma_\flat \to \mathcal{B}^+(D \times Q' \times \Gamma_\flat^*)$. Before giving the formal definition, we show an example. Consider, a transition from the configuration $(\langle p, \forall X\psi, \exists, p_e \rangle, \gamma \cdot \alpha)$, where $(p, \gamma) \in Env$. First, if the transition to $(p, \gamma \cdot \alpha)$ is disabled (that is, the automaton reads $\bot$), then, as the current mode is existential, the run is rejecting. If the transition to $(p, \gamma \cdot \alpha)$ is enabled, then the successors of $(p, \gamma \cdot \alpha)$ that are enabled should satisfy ψ. Note that all the successors of $(p, \gamma \cdot \alpha)$ that are indistinguishable by the environment are sent by the automaton to the same direction v. This guarantees that either all these successors are enabled by the strategy (in case the letter to be read in direction v is $\top$) or all are disabled (in case the letter in direction v is $\bot$). In addition, since the requirement to satisfy ψ concerns only successors of $(p, \gamma \cdot \alpha)$ that are enabled, the mode of the new states is universal. The copies of $\mathcal{A}_{S,\varphi}$ that check the composition with the strategy to be deadlock-free guarantee that at least one successor of $(p, \gamma \cdot \alpha)$ is enabled. As noted earlier, the enable/disable instructions of the strategy are ignored in every configuration $(p, \gamma \cdot \alpha)$ that is a successor of a system configuration. Also note that since we assume that no configuration in $\mathcal{M}_S$ has no successors, the conjunctions and disjunctions in δ' cannot be empty.

  We now formally define the transition function δ'. For $(p, \gamma \cdot \alpha) \in W$, we define the set of successors of $(p, \gamma \cdot \alpha)$ in $\mathcal{M}_S$, to be $s(p, \gamma) = \{(q, \beta) \mid ((p, \gamma), (q, \beta)) \in \delta\}$. The transition function $\delta' : Q' \times \Sigma \times \Gamma_\flat \to \mathcal{B}^+(D \times Q' \times \Gamma_\flat^*)$ is defined as follows. In the rules below, for the sake of succinctness, we consider $m \in \{\exists, \forall\} \times \{p_e, p_s\}$, $h \in AP \cup \{\mathbf{true}, \mathbf{false}\}$. Also, given a transition from $(\langle p, \psi, m\rangle, \top, \gamma)$, we let $p_x = p_e$ if $(p, \gamma) \in Env$ and $p_x = p_s$, otherwise.

- $\delta'(q_0', \bot, \flat) = \mathbf{false}$ and $\delta'(q_0', \top, \flat) = \delta'(\langle q_0, p_\top, \exists, p_s\rangle, \top, \flat) \wedge \delta'(\langle q_0, \varphi, \exists, p_s\rangle, \top, \flat)$.
- For all p, ψ, and γ, we have $\delta'(\langle p, \psi, \forall, p_e\rangle, \bot, \gamma) = \mathbf{true}$ and $\delta'(\langle p, \psi, \exists, p_e\rangle, \bot, \gamma) = \mathbf{false}$.
- For all p, ψ, and γ, we have $\delta'(\langle p, \psi, \forall, p_s\rangle, \bot, \gamma) = \delta'(\langle p, \psi, \forall, p_s\rangle, \top, \gamma)$ and $\delta'(\langle p, \psi, \exists, p_s\rangle, \bot, \gamma) = \delta'(\langle p, \psi, \exists, p_s\rangle, \top, \gamma)$.
- $\delta'(\langle p, p_\top, m\rangle, \top, \gamma) = (\bigvee_{(q,\beta)\in s(p,\gamma)}(vis(q,\beta), \langle q, p_\top, \exists, p_x\rangle, \beta)) \wedge (\bigwedge_{(q,\beta)\in s(p,\gamma)}(vis(q,\beta), \langle q, p_\top, \forall, p_x\rangle, \beta))$.
- $\delta'(\langle p, h, m\rangle, \top, \gamma) = \mathbf{true}$ if $h \in \eta((p,\gamma))$, or $h = \mathbf{true}$.
- $\delta'(\langle p, h, m\rangle, \top, \gamma) = \mathbf{false}$ if $h \notin \eta((p,\gamma))$, or $h = \mathbf{false}$.
- $\delta'(\langle p, \neg h, m\rangle, \top, \gamma) = \mathbf{true}$ if $h \notin \eta((p,\gamma))$, or $h = \mathbf{false}$.
- $\delta'(\langle p, \neg h, m\rangle, \top, \gamma) = \mathbf{false}$ if $h \in \eta((p,\gamma))$, or $h = \mathbf{true}$.
- $\delta'(\langle p, \psi_1 \wedge \psi_2, m\rangle, \top, \gamma) = \mathsf{split}(\delta'(\langle p, \psi_1, m\rangle, \top, \gamma) \wedge \delta'(\langle p, \psi_2, m\rangle, \top, \gamma))$.
- $\delta'(\langle p, \psi_1 \vee \psi_2, m\rangle, \top, \gamma) = \mathsf{split}(\delta'(\langle p, \psi_1, m\rangle, \top, \gamma) \vee \delta'(\langle p, \psi_2, m\rangle, \top, \gamma))$.
- $\delta'(\langle p, [n]\psi, m\rangle, \top, \gamma) = \mathsf{split}(\bigvee_{Y\subseteq s(p,\gamma)\wedge |Y| = |s(p,\gamma)|-n} \bigwedge_{(q,\beta)\in Y}(vis(q,\beta), \langle q, \psi, \forall, p_x\rangle, \beta))$.
- $\delta'(\langle p, \langle n\rangle\psi, m\rangle, \top, \gamma) = \mathsf{split}(\bigvee_{Y\subseteq s(p,\gamma)\wedge |Y| = n+1} \bigwedge_{(q,\beta)\in Y}(vis(q,\beta), \langle q, \psi, \exists, p_x\rangle, \beta))$.
- $\delta'(\langle p, \mu y.\varphi(y), m\rangle, \top, \gamma) = \mathsf{split}(\delta'(\langle p, \varphi(\mu y.\varphi(y)), m\rangle, \top, \gamma))$.
- $\delta'(\langle p, \nu y.\varphi(y), m\rangle, \top, \gamma) = \mathsf{split}(\delta'(\langle p, \varphi(\nu y.\varphi(y)), m\rangle, \top, \gamma))$.

The definition of the function $\mathsf{split} : \mathcal{B}^+(D \times Q' \times \Gamma_\flat^*) \to \mathcal{B}^+(D \times Q' \times \Gamma_\flat^*)$ is a simple adaptation of the definition found in [KVW00]. For every $d \in D, q \in Q, m \in \{\exists, \forall\} \times \{p_e, p_s\}$ and $\beta \in \Gamma_\flat^*$ we have the following:

- $\mathsf{split}(\mathbf{true}) = \mathbf{true}, \mathsf{split}(\mathbf{false}) = \mathbf{false}$.
- $\mathsf{split}(\theta_1 \vee \theta_2) = \mathsf{split}(\theta_1) \vee \mathsf{split}(\theta_2)$ and $\mathsf{split}(\theta_1 \wedge \theta_2) = \mathsf{split}(\theta_1) \wedge \mathsf{split}(\theta_2)$.
- If $\psi \in cl(\varphi)$ is of the form $p, \neg p, [n]\psi', \langle n\rangle\psi', \mu y.\psi'(y)$ or $\nu y.\psi'(y)$, then $\mathsf{split}(d, \langle p, \psi, m\rangle, \beta) = (d, \langle p, \psi, m\rangle, \beta)$.
- $\mathsf{split}(d, \langle p, \psi_1 \vee \psi_2, m\rangle, \beta) = \mathsf{split}(d, \langle p, \psi_1, m\rangle, \beta) \vee \mathsf{split}(d, \langle p, \psi_2, m\rangle, \beta)$.
- $\mathsf{split}(d, \langle p, \psi_1 \wedge \psi_2, m\rangle, \beta) = \mathsf{split}(d, \langle p, \psi_1, m\rangle, \beta) \wedge \mathsf{split}(d, \langle p, \psi_2, m\rangle, \beta)$.

- It remains to define the acceptance condition F. Let d be the maximal alternation level of (fixpoint) sub-formulas of φ. Denote by G_i the set of all ν-formulas in $cl(\varphi)$ of alternation level i. Denote by B_i the set of all μ-formulas in $cl(\varphi)$ of alternation depth less than or equal to i. Now, $F = \{F_0, F_1, \ldots, F_{2d}\}$, where $F_0 = \emptyset$ and for every $1 \le i \le d$ we have $F_{2i-1} = F_{2i-2} \cup (Q \times B_i \times \{\forall, \exists\} \times \{p_e, p_s\})$, and $F_{2i} = F_{2i-1} \cup (Q \times G_i \times \{\forall, \exists\} \times \{p_e, p_s\})$. Clearly, $F_0 \subseteq F_1 \subseteq F_2 \subseteq \ldots \subseteq F_{2d}$. Since by the definition of PD-SPT a path π of a run r is accepting if the minimal i with $\mathsf{Inf}(\pi) \cap F_i \neq \emptyset$ is even, by our definition of F, such an index i corresponds to the outermost fixpoint formula that was visited infinitely often. Thus, the acceptance condition makes sure that, if a fixpoint formula is visited infinitely often, then this is a greatest fixpoint formula, and that all of its least fixpoint super-formulas are visited only finitely many times.

Let us now discuss the size of $\mathcal{A}_{\mathcal{S},\varphi}$. It is easy to see that $|Q'| = O(|Q| * |\varphi|)$, and $|\delta'| = O(|\delta| * |\varphi|)$. Hence, the size of $\mathcal{A}_{\mathcal{S},\varphi}$ is $O(|\mathcal{S}| * |\varphi|)$.

Finally, we show that $\mathcal{A}_{\mathcal{S},\varphi}$ is semi-alternating. It is sufficient to show that for every $(t,\beta) \in D$, $\sigma \in \Sigma$, $p,p' \in Q'$, and $\gamma \in \Gamma$, if $((t,\beta),p',\beta')$ appears in $\delta'(p,\sigma,\gamma)$ then $\beta = \beta'$. To see that, notice that $((t,\beta),p',\beta')$ appears in $\delta'(p,\sigma,\gamma)$ only if $vis(q,\beta') = (t,\beta)$, for some $q \in Q$. Since by definition (because the pushdown store is completely visible) we have that $vis(q,\beta') = (vis(q),\beta')$, and we are done. □

Theorem 2 implies that $\mathcal{M}_{\mathcal{S}} \models_r \psi$ iff the language of the automaton $\mathcal{A}_{\mathcal{S},\neg\psi}$ is empty. We can now show the main result of the paper.

Theorem 3. *Given an* OPD $\mathcal{S}$ *and a formula* φ*, the pushdown module checking problem with imperfect state information is* 2EXPTIME*-complete if* φ *is a propositional or a graded* μ*-calculus formula, and* 3EXPTIME*-complete if* φ *is a* CTL* *formula.*

Proof. The lower bound follows from the known bound for pushdown module checking with perfect information (see [FMP07] for propositional and graded μ-calculus, and [BMP05] for *CTL**). For the upper bound, by Theorem 2, it is enough to check that $\mathcal{A}_{\mathcal{S},\neg\varphi}$ is empty. Recall that when considering propositional and graded μ-calculus (resp. *CTL**) $\mathcal{A}_{\mathcal{S},\neg\varphi}$ is a PD-SPT with $n = O(|\mathcal{S}| * |\varphi|)$ (resp. $n = O(|\mathcal{S}| * 2^{|\varphi|})$) states, and index $k = O(|\varphi|)$. Let $\mathcal{M}_{\mathcal{S}} = \langle AP, W_s, W_e, w_0, R, L, \cong \rangle$ be the module induced by $\mathcal{S}$. Observe that the set of directions of the strategy trees that are the input of $\mathcal{A}_{\mathcal{S},\neg\varphi}$ is $D = \{(vis(q),\beta) \mid ((p,\alpha),(q,\beta)) \in R$ for some p,q,α and $\beta\}$, and it is bounded from above by $|S|$. By Corollary 1, the emptiness of $\mathcal{A}_{\mathcal{S},\neg\varphi}$ can be decided in time double exponential in $|D|nk$. Thus, deciding if $\mathcal{M}_S \models_r \varphi$ can be done in time double-exponential in $|\mathcal{S}| * |\varphi|$ when considering propositional and graded μ-calculus, and triple-exponential in $|\mathcal{S}| * |\varphi|$ when considering CTL^*. □

Acknowledgment. The first and third author wish to thank Nir Piterman for useful discussions.

References

[AMV07] B. Aminof, A. Murano, and M.Y. Vardi. Pushdown module checking with imperfect information. In *CONCUR '07*, LNCS 4703, pages 461–476. Springer-Verlag, 2007.

[BLMV06] P.A. Bonatti, C. Lutz, A. Murano, and M.Y. Vardi. The complexity of enriched μ-calculi. In *ICALP'06*, LNCS 4052, pages 540-551, 2006.

[BMP05] Laura Bozzelli, Aniello Murano, and Adriano Peron. Pushdown module checking. In *LPAR'05*, LNCS 3835, pages 504–518. Springer-Verlag, 2005.

[CE81] E.M. Clarke and E.A. Emerson. Design and verification of synchronization skeletons using branching time temporal logic. In *Proceedings of Workshop on Logic of Programs*, LNCS 131, pages 52–71. Springer-Verlag, 1981.

[EH86] E.A. Emerson and J.Y. Halpern. Sometimes and not never revisited: On branching versus linear time. *J. of the ACM*, 33(1):151–178, 1986.

[FM07] A. Ferrante and A. Murano. Enriched μ–calculus module checking. In *FOSSACS'07*, volume 4423 of *LNCS*, pages 183-197, 2007.

[FMP07] A. Ferrante, A. Murano, and M. Parente. Enriched μ–calculus pushdown module checking. In *LPAR'07*, volume 4790 of *LNAI*, pages 438–453, 2007.

[Hoa85] C.A.R. Hoare. *Communicating Sequential Processes.* Prentice-Hall, 1985.

[HP85] D. Harel and A. Pnueli. On the development of reactive systems. In *Logics and Models of Concurrent Systems*, volume F-13 of *NATO Advanced Summer Institutes*, pages 477–498. Springer-Verlag, 1985.

[HU79] J. E. Hopcroft and J. D. Ullman. *Introduction to Automata Theory, Languages and Computation.* Addison-Wesley, 1979.

[Koz83] D. Kozen. Results on the propositional mu–calculus. *Theoretical Computer Science*, 27:333–354, 1983.

[KPV02] O. Kupferman, N. Piterman, and M.Y. Vardi. Pushdown specifications. In *LPAR'02*, LNCS 2514, pages 262–277. Springer-Verlag, 2002.

[KSV02] O. Kupferman, U. Sattler, and M.Y. Vardi. The complexity of the graded μ-calculus. In *CADE'02*, LNAI 2392, pages 423-437, 2002.

[KV96] O. Kupferman and M.Y. Vardi. Module checking. In *CAV'96*, LNCS 1102, pages 75–86. Springer-Verlag, 1996.

[KV97] O. Kupferman and M. Y. Vardi. Module checking revisited. In *Proc. 9th International Computer Aided Verification Conference*, LNCS 1254, pages 36–47. Springer-Verlag, 1997.

[KVW00] O. Kupferman, M.Y. Vardi, and P. Wolper. An Automata-Theoretic Approach to Branching-Time Model Checking. *J. of the ACM*, 47(2):312–360, 2000.

[KVW01] O. Kupferman, M.Y. Vardi, and P. Wolper. Module Checking. *Information and Computation*, 164(2):322–344, 2001.

[MS87] D.E. Muller and P.E. Schupp. Alternating automata on infinite trees. *Theoretical Computer Science*, 54:267–276, 1987.

[QS81] J.P. Queille and J. Sifakis. Specification and verification of concurrent programs in Cesar. In *Proceedings of the Fifth International Symposium on Programming*, LNCS 137, pages 337–351. Springer-Verlag, 1981.

From Formal Proofs to Mathematical Proofs: A Safe, Incremental Way for Building in First-order Decision Procedures

Fréderic Blanqui[1], Jean-Pierre Jouannaud[2], and Pierre-Yves Strub[2]

[1] INRIA & LORIA, Campus Scientifique, BP 239, 54506 Vandoeuvre-lès-Nancy Cedex, France, blanqui@loria.fr

[2] LIX, UMR 7161, Project INRIA TypiCal, École Polytechnique, 91128 Palaiseau, France, jouannaud, strub@lix.polytechnique.fr

Abstract. We investigate here a new version of the Calculus of Inductive Constructions (CIC) on which the proof assistant Coq is based: the Calculus of Congruent Inductive Constructions, which truly extends CIC by building in arbitrary first-order decision procedures: deduction is still in charge of the CIC kernel, while computation is outsourced to dedicated first-order decision procedures that can be taken from the shelves provided they deliver a proof certificate. The soundness of the whole system becomes an incremental property following from the soundness of the certificate checkers and that of the kernel. A detailed example shows that the resulting style of proofs becomes closer to that of the working mathematician.

1 Introduction

Proof assistants based on the Curry-Howard isomorphism such as Coq [9] allow to build the proof of a given proposition by applying appropriate proof tactics available from existing libraries or that can otherwise be developed for achieving a specific task. These tactics generate a proof term that can be checked with respect to the rules of logic. The proof-checker, also called the *kernel* of the proof assistant, implements the deduction rules of the logic on top of a term manipulation layer. In this model, the mathematical correctness of a proof development relies entirely on the kernel. Trusting the kernel is therefore vital.

The (intuitionist) logic on which Coq is based is the Calculus of Constructions (CC) of Coquand and Huet [10], an impredicative type theory incorporating polymorphism, dependent types and type constructors. Unlike logics without dependent types, CC enjoys a powerful type-checking rule, called *conversion*, which incorporates computations within deductions, making decidability of type-checking a non-trivial property of the calculus.

In CC, computation reduces to (pure) functional evaluation in the underlying lambda calculus. The notion of computation is richer in the Calculus of Inductive Constructions of Coquand and Paulin (CIC), obtained from CC by adding inductive types and the corresponding rules for higher-order primitive recursion [11]. The recent versions of Coq are based on a slight generalization of this calculus [15]. Still, such a simple function as *reverse* of a *dependent list* cannot be defined in CIC as

Please use the following format when citing this chapter:

Blanqui, F., Jouannaud, J.-P. and Strub, P.-Y., 2008, in IFIP International Federation for Information Processing, Volume 273; *Fifth IFIP International Conference on Theoretical Computer Science*; Giorgio Ausiello, Juhani Karhumäki, Giancarlo Mauri, Luke Ong; (Boston: Springer), pp. 349–365.

one would expect, because $(reverse\ l :: l')$ and $(reverse\ l') :: (reverse\ l)$, assuming :: is list concatenation, have non-convertible types $list(n+m)$ and $list(m+n)$, assuming $(reverse\ l)$ has for type the type of its argument l. This is so because the usual definition of $+$ by induction on one of its arguments does not reduce the proof of $m+n=n+m$ to a computation.

We do believe that scaling up the proof development process requires being able to mimic the mathematician when replacing the proof of a proposition P by the proof of an equivalent proposition P' obtained from P thanks to possibly complex calculations in which *easy steps* are hidden away. It is our program to make this view a reality.

A way to incorporate decision procedures to Coq is by developing a tactic and then use a reflexion technique to omit checking the proof term being built by proving the decision procedure itself. But the soundness of the entire mechanism cannot be guaranteed in general [12]. Further, this does not answer the question of hiding easy steps away.

A first attempt towards our goal is the Calculus of Algebraic Constructions (CAC), obtained by adding to CC user-defined computations as rewrite rules [5, 3]. Although conceptually quite powerful since CAC captures CIC [4], this paradigm does not yet fulfill all needs. In particular, the user needs to hide away the easy steps by himself, that is by giving the necessary rewrite rules and by verifying that they satisfy the assumptions of the *general schema* [5, 3].

The proof assistant PVS uses a potentially stronger paradigm than Coq by combining its deduction mechanism with a notion of computation based on the powerful Shostak's method for combining decision procedures [20], a framework dubbed *little proof engines* by Shankar [19]. Indeed, the little engines of proof hide away the easy computational steps, without any user assistance. Unfortunately, proof-checking is not decidable in PVS. Further, since the little engines of proofs involve complex coding, as well as Shostak's algorithm itself, one can only *believe* a PVS proof, while one can *check* and *trust* a Coq proof.

Two steps in the direction of integrating decision procedures into CC are Stehr's Open Calculus of Constructions (OCC) [21] and Oury's Extensional Calculus of Constructions (ECC) [17]. Implemented in Maude, OCC allows for the use of an arbitrary equational theory in conversion. ECC can be seen as a particular case of OCC in which all provable equalities can be used in conversion, which can also be achieved by adding the extensionality and Streicher's axioms to CC [22], hence the name of this calculus. Unfortunately, strong normalization and decidability of type checking are then lost, which shows that we should seek for more restrictive extensions.

In a preliminary work, we designed a new, quite restrictive framework, the Calculus of Congruent Constructions (CCC), which incorporates the congruence closure algorithm in CC's conversion [7], while preserving the good properties of CC, including the decidability of type checking. In [6], we have described $CC_{\mathbb{N}}$, in which the decision procedure was Presburger arithmetic and strong elimination ruled out. The present work is a continuation of the latter.

Theoretical contribution. Our main theoretical contribution is the definition and the meta-theoretical investigation of the Calculus of Congruent Inductive Constructions (CCIC), which incorporates arbitrary *first-order theories* for which entailment is decidable into deductions via an abstract conversion rule of the calculus. A major technical innovation of this work lies in the computation mechanism: goals are sent to the decision procedure together with the set of user hypotheses available from the current context. Our main result shows that this extension of CIC does not compromise its properties: confluence, strong normalization, coherence and decidability of proof-checking are all preserved.

Unlike previous calculi, the difficulty with CCIC is not strong normalization, for which we have reused the strong normalization proof of CAC [3]. A major difficulty was a traditional step towards subject-reduction: compatibility of conversion with products. Decidability of type checking required restricting conversions below recursors [23].

Practical contribution. We give several examples showing the usefulness of this new calculus, in particular for using dependent types such as dependent lists, which has been an important weakness of Coq until now. Further studies are needed to explore other potential applications, to match inductive definition-by-case modulo theories of constructors-destructors, another very different weakness of Coq. A detailed example shows that the resulting style of proofs becomes closer to that of the working mathematician.

Methodological contribution. The safety of proof assistants is based on their kernel. In the early days of Coq, the safety of its kernel relied on its small size and its clear structure reflecting the inference rules of the intuitionist type theory, CC, on which it was based. The slogan was that of a *readable kernel*. Moving later to CIC allowed to ease the specification tasks, making the system very popular among proof developers, but resulted in a more complex kernel that can now hardly be read except by a few specialists. The slogan changed to a *provable kernel*, and indeed one version of it was once proved with an earlier version (using strong normalization as an assumption), and a new safe kernel extracted from that proof.

Of course, there has been many changes in the kernel since then, and its correctness proof was not maintained. This is a first weakness with the *readable kernel* paradigm: it does not resist changes. There is a second which relates directly to CCIC: there is no guarantee that a decision procedure taken from the shelf implements correctly the complex mathematical theorem on which it is based, since carrying out such a proof may require an entire PhD work. Therefore, these procedures *cannot* be part of the kernel.

Our solution to these problems is a new shift of paradigm to that of an *incremental kernel*. The calculus on which a proof assistant is based should come in two parts: a stable calculus implementing deduction, CIC in our case, which should satisfy the *readable* or *provable kernel* paradigm; a collection of independent decision procedures implementing computations, that produce checkable proof certificates. The certificate checker should of course itself satisfy the *readable* or *provable code* paradigm. Note that a Coq proof is a particular case of a checkable certificate.

This paradigm has many advantages. First, it allows for a modular, cooperative development of the system, by separating the development of the kernel from that of the decision procedures. Second, it allows for an *unsafe mode* in case a decision procedure is used that does not have a certificate generator yet. Third, it allows to better trace errors in case the system rejects a proof, by using decision procedures that output *explanations* when they fail. Last, it allows the user to use any decision procedure she needs by simply hooking it to the system, possibly in unsafe mode.

This incremental schema is quite flexible, assuming that decision procedures come one by one. However, even so, they are not independent, they must be combined. Combining first-order decision procedures is not a new problem, it was considered in the early 80's by Nelson and Oppen on the one hand, by Shostak on the other hand, and has generated much work since then. There are several possibilities to build in this mechanism: in the kernel, via a certificate generator and checker again, or by reflection. This design decision has not been made yet.

2 Congruent Inductive Constructions

The Calculus of Congruent Inductive Constructions (CCIC) is an extension of CIC which embeds in its conversion rule the validity entailment of a fixed first order theory. First, we recall the basics of CIC before to introduce parametric multi-sorted algebras and then embed these first-order algebras into CIC. We are then able to define our calculus relative to a specific congruence that is defined last. For simplicity, we will only consider here the particular case of parametric lists and that of the natural numbers equipped with Presburger arithmetic. This simple case allows us to build lists of natural numbers, as well as lists of lists of natural numbers, and so on. It indeed has the complexity of the whole calculus, which is not at all the case when natural numbers only are considered as in [6].

2.1 Calculus of Inductive Constructions

Terms. We start our presentation by first describing the terms of CIC.

CIC uses two *sorts*: $\star$ (or Prop, or *object level universe*), $\Box$ (or Type, or *predicate level universe*) and $\triangle$. We denote $\{\star, \Box, \triangle\}$, the set of CIC sorts, by $\mathscr{S}$.

Following the presentation of *Pure Type Systems* (PTS) [14], we use two classes of variables: $\mathscr{X}^\star$ and $\mathscr{X}^\Box$ are countably infinite sets of *term variables* and *predicate variables* such that $\mathscr{X}^\star$ and $\mathscr{X}^\Box$ are disjoint. We write $\mathscr{X}$ for $\mathscr{X}^\star \cup \mathscr{X}^\Box$.

We shall use $\overline{u}$ for a list $(u_1, \ldots, u_n)$, s for a sort in $\mathscr{S}$, $x, y, \ldots$ for variables in $\mathscr{X}^\star$, $X, Y, \ldots$ for variables in $\mathscr{X}^\Box$.

Definition 1 (Pseudo-terms). The algebra $\mathscr{L}$ of *pseudo-terms* of CIC is defined by:

$$t,u,T,U,\ldots := s \in \mathscr{S} \mid x \in \mathscr{X} \mid \forall(x:T).t \mid \lambda[x:T].t$$
$$\mid\ t\,u \mid \mathrm{Ind}(X:t)\{\overline{T_i}\} \mid t^{[n]} \mid \mathrm{Elim}(t:T\,[\overline{u_i}] \to U)\{\overline{w_j}\}$$

The notion of free variables is as usual - the binders being λ, $\forall$ and Ind (in $\mathrm{Ind}(X : t)\{\overline{T_i}\}$, X is bound in the T_i's). We write $\mathrm{FV}(t)$ for the set of free variables of t. We say that t is closed if $\mathrm{FV}(t) = \emptyset$. A variable x *freely occurs* in t if $x \in \mathrm{FV}(t)$.

Inductive types. The novelty of CIC was to introduce inductive types, denoted by $I = \mathrm{Ind}(X : T)\{\overline{C_i}\}$ where the $\overline{C_i}$'s describe the types of the *constructors* of I, and T the type (or *arity*) of I which must be of the form $\forall(\overline{x_i : T_i}).\star$. The k-th constructor of the inductive type I, of type $C_k\{X \mapsto I\}$, will be denoted by $I^{[k]}$.

As an easy first example, we define natural numbers: $\mathbf{nat} := \mathrm{Ind}(X : \star)\{X, X \to X\}$. We shall use $\mathbf{0}$ and $\mathbf{S}$ as constructors for natural numbers, of respective types $\mathbf{nat}$ and $\mathbf{nat} \to \mathbf{nat}$, obtained by replacing X by $\mathbf{nat}$ in the above two expressions X and $X \to X$. Elimination rules for $\mathbf{nat}$ are as follows:

$$\begin{array}{l} \mathrm{Elim}\mathbb{N}(\mathbf{0}, Q)\{v_{\mathbf{0}}, v_{\mathbf{S}}\} \xrightarrow{\iota} v_{\mathbf{0}} \\ \mathrm{Elim}\mathbb{N}(\mathbf{S}x, Q)\{v_{\mathbf{0}}, v_{\mathbf{S}}\} \xrightarrow{\iota} v_{\mathbf{S}}\, x\, (\mathrm{Elim}\mathbb{N}(x, Q)\{v_{\mathbf{0}}, v_{\mathbf{S}}\}) \text{ with } Q : \mathbf{nat} \to s,\ \in \mathscr{S}. \end{array}$$

Similarly, we now define parametric lists: $\mathbf{list} := \lambda[T : \star].\,\mathrm{Ind}(X : \star)\{X, T \to X \to X\}$. We shall use $\mathbf{nil}$ and $\mathbf{cons}$ as constructors for parametrized lists, of respective types $\forall(T : \star).\,\mathbf{list}(T)$ and $\forall(T : \star).\,T \to \mathbf{list}(T) \to \mathbf{list}(T)$. Elimination rules for $\mathbf{list}$ are:

$$\begin{array}{ll} \mathrm{Elim}\mathbb{L}(\mathbf{nil}, Q)\{v_{\mathbf{nil}}, v_{\mathbf{cons}}\} & \xrightarrow{\iota} v_{\mathbf{nil}} \\ \mathrm{Elim}\mathbb{L}(\mathbf{cons}\, x\, l, Q)\{v_{\mathbf{nil}}, v_{\mathbf{cons}}\} & \xrightarrow{\iota} v_{\mathbf{cons}}\, x\, l\ \mathrm{Elim}\mathbb{L}(l, Q)\{v_{\mathbf{nil}}, v_{\mathbf{cons}}\}) \end{array}$$

Finally, we define dependent words over an alphabet A:

$$\mathbf{word} = \mathrm{Ind}(X : \mathbf{nat} \to \star)\{X\,\mathbf{0}, A \to X\ (\mathbf{S}\,\mathbf{0}), \forall(y, z : \mathbf{nat}).\,X\,y \to X\,z \to X(y+z)\}$$

We shall use ε, $\mathbf{char}$ and $\mathbf{app}$ for its three constructors, of respective types $\mathbf{word}\,\mathbf{0}$, $A \to \mathbf{word}\,(\mathbf{S}\,\mathbf{0})$, and $\forall(n, m : \mathbf{nat}).\,\mathbf{word}\,n \to \mathbf{word}\,m \to \mathbf{word}\,(n+m)$ obtained as previously by replacing X by $\mathbf{word}$ in the three expressions $X\,\mathbf{0}, A \to X\ (\mathbf{S}\,\mathbf{0})$, and $\forall(y, z : \mathbf{nat}).\,X\,y \to X\,z \to X(y+z)$. Elimination rules for dependent words are:

$$\begin{array}{ll} \mathrm{Elim}\mathbb{W}(\varepsilon, Q)\{v_\varepsilon, v_{\mathbf{char}}, v_{\mathbf{app}}\} & \xrightarrow{\iota} v_\varepsilon \\ \mathrm{Elim}\mathbb{W}(\mathbf{char}\, x, Q)\{v_\varepsilon, v_{\mathbf{char}}, v_{\mathbf{app}}\} & \xrightarrow{\iota} v_{\mathbf{char}}\, x \\ \mathrm{Elim}\mathbb{W}(\mathbf{app}\, n\, m\, l\, l', Q)\{v_\varepsilon, v_{\mathbf{char}}, v_{\mathbf{app}}\} & \xrightarrow{\iota} v_{\mathbf{app}}\, n\, m\, l\, l'\ (\mathrm{Elim}\mathbb{W}(l, Q)\{v_\varepsilon, v_{\mathbf{char}}, v_{\mathbf{app}}\}) \\ & \qquad (\mathrm{Elim}\mathbb{W}(l', Q)\{v_\varepsilon, v_{\mathbf{char}}, v_{\mathbf{app}}\}) \end{array}$$

Definitions by induction. We can now define functions by induction over natural numbers, lists or words. Since using the CIC syntax is a bit painful, we give only a quite simple example defining append (written @) for lists of natural numbers, of type $\forall(T : \star).\,\mathbf{list}(T) \to \mathbf{list}(T) \to \mathbf{list}(T)$:

$$@ := \lambda[l : \mathbf{list}\,\mathbf{nat}][l' : \mathbf{list}\,\mathbf{nat}].\,\mathrm{Elim}\mathbb{L}(l, Q)\left\{\begin{array}{c} \lambda[x : \mathbf{nat}][l'' : \mathbf{list}\,\mathbf{nat}]. \\ l',\quad \lambda[l1 : \mathbf{list}\,\mathbf{nat}][l2 : \mathbf{list}\,\mathbf{nat}]. \\ \lambda[L : Q\,l1\,l2].\,\mathbf{cons}\,x\,L \end{array}\right\}$$

Strong and Weak reductions. CIC distinguishes *strong* ι-elimination when the type Q of terms constructed by induction is at predicate level, from weak ι-elimination when Q is at object level. Strong elimination is restricted to *small* inductive types to ensure logical consistency [24].

Typing judgments. A *typing environment* Γ is a sequence of pairs $x_i : T_i$ made of a variable x_i and a term T_i (we say that Γ binds x_i to the type T_i), such that Γ does not bind a variable twice. The typing judgments are classically written $\Gamma \vdash t : T$, meaning that the *well formed term* t is a proof of the proposition T (has type T) under the *well formed environment* Γ. $x\Gamma$ will denote the type associated to x in Γ, and we write $\mathrm{dom}(\Gamma)$ for the domain of Γ as well.

The typing rules of CIC given in 1 are made of the typing rules for CC and the typing rules for inductive types, given for the particular case of **nat** and **list**.

$$\frac{}{\vdash \star : \Box}\ [\textsc{Ax-1}] \qquad \frac{}{\vdash \Box : \triangle}\ [\textsc{Ax-2}] \qquad \frac{\Gamma \vdash V : s \quad \Gamma \vdash t : T \quad s \in \{\star, \Box\} \quad x \in \mathscr{X}^s - \mathrm{dom}(\Gamma)}{\Gamma, [x : V] \vdash t : T}\ [\textsc{Weak}]$$

$$\frac{\Gamma \vdash T : s_T \quad \Gamma, [x : T] \vdash U : s_U}{\Gamma \vdash \forall(x : T).U : s_U}\ [\textsc{Prod}] \qquad \frac{x \in \mathrm{dom}(\Gamma) \cap \mathscr{X}^{s_x} \quad \Gamma \vdash x\Gamma : s_x}{\Gamma \vdash x : x\Gamma}\ [\textsc{Var}]$$

$$\frac{\Gamma \vdash \forall(x : T).U : s \quad \Gamma, [x : T] \vdash u : U}{\Gamma \vdash \lambda[x : T].u : \forall(x : T).U}\ [\textsc{Abs}] \qquad \frac{\Gamma \vdash t : T \quad \Gamma \vdash T' : s' \quad T \overset{\beta\iota}{\longleftrightarrow}_* T'}{\Gamma \vdash t : T'}\ [\textsc{Conv}]$$

$$\frac{\Gamma \vdash t : \forall(x : U).V \quad \Gamma \vdash u : U}{\Gamma \vdash t\,u : V\{x \mapsto u\}}\ [\textsc{App}]$$

$$\frac{\vdash \tau_f : s \in \{\star, \Box\}}{\vdash f : \tau_f}\ [\textsc{Symb}] \qquad \frac{\begin{array}{c}\Gamma \vdash T : \star \quad \Gamma \vdash p : \mathbf{nat} \quad \Gamma \vdash l : \mathbf{list}\,T\,p \\ \Gamma \vdash Q : \forall(n : nat).\,\mathbf{list}\,T\,n \to s \in \{\star, \Box\} \\ \Gamma \vdash v_{\mathbf{nil}} : Q\,\mathbf{0}\,(\mathbf{nil}\,T) \\ \Gamma \vdash v_{\mathbf{cons}} : \begin{array}{c}\forall(x : T)(n : \mathbf{nat})(l : \mathbf{list}\,T\,n). \\ Q\,n\,l \to Q\,(\mathbf{S}n)(\mathbf{cons}\,T\,x\,n\,l)\end{array}\end{array}}{\mathrm{Elim}\mathbb{L}(l, Q)\{v_{\mathbf{0}}, v_{\mathbf{S}}\} : Q\,p\,l}\ [\textsc{Elim}]$$

$$\frac{\begin{array}{c}\Gamma \vdash Q : \mathbf{nat} \to s \in \{\star, \Box\} \\ \Gamma \vdash n : \mathbf{nat} \quad \Gamma \vdash v_{\mathbf{0}} : Q\,\mathbf{0} \\ \Gamma \vdash v_{\mathbf{S}} : \forall(p : \mathbf{nat}).\,Q\,p \to Q\,(\mathbf{S}\,p)\end{array}}{\mathrm{Elim}\mathbb{N}(n, Q)\{v_{\mathbf{0}}, v_{\mathbf{S}}\} : Q\,n}\ [\textsc{Elim}]$$

Fig. 1 CIC typing rules for **nat** and **list**

We did not give the general typing elimination rule for arbitrary inductive types, which is quite complicated. Instead, we gave the elimination rules obtained for our three inductive types **nat**, **list** and **word**. We refer to [18, 24] for the general case, and for the precise typing rule of $\mathrm{Elim}\mathbb{W}$.

2.2 Parametric sorted algebras

Parametric sorted signature. Order-sorted algebras were introduced as a formal framework for the OBJ language in [13], before to be generalized as *membership equational logic* in [8]. We use here a polymorphic version of a restriction of the latter, by assuming given a signature (Λ, Σ), Λ for the sort constructors, and Σ for the function symbols made of a set of constructors for each sort constructor, and of a set of defined symbols. We shall use the notation $f : \forall\overline{\alpha}.\sigma_1 \times \cdots \times \sigma_n \to \tau$ for symbol declarations. As an example, we describe natural numbers and parametric (non-dependent) list using an OBJ-like syntax. We rule out here partiality, as introduced in practice by destructor symbols, for sake of clarity.

We shall use $\mathscr{V} = \{\alpha, \beta, \ldots\}$ for the set of sort variables, and $\mathscr{T}(\Sigma, \mathscr{V}) = \{\sigma, \tau, \ldots\}$ for the set of sort expressions.

`sort` **nat** : $*$	`cons` **S**	:	**nat** $\rightarrow$ **nat**
`sort` **list** : $* \rightarrow *$	`fun` $+$	:	**nat** $\times$ **nat** $\rightarrow$ **nat**
`svar` α : $*$	`cons` **nil**	:	**list**(α)
`cons` **0** : **nat**	`cons` **cons**	:	$\alpha \times$ **list**$(\alpha) \rightarrow$ **list**(α)
	`fun` @	:	**list**$(\alpha) \times$ **list**$(\alpha) \rightarrow$ **list**(α)

Definition 2 (Terms). For any sort σ, let $\mathscr{X}^\sigma$ be a countably infinite set of *variables of sort* σ, s.t. all the $\mathscr{X}^\sigma$'s are pairwise disjoint. Let $\mathscr{X} = \bigcup_\sigma \mathscr{X}^\sigma$. For any $x \in \mathscr{X}$, we say that x has sort σ if $x \in \mathscr{X}^\sigma$. For any sort σ, the set $\mathscr{T}_\sigma(\Sigma, \mathscr{X})$ of *terms of sorts* σ *with variables* $\mathscr{X}$ is the smallest set s.t.:

1. if $x \in \mathscr{X}^\tau$, then $x \in \mathscr{T}_\tau(\Sigma)$,
2. if $t_1, \cdots, t_n \in \mathscr{T}_{\sigma_1 \xi}(\Sigma, \mathscr{X}) \times \cdots \times \mathscr{T}_{\sigma_2 \xi}(\Sigma, \mathscr{X})$ where $f : \forall \overline{\alpha}. \sigma_1 \times \cdots \times \sigma_n \rightarrow \tau$ and ξ is a sort substitution, then $f(t_1, \ldots, t_n) \in \mathscr{T}_{\tau\xi}(\Sigma, \mathscr{X})$.

Let $\mathscr{T}(\Sigma, \mathscr{X}) = \bigcup_\sigma (\mathscr{T}_\sigma(\Sigma, \mathscr{X}))$. A term t has sort σ if $t \in \mathscr{T}_\sigma(\Sigma, \mathscr{X})$.

Note that the sets $\mathscr{X}^\sigma$ play the role of a typing context.

Example 1. Assuming that x is a variable of sort **nat**, then 0 and $0 + x$ are of sort **nat**, while **nil** is of sort **list**(α), **list**$($**nat**$)$, **list**$($**list**$($**nat**$))$, etc.

Definition 3 (Equations). Equations $t =^\sigma u$ are pairs of terms of the same sort σ.

Example 2. Assuming x of sort **nat** and l of sort **list**$($**list**$(($**nat**$))$, $x + 0 =^{\mathbf{nat}} x$ is an equation of sort **nat** and $cons(x, nil) =^{list(\mathbf{nat})} car(l)$ is an equation of sort **list**$($**nat**$)$.

We can therefore as usual build parametrized algebras for **list**, algebras for **nat** and therefore get algebras for **nat**, **list**$($**nat**$)$, etc. Satisfaction of an equation in these algebras is defined as usual. In practice, type superscripts may be omitted when they can be infered from the context.

2.3 Embedding parametric algebras in CIC

Our purpose here is to embed parametric multi-sorted algebra into CIC. As a result, two different, but related kinds of symbols will coexist, in CIC and in the embedded algebraic sub-world. We shall distinguish them by underlying symbols in CIC.

The first step of the translation maps, respectively sort constructors and constructor symbols to CIC inductive types and constructors. We start with natural numbers and its sort constructor **nat**. Constructor symbols of **nat** are simply all the constructors symbols whose codomain is **nat**, i.e. here **0** and **S**. We thus define $\underline{\mathbf{nat}}$ (the CIC inductive type attached to **nat**) as an inductive type with two constructor types (one for **0**, and one for **S**): $\underline{\mathbf{nat}} := \mathrm{Ind}(X : \star)\{C_1(X), C_2(X)\}$.

The constructor types of $\underline{\mathbf{nat}}$ are simply the arities of **0** and **S** where **nat** is replaced with the constructor type variable: $C_1(X) = X$ and $C_2(X) = X \rightarrow X$. As expected, we obtain here the standard inductive definition of natural numbers given in Section 2.1: $\mathrm{Ind}(X:\star)\{X, X \to X\}$. The translation $\underline{\mathbf{0}}$ of $\mathbf{0}$ (resp. $\underline{\mathbf{S}}$ of $\mathbf{S}$) is then simply $\underline{\mathbf{nat}}^{[1]}$ (resp. $\underline{\mathbf{nat}}^{[2]}$).

Translating **list** is not very different. Being of arity 1, with two associated constructor symbols (**nil** and **cons**), **list** is mapped to the already seen parametrized inductive type $\underline{\mathbf{list}} = \lambda[A:T].\mathrm{Ind}(\star)\{X, A \to X \to X\}$. Translation of constructors is done the same way. We just need to care about curryfication of symbols, and to replace sort variables with CIC type variables.

Finally, defined symbols are mapped to CIC defined symbols, after translating their type appropriately.

2.4 Building in a first-order theory

We now start describing our new calculus CCIC.

Terms. CCIC uses the same set of sorts $\mathscr{S} = \{\star, \Box, \triangle\}$ and sets of variables $\mathscr{X} = \mathscr{X}^\star \cup \mathscr{X}^\Box$ of CIC. For any sort $\sigma \in \Lambda$, let $\mathscr{X}_\sigma \subseteq \mathscr{X}^\star$ a infinite set of variables of sort σ s.t. $\{\mathscr{X}_\sigma\}_\sigma$ is a family of pairwise disjoint sets. We also assume that $\mathscr{X} - \bigcup_\sigma \mathscr{X}_\sigma$ is infinite.

Let $\mathscr{A} = \{\mathrm{r}, \mathrm{u}\}$ a set of two constants, called *annotations*, totally ordered by $\mathrm{u} \prec_{\mathscr{A}} \mathrm{r}$, where r stands for *restricted* and u for *unrestricted*. We use a for an arbitrary annotation. The role of annotations will be explained later.

Definition 4 (Pseudo-terms of CCIC). Given a parametric sorted signature (Λ, Σ), the algebra $\mathscr{L}$ of *pseudo-terms* of CCIC is defined as:

$$t, u, T, U, \ldots := s \in \mathscr{S} \mid x \in \mathscr{X} \mid \forall (x :^a T).t \mid \lambda[x :^a T].t \mid t\,u \mid f \in \Sigma \mid \sigma \in \Lambda$$
$$\mid \doteq \mid \mathrm{Eq}_T(t) \mid \mathrm{Ind}(X:t)\{\overline{T_i}\} \mid t^{[n]} \mid \mathrm{Elim}(t : T\,[\overline{u_i}] \to U)\{\overline{w_j}\}$$

In order to make definitions more convenient, we assume in the following that Λ contains the symbols $\doteq$, **nat** and **list**, and that Σ contains the symbols $\mathbf{0}, \mathbf{S}$ and Eq.

Compared with CIC, the differences are:

- the internalization of the first-order symbols,
- the internalization of the equality predicate:
  - $t \doteq_T u$ denotes the equality of the two terms (of type T) t and u,
  - $\mathrm{Eq}_T(t)$ represents the reflexivity proof of $t \doteq_T t$.
- annotations in products and abstractions are used to control the formation of applications as it can be seen from the new [APP] rule given at Figure 2.

Notation 2.1 *When x is not free in t, $\forall(x :^a T).t$ is written $T \to^a t$. The default annotation, when not specified in a product or abstraction, is the* unrestricted *one.*

As usual, there is a layered set of syntactic classes for $\mathscr{L}$:

Definition 5 (Syntactic classes). The pairwise disjoint syntactic classes of CCIC called *objects* ($\mathscr{O}$), *predicates* ($\mathscr{P}$), *kinds* ($\mathscr{K}$), *kinds predicates* ($\mathscr{M}$), and $\triangle$ are defined as usual:

- $\mathscr{O} ::= \mathscr{X}^\star \mid f \in \Sigma \mid \mathscr{O}\,\mathscr{O} \mid \mathscr{O}\,\mathscr{P} \mid \lambda[x^\star :^a \mathscr{P}].\mathscr{O} \mid \lambda[x^\square :^a \mathscr{K}].\mathscr{O} \mid \mathrm{Elim}(\mathscr{O} : \mathscr{P}[\overline{\mathscr{O}}] \to \mathscr{O})\{\overline{\mathscr{O}}\}$
- $\mathscr{P} ::= \mathscr{X}^\square \mid \sigma \in \Lambda \mid \mathscr{P}\,\mathscr{O} \mid \mathscr{P}\,\mathscr{P} \mid \lambda[x^\star :^a \mathscr{P}].\mathscr{P} \mid \lambda[x^\square :^a \mathscr{K}].\mathscr{P}$
  $\mid \mathrm{Elim}(\mathscr{O} : \mathscr{P}[\overline{\mathscr{O}}] \to \mathscr{P})\{\overline{\mathscr{P}}\} \mid \forall(x^\star :^a \mathscr{P}).\mathscr{P} \mid \forall(x^\square :^a \mathscr{K}).\mathscr{P}$
- $\mathscr{K} ::= \star \mid \mathscr{K}\,\mathscr{O} \mid \mathscr{K}\,\mathscr{P} \mid \lambda[x^\star :^a \mathscr{P}].\mathscr{K} \mid \lambda[x^\square :^a \mathscr{K}].\mathscr{K} \mid \forall(x^\star :^a \mathscr{P}).\mathscr{K} \mid \forall(x^\square :^a \mathscr{K}).\mathscr{K}$
- $\mathscr{M} ::= \square \mid \forall(x^\star :^a \mathscr{P}).\mathscr{M} \mid \forall(x^\square :^a \mathscr{K}).\mathscr{M}$
- $\triangle ::= \triangle$

This enumeration defines a successor function $+1$ on classes ($\mathscr{O}+1 = \mathscr{P}$, $\mathscr{P}+1 = \mathscr{K}$, $\mathscr{K}+1 = \mathscr{M}$, $\mathscr{M}+1 = \triangle$). We also define $\mathrm{Class}(t) = \mathscr{D}$ if $t \in \mathscr{D}$ and $\mathscr{D} \in \{\mathscr{O}, \mathscr{P}, \mathscr{K}, \mathscr{M}, \triangle\}$.

From now on, we only consider *well-constructed terms* (i.e. terms whose class is not $\perp$) and *well-constructed substitution* (i.e. substitutions s.t. $\mathrm{Class}(x) = \mathrm{Class}(x\theta)$ for any x in its domain). It is easy to check that if t is a well-constructed term and θ a well-constructed substitution, then $\mathrm{Class}(t) = \mathrm{Class}(t\theta)$. It is also well-known that $\xrightarrow{\beta\iota}$-reduction preserves term classes.

Definition 6 (Pseudo-contexts of CCIC). The typing environments of CIC are defined as $\Gamma, \Delta ::= [] \mid \Gamma, [x :^a T]$ s.t. a variable cannot be declared twice. We use $\mathrm{dom}(\Gamma)$ for the domain of Γ and $x\Gamma$ for the type associated to x in Γ.

The rules defining the CCIC typing judgment $\Gamma \vdash t : T$ are the same as for CIC except the rules for application and conversion given at Figure 2.

$$\frac{\Gamma \vdash t : \forall(x :^a U).V \quad \Gamma \vdash u : U \quad \text{if } a = \mathrm{r} \text{ and } U \xrightarrow{\beta}_* t_1 \doteq_T t_2 \text{ with } t_1, t_2 \in \mathscr{O} \text{ then } t_1 \sim_\Gamma t_2 \text{ must hold}}{\Gamma \vdash t\,u : V\{x \mapsto u\}}\ [\textsc{App}]$$

$$\frac{\Gamma \vdash t : T \quad \Gamma \vdash T' : s' \quad T \sim_\Gamma T'}{\Gamma \vdash t : T'}\ [\textsc{Conv}]$$

Fig. 2 CCIC modified typing rules

2.5 Conversion

We are now left with defining the conversion relation $\sim_\Gamma$, whose definition needs some preparation, since:

- conversion is defined on CCIC terms, but the first-order decision procedures operate on algebraic terms. We therefore need to translate CCIC terms into algebraic terms, a process we call *algebraisation*.
- conversion will operate on weak terms only, a notion introduced in Section 2.5. Non-weak terms will be converted with $\beta\iota$-reduction only, to forbid lifting up inconsistencies from the object level to the type level. This is crucial to avoid breaking strong normalization, and therefore decidability of type-checking in presence of inconsistent user's assumptions.

Algebraisation. Our calculus has a complex notion of computation reflecting its rich structure made of three ingredients: the typed lambda calculus, the inductive types with their recursors and the integration of the first order theory $\mathscr{T}$ in its conversion. To achieve this integration, goals are sent to the first order theory $\mathscr{T}$ together with a set of proof hypotheses extracted from the current context.

Algebraisation is the first step of this extraction: it allows transforming a CCIC term into its first-order counterpart. We illustrate this with an example, $\mathscr{T}$ being Presburger's arithmetic.

We begin by the simplest case, directly taken from $\mathrm{CC}_{\mathbb{N}}$, the extraction of pure algebraic, non parametric, equations. Suppose that the proof environment contains equations of the form $c \doteq 1+d$ and $d \doteq 2$ with c and d variables of sort **nat**. What is expected is that the set of hypotheses sent to the theory $\mathscr{T}$ contains the two well formed $\mathscr{T}$-formulas $c = 1+d$ and $d = 2$. This leads to a first definition of equations extraction:

1. a term is algebraic if it is of the form 0, or St, or $t+u$, or $x \in \mathscr{X}_{\mathbb{N}}$. The *algebraisation* $\mathscr{A}(t)$ of an algebraic term is then defined by induction: $\mathscr{A}(0) = 0$, $\mathscr{A}(St) = S(\mathscr{A}(t))$, $\mathscr{A}(t+u) = \mathscr{A}(t) + \mathscr{A}(u)$ and $\mathscr{A}(x_{\mathbb{N}}) = x_{\mathbb{N}}$,
2. a term is an extractable equation if it is of the form $t \doteq u$ with t and u algebraic terms. The extracted equation is then $\mathscr{A}(t) = \mathscr{A}(u)$.

The definition becomes harder for parametric signatures. The theory of lists gives us a paradigmatic example. From the definition of embedding a polymorphic multi-sorted algebra into CIC, we know that the symbol @ has $\forall(T:\star).\, \mathbf{list}\, T \to \mathbf{list}\, T \to \mathbf{list}\, T$ for type. Thus, a fully applied, well formed term having the symbol @ at head position must be of the form $(@\, T\, l1\, l2)$, T being the type of the elements of the lists $l1$ and $l2$. Algebraisation of such a term will erase all type parameters: in our example, $\mathscr{A}(@\, T\, l1\, l2) = @(\mathscr{A}(l1), \mathscr{A}(l2))$.

Algebraisation of non-pure algebraic terms is done by abstracting non-algebraic subterms with fresh variables. For example, algebraisation of $1+t$ with t non-algebraic will lead to $1 + x_{\mathbf{nat}}$ where $x_{\mathbf{nat}}$ is an abstraction variable of sort **nat** for t. Of course, if the proof context contains two equations of the form $c \doteq 1+t$ and $d \doteq 1+u$ with t and u $\beta\iota$-convertible, t and u should be abstracted by a unique variable so that $c = d$ can be deduced in $\mathscr{T}$ from $c = 1 + y_{\mathbf{nat}}$ and $d = 1 + y_{\mathbf{nat}}$. The problem is harder for:

- *parametric symbols*: in $(\mathbf{cons}\, T\, t\, (\mathbf{nil}\, U))$ with t non algebraic, should t be abstracted by a variable of sort **nat** or **list**(**nat**) ?
- *ill-formed terms*: should $(\mathbf{cons}\, T\, 0\, (\mathbf{cons}\, T\, (\mathbf{nil}\, U)\, (\mathbf{nil}\, T)))$ be abstracted as a list of natural numbers or as a list of lists ?

Our solution is to postpone decisions: $\mathscr{A}(t)$ will be a function from Λ to the terms of $\mathscr{T}$ s.t. $\mathscr{A}(t)(\sigma)$ is the algebraisation of t under the condition that t is a CCIC representation of a first order term of sort σ.

We now give the formal definition of $\mathscr{A}(\cdot)$. We assume:

- a Λ-sorted family $\{\mathscr{Y}_\sigma\}_\sigma$ of pairwise disjoint countable infinite sets of variables of sort σ. Let $\mathscr{Y} = \bigcup_\sigma \mathscr{Y}_\sigma$;

- for any equivalence relation $\mathscr{R}$ and sort $\sigma \in \Lambda$, we assume a function $\pi^\sigma_{\mathscr{R}}$: $\mathrm{CCIC}(\mathscr{X}) \to \mathscr{Y}_\sigma$ s.t. $\pi^\sigma_{\mathscr{R}}(t) = \pi^\sigma_{\mathscr{R}}(u)$ if and only if $t\, \mathscr{R}\, u$ (i.e. $\pi^\sigma_{\mathscr{R}}(t)$ is the element of $\mathscr{Y}_\sigma$ representing the class of t modulo $\mathscr{R}$).

Definition 7 (Well applied term). A term is well applied if it is of the form $f\,[\overline{T_\alpha}]_{\alpha\in\overline{\alpha}}\,t_1\cdots t_n$ with $f:\forall\overline{\alpha}.\,\sigma_1\times\cdots\times\sigma_n\rightarrow\sigma$.

Example 3. Example of well applied terms are 0, $S\,t$, or $\mathbf{cons}\,T\,x\,l$, T being the type parameter here. Note that we do not require the term to be well formed.

In case of partial symbols, such as **car** for lists, this definition must be changed slightly by adding a new argument, the proof that the input satisfies the appropriate guard, here that it is not **nil**.

Definition 8 (Algebraisation). The *algebraisation of $t\in$ CCIC modulo an equivalence relation $\mathscr{R}$* is the function $\mathscr{A}_{\mathscr{R}}(t):\Lambda\rightarrow\mathscr{T}(\mathscr{X}^\star\cup\mathscr{Y})$ defined by:

$$\begin{aligned}\mathscr{A}_{\mathscr{R}}(x_\sigma)(\sigma) &= x_\sigma\\ \mathscr{A}_{\mathscr{R}}(f\overline{T}\,[u_i]_{i\in n})(\tau\xi) &= f(\mathscr{A}_{\mathscr{R}}(u_1)(\sigma_1\xi),\ldots,\mathscr{A}_{\mathscr{R}}(u_n)(\sigma_n\xi))\\ \mathscr{A}_{\mathscr{R}}(t)(\tau) &= \pi^\tau_{\mathscr{R}}(t)\quad\text{otherwise}\end{aligned}$$

where $f:\forall\overline{\alpha}.\,\sigma_1\times\cdots\sigma_n\rightarrow\sigma$, $f\overline{T}\,[u_i]_{i\in n}$ is well applied, and ξ is a Λ-substitution.

For any relation R, $\mathscr{A}_R$ is defined as $\mathscr{A}_{\mathscr{R}}$ where $\mathscr{R}$ is the smallest equivalence relation containing R. We call *σ-alien* (or *alien* when the context is clear) a subterm of t abstracted by a variable in $\mathscr{Y}_\sigma$, and say that t is *algebraic* w.r.t. σ if contains no σ-alien. We denote by $\mathscr{A}\mathrm{lg}_\sigma$ the set of algebraic terms w.r.t. σ, and by $\mathscr{A}\mathrm{lg}=\bigcup_{\sigma\in\Lambda}\mathscr{A}\mathrm{lg}_\sigma$ the set of algebraic terms.

Example 4. Let $t\equiv\mathbf{cons}\,T\,\mathbf{0}\,(\mathbf{cons}\,U\,(\mathbf{nil}\,V)\,(\mathbf{nil}\,U))$, R be a relation on CCIC terms, $\sigma=\mathbf{list}(\mathbf{nat})$, and $x_{\mathbf{nat}},y_{\mathbf{list}},z_{\mathbf{nat}},x_\alpha$ and y_α be abstraction variables. Then:

$$\begin{aligned}\mathscr{A}_R(t)(\sigma) &= \mathbf{cons}(\mathscr{A}_R(\mathbf{0})(\mathbf{nat}),\mathscr{A}_R(\mathbf{cons}\,U\,(\mathbf{nil}\,V)\,(\mathbf{nil}\,U))(\sigma))\\ &= \mathbf{cons}(0,\mathbf{cons}(\mathscr{A}_R(\mathbf{nil}\,V)(\mathbf{nat}),\mathscr{A}_R(\mathbf{nil}\,U)(\sigma)))=\mathbf{cons}(0,\mathbf{cons}(x_{\mathbf{nat}},\mathbf{nil}))\\ \mathscr{A}_R(t)(\mathbf{list}(\sigma)) &= \mathbf{cons}(\mathscr{A}_R(\mathbf{0})(\sigma),\mathscr{A}_R(\mathbf{cons}U\,(\mathbf{nil}\,V)\,(\mathbf{nil}\,U))(\mathbf{list}(\sigma)))\\ &= \mathbf{cons}(y_{\mathbf{list}},\mathbf{cons}(\mathscr{A}_R(\mathbf{nil}\,V)(\sigma),\mathscr{A}_R(\mathbf{nil}\,U)(\mathbf{list}(\sigma))))=\mathbf{cons}(y_{\mathbf{list}},\mathbf{cons}(\mathbf{nil},\mathbf{nil}))\end{aligned}$$

$\mathscr{A}_R(t)(\mathbf{list}(\alpha))=\mathbf{cons}(x_\alpha,\mathbf{cons}(y_\alpha,\mathbf{nil}))$ and $\mathscr{A}_R(t)(\mathbf{nat})=z_{\mathbf{nat}}$.

It is clear from the above example that the algebraisation of a term depends on the expected sort of the result: when abstracting the (heterogeneous and ill-formed) list $0::\mathbf{nil}::\mathbf{nil}$ as a list of lists, 0 is seen as an alien which must be abstracted. When this list is abstracted as a list of natural numbers or as a polymorphic list, 0 is considered algebraic and the first occurrence of **nil** as an alien to be abstracted. Finally, if the list is algebraised as a natural number, it is abstracted by a variable.

Weak terms. We first distinguish a class of terms called *weak*. This class of terms will play an important role in the following as they restrict the interaction between the conversion at object level and the strong ι-reduction.

An example of non weak term is

$$t=\lambda[x:\mathbf{nat}].\,\mathrm{Elim}^{\mathscr{S}}(x:\mathbf{nat}\,[]\rightarrow Q)\{\mathbf{nat},\lambda[x:\mathbf{nat}][T:Q\,x].\,\mathbf{nat}\rightarrow\mathbf{nat}\}$$

Such a term is problematic in the sense that when applied to convertible terms, it can $\beta\iota$-reduce to type-level terms that are not $\beta\iota$-convertible. Suppose that the conversion relation is canonically extended to CCIC. Assume a typing environment Γ s.t. $\mathbf{0}\sim_\Gamma\mathbf{S0}$, and hence, by congruence, $t\,\mathbf{0} \sim_\Gamma t\,(\mathbf{S}\,\mathbf{0})$. Now, it is easy to check that $t\,\mathbf{0} \xrightarrow{\beta\iota}_* \mathbf{nat}$ and $t\,(\mathbf{S}\,\mathbf{0}) \xrightarrow{\beta\iota}_* (\mathbf{nat} \to \mathbf{nat})$. Strong normalization of β-reduction is then broken by encoding the term $\omega = \lambda[x : \mathbf{nat}].xx$.

In contrast, *weak* terms lift no inconsistencies from object level to a higher level:

Definition 9 (Weak terms). A term is *weak* if it contains no i) applied type-level variable, and ii) term of the form $\mathrm{Elim}(t : I\,[\overline{u}] \to Q)\{\overline{f}\}$ with t open.

Extractable terms. From now on, let $\mathscr{O}^+$ be an arbitrary set of CCIC terms. This set will be used in the conversion definition to restrict the set of *extractable equations* of a given environment: only equation of the form $t \doteq u$ with t and u in $\mathscr{O}^+$ will be considered.

At the moment, we only require $\mathscr{O}^+$ to be a subset of $\mathscr{O}$. Note that taking $\mathscr{O}^+ = \mathscr{O}$ does not compromise the standard calculus properties (subject reduction, type unicity, strong normalization of $\beta\iota$-reduction, ...) but the decidability. E.g., if $\mathscr{T}$ is the Presburger arithmetic, allowing the extraction of

$$\lambda[x :^a \mathbf{nat}].\,f\,x \doteq \lambda[x :^a \mathbf{nat}].\,f\,(x \dot{+} 2)$$

would require - for checking conversion - to decide any statement of the form

$$\mathscr{T} \models (\forall x.\, f(x) = f(x+2)) \to t = u,$$

which is well known to be impossible.

Conversion relation. We have now all necessary ingredients to define our conversion relation $\sim_\Gamma$:

Definition 10 (Conversion relation). Rules of Figure 3 define a family $\{\sim_\Gamma\}$ of CCIC binary relations indexed by a (non-necessarily well-formed) context Γ.

Note that the rule [DED] performing deductions in the first order theory, here Presburger arithmetic, outputs a certificate $[\_,\_,\_]$ made of the environment and the two terms to be proved equivalent under this environment, each time it is called. While this certificate must depend on these three data, it may of course carry additional information depending on the considered first-order theory.

The main differences with the calculus $\mathrm{CC}_\mathbb{N}$ defined in [6] are the following:

- The [APP] rule has been split into two rules: [APP$^\mathscr{S}$] and [APP$^\mathscr{W}$]. Conversion for strong terms is restricted to $\beta\iota$-conversion.
- Conversion for the first argument of an Elim is restricted to $\beta\iota$-conversion.
- The rules for transitivity and symmetry have been removed, which eases the proofs, notably that the deduction part of the conversion relation works at object level only. We prove later that the conversion relation is transitive and symmetric on well formed terms, thus recovering type unicity.
- The rules for $\beta\iota$-conversion perform one reduction step only, which also eases proofs. Therefore $u \xleftrightarrow{\beta\iota}_* v$ should be understood as $\exists w$ s.t. $u \xrightarrow{\beta\iota} w$ and $v \xrightarrow{\beta\iota} w$.

$$\frac{}{t \sim_\Gamma t}\ [\text{REFL}] \qquad \frac{[x :^{\mathrm{r}} T] \in \Gamma \quad T \xrightarrow{\beta\iota}_* t \doteq u \quad t,u \in \mathscr{O}^+}{t \sim_\Gamma u}\ [\text{EQ}]$$

$$\frac{T \sim_\Gamma U \quad t \sim_{\Gamma,[x:^a T]} u}{\lambda[x :^a T].t \sim_\Gamma \lambda[x :^a U].u}\ [\text{LAM}] \qquad \frac{T \sim_\Gamma U \quad t \sim_{\Gamma,[x:^a T]} u}{\forall(x :^a T).t \sim_\Gamma \forall(x :^a U).u}\ [\text{PROD}]$$

$$\frac{t \xrightarrow{\beta\iota} t' \quad t' \sim_\Gamma u}{t \sim_\Gamma u}\ [\beta\iota\text{-LEFT}] \qquad \frac{\begin{array}{c} t,t',f,f' \text{ are weak} \\ t \xleftrightarrow{\beta\iota}_* t' \quad I \sim_\Gamma I' \quad Q \sim_\Gamma Q' \quad \overline{v} \sim_\Gamma \overline{v'} \quad \overline{f} \sim_\Gamma \overline{f'} \end{array}}{\mathrm{Elim}(t : I[\overline{v}] \to Q)\{\overline{f}\} \sim_\Gamma \mathrm{Elim}(t' : I'[\overline{v'}] \to Q')\{\overline{f'}\}}$$

$$\frac{u \xrightarrow{\beta\iota} u' \quad t \sim_\Gamma u'}{t \sim_\Gamma u}\ [\beta\iota\text{-RIGHT}] \qquad \frac{t_1 \sim_\Gamma u_1 \quad t_2 \sim_\Gamma u_2 \quad t_i, u_i \text{ are weak}}{t_1 t_2 \sim_\Gamma u_1 u_2}\ [\text{APP}^{\mathscr{W}}]$$

$$\frac{\begin{array}{c} E \vDash \mathscr{A}_{\sim_\Gamma}(t)(\tau) = \mathscr{A}_{\sim_\Gamma}(u)(\tau) \quad t,u \in \mathscr{O}^+ \\ E = \{\mathscr{A}_{\sim_\Gamma}(w_1)(\sigma) = \mathscr{A}_{\sim_\Gamma}(w_2)(\sigma) \\ \mid w_1 \sim_\Gamma w_2, \sigma \in \Lambda, w_1, w_2 \in \mathscr{O}^+\} \end{array}}{t \sim_\Gamma u \quad [\Gamma,t,u]}\ [\text{DED}]$$

Fig. 3 CCIC conversion relation

2.6 Decidability of type-checking

CCIC enjoys all needed meta-theoretical properties (strong normalization, confluence, subject reduction), and therefore consistency follows:

Theorem 1. *There is no proof of* $\forall(x : \star).x$ *in the empty environment.*

All proofs are similar to those made for PTSs with the same succession of meta-theoretical lemmas, but need more preparation. This is in particular the case with the substitution lemma which is much harder than usual.

As said, type-checking in a dependent type theory is non-trivial, since the rule [CONV] is not syntax-oriented. The classical solution to this problem is to eliminate [CONV] and replace [APP] by the following rule.The proof is not difficult.

$$\frac{\begin{array}{c} \Gamma \vdash t : \forall(x :^a U).V \quad \Gamma \vdash u : U' \quad U \sim_\Gamma U' \\ \text{if } a = \mathrm{r} \text{ and } U \xrightarrow{\beta}_* t_1 \doteq_\Gamma t_2 \text{ with } t_1, t_2 \in \mathscr{O} \text{then } t_1 \sim_\Gamma t_2 \text{ must hold} \end{array}}{\Gamma \vdash tu : V\{x \mapsto u\}}\ [\text{APP}]$$

Decidability of type-checking in *CCIC* therefore reduces to decidability of $\sim_\Gamma$, the environment Γ being arbitrary, possibly containing ill-formed terms or even being inconsistent. To show that $\sim_\Gamma$ is decidable, we proceed as previously, by modifying the definition in order to make it syntax-oriented: we show that two arbitrary terms are convertible iff their $\beta\iota$-normal forms are convertible by the syntax-oriented *weak convertibility* relation $\approx_\Gamma$ given at Figure 4, in which, to any environment Γ, we associate the set $\mathrm{Eq}(\Gamma) = \{t = u \mid [x :^{\mathrm{u}} T] \in \Gamma, x\Gamma \to_* t \doteq u, t, u \in \mathscr{A}\}$.

Lemma 1. *Given Γ an environment and t,u two terms, $t \sim_\Gamma u$ iff $t \downarrow_{\beta\iota} \approx_\Gamma u \downarrow_{\beta\iota}$.*

This is the main technical result of the decidability proof, which proceeds by induction on the definition of $\sim_\Gamma$. Note that the numerous conditions of the form $\mathscr{T},\mathrm{Eq}(\Gamma) \not\models 0 = 1$ in the rules defining $\approx_\Gamma$ are required to make them mutually exclusive.

$$\frac{}{\star \approx_\Gamma \star}\,[\textsc{Refl-}\star] \qquad \frac{}{\square \approx_\Gamma \square}\,[\textsc{Refl-}\square] \qquad \frac{x \in \mathscr{X} \quad \mathscr{T},\mathrm{Eq}(\Gamma) \not\models 0 = 1 \text{ or } x \notin \mathscr{X}^\star}{x \approx_\Gamma x}\,[\textsc{Refl-}\mathscr{X}]$$

$$\frac{t,u \in \mathscr{O} \quad \mathscr{T},\mathrm{Eq}(\Gamma) \models 0 = 1}{t \approx_\Gamma u}\,[\textsc{Unsat}] \qquad \frac{\begin{array}{c} T \approx_\Gamma U \quad t \approx_{\Gamma,[x:^a T]} u \\ \mathscr{T},\mathrm{Eq}(\Gamma) \not\models 0 = 1 \text{ or} \\ \lambda[x:^a T].t \text{ and } \lambda[x:^a U].u \text{ not in } \mathscr{O} \end{array}}{\lambda[x:^a T].t \approx_\Gamma \lambda[x:^a U].u}\,[\textsc{Lam}]$$

$$\frac{T \approx_\Gamma U \quad t \approx_{\Gamma,[x:^a T]} u}{\forall(x:^a T).t \approx_\Gamma \forall(x:^a U).u}\,[\textsc{Prod}] \qquad \frac{\begin{array}{c} t = t' \quad I \approx_\Gamma I' \quad Q \approx_\Gamma Q' \quad \overline{v} \approx_\Gamma \overline{v'} \quad \overline{f} \approx_\Gamma \overline{f'} \\ t,t',\overline{f},\overline{f'} \text{ are weak} \mathscr{T},\mathrm{Eq}(\Gamma) \not\models 0 = 1 \text{ or} \\ \mathrm{Elim}(t,\ldots)\{\ldots\} \text{ and } \mathrm{Elim}(t',\ldots)\{\cdots\} \text{ not in } \mathscr{O} \end{array}}{\mathrm{Elim}(t : I[\overline{v}] \to Q)\{\overline{f}\} \approx_\Gamma \mathrm{Elim}(t' : I'[\overline{v'}] \to Q')\{\overline{f'}\}}\,[\mathscr{W}]$$

$$\frac{\begin{array}{c} t_1 \equiv u_1 \quad t_2 \equiv u_2 \\ \mathscr{T},\mathrm{Eq}(\Gamma) \not\models 0 = 1 \text{ or} \\ t_1\, t_2 \text{ and } u_1\, u_2 \text{ not in } \mathscr{O} \\ t_1\, t_2 \text{ or/and } u_1\, u_2 \text{ is not weak} \end{array}}{t_1\, t_2 \approx_\Gamma u_1\, u_2}\,[\textsc{App}^{\mathscr{S}}] \qquad \frac{\begin{array}{c} t_1 \approx_\Gamma u_1 \quad t_2 \approx_\Gamma u_2 \quad t_i, u_i \text{ weak} \\ \mathscr{T},\mathrm{Eq}(\Gamma) \not\models 0 = 1 \text{ or} \\ t_1\, t_2 \text{ and } u_1\, u_2 \text{ not in } \mathscr{O} \end{array}}{t_1\, t_2 \approx_\Gamma u_1\, u_2}\,[\textsc{App}^{\mathscr{W}}]$$

$$\frac{\begin{array}{c} \mathscr{T},\mathrm{Eq}(\Gamma) \not\models 0 = 1) \\ t = C_t[a_1,\ldots,a_k] \quad u = C_u[a_{k+1},\ldots,a_{k+l}] \\ C_t \text{ or } C_u \text{ is a non-empty algebraic context} \\ \text{all the } a_i\text{'s have empty algebraic caps} \\ \text{the } c_i\text{'s are fresh constants s.t. } c_i = c_j \text{ iff } a_i \approx_\Gamma b_j \\ \mathscr{T},\mathrm{Eq}(\Gamma) \models C_t[c_1,\ldots,c_k] = C_u[c_{k+1},\ldots,c_{k+l}] \end{array}}{t \approx_\Gamma u}\,[\textsc{Ded}]$$

Fig. 4 CCIC syntax-oriented conversion

Example 5. Let $\Gamma = [c : \mathbf{nat}], [p :^r (\lambda[x : \mathbf{nat}].x)\,\mathbf{0} \doteq c]$. Then $(\lambda[x : \mathbf{nat}].x + x)\,\mathbf{0} \approx_\Gamma c$ and $(\lambda[x : \mathbf{nat}].x + x)\,\mathbf{0} \approx_\Gamma c$, using congruence and deduction of $\sim_\Gamma$ and $\approx_\Gamma$.

In contrast, β-reducing $(\lambda[x : \mathbf{nat}].x + x)\,\mathbf{0}$ yields $\mathbf{0} \dotplus \mathbf{0} \sim_\Gamma c$, but not $\mathbf{0} \dotplus \mathbf{0} \approx_\Gamma c$. Indeed, $(\lambda[x : \mathbf{nat}].x \dotplus x)\,\mathbf{0}$ and $\mathbf{0} \dotplus \mathbf{0}$ are no more $\approx_\Gamma$-convertible, a direct consequence of removing $\beta\iota$-reduction from $\sim_\Gamma$: the equation $(\lambda[x : \mathbf{nat}].x)\,\mathbf{0} \doteq c$ cannot be used anymore, since $\mathbf{0} \dotplus \mathbf{0}$ is not $\approx_\Gamma$ convertible to $(\lambda[x : \mathbf{nat}].x)\,\mathbf{0})$.

Now, normalizing all terms as well as the environment Γ, we can recover convertibility for $\approx$: $\mathbf{0} \dotplus \mathbf{0} \approx_{\Gamma_{\downarrow\beta\iota}} c$, the extractable equation of $\Gamma_{\downarrow\beta\iota}$ being now $\mathbf{0} \doteq c$.

As a consequence, we obtain:

Theorem 2. *$\sim_\Gamma$ is decidable for any environment Γ when taking for $\mathscr{O}^+$ the set of terms that are reducible to an algebraic terms.*

and therefore, our main result follows:

Theorem 3. *The type-checking relationship $\Gamma \vdash t : T$ is decidable in CCIC.*

3 Using CCIC

We give here a detailed example illustrating the advantages of CCIC, based on the inductive type of words introduced in Section 2.1.

In Coq. First, we give a development in Coq, therefore based on CIC.

```
Variable T : Set.

Inductive word : nat -> Set :=
| epsilon : word 0
| char : T -> word 1
| append : forall n p, word n -> word p -> word (n+p).

Lemma plus_n_0_transparent : forall n, n+0=n.
Proof. induction n as [| n IHn]; simpl;
  [idtac | rewrite -> IHn]; trivial. Defined.

Lemma plus_n_Sm_transparent: forall n m, n+(S m)=S(n+m).
Proof. intros n m; induction n as [| n IHn];
  simpl; [idtac | rewrite -> IHn]; trivial. Defined.

Lemma plus_assoc_transparent: forall n p q, (n+p)+q=n+(p+q).
Proof. intros n p q; elim n; [trivial | intros k].
  simpl; intros H; rewrite -> H; trivial. Defined.

Definition reverse_acc : forall n, word n -> forall p, word p -> word (p+n).
Proof. intros n wn; induction wn as [| c | n p wn IHwn wp IHwp];
  intros k wk. rewrite plus_n_0_transparent; exact wk.
  rewrite plus_n_Sm_transparent; rewrite plus_n_0_transparent;
    exact (append (char c) wk).
  rewrite <- plus_assoc_transparent; exact (IHwp _ (IHwn _ wk)). Defined.

Fixpoint reverse n (w : word n) {struct w} : word n :=
match w in word k return word k with
| epsilon => epsilon
| char c => char c
| append n1 n2 w1 w2 => reverse_acc w2 w1 end.
```

The example of *palindromes* as words satisfying the property `word_eq m reverse m` is carried out in Strub's thesis (see his website). It yields a much more complex Coq development than the above, since it involves the equality over (quotients) of words.

In CCIC. We now make the similar development in CCIC, using a self-explanatory syntax. The definition of `reverse` reduces then to:

```
Fixpoint reverse n (w : word n) {struct w} : word n := match w with
| epsilon          => epsilon
| char c           => char c
| append _ _ w1 w2 => append (reverse w2) (reverse w1) end.
```

Typing of the third clause of reverse will use here Presburger's arithmetic, since `append n1 n2 w1 w2` has type `word (n1 + n2)`, while `append n2 n1 w2 w1` has type `word (n2 + n1)`, two types that are not convertible in CIC, but which become convertible in CCIC. We can easily see with this example the immense benefit brought by internalizing Presburger's arithmetic. Note that a single certificate is generated for this conversion:

```
[n1 : nat, n2: nat, w1 : word n1, w2: word n2, n1 + n2, n2 + n1]
```

4 Conclusion

CCIC is an extension of CIC by arbitrary first-order decision procedures for equality. We have shown here with a detailed example using Presburger's arithmetic the benefit of the approach with respect to the current implementation of Coq based on CIC: more terms can be typed especially in presence of types such as dependent lists which become easy to use; many proofs get automated, making the life of the user easier (developing the example of reverse for dependent lists in the currently distributed version of Coq took us a day of work, and we don't believe this can be shrinked to one hour); and proofs are much smaller, some seemingly complex proofs becoming simple reflexivity proofs. We believe that the resulting style of proofs becomes much closer to that of the working mathematician.

We have also explained the advantage of the approach insofar as it allows to clearly separate computation from deduction, therefore allowing for an incremental development of the kernel of the system.

So far, we have considered only decidable -equality- theories. However, thanks to the decidability assumption, a decidable non-equality theory can always be transformed into a decidable equality theory over the type Bool of truth values equipped with its usual operations.

There are still many directions to be investigated. A first is to embed membership equational logic in CIC along the lines of the simpler embedding described here. A second is to consider the case of dependent algebras instead of the simpler parametric algebras. This is a much more difficult question, which requires using a stronger notion of conversion in the main argument of an elimination, but would further help us addressing other weaknesses of Coq.

Finally, we strongly believe that the use of decision procedures outputing certificates when they succeed and explanations when they fail will change our way of making formal, and enlarge the audience of proof assistants.

Acknowledgement. We thank the Coq group for many useful discussions and suggestions, and the referees for their useful remarks.

References

1. H. Barendregt. Lambda calculi with types. In S. Abramski, D. Gabba, and T. Maibaum, editors, *Handbook of Logic in Computer Science*, volume 2. Oxford University Press, 1992.
2. B. Barras. *Auto-validation d'un système de preuves avec familles inductives*. PhD thesis, University of Paris VII, 1999.
3. F. Blanqui. Definitions by rewriting in the calculus of constructions. *Mathematical Structures in Computer Science*, 15(1):37–92, 2005. Journal version of LICS'01.
4. F. Blanqui. Inductive types in the calculus of algebraic constructions. *Fundamenta Informaticae*, 65(1-2):61–86, 2005. Journal version of TLCA'03.
5. F. Blanqui, J.-P. Jouannaud, and M. Okada. The Calculus of Algebraic Constructions. In RTA, *Lecture Notes in Computer Science* 1631:301–316. Springer-Verlag, 1999.
6. F. Blanqui, J. Jouannaud, and P. Strub. Building decision procedures in the calculus of inductive constructions. In *Proceedings 16th CSL 2007. LNCS 4646*, 2007.
7. F. Blanqui, J.-P. Jouannaud, and P.-Y. Strub. A Calculus of Congruent Constructions. Unpublished draft, 2005.

8. A. Bouhoula, J.-P. Jouannaud, and J. Meseguer. Specification and proof in membership equational logic. *Theoretical Comput. Sci.*, 236:35–132, 2000.
9. Coq-Development-Team. *The Coq Proof Assistant Reference Manual - Version 8.0.* INRIA, INRIA Rocquencourt, France, 2004. `http://coq.inria.fr/`.
10. T. Coquand and G. Huet. The Calculus of Constructions. *Information and Computation*, 76(2-3):95–120, 1988.
11. T. Coquand and C. Paulin-Mohring. Inductively defined types. *Colog'-88, International Conference on Computer Logic*, volume 417 of *LNCS*, pages 50–66. Springer-Verlag, 1990.
12. P. Corbineau. *Démonstration automatique en Théorie des Types.* PhD thesis, University of Paris IX, 2005.
13. K. Futatsugi, J. Goguen, J.-P. Jouannaud, and J. Meseguer. Principles of OBJ2. *Proceedings of 12th ACM Conference on Principles of Programming Languages*, 1985.
14. J. H. Geuvers and M. Nederhof. A modular proof of strong normalization for the calculus of constructions. *J. of Functional programming*, 1,2:155–189, 1991.
15. E. Giménez. Structural recursive definitions in type theory. In *Proceedings of ICALP'98*, volume 1443 of *LNCS*, pages 397–408, July 1998.
16. G. Gonthier. The four color theorem in Coq. In *TYPES 2004 International Workshop*, 2004.
17. N. Oury. Extensionality in the calculus of constructions. In *Proceedings 18th TPHOL, Oxford, UK. LNCS 3603*, 2005.
18. C. Paulin-Mohring. Inductive definitions in the system COQ. In *Typed Lambda Calculi and Applications*, pages 328–345. Springer Verlag, 1993. LNCS 664.
19. N. Shankar. Little engines of proof. In G. Plotkin, editor, *Proceedings of the Seventeenth Annual IEEE Symp. on Logic in Computer Science*. IEEE Computer Society Press, 2002.
20. R. E. Shostak. An efficient decision procedure for arithmetic with function symbols. *J. of the Association for Computing Machinery*, 26(2):351–360, 1979.
21. M. Stehr. The Open Calculus of Constructions: An equational type theory with dependent types for programming, specification, and interactive theorem proving (part I and II). *Fundamenta Informaticae* 68(1-2), p. 131-174, 2005.
22. T. Streicher. Investigations into intensional type theory, Habilitation, Münich University, 1993.
23. P.-Y. Strub. *The Calculus of Congruent Inductive Constructions.* PhD thesis, École Polytechnique, 2008.
24. B. Werner. *Une Théorie des Constructions Inductives.* PhD thesis, University Paris VII, 1994.

On Traits and Types in a Java-like Setting

Viviana Bono, Ferruccio Damiani, and Elena Giachino

Dipartimento di Informatica, Università di Torino
{bono,damiani,giachino}@di.unito.it

Abstract. Both single and multiple class-based inheritance are often inappropriate as a reuse mechanism, because classes play two competing roles. Namely, a class is both a *generator of instances* and a *unit of reuse*. *Traits* are composable pure units of behavior reuse, consisting only of methods, that have been proposed as an add-on to single class-based inheritance in order to improve reuse. However, adopting traits as an add-on to traditional class-based inheritance is not enough: classes, besides their primary role of generators of instances, still play the competing role of units of reuse. Therefore, a style of programming oriented to reuse is not enforced by the language, but left to the programmer's skills. Traits have been originally proposed in the setting of dynamically typed language. When static typing is also taken into account, the role of unit of reuse and the role of type are competing, too.
We argue that, in order to support the development of reusable program components, object oriented programming languages should be designed according to the principle that *each software structuring construct must have exactly one role*. We propose a realignment of the class-based object-oriented paradigm by presenting programming language features that separate completely the declarations of object *type*, *behavior* and *generator*. We illustrate our proposal through a core calculus and prove the soundness of the type system w.r.t. the operational semantics.

Key words: Type System, Inheritance, Composition, Flattening.

1 Introduction

It is common opinion that standard class-based inheritance does not support low coupling and, therefore, does not support well code reuse. This phenomenon is often described as the *fragile base-class problem* and it is well-described in the work by Mikhajlov and Sekerinski [20]. A well-known technique to circumvent the fragile base-class problem is to promote the use of interface-based polymorphism. This idea is also present in most of the design patterns, such as the GoF design patterns [14], in order to make the patterns as higher-level as possible with respect to the implementation details.

Class-based inheritance was criticized again recently by Schärli et al. [25, 10], by pointing out that both single and multiple class-based inheritance are often inappropriate as a reuse mechanism. They identify the problem in the fact that classes play two competing roles. Namely, a class is both a *generator of instances* (hence it must provide a *complete* set of basic features) and a *unit of reuse*

Please use the following format when citing this chapter:

Bono, V., Damiani, F. and Giachino, E., 2008, in IFIP International Federation for Information Processing, Volume 273; *Fifth IFIP International Conference on Theoretical Computer Science*; Giorgio Ausiello, Juhani Karhumäki, Giancarlo Mauri, Luke Ong; (Boston: Springer), pp. 367–382.

(hence it should provide a *minimal* set of sensibly reusable features). Schärli et al. also observed that *mixins* [7, 17, 13, 3], which are subclasses parameterized over their superclasses, are not necessarily appropriate for composing units of reuse. The problem is due to the fact that, being based on the ordinary single inheritance operator, mixing composition is linear. Indeed, the formulation of mixins given by Bracha in JIGSAW [6] does not suffer of this problem,[1] but most of the subsequent formulations of the mixin construct do.

To overcome these problems, Schärli et al. proposed *traits*, composable pure units of behavior reuse consisting only of methods, that can be composed in an arbitrary order via operations ensuring that the composite unit (trait or class) has complete control over the composition and must resolve conflicts explicitly. However, both in the original proposal and (to the best of our knowledge) in all the trait-based approaches that can be found in the literature (with the exception of the FORTRESS language proposal [1], currently under development), traits live together with the traditional class-based inheritance. Therefore, besides their primary role of generators of instances, classes can still play the competing role of units of reuse, and a style of programming oriented to reuse is not enforced by the language, but left to the programmer's skills.

The original proposal of Schärli et al. does not address typing issues. Various proposals for using traits in connection with static typing can be found in the literature (we refer to [21] for a brief overview). In some of these proposals (notably in the SCALA [22] and in the FORTRESS [1] languages) each trait, like each class, also defines a type. However, as a matter of fact, the role of unit of reuse and the role of type are competing. For instance, in order be able to define the subtyping relation on traits in such a way that a trait (or a class) is always a subtype of the component traits, SCALA and FORTRESS rule out operations on traits such as method exclusion and renaming, limiting the reuse potential of traits. The distinction between the role of type and the role of unit of reuse, described in terms of type and class, dates back at least to Snyder [27] (see also Cook et al. [9]).

Having in mind the need of promoting interface-based polymorphism and arbitrarily composable units of reuse, we would like to go further and give classes the role of object generators only.

We argue that, in order to support the development of reusable program components, object oriented programming languages should be designed according to the principle that *each software structuring construct must have exactly one role.* We propose programming language features that separate completely the declarations of object *type*, *behavior* and *generator*. Namely, we consider:

- *Interfaces*, as pure types.
- *Traits*, as pure units of behavior reuse.
- *Classes*, as pure generators of instances.

[1] JIGSAW introduces very general operators for module manipulation. Some of them have been later, independently, developed for traits.

Interfaces can be defined by extending other interfaces (the interface hierarchy induces subtyping). Traits can be defined by composing other traits. Classes are defined by composing traits, implementing interfaces, and defining fields.

Note that there are no hierarchical dependencies among classes. Therefore, a first outcome of the complete role separation is that problems of fragility in a class hierarchy (that arise with class-based and mixin-based inheritance) are avoided *a priori*: there is no class hierarchy. Since traits and classes do not define types, another outcome of the complete role separation is that the use of operations like method exclusion and renaming is not limited by the need of ensuring that each trait (or class) is a subtype of the composing traits (see Sect. 2).

Recently, Bergel et al. [4] pointed out several limitations of the trait model. In order to overcome these limitations, they propose (in a SMALLTALK-like setting) to make traits *stateful* by allowing traits to have private fields that, through a variable access operator, may be accessed from the clients possibly under a new name, and possibly merged with other variables. Our proposal provides (in a JAVA-like setting) an alternative solution to the limitations of the stateless trait model. Also, Bergel et al. observed that: "An open question for further study is whether trait composition can subsume class-based inheritance, leading to a programming language based on composition rather than inheritance as the primary mechanism for structuring code following JIGSAW [6] design." Our investigation addresses the previous question by providing a foundation for a realignment of the class-based object-oriented paradigm to support the systematic structuring of code in "single-role" reusable units. Besides their power of reuse, traits have attracted a great deal of attention in the programming language research community because of their simple semantics. We believe that our proposal is a step forward towards simplicity.

A preliminary version of the results presented in this paper appeared as [5].

Organization of the Paper. Section 2 illustrates our proposal through an example. Section 3 presents the syntax of FRJ (a core calculus for reusable units based on the constructs introduced above), outlines its type system and its operational semantics, and states a type soundness result. We conclude by discussing some related work and outlining possible directions for further work.

2 An Example

In this section we provide a simple example of code that cannot lead to unanticipated reuse both in traditional class-based languages and in trait-based languages where a composite trait is a subtype of the component traits, but that can be reused in an unanticipated way in a language based on our proposal. We exploit a standard Java-like notation, in particular we use a more general syntax for constructors than the one that will be presented in Section 3.1.

Consider the task of developing a class `Stack` that implements the interface:

```
interface IStack { boolean isEmpty(); void push(Object o); Object pop(); }
```

In a traditional class-based language (like, e.g., JAVA) it is natural to write a class like:

```
class Stack implements IStack { List l; Stack() { l=new LinkedList(); }
   boolean isEmpty() { return (l.size() == 0); }
   void push(Object o) { l.addFirst(); }
   Object pop() { Object o=l.getFirst(); l.removeFirst(); return o; }
}
```

Suppose that later on it becomes necessary to develop a class `Stack'` that implements the interface:

```
interface IStack' { Boolean isEmpty(); void push(Object o); void pop();
                    Object top(); }
```

In a traditional class-based language there is no straightforward way to reuse the code in class `Stack` and the simplest thing to do is to write a class like:

```
class Stack' implements IStack' { List l; Stack'() { l=new LinkedList(); }
   boolean isEmpty() { return (l.size() == 0); }
   void push(Object o) { l.addFirst(); }
   void pop() { l.removeFirst(); }
   Object top() { return l.getFirst(); }
}
```

To illustrate our proposal, we exploit a JAVA-like syntax (we still do not have an implementation). In a class the only (implicitly) public methods are those declared in the interfaces implemented by the class. All the other methods and the fields are (implicitly) private. All the constructors must be declared and are (implicitly) public. Moreover, for every library class (such as `Object`, `Integer`, etc.) we assume an interface and a trait. The same name can be used to denote the interface, the trait and the class. The `Object` interface is implicity extended by any interface and the `Object` trait is implicity used by any class.

A class `Stack`, whose instance type is the interface `IStack`, can be naturally written by defining separately instance behaviour and generation as follows:

```
trait TStack is { List l;
   boolean isEmpty() { return (l.size() == 0); }
   void push(Object o) { l.addFirst(); }
   Object pop() { Object o=l.getFirst(); l.removeFirst(); return o; } }

class Stack implements IStack by TStack
   { List l; Stack() { l=new LinkedList(); } }
```

A class `Stack'` that implements the interface `IStack'` can be straightforwardly written as follows by defining a trait `TStack'` that reuses the trait `TStack`:

```
trait TStack' is (TStack exclude pop)
               + { List l;  void pop() { l.removeFirst(); }
                            Object top() { return l.getFirst(); } }

class Stack' implements IStack' by TStack'
   { List l; Stack'() { l=new LinkedList(); } }
```

```
ID ::= interface I extends Ī { S̄; }
S  ::= I m (Ī x̄)

TD ::= trait T is TE
TE ::= {F̄; S̄; M̄} | T | TE + TE | TE exclude m | TE alias m as m
        | TE duplicate m as m | TE rename m to m | TE rename f to f
F  ::= I f
M  ::= S { return e; }
e  ::= x | e.f | e.m(ē) | new C(ē) |(I)e

CD ::= class C implements Ī by TE { F̄; K }
K  ::= C(Ī f̄) { this.f̄ = f̄; }
```

Fig. 1 FRJ: Syntax

The trait `TStack'` above can be alternatively defined as follows:

```
trait TStack' is (TStack rename pop to poptop)
                 + { Object poptop(); void push(Object);
                     void pop() { poptop(); }
                     Object top() { Object o=poptop(); push(o); return o; } }
```

Note that, if traits were types and composed traits were subtypes of the component traits, both the declarations of the trait `TStack'` would not typecheck.

3 FRJ: a Calculus for Reusable Units

In this section we provide a formal account of our idea by presenting FRJ (FEATHERWEIGHT REUSABLE JAVA), a minimal core calculus for interfaces, traits and classes, in the spirit of FJ (FEATHERWEIGHT JAVA) [15].

3.1 Syntax

The syntax of our calculus, FRJ, is presented in Fig. 1. We also consider a calculus, FFRJ (FLAT FRJ), obtained by removing the portions of the syntax highlighted in grey.

We use the overbar sequence notation according to [15]. For instance: "$\bar{\mathtt{f}}$" denotes the possibly empty sequence "$\mathtt{f}_1, ..., \mathtt{f}_n$", the pair "$\bar{\mathtt{I}}\,\bar{\mathtt{x}}$" stands for "$\mathtt{I}_1\,\mathtt{x}_1, ..., \mathtt{I}_n\,\mathtt{x}_n$", "$\bar{\mathtt{I}}\,\bar{\mathtt{f}};$" stands for "$\mathtt{I}_1\,\mathtt{f}_1; ...; \mathtt{I}_n\,\mathtt{f}_n;$", and the assignment "$\mathtt{this}.\bar{\mathtt{f}} = \bar{\mathtt{f}};$" stands for "$\mathtt{this.f}_1 = \mathtt{f}_1; ...; \mathtt{this.f}_n = \mathtt{f}_n;$". The empty sequence is denoted by "$\bullet$".

Sequences of named elements (e.g., methods signatures, fields declarations,...) are assumed to contain no duplicate names, the sequence of the names of the elements of $\bar{\mathtt{S}}$ is denoted by *names*($\bar{\mathtt{S}}$), the subsequence of the elements of $\bar{\mathtt{S}}$ with the names $\bar{\mathtt{n}}$ is denoted by *extract*($\bar{\mathtt{n}}, \overline{\mathtt{S}}$), and *discard*($\bar{\mathtt{n}}, \overline{\mathtt{S}}$) denotes the sequence obtained from $\overline{\mathtt{S}}$ by removing the elements with the names $\bar{\mathtt{n}}$. Following [15], we use a set-based notation for operators over sequences of named elements. For instance, $\mathtt{M} = \mathtt{I\,m\,(\bar{I}\,x)\{return\ e\}} \in \bar{\mathtt{M}}$ means that the method declaration $\mathtt{M}$ occurs in $\bar{\mathtt{M}}$. In the union and in the intersection of sequences of named elements, denoted by $\overline{\mathtt{S}} \cup \overline{\mathtt{Z}}$ and $\overline{\mathtt{S}} \cap \overline{\mathtt{Z}}$, respectively, it is assumed that if $\mathtt{n} \in$ *names*($\overline{\mathtt{S}}$) and $\mathtt{n} \in$ *names*($\overline{\mathtt{Z}}$) then *extract*($\mathtt{n}, \overline{\mathtt{S}}$) = *extract*($\mathtt{n}, \overline{\mathtt{Z}}$).

The concatenation of two sequences $\overline{\mathtt{S}}$ and $\overline{\mathtt{Z}}$ is denoted by $\overline{\mathtt{S}}\cdot\overline{\mathtt{Z}}$, where, if $\overline{\mathtt{S}}$ and $\overline{\mathtt{Z}}$ are sequences of named elements, it is assumed that *names*($\overline{\mathtt{S}}$) $\cap$ *names*($\overline{\mathtt{Z}}$) $= \emptyset$.

A class table CT is a map from class names to class declarations. Similarly, an interface table IT and a trait table TT map interface and trait names to interface and trait declarations, respectively. A FRJ program is a 4-tuple (IT, TT, CT, e). In presenting the type system and the flattening translation we assume fixed, global tables IT, TT, and CT. We also assume that these tables are *well-formed*, i.e., they contain an entry for each interface/trait/class mentioned in the program, and the interface subtyping and trait reuse graphs are acyclic.

The distinguishing features of FRJ w.r.t. the original trait proposal [10] and to other proposals of traits for JAVA-like setting [26, 18, 21] are the following:

- Classes and traits are not types and class-based inheritance is not present.
- Traits (and classes) can be typechecked in isolation (as in *Chai*$_2$ [26]).
- A basic trait expression $\{\,\bar{\mathtt{F}};\ \overline{\mathtt{S}};\ \bar{\mathtt{M}}\,\}$ provides the methods $\bar{\mathtt{M}}$ and declares the type of the required fields $\bar{\mathtt{F}}$ and methods $\overline{\mathtt{S}}$ (that can can be directly accessed by the bodies of the methods $\bar{\mathtt{M}}$).[2]
- In the *symmetric sum* operation (that merges two traits to form a new trait) we require that the summed traits must be disjoint (that is, they must not provide identically named methods).[3]
- The operation *exclude*, that forms a new trait by removing a method from an existing trait, is the usual one (i.e., as in [10, 26, 18, 21]).
- We have the operations *alias* and *duplicate* that form a new trait by giving a new name to an existing method. The two operations are identical on non-recursive methods. When a recursive method is aliased, its recursive invocation refers to the original method (as in [10]). When a recursive method is duplicated, its recursive invocation refers to the duplicate (as in the interpretation of aliasing proposed in [18]).
- We have the operation *rename* that creates a new trait by renaming all the occurrences of a required field name or of a required/provided method name from an existing trait.[4]

[2] Field requirements were not present in [10] and in [26, 18, 21]. They have been introduced in [12] in the setting of ML-like languages.

[3] According to other proposals, two methods with the same name do not conflict if they are syntactically equal [10, 21] or if they originate from the same subtrait [18].

[4] Method renaming is not present in [10] and in [26, 18, 21]. It has been introduced in [23] in the setting of ML-like languages. At the best of our knowledge, required field renaming is new.

- The *override* operation, that layers additional methods over an existing trait, is not present. It can be simulated by exclusion and symmetric sum.
- We use *interfaces* to explicitly declare the public methods of a class.

3.2 Typing

The FRJ type system combines nominal and structural typing. Within a basic trait expression, the uses of method parameters are type-checked according to the nominal notion of typing defined by the interface hierarchy, while the uses of the `this` metavariable are type-checked according to a structural notion of typing that takes into account the field and methods *required* by the trait and the methods *provided* by the trait.

3.2.1 Types, Constraints and Subtying. *Pure signatures*, ranged over by σ and ζ, are method signatures deprived of parameter names. For instance, the pure signature associated to the signature $\mathtt{I\,m(I_1\,x_1, ..., I_n\,x_n)}$ is $\mathtt{I\,m(I_1, ..., I_n)}$.

The syntax of *nominal types* is as follows: $\boxed{\eta \ ::= \ \mathtt{C} \mid \mathtt{I}}$. (I.e., a nominal type is either a class name or an interface name.) The syntax of *types for expressions* is as follows: $\boxed{\theta \ ::= \ \langle\, \bar{\mathtt{F}} \mid \bar{\sigma}\, \rangle \mid \eta}$. The type of the expression `this` is a pair $\langle\, \bar{\mathtt{F}} \mid \bar{\sigma}\, \rangle$, specifying that `this` has the fields $\bar{\mathtt{F}}$ and methods with (pure) signatures $\bar{\sigma}$. The type of an object creation expression $\mathtt{new\,C(\cdots)}$ is the class `C`. The type of any other expressions `e` is an interface name.

Besides assigning to each expression `e` a type describing the object yielded by the evaluation of `e`, the FRJ type system infers also the constraints on `this` imposed by its use within `e`. *Constraints*, ranged over by γ, are triples $\langle\, \bar{\mathtt{F}} \mid \bar{\sigma} \mid \bar{\mathtt{I}}\, \rangle$ specifying that the expression `e` selects the fields $\bar{\mathtt{F}}$ and the methods $\bar{\sigma}$ on `this`, and requires that `this` has the nominal types (interfaces) $\bar{\mathtt{I}}$. In particular, the interfaces in $\bar{\mathtt{I}}$ are the types of the method formal parameters to which `this` is passed inside the expression `e`. We recall that `this` will assume a meaning according to the class where the traits will be used. The typing rule for classes will check that such a class satisfies the constraints inferred for the bodies of the methods declared in the composing traits.

The *subtyping relation* for nominal types is the reflexive and transitive closure of the interface implementatation/extension relation declared by the `implements` clauses in the class table `CT` and by the `extends` clauses in the interface table `IT`. It is formalized by the judgement $\boxed{\eta_1 <: \eta_2}$ to be read: "η_1 is a subtype of η_2".

3.2.2 Typing Rules. An environment Γ is either a finite mapping form variable names (including `this`) to types, written "$\bar{\mathtt{x}} : \bar{\mathtt{I}}, \mathtt{this} : \langle\, \bar{\mathtt{F}} \mid \bar{\sigma}\, \rangle$", or the empty mapping, written "$\bullet$". The typing rules for interface declarations, expressions, method declarations, trait declarations and class declarations are syntax directed, with one rule for each form of term, except that (following [15]) there are three different rules for casts (to distinguish between *upcasts*, *downcasts*, and *stupid casts*). The typing judgements are the following:

- $\boxed{\vdash \texttt{interface I extends } \bar{\texttt{I}}\ \{\bar{\texttt{S}};\}\ \ \text{OK}}$ to be read: "the declaration of the interface I is well-typed".
- $\boxed{\Gamma \vdash \texttt{e} : \theta \mid \gamma}$ to be read: "under the assumption in Γ, the expression e is well-typed with type θ and constraints γ".
- $\boxed{\texttt{this} : \langle\, \bar{\texttt{F}} \mid \bar{\sigma}\,\rangle \vdash \texttt{I m}\,(\bar{\texttt{I}}\ \bar{\texttt{x}})\{\texttt{return e};\} : \mu}$ where $\mu = \zeta \mid \gamma$. To be read: "under the assumption that this has fields $\bar{\texttt{F}}$ and methods $\bar{\sigma}$, the declaration of method m is well-typed with type μ". I.e., the method m has signature ζ and its body enforces the constraints γ.
- $\boxed{\vdash \texttt{TE} : \bar{\mu}}$ where $\bar{\mu} = \mu_1 ... \mu_n \quad (n \geq 0)$. To be read: "the trait expression TE is well-typed with type $\bar{\mu}$". I.e., TE provides n methods with types $\mu_1, ..., \mu_n$, respectively.
- $\boxed{\vdash \texttt{trait T is TE} : \bar{\mu}}$ to be read: "the declaration of trait T is well-typed with type $\bar{\mu}$".
- $\boxed{\vdash \texttt{class C implements } \bar{\texttt{I}} \texttt{ by TE}\ \{\bar{\texttt{F}};\ \texttt{K}\}\ \ \text{OK}}$ to be read: "the declaration of the class C is well-typed".

Note that, within a basic trait expression $\{\bar{\texttt{F}};\ \bar{\texttt{S}};\ \bar{\texttt{M}}\}$, we ask the programmer to declare exactly the fields $\bar{\texttt{F}}$ and the methods $\bar{\texttt{S}}$ selected on this within the method bodies in $\bar{\texttt{M}}$. Declaring the types of fields and methods has two benefits: (*i*) it provides a form of documentation that enforces awareness of what it is actually used in a program; (*ii*) it simplifies the inferred constraints. We decided not to ask to declare the name of the interfaces that are used as types of this within the method bodies in $\bar{\texttt{M}}$, as this would not introduce any benefits to counterbalance the overhead.

3.2.3 Well-typed* FRJ *programs. We write $\vdash_{\text{FRJ}} (\texttt{IT}, \texttt{TT}, \texttt{CT}, \texttt{e}) : \eta$, to be read: "the program (IT, TT, CT, e) is well-typed with type η", to mean that the interfaces in IT, the traits in TT and the classes in CT are well-typed, and the expression e is well typed with type η and empty constraints under the empty set of assumptions (i.e., the judgement $\bullet \vdash \texttt{e} : \eta \mid \langle \bullet \mid \bullet \mid \bullet \rangle$ holds).

3.3 Flattening and Reduction

Our traits enjoy the *flattening property* [21], i.e., when a class uses a trait the semantics of the methods defined within the trait declaration is the same as if the methods were defined within the class declaration.[5] The semantics of FRJ is specified by means of a flattening translation that maps a FRJ program into a FFRJ program and of a reduction semantics for FFRJ programs.

[5] Flattening just aims to provide a canonical semantics to traits, it is not an especially effective implementation technique.

$$\begin{array}{rcl}
[\![\texttt{class C implements }\bar{\texttt{I}}\texttt{ by TE }\{\,\bar{\texttt{F}};\ \texttt{K}\,\}]\!] & \stackrel{\text{def}}{=} & \texttt{class C implements }\bar{\texttt{I}}\texttt{ by }\{\,\bar{\texttt{F}};\ \bullet;\ [\![\texttt{TE}]\!]\,\}\ \{\,\bar{\texttt{F}};\ \texttt{K}\,\}\\
[\![\{\,\bar{\texttt{F}};\ \bar{\texttt{S}};\ \bar{\texttt{M}}\,\}]\!] & \stackrel{\text{def}}{=} & \bar{\texttt{M}}\\
[\![\texttt{T}]\!] & \stackrel{\text{def}}{=} & [\![\texttt{TE}]\!]\quad \textbf{if}\ \texttt{TT(T)} = \texttt{trait T is TE}\\
[\![\texttt{TE}_1 + \texttt{TE}_2]\!] & \stackrel{\text{def}}{=} & [\![\texttt{TE}_1]\!]\cdot[\![\texttt{TE}_2]\!]\\
[\![\texttt{TE exclude m}]\!] & \stackrel{\text{def}}{=} & \mathit{discard}(\texttt{m}, [\![\texttt{TE}]\!])\\
[\![\texttt{TE alias m as m}']\!] & \stackrel{\text{def}}{=} & \bar{\texttt{M}}\cdot(\texttt{I m}'(\bar{\texttt{I}}\ \bar{\texttt{x}})\{\texttt{return e};\})\\
 & & \textbf{if}\ [\![\texttt{TE}]\!] = \bar{\texttt{M}}\ \textbf{and}\ \texttt{I m}(\bar{\texttt{I}}\ \bar{\texttt{x}})\{\texttt{return e};\} \in \bar{\texttt{M}}\\
[\![\texttt{TE duplicate m as m}']\!] & \stackrel{\text{def}}{=} & \bar{\texttt{M}}\cdot(\texttt{I m}'(\bar{\texttt{I}}\ \bar{\texttt{x}})\{\texttt{return e}[^{\texttt{this.m}'}/_{\texttt{this.m}}];\})\\
 & & \textbf{if}\ [\![\texttt{TE}]\!] = \bar{\texttt{M}}\ \textbf{and}\ \texttt{I m}(\bar{\texttt{I}}\ \bar{\texttt{x}})\{\texttt{return e};\} \in \bar{\texttt{M}}\\
[\![\texttt{TE rename f to f}']\!] & \stackrel{\text{def}}{=} & [\![\texttt{TE}]\!][^{\texttt{f}'}/_{\texttt{f}}]\\
[\![\texttt{TE rename m to m}']\!] & \stackrel{\text{def}}{=} & \mathit{mR}([\![\texttt{TE}]\!], \texttt{m}, \texttt{m}')\\
\\
\mathit{mR}(\texttt{I n}(\bar{\texttt{I}}\ \bar{\texttt{x}})\{\texttt{return e};\}, \texttt{m}, \texttt{m}') & \stackrel{\text{def}}{=} & \texttt{I n}[^{\texttt{m}'}/_{\texttt{m}}](\bar{\texttt{I}}\ \bar{\texttt{x}})\{\texttt{return e}[^{\texttt{this.m}'}/_{\texttt{this.m}}];\}\\
\mathit{mR}(\texttt{M}_1\cdot\ldots\cdot\texttt{M}_n, \texttt{m}, \texttt{m}') & \stackrel{\text{def}}{=} & (\mathit{mR}(\texttt{M}_1, \texttt{m}, \texttt{m}'))\cdot\ldots\cdot(\mathit{mR}(\texttt{M}_n, \texttt{m}, \texttt{m}'))
\end{array}$$

Fig. 2 Flattening FRJ to FFRJ

***3.3.1 Flattening Translation for* FRJ.** A FFRJ program is a FRJ program with an empty trait table. The translation removes the trait table and replaces the class table with a suitable one containing only FFRJ classes. The translation is specified through the function $[\![\cdot]\!]$, given in Fig. 2, that maps a FRJ class declaration to a FFRJ class declaration and a trait expression to a sequence of method declarations. We will write $[\![\texttt{CT}]\!]$ to denote the class table containing the translation of all the classes in CT. The clauses in Fig. 2 are self-explanatory. Note that the clause for field renaming is simpler than the clause for method renaming (which uses the auxiliary function *mR*); this is due to the fact that fields can be accessed only on `this`.

***3.3.2 Reduction for* FFRJ.** A FFRJ program is a 4-tuple $(\texttt{IT}, \bullet, \texttt{CT}, \texttt{e})$. A FFRJ class "$\texttt{class C implements }\bar{\texttt{I}}\texttt{ by }\{\,\bar{\texttt{F}};\ \bullet;\ \bar{\texttt{M}}\,\}\ \{\,\bar{\texttt{F}};\ \texttt{K}\,\}$" can be understood as the JAVA class "$\texttt{class C implements }\bar{\texttt{I}}\ \{\,\bar{\texttt{F}};\ \texttt{K}\ \bar{\texttt{M}}\,\}$". Following FJ [15], we give the semantics of FFRJ by means of a reduction relation of the form $\texttt{e} \rightarrow \texttt{e}'$, to be read "expression e reduces to expression e′ in one step". We write $\rightarrow^\star$ to denote the reflexive and transitive of $\rightarrow$. Values are defined by the following syntax: $\boxed{\texttt{v} ::= \texttt{new C}(\bar{\texttt{v}})}$.

3.4 Properties

The flattening translation preserves the type of programs.

Theorem 1 (Flattening Preserves the Type of Programs).
If $\vdash_{\text{FRJ}} (\texttt{IT}, \texttt{TT}, \texttt{CT}, \texttt{e}) : \eta$, *then* $\vdash_{\text{FRJ}} (\texttt{IT}, \bullet, [\![\texttt{CT}]\!], \texttt{e}) : \eta$.

To prove the type soundness result for FFRJ we need to consider a suitable notion of typing for runtime expressions. As for FJ [15], the syntax of runtime expressions is the same of expressions. Constraints are not needed to prove the type soundness for FFRJ, therefore the typing for runtime expressions do not consider constraints. An environment for runtime expressions Δ is either a finite mapping from variable names (including `this`) to types, written "$\bar{\mathtt{x}} : \bar{\mathtt{I}}, \mathtt{this} : \mathtt{C}$", or the empty mapping, written "$\bullet$". The typing judgement for runtime expressions is $\boxed{\Delta \vdash' \mathtt{e} : \eta}$ to be read: "under the assumption in Δ, the runtime expression $\mathtt{e}$ is well-typed with type η".

The type soundness result comes in two parts: first it relates the typing of expressions with the typing of runtime expressions, then it proves the type soundness with respect to the runtime expression typing.

Theorem 2 (Well-typed FFRJ expressions are well-typed runtime expression with a more specific type).
If $\bullet \vdash \mathtt{e} : \eta$, *then* $\bullet \vdash' \mathtt{e} : \eta'$ *for some* η' *such that* $\eta' <: \eta$.

The following theorem can be proved by using the standard technique of subject reduction and soundness theorems.

Theorem 3 (FFRJ Type Soundness).
If $\bullet \vdash' \mathtt{e} : \eta$ *and* $\mathtt{e} \rightarrow^\star \mathtt{e}'$ *with* $\mathtt{e}'$ *a normal form, then* $\mathtt{e}'$ *is: Either a value* $\mathtt{v}$ *with* $\bullet \vdash' \mathtt{v} : \mathtt{C}$ *and* $\mathtt{C} <: \eta$*; Or an expression containing* $(\mathtt{I})\mathtt{new\,C}(\bar{\mathtt{e}})$ *where* $\mathtt{C} \not<: \mathtt{I}$.

Following FJ [15], we say that a well-typed program $(\mathtt{IT}, \mathtt{TT}, \mathtt{CT}, \mathtt{e})$ is *cast-safe* if the type derivations involved in $\vdash_{\mathrm{FRJ}} (\mathtt{IT}, \mathtt{TT}, \mathtt{CT}, \mathtt{e}) : \eta$ include no downcasts or stupid casts. The following results hold.

Theorem 4 (Flattening Preserves Cast-Safeness).
If $(\mathtt{IT}, \mathtt{TT}, \mathtt{CT}, \mathtt{e})$ *is cast-safe, then* $(\mathtt{IT}, \bullet, [\![\mathtt{CT}]\!], \mathtt{e})$ *is cast-safe.*

Theorem 5 (No Typecast Errors in Cast-Safe FFRJ Programs).
If $(\mathtt{IT}, \bullet, [\![\mathtt{CT}]\!], \mathtt{e})$ *is cast-safe and* $\mathtt{e} \rightarrow^\star \mathtt{e}'$ *with* $\mathtt{e}'$ *a normal form, then* $\mathtt{e}'$ *is a value.*

4 Conclusions, Related and Further Work

The competing roles played by the same software structuring construct complicate the semantics and limits the reuse potential in mainstream object-oriented class-based programming languages.

To the best of our knowledge, the conflict between the roles of *unit of reuse* and *generator of instances* was firstly described by Schärli et al. [25, 10]. We claim also that the roles of *unit of reuse* and *type* are competing (see Sect. 1). In this respect, we propose to increase both the simplicity and the flexibility of the object-oriented paradigm by adopting programming language features that separate completely the declarations of object *type*, *behavior*, and *generator*. We developed a hybrid nominal/structural type system that allows to typecheck traits in isolation and proved its soundness.

The literature related to our proposal has been partially quoted through the paper. We add here comparisons and remarks concerning the type system and the recent proposals on *trait-based metaproprograming* [24] and *stateful traits* [4].

Sophisticated hybrid nominal/structural type systems have been already proposed [11, 19, 24]. In particular, in [24], the combination of nominal and structural types is conceptually similar to ours, but exploited at a different level. Namely, it is exploited to type trait functions, that provide a mechanism (termed trait-based metaproprograming) to obtain reusable class-member-level patterns. Another important difference between our proposal and the one in [24] is that, in the latter, traits play also the competing role of type, which instead we want to avoid.

Stateful traits [4] were introduced (in the setting of SMALLTALK-like languages) to avoid duplication of code connected directly with field initialization and manipulation. Our traits are stateless, however, since they can have required fields, it is possible to avoid the same kind of duplication of code that motivated the introduction of stateful traits. Moreover required fields names are unimportant because we provide a field rename operation. As byproducts, since required field renaming works synergically with method renaming, exclusion, aliasing, and duplication, we obtain more reuse potential.

In further work, we would like to formulate our proposal in a SMALLTALK-like setting (this would allow a careful comparison with the stateful trait proposal), to extend our type system to deal with generics, and to adapt our proposal to deal with *dynamic trait substitution* (see *Chai*$_3$ [26]). We also plan to develop prototypical implementations.

A special form of *reuse* is at the base of the contemporary *agile* software development methodologies [2]. Such methodologies are based on an iterative approach, where each iteration may include all of the phases necessary to release a small increment of a new functionality: planning, requirements analysis, design, coding, testing, and documentation. While an iteration may not add enough functionality to guarantee the release of a final product, an agile software project intends to be capable of releasing new software at the end of every iteration, but this means that the next iteration will *reuse* the software produced in the previous ones. We believe that an interesting future research direction is to investigate whether the programming language features proposed in this paper may help in writing software following an agile methodology. In this respect, we plan both to develop a trait-oriented agile methodology, suitable to be used directly within trait-based languages, and some trait-mining strategies, in order to re-engineer class-based libraries into trait-based ones. Some work in this respect was already done in the SMALLTALK-like setting [8, 16].

4.0.1 Acknowledgements. We thank Davide Ancona, Stéphane Ducasse, Paola Giannini, Oscar Nierstrasz, Betti Vennneri, Elena Zucca and the anonymous referees for comments, suggestions, and bibliographic references.

References

1. The Fortress language specification. http://research.sun.com/projects/plrg/fortress.pdf.
2. The Agile Alliance. Manifesto for Agile Software Development. `http://agilemanifesto.org/`.
3. D. Ancona, G. Lagorio, and E. Zucca. Jam—designing a Java extension with mixins. *ACM TOPLAS*, 25(5):641–712, September 2003.
4. A. Bergel, S. Ducasse, O. Nierstrasz, and R. Wuyts. Stateful traits. In *Advances in Smalltalk — Proceedings of 14th International Smalltalk Conference (ISC 2006)*, volume 4406 of *LNCS*, pages 66–90. Springer, 2007.
5. V. Bono, F. Damiani, and E. Giachino. Separating Type, Behavior, and State to Achieve Very Fine-grained Reuse. In *Electronic proceedings of FTfJP'07 (http://www.cs.ru.nl/ftfjp/)*, 2007.
6. G. Bracha. *The Programming Language JIGSAW: Mixins, Modularity and Multiple Inheritance*. PhD thesis, Department of Comp. Sci., Univ. of Utah, 1992.
7. G. Bracha and W. Cook. Mixin-based inheritance. In *ACM Symp. on Object-Oriented Programming: Systems, Languages and Applications 1990*, volume 25(10) of *SIGPLAN Notices*, pages 303–311. ACM Press, October 1990.
8. D. Cassou, S. Ducasse, and R. Wuyts. Redesigning with traits: the nile stream trait-based library. In *Proc. International Conference on Dynamic Languages*, pages 50–79, 2007.
9. W.R. Cook, W.L. Hill, and P.S. Canning. Inheritance is not subtyping. In *ACM Symp. on Principles of Programming Languages 1990*, pages 125–135. ACM Press, 1990.
10. S. Ducasse, O. Nierstrasz, N. Schärli, R. Wuyts, and A. Black. Traits: A mechanism for fine-grained reuse. *ACM TOPLAS*, 28(2):331–388, 2006.
11. K. Fisher and J. Reppy. Inheritance-based subtyping. *Information and Computation*, 177(1):28–55, 2002.
12. K. Fisher and J. Reppy. A typed calculus of traits. In *Electronic proceedings of FOOL 2004*, 2004.
13. M. Flatt, S. Krishnamurthi, and M. Felleisen. Classes and mixins. In *ACM Symp. on Principles of Programming Languages 1998*, pages 171–183. ACM Press, 1998.
14. E. Gamma, R. Helm, R. Johnson, and J. Vlissides. *Design Patterns: Elements of Reusable Object-Oriented Software*. Addison-Wesley, 1995.
15. A. Igarashi, B. Pierce, and P. Wadler. Featherweight Java: A minimal core calculus for Java and GJ. *ACM TOPLAS*, 23(3):396–450, 2001.
16. A. Lienhard, S. Ducasse, and G. Arévalo. Identifying traits with formal concept analysis. In *Proc. 20th Conference on Automated Software Engineering (ASE'05)*, pages 66–75. IEEE Computer Society, 2005.
17. M. Van Limberghen and T. Mens. Encapsulation and composition as orthogonal operators on mixins: A solution to multiple inheritance problems. *Object Oriented Systems*, 3(1):1–30, 1996.
18. L. Liquori and A Spiwack. Feathertrait: A modest extension of featherweight java. *ACM TOPLAS*. To appear.
19. D. Maleyery and J. Aldrich. Combining structural subtyping and external dispatch. In *Electronic proceedings of FOOL/WOOD 2007*, 2007.
20. L. Mikhajlov and E. Sekerinski. A Study of the Fragile Base Class Problem. In *Proc. ECOOP '98*, volume 1445 of *LNCS*, pages 355–382. Springer-Verlag, 1998.

21. O. Nierstrasz, S. Ducasse, and N. Schärli. Flattening traits. *JOT (www.jot.fm)*, 5(4):129–148, 2006.
22. M. Odersky. The Scala Language Specification, version 2.4. Technical report, Programming Methods Laboratory, EPFL, Switzerland, 2007.
23. J. Reppy and A. Turon. A foundation for trait-based metaprogramming. In *Electronic proceedings of FOOL/WOOD 2006*, 2006.
24. J. Reppy and A. Turon. Metaprogramming with traits. In *ECOOP 2007*, volume 4609 of *LNCS*, pages 373–398. Springer, 2007.
25. N. Schärli, S. Ducasse, O. Nierstrasz, and A. Black. Traits: Composable units of behavior. In *ECOOP 2003*, volume 2743 of *LNCS*, pages 248–274. Springer, 2003.
26. C. Smith and S. Drossopoulou. *Chai*: Traits for java-like languages. In *ECOOP'05*, LNCS 3586, pages 453–478. Springer, 2005.
27. A. Snyder. Encapsulation and inheritance in object-oriented programming languages. In *ACM Symp. on Object-Oriented Programming: Systems, Languages and Applications 1986*, volume 21(11) of *SIGPLAN Notices*, pages 38–45. ACM Press, 1986.

5 FRJ Typing Rules

The typing rules use the auxiliary functions given in Fig. 3: *fields* (that returns the sequence of the fields declared in a class C), *interfaces* (that when applied to an interface name I returns the name I itself, and when applied to a class name C returns the sequence of the interface names implemented by the class C), *methods* (that returns the sequence of the methods declared in a class C) and *mPSig* (that returns the sequence of the pure signatures of the methods associated to a sequence of method non-pure signatures, or interfaces, or method declarations).

Fields lookup (function *fields*)

$$\textit{fields}(\texttt{C}) = \bar{\texttt{F}} \qquad \textbf{if}\ \texttt{CT(C)} = \texttt{class C} \cdots \{\ \bar{\texttt{F}};\ \texttt{C}(\bar{\texttt{F}})\{\cdots\}\ \}$$

Interfaces lookup (function *interfaces*)

$$\textit{interfaces}(\texttt{C}) = \bar{\texttt{I}} \qquad \textbf{if}\ \texttt{CT(C)} = \texttt{class C implements}\ \bar{\texttt{I}}\ \texttt{by} \cdots$$
$$\textit{interfaces}(\texttt{I}) = \texttt{I}$$

Methods lookup (function *methods*)

$$\textit{methods}(\texttt{C}) = \bar{\texttt{M}} \qquad \textbf{if}\ \texttt{CT(C)} = \texttt{class C} \cdots \texttt{by}\ \{\cdots;\ \bullet;\ \bar{\texttt{M}}\}\ \{\cdots\}$$

Method pure signatures lookup (function *mPSig*)

$$\textit{mPSig}(\texttt{I m}\ (\bar{\texttt{I}}\ \bar{\texttt{x}})) = \texttt{I m}\ (\bar{\texttt{I}})$$
$$\textit{mPSig}(\texttt{S}_1; ...; \texttt{S}_n;) = \textit{mPSig}(\texttt{S}_1) \cdot ... \cdot \textit{mPSig}(\texttt{S}_n)$$
$$\textit{mPSig}(\texttt{I}) = \textit{mPSig}(\bar{\texttt{I}}) \cup \textit{mPSig}(\bar{\texttt{S}};) \qquad \textbf{if}\ \texttt{IT(I)} = \texttt{interface I extends}\ \bar{\texttt{I}}\ \{\ \bar{\texttt{S}};\ \}$$
$$\textit{mPSig}(\texttt{I}_1, ..., \texttt{I}_n) = \textit{mPSig}(\texttt{I}_1) \cup ... \cup \textit{mPSig}(\texttt{I}_n)$$
$$\textit{mPSig}(\texttt{S}\ \{\texttt{return e};\}) = \textit{mPSig}(\texttt{S})$$
$$\textit{mPSig}(\texttt{M}_1 ... \texttt{M}_n) = \textit{mPSig}(\texttt{M}_1) \cdot ... \cdot \textit{mPSig}(\texttt{M}_n)$$
$$\textit{mPSig}(\texttt{C}) = \textit{mPSig}(\textit{methods}(\texttt{C}))$$

Fig. 3 FRJ: Auxiliary function *fields*, *interfaces*, *methods* and *mPSig*

The typing rule for interface declarations, the typing rules for expressions and method declarations, the typing rules for trait expressions and trait declarations, and the typing rule for class declarations are given in Fig.s 6, 4, 5 and 7. Some of the typing rules use assumptions of the form "E **ok**" to mean that the expression E (involving operations on sequences of named elements) yields a (well defined) sequence of named elements. For instance, the assertion "$(\overline{\sigma} \cup \overline{\zeta})$ **ok**" holds if and only if $\mathtt{n} \in \mathit{names}(\overline{\sigma})$ and $\mathtt{n} \in \mathit{names}(\overline{\zeta})$ imply $\mathit{extract}(\mathtt{n}, \overline{\sigma}) = \mathit{extract}(\mathtt{n}, \overline{\zeta})$.

Expression typing:

$$\Gamma \vdash \mathtt{x} : \Gamma(\mathtt{x}) \mid \langle \bullet \mid \bullet \mid \bullet \rangle \qquad \text{(T-VAR)}$$

$$\frac{\Gamma \vdash \mathtt{this} : \langle\, \bar{\mathtt{F}} \mid \ldots \,\rangle \mid \langle \bullet \mid \bullet \mid \bullet \rangle \qquad \mathit{extract}(\mathtt{f}, \bar{\mathtt{F}}) = \mathtt{I\,f}}{\Gamma \vdash \mathtt{this.f} : \mathtt{I} \mid \langle\, \mathtt{I\,f} \mid \bullet \mid \bullet \,\rangle} \qquad \text{(T-FIELD)}$$

$$\frac{\begin{array}{l} \Gamma \vdash \mathtt{e} : \theta \mid \langle\, \bar{\mathtt{F}}^{(0)} \mid \overline{\sigma}^{(0)} \mid \bar{\mathtt{I}}^{(0)} \,\rangle \\ \theta = \Gamma(\mathtt{this}) = \langle \ldots \mid \overline{\sigma} \,\rangle \ \textbf{implies}\ \mathtt{I\,m}\,(\mathtt{I}_1, \ldots, \mathtt{I}_n) = \mathit{extract}(\mathtt{m}, \overline{\sigma}) \\ \theta \neq \Gamma(\mathtt{this})\ \textbf{implies}\ \mathtt{I\,m}\,(\mathtt{I}_1, \ldots, \mathtt{I}_n) = \mathit{extract}(\mathtt{m}, \mathit{mPSig}(\mathit{interfaces}(\theta))) \\ \forall i \in 1..n, \qquad \Gamma \vdash \mathtt{e}_i : \theta_i \mid \langle\, \bar{\mathtt{F}}^{(i)} \mid \overline{\sigma}^{(i)} \mid \bar{\mathtt{I}}^{(i)} \,\rangle \\ \mathcal{T} = \{ i \mid i \in 1..n \textbf{ and } \theta_i = \Gamma(\mathtt{this}) \} \\ \forall i \in 1..n - \mathcal{T}, \quad \theta_i <: \mathtt{I}_i \qquad \forall i \in \mathcal{T}, \quad \overline{\sigma} \cup \mathit{mPSig}(\mathtt{I}_i)\ \textbf{ok} \end{array}}{\Gamma \vdash \mathtt{e.m}(\mathtt{e}_1, \ldots, \mathtt{e}_n) : \mathtt{I} \mid \langle\, \cup_{i \in 0..n} \bar{\mathtt{F}}^{(i)} \mid \cup_{i \in 0..n} \overline{\sigma}^{(i)} \mid (\cup_{i \in 0..n} \bar{\mathtt{I}}^{(i)}) \cup (\cup_{i \in \mathcal{T}} \mathtt{I}_i \,) \rangle} \qquad \text{(T-INVK)}$$

$$\frac{\begin{array}{l} \mathit{fields}(\mathtt{C}) = \mathtt{I}_1\,\mathtt{f}_1; \ldots; \mathtt{I}_n\,\mathtt{f}_n; \qquad \forall i \in 1..n, \quad \Gamma \vdash \mathtt{e}_i : \theta_i \mid \langle\, \bar{\mathtt{F}}^{(i)} \mid \overline{\sigma}^{(i)} \mid \bar{\mathtt{I}}^{(i)} \,\rangle \\ \mathcal{T} = \{ i \mid i \in 1..n \textbf{ and } \theta_i = \Gamma(\mathtt{this}) = \langle \ldots \mid \overline{\sigma} \,\rangle \} \\ \forall i \in 1..n - \mathcal{T}, \quad \theta_i <: \mathtt{I}_i \qquad \forall i \in \mathcal{T}, \quad \overline{\sigma} \cup \mathit{mPSig}(\mathtt{I}_i)\ \textbf{ok} \end{array}}{\Gamma \vdash \mathtt{new\,C}(\mathtt{e}_1, \ldots, \mathtt{e}_n) : \mathtt{C} \mid \langle\, \cup_{i \in 1..n} \bar{\mathtt{F}}^{(i)} \mid \cup_{i \in 1..n} \overline{\sigma}^{(i)} \mid (\cup_{i \in 1..n} \bar{\mathtt{I}}^{(i)}) \cup (\cup_{i \in \mathcal{T}} \mathtt{I}_i \,) \rangle} \qquad \text{(T-NEW)}$$

$$\frac{\Gamma \vdash \mathtt{e} : \eta \mid \gamma \qquad \eta <: \mathtt{I}}{\Gamma \vdash (\mathtt{I})\mathtt{e} : \mathtt{I} \mid \gamma} \qquad \text{(T-UCAST)}$$

$$\frac{\Gamma \vdash \mathtt{e} : \mathtt{J} \mid \gamma \qquad \mathtt{I} <: \mathtt{J} \qquad \mathtt{I} \neq \mathtt{J}}{\Gamma \vdash (\mathtt{I})\mathtt{e} : \mathtt{I} \mid \gamma} \qquad \text{(T-DCAST)}$$

$$\frac{\Gamma \vdash \mathtt{e} : \eta \mid \gamma \qquad \eta \not<: \mathtt{I} \qquad \mathtt{I} \not<: \eta \qquad \textbf{stupid warning}}{\Gamma \vdash (\mathtt{I})\mathtt{e} : \mathtt{I} \mid \gamma} \qquad \text{(T-SCAST)}$$

Method declaration typing:

$$\frac{\begin{array}{l} \mathtt{this} : \langle\, \bar{\mathtt{F}} \mid \overline{\sigma} \,\rangle, \bar{\mathtt{x}} : \bar{\mathtt{J}} \vdash \mathtt{e} : \theta \mid \langle\, \bar{\mathtt{F}}' \mid \overline{\sigma}' \mid \bar{\mathtt{I}} \,\rangle \\ \theta = \langle\, \bar{\mathtt{F}} \mid \overline{\sigma} \,\rangle \ \textbf{implies}\ (\, \overline{\sigma} \cup \mathit{mPSig}(\mathtt{J})\ \textbf{ok and}\ \bar{\mathtt{I}}' = \bar{\mathtt{I}} \cup \mathtt{J} \,) \\ \theta \neq \langle\, \bar{\mathtt{F}} \mid \overline{\sigma} \,\rangle \ \textbf{implies}\ (\, \theta <: \mathtt{J}\ \textbf{and}\ \bar{\mathtt{I}}' = \bar{\mathtt{I}} \,) \end{array}}{\mathtt{this} : \langle\, \bar{\mathtt{F}} \mid \overline{\sigma} \,\rangle \vdash \mathtt{J\,m}\,(\bar{\mathtt{J}}\,\bar{\mathtt{x}})\{\mathtt{return\ e;}\} : \mathtt{J\,m}\,(\bar{\mathtt{J}}) \mid \langle\, \bar{\mathtt{F}}' \mid \overline{\sigma}' \mid \bar{\mathtt{I}}' \,\rangle} \qquad \text{(M-OK)}$$

Fig. 4 FRJ: Typing rules for expressions and method declarations

Trait expression typing:

$$\frac{\begin{array}{l} mPSig(\overline{\mathtt{S}}) = \overline{\sigma} \qquad mPSig(\mathtt{M}_1 ... \mathtt{M}_p) = \zeta_1 ... \zeta_p \qquad p \geq 0 \\ \forall i \in 1..p, \quad \mathbf{this} : \langle\, \bar{\mathtt{F}} \mid \overline{\sigma} \cdot \zeta_1 ... \zeta_p \,\rangle \vdash \mathtt{M}_i : \mu_i \qquad \mu_i = \zeta_i \mid \langle\, \bar{\mathtt{F}}^{(i)} \mid \overline{\zeta}^{(i)} \mid \bar{\mathtt{I}}^{(i)} \,\rangle \\ \bar{\mathtt{F}} = \cup_{i \in 1..p} \bar{\mathtt{F}}^{(i)} \qquad \overline{\sigma} = \mathit{discard}(\mathit{names}(\zeta_1 ... \zeta_p), (\cup_{i \in 1..p} \overline{\zeta}^{(i)})) \\ \zeta_1 ... \zeta_p \cup (\cup_{i \in 1..p} \overline{\zeta}^{(i)}) \cup mPSig(\cup_{i \in 1..p} \bar{\mathtt{I}}^{(i)}) \ \mathbf{ok} \end{array}}{\vdash \{\, \bar{\mathtt{F}};\ \overline{\mathtt{S}};\ \mathtt{M}_1 ... \mathtt{M}_p \,\} : \mu_1 ... \mu_p} \quad \text{(T-TEBASIC)}$$

$$\frac{\vdash \mathtt{trait\ T} \cdots : \bar{\mu}}{\vdash \mathtt{T} : \bar{\mu}} \quad \text{(T-TE)}$$

$$\frac{\begin{array}{l} \vdash \mathtt{TE}_1 : \mu_1 ... \mu_p \qquad \vdash \mathtt{TE}_2 : \mu_{p+1} ... \mu_{p+q} \\ p, q \geq 1 \qquad \forall i \in 1..p+q, \quad \mu_i = \zeta_i \mid \langle\, \bar{\mathtt{F}}^{(i)} \mid \overline{\sigma}^{(i)} \mid \bar{\mathtt{I}}^{(i)} \,\rangle \\ \cup_{i \in 1..p+q} \bar{\mathtt{F}}^{(i)} \ \mathbf{ok} \qquad \zeta_1 ... \zeta_{p+q} \cup (\cup_{i \in 1..p+q} \overline{\sigma}^{(i)}) \cup mPSig(\cup_{i \in 1..p+q} \bar{\mathtt{I}}^{(i)}) \ \mathbf{ok} \end{array}}{\vdash \mathtt{TE}_1 + \mathtt{TE}_2 : \mu_1 ... \mu_{p+q}} \quad \text{(T-TESUM)}$$

$$\frac{\vdash \mathtt{TE} : \bar{\mu} \cdot \mu \cdot \bar{\mu}' \qquad \mathit{names}(\mu) = \mathtt{m}}{\vdash \mathtt{TE\ exclude\ m} : \bar{\mu} \cdot \bar{\mu}'} \quad \text{(T-TEEX)}$$

$$\frac{\begin{array}{l} \vdash \mathtt{TE} : \mu_1 ... \mu_n \qquad n \geq p \geq 1 \qquad \forall i \in 1..n, \quad \mu_i = \zeta_i \mid \langle\, \bar{\mathtt{F}}^{(i)} \mid \overline{\sigma}^{(i)} \mid \bar{\mathtt{I}}^{(i)} \,\rangle \\ \mathit{names}(\zeta_p) = \mathtt{m} \qquad \mathtt{m}' \notin \mathit{names}(\zeta_1 ... \zeta_n) \qquad \zeta_p[^{\mathtt{m}'}/_{\mathtt{m}}] \cup (\cup_{i \in 1..n} \overline{\sigma}^{(i)}) \ \mathbf{ok} \\ \mu = \zeta_p[^{\mathtt{m}'}/_{\mathtt{m}}] \mid \langle\, \bar{\mathtt{F}}^{(p)} \mid \overline{\sigma}^{(p)} \mid \bar{\mathtt{I}}^{(p)} \,\rangle \end{array}}{\vdash \mathtt{TE\ alias\ m\ as\ m}' : \mu_1 ... \mu_n \mu} \quad \text{(T-TEAL)}$$

$$\frac{\begin{array}{l} \vdash \mathtt{TE} : \mu_1 ... \mu_n \qquad n \geq p \geq 1 \qquad \forall i \in 1..n, \quad \mu_i = \zeta_i \mid \langle\, \bar{\mathtt{F}}^{(i)} \mid \overline{\sigma}^{(i)} \mid \bar{\mathtt{I}}^{(i)} \,\rangle \\ \mathit{names}(\zeta_p) = \mathtt{m} \qquad \mathtt{m}' \notin \mathit{names}(\zeta_1 ... \zeta_n) \qquad \zeta_p[^{\mathtt{m}'}/_{\mathtt{m}}] \cup (\cup_{i \in 1..n} \overline{\sigma}^{(i)}) \ \mathbf{ok} \\ \mu = \zeta_p[^{\mathtt{m}'}/_{\mathtt{m}}] \mid \langle\, \bar{\mathtt{F}}^{(p)} \mid \overline{\sigma}^{(p)}[^{\mathtt{m}'}/_{\mathtt{m}}] \mid \bar{\mathtt{I}}^{(p)} \,\rangle \end{array}}{\vdash \mathtt{TE\ duplicate\ m\ as\ m}' : \mu_1 ... \mu_n \mu} \quad \text{(T-TEDU)}$$

$$\frac{\begin{array}{l} \vdash \mathtt{TE} : \mu_1 ... \mu_n \qquad n \geq 1 \qquad \forall i \in 1..n, \quad \mu_i = \zeta_i \mid \langle\, \bar{\mathtt{F}}^{(i)} \mid \overline{\sigma}^{(i)} \mid \bar{\mathtt{I}}^{(i)} \,\rangle \\ \overline{\zeta} = \zeta_1 ... \zeta_n \qquad \overline{\sigma} = \overline{\sigma}^{(1)} \cup ... \cup \sigma^{(n)} \qquad \mathtt{m} \in \mathit{names}(\overline{\zeta} \cup \overline{\sigma}) \qquad \mathtt{m}' \notin \mathit{names}(\overline{\zeta}) \\ (\overline{\zeta} \cup \overline{\sigma})[^{\mathtt{m}}/_{\mathtt{m}'}] \cup mPSig(\cup_{i \in 1..n} \bar{\mathtt{I}}^{(i)}) \ \mathbf{ok} \\ \forall i \in 1..n, \quad \mu'_i = \zeta_i[^{\mathtt{m}'}/_{\mathtt{m}}] \mid \langle\, \bar{\mathtt{F}}^{(i)} \mid \overline{\sigma}^{(i)}[^{\mathtt{m}'}/_{\mathtt{m}}] \mid \bar{\mathtt{I}}^{(i)} \,\rangle \end{array}}{\vdash \mathtt{TE\ rename\ m\ to\ m}' : \mu'_1 ... \mu'_n} \quad \text{(T-TEREM)}$$

$$\frac{\begin{array}{l} \vdash \mathtt{TE} : \mu_1 ... \mu_n \qquad n \geq 1 \qquad \forall i \in 1..n, \quad \mu_i = \zeta_i \mid \langle\, \bar{\mathtt{F}}^{(i)} \mid \overline{\sigma}^{(i)} \mid \bar{\mathtt{I}}^{(i)} \,\rangle \\ \bar{\mathtt{F}} = \bar{\mathtt{F}}^{(1)} \cup ... \cup \bar{\mathtt{F}}^{(n)} \qquad \mathtt{f} \in \mathit{names}(\bar{\mathtt{F}}) \qquad \bar{\mathtt{F}}[^{\mathtt{f}'}/_{\mathtt{f}}] \ \mathbf{ok} \\ \forall i \in 1..n, \quad \mu'_i = \zeta_i \mid \langle\, \bar{\mathtt{F}}^{(i)}[^{\mathtt{f}'}/_{\mathtt{f}}] \mid \overline{\sigma}^{(i)} \mid \bar{\mathtt{I}}^{(i)} \,\rangle \end{array}}{\vdash \mathtt{TE\ rename\ f\ to\ f}' : \mu'_1 ... \mu'_n} \quad \text{(T-TEREF)}$$

Trait declaration typing:

$$\frac{\vdash \mathtt{TE} : \bar{\mu}}{\vdash \mathtt{trait\ T\ is\ TE} : \bar{\mu}} \quad \text{(T-OK)}$$

Fig. 5 FRJ: Typing rules for trait expressions and trait declarations

Interface declaration typing:

$$\frac{mPSig(\mathtt{I})\ \mathbf{ok}}{\vdash\ \mathtt{interface\ I\ extends\ \bar{J}}\ \{\ \bar{\mathtt{S}}\ \}\ \ \mathrm{OK}}\quad (\text{I-OK})$$

Fig. 6 FRJ: Typing rule for interface declarations

Class declaration typing:

$$\frac{\begin{array}{lll}\vdash \mathtt{TE} : \mu_1 ... \mu_p & p \geq 0 & \forall i \in 1..p,\quad \mu_i = \zeta_i \mid \langle\ \bar{\mathtt{F}}^{(i)} \mid \bar{\sigma}^{(i)} \mid \bar{\mathtt{I}}^{(i)}\ \rangle \\ \cup_{i \in 1..p} \bar{\mathtt{F}}^{(i)} = \bar{\mathtt{J}}\ \bar{\mathtt{g}} & & \zeta_1 ... \zeta_p \supseteq ((\cup_{i \in 1..p} \bar{\sigma}^{(i)}) \cup mPSig(\bar{\mathtt{I}})) \\ \forall \mathtt{I}' \in \cup_{i \in 1..p} \bar{\mathtt{I}}^{(i)},\ \exists \mathtt{I} \in \bar{\mathtt{I}},\quad \mathtt{I} <: \mathtt{I}' & & \end{array}}{\vdash\ \mathtt{class\ C\ implements}\ \bar{\mathtt{I}}\ \mathtt{by\ TE}\ \{\ \bar{\mathtt{J}}\ \bar{\mathtt{g}};\ \ \mathtt{C}(\bar{\mathtt{J}}\ \bar{\mathtt{g}})\ \{\ \mathtt{this}.\bar{\mathtt{g}} = \bar{\mathtt{g}};\ \}\ \}\ \ \mathrm{OK}}\quad (\text{C-OK})$$

Fig. 7 FRJ: Typing rule for class declarations

6 FFRJ Reduction Rules

The FFRJ reduction rules are given in Fig. 8 (the auxiliary functions *fields*, *methods* and *interfaces* are given in Fig. 3).

Evaluation contexts and redexes:

$$\begin{array}{l} E ::= [\,] \mid E.\mathtt{f} \mid E.\mathtt{m}(\bar{\mathtt{e}}) \mid \mathtt{v.m}(\bar{\mathtt{v}}, E, \bar{\mathtt{e}}) \mid (\mathtt{I})E \mid \mathtt{new\,C}(\bar{\mathtt{v}}, E, \bar{\mathtt{e}}) \\ r ::= (\mathtt{new\,C}(\bar{\mathtt{v}})).\mathtt{f} \mid (\mathtt{new\,C}(\bar{\mathtt{v}})).\mathtt{m}(\bar{\mathtt{v}}) \mid (\mathtt{I})(\mathtt{new\,C}(\bar{\mathtt{e}})) \end{array}$$

Reduction rules:

$$\frac{fields(\mathtt{C}) = \mathtt{I}_1\,\mathtt{f}_1; ...; \mathtt{I}_n\,\mathtt{f}_n}{E[(\mathtt{new\,C}(\mathtt{v}_1, ..., \mathtt{v}_n)).\mathtt{f}_i] \rightarrow E[\mathtt{v}_i]}\quad (\text{R-FIELD})$$

$$\frac{\mathtt{I\,m}\,(\bar{\mathtt{I}}\,\bar{\mathtt{x}})\ \{\,\mathtt{return\,e};\ \} \in methods(\mathtt{C})}{E[(\mathtt{new\,C}(\bar{\mathtt{v}})).\mathtt{m}(\bar{\mathtt{u}})] \rightarrow E[\mathtt{e}[{}^{\bar{\mathtt{u}}}/_{\bar{\mathtt{x}}}, {}^{\mathtt{new\,C}(\bar{\mathtt{v}})}/_{\mathtt{this}}]]}\quad (\text{R-INVK})$$

$$\frac{\exists \mathtt{J} \in interfaces(\mathtt{C}),\quad \mathtt{J} <: \mathtt{I}}{E[(\mathtt{I})(\mathtt{new\,C}(\bar{\mathtt{e}}))] \rightarrow E[\mathtt{new\,C}(\bar{\mathtt{e}})]}\quad (\text{R-CAST})$$

Fig. 8 FFRJ: Reduction rules

Canonical Sequent Proofs via Multi-Focusing

Kaustuv Chaudhuri[1], Dale Miller[2], and Alexis Saurin[3]

[1] INRIA Saclay – Île-de-France, Kaustuv.Chaudhuri@inria.fr
[2] INRIA Saclay – Île-de-France & LIX, École Polytechnique, Dale.Miller@inria.fr
[3] INRIA Saclay – Île-de-France & LIX, École Polytechnique, Alexis.Saurin@inria.fr

Abstract. The sequent calculus admits many proofs of the same conclusion that differ only by trivial permutations of inference rules. In order to eliminate this "bureaucracy" from sequent proofs, deductive formalisms such as proof nets or natural deduction are usually used instead of the sequent calculus, for they identify proofs more abstractly and geometrically. In this paper we recover permutative canonicity directly in the cut-free sequent calculus by generalizing focused sequent proofs to admit multiple foci, and then considering the restricted class of *maximally multi-focused proofs*. We validate this definition by proving a bijection to the well-known proof-nets for the unit-free multiplicative linear logic, and discuss the possibility of a similar correspondence for larger fragments.

1 Introduction

Sequent calculus proofs are much less proof objects than they are traces of the computation of a more abstract proof object. In particular, the infernece rules of the sequent calculus are minute and there are many choices in the order of their application that seem equivalent although, formally, they result in different sequent proofs. One way to get a more abstract notion of proof is to declare that two cut-free proofs are *equivalent* if it is possible to permute the inference rules in one to get the other. Such equivalence classes are unsatisfactory for at least two reasons. First, computing permutations of inference rules might require examining and reorganizing arbitrary parts of a proof: attempting to move a given inference rule to the bottom of a proof could cause changes to many parts of the proof. Second, since equivalence classes are not, themselves, inductive structures, familiar arguments involving inductive reasoning over proof structures cannot be applied easily to equivalence classes. Many people working in proof theory and particularly those interested in the problem of the *identity of proofs* discard sequent proofs for more abstract proof structures like natural deduction proofs or proof nets. In these later objects, a more geometric structure of proofs requires less sequentialization of inference rules and allows one to work on proofs more abstractly.

We shall argue in this paper that one does not need to discard the sequent calculus in order to factor out many of these irrelevant sequentializations of inference rules. We shall show that there are, in fact, normal forms of sequent proofs that provide unique representatives of their permutative equivalence classes. To be concrete, we shall assume a setting of the standard cut-free sequent calculus for multiplicative-additive linear logic (MALL), including units and literals. Motivating the construction of canonical representatives is as follows. A first step is to consider only *focused proofs* [2],

Please use the following format when citing this chapter:

Chaudhuri, K., Miller, D. and Saurin, A., 2008, in IFIP International Federation for Information Processing, Volume 273; *Fifth IFIP International Conference on Theoretical Computer Science*; Giorgio Ausiello, Juhani Karhumäki, Giancarlo Mauri, Luke Ong; (Boston: Springer), pp. 383–396.

with a strict alternation of negative (invertible) and positive (focused) phases. Focused proofs systems can be used to distinguish between *micro* rules, *i.e.*, introduction rules in the ordinary sequent calculus, and the *macro* rules that comprise an entire focusing phases and correspond to the introduction of *synthetic connectives* [5]. A first abstraction is then to consider proofs as built up from macro rules introducing synthetic connectives. Unfortunately, this layer of abstraction does not yield canonical representatives of equivalence classes since the selection of foci is still sequentialized even when the selection order is irrelevant. Such parallelism can be captured by the addition of the *multi-focus* rule that permits focusing on several formulas within one phase. If we then require that such multi-focus inference rules select a "maximal focus" then, as we show in Section 4, we have achieved canonical representatives of equivalence classes of proofs.

Proof nets for MLL and MALL have been used also as abstractions of the class of cut-free proofs under the equivalence of permuting inference rules. We show that maximally multi-focused sequent proofs (modulo the weak "iso-polar" equivalence) are in one-to-one correspondence with MLL proof nets [9]: we show how to uniquely associate a maximally multi-focused proof to an MLL proof net. We also discuss proof nets in MALL without units [10, 12] and for other fragments of linear logic: maximal multi-focusing proofs should also be applicable in various other richer logics where the nature of proof nets is less well developed or satisfying, such as linear logic with units and exponentials.

This paper is organized as follows: in Sec. 2 we recall the sequent calculus for MALL. In Sec. 3 we present our multi-focal generalization of Andreoli's focusing calculus. In Sec. 4 we define the notion of *maximality* and prove the key canonicity result (Theorem 7). In Sec. 5 we exhibit a one-to-one correspondence between maximally multi-focused proofs and proof-nets for MLL without units.

2 Sequent calculus for MALL

MALL formulas are defined by the following grammar:

$$A, B, \ldots ::= a \mid a^\perp \mid A \otimes B \mid \mathbf{1} \mid A \mathbin{\wp} B \mid \perp \mid A \mathbin{\&} B \mid \top \mid A \oplus B \mid \mathbf{0}$$

A *literal* is either an atomic formula, written using minuscule scheme variables $(a, b, \ldots)$, or it is a negated atom $(a^\perp, b^\perp, \ldots)$. As usual, MALL formulas are assumed to be in negation-normal form, and the pairs $(\otimes, \wp)$, $(\mathbf{1}, \perp)$, $(\&, \oplus)$, and $(\top, \mathbf{0})$ are de Morgan duals, *i.e.*, $(A \otimes B)^\perp = A^\perp \mathbin{\wp} B^\perp$, *etc.* The sequent calculus for MALL uses one-sided sequents of the form $\vdash \Gamma$, where the context Γ is a multiset of formulas. Figure 1 contains the standard proof rules for such sequents [9].

Script majuscule letters $\mathcal{D}, \mathcal{E}, \ldots$ are used to denote proofs and the expression $\mathcal{D} \vdash \Gamma$ signifies that $\mathcal{D}$ is a proof of $\vdash \Gamma$. It is well-known that the following cut and (non-atomic) initial rules are admissible.

$$\frac{}{\vdash a, a^\perp}\ \mathrm{I} \qquad \frac{\vdash \Gamma, A \qquad \vdash \Delta, B}{\vdash \Gamma, \Delta, A \otimes B}\ \otimes \qquad \frac{}{\vdash \mathbf{1}}\ \mathbf{1} \qquad \frac{\vdash \Gamma, A, B}{\vdash \Gamma, A \parr B}\ \parr \qquad \frac{\vdash \Gamma}{\vdash \Gamma, \bot}\ \bot$$

$$\frac{\vdash \Gamma, A \qquad \vdash \Gamma, B}{\vdash \Gamma, A \,\&\, B}\ \& \qquad \frac{}{\vdash \Gamma, \top}\ \top \qquad \frac{\vdash \Gamma, A_i}{\vdash \Gamma, A_1 \oplus A_2}\ \oplus_i$$

Fig. 1 Sequent calculus for MALL. In the $\oplus_i$ rule, $i \in \{1,2\}$.

$$\frac{\vdash \Gamma, A \qquad \vdash \Delta, A^\perp}{\vdash \Gamma, \Delta}\ \mathrm{C} \qquad \text{and} \qquad \frac{}{\vdash A, A^\perp}\ \mathrm{I}^*$$

Local permutations of inference rules form a natural relation between cut-free proofs [13]. For example, in a proof of the form

$$\frac{\mathcal{D} \vdash \Gamma, A \qquad \dfrac{\mathcal{E} \vdash \Delta, B, C \qquad \mathcal{F} \vdash \Delta, B, D}{\vdash \Delta, B, C \,\&\, D}\ \&}{\vdash \Gamma, \Delta, A \otimes B, C \,\&\, D}\ \otimes, \tag{1}$$

the order of the $\otimes$ and & rules may be locally switched to yield the proof

$$\frac{\dfrac{\mathcal{D} \vdash \Gamma, A \qquad \mathcal{E} \vdash \Delta, B, C}{\vdash \Gamma, \Delta, A \otimes B, C}\ \otimes \qquad \dfrac{\mathcal{D} \vdash \Gamma, A \qquad \mathcal{F} \vdash \Delta, B, D}{\vdash \Gamma, \Delta, A \otimes B, D}\ \otimes}{\vdash \Gamma, \Delta, A \otimes B, C \,\&\, D}\ \&. \tag{2}$$

This switching causes the proof $\mathcal{D}$ to be duplicated in (2), but does not alter the constituent sub-proofs $\mathcal{D}$, $\mathcal{E}$ and $\mathcal{F}$. We denote a site of a local permutation, *i.e.*, a pair of neighbouring inference rules r_1 followed by r_2 as r_1/r_2; for example, (1) ends with a $\&/\otimes$ along the right branch of the final rule.

Consider, instead, the following proof figures.

$$\frac{\mathcal{D} \vdash \Gamma, A \qquad \dfrac{}{\vdash \Delta, B, \top}\ \top}{\vdash \Gamma, \Delta, A \otimes B, \top}\ \otimes \qquad \frac{}{\vdash \Gamma, \Delta, A \otimes B, \top}\ \top \tag{3}$$

Moving from left-to-right can be seen as moving the $\top$ inference rule below the $\otimes$ rule: in the process the entire proof $\mathcal{D}$ is deleted. Since we wish to establish an equivalence based on permutations, moving from right-to-left can be seen as "creating" the proof $\mathcal{D}$. While deletion of proofs can be seen as problematic when one is attempting to capture the "essence" of proofs, creation is certainly problematic in this sense. Thus, we introduce the following restriction on permutations to avoid this kind of proof creation within equivalent proofs.

Definition 1 *Two proofs $\mathcal{D}$ and $\mathcal{E} \vdash \Gamma$ are* iso-initial, *written $\mathcal{D} \simeq \mathcal{E}$, if each can be rewritten to the other using local permutations* and *the set of initial sequents in both $\mathcal{D}$ and $\mathcal{E}$ are the same. The sets under consideration are of pairs of formula occurrences.*

The additional restriction on the sets of initial sequents allows the deletion and creation of subproofs during permutation only when such proofs are without initial rules.

For the $\top$-free fragment of MALL, this restriction is trivial, as all permutations preserve the set of initial sequents. However, because $\top$ can arbitrarily rewrite a branch of a proof, allowing all permutations with $\top$ would identify too many proofs. This restriction is further motivated by the observation from unit-free multiplicative proof nets, where the axiom links (which correspond to the initial sequents) contain the essential dynamics of a proof. These dynamics should not be suppressed by trivial permutations. Note that because we don't allow all permutations of $\top$, we are decidedly not equating all proofs that are equated in the standard categorical model of MALL proofs; *i.e.*, $\top$ is no longer a terminal object in a suitable $\star$-autonomous category where & is the Cartesian product.

3 Multi-focusing for MALL

In the remainder of this paper, we shall consider only cut-free proofs.

The formulas of MALL can be classified, based on their permutative affinities or *polarity*, into the following two classes.

$$\begin{array}{llll} \textit{(positive)} & P, Q, \ldots & ::= & a \mid A \otimes B \mid \mathbf{1} \mid A \oplus B \mid \mathbf{0} \\ \textit{(negative)} & N, M, \ldots & ::= & a^{\perp} \mid A \mathbin{⅋} B \mid \perp \mid A \,\&\, B \mid \top \end{array}$$

A logical rule that applies to a positive (resp. negative) formula will henceforth be called a positive (resp. negative) rule. If r_1 is a positive rule and r_2 is a negative rule, then r_1/r_2 is an instance of the local permutation class pos/neg; similarly for pos/pos, neg/neg, and neg/pos. All pos/pos and neg/neg permutations are valid. Furthermore, neg/pos permutations are also valid since the negative rules are invertible and, hence, may be applied arbitrarily early (reading bottom-up). From a proof-search perspective, the negative rules are, therefore, *asynchronous* since their application does not depend on the structure of the side contexts. The positive rules, on the other hand, are non-invertible and, therefore, *synchronous*: their application depends on the structure of the remaining context and the sequence of rules that have been applied lower in the proof.

Andreoli [2] presented a *focused* proof system (for all of first-order linear logic) in which proofs have two phases. When reading proofs from the conclusion to the premises, a *focal* phase begins by granting focus to a positive formula from the available positive formulas: this focus can be indicated explicitly in the sequents by writing them as $\vdash \Gamma \Downarrow A$ where A is under focus. Once the focused formula becomes negative, *i.e.*, the sequent is of the form $\vdash \Gamma \Downarrow N$, the focus is *released* and the search enters the negative (asynchronous) phase where the negative connectives are decomposed; this phase is indicated in sequents of the form $\vdash \Gamma \Uparrow \Delta$. This phase separation is complete for cut-free proofs, *i.e.*, every provable sequent has a focused proof [2, 16].

In this paper, we generalize this usual focusing strategy further in the following way: when deciding to focus, we may focus on more than one positive formula at a time, *i.e.*, our positive sequents are now of the form $\vdash \Gamma \Downarrow \Delta$ (with Δ non-empty). All the formulas under focus are decomposed until only negative formulas remain in focus;

$$\frac{\vdash \Gamma_1 \Downarrow A, \Delta_1 \quad \vdash \Gamma_2 \Downarrow B, \Delta_2}{\vdash \Gamma_1, \Gamma_2 \Downarrow A \otimes B, \Delta_1, \Delta_2}\,[\otimes] \qquad \frac{}{\vdash \cdot \Downarrow \mathbf{1}}\,[\mathbf{1}] \qquad \frac{\vdash \Gamma \Downarrow A_i, \Delta}{\vdash \Gamma \Downarrow A_1 \oplus A_2, \Delta}\,[\oplus_i]$$

$$\frac{\vdash \Gamma \Uparrow A, \Delta \quad \vdash \Gamma \Uparrow B, \Delta}{\vdash \Gamma \Uparrow A \,\&\, B, \Delta}\,[\&] \qquad \frac{}{\vdash \Gamma \Uparrow \top, \Delta}\,[\top] \qquad \frac{\vdash \Gamma \Uparrow A, B, \Delta}{\vdash \Gamma \Uparrow A \parr B, \Delta}\,[\parr] \qquad \frac{\vdash \Gamma \Uparrow \Delta}{\vdash \Gamma \Uparrow \bot, \Delta}\,[\bot]$$

$$\frac{}{\vdash a^\perp \Downarrow a}\,[\mathrm{I}] \qquad \frac{\vdash \Gamma \Downarrow \Delta}{\vdash \Gamma, \Delta \Uparrow \cdot}\,[\mathrm{MF}] \qquad \frac{\vdash \Gamma, A \Uparrow \Delta}{\vdash \Gamma \Uparrow A, \Delta}\,[\mathrm{R}{\Uparrow}] \qquad \frac{\vdash \Gamma \Uparrow \Delta}{\vdash \Gamma \Downarrow \Delta}\,[\mathrm{R}{\Downarrow}]$$

Fig. 2 Multi-focusing sequent calculus, MF. The contexts on the left of $\Downarrow$ and $\Uparrow$ contain only positive formulas or negated atoms. In the [MF] rule, Δ contains at least one positive formula. In the [R⇑] rule, A is positive or a negated atom. In the [R⇓] rule, Δ is all negative. In $[\oplus_i]$, $i \in \{1, 2\}$.

then, the focus is released and the negative formulas are decomposed in the negative phase. The rules of this calculus of *multi-focused* proofs are presented in Figure 2.

Definition 2 *If $\mathcal{D} \vdash \Gamma \Uparrow \Delta$ or $\mathcal{D} \vdash \Gamma \Downarrow \Delta$, then we write $\lfloor \mathcal{D} \rfloor$ for that proof of $\vdash \Gamma, \Delta$ that replaces every sequent of the form $\vdash \Gamma' \Uparrow \Delta'$ or $\vdash \Gamma' \Downarrow \Delta'$ in $\mathcal{D}$ with $\vdash \Gamma', \Delta'$, elides all instances of* [R⇑], [R⇓] *and* [MF], *and renames all other rules to their unbracketed forms* ([⊗] *to* ⊗, *etc).*

Theorem 3 (Correctness of multi-focusing)

1. *If $\mathcal{D} \vdash \Gamma \Downarrow \Delta$ or if $\mathcal{D} \vdash \Gamma \Uparrow \Delta$, then $\lfloor \mathcal{D} \rfloor \vdash \Gamma, \Delta$ (soundness).*
2. *If $\vdash \Gamma$, then $\vdash \cdot \Uparrow \Gamma$ (completeness).*

Proof. Soundness is immediate. Completeness follows by observing that Andreoli's focusing calculus for MALL is recovered in MF by restricting the context Δ in [MF] to a singleton, and then using the analogous completeness theorem there [2, 16]. Note that the proof in [2] is for full first-order, multiplicative-additive-exponential linear logic. □

Given the phase separation induced by focusing, we define the following primitive equivalence on proofs that identifies proofs that differ from each other only inside a phase.

Definition 4 *Two proofs $\mathcal{D}$ and $\mathcal{D}' \vdash \Gamma \Updownarrow \Delta$ are* iso-polar, *written $\mathcal{D} \approx \mathcal{D}'$, if they are equal up to permutations restricted to the pos/pos and neg/neg types.*

This equivalence seems natural because the interchange of the pos/pos and neg/neg inference rules are truly parallel and non-interacting. Indeed, two iso-polar proofs have the same synthetic inference rules, *i.e.*, the derived rules where the details of the positive and negative phases are elided, and only [I] and the phase transitions [R⇓] and [MF] are noted. For example, one proof of $\vdash a^\perp, a \otimes (b \,\&\, c), d \oplus \top \Uparrow \cdot$ using only synthetic rules is:

$$\frac{\dfrac{}{\vdash a^\perp \Downarrow a}\,[\mathrm{I}] \quad \dfrac{\dfrac{}{\vdash \cdot \Uparrow b \,\&\, c, \top}}{\vdash \cdot \Downarrow b \,\&\, c, \top}\,[\mathrm{R}{\Downarrow}]}{\vdash a^\perp, a \otimes (b \,\&\, c), d \oplus \top \Uparrow \cdot}\,[\mathrm{MF}]$$

The instance of [MF] focuses on $a \otimes (b \mathbin{\&} c)$ and $d \oplus \top$, but the instances of [$\otimes$] and [$\oplus$] above it are elided, as are any [&] and [$\top$] rules used above the instance of [R$\Downarrow$].

A single representative of the $\approx$-classes can be constructed by treating the contexts Δ to the right of $\Uparrow$ and $\Downarrow$ in MF as ordered contexts, similar to Andreoli's original focusing proof system [2]. This order on the context induces a fixed but arbitrary order of the pos/pos and neg/neg rules.

4 Maximality and canonicity

We now revisit the question of permutations of the synthetic inference rules induced by focusing. In the unfocused calculus, it is easy to see that the synthetic rule for a negative synthetic connective, which is a sequence of negative rules for the constituents of the synthetic connective, permutes with that of another synthetic negative connective: it is a simple matter of sequencing permutations. Similarly, the positive synthetic rules commute with other positive synthetic rules, and likewise for a neg/pos permutation of synthetic rules. As before, the only disallowed permutations in general are the pos/neg permutations.

Definition 5 *Suppose* $\mathcal{D} = \dfrac{\mathcal{D}' \vdash \Gamma \Downarrow \Delta}{\vdash \Gamma, \Delta \Uparrow \cdot}$ [MF]. *Then, Δ are called the* roots *of $\mathcal{D}$, written* roots($\mathcal{D}$).

We intend to show that every member of an iso-initial class of proofs of $\vdash \Gamma$ is equivalent to a unique proof (upto iso-polarity) of $\vdash \cdot \Uparrow \Gamma$. In fact, we shall call these representatives of the iso-initial equivalence class the *maximally multi-focused proofs*.

Definition 6 *A proof $\mathcal{D}$ of $\vdash \Gamma \Updownarrow \Delta$ is* maximal *if for every sub-proof $\mathcal{E} \vdash \Gamma' \Uparrow \cdot$ of $\mathcal{D}$, it is the case for any $\mathcal{E}' \simeq \mathcal{E} \vdash \Gamma' \Uparrow \cdot$ that* roots($\mathcal{E}'$) $\subseteq$ roots($\mathcal{E}$).

Our goal with maximal proofs is the following canonicity result:

Theorem 7 (canonicity) *If $\mathcal{D} \simeq \mathcal{E} \vdash \Gamma \Uparrow \cdot$ are both maximal, then $\mathcal{D} \approx \mathcal{E}$.*

The proof of this theorem will require considering permutations of entire synthetic connectives. Following Andreoli [2], we call a neighbouring pair of phases, with the bottom phase having a positive synthetic connective as its principal formula, and the top phase being its corresponding negative synthetic rules, a *bipole*. Consider two neighbouring bipoles: if the positive phase of the top bipole permutes with the negative phase of the bottom bipole, then in an unfocused form we can perform the permutation and merge the two bipoles by uniting their positive and negative phases, obtaining another (multi-)focused proof.

The MF rules are, however, too rigid to express any but the final points of the permutation. Thus, in this section we shall consider a comparitively more relaxed focusing calculus where a negative phase (of the bottom bipole) can be "carried through" the positive phase (of the top bipole). The bottom negative phase is first (temporarily) preempted by the top positive phase; for this, we use sequents of the form $\vdash \Gamma \downarrow \Delta \,;\, \Xi$

$$\frac{\vdash \Gamma_1 \downarrow A, \Delta_1 ; \Xi_1 \qquad \vdash \Gamma_2 \downarrow B, \Delta_2 ; \Xi_2}{\vdash \Gamma_1, \Gamma_2 \downarrow A \otimes B, \Delta_1, \Delta_2 ; \Xi_1, \Xi_2} [\otimes] \qquad \frac{}{\vdash \cdot \downarrow \mathbf{1} ; \cdot} [\mathbf{1}] \qquad \frac{\vdash \Gamma \downarrow A_i, \Delta ; \Xi}{\vdash \Gamma \downarrow A_1 \oplus A_2, \Delta ; \Xi} [\oplus_i]$$

$$\frac{\vdash \Gamma \Uparrow A, \Delta \qquad \vdash \Gamma \Uparrow B, \Delta}{\vdash \Gamma \Uparrow A \,\&\, B, \Delta} [\&] \qquad \frac{}{\vdash \Gamma \Uparrow \top, \Delta} [\top] \qquad \frac{\vdash \Gamma \Uparrow A, B, \Delta}{\vdash \Gamma \Uparrow A \parr B, \Delta} [\parr] \qquad \frac{\vdash \Gamma \Uparrow \Delta}{\vdash \Gamma \Uparrow \bot, \Delta} [\bot]$$

$$\frac{}{\vdash a^\perp \downarrow a ; \cdot} [\mathrm{I}] \qquad \frac{\vdash \Gamma \downarrow \Delta ; \Xi}{\vdash \Gamma, \Delta \Uparrow \Xi} [\mathrm{PMF}_1] \qquad \frac{\vdash \Gamma \downarrow \Delta, \Psi ; \Xi}{\vdash \Gamma, \Delta \downarrow \Psi ; \Xi} [\mathrm{PMF}_2]$$

$$\frac{\vdash \Gamma, A \Uparrow \Delta}{\vdash \Gamma \Uparrow A, \Delta} [\mathrm{R}\Uparrow] \qquad \frac{\vdash \Gamma \downarrow \Delta ; N, \Xi}{\vdash \Gamma \downarrow \Delta, N ; \Xi} [\mathrm{R}\downarrow] \qquad \frac{\vdash \Gamma \Uparrow \Xi}{\vdash \Gamma \downarrow \cdot ; \Xi} [\mathrm{R}]$$

Fig. 3 Rules of the pre-emptive multi-focusing calculus, PMF. All side conditions from MF (Fig. 2) are carried over; in particular, for $[\mathrm{PMF}_1]$ and $[\mathrm{PMF}_2]$, the context Δ is non-empty.

where Δ is under focus, and Ξ is a suspended context. Later, when the positive phase has permuted down, the negative phases are awakened into active sequents of the form $\vdash \Gamma \Uparrow \Delta$. The rules of this *pre-emptive multi-focusing* calculus, called PMF, are in Figure 3. A straightforward injection $(-)^\#$ from MF to PMF derivations is assumed.

Fact 8 *The following are seen by straightforward induction.*

1. *If* $\vdash_{\mathrm{MF}} \Gamma \Downarrow \Delta$, *then* $\vdash_{\mathrm{PMF}} \Gamma \downarrow \Delta ; \cdot$.
2. *If* $\vdash_{\mathrm{PMF}} \Gamma \downarrow \Delta ; \Xi$, *then* $\vdash_{\mathrm{MF}} \Gamma \Downarrow \Delta, \Xi$.
3. $\vdash_{\mathrm{MF}} \Gamma \Uparrow \Delta$ *if and only if* $\vdash_{\mathrm{PMF}} \Gamma \Uparrow \Delta$.

Because both positive and negative phases can be pre-empted using the $[\mathrm{PMF}_i]$ rules, we can explicitly sequence two positive phases by introducing new instances of $[\mathrm{PMF}_2]$. Note that focus, once granted, cannot be removed until the formula becomes negative; thus, PMF does not destroy synthetic positive connectives, which are the essential innovation of focusing. After the positive phase of the top bipole has permuted through the negative phase of the bottom bipole, the suspended negative phases are awakened, which might give rise to a number of different sub-derivations (due to &). If $\mathbf{D}$ is this multiset of sub-derivations, then we indicate that it finishes with the negative phase for Ξ as $\mathbf{D} / \Xi$.

Definition 9

1. $(\mathbf{D} / \Xi) \vdash \Gamma \downarrow \Delta ; \Xi$, *where* $\mathbf{D}$ *is a multiset of derivations, has one of the following forms:*

$$\frac{(\mathbf{D} / N, \Xi) \vdash \Gamma \downarrow \Delta ; N, \Xi}{\vdash \Gamma \downarrow \Delta, N ; \Xi} [\mathrm{R}\downarrow] \qquad \frac{(\mathbf{D} / \Xi) \vdash \Gamma \Uparrow \Xi}{\vdash \Gamma \downarrow \cdot ; \Xi} [\mathrm{R}] \qquad \frac{(\mathbf{D} / \Xi) \vdash \Gamma \downarrow \Delta, A_i ; \Xi}{\vdash \Gamma \downarrow \Delta, A_1 \oplus A_2 ; \Xi} [\oplus_i]$$

$$\frac{(\mathbf{D} / \Xi) \vdash \Gamma_1 \downarrow \Delta_1, A ; \Xi \qquad \mathcal{E} \vdash \Gamma_2 \downarrow \Delta_2, B ; \cdot}{\vdash \Gamma_1, \Gamma_2 \downarrow \Delta_1, \Delta_2, A \otimes B ; \Xi} [\otimes] \qquad \frac{(\mathbf{D} / \Xi) \vdash \Gamma \downarrow \Delta, \Delta' ; \Xi}{\vdash \Gamma, \Delta' \downarrow \Delta ; \Xi} [\mathrm{PMF}_2]$$

(And the symmetric case for $[\otimes]$.*)*

2. $(\mathbf{D} / \Xi) \vdash \Gamma \Uparrow \Delta, \Xi$ *where* $\mathbf{D}$ *is a multiset of derivations, has one of the following forms:*

$$\frac{(\mathbf{D}_1 / \Xi', A) \vdash \Gamma \Uparrow \Delta, \Xi', A \qquad (\mathbf{D}_2 / \Xi', B) \vdash \Gamma \Uparrow \Delta, \Xi', B}{\vdash \Gamma \Uparrow \Delta, \Xi', A \,\&\, B} [\&] \qquad \ldots \text{and } \mathbf{D} = \mathbf{D}_1, \mathbf{D}_2$$

$$\dfrac{}{\vdash \Gamma \Uparrow \Delta, \Xi', \top}\,[\top] \qquad \ldots \text{ and } \mathbf{D} = \cdot$$

$$\dfrac{(\mathbf{D} / \Xi', A, B) \vdash \Gamma \Uparrow \Delta, \Xi', A, B}{\vdash \Gamma \Uparrow \Delta, \Xi', A \parr B}\,[\parr] \qquad \dfrac{(\mathbf{D} / \Xi') \vdash \Gamma \Uparrow \Delta, \Xi'}{\vdash \Gamma \Uparrow \Delta, \Xi', \bot}\,[\bot]$$

$$\dfrac{(\mathbf{D} / \Xi') \vdash \Gamma, P \Uparrow \Delta, \Xi'}{\vdash \Gamma \Uparrow \Delta, \Xi', P}\,[R\Uparrow]$$

with $\Xi = \Xi', F$ *for* F *being* $A \parr B$, $A \,\&\, B$, $\top$, $\bot$, *or* P. *Additionally,* $(\mathcal{D} / \cdot) = \mathcal{D}$.

We define the merge operation in terms of a rewrite $\longrightarrow$ between PMF proofs such that in each case of the rewrite at least one root of a [PMF_1] is permuted lower in the derivation. Eventually, this will bring two instances of [PMF_i] next to each other, at which point they are merged. All negative rules encountered during the rewrite are immediately suspended, causing them to permute above the positive phase rooted at the [PMF_i] being permuted. To obtain confluence globally, we must first split the roots to obtain the subset that can merge with the roots of the bottom bipole; otherwise, we might merge bipoles in the wrong order and block possible merges.

Definition 10 *The rewrite* $\longrightarrow$ *between* PMF *proofs has the following rules.*

$$\dfrac{\mathcal{D} \vdash \Gamma \downarrow \Delta, \Delta' ; \Xi}{\vdash \Gamma, \Delta, \Delta' \Uparrow \Xi}\,[\text{PMF}_1] \quad \longrightarrow \quad \dfrac{\dfrac{\mathcal{D} \vdash \Gamma \downarrow \Delta, \Delta' ; \Xi}{\vdash \Gamma, \Delta \downarrow \Delta' ; \Xi}\,[\text{PMF}_2]}{\vdash \Gamma, \Delta, \Delta' \Uparrow \Xi}\,[\text{PMF}_1]$$

$$\dfrac{\dfrac{(\mathbf{D} / \Xi) \vdash \Gamma, P \downarrow \Delta ; \Xi}{\vdash \Gamma, P, \Delta \Uparrow \Xi}\,[\text{PMF}_1]}{\vdash \Gamma, \Delta \Uparrow \Xi, P}\,[R\Uparrow] \quad \longrightarrow \quad \dfrac{(\mathbf{D} / \Xi, P) \vdash \Gamma \downarrow \Delta ; P, \Xi}{\vdash \Gamma, \Delta \Uparrow \Xi, P}\,[\text{PMF}_1]$$

$$\dfrac{\dfrac{(\mathbf{D}_1 / \Xi, C) \vdash \Gamma \downarrow \Delta ; \Xi, C}{\vdash \Gamma, \Delta \Uparrow \Xi, C}\,[\text{PMF}_1] \quad \dfrac{(\mathbf{D}_2 / \Xi, D) \vdash \Gamma \downarrow \Delta ; \Xi, D}{\vdash \Gamma, \Delta \Uparrow \Xi, D}\,[\text{PMF}_1]}{\vdash \Gamma, \Delta \Uparrow \Xi, C \,\&\, D}\,[\&]$$

$$\longrightarrow \quad \dfrac{(\mathbf{D}_1, \mathbf{D}_2 / \Xi, C \,\&\, D) \vdash \Gamma \downarrow \Delta ; \Xi, C \,\&\, D}{\vdash \Gamma, \Delta \Uparrow \Xi, C \,\&\, D}\,[\text{PMF}_1]$$

$$\dfrac{\dfrac{(\mathbf{D} / \Xi, C, D) \vdash \Gamma \downarrow \Delta ; \Xi, C, D}{\vdash \Gamma, \Delta \Uparrow \Xi, C, D}\,[\text{PMF}_1]}{\vdash \Gamma, \Delta \Uparrow \Xi, C \parr D}\,[\parr] \quad \longrightarrow \quad \dfrac{(\mathbf{D} / \Xi, C \parr D) \vdash \Gamma \downarrow \Delta ; \Xi, C \parr D}{\vdash \Gamma, \Delta \Uparrow \Xi, C \parr D}\,[\text{PMF}_1]$$

$$\dfrac{\dfrac{(\mathbf{D} / \Xi) \vdash \Gamma \downarrow \Delta ; \Xi}{\vdash \Gamma, \Delta \Uparrow \Xi}\,[\text{PMF}_1]}{\vdash \Gamma, \Delta \Uparrow \Xi, \bot}\,[\bot] \quad \longrightarrow \quad \dfrac{(\mathbf{D} / \Xi, \bot) \vdash \Gamma \downarrow \Delta ; \Xi, \bot}{\vdash \Gamma, \Delta \Uparrow \Xi, \bot}\,[\text{PMF}_1]$$

$$\dfrac{\dfrac{(\mathbf{D} / N, \Xi) \vdash \Gamma \downarrow \Delta, \Psi ; N, \Xi}{\vdash \Gamma, \Delta \downarrow \Psi ; N, \Xi}\,[\text{PMF}_1]}{\vdash \Gamma, \Delta \downarrow \Psi, N ; \Xi}\,[R\downarrow] \quad \longrightarrow \quad \dfrac{\dfrac{(\mathbf{D} / N, \Xi) \vdash \Gamma \downarrow \Delta, \Psi ; N, \Xi}{\vdash \Gamma \downarrow \Delta, \Psi, N ; \Xi}\,[R\downarrow]}{\vdash \Gamma, \Delta \downarrow \Psi, N ; \Xi}\,[\text{PMF}_1]$$

$$\dfrac{\dfrac{(\mathbf{D} / \Xi) \vdash \Gamma \downarrow \Delta ; \Xi}{\vdash \Gamma, \Delta \Uparrow \Xi}\,[\text{PMF}_1]}{\vdash \Gamma, \Delta \downarrow \cdot ; \Xi}\,[R] \quad \longrightarrow \quad \dfrac{(\mathbf{D} / \Xi) \vdash \Gamma \downarrow \Delta ; \Xi}{\vdash \Gamma, \Delta \downarrow \cdot ; \Xi}\,[\text{PMF}_2]$$

$$\dfrac{\dfrac{\mathcal{D} \vdash \Gamma_1 \downarrow \Psi, \Delta_1, A \,;\, \Xi_1}{\vdash \Gamma_1, \Psi \downarrow \Delta_1, A \,;\, \Xi_1}\,[\mathrm{PMF}_2] \qquad \mathcal{E} \vdash \Gamma_1 \downarrow \Delta_2, B \,;\, \Xi_2}{\vdash \Gamma_1, \Gamma_2, \Psi \downarrow \Delta_1, \Delta_2, A \otimes B \,;\, \Xi_1, \Xi_2}\,[\otimes]$$

$$\longrightarrow \quad \dfrac{\dfrac{\mathcal{D} \vdash \Gamma_1 \downarrow \Psi, \Delta_1, A \,;\, \Xi_1 \qquad \mathcal{E} \vdash \Gamma_1 \downarrow \Delta_2, B \,;\, \Xi_2}{\vdash \Gamma_1, \Gamma_2 \downarrow \Psi, \Delta_1, \Delta_2, A \otimes B \,;\, \Xi_1, \Xi_2}\,[\otimes]}{\vdash \Gamma_1, \Gamma_2, \Psi \downarrow \Delta_1, \Delta_2, A \otimes B \,;\, \Xi_1, \Xi_2}\,[\mathrm{PMF}_2]$$

$$\dfrac{\dfrac{\mathcal{D} \vdash \Gamma \downarrow \Psi, \Delta, A_i \,;\, \Xi}{\vdash \Gamma, \Psi \downarrow \Delta, A_i \,;\, \Xi}\,[\mathrm{PMF}_2]}{\vdash \Gamma, \Psi \downarrow \Delta, A_1 \oplus A_2 \,;\, \Xi}\,[\oplus_i] \quad \longrightarrow \quad \dfrac{\dfrac{\mathcal{D} \vdash \Gamma \downarrow \Psi, \Delta, A_i \,;\, \Xi}{\vdash \Gamma \downarrow \Psi, \Delta, A_1 \oplus A_2 \,;\, \Xi}\,[\oplus_i]}{\vdash \Gamma, \Psi \downarrow \Delta, A_1 \oplus A_2 \,;\, \Xi}\,[\mathrm{PMF}_2]$$

$$\dfrac{\dfrac{\mathcal{D} \vdash \Gamma \downarrow \Psi_1, \Psi_2, \Delta \,;\, \Xi}{\vdash \Gamma, \Psi_1 \downarrow \Delta, \Psi_2 \,;\, \Xi}\,[\mathrm{PMF}_2]}{\vdash \Gamma, \Psi_1, \Psi_2 \downarrow \Delta \,;\, \Xi}\,[\mathrm{PMF}_2] \quad \longrightarrow \quad \dfrac{\mathcal{D} \vdash \Gamma \downarrow \Psi_1, \Psi_2, \Delta \,;\, \Xi}{\vdash \Gamma, \Psi_1, \Psi_2 \downarrow \Delta \,;\, \Xi}\,[\mathrm{PMF}_2]$$

$$\dfrac{\dfrac{\mathcal{D} \vdash \Gamma \downarrow \Psi_1, \Psi_2, \Delta \,;\, \Xi}{\vdash \Gamma, \Psi_1 \downarrow \Psi_2 \,;\, \Xi}\,[\mathrm{PMF}_2]}{\vdash \Gamma, \Psi_1, \Psi_2 \Uparrow \Xi}\,[\mathrm{PMF}_1] \quad \longrightarrow \quad \dfrac{\mathcal{D} \vdash \Gamma \downarrow \Psi_1, \Psi_2 \,;\, \Xi}{\vdash \Gamma, \Psi_1, \Psi_2 \Uparrow \Xi}\,[\mathrm{PMF}_1]$$

The symmetric cases for [PMF$_1$] / [$\otimes$] *and* [PMF$_1$] / [$\oplus$]$_i$ *are elided.*

The rewrite in defn. 10 is a permutation on MF derivations modulo the injection into PMF. The intermediate points of the permutation after the injection are not interesting, but the reflexive-transitive closure of the PMF rewrite also defines the following MF rewrite.

Definition 11 *If* $\mathcal{D}, \mathcal{E} \vdash_{\mathrm{MF}} \Gamma \Updownarrow \Delta$, *and* $\mathcal{D}^\# \longrightarrow^* \mathcal{E}^\#$, *then* $\mathcal{D} \longrightarrow \mathcal{E}$.

We shall show that this rewrite on MF derivations will generate the maximal proofs. The proof itself will be a trivial consequence of two decomposition lemmas. The left-decomposition lemma below shows that the maximal proofs are $\longrightarrow$-normal upto iso-polarity.

Lemma 12 (left decomposition)
If $\mathcal{D} \vdash \Gamma \Updownarrow \Delta$ *is maximal and* $\mathcal{D} \longrightarrow \mathcal{E}$, *then* $\mathcal{D} \approx \mathcal{E}$.

Proof. Note that in every case of the rewrite $\longrightarrow$ on PMF derivations, an instance of [PMF$_1$] is brought closer to the root of the derivation. Therefore, the rewrite $\longrightarrow$ on MF proofs can only enlarge the lowermost roots in $\mathcal{D}$. But, $\mathcal{D}$ is already maximal. So $\mathcal{E}$ has the same instances of [MF] as $\mathcal{D}$, *i.e.*, $\mathcal{D} \approx \mathcal{E}$. □

The second key lemma is a right-decomposition that establishes that the maximal proofs are reachable by $\longrightarrow$.

Lemma 13 (right decomposition)
If $\mathcal{D} \simeq \mathcal{E} \vdash_{\mathrm{MF}} \Gamma \Updownarrow \Delta$ *and* $\mathcal{E}$ *is maximal, then* $\mathcal{D} \longrightarrow \mathcal{E}$.

Proof (Sketch). We have to show that all ways of permuting a root downwards in a proof can be generated by $\longrightarrow$. But this is easily seen because the $\longrightarrow$ is allowed to divide the roots and permute only the necessary fragment downwards. For a representative example, suppose the following is a sub-derivation of $\mathcal{D}^\#$:

$$\mathcal{F} = \cfrac{\cfrac{\cfrac{\mathcal{F}' \vdash \Gamma, P \downarrow \Delta, Q \,;\, \cdot}{\vdash \Gamma, P, Q, \Delta \Uparrow \cdot}\,[\mathrm{PMF}_1]}{\vdash \Gamma, \Delta \Uparrow P, Q}\,[\mathrm{R}\Uparrow]^2}{\vdash \Gamma, \Delta \Uparrow P \parr Q}\,[\parr]$$

Of the roots Δ, Q, only Δ can possibly permute below $P \parr Q$, because Q is one of its sub-formulas. According to the rewrite rules, we first remove Q from the roots of the [PMF] rule by inserting another [PMF]. The permutation can now proceed (for some $\mathbf{F} \,/\, P, Q \simeq \mathcal{F}''$):

$$\cfrac{\cfrac{\cfrac{\mathcal{F}'' = \cfrac{\mathcal{F}' \vdash \Gamma, P \downarrow \Delta, Q \,;\, \cdot}{\vdash \Gamma, P, Q \downarrow \Delta \,;\, \cdot}\,[\mathrm{PMF}_2]}{\vdash \Gamma, P, Q, \Delta \Uparrow \cdot}\,[\mathrm{PMF}_1]}{\vdash \Gamma, \Delta \Uparrow P, Q}\,[\mathrm{R}\Uparrow]^2}{\vdash \Gamma, \Delta \Uparrow P \parr Q}\,[\parr] \longrightarrow \cfrac{\cfrac{(\mathbf{F} \,/\, P, Q) \vdash \Gamma \downarrow \Delta \,;\, P, Q}{\vdash \Gamma, \Delta \Uparrow P, Q}\,[\mathrm{PMF}_1]}{\vdash \Gamma, \Delta \Uparrow P \parr Q}\,[\parr]$$

$$\longrightarrow \cfrac{(\mathbf{F} \,/\, P \parr Q) \vdash \Gamma \downarrow \Delta \,;\, P \parr Q}{\vdash \Gamma, \Delta \Uparrow P \parr Q}\,[\mathrm{PMF}_1]$$

The instance of $[\mathrm{PMF}_1]$ that permutes down is free of the disallowed root Q. □

Proof (of theorem 7). Let $\mathcal{D} \simeq \mathcal{E} \vdash_{\mathrm{MF}} \Gamma \Updownarrow \Delta$ be given such that both $\mathcal{D}$ and $\mathcal{E}$ are maximal. By lemma 13, $\mathcal{D} \longrightarrow \mathcal{E}$; hence, by lemma 12, $\mathcal{D} \approx \mathcal{E}$. □

5 Multi-focusing and proof nets

The usual approach to the proof identity problem in linear logic (and to providing a canonical representation of proofs) consists in using proof nets which were first introduced by Girard [9]. Since we proved that maximally multi-focused proofs also provide such a canonical approach to proofs it is natural to compare our approach with proof nets. This is the aim of the present section where we deal with a restricted fragment of MALL proofs, the unit-free cut-free multiplicative fragment, MLL^-, for which proof nets are especially well-behaved: we shall provide a direct proof that maximally multi-focused proofs in MLL^- are in a one-to-one correspondence with cut-free MLL^- proof nets.

The previous results of the paper already ensure that such a result is true but we shall now give a direct evidence of this fact by actually building the class of iso-polar maximally multi-focused proofs corresponding to a given proof net. The converse, namely that two iso-polar maximally multi-focused proofs correspond to the same proof nets is trivial.

Proof nets are structures that do not retain all the unnecessary ordering information contained in a sequent proof. A MLL^- proof structure is thus a graph structure consisting in the formula tree of the sequent $\vdash \Gamma$ together with some more structure representing the initial rules:

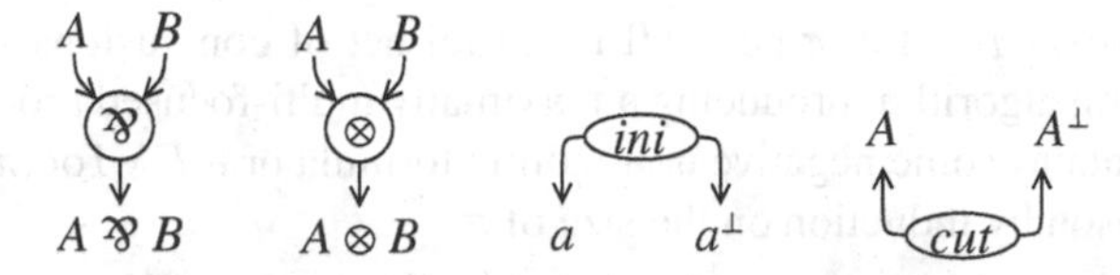

Fig. 4 Unit-free cut-free MLL proof nets. *ini* is restricted to the atomic formulas.

Definition 14 (MLL⁻ proof structure) *A MLL⁻ proof structure on $\vdash \Gamma$ is a graph made of cells represented in Figure 4 which are linked by edges labeled with MLL⁻ formulas. There is one pending edge for each formula F in $\vdash \Gamma$ which is labeled with F and which is called a conclusion.*

Additional conditions are imposed in order to ensure that this *proof structure* is actually a logical object and represents a proof:

Definition 15 (MLL⁻ proof net) *A MLL⁻ proof net on $\vdash \Gamma$ is a proof structure that results from the desequentialization of a sequent proof π of $\vdash \Gamma$ by forgetting the inference rule ordering*[1].

The previous definition does not provide a convenient criterion that can be helpful to check that a given proof structure is indeed a proof net. Many more satisfying criteria have been provided to characterize proof nets, they all have in common not to be inductive but geometric criteria (they deal with the structure as a whole, not as made of elementary components). In the following, we shall only consider cut-free MLL⁻ proof structures.

As already mentioned, we shall now be interested in providing a direct proof of the following theorem:

Theorem 16 *Two maximally multi-focused MLL⁻ proofs of $\vdash \cdot \Uparrow \Gamma$ are iso-polar iff they have the same MLL⁻ proof net.*

The theorem will be proved by showing that for every proof net there is a unique maximally multi-focused proof (up to iso-polarity) associated with it. We first recall two definitions from [1] which develops a focused sequentialization algorithm for MLL⁻ proof nets:

Definition 17 (split(π), foc(π), from [1]) *Let π be an MLL⁻ proof net.*

1. *split(π) is the set of positive conclusions P of π such that removing the concluding ⊗-link of P disconnects π in two proof nets π_1 and π_2.*
2. *foc(π) is the set of conclusions F of π such that F is a positive atom and π is just an ini link; or $F \in$ split(π) and its premisses A and B are conclusions of the two sub-nets π_1 and π_2 where A (resp. B) is negative or $A \in$ foc(π_1) (resp. $B \in$ foc(π_2)).*

[1] A MLL⁻ inference rule is turned to the corresponding cell of Figure 4 and the cells are combined by tracing the formulas occurrences in the sequent proof.

Proof (of Theorem 16). Let π be a MLL$^-$ proof net of conclusions Γ. We outline a sequentialization algorithm producing a maximally multi-focused proof of conclusion $\vdash \cdot \Uparrow \Gamma$ if Γ contains some negative non-atomic formula or $\vdash \Gamma \setminus \mathtt{foc}(\pi) \Downarrow \mathtt{foc}(\pi)$ otherwise. We reason by induction on the size of π.

Case Γ *contains at least one negative formula.* We remove all negative cells (that is, the $⅋$ cells) of π up to reaching a positive cell or an initial cell. The resulting proof structure is a proof net π' and its conclusions Γ' are positive. By induction hypothesis, we can sequentialize it into a maximally multi-focused proof $\mathcal{D}'$ of conclusion $\vdash \Gamma' \setminus \mathtt{foc}(\pi') \Downarrow \mathtt{foc}(\pi')$ by sequentializing in an arbitrary order (the different possibilities give rise to iso-polar proofs) the negative rules that have been removed in the previous step, we obtain a proof $\mathcal{D}$ of the form:

$$\dfrac{\dfrac{\dfrac{\mathcal{D}' \vdash \Gamma' \setminus \mathtt{foc}(\pi') \Downarrow \mathtt{foc}(\pi')}{\vdash \Gamma' \Uparrow \cdot}\,[\mathrm{MF}]}{\vdots}\,[⅋]}{\mathcal{D} \vdash \cdot \Uparrow \Gamma}\,[⅋]$$

Case Γ *contains only positive formulas.* Since π is a proof net, $\mathtt{foc}(\pi) \neq \emptyset$. Consider the formulas in $[()\pi)$ and remove the top-most positive connectives of every $F \in \mathtt{foc}(\pi)$. The resulting proof structure is not a proof net since it is not connected; however, each of its connected components is. Let them be $\pi_1, \ldots, \pi_n$. For $1 \leq i \leq n$, π_i has conclusions Γ_i which has at least one negative formula or which is reduced to an axiom link. In the first case, one can inductively sequentialize it into of maximally multi-focused proof $\mathcal{D}_i$. In order to conclude, we only need to show that one can obtain a proof of $\vdash \cdot \Uparrow \Gamma$ from the $\mathcal{D}_i$ and the positive cells of the formulas of $\mathtt{foc}(\pi)$, which follows from the fact that the formulas in $\mathtt{foc}(\pi)$ are hereditarily splitting: applying these formulas in any order (as long as the sub-formula priority is maintained), gives rise to a way to sequentialize π.

We finally need to check that the proof obtained with this process is indeed maximal, but this is done very easily: let F be a formula that could potentially enlarge the set of foci and let us consider a proof $\mathcal{D}_F$ that witnesses this fact ($\mathcal{D}_F$ focuses on F). By desequentializing $\mathcal{D}_F$, we get a proof net π and since $\mathcal{D}_F$ is a sequentialization of π that focuses on F which is positive, then F is hereditarily splitting, that is $F \in \mathtt{foc}(\pi)$, so $\mathtt{foc}(\pi)$ is maximal.

The process considered in this proof is non-deterministic (within a negative or positive phase, we sequentialize *in any order*) and we can check that the different proofs that can result from this process are exactly all the iso-polar maximally multi-focused proofs of the iso-polarity class corresponding to proof net π. □

We showed in this section that there is a bijection between MLL$^-$ proof nets and classes of iso-polar maximally multi-focused proofs. MLL$^-$ proof nets are certainly the most concise canonical structures for this fragment. There are candidates to extend MLL$^-$ proof nets to broader fragments (MLL with units [14], MALL [12] or MELL) but they are not as satisfactory as for MLL$^-$. The analysis we just made could be carried to MALL proof nets as introduced by Hughes and van Glabbeek [12] for the appropriate extension of definition 17 (in particular to take into account the fact that with MALL proof nets there is not only one linking but a set of linkings corresponding to the additive slices of the proof net).

The problem of proof-nets for MALL with units is still open. Yet, these fragments have standard sequent calculi with well understood focusing systems. We expect that an analysis of the maximally multi-focused sequent proofs would yield a better understanding of proof net-like structures for such fragments.

6 Conclusion

The contributions of this paper are three-fold: *(i)* we extend Andreoli's definition of focused proofs to multi-focused proofs, for which we define a notion of *maximality*; *(ii)* we show that the maximally multi-focused proofs are representatives of their $\simeq$-equivalence class upto a trivial iso-polar equivalence; and *(iii)* we prove that unit-free multiplicative proof nets are in bijective correspondence with maximal multi-focused proofs for unit-free multiplicative linear logic.

The notion of multi-focusing in this paper was first considered by Saurin and Miller [16] as naturally arising in the structure of *focalization graphs* to prove the focalization theorem. Multi-focusing was subsequently also used by Delande and Miller [7] as a necessary generalization of Andreoli's asymmetric treatment of the positive formulas. Andreoli studied focusing in proof nets [1, 3] and defined a notion of "multi-focus" [3] with a different meaning: there, it refers to a part of the context which is needed in order to apply the decide rule. He also investigates the use of focusing to construct proof nets for a restricted fragment of MLL^{-}.

Faggian *et al* [8, 6] introduced L-nets as a generalization of *designs* from Girard's ludics [11]: L-nets can be seen as designs with a flexible degree of sequentiality, falling between sequent proofs and proof nets. This appears similar to multi-focusing which covers the spectrum from singly focused proofs to maximally multi-focused proofs, and thus exhibits some flexibility about the degree of sequentiality. This flexibility is also observed in [7] which presents the search for proofs and refutations as a pair of mutually normalizing interpretations of a neutral procedure for the cut and atom-free MALL. Relating these diverse approaches is an important matter for future work.

Several other open questions remain about multi-focused proofs. Firstly, we lack a cut-elimination theorem for multi-focused proofs that generalizes similar theorems for singly focused proofs (see, *eg*. [4, 15]). Moreover, it is considerably unclear how maximality interacts with cut-elimination, for the standard procedure would not preserve maximality. In terms of larger fragments of linear logic, multi-focusing generalizes easily to admit the exponentials and first-order quantification; however, the respective notions of maximality remain to be developed for these fragments.

References

1. J.-M. Andreoli and R. Maieli. Focusing and proof nets in linear and noncommutative logic. In *International Conference on Logic for Programming and Automated Reasoning (LPAR)*, volume 1581 of *LNAI*. Springer, 1999.
2. Jean-Marc Andreoli. Logic programming with focusing proofs in linear logic. *J. of Logic and Computation*, 2(3):297–347, 1992.
3. Jean-Marc Andreoli. Focussing proof-net construction as a middleware paradigm. In Andrei Voronkov, editor, *18th Conference on Automated Deduction (CADE)*, number 2392 in LNAI, pages 501–516. Springer, 2002.
4. Kaustuv Chaudhuri, Frank Pfenning, and Greg Price. A logical characterization of forward and backward chaining in the inverse method. *J. of Automated Reasoning*, 40(2-3):133–177, March 2008.
5. Pierre-Louis Curien. Introduction to linear logic and ludics, Part I. *Advances in Mathematics (China)*, 34(5):513–544, January 2005.
6. Pierre-Louis Curien and Claudia Faggian. L-nets, strategies and proof-nets. In C.-H. Luke Ong, editor, *CSL 2005: Computer Science Logic*, volume 3634 of *LNCS*, pages 167–183. Springer, 2005.
7. Olivier Delande and Dale Miller. A neutral approach to proof and refutation in MALL. In F. Pfenning, editor, *23th Symp. on Logic in Computer Science*. IEEE Computer Society Press, 2008.
8. Claudia Faggian and François Maurel. Ludics nets, a game model of concurrent interaction. In *20th Symp. on Logic in Computer Science*, pages 376–385. IEEE Computer Society, 2005.
9. Jean-Yves Girard. Linear logic. *Theoretical Computer Science*, 50:1–102, 1987.
10. Jean-Yves Girard. Proof-nets: the parallel syntax for proof-theory. In Aldo Ursini and Paolo Agliano, editors, *Logic and Algebra*, volume 180 of *Lecture Notes In Pure and Applied Mathematics*, pages 97–124, New York, 1996. Marcel Dekker.
11. Jean-Yves Girard. Locus solum. *Mathematical Structures in Computer Science*, 11(3):301–506, June 2001.
12. Dominic Hughes and Rob Van Glabbeek. Proof nets for unit-free multiplicative-additive linear logic. *ACM Trans. on Computational Logic*, 6:784–842, 2005.
13. Stephen Cole Kleene. Permutabilities of inferences in Gentzen's calculi LK and LJ. *Memoirs of the American Mathematical Society*, 10:1–26, 1952.
14. François Lamarche and Lutz Straßburger. From proof nets to the free *-autonomous category. *Logical Methods in Computer Science*, 2(4:3):1–44, 2006.
15. Chuck Liang and Dale Miller. Focusing and polarization in intuitionistic logic. In J. Duparc and T. A. Henzinger, editors, *CSL 2007: Computer Science Logic*, volume 4646 of *LNCS*, pages 451–465. Springer, 2007.
16. Dale Miller and Alexis Saurin. From proofs to focused proofs: a modular proof of focalization in linear logic. In J. Duparc and T. A. Henzinger, editors, *CSL 2007: Computer Science Logic*, volume 4646 of *LNCS*, pages 405–419. Springer, 2007.

Universal Coinductive Characterisations of Process Semantics

David de Frutos Escrig* and Carlos Gregorio Rodríguez**

Department of Sistemas Informáticos y Computación
Universidad Complutense de Madrid
defrutos,cgr@sip.ucm.es

Abstract. We present a theoretical framework which allows to define in a uniform way coinductive characterisations of nearly any semantic preorder or equivalence between processes, by means of *simulations up-to* and *bisimulations up-to*. In particular, all the semantics in the linear time-branching time spectrum are covered. *Constrained simulations*, that generalise plain simulations by including a constraint that all the pairs of related processes must satisfy, are the key to obtain such a general framework. We provide a simple axiomatisation of any constrained simulation preorder and also for the corresponding equivalence. These axiomatizations allow us to prove in a uniform way that each constrained simulation preorder (equivalence) defines a class of process preorders (equivalences) which share commons properties, like the possibility of giving coinductive characterisations for all of them, or the existence of a canonical preorder inducing each of these equivalences.

1 Introduction and Related Work

One of the essential decisions that should be taken when defining a process algebra is to settle in the most adequate way its underlying semantics. Occasionally the semantics is directly determined by an equivalence relation but more often it is based on a preorder relation, although certainly every preorder induces an equivalence by means of its kernel; besides, the ordering relations can be used to compare non-equivalent processes or to define continuous domains in order to apply fix point arguments to define the behaviour of recursive processes.

Every semantics sets forth a level of abstraction that determines which aspects of the behaviour of processes are of importance and which are not. Mainly because of the generality and diversity of the applications of process algebras, there is no prevailing semantic notion, but rather a number of different proposals have arisen from diverse approaches, contexts and applications.

We consider that this variety of process semantics is a good sign of the applicability of process algebras just proving the healthiness of the formalism. However, this plurality of semantics becomes a hindrance when the goal is to

* Partially supported by the Spanish MEC project DESAFIOS TIN2006-15660-C02-01 and the project PROMESAS-CAM S-0505/TIC/0407.

** Partially supported by the Spanish MEC project WEST/FAST TIN2006-15578-C02-01.

Please use the following format when citing this chapter:

de Frutos Escrig, D. and Rodríguez, C.G., 2008, in IFIP International Federation for Information Processing, Volume 273; *Fifth IFIP International Conference on Theoretical Computer Science*; Giorgio Ausiello, Juhani Karhumäki, Giancarlo Mauri, Luke Ong; (Boston: Springer), pp. 397–412.

study general properties for all of them, to compare different semantics or to determine what semantics suits a given application better.

That is why it would be nice to have a unified model that could provide us with a general and uniform approach to the different semantics. Our work targets this goal and looks for a common framework in which to include the semantics for processes. Bisimulation semantics [Par81,Mil89] is one of the most elegant and powerful equivalences defined for processes and it was our starting point to achieve this goal of uniformity. In [dFG05] we showed how to weaken the notion of bisimulation defining our bisimulations up-to that characterise many other interesting equivalences. In [dFG07] we continued that work, extending our results by considering process preorders instead of equivalences, and we have found out that this approach is indeed even more general, giving rise to a richer and more elegant theory. In both cases ready simulation [BIM95] was our main support, and that meant that we could only apply our results to preorders that were coarser than ready simulation. This restriction also appear in other related works such as [AFI07].

However, it was not clear that only the semantics coarser than ready simulation would satisfy our results. In fact, we had already presented in [dFG05] a result (Theorem 2, there) proving that we could also get coinductive characterisations of some equivalences finer than the ready simulation equivalence. To prove that theorem we required a quite ad-hoc property, that we called Hoare-Equivalence, but in [dFG07] we did not find the way to transfer these results to the framework of semantic preorders.

This paper focuses on the generalisation of the simulations up-to, and provides a general coinductive characterisation of a great variety of semantics preorders, either coarser or finer than the ready simulation preorder, in particular, this characterisation can be applied to all the semantics in the linear time-branching time spectrum [Gla01]. This generalisation has been possible after the observation that ready simulation was just a significant example of what we have called *constrained simulations*. This kind of simulations preserve the properties we need in order to prove the generalisation of previous results.

The rest of the paper is structured as follows. In Section 2 we introduce the basic definitions and notations on processes and preorders, and we recall some results from our previous works [dFG05, dFG07]. In Section 3 we define the family of *constrained simulations*, where simulations are constrained by the obligation to relate processes that satisfy some adequate condition. We provide a sound and complete axiomatization for the preorders and the induced equivalence relations (see Theorem 4 and 5). These axiomatizations are one of the key points in the proofs of the main results of the paper that follow in the next sections.

The core of our results is collected in Sections 4 and 5, where we define the notion of *constrained simulation up-to a preorder*; we develop our theory through a number of results that provide characterisations of the semantic preorders and equivalences in terms of constrained simulations up-to (see Theorem 6, 7, 8 and 9). Some additional results that illustrate the applications of the theory are also included (see Theorem 10). Finally in Section 6 we conclude by discussing some research lines for future work.

2 Preliminaries and Previous Work

The behaviour of processes is usually described using the well-established formalism of *labelled transition systems* [Plo81] or lts for short.

Definition 1. A labelled transition system is a structure $\mathcal{T} = (\mathcal{P}, Act, \rightarrow)$ where

- $\mathcal{P}$ is a set of processes, agents or states,
- *Act* is a set of actions, and
- $\rightarrow \subseteq \mathcal{P} \times Act \times \mathcal{P}$ is a transition relation.

A rooted lts is a pair $(\mathcal{T}, p_0)$ with $p_0 \in \mathcal{P}$.

Act is the set of actions that processes can perform and the relation $\rightarrow$ describes the process transitions after the execution of actions. The triple $\langle p, a, q\rangle$ is represented by $p \xrightarrow{a} q$, indicating that process p performs action a evolving to process q. A rooted lts describes the semantics of a process: that corresponding to its initial state p_0.

Some usual notations on lts are used. We write $p \xrightarrow{a}$ if there exists a process q such that $p \xrightarrow{a} q$. The function I calculates the set of initial actions of a process, $I(p) = \{a \mid a \in Act \text{ and } p \xrightarrow{a}\}$.

Lts for finite processes are just directed graphs which become finite trees if expanded. These finite trees can be syntactically described by the basic process algebra BCCSP, which was also used, for instance, in [Gla01, dFG05].

Definition 2. Given a set of actions *Act*, the set of BCCSP processes is defined by the following BNF-expression:

$$p ::= \mathbf{0} \mid ap \mid p + q$$

where $a \in Act$. $\mathbf{0}$ represents the process that performs no action; for every action in *Act*, there is a prefix operator; and $+$ is a choice operator.

All the definitions we present in the paper are valid for arbitrary processes, that is, for arbitrary rooted lts, either finite or infinite. We are going to prove the results in this paper mainly by induction on the depth of BCCSP processes. Then, by using continuity arguments (in a similar way as we did in [dFG05]) these results can be extended to arbitrary finitely branching transition systems, since by unfolding any of them we can get an equivalent finitary tree process.

The operational semantics for BCCSP terms is defined in Fig. 1. The depth of a BCCSP process is the depth of the tree it denotes.

$$ap \xrightarrow{a} p \qquad \frac{p \xrightarrow{a} p'}{p+q \xrightarrow{a} p'} \qquad \frac{q \xrightarrow{a} q'}{p+q \xrightarrow{a} q'}$$

Fig. 1 Operational Semantics for BCCSP Terms

As usual, trailing occurrences of the constant $\mathbf{0}$ are omitted: we write a instead of $a\mathbf{0}$. By using $\sum$ as a shorthand for multiple choice (which is commutative and associative) we can define any process as $\sum_i \sum_j a_i p_{ij}$. A process aq' is a summand of the process q if and only if $q \xrightarrow{a} q'$. Given $a \in Act$ we define $p|_a$ as the (sub)process we get by adding all the a-summands of p. That is, if $p = \sum_i \sum_j a_i p_{ij}$, then $p|_{a_i} = \sum_j a_i p_{ij}$.

Preorders, that we represent by $\sqsubseteq$, are reflexive and transitive relations. We use the symbol $\sqsupseteq$ to represent the preorder relation $\sqsubseteq^{-1}$. Every preorder induces an equivalence relation that we denote by $\equiv$, that is $p \equiv q$ if and only if $p \sqsubseteq q$ and $q \sqsubseteq p$. Finally, bisimulation equivalence is denoted by $=_B$.

Definition 3. A preorder relation $\sqsubseteq$ over processes is a *behaviour preorder* if it is coarser than the bisimulation equivalence, i.e. $p =_B q \Rightarrow p \sqsubseteq q$, and it is a precongruence with respect to the prefix and choice operators, i.e. if $p \sqsubseteq q$ then $ap \sqsubseteq aq$ and $p + r \sqsubseteq q + r$. Besides, if the relation is symmetric, i.e. is an equivalence relation, we say that it is a *behaviour equivalence.*

In [dFG05] we introduced bisimulations up-to a preorder (that we denote by $\approx_{\sqsubseteq}$) in order to weaken the definition of bisimulations in such a way that weaker equivalences could be captured by a coinductive definition.

Definition 4. Let $\sqsubseteq$ be a behaviour preorder. Then a binary relation S over processes is a *bisimulation up-to* $\sqsubseteq$, if pSq implies that:

- For every a, if $p \xrightarrow{a} p'_a$, then there exist q' and q'_a, $q \sqsupseteq q' \xrightarrow{a} q'_a$ and $p'_a S q'_a$;
- For every a, if $q \xrightarrow{a} q'_a$, then there exist p' and p'_a, $p \sqsupseteq p' \xrightarrow{a} p'_a$ and $p'_a S q'_a$.

Two processes are *bisimilar up-to* $\sqsubseteq$, written $p \approx_{\sqsubseteq} q$, if there exists a bisimulation up-to $\sqsubseteq$, S, such that pSq.

The added capability introduced by the $\sqsupseteq$-reduction generalises the original definition of bisimulation, so that we have now more chances to prove the equivalence of two processes. When the behaviour preorder is just the identity relation we get the bisimulation equivalence, but, as we proved in [dFG05], we get other interesting semantics (traces, failures, ready simulation and so on) by considering other behaviour preorders. One of the main results in that paper (see Theorem 1 below) required the preorders to satisfy the axiom (RS), $ax \sqsubseteq ax + ay$ (that characterises the ready simulation preorder [BIM95]) so that it could only be applied to semantics coarser than the ready simulation.

Definition 5. A behaviour preorder $\sqsubseteq$ is *initials preserving* when $p \sqsubseteq q$ implies $I(p) \subseteq I(q)$. It is *action factorised* (or just *factorised*) when $p \sqsubseteq q$ implies $p|_a \sqsubseteq q|_a$, for all $a \in I(p)$.

Theorem 1 ([dFG05]). *For every behaviour preorder $\sqsubseteq$ that is initials preserving, action factorised and satisfying the axiom (RS), we have that $p \eqsim_{\sqsubseteq} q$ if and only if $p \equiv q$.*

This theorem provides a symmetric, coinductive, bisimulation-like characterisation for any equivalence in the linear time-branching time spectrum from trace equivalence to ready simulation equivalence.

Once we had coinductive characterisations for many semantic equivalences we shifted the focus from equivalences to preorders. In [dFG07] we first achieved characterisations of some semantic preorders in terms of simulations up-to.

Definition 6. Let $\sqsubseteq$ be a behaviour preorder, we say that a binary relation S over processes is a *simulation up-to* $\sqsubseteq$, if pSq implies that:

- For every a, if $p \xrightarrow{a} p'_a$ there exist q' and q'_a, $q \sqsupseteq q' \xrightarrow{a} q'_a$ and $p'_a S q'_a$.

We say that process p is simulated up-to $\sqsubseteq$ by process q, or that q simulates p up-to $\sqsubseteq$, written $p \sqsubseteq_{\sqsubseteq} q$, if there exists a simulation up-to $\sqsubseteq$, S, such that pSq.

Theorem 2 ([dFG07]). *For every behaviour preorder $\sqsubseteq$ that satisfies the axiom (S) $x \sqsubseteq x + y$, we have $p \sqsubseteq_{\sqsubseteq} q$ if and only if $p \sqsubseteq q$.*

This result only applies to preorders coarser than the simulation preorder and therefore it falls short of the generality we got in Theorem 1. In order to regain this generality we needed to strengthen the simulation relation to achieve a greater discriminating power. Ready simulation was again called to play an essential role.

Definition 7. Let I be the binary relation that captures the equivalence of initial actions and is defined over pairs of processes by $pIq \Leftrightarrow I(p) = I(q)$. Let $\sqsubseteq$ be a behaviour preorder, we say that a binary relation S over processes is an *I-simulation up-to* $\sqsubseteq$, if $S \subseteq I$ (that is, $pSq \Rightarrow pIq$), and S is a simulation up-to $\sqsubseteq$. Or, equivalently, in a coinductive way, whenever we have pSq we also have:

- For every a, if $p \xrightarrow{a} p'_a$ there exist q', q'_a such that $q \sqsupseteq q' \xrightarrow{a} q'$ and $p'_a S q'_a$;
- pIq;

We say that process p is I-simulated up-to $\sqsubseteq$ by process q, or that q I-simulates p up-to $\sqsubseteq$, written $p \sqsubseteq^I_{\sqsubseteq} q$, if there exists an I-simulation up-to $\sqsubseteq$, S, such that pSq.

By using this definition we were able to characterise all the preorders finer than failures and coarser than ready simulation.

Theorem 3 ([dFG07]). *For every behaviour preorder $\sqsubseteq$ that satisfies the axiom (RS), and $\sqsubseteq \subseteq I$, we have that $p \precsim^{I}_{\sqsubseteq} q$ if and only if $p \sqsubseteq q$.*

Once again, we needed the axiom (RS) to prove the theorem above. In the rest of the paper we will generalise our results by taking as starting point a more general class of simulations that we have called *constrained simulations.*

3 Constrained Simulations

C-constrained simulations are just plain simulations to which we impose that their pairs should also be related by the constraint C.

Definition 8. Given a relation C over BCCSP processes, a relation S_C is a *C-constrained simulation*, if pS_Cq implies:

- For every a, if $p \xrightarrow{a} p'$ there exists q', $q \xrightarrow{a} q'$ and $p'S_Cq'$, and
- pCq.

We say that process p is C-simulated by process q, or that q C-simulates p, written $p \sqsubseteq_{\rightarrow}^{C} q$, whenever there exists a C-constrained simulation S_C, such that pS_Cq.

Since we want to characterise behaviour preorders by using C-simulations it is reasonable to impose on these simulations the condition of being themselves behaviour preorders; that is guaranteed whenever the constraints are also behaviour preorders. Given that the operators in our basic algebra BCCSP are those generating finite trees, this condition is quite natural and the results we will prove based on it are indeed rather general.

Example 1. Let us briefly present several examples of constrained simulations, all of them corresponding to relations being behaviour preorders.

- Ordinary simulation is a constrained simulation taking as C the universal relation, xCy for every x and y.
- Ready simulation is just the I-constrained simulation, where $pIq \Leftrightarrow I(p) = I(q)$.
- Ready simulation is perhaps the most important C-constrained simulation but we can also achieve a greater discriminating power. Let us consider, for instance, the simulation preorder $\sqsubseteq_S$ and $C = \sqsubseteq_S^{-1}$; then 2-nested simulations [GV92] are just the corresponding class of C-constrained simulations.

One could argue that, whenever we admit a nearly arbitrary constraint when defining the constrained simulations, we are vitiating the local character of the notion of simulation, thus spoiling the coinductive nature of the generalisation. We do not agree with such an opinion for several reasons. First, we can still consider local constraints that provide interesting results, such as I, as we are going to see. More in general, this family of constrained simulations will allow us to set a benchmark to compare and classify the complexity of process semantics, most of these semantics can be characterised with a local constraint, but others, such as 2-nested simulation, are intrinsically non-local (see [AFGI04]). To have a common framework to unify all these simulation semantics is very useful because we can prove general results for all of then as we present in the rest of this section.

C-constrained similarity, $\underset{\rightarrow}{\sqsubseteq}^C$, can be conditionally axiomatized in a simple way. For any constraint C we just need to consider the axiom

$$(P_C) \qquad xCy \Rightarrow x \sqsubseteq x + y$$

We define the axiomatization $\mathcal{P}_C$ as the set of axioms obtained by adding the axiom P_C to the set of axioms that characterises bisimulation equivalence (Figure 2), $\mathcal{P}_C = \{B_1, B_2, B_3, B_4, P_C\}$. As usual, we write $\mathcal{P}_C \vdash p \sqsubseteq q$ when the relation $p \sqsubseteq q$ is provable from $\mathcal{P}_C$ using the rules of inequational logic. $\mathcal{P}_C$ is sound and complete with respect to $\underset{\rightarrow}{\sqsubseteq}^C$.

$$\begin{array}{llll} (B_1) & x + y = y + x & (B_3) & x + x = x \\ (B_2) & (x + y) + z = x + (y + z) & (B_4) & x + \mathbf{0} = x \end{array}$$

Fig. 2 Axiomatisation for the (Strong) Bisimulation Equivalence

Theorem 4. *For every constraint C being a behaviour preorder, we have that*

$$\mathcal{P}_C \vdash p \sqsubseteq q \Longleftrightarrow p \underset{\rightarrow}{\sqsubseteq}^C q$$

Proof. Soundness. Bisimilarity axioms are sound for both the relation C and for the C-constrained simulation preorder. Therefore, we only need to prove that the axiom (P_C) is also sound. Process $p + q$ can obviously simulate p and since we have pCq and C is a congruence with respect to choice, we also have $pC(q + p)$ and we conclude that $p \underset{\rightarrow}{\sqsubseteq}^C p + q$.

Completeness. By induction on the depth of processes. If $p = \mathbf{0}$ then $\mathbf{0}Cq$ and, applying $(P_C), (B_1)$ and (B_4), $\mathcal{P}_C \vdash \mathbf{0} \sqsubseteq q$. Consider now the general case $p = \sum a_i p_i$. On the one hand, if $p \underset{\rightarrow}{\sqsubseteq}^C q$ then pCq and we can use (P_C) to prove that $\mathcal{P}_C \vdash p \sqsubseteq p + q$. On the other hand, whenever $p \xrightarrow{a_i} p_i$ then $q \xrightarrow{a_i} q_{j_i}$ with $p_i \underset{\rightarrow}{\sqsubseteq}^C q_{j_i}$; by induction hypothesis we have $\mathcal{P}_C \vdash p_i \sqsubseteq q_{j_i}$, therefore we have $\mathcal{P}_C \vdash q + \sum a_i p_i \sqsubseteq q + \sum a_i q_{j_i}$, equivalently $\mathcal{P}_C \vdash q + p \sqsubseteq q$. Combining both cases, $\mathcal{P}_C \vdash p \sqsubseteq q + p \sqsubseteq q$.

We next study the axiomatization of the equivalence relation associated to the C-constrained simulation, $\rightleftarrows^C = \underset{\rightarrow}{\sqsubseteq}^C \cap \underset{\leftarrow}{\sqsupseteq}^C$. We propose the following axiom for each constraint C:

$$(E_C) \qquad xCy \Rightarrow a(x+y) = a(x+y) + ay$$

We define the set $\mathcal{E}_C = \{B_1, B_2, B_3, B_4, E_C\}$, containing the axioms that characterise bisimulation equivalence (Figure 2) and the axiom E_C. We write $\mathcal{E}_C \vdash p = q$ when the equation $p = q$ is provable from $\mathcal{E}_C$.

$\mathcal{E}_C$ is sound and complete with respect to $\overset{C}{\rightleftarrows}$. However, in this case, to prove this result the constraint has to be symmetric, that is, it has to be a behaviour equivalence; below we will comment more on this subject.

Theorem 5. *For every constraint C being a behaviour equivalence, we have that*

$$\mathcal{E}_C \vdash p = q \Leftrightarrow p \overset{C}{\rightleftarrows} q$$

Proof. Soundness. Let us just prove that (E_C) is sound. Whenever pCq we also have $\mathcal{E}_C \vdash a(p+q) = a(p+q)+aq$. We have to prove that $a(p+q) \underset{\rightarrow}{\sqsubseteq}^C a(p+q)+aq$ and $a(p+q)+aq \underset{\rightarrow}{\sqsubseteq}^C a(p+q)$. Let us start by proving $a(p+q) \underset{\rightarrow}{\sqsubseteq}^C a(p+q)+aq$, processes in both sides of the relation can trivially simulate each other and taking into account that C is a behaviour preorder, from pCq we derive $a(p+q)\ C\ a(p+q)+aq$ and, we immediately conclude that we have a C-simulation.

Let us prove now that $a(p+q)+aq \underset{\rightarrow}{\sqsubseteq}^C a(p+q)$, as before, processes in both sides can simulate each other, to have a C-simulation we just need to prove $a(p+q)+aq\ C\ a(p+q)$. As before, from pCq we derive $a(p+q)\ C\ a(p+q)+aq$ and, since C is symmetric, we conclude $a(p+q)+aq\ C\ a(p+q)$.

Completeness. The proof of the completeness of the axiomatization of the simulation equivalence in [Gla01] (Section 17.2) can be transferred without any changes just checking the additional proof obligations imposed by the condition in the axiom (E_C).

It is interesting to note that the cases in which C is not symmetric are not completely excluded from the result above. This can be concluded from the following results.

Definition 9. We say that two constraints C_1 and C_2 are *cs-equivalent*, which we denote by $C_1 \sim C_2$, iff they define the same C-constrained similarity relation, that is $\underset{\rightarrow}{\sqsubseteq}^{C_1} = \underset{\rightarrow}{\sqsubseteq}^{C_2}$.

Next proposition is just a snapshot of a nice algebraic theory that can be developed around constrained simulations and cs-equivalence.

Proposition 1. *For any behaviour preorders C, C_1 and C_2 we have:*

1. *$C_1 \sim C_2 \Rightarrow (C_1 \cap C_2) \sim C_1$.*
2. *$C \sim \underset{\rightarrow}{\sqsubseteq}^C$ and $\underset{\rightarrow}{\sqsubseteq}^C$ is the smallest C-simulation that is cs-equivalent to C.*
3. *If $C_1 \sim C_2$ and $C_1 \supseteq C \supseteq C_2$, then $C \sim C_1$.*

4. For the simulation preorder $\sqsubseteq_S$ we have that $C \sim (C \cap \sqsubseteq_S)$.

Example 2. Next we show some illustrative examples of cs-equivalent constraints.

- Let us consider the classical simulation preorder, $\sqsubseteq_S$. $\sqsubseteq_S = \underset{\rightarrow}{\sqsubseteq}^U$ where U is the universal relation, xUy for every x and y. On the other hand, if we use $\sqsubseteq_S$ as constraint, it is immediate to see that $\sqsubseteq_S = \underset{\rightarrow}{\sqsubseteq}^{\sqsubseteq_S}$ and therefore $U \sim \sqsubseteq_S$, but while U is symmetric, $\sqsubseteq_S$ is not.
- Taking the constraint $I_{\supseteq}$ given by $pI_{\supseteq}q \Leftrightarrow I(p) \supseteq I(q)$, we have that the ready simulation preorder is $\underset{\rightarrow}{\sqsubseteq}^{I_{\supseteq}}$, as it was originally defined in [BIM95]. It is well known that the ready simulation preorder also coincides with $\underset{\rightarrow}{\sqsubseteq}^I$, see for instance [Gla01]. Again, I is a symmetric relation, $I_{\supseteq}$ is not, and $I \sim I_{\supseteq}$.
- In a similar way, we can define the 2-nested simulation as a constrained simulation using either $\sqsubseteq_S^{-1}$ or the equivalence relation $\sqsubseteq_S^{-1} \cap \sqsubseteq_S$.

From the examples above, one could guess that any constraint might be cs-equivalent to some symmetric one. This is indeed the case for any "interesting" constraint we have found, but in general it is not true, as the following counterexample shows.

Example 3. If we consider the behaviour preorder $\sqsubseteq$ defined by the axioms of bisimulation equivalence (Figure 2) together with the axiom $x \sqsubseteq x + aa$, where a represents any arbitrary action in *Act*, it can be checked that there is no symmetric constraint cs-equivalent to $\sqsubseteq$.

4 Constrained Simulations Up-to a Preorder

Starting from constrained simulations we define a general notion of constrained simulation up-to a preorder that will allow us to provide simulation-like characterisations for behaviour preorders.

Definition 10. Let $\sqsubseteq$ be a behaviour preorder, and C a relation over processes. We say that a binary relation S over processes is a *C-simulation up-to* $\sqsubseteq$, if $S \subseteq C$ (that is, $pSq \Rightarrow pCq$), and S is a simulation up-to $\sqsubseteq$. Or, equivalently, in a coinductive way, whenever we have pSq, we also have:

- For every a, if $p \xrightarrow{a} p'_a$ there exist q', q'_a such that $q \sqsupseteq q' \xrightarrow{a} q'_a$ and $p'_a S q'_a$;
- pCq.

We say that process p is C-simulated up-to $\sqsubseteq$ by process q, or that process q C-simulates process p up-to $\sqsubseteq$, written $p \lesssim_{\sqsubseteq}^{C} q$, if there exists a C-simulation up-to $\sqsubseteq$, S, such that pSq.

We often just write $\underset{\sim}{\sqsubset}^{C}$, instead of $\underset{\sim}{\sqsubset}^{C}_{\sqsubseteq}$, when the behaviour preorder is clear from the context.

The following proposition highlights the tight relation between a behaviour preorder and its kernel equivalence.

Proposition 2. *Given a behaviour preorder $\sqsubseteq$ and a behaviour equivalence C such that $\underset{\rightarrow}{\sqsubset}^{C} \subseteq \sqsubseteq \subseteq C$, we have $p \sqsubseteq q \Longleftrightarrow q \equiv q + p \wedge pCq$.*

Proof. *First we prove the right to left implication. Given that processes p and q satisfy the constraint C, we can apply the axiomatic characterisation for $\underset{\rightarrow}{\sqsubset}^{C}$ in Theorem 4, and using the axiom (P_C) we obtain $p \underset{\rightarrow}{\sqsubset}^{C} p + q$. Now, since $\underset{\rightarrow}{\sqsubset}^{C} \subseteq \sqsubseteq$, we also have $p \sqsubseteq p + q \equiv q$ and therefore $p \sqsubseteq q$.*

We now prove the left to right implication. On the one hand, since $\sqsubseteq$ is a behaviour preorder we have $p \sqsubseteq q \Rightarrow p + q \sqsubseteq q$. On the other hand, if $p \sqsubseteq q$ then pCq and also qCp, since C is symmetric. As before, we use the axiom for C-similarity to obtain $q \underset{\rightarrow}{\sqsubset}^{C} q + p$. Since $\underset{\rightarrow}{\sqsubset}^{C} \subseteq \sqsubseteq$, we conclude $q \sqsubseteq q + p$.

Next we show that whenever p is C-simulated up-to $\sqsubseteq$ by q we also have $q \equiv q + p$. Using this lemma and the previous proposition we will prove later our Theorem 6.

Lemma 1. *For every behaviour preorder $\sqsubseteq$ and every behaviour equivalence C, such that $\underset{\rightarrow}{\sqsubset}^{C} \subseteq \sqsubseteq \subseteq C$, we have $p \underset{\sim}{\sqsubset}^{C}_{\sqsubseteq} q \Rightarrow q \equiv q + p$*

Proof. *We prove it by induction on the depth of process p. Since $\equiv$ is a congruence with respect to the choice operator, it is enough to show that $p \underset{\sim}{\sqsubset}^{C}_{\sqsubseteq} q \Rightarrow q \equiv q + p|_a$ for every $a \in I(p)$. Whenever $p \xrightarrow{a} p'_a$ then there exist q_a and q'_a such that $q \sqsupseteq q_a \xrightarrow{a} q'_a$ with $p'_a \underset{\sim}{\sqsubset}^{C}_{\sqsubseteq} q'_a$, so that $p'_a C q'_a$ and, applying the induction hypothesis, $q'_a \equiv q'_a + p'_a$. Since $\underset{\rightarrow}{\sqsubset}^{C} \subseteq \sqsubseteq$, we can use the axiomatization given in Theorem 5 and apply the axiom (E_C) to obtain $p'_a C q'_a \Rightarrow aq'_a \equiv a(q'_a + p'_a) + ap'_a$; therefore $aq'_a \equiv aq'_a + ap'_a$. Hence we get $\sum aq'_a \equiv \sum aq'_a + p|_a$. On the other hand, $q_a = aq'_a + r_a$ and since $\equiv$ is a congruence with respect to choice $\sum aq'_a + \sum r_a \equiv \sum aq'_a + \sum r_a + p|_a$, that is, $\sum q_a \equiv \sum q_a + p|_a$. We can also add q in both sides, getting $\sum q_a + q \equiv \sum q_a + q + p|_a$. Now, applying Proposition 2 to $\sqsubseteq$, since for every q_a we have $q_a \sqsubseteq q$, we conclude $q \equiv q + q_a$, and therefore $q \equiv q + p|_a$, as we wanted to prove.*

Theorem 6. *For every behaviour preorder $\sqsubseteq$ and every behaviour equivalence C such that $\underset{\rightarrow}{\sqsubset}^{C} \subseteq \sqsubseteq \subseteq C$, we have $p \underset{\sim}{\sqsubset}^{C}_{\sqsubseteq} q \Leftrightarrow p \sqsubseteq q$.*

Proof. *The right to left implication is obvious: we have pCq and if $p \xrightarrow{a} p'_a$ then we can take $q \sqsupseteq p \xrightarrow{a} p'_a$. To prove the left to right implication we use Lemma 1, $p \underset{\sim}{\sqsubset}^{C}_{\sqsubseteq} q \Rightarrow q \equiv q + p$ and, since pCq is also satisfied, we can now apply Proposition 2, to conclude $p \underset{\sim}{\sqsubset}^{C}_{\sqsubseteq} q \Rightarrow p \sqsubseteq q$.*

The condition of symmetry imposed to C is necessary, as the following counterexample shows.

Example 4. Let us consider the non-symmetric constraint $I_{\supseteq}$ defined in Example 2. Let us consider the preorder $\sqsubseteq$ defined by the axioms that define the ready simulation preorder (that is, the axioms that characterise bisimulation equivalence plus the axiom $ax \sqsubseteq ax + ay$), together with the axiom $x + bb \sqsubseteq x$. The constrained simulation $\underset{\rightarrow}{\sqsubset}^{I_{\supseteq}}$ is the ready simulation preorder and thus it is immediate to check both $\underset{\rightarrow}{\sqsubset}^{I_{\supseteq}} \subseteq \sqsubseteq$ and $\sqsubseteq \subseteq I_{\supseteq}$. However this preorder does not satisfy the thesis of Theorem 6, because if we consider the processes $p = a + bb$ and $q = a + b$, we have $p \not\sqsubseteq q$ but $p \underset{\sim_{\sqsubseteq}}{\sqsubset}^{I_{\supseteq}} q$, because $I(p) = I(q)$ and if $p \xrightarrow{a} \mathbf{0}$ then $q \xrightarrow{a} \mathbf{0}$, while for $p \xrightarrow{b} b$ we can take $q \sqsupseteq a + bb + b \xrightarrow{b} b$.

Theorem 6 generalises Theorem 2 and 3 (and also Theorem 44 in [dFG08]). And what is even more important, it provides a uniform framework that allows to better understand the role of the premises in the previous results we had on simulations up-to. Thanks to the use of constrained simulations up-to, now we can clearly see that a great deal of semantic preorders both coarser and finer than the ready simulation preorder can be characterised by using simulations up-to. In particular, we can characterise not only all the preorders coarser than the ready simulation that appear in Van Glabbeek's linear time-branching time spectrum but also the rest of semantics there. Next example illustrates the case of possible-futures semantics that is not coarser than the ready simulation and therefore falls outside the scope of Theorem 3.

Example 5. If we denote by $\sqsubseteq_{PF}$ the possible-futures preorder [RB81, Gla01], then, taking the constraint $pTq \Leftrightarrow \text{traces}(p) = \text{traces}(q)$, we have that T is a behaviour equivalence and that $\underset{\rightarrow}{\sqsubset}^{T} \subseteq \sqsubseteq_{PF} \subseteq T$. Thus we are under the hypothesis of Theorem 6 and therefore $\underset{\sim_{\sqsubseteq_{PF}}}{\sqsubset}^{T}$ and $\sqsubseteq_{PF}$ are the same relation.

Once again, the next result shows the interplay between preorders and the induced equivalences. For a given preorder, we could use the induced equivalence relation $\equiv$ to characterise the preorder by means of a C-simulation up-to $\equiv$.

Theorem 7. *For every behaviour preorder $\sqsubseteq$ and its induced equivalence relation $\equiv$, for every behaviour equivalence C such that $\underset{\rightarrow}{\sqsubset}^{C} \subseteq \sqsubseteq \subseteq C$, we have $p \underset{\sim_{\sqsubseteq}}{\sqsubset}^{C} q \Leftrightarrow p \underset{\sim_{\equiv}}{\sqsubset}^{C} q$.*

Proof. The right to left implication is obvious. The left to right implication is a consequence of Proposition 2; whenever q would reduce into q' by applying $q \sqsupseteq q'$ we could also reduce it by $\equiv$, by applying $q \equiv q + q'$, and then we could execute all the transitions of q'.

5 Constrained Simulations up-to an Equivalence

The previous section was devoted to the study of constrained simulations up-to a preorder. The starting point there was a given preorder, instead in this section we show that the theory of constrained simulations up-to can be developed even if we do not have such a preorder to start from. It is true that equivalence relations are particular cases of preorders but equivalences are symmetric relations and cannot be characterised by means of proper simulations that are intrinsically non-symmetric. An interesting result that we present in this section is how to use our up-to technique to build up an adequate preorder for a given equivalence relation.

Lemma 2. *For every behaviour equivalence* $\equiv$, *and for every constraint* C *that is a behaviour equivalence such that* $\rightleftarrows^C \subseteq \equiv$, *we have* $p \sqsubseteq^C_{\approx_\equiv} q \Rightarrow q \equiv q + p$.

Proof. *The proof uses the same notations and follows similar arguments to those in the proof of Lemma 1. We use induction on the depth of process* p. *Since* $\equiv$ *is a congruence with respect to the choice operator, it is enough to show that* $p \sqsubseteq^C_{\approx_\equiv} q \Rightarrow q \equiv q + p|_a$ *for every* $a \in I(p)$. *Whenever* $p \xrightarrow{a} p'_a$ *then* $q \equiv q_a \xrightarrow{a} q'_a$ *and* $p'_a \sqsubseteq^C_{\approx_\equiv} q'_a$, *and by applying the induction hypothesis we obtain* $q'_a \equiv q'_a + p'_a$. *On the other hand, since* $\rightleftarrows^C \subseteq \equiv$, *we can use the axiomatic characterisation given in Theorem 5 and apply the axiom* (E_C) *to obtain* $aq'_a \equiv a(q'_a + p'_a) + ap'_a$; *therefore* $aq'_a \equiv a(q'_a + p'_a) + ap'_a \equiv aq'_a + ap'_a$. *Adding all up* $\sum aq'_a \equiv \sum aq'_a + p|_a$. *Since* $\equiv$ *is a congruence with respect to the choice operator we can add subterms in both sides of the equivalence, in particular, every* $q_a = aq'_a + r_a$ *and therefore* $\sum aq'_a + \sum r_a \equiv \sum aq'_a + \sum r_a + p|_a$, *that is,* $\sum q_a \equiv \sum q_a + p|_a$. *Since for all* q_a *we have* $q \equiv q_a$, *then we conclude that* $q \equiv q + p|_a$.

Theorem 8. *For every behaviour equivalence* $\equiv$, *and every constraint* C *that is a behaviour equivalence such that* $\rightleftarrows^C \subseteq \equiv \subseteq C$, *we have* $p \sqsubseteq^C_{\approx_\equiv} q \wedge p \sqsupseteq^C_{\approx_\equiv} q \Leftrightarrow p \equiv q$.

Proof. *The right to left implication is obvious, just considering that* $\equiv \subseteq C$. *We prove the left to right implication. Since* $p \sqsubseteq^C_{\approx_\equiv} q$ *then, by Lemma 2, we have that* $q \equiv q + p$; *symmetrically, from* $q \sqsubseteq^C_{\approx_\equiv} p$ *we get* $p \equiv q + p$ *and therefore* $p \equiv q$.

As desired, for any equivalence relation fulfilling the hypothesis of the theorem above, we get a preorder such that its kernel is the original equivalence. Moreover, this preorder satisfies some interesting properties.

Proposition 3. *For every behaviour equivalence* $\equiv$, *and for every constraint* C *that is a behaviour equivalence such that* $\rightleftarrows^C \subseteq \equiv \subseteq C$, *we have that* $\sqsubseteq^C_{\approx_\equiv}$ *is a behaviour preorder and* $\sqsubseteq^C_{\rightarrow} \subseteq \sqsubseteq^C_{\approx_\equiv} \subseteq C$.

Proof. *That $\sqsubseteq^C_{\sim\equiv}$ is a precongruence with respect to the choice operator follows from the congruence with respect to the choice of $\equiv$. The rest of the properties are immediate.*

Now we can say that for any behaviour equivalence fulfilling the hypothesis of Theorem 8, the preorder $\sqsubseteq^C_{\sim\equiv}$ is canonical in the sense specified in the following result.

Theorem 9. *For every behaviour equivalence $\equiv$, and for every constraint C that is a behaviour equivalence, such that $\rightleftarrows^C \subseteq \equiv \subseteq C$, the preorder $\sqsubseteq^C_{\sim\equiv}$ is the only behaviour preorder that satisfies $\sqsubseteq^C_{\rightarrow} \subseteq \sqsubseteq^C_{\sim\equiv} \subseteq C$, and whose kernel is $\equiv$. Therefore, it can be said to be the* canonical preorder under the constraint C *that induces the equivalence $\equiv$.*

Proof. *Proposition 3 says that $\sqsubseteq^C_{\sim\equiv}$ satisfies the hypothesis of the results in Section 4. On the other hand, if there is any other behaviour preorder $\sqsubseteq$ such that $\equiv = \sqsubseteq \cap \sqsupseteq$, then by applying Theorem 6 and 7 we conclude $\sqsubseteq \Leftrightarrow \sqsubseteq^C_{\sim\sqsubseteq} \Leftrightarrow \sqsubseteq^C_{\sim\equiv}$.*

The canonicity of the preorder $\sqsubseteq^C_{\sim\equiv}$ is bounded by the constraint C appearing in its definition. For instance, trace equivalence $\equiv_T$ satisfies $\rightleftarrows^U \subseteq \equiv_T$, and therefore we can obtain the U-canonical preorder $\sqsubseteq^U_{\sim\equiv_T}$, which is just the classic preorder $\sqsubseteq_T$. But we also have $\rightleftarrows^I \subseteq \equiv_T \subseteq I$, and then we also consider the I-canonical preorder $\sqsubseteq^I_{\sim\equiv_T}$ whose kernel is also trace equivalence, but it is strictly finer than $\sqsubseteq_T$. Then we could even conclude that we should not call canonical to these generated preorders. We have decided to maintain this term because it is true that given both the equivalence $\equiv$ and the constraint C, then the generated preorder $\sqsubseteq^C_{\sim\equiv}$ is unique indeed.

Anyway, if we want to associate to an equivalence a unique canonical preorder we can define it as the C-canonical preorder with the coarsest constraint C. That is, the coarsest C such that $\rightleftarrows^C \subseteq \equiv \subseteq C$ that generates the preorder $\sqsubseteq^C_{\sim\equiv}$. We have not been able to prove the existence of such a coarser constraint for any arbitrary behaviour equivalence. However, if we restrict ourselves to the semantics in the linear time-branching time spectrum, it is easy to see that such a coarsest constraint exists.

Proposition 4. *Let $\mathcal{O} \in \{T, S, CT, CS, F, R, FT, RT, PW, RS, 2N, PF\}$ be any of the semantics in the linear time-branching time spectrum [Gla01]. Then, there exists a coarsest constrained $C_\mathcal{O}$ such that $\rightleftarrows^{C_\mathcal{O}} \subseteq \equiv_\mathcal{O} \subseteq C_\mathcal{O}$ and that is defined in Table 1.*

Proof. *We prove some of the results, the rest can be proved in a similar way.*

T *Since $\rightleftarrows^U \subseteq \equiv_T$, we have immediately $C_T = U$.*

	T	S	CT	CS	F	R	FT	RT	PW	RS	PF	$2N$
$C_{\mathcal{O}}$	U	U	V	V	I	I	I	I	I	I	W	X

$$pUq \; \forall p, q \qquad\qquad pWq \iff p \equiv_T q$$
$$pVq \iff (p = \mathbf{0} \Leftrightarrow q = \mathbf{0}) \qquad pXq \iff p \equiv_S q$$
$$pIq \iff I(p) = I(q)$$

Table 1 Coarsest Constraints for the Semantics in the ltbt Spectrum

F *We have* $\rightleftarrows^I \subseteq \equiv_F \subseteq I$. *We know that* $\rightleftarrows^I$ *can be axiomatized by* $pIq \Rightarrow a(p+q) \equiv a(p+q) + ap$. *If we consider any other* $\rightleftarrows^C \subseteq \equiv_F \subseteq C$ *with* $\rightleftarrows^I \not\subseteq \rightleftarrows^C$ *we should have* $I \not\subseteq C$ *and therefore there should be some processes such that* $p'Cq'$ *and some action* $b \in I(q') - I(p')$. *From the axiomatization of* $\rightleftarrows^C$ *we would obtain* $a(p'+q') \equiv a(p'+q') + ap'$ *but these two processes are not failure equivalent, since the right one can reject* $\{b\}$ *after executing* a, *and the leftone cannot.*

2N *As for the other simulation semantics,* $\mathcal{O}_{2N}$ *is just the constraint used in the definition.*

PF *We have* $\rightleftarrows^T \subseteq \equiv_{PF} \subseteq T$. *As above, if we have another constraint* $\rightleftarrows^T \not\subseteq \rightleftarrows^C$ *we should have some processes such that* $p'Cq'$, *and therefore* $a(p'+q') \equiv a(p'+q') + ap'$, *but* $T(p) \neq T(q)$. *This is not possible because a possible future for the process in the right side after executing* a *is* $T(p')$, *that is not a possible future for the process on the left side.*

As a byproduct, we have detected a new simulation semantics that does not appear in the linear time-branching time spectrum, the T-constrained simulation, which we could call trace equivalence simulation semantics. This new semantics should be added in the spectrum between ready simulation and 2-nested simulation, and above possible futures.

We can take advantage of the close relation between any behaviour equivalence and the corresponding canonical preorder by turning an axiomatic characterisation of the former into an axiomatization of the latter.

Theorem 10. *Let* $\equiv$ *be a behaviour equivalence and* C *a constraint that is a behaviour equivalence such that* $\rightleftarrows^C \subseteq \equiv \subseteq C$. *If* A_E *is an axiomatization of the equivalence* $\equiv$, *taking the axiom* (P_C) *to be* $xCy \Rightarrow x \sqsubseteq x + y$, *we have that* $A_P = A_E \cup \{P_C\}$ *is an axiomatization of the relation* $\underset{\sim_\equiv}{\sqsubset}^C$.

Proof. As usual we write $A_P \vdash p \sqsubseteq q$ *when the inequality* $p \sqsubseteq q$ *is provable from the set of axioms* A_P. *We prove that* $A_P \vdash p \sqsubseteq q$ *iff* $p \underset{\sim_\equiv}{\sqsubset}^C q$.

Soundness. On the one hand, $\equiv \subseteq \underset{\sim_\equiv}{\sqsubset}^C$; *on the other hand, from Proposition 3 we know that* $\underset{\to}{\sqsubset}^C \subseteq \underset{\sim_\equiv}{\sqsubset}^C \subseteq C$; *besides, from Theorem 4 we have that* P_C *is one of the axioms that characterise* $\underset{\to}{\sqsubset}^C$.

Completeness. By Lemma 2 we know that $p \sqsubseteq^C_{\sim_\equiv} q \Rightarrow q \equiv q + p$ *and therefore* $A_P \vdash q \equiv q + p$. *On the other hand, we have also* pCq *and then, using* P_C, *we get* $A_P \vdash p \sqsubseteq p + q$. *All together,* $A_P \vdash p \sqsubseteq p + q \equiv q$.

With this result, the correctness and completeness of the axiomatization for the preorders are proved once and for all for a great variety of semantics.

6 Conclusions and Future Work

In this paper we have universalised the presentation of our theory of simulation up-to by means of which we provide coinductive characterisations for a great variety of semantics either coarser or finer than the ready simulation, including those in the linear time-branching time spectrum. Constrained simulations have played an essential role in our development; we have provided an axiomatization of the preorders defined by them and also of the induced equivalences.

An interesting result was that any behaviour equivalence induces a canonical preorder whose kernel is the given equivalence relation. It is nice to find that for all the semantics in the linear time-branching time spectrum the so obtained canonical preorder coincides with the one we already knew from the literature. As a consequence of the canonicity, some properties can be proved in general, once and for all. We have illustrated this fact by giving a general axiomatization of the preorders in terms of the axiomatization of the equivalences and the axiomatization of the constrained simulations.

There are several directions in which we plan to continue the study of the relations between preorders and equivalences. For instance, the axiomatization of the preorders that we have obtained is conditional. It would be interesting to know in which cases the axiom (P_C) can be turned into an equivalent finite collection of equational axioms. This is in fact the case for the semantics in the linear time-branching time spectrum coarser than the ready simulation, for which there exists an equational axiomatization equivalent to our conditional axiomatization.

In [AFIL05] you can find a review of (in)axiomatizability results, that its authors have recently completed giving other new results on the subject. We think that our characterisation of the semantics using constrained simulations up-to will be useful to get other new (in)axiomatizability general results. For instance, we conjecture that any interesting preorder stronger than the ready simulation is not finitely axiomatizable (see [AFGI04] for the seed results supporting this conjecture).

References

[AFGI04] Luca Aceto, Wan Fokkink, Rob van Glabbeek, and Anna Ingólfsdóttir. Nested semantics over finite tree are equationally hard. *Information and Computation*, 191(2):203–232, 2004.

[AFI07] Luca Aceto, Wan Fokkink, and Anna Ingólfsdóttir. Ready to preorder: get your BCCSP axiomatization for free! In *CALCO'07*, volume 4624 of *Lecture Notes in Computer Science*, pages 65–79. Springer, 2007.

[AFIL05] Luca Aceto, Wan Fokkink, Anna Ingólfsdóttir, and Bas Luttik. Finite equational bases in process algebra: Results and open questions. In *Processes, Terms and Cycles*, volume 3838 of *LNCS*, pages 338–367. Springer, 2005.

[BIM95] Bard Bloom, Sorin Istrail, and Albert R. Meyer. Bisimulation can't be traced. *Journal of the ACM*, 42(1):232–268, 1995.

[dFG05] David de Frutos-Escrig and Carlos Gregorio-Rodríguez. Bisimulations up-to for the linear time-branching time spectrum. In *CONCUR 2005*, volume 3653 of *Lecture Notes in Computer Science*, pages 278–292. Springer, 2005.

[dFG07] David de Frutos-Escrig and Carlos Gregorio-Rodríguez. Simulations up-to and canonical preorders (extended abstract). In *SOS 2007*, volume 192 of *ENTCS*, pages 13–28. Elsevier, 2007.

[dFG08] David de Frutos-Escrig and Carlos Gregorio-Rodríguez. (Bi)simulations up-to characterise process semantics. *Information and Computation*, (to appear), 2008.

[Gla01] Rob J. van Glabbeek. *Handbook of Process Algebra*, chapter The Linear Time - Branching Time Spectrum I: The Semantics of Concrete, Sequential Processes, pages 3–99. Elsevier, 2001.

[GV92] Jan Friso Groote and Frits Willem Vaandrager. Structured operational semantics and bisimulations as a congruence. *Information and Computation*, 100(2):202–260, 1992.

[Mil89] Robin Milner. *Communication and Concurrency*. Prentice Hall, 1989.

[Par81] David M.R. Park. Concurrency and automata on infinite sequences. In *Theoretical Computer Science, 5th GI-Conference*, volume 104 of *Lecture Notes in Computer Science*, pages 167–183. Springer, 1981.

[Plo81] Gordon D. Plotkin. A structural approach to operational semantics. Tech. Report DAIMI FN-19, Comp. Sci. Dept., Aarhus University, 1981.

[RB81] William Rounds and Stephen Brooks. Possible futures, acceptances, refusals, and communicating processes. In *22nd Foundations of Computer Science Annual Symposium*, pages 140–149. IEEE, 1981.

Static and dynamic typing for the termination of mobile processes

Romain Demangeon[1], Daniel Hirschkoff[1], and Davide Sangiorgi[2]

[1] ENS Lyon, Université de Lyon, CNRS, INRIA – France
[2] Dipartimento di Informatica, Università di Bologna – Italia

Abstract. A process terminates if all its reduction sequences are finite. We propose two type systems that ensure termination of π-calculus processes.
Our first type system is purely static. It refines previous type systems by Deng and Sangiorgi by taking into account certain partial order information on names so to enhance the techniques from term rewriting (based on lexicographic and multiset orderings) that underpin the proof of termination. The second system is mixed, in that it combines a static and a dynamic analysis. During the static analysis, processes are annotated with assertions. These are then used at run time to monitor the execution of processes. An exception may be raised if certain conditions that may lead to divergence are met.
We illustrate the expressiveness of the solutions proposed with a few examples of programming idioms that were beyond reach for previous type systems.

1 Termination of Concurrent Processes

Following the introduction of the π-calculus, a lot of research has been put into the study of languages for process mobility in which computing is exchange of messages between processes. Programs of these languages often produce dynamic recursive structures, that is, systems consisting of a variable number of components (at run time, new components may be created, and existing ones may be removed).

In this paper we study the problem of termination for mobile processes. We focus on the π-calculus, the commonly accepted model for them. A process terminates if all its internal runs are finite; that is, the process has no infinite sequence of reductions. Termination is a fundamental property in sequential languages. It is also important in concurrency. For instance, termination can be used to guarantee that interaction with a resource will eventually end (avoiding denial of service situations), or to ensure that the participants in a transaction will eventually reach an agreement. Unfortunately, termination is also a hard property to ensure, both in functional languages and in concurrency. Termination is particularly hard in the π-calculus, due to the expressiveness of this formalism. A number of programming language features can be encoded into the π-calculus, including functions, objects, and state (in the sense of imperative languages) [8]. Thus the notoriously hard problems of termination for these features hit the π-calculus too.

Previous work on termination in the π-calculus relies on *type systems* to guarantee the property. Languages of terminating processes are proposed in [10] and [7]. In

Please use the following format when citing this chapter:

Demangeon, R., Hirschkoff, D. and Sangiorgi, D., 2008, in IFIP International Federation for Information Processing, Volume 273; *Fifth IFIP International Conference on Theoretical Computer Science*; Giorgio Ausiello, Juhani Karhumäki, Giancarlo Mauri, Luke Ong; (Boston: Springer), pp. 413–427.

both cases, the proofs of termination make use of logical relations, a well-known technique from functional languages. The languages of terminating processes so obtained are however rather 'functional', in that the structures allowed are similar to those derived when encoding functions as processes. For this reason, subsequent work [3, 2] has explored type systems in which termination is proved using techniques from term rewriting systems, essentially defining a measure (a 'weight') that decreases with reductions. These type systems maintain however limitations on the form of recursive structures handled.

To explain the kind of problems encountered, consider a tree-like data structure, where search along the tree may involve recursive calls on all the subtrees of a given node. Termination of a call to the search procedure should intuitively follow from the acyclicity of the data structure. However, the tree cannot be type checked in [3], for the following reason. The type systems in [3] are based on an assignment of *levels* to π-calculus names, where a level is a positive integer. Since all nodes in a tree play the same role, names used in different nodes must have the same type and hence also the same level. As a consequence, when a search at a given node triggers several searches in subtrees of the node, the weight of the process (roughly, the multiset of the levels of the names in the "active" nodes) may increase (since the number of active nodes may be bigger). This breaks the reasoning needed in the proof of termination.

The contribution of the present paper is twofold. First, we refine the type systems in [3] so to be able to handle more complex recursive structures. The main improvement is given by the addition, into the type system, of a partial order that is used to compare names with the same level. While this possibility was already suggested in [3], only a very restrictive form of it had been investigated. In the present paper the idea is explored in depth. As we illustrate in Section 3, setting the balance between partial order and levels (or weight) is delicate, as counterexamples easily arise as soon as we abandon a purely lexicographical ordering between the weight and partial order information. Major problems are updating the partial order when new names are created, and ensuring that only well-founded partial orders are generated (a non-terminating computation could be produced following the step-by-step generation of an infinite descending path).

All type systems mentioned above are purely static: the whole type analysis is made before the processes are run. There is a tradeoff between expressiveness and complexity of the type systems (here complexity refers both to the intricacy of the typing rules and to the actual complexity of the type inference problem). We think that our new static type system is justified by the gain in expressiveness. We do not present, in contrast, other static type systems that we have examined, as the overall gain is more dubious.

We discuss, instead, as the second main contribution of the paper, an alternative approach: a simple and efficient static type system enhanced with dynamic (i.e., run time) checks. The static type system we adopt only exploits information about the level of names; however it annotates the positions where extra information (a partial order), is needed to justify termination. At runtime, these annotations are used to perform dynamic checks. Correctness of the resulting *mixed* type system is stated as follows:

if the first phase succeeds on a process, then the resulting annotated process cannot exhibit an infinite computation: its execution either terminates or raises an exception.

The advantage of dynamic typing is that only the parts of a process that are actually executed need to be analysed. This may considerably reduce the type constraints generated, especially in computations with data dependencies and/or non-determinism. For instance, in our case, dynamic typing reduces the risk of generating non-well-founded paths in the partial order over the names. We illustrate the expressive power of the mixed type system on some non-trivial examples. They include recursive structures created through merges of smaller structures, like, for instance, trees created from smaller trees via append operations (where a tree may be appended onto one or more leaves of another tree). In previous type systems, manipulations of this kind are not allowed, intuitively because all names connecting components of a recursive structure must be created locally. This forbids extensions of the structure with components (or just names) coming from the outside.

The price to pay with dynamic typing is the time for the additional checks at run time and the space for the data structure needed in the checks. In our mixed type system, the time for a dynamic check is at most linear in the number of annotated names created (the annotated names are those in positions that have been marked during the initial phase of static typing). The mixed system could be used in cases where the termination property is important and other, purely static, type systems have failed.

In type systems, the idea of extending static analyses with forms of dynamic checking is certainly not new. Works on the addition of dynamic types to statically typed languages include [1, 4]. We can also mention stack inspection, as, e.g., in [5], where checks on the access to resources are made at run time. Similar mechanisms are also employed in incremental garbage collectors, through read- or write-barriers, to prevent the program from accessing data that need to be processed by the collector (see [9]). In the present work, we use the phrase 'dynamic typing' by analogy with the aforementioned approaches, although the analysis that we make at runtime boils down to some lightweight sanity checks on partial orders.

2 The π-Calculus

We let $a,b,c,\dots,p,q,\dots,x,y,z$ range over an infinite set of *names*. Processes, ranged over using P,Q, are described by the following grammar (we use notation $\widetilde{n}$ to range over possibly empty tuples of names):

$$P ::= \mathbf{0} \mid P_1|P_2 \mid (\nu c)P \mid P_1+P_2 \mid a(\widetilde{x}).P \mid \overline{a}\langle\widetilde{v}\rangle.P \mid !a(\widetilde{x}).P\ .$$

The constructs of input, replicated input and restriction are binding. We sometimes call a bound name a *variable* and a free name a *channel*. We implicitly suppose that in all processes bound names are pairwise distinct and distinct from all free names. In an input $a(\widetilde{x}).P$ and an output $\overline{a}\langle\widetilde{v}\rangle.P$ we call a the *subject* name. As usual, trailing occurrences of $\mathbf{0}$ are omitted, and emissions and receptions of empty tuples of names along a are respectively abbreviated $a.P$ and $\overline{a}.P$. The reduction relation of the calculus is standard (see Appendix 6).

Strict partial orders. As the rules of our type system heavily rely on partial orders, we first introduce some notations for partial orders on names. We use $\mathscr{R}$ to range over *strict* partial orders on names, and $\mathrm{dom}(\mathscr{R})$ is the domain of the order (the set of related elements). In the sequel, the phrase 'partial order' will always be used to denote a strict partial order, since we are only interested in these. Partial orders will be represented as the set of all pairs of related elements. However, for writing convenience, we usually indicate only a subset of the pairs, namely a subset whose transitive closure gives the induced partial order. For instance, the set $\{(a,b),(b,c)\}$ stands for the (strict) partial order $\{(a,b),(b,c),(a,c)\}$.

3 A purely static type system

3.1 Previous Type Systems: a Motivating Example

We recall here the basic ideas behind the type systems of [3, 2], using an example that also illustrates some of the limitations of these systems on recursive structures. Here and in the sequel, we shall use extensions of the π-calculus of Section 2 for the presentation of examples: the additional constructs are standard, and do not raise any particular difficulty for type checking termination. The example is about the implementation of a symbol table as a binary tree. Each node in the tree is a simple π-calculus process. The process T_0 below is the generator of nodes. An output $\overline{node}\langle a,l,r,s,e\rangle$ produces a node that stores a string s whose key is e, that is connected to its parent node (or to the environment, in case of the root node) with name a, and to its children nodes with names l and r. A tree at a (that is, a tree whose root uses a for interactions with the outside) is searched for a value v via requests of the form $\overline{a}\langle \mathsf{search}, v, ans\rangle$ where ans is a return channel. When the search reaches a node, if the value is found in the node, then the corresponding key is sent back on ans; otherwise the request is concurrently propagated to both subtrees of the node. (We omit the details of a search operation that fails.)

$$\begin{array}{lll} T_0 & \stackrel{\mathrm{def}}{=} & !node(a,l,r,s,e).a(mode,v,ans). \\ & & \quad \text{if } mode = \mathsf{search} \text{ then} \\ & & \qquad \text{if } v = s \text{ then } \ \overline{ans}\langle e\rangle \mid \overline{node}\langle a,l,r,s,e\rangle \\ & & \qquad \text{else } \ \overline{l}\langle mode,v,ans\rangle \mid \overline{r}\langle mode,v,ans\rangle \mid \overline{node}\langle a,l,r,s,e\rangle \\ & & \quad \text{else } \ldots \end{array}$$

The type systems in [3, 2] recognise a system as terminating if the continuations activated in an interaction (i.e., the processes underneath the interacting prefixes) have a smaller "weight" than that of the output that has been consumed to trigger the interaction. This notion of weight is formalised with an assignment of levels (positive

$$\frac{}{\mathscr{R} \vdash \mathbf{0}} \qquad \frac{\mathscr{R}_1 \vdash P_1 \qquad \mathscr{R}_2 \vdash P_2}{\mathscr{R}_1 + \mathscr{R}_2 \vdash P_1 \mid P_2} \qquad \frac{\mathscr{R}_1 \vdash P_1 \qquad \mathscr{R}_2 \vdash P_2}{\mathscr{R}_1 + \mathscr{R}_2 \vdash P_1 + P_2} \qquad \frac{\mathscr{R} \vdash P}{\mathscr{R} \Downarrow_c \vdash (\nu c) P}$$

$$\frac{\mathscr{R} \vdash P \qquad \vdash a : \sharp^{l_a}_{SS} \widetilde{T} \qquad \vdash \widetilde{v} : \widetilde{T} \qquad SS * \widetilde{v} \subseteq \mathscr{R}}{\mathscr{R} \vdash \overline{a} \langle \widetilde{v} \rangle . P} \qquad \frac{\mathscr{R} + (SS * \widetilde{x}) \vdash P \qquad \vdash a : \sharp^{l_a}_{SS} \widetilde{T} \qquad \vdash \widetilde{x} : \widetilde{T}}{\mathscr{R} \vdash a(\widetilde{x}).P}$$

Fig. 1 Static system: typing rules (see main text for the typing of replication)

integers) to the types of the names. Now, consider the system composed by a tree at a and a search request $\overline{a}\langle \mathsf{search}, v, ans \rangle$. Names a, l and r play the same role in the structure, and therefore must have the same level. As the consumption of the output at a may produce outputs at l and r (the 'else' branch in T_0), the overall weight of the system increases (indeed, ensuring termination essentially boils down to controlling the outputs that can be generated along computation, since outputs may be used to trigger new copies of replicated processes). Due to this increase, T_0 is not typable. (The systems of [3, 2] allow the weight of the derivatives of an interaction to be at most the same as that of the initial process, and for this they rely on a rudimentary partial order information on names; however, the weight may never increase, as is instead the case for T_0.)

In the new type system that we propose below, replications in which the weight increases may be typed (indeed T_0 is typable, see Section 3.3). The greater expressiveness is achieved by enforcing a tight coupling between weight and a well-founded partial order. Increases in weight through reductions are possible, provided they are appropriately compensated in the partial order. This schema, while intuitively simple, is rather delicate. As an example of the possible problems (other examples will be given later in the paper), consider the system

$$T_1 \stackrel{\text{def}}{=} \overline{u} \,|\, \overline{v} \,|\, U_1 \,|\, U_2 \qquad \text{with} \qquad \begin{cases} U_1 \stackrel{\text{def}}{=} !p(a,b,c).a.(\overline{b} \mid \overline{c}) \\ U_2 \stackrel{\text{def}}{=} !u.v.(\overline{w} \mid \overline{p}\langle w,u,v \rangle) \end{cases}$$

where names w, u, v have the same level k and p has level $k' < k$ (this can be imposed, e.g., by adding extra processes in parallel). In U_1, the weight increases underneath the initial inputs at p and a; but the new outputs are smaller in the partial order, if we set a above b and c. In U_2, the weight decreases underneath the top two inputs. The system seems to meet the termination conditions; however, it does not terminate (the outputs at u and v trigger U_2, which in turn triggers U_1 and we are back to T_1).

3.2 The Type System

Figure 1 presents the typing rules for our new system. As in [3, 2], the type system follows the Church style, in the sense that each name is assigned a type a priori. We write $\vdash a : T$ if T is the type so assigned to name a. We add the termination analysis on top of the simply-typed polyadic π-calculus. Accommodating other standard type constructs would be straightforward; indeed, in examples, we sometimes use primitive types for values (integers and booleans, with the related if-then-else operator in the syntax of processes) and, in one case, recursive types. The grammar for types is given by

$$T \quad ::= \quad \sharp^l_{SS}\widetilde{T} \ .$$

In $\vdash p : \sharp^l_{SS}\widetilde{T}$, integer l is called the *level* of p, written $\mathrm{lvl}(p) = l$; and SS is the partial order associated to tuples of names carried along p, in which the i-th component is represented by integer i. For instance, if $SS = \{(2,3)\}$, then the second component of a tuple should be above the third one; thus an output $\overline{p}\langle u,v,w\rangle.P$ is typable only if v is above w in the partial order with which the output is typed. We let SS range over partial orders on integers of this kind, and use operator $*$ to 'project' them onto a relation on names. For example, if $SS = \{(1,2),(4,3)\}$, then, if $\mathscr{R} = SS * (u,v,w,t)$, we have $u\mathscr{R}v$ and $t\mathscr{R}w$.

We need some further notations to define the type system. Two partial orders $\mathscr{R}_1$ and $\mathscr{R}_2$ are compatible if $\mathscr{R}_1 \cup \mathscr{R}_2$ yields a partial order. If $\mathscr{R}_1$ and $\mathscr{R}_2$ are compatible then $\mathscr{R}_1 + \mathscr{R}_2$ is the partial order (induced by) $\mathscr{R}_1 \cup \mathscr{R}_2$; if they are not compatible, then $\mathscr{R}_1 + \mathscr{R}_2$ is undefined. $\mathscr{R} \Downarrow_c$ stands for the relation obtained by removing all pairs involving c after closing $\mathscr{R}$ by transitivity.

The typing judgements for processes are of the form $\mathscr{R} \vdash P$, where $\mathscr{R}$ is a partial order. They are defined by the rules of Figure 1, plus the rules for replication below. The rules of Figure 1 are similar to those in [3]. The typing of replication, however, is different. We comment on the main typing rules. In the rule for output, the partial order $\mathscr{R}$ must include $SS * \widetilde{v}$, which is the partial order derived from the type of the subject a. Similarly, in the rule for input, the partial order is extended with constraints on bound names of the input as derived from the type of its subject. We now present the two rules for replication. They are defined on processes of the form $!\kappa.P$, where κ is a sequence of inputs, such as $a_1(\widetilde{x_1}).a_2(\widetilde{x_2})\ldots a_n(\widetilde{x_n})$; moreover the sequence is maximal, in the sense that the outermost process operator in P is not an input. If $\kappa = a_1(\widetilde{x_1}).a_2(\widetilde{x_2})\ldots a_n(\widetilde{x_n})$, then M_κ is the multiset of the names $a_1,\ldots,a_n$ that occur in subject position in κ. Moreover, if $\sharp^{l_i}_{SS_i}\widetilde{T_i}$ is the type of a_i, for $i = 1,..,n$, then $\mathscr{R}_\kappa$ stands for $(SS_1 * \widetilde{x_1}) \cup \cdots \cup (SS_n * \widetilde{x_n})$. For a given multiset M of names, $M|_l$ is the multiset of names in M whose level is equal to l, and $\mathrm{card}(M)$ is the cardinality of M.

$$[\mathtt{Rep1}]\ \frac{\mathscr{R} \vdash \kappa.P \qquad \exists l > 0 \text{ s.t. } \begin{cases} (i)\ \forall j > l, M_\kappa|_j = os(P)|_j \\ (ii)\ \forall j \geq l, rs(P)|_j = \emptyset \\ (iii)\ os(P)|_l \subsetneq M_\kappa|_l \end{cases}}{\mathscr{R} \vdash !\kappa.P}$$

$$[\texttt{Rep2}]\ \frac{\mathscr{R} \vdash \kappa.P \qquad \exists l > 0 \text{ s.t.} \begin{cases} (i)\ \forall j > l, M_\kappa|_j = os(P)|_j \\ (ii)\ \forall j \geq l, rs(P)|_j = \emptyset \\ (iii)\ \text{card}(M_\kappa|_l) \leq \text{card}(os(P)|_l) \\ (iv)\ M_\kappa|_l\,(\mathscr{R}_\kappa)_{\text{mul}}\, os(P)|_l \end{cases}}{\mathscr{R} \vdash !\kappa.P}$$

Although the definition of rules [Rep1] and [Rep2] is complex, the checks made are fairly simple. The rules differ only in conditions (iii) and (iv). Besides the expected condition on the typing of $\kappa.P$, the most important aspect is the comparison between M_κ, the multiset of the subjects of the inputs in κ, and $os(P)$, the multiset of the subjects of the outputs in P not occurring under a replication. In the two rules, l is the maximal level on which the weights of M_κ and $os(P)$ differ (at higher levell they are the same: condition (i)). In [Rep1], intuitively, levels are sufficient to guarantee termination. We indeed check in *(iii)* that, at level l, M_κ *strictly* contains $os(P)$, as a multiset. This condition enforces two properties: first, the weight decreases at level l when consuming the sequence κ; second, the partial order cannot be used to produce diverging computations, by compensating the loss in weight with an increase in the partial order.

In rule [Rep2] (that uses the same notations as [Rep1]), condition (iii) says that at level l the weight of M_κ is not bigger. Hence weight alone is not sufficient to guarantee termination, and the partial order becomes crucial: we check in (iv) that, at level l, M_κ dominates $os(P)$ according to the strict partial order associated to the multiset extension of $\mathscr{R}_\kappa$. Precisely, $M_\kappa|_l\,(\mathscr{R}_\kappa)_{\text{mul}}\, os(P)|_l$ holds if $M_\kappa|_l \neq os(P)|_l$, and there is a multiset C included in M_κ and $os(P)$ s.t. for all $b \in os(P)|_l \setminus C$, there is $a \in M_\kappa|_l \setminus C$ with $a\mathscr{R}b$.

In both rules, the remaining condition (ii) ensures that no name is created at level l or higher; the need for this technical condition will be shown in the second example of Section 3.3.

3.3 Examples

We present two examples that illustrate some of the technicalities of the type system. The first example explains the need for condition (iv) in rule [Rep1]. Consider the system T_1 in Section 3.1. The system diverges. We explain why it is rejected by our system, supposing, as we did in Section 3.1, that we must have $\text{lvl}(a) > \text{lvl}(p)$; e.g., p has level 1 and all other names have level 2. This way, we can type U_2 using [Rep1] (two inputs at level 2 'weight more' than one at level 2 and one at level 1). For U_1, since the weight is increasing, we must resort to the partial order, and impose that the first component of tuples transmitted on p dominates the two other components (so that name a dominates b and c), and we can use [Rep2]. However, this renders the typing of U_2 invalid, because condition (iv) of [Rep1] is not satisfied: $M_\kappa|_2 = \{u,v\}$ does not contain $os(P)|_2 = \{w\}$. It can be shown, more generally, that for any assignment of levels to names, T_1 cannot be typed.

Another delicate aspect of the type system is the control of the creation of new names. In a replication $!\kappa.P$, a new name that is created should have a level smaller than the maximal level that decreases when moving from κ to P (this is imposed by condition (ii) in [Rep1] and [Rep2]). The need for this constraint is illustrated by the following process.

$$T_2 \stackrel{\text{def}}{=} !p(a,e,f).a.(\overline{e} \mid \overline{f} \mid \overline{p}\langle a,e,f\rangle) \mid !p(a,e,f).e.f.(\nu c)(\overline{c} \mid \overline{p}\langle c,e,f\rangle) \ .$$

Without the constraint on the creation of new names, T_2 could be typed, by setting the level of p to 1, the level of all the other names to 2, and annotating the type of p with a partial order that forces a to be above e and f. But T_2, when put in parallel with $\overline{u} \mid \overline{p}\langle u,v,w\rangle$ (which would also be typable), diverges:

$$T_2 \mid \overline{u} \mid \overline{p}\langle u,v,w\rangle \longrightarrow\longrightarrow T_2 \mid \overline{v} \mid \overline{w} \mid \overline{p}\langle u,v,w\rangle \longrightarrow\longrightarrow (\nu c)\left(T_2 \mid \overline{c} \mid \overline{p}\langle c,v,w\rangle\right)$$

At the end, c plays the role played by u in the initial state. In the second replication in T_2, where the new name c is created, the maximal level that decreases is 2 (at level 2, two outputs are consumed to reach the body of the replication, namely e and f, and only one output is produced, namely c). The newly created name has precisely level 2, hence typing fails.

Process T_0 presented in Section 3.1 can be type-checked, by assigning type T_a to names a,l,r, type T_{ans} to *ans*, and type T_{node} to *node*, with

$$T_a = \sharp^3(\mathsf{M},\mathsf{S},T_{ans}), \quad T_{ans} = \sharp^2(\mathsf{K}), \quad T_{node} = \sharp^1_{\{(1,2),(1,3)\}}(T_a,T_a,T_a,\mathsf{S},\mathsf{K}),$$

where S is the type of the value v (strings in the example), K the type of the key associated to a value, M the type of tags indicating the method that is invoked on the tree. In the typing, the critical part is the 'else' branch in T_0; here the input on a at level 1 is traded for two outputs, on names l and r, at the same level, and we rely on the partial order derived from p to conclude the typing ([Rep2] – a dominates both l and r). Note that at the higher level, level 3, the weight does not change, as the input at *node* is followed by an output on the same channel.

3.4 Soundness of the Type System

Theorem 1. *If $\mathscr{R} \vdash P$ then P terminates.*

Proof (Sketch). Suppose P is non-terminating, i.e. there exists an infinite derivation $\mathscr{D} : P_1 = P \to P_2 \to \ldots$ We write $\kappa_1,\ldots,\kappa_n$ for the (finitely many) prefix sequences occurring in P. The typing for P determines, for each κ_i, a level, written $\mathrm{lvl}(\kappa_i)$, which is the maximal level at which either the order or the weight decreases (this is integer l in [Rep1] and [Rep2]).

Some steps in $\mathscr{D}$ correspond to a communication that erases the last input prefix of one of the κ_is – we call such steps *gaps*; there are necessarily infinitely many gaps, otherwise no divergence could arise. Since there are finitely many κ_is, at least one of the κ_i is involved in an infinite number of gaps in $\mathscr{D}$. We let $k = \max\{\mathrm{lvl}(\kappa_i).\ \kappa_i$ is fired an infinite number of times in $\mathscr{D}\}$. We focus on reductions that involve gaps at level k to derive a contradiction.

By definition of k, and because P is typable, there exists a step in $\mathscr{D}$ after which: (i) no new name is created at a level $\geq k$ (and hence the support of the partial order involving free names remains the same at level k); (ii) no output occurring at a level strictly greater than k is triggered. After that step, there are necessarily infinitely many gaps involving some κ at level k along which the order decreases: if this was not the case, there would exist a step after which all such gaps would correspond to a strictly decreasing weight, which is impossible. Since for such gaps the partial order cannot grow (condition (iv) in [Rep1]), and since the support of the partial order remains the same, we derive a contradiction ($\mathscr{R}_{\mathrm{mul}}$ is well-founded whenever R is). $\square$

The proof of Theorem 1 departs considerably from the correctness proof of the systems in [3]. The strategy of the latter proof is less robust, because it exploits additional syntactical hypotheses about prefixes in processes to rearrange reductions in an infinite computation. In the present proof, we extract some ordering information from a diverging computation in order to derive a contradiction.

4 The Mixed Type System

In this section we discuss another approach to typing termination. We present a mixed system in which the type checks are performed in two separated phases: a phase that precedes execution, and the phase of execution itself. Below, these two phases are referred to as *static* and *dynamic*, respectively; correspondingly we distinguish between static and dynamic typing.

The static typing, besides making the type checks, inserts into the processes *assertions* on names of the form $[a > b]$. We call a process with assertions an *annotated process*. The grammar for annotated processes is the same as that of ordinary processes in Section 2, with the addition of the production $[a > b]P$ for assertions. We use $A, B, \ldots$ to range over annotated processes.

The assertions are needed in the dynamic typing. Precisely, at run time we check that the transitive closure of the assertions encountered during execution is well-founded. Thus the operational semantics is defined on pairs $(A, \mathscr{R})$ where A is an annotated process and $\mathscr{R}$ a partial order (as usual, represented by a set of pairs whose reflexive and transitive closure induces the partial order).

Failure in the dynamic checks occurs when the addition of a new assertion introduces a cycle; in this case the special term $\perp$ is produced, meaning that an exception has been raised. We call $\perp$ and the pairs $(A, \mathscr{R})$ *configurations*. We first define the dynamic system, and then the static system.

4.1 The dynamic system

The operational semantics on ordinary processes is extended to configurations as expected, and we write $\mapsto$ for the reduction relation on configurations. The only new rule is the following (see Appendix 6.2 for a complete presentation).

$$[a > b]A, \mathscr{R} \mapsto \begin{cases} A, (\mathscr{R} \cup \{(a,b)\}) & \text{if } \mathscr{R} \cup \{(a,b)\} \text{ is a partial order} \\ \perp & \text{otherwise} \end{cases}$$

An annotated process A is *divergent* if there is an infinite sequence of reductions emanating from $(A, \emptyset)$ (where $\emptyset$ is the empty relation).

4.2 The static system

The static type system takes an ordinary process, performs some type checks on it, and returns an annotated process. Judgements for processes are of the form $\vdash P \rightsquigarrow A$, meaning that P is well typed and A is the annotated version of P that is produced.

The rules are presented in Figure 2. As in Figure 1, the main termination analysis is performed in the rule for replication. To type a replication $!a(\tilde{x}).P$, we insert an assertion whenever we encounter an output in P that is not under a replication and whose subject has the same level as a; in this situation, levels alone are not sufficient to guarantee termination, and further checks, via the assertions, are postponed at run time.

We explain the notations used in the rule for replication. If A is an annotated process and a a name, then $C(A,a)$ stands for the annotated process obtained from A by inserting an assertion $[a > b]$ in front of each output (not guarded by replication) whose subject name b has the same level as a. Intuitively, $[a > b]$ is a sanity check: a has to dominate b according to the partial order to guarantee that the process does not loop (see examples in Section 4.4). We write $\mathrm{lvl}(os(P))$ and $\mathrm{lvl}(rs(P))$ for the sets of the levels of the names in $os(P)$ and $rs(P)$, respectively. Thus $l_a \succeq \mathrm{lvl}(os(P))$ means that l_a is greater than, or equal to, the level of each name in $os(P)$; and $l_a \succ \mathrm{lvl}(rs(P))$ means that l_a is strictly greater than the level of each name in $rs(P)$.

Remark 1. In the rule for replication, only the initial input of the replication is examined. The system can be made more powerful by taking into account sequences of inputs, along the lines of the type system of Section 3 (where sequences are indicated by the κ prefix). We have not done so for simplicity of presentation and for efficiency: as discussed in Section 4.5, inference for the present system is polynomial. It would become NP-complete with sequences (as a matter of fact, it can be proved along the lines of [2] that type inference is NP-complete for the inference problem for the type system of Section 3).

Since we do not take sequences into account, the systems of Sections 3 and 4 are incomparable: none of them captures more processes than the other.

$$\frac{\vdash P \leadsto A \qquad \vdash a : \sharp^{l_a}\widetilde{T} \qquad \vdash \widetilde{v} : \widetilde{T}}{\vdash \overline{a}\langle\widetilde{v}\rangle.P \leadsto \overline{a}\langle\widetilde{v}\rangle.A} \qquad \frac{\vdash P \leadsto A \qquad \vdash a : \sharp^{l_a}\widetilde{T} \qquad \vdash \widetilde{x} : \widetilde{T}}{\vdash a(\widetilde{x}).P \leadsto a(\widetilde{x}).A}$$

$$\frac{\vdash P \leadsto A \qquad \vdash Q \leadsto B}{\vdash P \mid Q \leadsto A \mid B} \qquad \frac{\vdash P \leadsto A \qquad \vdash Q \leadsto B}{\vdash P+Q \leadsto A+B} \qquad \frac{\vdash P \leadsto A}{\vdash (\nu c)P \leadsto (\nu c)A} \qquad \frac{}{\vdash \mathbf{0} \leadsto \mathbf{0}}$$

$$\frac{\begin{array}{cccc}\vdash P \leadsto A & \vdash a : \sharp^{l_a}\widetilde{T} & \vdash \widetilde{x} : \widetilde{T} & a \notin os(A) \\ l_a \succeq \mathrm{lvl}(os(P)) & l_a \succ \mathrm{lvl}(rs(P)) & & A' = C(A,a)\end{array}}{\vdash !a(\widetilde{x}).P \leadsto !a(\widetilde{x}).A'}$$

Fig. 2 The static type analysis in the mixed system

4.3 Soundness

Theorem 2. *If $\vdash P \leadsto A$, then A has no diverging computation.*

For lack of space, we omit the proof of this result, that follows the same general strategy as the proof of Theorem 1.

The following proposition says that a process and its annotated version perform the same reductions, unless the annotated one raises an exception. Relation $\longrightarrow^*$ (resp. $\longmapsto^*$) is the reflexive transitive closure of $\longrightarrow$ (resp. $\longmapsto$). We write erase(A) for the process obtained by removing the assertions from A.

Proposition 1. *Suppose $\vdash P \leadsto A$. If $P \longrightarrow^* P'$, then either $A, \emptyset \longmapsto^* A', \mathscr{R}$ with* $\mathrm{erase}(A') = P'$ *for some $\mathscr{R}$, or $A, \emptyset \longmapsto^* \bot$. Conversely, if $A, \emptyset \longmapsto^* A', \mathscr{R}$, then $P \longrightarrow^* P'$ for some P' with* $\mathrm{erase}(A') = P'$.

4.4 Examples

The first example shows a divergent process that passes the static phase of the mixed system and produces a failure exception at run time. Let

$$R \stackrel{\mathrm{def}}{=} !p(a,b,c).(!a.\overline{b} \mid !b.\overline{c} \mid !c.\overline{a}) \mid \overline{p}\langle u,v,w\rangle \mid \overline{u} \ .$$

R is typable: we have a derivation for

$$\vdash R \leadsto A \stackrel{\mathrm{def}}{=} !p(a,b,c).(!a.[a>b].\overline{b} \mid !b.[b>c].\overline{c} \mid !c.[c>a].\overline{a}) \mid \overline{p}\langle u,v,w\rangle \mid \overline{u}$$

by assigning the same level to a,b,c. At run time we have the following (deterministic) sequence of reductions:

$$
\begin{array}{l}
!build(a,s_0,e_0).\ (\nu state)\ (\\
\quad \overline{state}\langle nil, nil, s_0, e_0\rangle \\
\quad \mid\ !a(chan, mode).state(l,r,s,e).chan(v,ans,n). \\
\qquad \text{if } mode = \text{merge then} \\
\qquad\quad \text{if } l = \text{nil then } \overline{state}\langle n,r,s,e\rangle \\
\qquad\quad \text{else if } r = \text{nil then } \overline{state}\langle l,n,s,e\rangle \\
\qquad\quad \text{else } (\nu chan')\ \ (\overline{l}\langle chan', \text{merge}\rangle.\overline{chan'}\langle v,ans,n\rangle.\overline{state}\langle l,r,s,e\rangle \\
\qquad\qquad\qquad\qquad + \overline{r}\langle chan', \text{merge}\rangle.\overline{chan'}\langle v,ans,n\rangle.\overline{state}\langle l,r,s,e\rangle\,) \\
\qquad \text{else} \quad \ldots)
\end{array}
$$

Fig. 3 Merging tree structures

$$
\begin{array}{rl}
(A,\emptyset) \rightarrow\rightarrow & (A \mid !u.[u>v].\overline{v} \mid !v.[v>w].\overline{w} \mid !w.[w>u].\overline{u}) \mid [u>v].\overline{v},\emptyset) \\
\rightarrow\rightarrow & (A \mid !u.[u>v].\overline{v} \mid !v.[v>w].\overline{w} \mid !w.[w>u].\overline{u}) \mid [v>w].\overline{w},\mathscr{R}_1) \\
\rightarrow\rightarrow & (A \mid !u.[u>v].\overline{v} \mid !v.[v>w].\overline{w} \mid !w.[w>u].\overline{u}) \mid [w>u].\overline{u},\mathscr{R}_2) \\
\rightarrow & \bot
\end{array}
$$

where $\mathscr{R}_1$ is $\{(u,v)\}$ and $\mathscr{R}_2$ is $\{(u,v),(v,w)\}$. Process A eventually produces $\bot$ as the three inner replications create a cycle in the relation.

The next example illustrates an advantage of dynamic typing on data-dependent or non-deterministic computations. Let

$$
Q \stackrel{\text{def}}{=} (Q_0 \mid \overline{p}\langle u,v\rangle \mid \overline{u} \mid \overline{g}) \qquad \text{where} \qquad Q_0 \stackrel{\text{def}}{=} !p(a,b).(!a.\overline{b} \mid (g.!b.\overline{a} + g.b)) \ .
$$

When the output at p is consumed we obtain the process

$$
Q' \stackrel{\text{def}}{=} Q_0 \mid !u.\overline{v} \mid (g.!v.\overline{u} + g.v) \mid \overline{u} \mid \overline{g} \ .
$$

If the output on g synchronises with the left summand, a loop is produced by the two replications. If the right summand is selected, the divergence is avoided. A static type system would necessarily reject Q, due to the potential loop in the two replications. In our mixed system, with appropriate choice of levels, Q passes the static analysis. At run time, one computation of Q will yield $\bot$, the other will not. We omit the details for lack of space.

We now discuss the typing of recursive structures as those in Section 1 and Section 3.1: trees with operations of *remote allocation*, that allow one to merge two trees by attaching the root of a tree to a leaf of another tree. To type the tree T_0 of Section 3.1, we need to take into account sequences of inputs in replications, that is, replications of the form $!\kappa.P$ as we do in the type system of Section 3. (Precisely, in the subterm $!node(a,\ldots).a(\ldots)\ldots$, we need to compare the sum of the levels of names *node* and a against the weight of the continuation.) This can be easily done, as discussed in Remark 1, by strengthening the typing rule for replication in the static phase of the mixed system. Alternatively, we can keep the present typing rules and make some modifications to the programs. We discuss this solution in the remainder. Figure 3 presents the modified tree structure. The topmost replication, $!build(a,s_0,e_0)$, acts as a constructor, invoked for the creation of a new node; this new node carries values e_0, s_0, and interacts with the parent node via channel a. The state of this node is represented by the floating message on *state* (in which the first two components are the names for accessing the children, and are set to the special value nil if the node is a leaf). We only show the code for the merge operation: the code for a search can be adapted from the example in Section 3.1. When merge is invoked, the transmitted channel should be attached to a leaf; if there is room, this happens in the current node; otherwise the merge is nondeterministically delegated to one of the children. (This is a simplified version of merge: the new tree is attached anywhere in the tree, without, for instance, ensuring that the tree remains well-balanced.) The code above is accepted by the static analysis of the mixed type system[1] modulo the insertion of just a few annotations: the highest level is affected to names a, l, r, and an annotation $[a > l]$ (resp. $[a > r]$) is inserted before the output at l (resp. at r). The resulting annotated process does not lead to failure exceptions at run time.

The mixed system remains, of course, incomplete — there are terminating processes whose annotated version yields $\perp$ — as the problem of the termination of a process is not decidable.

4.5 Efficiency

The static analysis of the mixed system can be made in time that is polynomial w.r.t. the size of the process being checked, by adapting the type inference algorithms in [2] (the modifications are mild). (Our system is more flexible than the one of [2], in that, e.g., we allow constraints relating a name received in some input prefix and a name defined above that prefix, but this does not affect the overall inference procedure for levels, which remains polynomial.) With such an algorithm, the static analysis introduces only the necessary assertions. More precisely, if the termination of a process can be proved by only relying on levels and weights (without referring to a partial order), then the static analysis will introduce no assertions and there will be no dynamic checks at run time.

Note that a trivial (and linear) static analysis would assign the same level to all names, and add assertions in front of all outputs prefixes. This would however mean that: all type checks are performed at run time; useful weight information is lost, so that the final termination analysis is rather rough.

The inference problem would become NP-complete if the system were refined by taking into account sequences of inputs underneath a replication as suggested in Remark 1 (this is proved by adapting the NP-completeness result for the system with sequences of inputs in [2]).

Concerning the efficiency of dynamic checks, each time a new constraint is added to the $\mathscr{R}$ component of a configuration, we have to check for acyclicity of the resulting relation. This can be done via a depth-first traversal of $\mathscr{R}$, whose cost is linear in $\#\mathscr{R} + |\mathscr{R}|$, where $\#\mathscr{R}$ (resp. $|\mathscr{R}|$) stands for the size of dom($\mathscr{R}$) (resp. the number of

[1] Recursive types are needed for typing, independently from the termination analysis; as mentioned in Section 3.1, recursive types, as well as other common type constructs, are straightforward to accommodate.

pairs in $\mathscr{R}$). In [6], an online algorithm is shown, that allows one to perform the same task in linear amortised time in $\#\mathscr{R}$ only.

5 Conclusion

In this paper we have investigated type systems for termination in which techniques of previous systems based on a lexicographical measure are enhanced with partial orders on names. The first system is purely static, the second mixes static and dynamic typing.

We have illustrated the expressiveness of the mixed system on a remote allocation example in which a recursive structure is extended with names and substructures imported from the environment. It would be difficult to handle this kind of system using a purely static system, due to the mobility of the names involved. The reason is that, intuitively, one cannot statically predict the precise name that is received in an input, and therefore one must make a worst-case approximation, using a set of names, guided by the type system. Further, in this situation, if a name is exported, then one has also to foresee the possibility that the name is sent back, which can easily create cycles that break the partial order on the names.

Acknowledgements. This work has been supported by the french ANR projects "MoDyFiable" and "CHoCo", by European Project FET-GC II IST-2005-16004 SENSORIA, and by Italian MIUR Project n. 2005015785, "Logical Foundations of Distributed Systems and Mobile Code".

References

1. M. Abadi, L. Cardelli, B. C. Pierce, and G. D. Plotkin. Dynamic typing in a statically typed language. *ACM Trans. Program. Lang. Syst.*, 13(2):237–268, 1991.
2. R. Demangeon, D. Hirschkoff, N. Kobayashi, and D. Sangiorgi. On the complexity of termination inference for processes. In *Proceedings of TGC'07*, LNCS. Springer, 2008. to appear.
3. Y. Deng and D. Sangiorgi. Ensuring termination by typability. *Inf. Comput.*, 204(7):1045–1082, 2006.
4. F. Henglein. Dynamic typing. In *Proc. of ESOP'92*, volume 582 of *Lecture Notes in Computer Science*, pages 233–253. Springer, 1992.
5. T. Lindholm and F. Yellin. *The Java^{TM} Virtual Machine Specification.* Addison Wesley, 1997.
6. A. Marchetti-Spaccamela, U. Nanni, and H. Rohnert. Maintaining a topological order under edge insertions. *Inf. Process. Lett.*, 59(1):53–58, 1996.
7. D. Sangiorgi. Termination of processes. *Mathematical Structures in Computer Science*, 16(1):1–39, 2006.
8. D. Sangiorgi and D. Walker. *The pi-calculus: a Theory of Mobile Processes.* Cambridge University Press, 2001.
9. P. R. Wilson. Uniprocessor garbage collection techniques. In *Proc. of IWMM'92*, volume 637 of *Lecture Notes in Computer Science*, pages 1–42. Springer, 1992.
10. N. Yoshida, M. Berger, and K. Honda. Strong normalisation in the pi-calculus. *Information and Computation*, 191(2):145–202, 2004.

6 Formal Definition of the Operational Semantics

6.1 π-calculus Processes

Structural congruence for the π-calculus is the least equivalence relation that is closed under α-conversion, satisfies the laws of an abelian monoid for $\mid$ and $+$ (with $\mathbf{0}$ as neutral element), and moreover validates the following axioms:

$$(\nu x)(\nu y)P \equiv (\nu y)(\nu x)P \qquad !P \mid P \equiv !P \qquad (\nu z)\mathbf{0} \equiv \mathbf{0} \qquad P+P \equiv P$$

$$(\nu x)(P \mid Q) \equiv (\nu x)P \mid Q \quad \text{if } x \text{ is not a free name of } Q$$

We moreover let $\equiv$ be closed by parallel composition, restriction, and replication, but *not* under prefixes (this is due to a technical reason related to the handling of κ in Section 3). The reduction relation is defined as follows.

$$\frac{}{(\overline{x}\langle\widetilde{v}\rangle.P_1 + M_1) \mid (x(\widetilde{z}).P_2 + M_2) \rightarrow P_1 \mid P_2\{\widetilde{v}/\widetilde{z}\}}$$

$$\frac{P_1 \rightarrow P_1'}{P_1 \mid P_2 \rightarrow P_1' \mid P_2} \qquad \frac{P \rightarrow P'}{(\nu c)\, P \rightarrow (\nu c)\, P'} \qquad \frac{P_1 \equiv P_2 \rightarrow P_2' \equiv P_1'}{P_1 \rightarrow P_1'}$$

6.2 Annotated Processes

Structural congruence for annotated processes is defined as above. If $\mathscr{R}$ is a relation on names, we write $\mathtt{ok}(\mathscr{R})$ if $\mathscr{R}$ induces a partial order, and $\neg\mathtt{ok}(\mathscr{R})$ otherwise. The reduction relation for configurations, written $\mapsto$, is defined by the following rules.

$$\frac{}{(\overline{x}\langle\widetilde{v}\rangle.A_1 + M_1) \mid (x(\widetilde{z}).A_2 + M_2), \mathscr{R} \mapsto A_1 \mid A_2\{\widetilde{v}/\widetilde{z}\}, \mathscr{R}}$$

$$\frac{A_1, \mathscr{R} \mapsto A_1', \mathscr{R}}{A_1 \mid A_2, \mathscr{R} \mapsto A_1' \mid A_2, \mathscr{R}} \qquad \frac{A, \mathscr{R} \mapsto A', \mathscr{R}}{\nu c\, A, \mathscr{R} \mapsto \nu c\, A', \mathscr{R}}$$

$$\frac{A_1 \equiv A_2 \qquad A_2, \mathscr{R} \mapsto A_2', \mathscr{R} \qquad A_2' \equiv A_1'}{A_1, \mathscr{R} \mapsto A_1', \mathscr{R}}$$

$$\frac{\mathtt{ok}(\mathscr{R} \cup \{(a,b)\})}{[a > b]A, \mathscr{R} \mapsto A, \mathscr{R} \cup \{(a,b)\}} \qquad \frac{\neg\mathtt{ok}(\mathscr{R} \cup \{(a,b)\})}{[a > b]A, \mathscr{R} \mapsto \bot}$$

Regular n-ary Queries in Trees and Variable Independence

Emmanuel Filiot and Sophie Tison

University of Lille 1, LIFL, UMR 8022 of CNRS, INRIA Lille - Nord Europe
Emmanuel.Filiot@lifl.fr Sophie.Tison@lifl.fr

Abstract. Regular n-ary queries in trees are queries which are definable by an MSO formula with n free first-order variables. We investigate the variable independence problem – originally introduced for databases – in the context of trees. In particular, we show how to decide whether a regular query is equivalent to a union of cartesian products, independently of the input tree. As an intermediate step, we reduce this problem to the problem of deciding whether the number of answers to a regular query is bounded by some constant, independently of the input tree. As a (non-trivial) generalization, we introduce variable independence w.r.t. a dependence forest between blocks of variables, which we prove to be decidable.

1 Introduction

Querying a tree consists of selecting nodes of its domain. This task has received a special interest from the XML community as it is fundamental to information extraction and document transformation. Several formalisms have been proposed to express unary queries [13], but less work has been done on n-ary queries, *ie* the selection of n-tuples of nodes [1, 10, 15]. Nevertheless, n-ary queries are of special interest, for instance to select tuples of the form *(name, addr, email, phone, fax)* in an XML document representing a directory. Since the arity of the query can easily get up to 10, efficiency becomes crucial to evaluate (n-ary) queries.

On the other hand, the notion of *variable independence* has been introduced in the context of (infinite) constraint databases [3, 5, 11, 12]. Query evaluation can be improved considerably when variables are independent. In particular, complexities of many evaluation algorithms on constraint databases are related to some independence between the components of the output tuples. For instance if the query can be decomposed into a Cartesian product of queries of lower arities, then all sub-queries of the product can be evaluated independently. *Orthographic dimension* has been proposed as a measure for variable independence [11]. It corresponds to the size of the largest block of dependent variables. It is shown in [5] that this notion is well-defined, since every pair of decompositions can be intersected into a decomposition of lower maximal block size.

Please use the following format when citing this chapter:

Filiot, E. and Tison, S., 2008, in IFIP International Federation for Information Processing, Volume 273; *Fifth IFIP International Conference on Theoretical Computer Science*; Giorgio Ausiello, Juhani Karhumäki, Giancarlo Mauri, Luke Ong; (Boston: Springer), pp. 429–443.

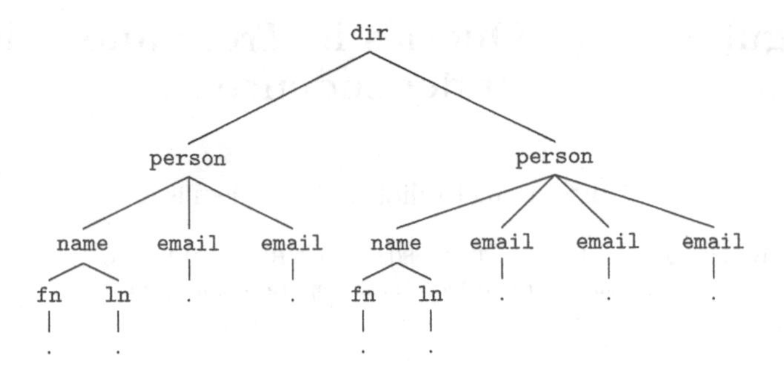

Fig. 1 A tree representing a directory

However, in the context of constraint databases, the structure is infinite but fixed. A natural question is whether these results carry over into the context of an infinite number of finite tree structures. The notion of variable independence is also closely related to the representation of the set of answers of an n-ary query. In particular, if the variables are independent, this allows to represent the set of answers as a Cartesian product of sets of (sub)answers of lower sizes. More generally, *aggregated answers* have been introduced in [14] as a compact representation of the set of answers. Basically, these are multipartite dags such that each part corresponds to a free variable, and the branching structure of the parts is a tree. This branching structure somehow reflects the structure of the input tree. Consider for instance the tree of Fig. 1 representing a directory. Data are omitted in this picture. Consider the ternary query q which selects all triples (x, y, z) where x is labeled `person`, y is the first name of this person, and z one of its emails. Once x is selected, then y and z are independent. The set of answers to this query can be represented as a 3-partite graph with sets of vertices $\{V_x, V_y, V_z\}$, where there is a directed edge from a node u of V_x to a node v of V_y, if v is the name of u. Similarly, edge relations between V_y and V_z correspond to the $(person, email)$ relations of the input tree. A person may have several emails, making this representation more compact than the set of all answers. Moreover, as argued in [14], this representation keeps the information on how the components of the output tuples are related in the tree. It is also particularly appropriate for post-processing tasks, such as answer searching, answer browsing, statistical computing, answer enumeration, and cascade-style querying. It raises the fundamental question of how compact this representation is. We answer this question by extending variable independence to *relative independence*. In particular, it allows more complex dependencies between variables, as emphasized by the previous example. We measure compactness in the settings of data-complexity, as we want to be independent of the query formalism.

To achieve the formal study of variable independence, we choose *Monadic Second Order Logic* to express n-ary queries, as it is often used as a yardstick logic in the context of trees [13].

We show that variable independence is decidable for MSO queries in trees, and that a decomposition is computable. We reduce this problem to testing whether a query is bounded, i.e. its number of answers is bounded by some constant independent of the input tree. We prove this problem to be decidable. We show that the notion of orthographic dimension is also well-defined in the context of trees. Finally, we introduce variable independence w.r.t. to a *dependence forest*, which introduces dependencies between blocks of variables.

Note that in the context of trees, a restricted notion of variable independence has been investigated in [15] and proved to be equivalent to non-ambiguity of tree automata.

Acknowledgments We are very grateful to the anonymous referees for their valuable comments, to Bruno Courcelle who pointed out some references and to Slawomir Staworko for his careful re-reading.

2 Preliminaries

Although XML documents are usually modeled as unranked trees [13], we consider finite binary trees only. All our results can easily be lifted to unranked trees via a binary encoding [13].

Binary trees We consider a finite alphabet Σ consisting of symbols ranged over by a, f, g. A *binary tree* t over Σ is inductively defined by the following grammar: $t ::= a \mid a(t,t) \quad a \in \Sigma$

The set of binary trees over Σ is denoted by T_Σ. The set of *nodes* N_t of a tree $t \in T_\Sigma$ is a set of words over $\{0,1\}$. We write ϵ for the empty word and $u.u'$ for the concatenation of u and u'. The set N_t is inductively defined by $N_a = \epsilon$ and $N_{a(t_0,t_1)} = \{\epsilon\} \cup \{b.u \mid b \in \{0,1\},\ u \in N_{t_b}\}$. Nodes $u \in N_t$ for which there is some $b \in \{0,1\}$ such that $u.b \in N_t$ are called *inner-nodes*. Other nodes are called *leaves*, and the node ϵ is called the *root* of t.

Let Σ' be another finite alphabet. Let $t \in T_\Sigma$ and $t' \in T_{\Sigma'}$ be two binary trees such that $N_t = N_{t'}$. The *product tree* $t \times t'$ is the tree over $\Sigma \times \Sigma'$ inductively defined by: $a \times b = (a,b)$, for all $a \in \Sigma, b \in \Sigma'$, and $f(t_0,t_1) \times g(t'_0,t'_1) = (f,g)(t_0 \times t'_0, t_1 \times t'_1)$, for all $f \in \Sigma, g \in \Sigma'$, $t_0, t_1 \in T_\Sigma$ and $t'_0, t'_1 \in T_{\Sigma'}$. More generally, we can define the product of n trees modulo associativity of $\times$.

Trees are also viewed as structures over the signature consisting of the binary successor symbols S_0 and S_1 and the unary symbols lab_a, for all $a \in \Sigma$, interpreted by their intuitive meanings.

MSO Monadic Second Order (MSO) logic extends first-order logic with quantification over sets. We consider first-order (resp. second-order) variables ranged over by x, y (resp. X, Y). MSO formulas consists of atomic formulas $lab_a(x)$, $S_0(x,y)$, $S_1(x,y)$ or $x \in X$, and are closed by boolean connectives $\wedge, \neg$ and quantification $\exists x, \exists X$. We let $t, \rho \models \phi$ denotes the satisfaction relation and say that the formula ϕ holds in the tree t under the variable assignment ρ. We refer the reader to [13] for more details about the semantics of MSO.

Definition 1 (regular n-ary queries). Let $n \geq 0$. An *n-ary query* q is a mapping from trees $t \in T_\Sigma$ into subsets of N_t^n. It is *regular* (or MSO-definable) if there is an MSO-formula $\phi(x_1, \dots, x_n)$ with n free first-order variables such that for all trees t, we have: $q(t) = \{(\rho(x_1), \dots, \rho(x_n)) \mid t, \rho \models \phi(x_1, \dots, x_n)\}$
If $n = 0$ (resp. $n = 1$, $n = 2$), q is called Boolean (resp. unary, binary). We also say that ϕ defines q and denote q by q_ϕ.

Those queries are called regular since there is a close correspondence between MSO-definable queries and tree automata. In particular, it is well-known that every MSO-definable n-ary query $\phi(x_1, \dots, x_n)$ on trees over Σ can be represented as a tree automaton A over the alphabet $\Sigma \times \{0,1\}^n$ [15, 19, 4]. Moreover, we can assume that A is *canonical*, i.e. for any tree $t \in L(A)$, and any $i \in \{1, \dots, n\}$, there is exactly one node of t such that the $(i+1)$-th component of its label is 1 [19, 4]. The following holds: for all its nodes $u_1, \dots, u_n$ of t, $t \models \phi(u_1, \dots, u_n)$ iff $t \times \chi_{u_1} \times \dots \times \chi_{u_n} \in L(A)$, where for all $i \in \{1, \dots, n\}$, χ_{u_i} is the tree on $\{0,1\}$ such that $N_t = N_{\chi_{u_i}}$ and all nodes except u_i are labeled 0.

3 Boundedness Properties of Regular Queries

In this section, we prove intermediate results which are useful for Section 4, but might be of independent interest.
Let $\phi(x_1, \dots, x_n)$ be an MSO formula whose free variables are first-order. We say that ϕ is *bounded* if there is $K \in \mathbb{N}$ such that for all $t \in T_\Sigma$, the number of assignments ρ of variables x_is into N_t such that $t, \rho \models \phi$ is bounded by K.

Lemma 1. *Given an MSO formula $\phi(x_1, \dots, x_n)$ with n free first-order variables $x_1, \dots, x_n$, it is decidable whether ϕ is bounded and, in this case, a bound is computable.*

Proof. As said at the end of Section 2, ϕ can be represented as a canonical tree automaton A over $\Sigma \times \{0,1\}^n$. Now, we can easily transform A in linear time into a bottom-up transducer T_A which takes trees t_1 over Σ as inputs and outputs trees t_2 over $\{0,1\}^n$ such that $N_{t_1} = N_{t_2}$ and $t_1 \times t_2 \in L(A)$. Since for all trees t, we have $|T_A(t)| = |q_\phi(t)|$, it suffices to test whether $|T_A(t)|$ is bounded by some constant, which is decidable in polynomial time (in the size of T_A). This is called finite valuedness in [17, 18]. Moreover, for all fixed $k \in \mathbb{N}$, one can decide in non-deterministic polynomial time whether the number of images by T_A is greater than k [17, 18] (with a constant factor which is several exponentials in the size of k). Moreover, the bound is lesser than $2^{2^{p(|T_A|)}}$ for some polynomial p independent of T_A. Hence, based on a dichotomy algorithm, one can compute the smallest upper bound. The time complexity however is several exponentials in the size of T_A. □

Concerning time complexity, it is known that the size of the tree automaton associated with ϕ might be non-elementary in the size of ϕ [19], making the whole procedure possibly non-elementary. However, if the query is given by a canonical tree automaton A, testing boundedness becomes polynomial in the size of A, since testing finite valuedness of a tree transducer can be done in polynomial time [18, 17].

Lemma 1 could also be deduced from a result of [2]. This paper considers an extension of MSO on infinite trees with bounding quantifiers. In particular, for any MSO formula $\psi(X)$, the bounding quantified formula $\mathbb{B}X.\psi(X)$ holds in an infinite tree t, if there is a bound $b \in \mathbb{N}$ such that the size of any subset of the set of nodes of t that satisfies $\psi(X)$ is bounded by b.
Two fragments are proved to be decidable: formulas of the form $\neg\mathbb{B}X.\psi(X)$, where ψ is in MSO, and formulas built from arbitrary MSO formulas and $\mathbb{B}, \exists, \vee$ and $\wedge$.
We can easily reduce our problem to satisfiability of some formula in the first fragment. First, boundedness of an n-ary query reduces to boundedness of all its projections. Hence, we only need to consider unary queries. Now, from a formula $\psi(x)$ in one free variable, we construct a closed formula γ such that $\psi(x)$ is bounded iff γ is unsatisfiable. The formula γ is a conjunction $\gamma = \gamma_1 \wedge \gamma_2$. The first formula γ_1 checks whether the model is a tree (possibly infinite) of the form $\#(t_1, \#(t_2, \#(t_3, \ldots)$, for some fresh symbol $\# \notin \Sigma$ and $t_1, t_2, t_3, \ldots \in T_\Sigma$ are finite trees over Σ. The second formula γ_2 has the form $\neg\mathbb{B}X.\gamma'(X)$, where $\gamma'(X)$ is an MSO formula which holds in $\#(t_1, \#(t_2, \#(t_3, \ldots)$ under some assignment ρ if there is $i \geq 1$ such that such that $\rho(X)$ corresponds to the set of nodes u of t_i such that $t_i \models \psi(u)$. The formula γ' is defined by first choosing some node x_0 labeled $\#$ and then relating $\psi(x)$ under x_0.
However, we cannot benefit from this reduction if the query is given by a tree automaton (in term of time complexity).

An *equivalence relation on n-tuples* is a $2n$-ary query , often denoted $\equiv$, such that for all trees t, $\equiv(t)$ is an equivalence relation on N_t^n. We let $\equiv_t$ stands for $\equiv(t)$. It is regular if $\equiv$ is regular. We say that $\equiv$ is of bounded index if for all trees t, the number of $\equiv_t$-equivalence classes is bounded by some constant which does not depend on the tree. We can define a regular query which selects the minimal representatives of the equivalence classes, for some MSO-definable order on tuples. Hence, as a corollary of Lemma 1, we get:

Corollary 1 (bounded index property). *Let $\equiv$ be a regular equivalence relation on n-ary tuples. It is decidable whether $\equiv$ is of bounded index.*

Proof. Let t be a tree, we define a total order $\leq_t$ on N_t^n. It suffices to start from a total order on N_t and to extend it to a lexicographic order $\leq_t$ on N_t^n. Take for instance the lexicographic order on words over $\{0,1\}$ which is a total order on N_t. We can easily show that the query $t \mapsto \leq_t$ is regular. Now, we define the n-ary query $q_{min} : t \mapsto \{\overline{u} \mid \forall \overline{u}' \in N_t^n,\ \overline{u} \equiv_t \overline{u}' \implies \overline{u} \leq_t \overline{u}'\}$. The query q_{min} is regular. Finally it suffices to verify boundedness of q_{min}, which is decidable by Lemma 1. □

Beyond Trees Deciding boundedness of an MSO formula can be done for classes of structures which are images of a regular set of trees by an MSO-transduction. We refer the reader to [6] for a definition of MSO-transductions. Given a set of integers $I \subseteq \mathbb{N}$, we say that I is *linear* if there are integers $\alpha_0, \ldots, \alpha_n \in \mathbb{N}$ such that $I = \{\alpha_0 + \sum_{i=1}^{n} \alpha_i x_i \mid x_1, \ldots, x_n \in \mathbb{N}\}$. It is *semi-linear* if it is a finite union of linear sets. Let σ be a signature and C be a class of σ-structures, θ an MSO-transduction from binary trees to σ-structures such that C is the image by θ of a set of binary trees. In [6], Courcelle proves[1] that given an MSO formula $\phi(X)$ over the signature σ with one free second-order variable X, the set $\{\#\rho(X) \mid M, \rho \models \phi(X),\ M \in C\}$ is semi-linear, and we can compute the coefficients of the polynomials if the transduction θ is known ($\#\rho(X)$ denotes the cardinality of $\rho(X)$). This is the case for instance for the class of graphs of clique-width less than k, for any fixed k [8]. To decide boundedness of an MSO formula $\phi(x)$ (where x is first-order), it suffices to compute the above coefficients for the formula $\Phi(X) = \forall x,\ x \in X \leftrightarrow \phi(x)$. The formula $\phi(x)$ is bounded iff the coefficients α_0 of the linear sets are the unique (possibly) non-null coefficients. Finally, boundedness of a formula $\phi(x_1, \ldots, x_n)$ reduces to boundedness of every projection of ϕ on a single variable x_i, for all $i \in \{1, \ldots, n\}$. Hence, boundedness is decidable, for instance, for structures of clique-width less than k, for any fixed k, or for unranked trees. However, to decide the bounded index property, we need an MSO-definable total order.

4 Variable Independence

The definition of variable independence was originally defined over a fixed structure [3, 12]. We state it over the class of binary trees. We let ϕ be an MSO formula with free variables $x_1, \ldots, x_n$, and $P = \{B_1, \ldots, B_k\}$ be a partition of $\{x_1, \ldots, x_n\}$. We write $\overline{x}_{B_i}$, $i = 1, \ldots, k$, to denote the tuple formed by variables of B_i given in order. We say that ϕ *conforms to* P, denoted $\phi \sim P$, if ϕ is equivalent to a formula of the form $\bigvee_{j=1}^{N} \phi_{j,1}(\overline{x}_{B_1}) \wedge \cdots \wedge \phi_{j,k}(\overline{x}_{B_k})$, where N is a natural and $\phi_{j,i}$ are MSO formulas with free variables in B_i. Note that if we require N to be equal to 1, the problem becomes easy, since it suffices to test whether ϕ is equivalent to the conjunction of the k projections of ϕ on the variables from each block B_i.

W.l.o.g., we assume that free variables of ϕ are ranked in order given by $B_1, \ldots, B_k$. In other words, we assume $x_1, \ldots, x_n = \overline{x}_{B_1}, \ldots, \overline{x}_{B_k}$ (modulo associativity). Now, for any $i \in \{1, \ldots, k\}$, and any tuples of variables $\overline{x}$, $\overline{y}$ such that $|\overline{x}| = |\overline{y}| = |B_i|$, we let $\psi_{\phi}^{i}(\overline{x}, \overline{y})$ be the formula defined by:

---

[1] Courcelle proves a more general result where several free variables are allowed

$$\psi^i_\phi(\overline{x},\overline{y}) = \forall \overline{x}_{B_1} \ldots \forall \overline{x}_{B_{i-1}} \forall \overline{x}_{B_{i+1}} \ldots \forall \overline{x}_{B_k} \;\; \phi(\overline{x}_{B_1},\ldots,\overline{x}_{B_{i-1}},\overline{x},\overline{x}_{B_{i+1}},\ldots,\overline{x}_{B_k}) \leftrightarrow \phi(\overline{x}_{B_1},\ldots,\overline{x}_{B_{i-1}},\overline{y},\overline{x}_{B_{i+1}},\ldots,\overline{x}_{B_k})$$

For any $i \in \{1,\ldots,k\}$, and any tree $t \in T_\Sigma$, we let R^t_i be the binary relation on $N_t^{|B_i|}$ defined by $R^t_i = \{(\overline{u},\overline{v}) \mid t \models \psi^i_\phi(\overline{u},\overline{v})\}$. Intuitively, $\overline{u}$ and $\overline{v}$ are equivalent if one can substitute $\overline{u}$ with $\overline{v}$, in any tuple selected by ϕ whose i-th block is $\overline{u}$, and conversely.

Lemma 2. *Let $i \in \{1,\ldots,k\}$ and $t \in T_\Sigma$. The following are true:*

1. *R^t_i is an equivalence relation on $N_t^{|B_i|}$;*
2. *if ϕ is equivalent to some formula of the form $\bigvee_{j=1}^N \phi_{j,1}(\overline{x}_{B_1}) \wedge \cdots \wedge \phi_{j,k}(\overline{x}_{B_k})$, then the number of R^t_i-equivalence classes is bounded by 2^N.*

Proof. We only prove the second point, as the first is easy. Given some natural $i \in \{1,\ldots,k\}$, some tree t and some node tuples $\overline{u},\overline{v}$ of length $|B_i|$, we let $\overline{u} \equiv^t_i \overline{v}$ if there is some set $F \subseteq \{1,\ldots,N\}$ (possibly empty) such that for all $j \in F$, we have $t \models \phi_{j,i}(\overline{u}) \wedge \phi_{j,i}(\overline{v})$, and for all $j \in \{1,\ldots,N\} \backslash F$, we have $t \models \neg\phi_{j,i}(\overline{u}) \wedge \neg\phi_{j,i}(\overline{v})$. We can easily prove that $\equiv^t_i$ is an equivalence relation on $N_t^{|B_i|}$ which has at most 2^N equivalence classes. We now prove that $\equiv^t_i$ is a refinement of R^t_i, which will be sufficient to conclude. Let $\overline{u},\overline{v}$ be two node tuples of length $|B_i|$ such that $\overline{u} \equiv^t_i \overline{v}$. Let $\overline{w}_1,\ldots,\overline{w}_{i-1},\overline{w}_{i+1},\ldots,\overline{w}_k$ be node tuples. We have $t \models \phi(\overline{w}_1,\ldots,\overline{w}_{i-1},\overline{u},\overline{w}_{i+1},\ldots,\overline{w}_k)$ iff $t \models \bigvee_{j=1}^N \phi_{j,1}(\overline{w}_1) \wedge \cdots \wedge \phi_{j,i}(\overline{u}) \wedge \cdots \wedge \phi_{j,k}(\overline{w}_k)$ iff (by definition of $\equiv^t_i$) $t \models \bigvee_{j=1}^N \phi_{j,1}(\overline{w}_1) \wedge \cdots \wedge \phi_{j,i}(\overline{v}) \wedge \cdots \wedge \phi_{j,k}(\overline{w}_k)$ iff $t \models \phi(\overline{w}_1,\ldots,\overline{w}_{i-1},\overline{v},\overline{w}_{i+1},\ldots,\overline{w}_k)$. Hence we get $(\overline{u},\overline{v}) \in R^t_i$. □

Lemma 3. *If for all $i \in \{1,\ldots,k\}$, there is some $m_i \in \mathbb{N}$ such that for all trees $t \in T_\Sigma$, the number of equivalence classes of R^t_i is bounded by m_i, then $\phi \sim P$.*

Proof. Let $i \in \{1,\ldots,n\}$. We define a successor relation between equivalence classes, then we introduce formulas $cl^l_i(\overline{x}), l = 1,\ldots,m_i$, to define the l-th equivalence class of R^t_i, in any tree $t \in T_\Sigma$.

As already seen in the proof of Corollary 1, there is an MSO-definable total order $\leq$ on node tuples. We now define a successor relation S^i_ϕ between the minimal representatives (for $\leq$) of the equivalence relation defined by ψ^i_ϕ. Now, let the formula $min^i_\phi(\overline{x})$ holds if $\overline{x}$ is the minimal representative of some equivalence class. It can be defined by $\forall\overline{y},\ \psi^i_\phi(\overline{x},\overline{y}) \rightarrow \overline{x} \leq \overline{y}$. The relation S^i_ϕ is now defined by the following MSO formula:

$$S^i_\phi(\overline{x},\overline{y}) = \overline{x} < \overline{y} \wedge min^i_\phi(\overline{x}) \wedge min^i_\phi(\overline{y}) \wedge \neg(\exists\overline{z}, \overline{x} < \overline{z} < \overline{y} \wedge min^i_\phi(\overline{z}))$$

We let $s_0(\overline{x})$ stand for $\neg\exists\overline{z}, S^i_\phi(\overline{z},\overline{x})$ and $s_l(\overline{x})$, $l \in \mathbb{N}$ stands for $\exists\overline{y},\ s_{l-1}(\overline{y}) \wedge S^i_\phi(\overline{y},\overline{x})$. We now define $cl^l_i(\overline{x})$ by $\exists\overline{y}, s_l(\overline{y}) \wedge \psi^i_\phi(\overline{x},\overline{y})$, for all $1 \leq l \leq m_i$. Intuitively $s_l(\overline{x})$ holds in t under some assignment ρ if $\rho(\overline{x})$ is the minimal

representative of the l-th equivalence class of R_i^t, while $cl_i^l(\overline{x})$ holds in t under ρ if $\rho(\overline{x})$ belongs to the l-th equivalence class of R_i^t.
Finally, we let L be the set of tuples of naturals $\overline{l} = (l_1, \ldots, l_k)$ such that $1 \leq l_i \leq m_i$, $i = 1, \ldots, k$, and we denote by $\beta^{\overline{l}}(\overline{x}_1)$ the formula $\exists \overline{x}_2 \ldots \overline{x}_k,\ \phi(\overline{x}_1, \overline{x}_2, \ldots, \overline{x}_k) \wedge \bigwedge_{j=1}^{k} cl_j^{l_j}(\overline{x}_j)$. We let $\phi^{\overline{l}}$ be the formula $\beta^{\overline{l}}(\overline{x}_1) \wedge cl_2^{l_2}(\overline{x}_2) \cdots \wedge cl_k^{l_k}(\overline{x}_k)$. We now prove that ϕ is equivalent to $\bigvee_{\overline{l} \in L} \phi^{\overline{l}}$.

Let $t \in T_\Sigma$ and $\overline{u}_1, \ldots, \overline{u}_k$ node tuples of t such that $t \models \phi(\overline{u}_1, \ldots, \overline{u}_k)$. For all $i \in \{1, \ldots, k\}$, we necessarily have $t \models \psi_\phi^i(\overline{u}_i, \overline{u}_i)$. Hence, there is some natural $l_i \in \{1, \ldots, m_i\}$ such that $t \models cl_i^{l_i}(\overline{u}_i)$, $i = 1, \ldots, k$. Let $\overline{l} = (l_1, \ldots, l_k)$. It is easy to see that $t \models \phi^{\overline{l}}(\overline{u}_1, \overline{u}_2, \ldots, \overline{u}_k)$.
Conversely, suppose there is some tuple $\overline{l} = (l_1, \ldots, l_k) \in L$ such that $t \models \phi^{\overline{l}}(\overline{u}_1, \ldots, \overline{u}_k)$. In particular, we have $t \models \beta^{\overline{l}}(\overline{u}_1)$, hence there are some node tuples $\overline{u}'_2, \ldots, \overline{u}'_k$, such that $t \models \phi(\overline{u}_1, \overline{u}'_2, \ldots, \overline{u}'_k)$, and for each $i \in \{2, \ldots, k\}$, $t \models cl_i^{l_i}(\overline{u}'_i)$. We now prove by induction that for all $p \in \{2, \ldots, k\}$, we have $t \models \phi(\overline{u}_1, \ldots, \overline{u}_p, \overline{u}'_{p+1}, \ldots, \overline{u}'_k)$. It is true for $p = 1$ by hypothesis. Suppose that it is true at rank $p > 1$. Since $t \models cl_{p+1}^{l_{p+1}}(\overline{u}_{p+1})$, we also have $t \models \psi_\phi^{p+1}(\overline{u}_{p+1}, \overline{u}'_{p+1})$. By induction hypothesis, we have $t \models \phi(\overline{u}_1, \ldots, \overline{u}_p, \overline{u}'_{p+1}, \ldots, \overline{u}'_k)$, and by definition of ψ_ϕ^{p+1}, we easily get $t \models \phi(\overline{u}_1, \ldots, \overline{u}_p, \overline{u}_{p+1}, \overline{u}'_{p+2}, \ldots, \overline{u}'_k)$. □

As a consequence of Lemma 2 and 3, and Corollary 1, we get the main result:

Theorem 1. *Given an MSO formula ϕ with free variables $x_1, \ldots, x_n$ and a partition P of $\{x_1, \ldots, x_n\}$, it is decidable whether $\phi \sim P$ holds or not. If it holds, a decomposition of ϕ is computable.*

Orthographic Dimension The notion of orthographic dimension has been introduced in [11] to measure the degree of independence between variables, over a fixed database. We define it for any tree structure. Given a formula $\phi(x_1, \ldots, x_n)$, the *orthographic dimension* d_ϕ of ϕ is defined by $d_\phi = min_{P\ :\ \phi \sim P}\ max_{B \in P} |B|$.
Theorem 1 gives us a naive algorithm to compute d_ϕ: for each partition P of $\{1, \ldots, n\}$, test whether $\phi \sim P$ and compute $max_{B \in P} |B|$. But as we next show, we can restrict the tests to 2-partitions.
Given two partitions P, P' of $\{1, \ldots, n\}$, we write $P \sqcap P'$ for the refinement of P and P'. Formally, we have $P \sqcap P' = \{B \cap B' \mid B \in P, B' \in P'\} - \varnothing$. Theorem 1 of [5] states well-definedness of the notion of orthographic dimension, for first-order logic on any vocabulary, over a fixed structure. In particular, it means that there is a unique partition whose largest block is equal to the orthographic dimension, such that the formula conforms to it. However, the proof given in [5] also works if the structure is not fixed. Moreover, it also works as soon as the logic contains first-order quantifiers, negations, and Boolean connectives. Hence, it also proves the following:

Theorem 2. *Let ϕ be an MSO formula in n free first-order variables $x_1, \ldots, x_n$, and P, P' be two partitions of $\{1, \ldots, n\}$. If $\phi \sim P$ and $\phi \sim P'$, then $\phi \sim P \sqcap P'$.*

Now, let $\mathcal{P} = \{P \mid P \text{ is a 2-partition of } \{1, \ldots, n\} \text{ and } \phi \sim P\}$. From Theorem 2, we can deduce that the orthographic dimension of ϕ is the size of the largest block of $\sqcap_{P \in \mathcal{P}} P$. Moreover, by Theorem 1, we can compute a decomposition which corresponds to the orthographic dimension.

Relation to the Answer Set Representation If $\phi \sim P$, then for any tree $t \in T_\Sigma$, we can represent $q_\phi(t)$ by an aggregated answer of size $O(n|t|^{d_\phi})$, computable in time $O(n|t|^{d_\phi})$ (ϕ is assumed to be fixed). Indeed, ϕ is equivalent to a (computable) formula of the form:$\bigvee_{i=1}^{N} \phi_{i,1}(x_{B_1}) \wedge \ldots \wedge \phi_{i,k}(x_{B_k})$ for some natural N. For every $i \in \{1, \ldots, N\}$, and $j \in \{1, \ldots, k\}$, we compute an automaton $A_{i,j}$ over $\Sigma \times \{0,1\}^{|B_j|}$ such that we can compute in time $O(|t|^{|B_j|}|A_{i,j}|)$ the set $q_{\phi_{i,j}}(t)$ [7]. The answer representation can be identified to the collection of k-tuples $(q_{\phi_{i,1}}(t), \ldots, q_{\phi_{i,k}}(t))$.

5 Relative Variable Independence

In this section, we generalize variable independence w.r.t. a partition to variable independence w.r.t. a dependence forest on free variables. Consider for instance the tree of Fig. 1, and let $\phi(x, y, z)$ be an MSO formula, where x denotes a person, y its first name and z one of its emails. Once the interpretation of x is fixed, then y and z are independent. We call this *relative independence.* Let T be the tree $x(y, z)$, we say that ϕ conforms to T, denoted $\phi \sim T$. We next show that relative independence is decidable as a consequence of Theorem 1.

As a slight generalization, we allow dependence forests to specify dependences between sets of variables instead of single variables. For example, if a formula $\phi(x, y, z, w)$ conforms to some dependence forest $\{x, y\}(\{z\}, \{w\})$, it means that once x and y are selected, then z and w are independent.
Formally, let $V = \{x_1, \ldots, x_n\}$ be a finite set of variables. A *dependence forest* F over V is a forest whose nodes are labeled by subsets of 2^V and such that the set of labels occurring in F form a partition of V. If F has only one root, then it is called a *dependence tree.* We often denote by $\{T_1, \ldots, T_k\}$ a dependence forest consisting of the dependence trees $T_1, \ldots, T_k$, and by $V'(F)$ a dependence tree consisting of a dependence forest F rooted by a set $V' \subseteq V$. Although we take a graph point of view, we use the same notations as for binary trees to denote the set of nodes of some forest F over V and its labeling function, respectively by N_F and $lab_F : N_F \rightarrow 2^V$. Finally, we confuse the set and tuple notations for variables, so that we sometimes write $V = \overline{x}$, for some tuple of variables $\overline{x}$.
Let $\phi(x_1, \ldots, x_n)$ be an MSO formula in n free first-order variables $x_1, \ldots, x_n$. Let μ be a mapping from N_F into MSO formulas. We say that μ is *admissible* for F if for all nodes $u, v \in N_F$, if u is the parent of v, then $\mu(v)$ is an MSO formula ψ whose free variables are $lab_F(u) \cup lab_F(v)$; if u is the root of F, then we require that $\mu(u)$ is an MSO formula whose free variables are $lab_F(u)$. If μ

is admissible, we naturally extend it to an MSO formula $\mu(F) = \bigwedge_{u \in N_F} \mu(u)$ in free variables $x_1, \ldots, x_n$.

Definition 2. We say that ϕ conforms to F, denoted $\phi \sim F$, if there is a finite sequence $\mu_1, \ldots, \mu_N$ of admissible mappings for F such that ϕ is equivalent to $\bigvee_{i=1}^{N} \mu_i(F)$.

We prove decidability of relative independence. We start by a base lemma (Lemma 4), for forests of the form $\overline{x}(\{\overline{y}, \overline{z}\})$. Then we give a recursive algorithm for the general case, that uses Lemma 4 and Theorem 1.

Lemma 4. *Given an MSO formula $\phi(\overline{x}, \overline{y}, \overline{z})$ with free variables $\overline{x}, \overline{y}, \overline{z}$, it is decidable whether ϕ is equivalent to some disjunction of the form $\bigvee_{i=1}^{n} \alpha_i(\overline{x}, \overline{y}) \wedge \beta_i(\overline{x}, \overline{z})$, for some natural n, and MSO formulas α_i, β_i for $i = 1, \ldots, n$. Moreover if it holds, a disjunction is computable.*

Proof. Intuitively, we fix the interpretation of $\overline{x}$ by extending the alphabet with Boolean tuples. This gives a formula $\phi_{\overline{x}}(\overline{y}, \overline{z})$, and we test whether $\phi_{\overline{x}}(\overline{y}, \overline{z}) \sim \{\overline{y}, \overline{z}\}$.

More formally, we first transform $\phi(\overline{x}, \overline{y}, \overline{z})$ into $\phi_{\overline{x}}(\overline{y}, \overline{z})$, interpreted on trees over the alphabet $\Sigma \times \{0,1\}^m$, where $m = |\overline{x}|$, such that the following property holds (P_1): for all trees t, all nodes $u_1, \ldots, u_m \in N_t$, and all node tuples $\overline{v}, \overline{w}$, we have $t \models \phi(u_1, \ldots, u_m, \overline{v}, \overline{w})$ iff $t \times \chi_{u_1} \times \cdots \times \chi_{u_m} \models \phi_{\overline{x}}(\overline{v}, \overline{w})$, where the trees χ_{u_i} are defined at the end of Section 2.
This can be done by repeating exhaustively the following transformation rule on ϕ: replace each atom of the form $P(\overline{x}_1, x, \overline{x}_2)$, where x is the i-th component of $\overline{x}$ and is a free occurrence in ϕ, by $\exists x, \bigvee_{(f,\overline{b}) \in \Sigma \times B_i} lab_{(f,\overline{b})}(x) \wedge P(\overline{x}_1, x, \overline{x}_2)$, where $B_i \subseteq \{0,1\}^m$ is the set of Boolean tuples whose i-th component is 1 for $i = 1, \ldots, m$. Hence, $\phi(\overline{x}, \overline{y}, \overline{z})$ rewrites to some formula $\phi'_{\overline{x}}(\overline{y}, \overline{z})$. We define $\phi_{\overline{x}}$ by $\phi'_{\overline{x}} \wedge \phi_{can}$, where ϕ_{can} is a sentence which ensures that all models $t \in T_{\Sigma \times \{0,1\}^m}$ of $\phi_{\overline{x}}$ are canonical (*ie* all nodes except one have their i-th component set to 0 $i = 1, \ldots, n$). We call R_1 the transformation from ϕ to $\phi_{\overline{x}}$.

Then it suffices to test whether $\phi_{\overline{x}}(\overline{y}, \overline{z}) \sim \{\overline{y}, \overline{z}\}$, which is decidable by Theorem 1. If it holds, then $\phi_{\overline{x}}(\overline{y}, \overline{z})$ is equivalent to some formula of the form $\psi_{\overline{x}}(\overline{y}, \overline{z}) = \bigvee_{i=1}^{n} \alpha_{i,\overline{x}}(\overline{y}) \wedge \beta_{i,\overline{x}}(\overline{z})$, for some MSO formulas $\alpha_{i,\overline{x}}, \beta_{i,\overline{x}}$. We next consider the following transformation rule R_2: replace each atom of the form $lab_{f,\overline{b}}(x)$ by $lab_f(x) \wedge \bigwedge_{b_i=1} x_i = x$, where b_i denotes the i-th component of $\overline{b}$, $i = 1, \ldots, m$. Suppose that $\overline{x} = x_1, \ldots, x_m$. Applying exhaustively this transformation rule on $\psi_{\overline{x}}$ leads to a formula $\psi(\overline{x}, \overline{y}, \overline{z})$ of the form $\bigvee_{i=1}^{n} \alpha_i(\overline{x}, \overline{y}) \wedge \beta_i(\overline{x}, \overline{z})$ interpreted on trees over Σ. We have the following property (P_2): for all trees t, all nodes $u_1, \ldots, u_m \in N_t$, and all node tuples $\overline{v}, \overline{w}$, we have $t \models \psi(u_1, \ldots, u_m, \overline{v}, \overline{w})$ iff $t \times \chi_{u_1} \times \cdots \times \chi_{u_m} \models \psi_{\overline{x}}(\overline{v}, \overline{w})$.

Finally, we prove this algorithm to be correct. Suppose that it returns a decomposition. By combining properties P_1 and P_2, we can prove correctness of this decomposition. Conversely, suppose that ϕ is equivalent to some formula of the form $\bigvee_{i=1}^{n} \alpha_i(\overline{x}, \overline{y}) \wedge \beta_i(\overline{x}, \overline{z})$. It is easy to see that $\phi_{\overline{x}}$ is equivalent (here

Algorithm 1 Testing Relative Independence

```
    procedure D(φ,F)
2:      case F is a leaf or is of the form x̄(ȳ):
        return φ
4:
        case F is of the form {x̄_1,...,x̄_k}:
6:      test whether φ ~ {x̄_1,...,x̄_k} as in the proof of Theorem 1 and return a decomposition. Otherwise breaks.

8:      case F is of the form x̄(ȳ,z̄):
        test whether φ ~ x̄(ȳ,z̄) as in the proof of Lemma 4 and return a decomposition. Otherwise breaks.
10:
        case F is of the form x̄(ȳ(F'),F''):
12:     z̄', z̄'' ← sets of variables occurring in F', F''
        ⋁_i α_i(x̄,ȳ,z̄') ∧ β_i(x̄,z̄'') ← D(φ, x̄({ȳ ∪ z̄', z̄''}))
14:     return ⋁_i D(α_i, ȳ(F',x̄)) ∧ D(β_i, x̄(F''))

16:     case F is of the form {T_1,...,T_k}:
        for i ∈ {1,...,k} do
18:         x̄_i ← set of variables occurring in T_i
        ⋁_{i=1}^n ⋀_{j=1}^k α_i^j(x̄_j) ← D(φ, {x̄_1,...,x̄_k})
20:     return ⋁_{i=1}^n D(α_i^1, T_1) ∧ ··· ∧ D(α_i^k, T_k)
```

we use canonicity of its models) to $\bigvee_{i=1}^{n} \alpha_{i,\overline{x}}(\overline{y}) \wedge \beta_{i,\overline{x}}(\overline{z})$, where $\phi_{\overline{x}}$, $\alpha_{i,\overline{x}}(\overline{y})$ and $\beta_{i,\overline{x}}(\overline{z})$ are obtained by applying the rewrite rule R_1 on respectively ϕ, $\alpha_i(\overline{x},\overline{y})$ and $\beta_i(\overline{x},\overline{z})$. Hence, $\phi_{\overline{x}} \sim \{\overline{y},\overline{z}\}$, and the proof follows since the algorithm of the proof of Theorem 1 is sound. □

Now, we extend the result of Lemma 4 to full independence forests:

Theorem 3. *Given a formula ϕ in free variables $V = \{x_1,\dots,x_n\}$ and a dependence forest F over V, it is decidable whether $\phi \sim F$ holds or not.*

Proof. Consider Algorithm 1. The inputs are a formula ϕ with free variables V and a dependence forest F over V. The symbols $T_1,\dots,T_k$ denote dependence trees while F', F'' denote (possibly empty) dependence forests.
First note that the algorithm terminates. Indeed, the number of nodes of the forest strictly decreases at each recursive call except for the 4th case when F'' is empty, but in this case the height of the forest strictly decreases.
Now, we can prove (inductively and by using Theorem 1 and Lemma 4 for the basic cases) that if this algorithm returns a formula, then it is a decomposition of the input formula ϕ w.r.t. the input forest F. It suffices to push up the disjunctive connectives to get a sequence of admissible mappings for F.
Conversely, let ϕ (resp. F) be an input formula (resp. an input dependence forest), such that we have $\phi \sim F$. We prove by induction that the algorithm outputs a decomposition. We use the fact that the decomposition is not arbitrary, but has a particular form, derived from the algorithms given in the proofs

of Theorem 1 and Lemma 4. The first case is obvious, and the two next cases have already been proved. First remark that we have the following property (*): let F be a dependence forest, γ_1, γ_2 two formulas, α a sentence and $\beta(\overline{x})$ a formula such that $\overline{x}$ is a label of F. If $\gamma_1 \sim F$ and $\gamma_2 \sim F$, we have $\gamma_1 \vee \gamma_2 \sim F$, $\gamma_1 \wedge \gamma_2 \sim F$, $\neg\gamma_1 \sim F$, $\alpha \wedge \gamma_1 \sim F$, and $\beta(\overline{x}) \wedge \gamma_1 \sim F$.

Suppose that F is of the form $\overline{x}(\overline{y}(F'), F'')$. Since $\phi \sim F$, in particular, $\phi \sim \overline{x}(\overline{y} \cup \overline{z}', \overline{z}'')$, where $\overline{z}', \overline{z}''$ are defined as in Algorithm 1. Now, we inspect the proof of Lemma 4. Let $\phi_{\overline{x}}$ be the result of applying the rewriting rule R_1 of this proof. It is clear, by hypothesis, that we have $\phi_{\overline{x}} \sim \{\overline{y} \cup \overline{z}', \overline{z}''\}$. Hence, algorithm of the proof of Theorem 1 outputs a decomposition of the form $\bigvee_{\overline{l} \in L} \beta_1^{\overline{l}}(\overline{y}, \overline{z}') \wedge cl_2^{l_2}(\overline{z}'')$, exactly as defined in the proof of Theorem 1. We let ψ^{-1} be the result of applying exhaustively the rewrite rule R_2 of the proof of Lemma 4 on ψ, for all formulas ψ. Hence, $\mathrm{D}(\phi, \overline{x}(\overline{y} \cup \overline{z}', \overline{z}''))$ returns the formula $\bigvee_{\overline{l} \in L} (\beta_1^{\overline{l}})^{-1}(\overline{x}, \overline{y}, \overline{z}') \wedge (cl_2^{l_2})^{-1}(\overline{x}, \overline{z}''))$. It remains to prove that formulas $(\beta_1^{\overline{l}})^{-1}$ and $(cl_2^{l_2})^{-1}$ satisfy $(\beta_1^{\overline{l}})^{-1} \sim \overline{y}(\overline{x}, F')$ and $(cl_2^{l_2})^{-1} \sim \overline{x}(F'')$. We only prove it for formulas $(cl_2^{l_2})^{-1}$, as the proof for formulas $(\beta_1^{\overline{l}})^{-1}$ is analogous. So let us fix some natural l_2. By going back to the definition of formula $cl_2^{l_2}$, we can prove that $(cl_2^{l_2})^{-1}$ is equivalent to a formula of the form $\Gamma = \exists \overline{z}_0,\ \gamma(\overline{x}, \overline{z}_0) \wedge \forall \overline{u}, \phi(\overline{x}, \overline{u}, \overline{z}'') \leftrightarrow \phi(\overline{x}, \overline{u}, \overline{z}_0)$, for some γ. Now, since $\phi \sim F$, it easy to see that ϕ is equivalent to a formula of the form $\Psi = \bigvee_{i=1}^{n} \epsilon_i^1(\overline{x}, \overline{y}, \overline{z}') \wedge \epsilon_i^2(\overline{x}, \overline{z}'')$, such that $\epsilon_i^1 \sim \overline{y}(\overline{x}, F')$ and $\epsilon_i^2 \sim \overline{x}(F'')$ for $i = 1, \ldots, n$. We replace in Γ the formula ϕ by Ψ, and, after a series of rewritings (by pushing up disjunctions and pushing down quantifiers), we can prove that $(cl_2^{l_2})^{-1}$ is equivalent to a formula of the form $\bigvee_{i=1}^{n} \gamma_i(\overline{x}) \epsilon_i^2(\overline{x}, \overline{z}'') \vee \bigvee_{P \subseteq \{1,\ldots,n\}} \gamma_P'(\overline{x}) \wedge \bigwedge_{i \in P} \neg\epsilon_i^2(\overline{x}, \overline{z}'')$, for formulas γ_i, γ_P' depending only on i and P. The conclusion follows by property (*) and the fact that every ϵ_i^2 satisfies $\epsilon_i^2 \sim \overline{x}(F'')$.

Suppose now that F is of the form $\{T_1, \ldots, T_k\}$, let $\overline{x}_i$ be the variables occurring in T_i for $i = 1, \ldots, k$, and let $P = \{\overline{x}_1, \ldots, \overline{x}_k\}$. Since $\phi \sim F$, in particular, $\phi \sim P$. Hence $\mathrm{D}(\phi, P)$ outputs a decomposition of the form $\bigvee_{\overline{l} \in L} \beta_1^{\overline{l}}(\overline{x}_1) \wedge cl_2^{l_2}(\overline{x}_2) \wedge \ldots \wedge cl_k^{l_k}(\overline{x}_k)$, exactly as defined in the proof of Theorem 1. We have to prove that for all $\overline{l} = (l_1, \ldots, l_k) \in L$, we have $\beta_1^{\overline{l}} \sim T_1$, and $cl_i^{l_i} \sim T_i$ for $i = 2, \ldots, k$. This is sufficient, since by induction hypothesis, D will output a decomposition of every $\beta_1^{\overline{l}}$ and $cl_i^{l_i}$. We only prove it for formulas $cl_i^{l_i}(\overline{x}_i)$, as it is similar for formulas $\beta_1^{\overline{l}}$. Let us fix some $\overline{l}$ and i. We come back to the definition of $cl_i^{l_i}$, and we can easily show that it is of the form $\exists \overline{y}, \gamma(\overline{y}) \wedge \psi_\phi^i(\overline{x}_i, \overline{y})$, for some formula γ which selects the minimal representatives of the l_i-th equivalence class of the relation defined by ψ_ϕ^i, where ψ_ϕ^i has been defined in Section 4. Now, since $\phi \sim F$, ϕ is equivalent to a formula of the form $\Psi = \bigvee_{p=1}^{n} \bigwedge_{j=1}^{k} \epsilon_p^j(\overline{x}_j)$ such that every ϵ_p^j satisfy $\epsilon_p^j \sim T_j$. Next, in $\exists \overline{y}, \gamma(\overline{y}) \wedge \psi_\phi^i(\overline{x}_i, \overline{y})$, we replace ϕ by Ψ (ψ_ϕ^i can be viewed as the result of applying the function $\psi_\cdot^i$ on ϕ, and we just replace $\psi_\phi^i(\overline{x}_i, \overline{y})$ by $\psi_\Psi^i(\overline{x}_i, \overline{y})$). We get a formula equivalent to $cl_i^{l_i}$ which, after a series of rewritings preserving equivalence (by moving up disjunctions and pushing down quantifiers), rewrites to a formula (equivalent to $cl_i^{l_i}$) of the form $\bigvee_{p=1}^{n} \phi_p \wedge \epsilon_p^i(\overline{x}_i) \vee \bigvee_{Q \subseteq \{1,\ldots,n\}} \psi_Q \wedge \bigwedge_{p \in Q} \neg \epsilon_p^i(\overline{x}_i)$, for some closed formulas ϕ_p, ψ_Q depending on p and Q. The conclusion follows by using property (*) and the fact that every ϵ_p^i satisfies $\epsilon_p^i \sim T_i$. □

Similarly as the case of variable independence, if $\phi(x_1, \ldots, x_n)$ conforms to F, then for any tree $t \in T_\Sigma$, $q_\phi(t)$ can be represented by an aggregated answer of size $O(n|t|^b)$ (ϕ is assumed to be fixed, and necessarily $|F| \leq n$). The parameter b denotes the maximal sum of the size of a label of F plus the size of the label of its father if it exists.

Note that variable independence w.r.t. a dependence forest subsumes variable independence w.r.t. a partition, since a partition can be viewed as a dependence forest consisting of a set of leaves. Moreover, as stated by the next theorem, there is an MSO formula ϕ such that there is no dependence forest F such that: (i) labels of F are singletons (ii) $\phi \sim F$. Nevertheless, we know that on trees, every MSO formula is equivalent to an existentially quantified Boolean combination of MSO formulas in two free variables [16, 10].

Theorem 4. *There is an MSO formula ϕ such that there is* **no** *dependence forest F whose labels are singletons and such that $\phi \sim F$.*

Proof. Let $x \preceq y$ be an MSO formula which holds in a tree if y is a descendant of x. It is well-known that it can easily be defined as the reflexive and transitive closure of $S_1 \vee S_2$, this closure being definable in MSO.
Now, let $\phi(x, y, z)$ be an MSO formula defined by:

$$\begin{aligned} \phi(x,y,z) = \exists \alpha \quad & \alpha \preceq x \wedge \alpha \preceq y \wedge \alpha \preceq z \\ & \wedge \forall \alpha' \alpha' \preceq x \wedge \alpha' \preceq y \implies \alpha' \preceq \alpha \\ & \wedge \forall \alpha' \; \alpha' \preceq x \wedge \alpha' \preceq z \implies \alpha' \preceq \alpha \end{aligned}$$

For all trees t and all nodes u, v, w of t, we have $t \models \phi(u, v, w)$ iff the least common ancestor of u and v is equal to the least common ancestor of u and w.

We now prove by applying algorithm 1 that there is no forest F whose labels are singletons such that $\phi \sim F$, by proving it for each forest over $\{x, y, z\}$. Let $n \geq 0$ and t_n be the tree over the alphabet $\{a\}$ inductively defined by $t_0 = a$ and $t_n = a(t_{n-1}, a)$. For all n, we denote by $v_0, v_1, \ldots, v_n$ the nodes $1^n, 1^{n-1}, \ldots, \epsilon$ respectively, and, if $n > 0$, by $w_1, \ldots, w_n$ the nodes $1^{n-1}.2, 1^{n-2}.2, \ldots, 2$ respectively.

1. $F = \{x, y, z\}$ or $F = \{x, y(z)\}$ or $F = \{x, z(y)\}$. We apply algorithm 1. The formula $\psi_\phi^1(x, x')$ is defined by $\forall y, z \; \phi(x, y, z) \leftrightarrow \phi(x', y, z)$. We prove that for all n, and all $i, j \leq n$, $v_i \neq v_j$ implies $t_n \not\models \psi_\phi^1(v_i, v_j)$. Indeed, if $v_i \neq v_j$ such that $desc(v_i, v_j)$, then we have $t \models \phi(v_i, v_i, v_j)$ but $t \not\models \phi(v_j, v_i, v_j)$. Hence, the number of classes of the equivalence relation defined by ψ_ϕ^1 is at least n, which is unbounded. So Algorithm 1 breaks;
2. $F = \{y, x(z)\}$ or $F = \{y, z(x)\}$. The formula $\psi_\phi^2(y, y')$ is defined by $\forall x, y, z \; \phi(x, y, z) \leftrightarrow \phi(x, y', z)$. We prove that for all n, and all $i, j \leq n$, $v_i \neq v_j$ implies $t_n \not\models \psi_\phi^2(v_i, v_j)$. Indeed, if $v_i \neq v_j$ such that $desc(v_i, v_j)$,

then we have $t \models \phi(v_j, v_i, v_i)$ but $t \not\models \phi(v_j, v_j, v_i)$. Hence, the number of classes of the equivalence relation defined by ψ_ϕ^2 is at least n, which is unbounded. So Algorithm 1 breaks;
3. $F = \{z, y(x)\}$ or $F = \{z, x(y)\}$. Those cases are symmetric to the previous ones.
4. $F = x(y, z)$. We let $\psi(y, y') = \forall z, \phi_x(y, z) \leftrightarrow \phi_x(y', z)$ where ϕ_x has been defined in the proof of Lemma 4. By definition of Algorithm 1, if the equivalence relation defined by ψ has an unbounded index, then the algorithm fails. This is what we next prove. Let $n \leq 0$. We fix x by a Boolean in the tree t_n: we let t'_n be the tree over $\{a\} \times \{0, 1\}$ such that $N_{t_n} = N_{t'_n}$ and all nodes are labeled $(a, 0)$ except v_0 which is labeled $(a, 1)$. It is easy to see that for all $i \geq 1$, we have $t'_n \models \phi_x(v_i, w_i)$ and for all $j > i$, we have $t'_n \not\models \phi_x(v_j, w_i$. Hence there are at least n equivalence classes for the relation defined by ψ. So Algorithm 1 breaks.
5. $F = y(x, z)$. Similarly to the previous case, we fix a variable. Let $\psi(z, z') = \forall x, \phi_y(x, z) \leftrightarrow \phi_y(x, z')$. Let t'_n be the tree defined in the previous case. Hence y is fixed to denote the node v_0. We can prove that for all $i > 0$, we have $t'_n \models \phi_y(v_i, w_i)$ but for all $j > i$, we have $t'_n \not\models \phi_y(v_i, w_j)$. Hence there are at least n equivalence classes for the relation defined by ψ. So Algorithm 1 breaks.
6. $F = z(x, y)$. This case is symmetric to the previous one. □

Further Extensions First note that all the results presented in the paper also hold for FO-queries, as we do not use second order variables in decompositions (but in this case we need to add in the tree structure a total order on the nodes). We would like to investigate independence problems for more general classes of structures C. Indeed, we can give two sufficient conditions for relative independence to be decidable on C: (i) boundedness of an MSO formula with first-order variables is decidable on C, (ii) there is a computable MSO-definable total order on the elements of the structures of C. The first point has already been detailed in Section 3, while the second point is studied in [9]. This is the case for instance for unranked tree structures, over the signature consisting of the first-child and next-sibling predicates, and predicates to test the labels.
Finally, we would like to extend independence w.r.t. a dependence forest to independence w.r.t. a dependence graph. The techniques presented here do not seem to be easily extendable to graphs, even for a clique of size 3 for instance. In particular, we cannot use an inductive proof based on Lemma 4 anymore.

References

1. A. Berlea. On-the-fly tuple selection for XQuery. In *International Workshop on XQuery Implementation, Experience and Perspectives*, June 2007.
2. M. Bojanczyk. A bounding quantifier. In *14th Annual Conference of the EACSL on Computer Science Logic*, 2004.

3. J. Chomicki, D. Goldin, G. Kuper, and D. Toman. Variable independence in constraint databases. *IEEE Transactions on Knowledge and Data Engineering*, 15(6):1422–1436, 2003.
4. H. Comon, M. Dauchet, R. Gilleron, C. Löding, F. Jacquemard, D. Lugiez, S. Tison, and M. Tommasi. Tree automata techniques and applications. Available on: http://www.grappa.univ-lille3.fr/tata, 2007.
5. S. Cosmadakis, G. Kuper, and L. Libkin. On the orthographic dimension of definable sets. *Inf. Process. Lett.*, 79(3):141–145, 2001.
6. B. Courcelle. Structural properties of context-free sets of graphs generated by vertex replacement. *Inf. Comput.*, 116(2):275–293, 1995.
7. B. Courcelle. Linear delay enumeration and monadic second-order logic. 2007. To appear in Discrete Applied Mathematics.
8. B. Courcelle and S. Oum. Vertex-minors, monadic second-order logic, and a conjecture by seese. *J. Comb. Theory Ser. B*, 97(1):91–126, 2007.
9. Bruno Courcelle. The monadic second-order logic of graphs x: linear orderings. *Theor. Comput. Sci.*, 160(1-2):87–143, 1996.
10. E. Filiot, J. Niehren, J.-M. Talbot, and S. Tison. Polynomial time fragments of xpath with variables. In *ACM Symposium on Principles of Database Systems*, 2007.
11. S. Grumbach, P. Rigaux, and L. Segoufin. On the orthographic dimension of constraint databases. In *7th International Conference on Database Theory*, pages 199–216. Springer-Verlag, 1999.
12. L. Libkin. Variable independence for first-order definable constraints. *TOCL*, 4(4):431–451, 2003.
13. L. Libkin. Logics over unranked trees: an overview. *Logical Methods in Computer Science*, 3(2):1–31, 2006.
14. Holger Meuss, Klaus U. Schulz, and François Bry. Towards aggregated answers for semistructured data. In *ICDT '01: Proceedings of the 8th International Conference on Database Theory*, pages 346–360, 2001.
15. J. Niehren, L. Planque, J.-M. Talbot, and S. Tison. N-ary queries by tree automata. In *10th International Symposium on Database Programming Languages*, volume 3774, pages 217–231, 2005.
16. T. Schwentick. On diving in trees. In *25th International Symposium on Mathematical Foundations of Computer Science*, pages 660–669, 2000.
17. H. Seidl. Ambiguity, valuedness and costs. 1992. Habilitation Thesis.
18. H. Seidl. Equivalence of finite-valued tree transducers is decidable. *Math. Syst. Theory*, 27(4):285–346, 1994.
19. J. W. Thatcher and J. B. Wright. Generalized finite automata with an application to a decision problem of second-order logic. 2:57–82, 1968.

Hamiltonicity of automatic graphs

Dietrich Kuske[1] and Markus Lohrey[2]

[1] Institut für Informatik, Universität Leipzig, kuske@informatik.uni-leipzig.de
[2] Institut für Informatik, Universität Leipzig, lohrey@informatik.uni-leipzig.de

Abstract. It is shown that the existence of a Hamiltonian path in a planar automatic graph of bounded degree is complete for Σ_1^1, the first level of the analytical hierarchy. This sharpens a corresponding result of Hirst and Harel for highly recursive graphs. Furthermore, we also show: (i) The Hamiltonian path problem for finite planar graphs that are succinctly encoded by an automatic presentation is NEXPTIME-complete, (ii) the existence of an infinite path in an automatic successor tree is Σ_1^1-complete, and (iii) an infinite version of the set cover problem is decidable for automatic graphs (it is Σ_1^1-complete for recursive graphs).

1 Introduction

The theory of *recursive structures* has its origins in computability theory. A structure is recursive, if its domain is a recursive set of naturals, and every relation is again recursive. Starting with the work of Manaster and Rosenstein [23] and Bean [1, 2], infinite variants of classical graph problems for finite graphs were studied for recursive graphs. It is not surprising that these problems are mostly undecidable for recursive graphs. This motivates the search for the precise level of undecidability. It turned out that some of the problems reside on low levels of the arithmetic hierarchy (e.g. the question whether a given recursive graph has an Eulerian path [3]), whereas others are complete for Σ_1^1 — the first level of the analytic hierarchy [21]. A classical example for the latter situation is the question whether a given recursive tree has an infinite path. With a technically quite subtle reduction from the latter problem, Harel proved in [13] that also the existence of a *Hamiltonian path* (i.e., a one-way infinite path that visits every node exactly once) in a recursive graph is Σ_1^1-complete. Σ_1^1-hardness holds already for highly recursive graphs, where a list of the neighbors of a node v can be computed effectively from v.

Hamiltonian paths in infinite graphs were also studied under a purely graph theoretic view. An important result of Dean, Thomas, and Yu [6] states that an infinite undirected graph G has a Hamiltonian path if it is (i) planar, (ii) 4-connected, and (iii) has only one end (see [7] for definitions). This extends a result of Tutte [27] for finite graphs.

In computer science, in particular in the area of automatic verification, focus has shifted in recent years from arbitrary recursive graphs to subclasses that have more amenable algorithmic properties. An important example for this is the class of *automatic graphs* [5, 16]. A graph is called automatic if it has an *automatic presentation*, which consists of a finite automaton that generates the set of nodes and a two-tape

Please use the following format when citing this chapter:

Kuske, D. and Lohrey, M., 2008, in IFIP International Federation for Information Processing, Volume 273; *Fifth IFIP International Conference on Theoretical Computer Science*; Giorgio Ausiello, Juhani Karhumäki, Giancarlo Mauri, Luke Ong; (Boston: Springer), pp. 445–459.

automaton with synchronously moving heads, which accepts the set of edges. One of the main motivations for investigating automatic graphs is the fact that every automatic graph has a decidable first-order theory [16], this result extends to first-order logic with infinity and modulo quantifiers [5, 19]. In contrast to these positive results, Khoussainov, Nies, and Rubin have shown that the isomorphism problem for automatic graphs is Σ_1^1-complete [17]. Results on the model theoretic complexity of automatic structures can be found in [15].

The main result of this paper states that the existence of a Hamiltonian path becomes Σ_1^1-complete already for a quite restricted subclass of recursive graphs, namely for automatic graphs, which are planar and of bounded degree. The latter means that there exists a constant c such that every node has at most c many neighbors. The proof of the Σ_1^1 lower bound (the non-trivial part) in Section 3 is based on a reduction from the *recurring tiling problem* [10, 12]. This is a variant of the classical tiling problem [29, 4] that asks whether a given finite set of tiles allows a tiling of the infinite quarter plane such that a distinguished color occurs infinitely often at the lower border. Harel proved that the recurring tiling problem is Σ_1^1-complete [10, 12]. In our reduction we use as building blocks some of the graph gadgets from the NP-hardness proof of the Hamiltonian path problem for finite planar graphs [9]. These gadgets have to be combined in a non-trivial way for the whole reduction.

The main purpose of automatic presentations is the finite representation of infinite structures. But automatic presentations can be also used as a tool for the succinct representation of large finite structures. An automatic presentation of size n may generate a finite graph of size $2^{O(n)}$. A straightforward adaptation of our proof for infinite automatic graphs shows that it is NEXPTIME-complete to check whether a finite planar graph given by an automatic presentation has a Hamiltonian path, see Section 4. Without the restriction to planar graphs, this result was already shown by Veith [28] in the slightly different context of graphs represented by ordered binary decision diagrams (OBDDs). The special OBDDs considered by Veith in [28] can be seen as automatic presentations of finite graphs.

Finally, in Section 5 we investigate some other graph problems in the automatic setting. Using a proof technique from [20, 15], we prove that the fundamental Σ_1^1-complete problem in recursion theory, namely the existence of an infinite path in a recursive tree remains Σ_1^1-complete if the input tree is automatic. For this result it is crucial that the tree is a *successor tree*, which means that it is an acyclic graph, where every node is reachable from a root node and every node except the root has exactly one incoming edge. If trees are given as particular partially ordered sets (order trees), then the existence of an infinite path is decidable for automatic trees [20].

From the above results, one might get the feeling that graph problems always have the same degree of undecidability in the recursive and in the automatic world. To the contrary, there are problems that are Σ_1^1-complete for recursive graphs [14] but decidable for automatic graphs. This applies to the existence of an infinite branch in an automatic *order tree* (i.e., the reflexive and transitive closure of a successor tree, Khoussainov, Rubin, and Stephan [20]) as well as to the existence of an infinite clique in an automatic graph (Rubin [25]). We show that also an infinite version of the set cover problem is decidable for automatic graphs. This result is achieved by providing a decision procedure for a fragment of second-order logic that allows to express the set cover problem as well as the two other decidable problems mentioned before.

Proofs, which are not included in this extended abstract will appear in the long version of this paper.

2 Preliminaries

Infinite graphs and Hamiltonian paths For details on graph theory see [7]. A *graph* is a pair $G = (V,E)$, where V is the (possibly infinite) set of nodes and $E \subseteq V \times V$ is the set of edges. It is *undirected* if $(u,v) \in E$ implies $(v,u) \in E$. The graph G has *degree at most c*, where $c \in \mathbb{N}$, if every node is contained in at most c many edges. If G has degree at most c for some constant c, then G has *bounded degree*. If it is only required that every node is involved in only finitely many edges then G is called *locally finite*. The graph G is *planar* if it can be embedded in the Euclidean plane without crossing edges and without accumulation points; any such embedding is a *plane graph*. A *finite path* in G is a sequence $[v_1, v_2, \ldots, v_n]$ of nodes such that $(v_i, v_{i+1}) \in E$ for all $1 \leq i \leq n$. The nodes v_1 and v_n are the end points of this path. The graph $G = (V,E)$ is *connected* if for all $u, v \in V$ there exists a finite path in the undirected graph $(V, E \cup \{(x,y) \mid (y,x) \in E\}$ with end points u and v. An *infinite path* in G is an infinite sequence $[v_1, v_2, \ldots]$ such that every initial segment is a finite path. A *Hamiltonian path* (or *spanning ray*) of an *infinite* graph G is an infinite path $[v_1, v_2, \ldots]$ in G that visits every node of G exactly once, i.e. the mapping $i \mapsto v_i$ ($i \in \mathbb{N}$) is a bijection between $\mathbb{N}$ and the set of nodes.

Recursive graphs and automatic graphs A *recursive graph* is a graph $G = (V,E)$ such that V and E are recursive subsets of $\mathbb{N}$ and $\mathbb{N} \times \mathbb{N}$, respectively. In case G is infinite, one can w.l.o.g. assume that $V = \mathbb{N}$. A recursive graph G is *highly recursive* if it is locally finite and for every node v a list of its finitely many neighbors can be computed from v. Harel [13] has shown the following result:

Theorem 1 ([13]). *It is Σ^1_1-complete to determine, whether a given highly recursive undirected graph of bounded degree has a Hamiltonian path.*

Recall that Σ^1_1 is the first level of the *analytic hierarchy* [21]. More precisely, it is the class of all subsets of $\mathbb{N}$ of the form $\{n \in \mathbb{N} \mid \exists A\, \varphi(A)\}$, where $\varphi(A)$ is a formula of first-order arithmetic. In Thm. 1, a recursive graph is encoded by a pair of Gödel numbers for machines for the node and edge set, respectively.

In [14], Hirst and Harel proved that for planar recursive graphs the existence of a Hamiltonian path is still Σ^1_1-complete. The aim of this paper is to extend the results from [13, 14] to the class of planar automatic graphs of bounded degree. We introduce this class of graphs briefly, more details can be found in [16, 5]

Let us fix $n \in \mathbb{N}$ and a finite alphabet Γ. Let $\# \notin \Gamma$ be an additional padding symbol. For words $w_1, \ldots, w_n \in \Gamma^*$ we define the *convolution* $w_1 \otimes w_2 \otimes \cdots \otimes w_n$, which is a word over the alphabet $(\Gamma \cup \{\#\})^n$, as follows: Let $w_i = a_{i,1} a_{i,2} \cdots a_{i,k_i}$ with $a_{i,j} \in \Gamma$ and $k = \max\{k_1, \ldots, k_n\}$. For $k_i < j \leq k$ define $a_{i,j} = \#$. Then $w_1 \otimes \cdots \otimes w_n =$

$(a_{1,1},\ldots,a_{n,1})\cdots(a_{1,k},\ldots,a_{n,k})$. Thus, for instance $aba \otimes bbabb = (a,b)(b,b)(a,a)(\#,b)(\#,b)$. An n-ary relation $R \subseteq (\Gamma^*)^n$ is called automatic if the language $\{w_1 \otimes \cdots \otimes w_n \mid (w_1,\ldots,w_n) \in R\}$ is a regular language.

Now let $\mathscr{A} = (A,(R_i)_{i\in J})$ be a relational structure with finitely many relations, where $R_i \subseteq A^{n_i}$. A tuple (Γ,L,h) is called an *automatic presentation* for $\mathscr{A}$ if (i) Γ is a finite alphabet, (ii) $L \subseteq \Gamma^*$ is a regular language, (iii) $h : L \to A$ is a surjective function, (iv) the relation $\{(u,v) \in L \times L \mid h(u) = h(v)\}$ is automatic, and (v) the relation $\{(u_1,\ldots,u_{n_i}) \in L^{n_i} \mid (h(u_1),\ldots,h(u_{n_i})) \in R_i\}$ is automatic for every $i \in J$. We say that $\mathscr{A}$ is *automatic* if there exists an automatic presentation for $\mathscr{A}$. In the rest of the paper we will mainly restrict to automatic graphs. Such a graph can be represented by an automaton for the node set and an automaton for the edge set. Clearly, a (locally finite) automatic graph is (highly) recursive.

In contrast to recursive graphs, automatic graphs have some nice algorithmic properties. In [16] it was shown that the first-order theory of an automatic structure is decidable. This result extends to first-order logic with infinity and modulo quantifiers [5, 19]. For general automatic structures, these logics do not allow elementary algorithms [5]. On the other hand, for automatic structures with a Gaifman graph of bounded degree first-order logic extended by a rather general class of counting quantifiers can be decided in triply exponential space [22].

In contrast to these positive results, several strong undecidability results show that algorithmic methods for automatic structures are quite limited. Since the configuration graph of a Turing machine is automatic, it follows easily that reachability in automatic graphs is undecidable. Khoussainov, Nies, and Rubin have shown that the isomorphism problem for automatic graphs is Σ_1^1-complete [17], whereas isomorphism of locally finite automatic graphs is Π_3^0-complete [24]. Our main result is the following:

Theorem 2. *It is Σ_1^1-complete to determine, whether a given planar automatic undirected graph of bounded degree has a Hamiltonian path.*

Note that the Σ_1^1 upper bound in Thm. 2 follows immediately from the corresponding result for general recursive graphs (Thm. 1). For the lower bound we use a special variant of the tiling problem [29, 4] that was introduced by Harel.

Tilings Our main tool for proving Σ_1^1-hardness of the existence of a Hamiltonian path in a planar automatic graph of bounded degree is the *recurring tiling problem* [10, 12]. An instance of the recurring tiling problem consists of (i) a finite set of *colors* $C = \{c_0, c_1, \ldots, c_n\}$, (ii) a distinguished color c_0, and (iii) a set $\mathscr{T} \subseteq C^4$ of *tile types*. For a tile type $t \in \mathscr{T}$ we write $t = (t_W, t_N, t_E, t_S)$ ("W" for west, "N" for north, "E" for east, and "S" for south); a visualization looks as follows:

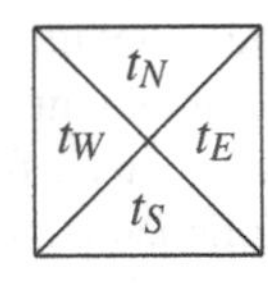

A mapping $f : \mathbb{N}^2 \to \mathscr{T}$ is a *tiling* if, for every $(i,j) \in \mathbb{N}^2$, we have $f(i,j)_N = f(i+1,j)_S$ and $f(i,j)_E = f(i,j+1)_W$. A *recurring tiling* is a tiling f such that for

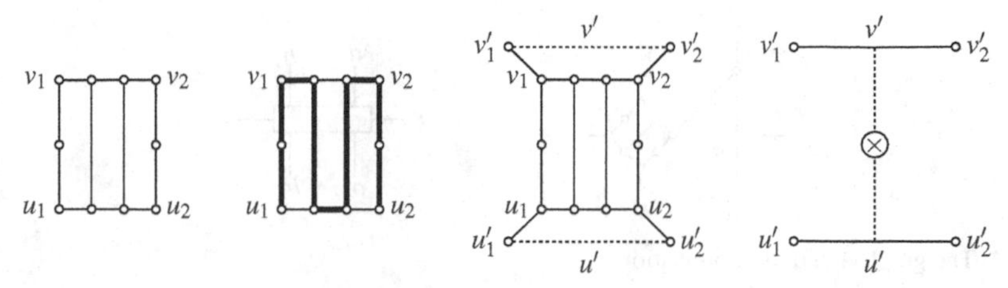

Fig. 1 The graph X, its use and abbreviation

infinitely many $j \in \mathbb{N}$, we have $f(0, j)_S = c_0$. Now the recurring tiling problem asks whether a given problem instance has a recurring tiling. Harel has shown the following result:

Theorem 3 ([10]). *The recurring tiling problem is Σ_1^1-complete.*

The recurring tiling problem turned out be very useful for proving Σ_1^1 lower bounds for certain satisfiability problems in logic [11].

3 Hamiltonicity for automatic graphs

In this section, we reduce the recurring tiling problem to the existence of a Hamiltonian path in a planar automatic graph of bounded degree. This proves Thm. 2 by Thm. 3.

3.1 Building blocks

Let us introduce several building blocks from which we assemble our final planar automatic graph of bounded degree. These building blocks are variants of graphs taken from the NP-hardness proof for the Hamiltonian path problem in finite planar graphs [9].

Exclusive or Consider the finite plane graph X in Fig. 1 (first picture). It has a Hamiltonian path from u_1 to u_2 (and similarly from v_1 to v_2) indicated in the second picture. Now suppose G' is some graph containing the edges u' and v'. Then we build a graph G as follows: in the disjoint union of G' and X, delete the edges u' and v' and connect their endpoints to u_1 and u_2 (to v_1 and v_2, resp., see Fig. 1, third picture). Now suppose H is a Hamiltonian path in G with no endpoint in X. Suppose u_1 is the first vertex from X in H. Then the restriction of H to X has to coincide with the Hamiltonian path from u_1 to u_2. Hence H gives rise to a Hamiltonian path in G' that coincides with H on G' but passes through the edge u' instead of taking the detour through X. Note that H' does not contain the edge v'. Conversely, every Hamiltonian path H' of G' that contains the edge u' but not the edge v' induces a Hamiltonian path H of G in a similar

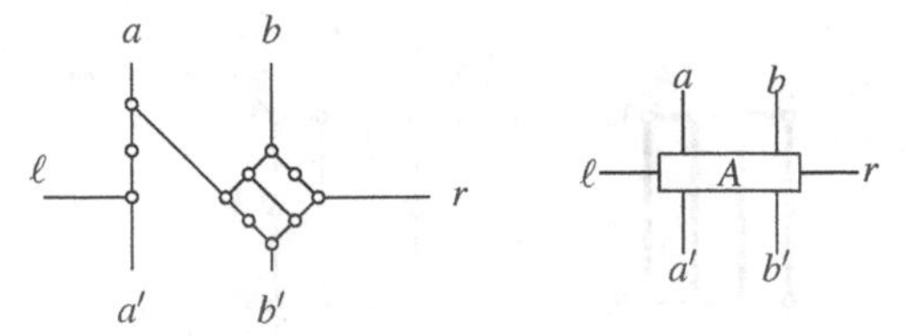

Fig. 2 The graph A and its abbreviation

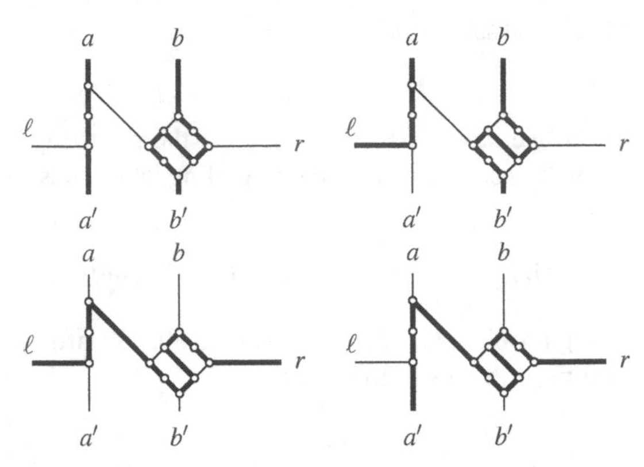

Fig. 3 Paths through the graph A

way. Joining X to the graph G' in this manner restricts the Hamiltonian paths to those that either contain the edge u' or the edge v', but not both. This also explains the name X: this graph acts as an "exclusive-or". Note that, if G' is planar and the two edges u' and v' belong to the same face, then also G can be constructed as a planar graph. Since we will make repeated use of this construction, we abbreviate it as in Fig. 1, fourth picture.

Boolean functions Let $f : \{0,1\}^n \to \{0,1\}$ be a Boolean function. In the NP-hardness proof of [9], a planar graph G together with distinguished edges $e_1, \dots, e_n$ is constructed such that $f(b_1, \dots, b_n) = 1$ iff G has a Hamiltonian cycle H with $b_i = 1 \Leftrightarrow e_i \in H$. We modify this construction slightly in order to place the edges e_i and two vertices u and v in a specified order at the boundary of the outer face.

Theorem 4. *There exists a constant c such that from given $k, \ell, n \in \mathbb{N}$ and $F \subseteq 2^{\{1,\dots,k+\ell+n\}}$, one can construct effectively in logspace a finite plane graph G_F of degree at most c such that:*

- *At the boundary of the outer face of G_F, we find (in this counter-clockwise order) edges $e_1, \dots e_k$, a vertex u, edges $e_{k+1}, \dots, e_{k+\ell}$, a vertex v, and edges $e_{k+\ell+1}, \dots, e_{k+\ell+n}$.*
- *For every $M \subseteq \{1, \dots, k+\ell+n\}$, $M \in F$ iff there is a Hamiltonian path H from u to v such that $M = \{i \mid e_i \text{ belongs to } H\}$.*

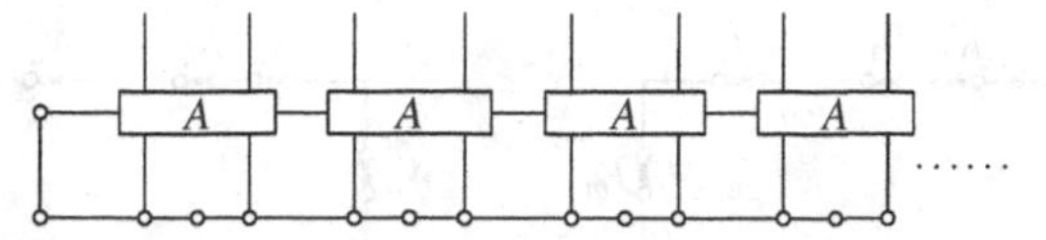

Fig. 4 The infinite graph L

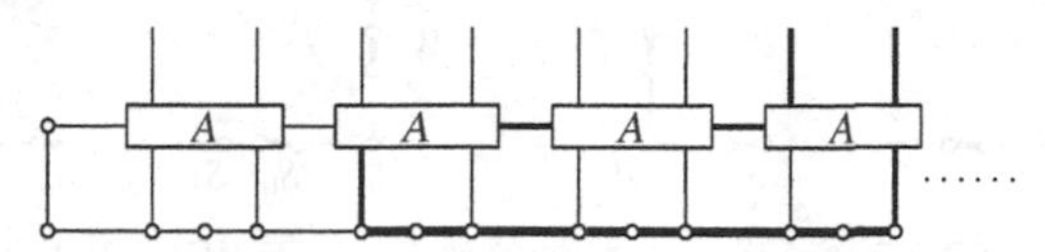

Fig. 5 A visit of a Hamiltonian path to the graph L

Infinity checking Next consider Fig. 2 – it depicts a graph A that is connected to some context via the edges ℓ, a, a', b, b', and r. If the complete graph has a Hamiltonian path, then locally, it has to be of one of the four forms depicted in Fig. 3.

Now consider Fig. 4 – it consists of infinitely many copies of the graph A arranged in a line, the edges a' and b' connect these copies of A with a line of nodes. Suppose the edges a and b of the copies of A are connected to some infinite graph G. Then, every Hamiltonian path H of the resulting graph has to enter and leave L infinitely often. Since the possibilities to pass A are restricted as shown in Fig.3, any such visit has to look as described in Fig. 5, i.e., the path enters from a into some copy of A, moves left to some copy of A (possibly without doing any step), moves down to the third line where it goes all the way back until it can enter the A-copy visited first via the edge b' and leave it via the edge b.

3.2 Assembling

From an instance of the recurring tiling problem, we construct in this section a planar automatic graph G of bounded degree that has an Hamiltonian path iff the instance of the recurring tiling problem admits a solution. So, we fix a finite set $C = \{c_0, c_1, \ldots, c_n\}$ of colors, a distinguished color c_0, and a set $\mathscr{T} \subseteq C^4$ of tile types. Next let

$$\mathscr{V} = \{W_0, W_1, \ldots, W_n, S_0, S_1, \ldots, S_n, \overline{N_0}, \overline{N_1}, \ldots, \overline{N_n}, \overline{E_0}, \overline{E_1}, \ldots, \overline{E_n}\}.$$

We will describe tile types by certain subsets of $\mathscr{V}$ where W_i expresses that the left color is c_i, and $\overline{N_i}$ denotes that the top color is *not* c_i (S_i and $\overline{E_i}$ refer to the bottom and right color and are to be understood similarly). More precisely, the tile $d = (c_i, c_j, c_k, c_\ell)$ is denoted by the set $\mathbb{S}_d = \{W_i\} \cup \{\overline{N_m} \mid m \neq j\} \cup \{\overline{E_m} \mid m \neq k\} \cup \{S_\ell\}$. Now let $F = \{\mathbb{S}_d \mid d \in \mathscr{T}\}$ be the descriptions of all the tile types d in $\mathscr{T}$. Then, by Thm. 4, there are finite plane graphs G_1, G_2, G_3, and G_4 with the following properties: (i) at the outer face, we find edges e for $e \in \mathscr{V}$ and nodes u and v in the order indicated in Fig. 6 and (ii) $M \in F$ iff there exists a Hamiltonian path H of G_x from u to v such that $M = \{v \in \mathscr{V} \mid v \text{ belongs to } H\}$ (for all $1 \leq x \leq 4$ and $M \subseteq \mathscr{V}$).

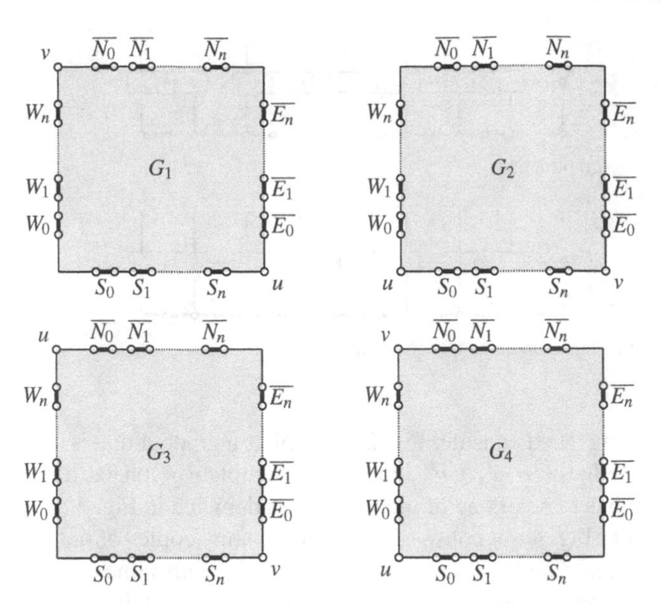

Fig. 6 The graphs G_x

Next we choose mutually disjoint graphs $G(k,\ell)$ (for $k,\ell \in \mathbb{N}$) such that

$$G(k,\ell) \cong \begin{cases} G_1 & \text{if } k+\ell \text{ is even and } k>0 \\ G_2 & \text{if } k+\ell \text{ is odd and } \ell=0 \\ G_3 & \text{if } k+\ell \text{ is odd and } \ell>0 \\ G_4 & \text{if } k+\ell \text{ is even and } k=0. \end{cases}$$

Then $u(k,\ell)$ and $v(k,\ell)$ refer to the nodes u and v of the graph $G(k,\ell)$; similarly, $e(k,\ell)$ for $e \in \mathscr{V}$ refers to the edge e of the graph $G(k,\ell)$. In the disjoint union of these graphs $G(k,\ell)$, we connect the node $v(k,\ell)$ by a new edge with the following node:

$$\begin{aligned} u(k+1,\ell) &\text{ for } k+\ell \text{ even and } \ell=0 \\ u(k+1,\ell-1) &\text{ for } k+\ell \text{ even and } \ell>0 \\ u(k-1,\ell+1) &\text{ for } k+\ell \text{ odd and } k>0 \\ u(k,\ell+1) &\text{ for } k+\ell \text{ odd and } k=0. \end{aligned}$$

The result G^1 of this construction is visualized in Fig. 7 where the vertices $u(k,\ell)$ are denoted by empty nodes and $v(k,\ell)$ by filled nodes. From G^1 we construct G^2 by replacing the edges $\overline{E_i}(k,\ell)$ and $W_i(k,\ell+1)$ as well as $\overline{N_i}(k,\ell)$ and $S_i(k+1,\ell)$ ($k,\ell \in \mathbb{N}$, $0 \le i \le n$) by a copy of the exclusive-or graph X, see Fig. 8. In a third step, we construct G^3 by adding to G^2 the graph L from Fig. 4. To connect L to G^2, the start node of the edges a and b, resp., of the ith copy of A in L is the left and right, resp., node of the

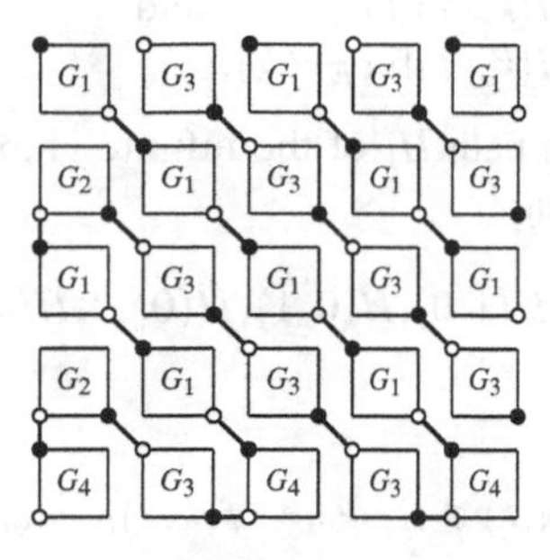

Fig. 7 First step in global construction - the graph G^1

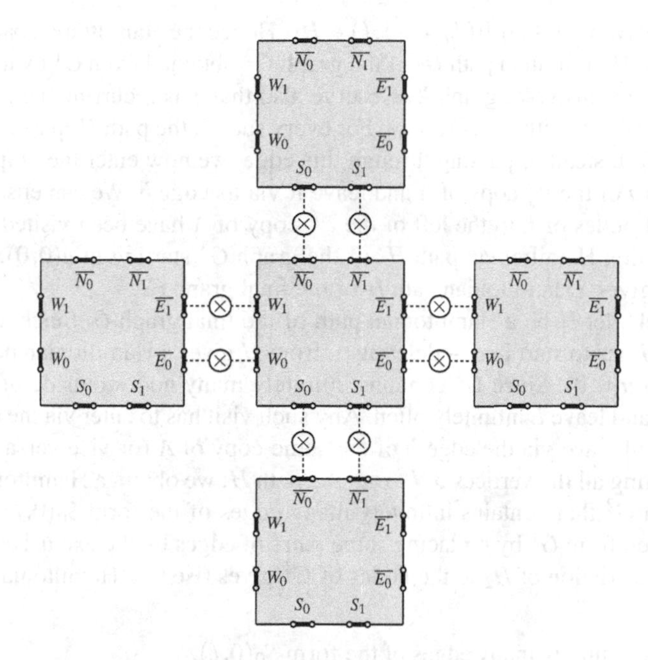

Fig. 8 Second step in global construction – the graph G^2 (for two colors c_0 and c_1)

edge $S_0(0,i)$. The final graph G is obtained from G^3 by adding a new node $\perp$ together with an edge between $\perp$ and $u(0,0)$.

Let us now prove that G has a Hamiltonian path iff $\mathscr{T}$ admits a recurring tiling. First suppose there is a recurring tiling $f : \mathbb{N} \times \mathbb{N} \to \mathscr{T}$. Let $k, \ell \in \mathbb{N}$ and $f(k,\ell) = (c_W, c_N, c_E, c_S)$. Then the graph $G(k,\ell) \in \{G_x \mid 1 \leq i \leq 4\}$ has a Hamiltonian path $H(k,\ell)$ from $u(k,\ell)$ to $v(k,\ell)$ such that for all $1 \leq i \leq n$

1. the edge S_i belongs to $H(k,\ell)$ iff $c_S = c_i$,
2. the edge W_i belongs to $H(k,\ell)$ iff $c_W = c_i$,

3. the edge $\overline{N_i}$ belongs to $H(k,\ell)$ iff $c_N \neq c_i$, and
4. the edge $\overline{E_i}$ belongs to $H(k,\ell)$ iff $c_E \neq c_i$.

Then we find a Hamiltonian path H_1 of the infinite graph G^1 in Fig. 7 by appending these Hamiltonian paths suitably:

$$H_1 = H(0,0), H(1,0), H(0,1), H(0,2), H(1,1), H(2,0)\ldots$$

Since f is a tiling, we get

$$\begin{aligned} \overline{E_i}(k,\ell) \notin H_1 &\iff f(k,\ell)_E = c_i \\ &\iff f(k,\ell+1)_W = c_i \\ &\iff W_i(k,\ell+1) \in H_1 \end{aligned}$$

and similarly $\overline{N_i}(k,\ell) \notin H_1$ iff $S_i(k+1,\ell) \in H_1$. Hence the Hamiltonian path H_1 can be extended to a Hamiltonian path H_2 of the graph G^2 obtained from G^1 by adding all the copies of the exclusive-or graph X. Observe also that f is recurring, i.e., there are infinitely many $\ell \in \mathbb{N}$ with $f(0,\ell)_S = c_0$. For every such ℓ, the path H_1 passes through the edge $S_0(0,\ell)$. Instead of passing through this edge, we now enter the graph L (Fig. 4) via the edge a of the ℓth copy of A and leave it via its edge b. We can ensure that after this visit, all nodes of L to the left of the ℓth copy of A have been visited (cf. Fig. 5). This results in a Hamiltonian path H_3 of the graph G^3 starting in $u(0,0)$. Prepending the node $\perp$ gives a Hamiltonian path H of the final graph G.

Conversely, let H be a Hamiltonian path of the final graph G. Since $\perp$ has degree 1, the path H has to start in $\perp$ – deleting $\perp$ from H gives a Hamiltonian path H_3 of G^3 that starts in $u(0,0)$. Since G^3 contains infinitely many nodes outside of L, this path has to enter and leave L infinitely often. Any such visit has to enter via the edge a some copy of A and leave via the edge b of the same copy of A (or vice versa, see Fig. 5). Hence, deleting all the vertices of L from the path H, we obtain a Hamiltonian path H_2 of the graph G^2 that contains infinitely many edges of the form $S_0(0,\ell)$. Recall that G^2 is obtained from G^1 by replacing some pairs of edges by the exclusive-or graph X. Hence, the restriction of H_2 to the nodes of G^1 gives rise to a Hamiltonian path H_1 of G^1 that

(a) contains infinitely many edges of the form $S_0(0,\ell)$,
(b) contains the edge $W_i(k,\ell+1)$ iff it does not contain the edge $\overline{E_i}(k,\ell)$, and
(c) contains the edge $S_i(k+1,\ell)$ iff it does not contain the edge $\overline{N_i}(k,\ell)$

for all $0 \leq i \leq n$ and $k,\ell \in \mathbb{N}$. Since H_1 has to pass through all the graphs $G(k,\ell)$, it has to be of the form

$$H(0,0), H(1,0), H(0,1), H(0,2), H(1,1), H(2,0)\ldots$$

where $H(k,\ell)$ is a Hamiltonian path of the graph $G(k,\ell)$ from $u(k,\ell)$ to $v(k,\ell)$. Now we are ready to define the mapping $f : \mathbb{N}^2 \to C^4$: set

(1) $f(k,\ell)_W = c_i$ iff $H(k,\ell)$ contains the edge $W_i(k,\ell)$,
(2) $f(k,\ell)_N = c_i$ iff $H(k,\ell)$ does not contain the edge $\overline{N_i}(k,\ell)$,

(3) $f(k,\ell)_E = c_i$ iff $H(k,\ell)$ does not contain the edge $\overline{E_i}(k,\ell)$, and
(4) $f(k,\ell)_S = c_i$ iff $H(k,\ell)$ contains the edge $S_i(k,\ell)$.

Since $H(k,\ell)$ is a Hamiltonian path of $G(k,\ell)$ from $u(k,\ell)$ to $v(k,\ell)$, we get $f(k,\ell) \in \mathscr{T}$ from the construction of the graphs G_1, G_2, G_3, G_4. By (1), (b), and (3), we have

$$\begin{aligned} f(k,\ell)_W = c_i &\iff W_i(k,\ell) \text{ belongs to } H(k,\ell) \\ &\iff \overline{E_i}(k,\ell+1) \text{ does not belong to } H(k,\ell+1) \\ &\iff f(k,\ell+1) = c_i \end{aligned}$$

and similarly $f(k,\ell)_N = f(k+1,\ell)_S$ follows from (2), (c), and (4). Thus, f is a tiling. Since H_1 contains infinitely many edges of the form $S_0(0,\ell)$, there are infinitely many $\ell \in \mathbb{N}$ such that $S_0(0,\ell)$ belongs to $H(0,\ell)$, i.e., $f(0,\ell)_S = c_0$.

Thus, we showed that indeed the graph G contains a Hamiltonian path iff the set of tiles $\mathscr{T}$ admits a recurring tiling.

Clearly, the undirected graph G is planar and has bounded degree. Thus, in order to finish the proof of Thm. 2, it remains to prove that G is automatic. Note that the graph G has a highly regular structure. It results from the infinite grid $\mathbb{N} \times \mathbb{N}$ by replacing each grid point by a finite graph and connecting these finite graphs in a regular pattern. It is not surprising that such a graph is automatic, in particular since the grid is automatic. Let us provide some more formal arguments for the automaticity of G.

Recall that G can be obtained from $\mathbb{N} \times \mathbb{N}$ by replacing every grid point $(k,\ell) \in \mathbb{N} \times \mathbb{N}$ by a finite graph $G'(k,\ell)$. This graph is a copy of one of the graphs G'_1, G'_2, G'_3, G'_4, where G'_i is the graph G_i together with copies of the XOR-graph X that connect $G(k,\ell)$ with $G(k+1,\ell)$ and $G(k,\ell+1)$. Whether $G'(k,\ell)$ is G'_i only depends on the parity of $k+\ell$ and whether k and ℓ are zero or non-zero, respectively.

The alphabet of our presentation consists of the elements of $\{0,1,\#\}^2 \setminus \{(\#,\#)\}$ and the nodes of the graphs $G'_1, \ldots, G'_4$. Then, the node set of G can be represented by the regular language

$$\{(\text{bin}(k) \otimes \text{bin}(\ell))v \mid k,\ell \geq 0,\ v \text{ is a node of } G'(k,\ell)\}, \tag{1}$$

where $\text{bin}(n)$ is the binary encoding of a number n (note that the parity of $k+\ell$ can be determined by a finite automaton from $\text{bin}(k) \otimes \text{bin}(\ell)$). Constructing from this node representation an automaton that recognizes the edge set of G is straightforward but tedious. This concludes the proof of Thm. 2.

There also exists the variant of two-way Hamiltonian paths in infinite graphs. A two-way Hamiltonian path in $G = (V,E)$ is a two-way infinite sequence $(v_i)_{i\in\mathbb{Z}}$ such that $(v_i, v_{i+1}) \in E$ for all $i \in \mathbb{Z}$ and for every node $v \in V$ there is exactly one $i \in \mathbb{Z}$ such that $v = v_i$. From the previous construction, it follows that also the existence of a two-way Hamiltonian path in a given planar automatic graph of bounded degree is Σ^1_1-complete. Take the disjoint union of two copies of our main graph G and connect the two $\perp$-nodes with an edge. The resulting graph G' has a two-way Hamiltonian path iff G has a (one-way) Hamiltonian path. Moreover, since G is automatic and the class of automatic graphs is closed under disjoint unions, G' is automatic as well.

4 Remarks about large finite graphs

The main purpose of automatic presentations is the finite representation of infinite structures. But automatic presentations can be also used as a tool for the succinct representation of large finite structures. Note that a finite automaton with n states can accept a finite language with $2^{O(n)}$ elements, which may serve as the domain of a finite structure.

In general, given an automatic presentation (Γ, L, h) for a *finite* graph (V, E) together with an automaton A for the node set language L, it is clear that $|V|$ is bounded by $|\Gamma|^n$, where n is the number of states of A. It follows that for every graph problem L in NP, the succinct version of L, where the input graph is given by an automatic presentation, belongs to NEXPTIME. In particular, the Hamiltonian path problem belongs to NEXPTIME for this succinct input representation.

For the lower bound, consider for $n \geq 1$ the finite planar graph G_n that results from our main infinite graph G by restricting it to the graphs $G(k,\ell)$ for $k+\ell \leq n$ and the connecting XOR-graphs between these graphs. Then G_n has a Hamiltonian path if and only if the finite set of tiles $\mathscr{T}$ admits a tiling of the "triangle" $D_n = \{(k,\ell) \in \mathbb{N} \times \mathbb{N} \mid k+\ell \leq n\}$ (tilings of finite parts of the grid $\mathbb{N} \times \mathbb{N}$ are defined analogously to tilings of the whole grid). Now we can use a result of Fürer [8]: It is NEXPTIME-complete (under logspace reductions) to check for a given number n (encoded in binary) and a finite set of tiles $\mathscr{T}$ whether $\mathscr{T}$ admits a tiling of D_n. Let us make a few remarks on Fürer's proof before continuing:

- Fürer proved NEXPTIME-completeness for tilings of the square $\{(k,\ell) \in \mathbb{N} \times \mathbb{N} \mid k, \ell \leq n\}$ instead of the triangle D_n. It is straightforward to adapt Fürer's proof for D_n.
- Fürer actually does not speak about NEXPTIME-completeness in his paper, but states explicit lower bounds. But in his proof he presents a generic reduction from the acceptance problem for nondeterministic exponential time Turing-machines to the problem of tiling $\{(k,\ell) \in \mathbb{N} \times \mathbb{N} \mid k, \ell \leq n\}$ for a given number coded in binary.
- Fürer states that all his construction can be carried out in polynomial time, but it is straightforward to check that they can be carried out even in logspace.

Finally, it is easy to construct from a number n coded in binary in logarithmic space an automatic presentation of the graph G_n. For this, we can basically use the automatic presentation of the infinite graph G, but restrict it to numbers of size at most n. Hence, we obtain:

Theorem 5. *It is NEXPTIME-complete under logspace reductions to check for a given automatic presentation of a finite planar graph, whether it has a Hamiltonian path.*

A variant of Thm. 5 was shown by Veith [28]. He considers finite structures that are represented by OBDDs (ordered binary decision diagrams). In this context, the node set of a graph is $\{0,1\}^n$ for some fixed n. The edge set is represented by an OBDD over variables $x_1, \ldots, x_n, y_1, \ldots, y_n$. Here the tuple $(x_1, \ldots, x_n) \in \{0,1\}^n$ represents the initial vertex of an edge, whereas $(y_1, \ldots, y_n) \in \{0,1\}^n$ represents the final node. The variable order of the OBDDs in [28] is fixed to the interleaved order $x_1, y_1, x_2, y_2, \ldots, x_n, y_n$. Under this variable order, OBDDs exactly correspond to deterministic acyclic automata that work on the convolution $(x_1 \cdots x_n) \otimes (y_1 \cdots y_n)$.

In [28], the following upgrading theorem was shown (here, only formulated for the classes NP and NEXPTIME): If a graph problem L is NP-complete under quantifier free first-order reductions then obdd(L) (the class of all OBDDs of the above form that encode a graph from L) is NEXPTIME-complete under polynomial time reductions. Since the Hamiltonian path problem (HP) is NP-compete under quantifier free first-order reductions [26], it follows that obdd(HP) is NEXPTIME-complete under polynomial time reductions. Thm. 5 strengthens this result in two points: we obtain NEXPTIME-completeness (i) under logspace reductions and (ii) for *planar* graphs. It is not clear for us, whether the *planar* Hamiltonian path problem is still NP-complete under quantifier free first-order reductions.

5 Further graph problems

An *order tree* is a partial order $(A, \preceq)$ with a least element such that the set $\{a \in A \mid a \preceq b\}$ is finite and linearly ordered for every $b \in A$, a *successor tree* is the covering relation of an order tree. It is decidable, whether an automatic order tree has an infinite path [20]. The following result is in sharp contrast to this positive result.

Theorem 6. *It is Σ_1^1-complete to determine whether a given automatic successor tree T has an infinite path.*

The proof idea is to transform a recursive successor tree into an automatic one by adding the computation (i.e., sequence of transitions) that verifies the edge (u, v) as a path between the nodes u and v; a similar idea was used in [20, 15].

Let us now present some graph problems which are Σ_1^1-complete for recursive graphs, but decidable in automatic graphs. For this, we introduce, inspired by [18, 25], a fragment $\mathrm{SO^r}$ of second-order logic, which extends first-order logic with the infinity quantifier and modulo quantifiers. Every relation that is definable in first-order logic with the infinity quantifier and modulo quantifiers has a regular set of representatives [16, 5, 19]. We will extend this result to $\mathrm{SO^r}$. The set of all formulas of $\mathrm{SO^r}$ is inductively defined as follows:

- Every atomic first-order formula is an $\mathrm{SO^r}$-formula.
- $X(x_1, \ldots, x_k)$ for $x_1, \ldots, x_k$ first-order variables and X a k-ary second-order variable is an $\mathrm{SO^r}$-formula.
- If φ and ψ are $\mathrm{SO^r}$-formulas, then also $\varphi \vee \psi$ is an $\mathrm{SO^r}$-formula.
- If φ is an $\mathrm{SO^r}$-formula, then also $\neg\varphi$, $\exists x\,\varphi$, $\exists^{\infty} x\,\varphi$ ("there are infinitely many x satisfying φ"), $\exists^{(k,p)} x\,\varphi$ for $0 \leq k < p \in \mathbb{N}$ ("the number of x satisfying φ is finite and congruent k modulo p") are $\mathrm{SO^r}$-formulas.
- If φ is an $\mathrm{SO^r}$-formula and X is a second-order variable of arity k such that for every k-tuple of first-order variables $x_1, \ldots, x_k$, φ contains the subformula $X(x_1, \ldots, x_k)$

only negatively (i.e. within an odd number of negations), then also $\exists X$ infinite : φ is an $\mathrm{SO^r}$-formula.

Note that the restriction on φ in the last point ensures that if φ is satisfied for some k-ary relation $X = R$ and $Q \subseteq R$, then φ is also satisfied for $X = Q$.

Using the proof ideas from [18] and [25], one can show the following two theorems. The first theorem refers to the decidability of the $\mathrm{SO^r}$-theory of every automatic structure, the second theorem implies that from a $\mathrm{SO^r}$-formula $\exists X$ infinite : $\alpha(X)$ true in an automatic structure $\mathscr{A}$, one can construct a regular witness to the validity of this formula.

Theorem 7. *From an automatic presentation (Γ, L, h) of an automatic structure $\mathscr{A}$ and an $\mathrm{SO^r}$-formula $\varphi(\overline{x})$ one can compute effectively an automaton for the convolution of the relation $\{(u_1,\ldots,u_n) \in L^n \mid \mathscr{A} \models \varphi(h(u_1),\ldots,h(u_n))\}$. Hence, if φ is an $\mathrm{SO^r}$-sentence, then $\mathscr{A} \models \varphi$ can be checked effectively.*

Theorem 8. *From an automatic presentation (Γ, L, h) of an automatic structure $\mathscr{A}$ and an $\mathrm{SO^r}$-sentence $\beta = \exists X$ infinite : $\alpha(X)$ with $\mathscr{A} \models \beta$, one can construct $H \subseteq L^n$ regular such that $h(H)$ is infinite and $\mathscr{A} \models \alpha(h(H))$.*

We use Thm. 7 and 8 to show that two problems, which are Σ^1_1-complete for recursive structures [14], are decidable for automatic structures. First, by taking the $\mathrm{SO^r}$-formula $\exists X$ infinite $\forall x, y : (x, y \in X \Rightarrow (x, y) \in E)$, we get:

Corollary 1 (cf. [25, Thm. 3.20]). *It is decidable whether a given automatic graph contains an infinite clique. If an infinite clique exists, a regular set of representatives of an infinite clique can be computed.*

The second problem is the infinite version of maximal set cover considered by Hirst and Harel [14]. It asks whether, given a set $X = \{X_i \mid i \in \mathbb{N}\}$ of sets $X_i \subseteq \mathbb{N}$, there exists $A \subseteq \mathbb{N}$ with $\bigcup_{a \in A} X_a = \mathbb{N}$ and $\mathbb{N} \setminus A$ infinite. Note that the collection X can be represented as a set of pairs E with $(i, j) \in E$ iff $j \in X_i$. Then there exists A as required iff the directed graph $(\mathbb{N}, E)$ satisfies $\exists B$ infinite $\forall j \exists i : i \notin B \wedge (i, j) \in E$ (then A is the complement of B). Hence we get:

Corollary 2. *The infinite version of maximal set cover is decidable if the collection X is given as an automatic set of pairs. In case a set cover as required exists, an infinite such can be computed.*

References

1. D. R. Bean. Effective coloration. *J. Symbolic Logic*, 41(2):469–480, 1976.
2. D. R. Bean. Recursive Euler and Hamilton paths. *Proc. Amer. Math. Soc.*, 55(2):385–394, 1976.
3. R. Beigel and W. I. Gasarch. unpublished results. 1986–1990.
4. R. Berger. The undecidability of the domino problem. *Mem. Am. Math. Soc.*, 66. AMS, 1966.
5. A. Blumensath and E. Grädel. Finite presentations of infinite structures: Automata and interpretations. *Theory Comput. Syst.*, 37(6):641–674, 2004.

6. N. Dean, R. Thomas, and X. Yu. Spanning paths in infinite planar graphs. *J. Graph Theory*, 23(2):163–174, 1996.
7. R. Diestel. *Graph Theory, Third Edition*. Springer, 2006.
8. M. Fürer. The computational complexity of the unconstrained limited domino problem (with implications for logical decision problems). In *Logic and machines: decision problems and complexity*, LNCS 171, pages 312–319. Springer, 1984.
9. M. Garey, D. Johnson, and R. E. Tarjan. The planar Hamiltonian circuit problem is NP-complete. *SIAM J. Comput.*, 5(4):704–714, 1976.
10. D. Harel. A simple undecidable domino problem (or, a lemma on infinite trees, with applications). In *Proc. Logic and Computation Conference*, Victoria, Australia, 1984. Clayton.
11. D. Harel. Recurring dominoes: making the highly undecidable highly understandable. *Ann. Discrete Math.*, 24:51–72, 1985.
12. D. Harel. Effective transformations on infinite trees, with applications to high undecidability, dominoes, and fairness. *J. Assoc. Comput. Mach.*, 33(1):224–248, 1986.
13. D. Harel. Hamiltonian paths in infinite graphs. *Israel J. Math.*, 76(3):317–336, 1991.
14. T. Hirst and D. Harel. Taking it to the limit: on infinite variants of NP-complete problems. *J. Comput. System Sci.*, 53:180–193, 1996.
15. B. Khoussainov and M. Minnes. Model theoretic complexity of automatic structures. In *Proceedings of TAMC 08*. Springer, 2008. to appear.
16. B. Khoussainov and A. Nerode. Automatic presentations of structures. In *LCC: International Workshop on Logic and Computational Complexity*, LNCS 960, pages 367–392, 1994.
17. B. Khoussainov, A. Nies, S. Rubin, and F. Stephan. Automatic structures: richness and limitations. *Log. Methods Comput. Sci.*, 3(2):2:2, 18 pp. (electronic), 2007.
18. B. Khoussainov, S. Rubin, and F. Stephan. On automatic partial orders. In *Proc. LICS 2003*, pages 168–177. IEEE Computer Society Press, 2003.
19. B. Khoussainov, S. Rubin, and F. Stephan. Definability and regularity in automatic structures. In *Proc. STACS 2004*, LNCS 2996, pages 440–451. Springer, 2004.
20. B. Khoussainov, S. Rubin, and F. Stephan. Automatic linear orders and trees. *ACM Trans. Comput. Log.*, 6(4):675–700, 2005.
21. D. Kozen. *Theory of Computation*. Springer, 2006.
22. D. Kuske and M. Lohrey. First-order and counting theories of omega-automatic structures. *J. Symbolic Logic*, 73:129–150, 2008.
23. A. B. Manaster and J. G. Rosenstein. Effective matchmaking (recursion theoretic aspects of a theorem of Philip Hall). *Proc. London Math. Soc. (3)*, 25:615–654, 1972.
24. S. Rubin. *Automatic Structures*. PhD thesis, University of Auckland, 2004.
25. S. Rubin. Automata presenting structures: A survey of the finite string case. *Bull. Symbolic Logic*, 2008. To appear.
26. I. A. Stewart. Using the Hamiltonian path operator to capture NP. *J. Comput. System Sci.*, 45(1):127–151, 1992.
27. W. T. Tutte. A theorem on planar graphs. *Trans. Amer. Math. Soc.*, 82:99–116, 1956.
28. H. Veith. How to encode a logical structure by an OBDD. In *Proc. 13th Annual Conf. Computational Complexity*, pages 122–131. IEEE Computer Society, 1998.
29. H. Wang. Proving theorems by pattern recognition. *Bell Syst. Tech. J.*, 40:1–41, 1961.

Marking the chops: an unambiguous temporal logic

Kamal Lodaya[1], Paritosh K. Pandya[2], and Simoni S. Shah[2]

[1] The Institute of Mathematical Sciences, CIT Campus, Chennai *600113*, India
[2] Tata Institute of Fundamental Research, Colaba, Mumbai *400005*, India
Correspondence pandya@tcs.tifr.res.in

Abstract. Interval Temporal Logic [11] is a highly expressive and succinct logic whose satisfiability over finite words is non-elementary in the number of alternations of chop and negation operators. All the sublogics of *ITL* with elementary decidability known to us restrict this alternation depth. In this paper, we define a sublogic of Interval Temporal Logic by replacing chops with marked chops but without any restriction on the alternation depth. We show that the resulting logic admits unique parsing of a word matching a formula, with the consequence that membership is in LOGDCFL and satisfiability is in PSPACE. As our first result, we give an *effective* model-preserving reduction from *UITL* to the partially ordered two-way deterministic finite automata of Schwentick, Thérien and Vollmer [14]. We show that the size of the resulting automaton is quadratic in the size of the formula. We also have an exponential converse reduction from *po2dfa* to *UITL*. It follows from the work of Schützenberger [13], Thérien and Wilke [19] that this unambiguous *ITL* has same expressive power as the first-order logic with two variables [10].

1 Introduction

Two-variable first-order logic FO^2 was first studied by Mortimer [10]. In recent years, a lot of research has centred around this logic, on words [5], data [1], Mazurkiewicz traces [7], trees [2], etc. In particular for words, the article by Tesson and Thérien [18] reveals the many facets of the class of languages defined by sentences of this logic. The logic was shown to be NEXPTIME-complete and equivalent to a natural fragment of linear temporal logic called *Unary TL* by Etessami, Vardi and Wilke [5] and to partially ordered two-way deterministic finite automata (henceforth *po2dfa*) by Schwentick, Thérien and Vollmer [14]. Thérien and Wilke showed [19] that it corresponds to the variety DA of unambiguous languages studied by Schützenberger [13]. Weis and Immerman [20] and Kufleitner and Weil [8] have recently examined the quantifier alternation hierarchy within $FO^2[<]$.

A proper treatment of syntax, we feel, is lacking. Tesson and Thérien's paper [18] does give a rudimentary syntax in terms of deterministic and co-deterministic products, which we close under boolean operations and call an deterministic or unambiguous subclass of propositional interval temporal logic *ITL* [11].

ITL is a highly succinct logic for specifying properties of finite words. The unconstrained chop operator (similar to concatenation of star-free expressions) leads to high decision complexity: the satisfiability of *ITL* is non-elementary in the number of

Please use the following format when citing this chapter:

Lodaya, K., Pandya, P.K. and Shah, S.S., 2008, in IFIP International Federation for Information Processing, Volume 273; *Fifth IFIP International Conference on Theoretical Computer Science*; Giorgio Ausiello, Juhani Karhumäki, Giancarlo Mauri, Luke Ong; (Boston: Springer), pp. 461–476.

alternations of the negation and chop operator [17]. Sublogics of *ITL* with elementary satisfiability have been obtained by constraining this alternation depth in some manner. *UITL* replaces chops with marked chops but without any restriction on their alternation depth with negation. Our first theorem is a consequence of the unique parsability: membership of a word w in the language of a formula is in LOGDCFL and nonemptiness is in NP.

Also exploiting exploiting this unique parsability, an effective quadratic translation from *UITL* to $FO^2[<]$ has been given by Shah [15]. From the work of Thérien and Wilke [19] it follows that *UITL* is expressively contained in in the unambiguous languages of Schützenberger [13]. That it was an open problem whether this *UITL* syntax matches the expressive power of FO^2 we learnt from [8], which was written concurrently and independently of this paper. We answer the question positively in this paper.

Our second theorem is an $O(n^2)$ translation from a formula of our logic to a *po2dfa* which accepts exactly the models of the formula. A partially ordered 2DFA [14] (also called linear by Löding and Thomas [9]) is a two-way DFA which has the property that once the automaton exits a state, it is never entered again. The translation from formulae to automata illustrates the difficulty of working with weak models such as *po2dfa*. To complete the characterisation of the expressive power, as our third theorem we construct for each *po2dfa* a formula exactly specifying its language. This solves the open problem mentioned above, as does the paper [8] using completely different techniques. The constructed formula is exponential in the size of the automaton.

$FO^2[<]$ and *Unary TL* are at a remove from the very deterministic notion of *po2dfa*. Our logic, which can be thought of as *ITL* but where the chop operator is forced to be deterministic, is much closer to the automata. As a consequence, satisfiability drops from nonelementary for *ITL* to PSPACE for our logic. In earlier work [6], we found that such unambiguity considerably improves the computational performance of a validity checking tool for *ITL*.

The idea of having deterministic temporal logics has been explored before. A "marked" operator in temporal logic *atnext* was studied by Borchert and Tesson [3]. Kufleitner simplified it to deterministic marked next and previous modalities X_a and Y_a to define a point-based linear temporal logic, and showed that it is expressively complete for $FO^2[<]$ over Mazurkiewicz traces (and hence also over words) [7]. However, a concrete exploitation of this to give explicit and efficient reduction from logic to automata seems new.

The rest of the paper is organized as follows. Section 2 defines the syntax and semantics of *UITL*. Section 3 discusses the partially ordered 2DFA and some expressions we use as a convenient notation for them. Section 4 gives the reduction from formulae of *UITL* to *po2dfa*. Section 5 gives the construction of a formula specifying the language accepted by a *po2dfa*. We end with some perspectives.

Acknowledgements The authors thank Manfred Kufleitner, Pascal Weil and Meena Mahajan for valuable inputs. The first author acknowledges the Indo-French project *Timed-Discoveri* for support. The second and the third authors acknowledge partial support from the Microsoft Research Grant for the project "formal specification and analysis of component-based designs.".

2 Unambiguous interval temporal logic: its syntax and semantics

We propose a fragment of *ITL* [11] where the chop operator is replaced by marked chop operators F_a and L_a. Our syntax derives from closing the $*_{n,k}$-expressions of Tesson and Thérien [18] under Boolean operations.

Fix an alphabet Σ. Let $a \in \Sigma$, $A \subseteq \Sigma$. Let D, D_1, D_2 range over formulas in *UITL*. The abstract syntax of *UITL* is given below.

$$\lceil\lceil A \rceil\rceil \mid \lceil\lceil A \rceil \mid \lceil A \rceil\rceil \mid \lceil A \rceil \mid D_1 \vee D_2 \mid \neg D \mid D_1 F_a D_2 \mid D_1 L_a D_2 \mid \oplus D \mid \ominus D$$

Let w be a nonempty finite word over Σ and let $pos(w) = \{1, \ldots, \#w\}$ be the set of positions. Let $INTV(w) = \{[i,j] \mid i,j \in pos(w), i \leq j\}$ be the set of intervals overs w. The satisfaction of a formula D is defined over intervals of a word model w as follows.

$$\begin{array}{l}
w,[i,j] \models \lceil\lceil A \rceil\rceil \text{ iff for all } k : i \leq k \leq j.\ \ w[k] \in A \\
w,[i,j] \models \lceil A \rceil \text{ iff for all } k : i < k < j.\ \ w[k] \in A \\
w,[i,j] \models \lceil\lceil A \rceil \text{ iff for all } k : i \leq k < j.\ \ w[k] \in A \\
w,[i,j] \models \lceil A \rceil\rceil \text{ iff for all } k : i < k \leq j.\ \ w[k] \in A \\
w,[i,j] \models D_1 F_a D_2 \text{ iff for some } k : i \leq k \leq j.\ \ w[k] = a \text{ and} \\
\qquad (\text{for all } m : i \leq m < k.\ w[m] \neq a) \text{ and} \\
\qquad w,[i,k] \models D_1 \text{ and } w,[k,j] \models D_2 \\
w,[i,j] \models D_1 L_a D_2 \text{ iff for some } k : i \leq k \leq j.\ \ w[k] = a \text{ and} \\
\qquad (\text{for all } m : k < m \leq j.\ w[m] \neq a) \text{ and} \\
\qquad w,[i,k] \models D_1 \text{ and } w,[k,j] \models D_2 \\
w,[i,j] \models \oplus D \text{ iff } i < j \text{ and } w,[i+1,j] \models D \\
w,[i,j] \models \ominus D \text{ iff } i < j \text{ and } w,[i,j-1] \models D
\end{array}$$

As usual, $w \models D$ iff $w,[1,\#w] \models D$ and $L(D) \stackrel{\text{def}}{=} \{w \mid w \models D\}$ is the language defined by D.

The proposition $\lceil\lceil A \rceil\rceil$ states that letters of all positions in the interval (including the endpoints) are in A. Similarly, $\lceil A \rceil$ says that all the strictly interior positions in an interval have only letters from A; thus it trivially holds for point (i.e. $[i,i]$) and unit (i.e. $[i,i+1]$) intervals. By similar reasoning, $\lceil\,\rceil \stackrel{\text{def}}{=} \lceil\lceil \emptyset \rceil\rceil$ holds only on point intervals, and $\neg\lceil\lceil \emptyset \rceil\rceil \wedge \lceil \emptyset \rceil$ only on unit intervals. The semantics of the "first" and "last" marked chops and the "next" and "previous" operators should be clear.

The derived operators $\wedge, \supset, \Leftrightarrow$ have their usual definitions. The constant $\top$ (denoting *true*) can be defined as $\lceil\lceil \Sigma \rceil\rceil$. We take both these to be of constant size, but in general the size of $\lceil\lceil A \rceil\rceil$ is $O(|A|)$. Conversely, $\lceil\lceil A \rceil\rceil \Leftrightarrow \bigwedge_{a \notin A} \neg(\top F_a \top)$. Similar equivalences can be given for $\lceil\lceil A \rceil$, $\lceil A \rceil\rceil$ and $\lceil A \rceil$. Negations can be pushed inwards to the level of literals using $\neg(D_1 F_a D_2) \Leftrightarrow (\lceil\lceil \Sigma \setminus \{a\} \rceil\rceil \vee (\top F_a \neg D_2) \vee (\neg D_1 F_a \top))$ and $\neg(\oplus D) \Leftrightarrow \lceil\,\rceil \vee \oplus(\neg D)$. All these translations are linear in the size of the formula. For later use in Section 5, we also designate as simple formulae those made of the atomic formulae and the marked chop operators F_a and L_a as well as operators $\ominus$ and $\oplus$ with the Boolean operators being disallowed.

Example 1. Consider the formula $D \stackrel{\text{def}}{=} (\top L_a((\neg(\top F_b \top))F_d \top))$ over the alphabet $\Sigma = \{a,b,c,d\}$. Intuitively, it states that between the last occurrence of a and subsequent first occurrence of d there is no occurrence of letter b. Thus it specifies the language $\Sigma^* ac^* d\{b,c,d\}^*$. □

Example 2. Formula $\lceil\ \rceil F_a(\oplus(\lceil\ \rceil F_a(\oplus(\lceil\ \rceil F_a \lceil\ \rceil))))$ holds exactly for the word aaa. Note that it is impossible to express this without using $\oplus$ or $\ominus$ operators. □

Following Kufleitner and Weil [8], we define two hierarchies of formulae. $R_1 = L_1$ consists of the formulae made up of the four kinds of atomic formulae and the Boolean operations, marked chops being disallowed. R_{n+1} extends L_n by allowing F_a operators (deterministic products) over formulas of L_n and closing under Boolean operations; symmetrically, L_{n+1} is the Boolean closure of L_a operators (co-deterministic products) over R_n. Thus $UITL = \bigcup_n R_n = \bigcup_n L_n$ is the full deterministic/co-deterministic hierarchy over the piecewise testable languages of Simon [16], which are characterized by R_1 (e.g. see the survey of Diekert, Gastin and Kufleitner [4]). R_1, R_2 and L_2 are known as J, R and L in the literature.

2.1 Unique parsing

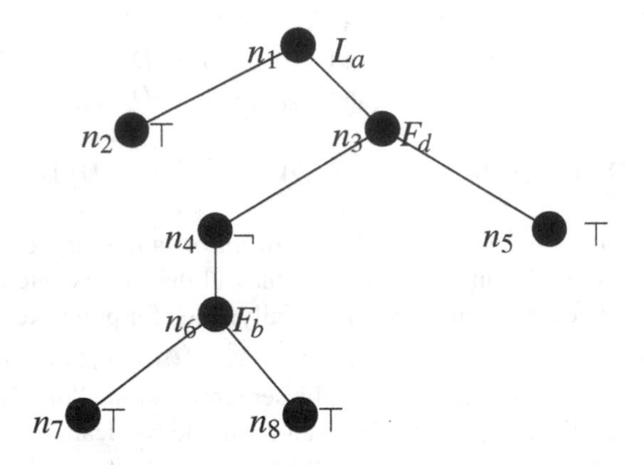

Fig. 1 Syntax Tree of Formula in Example 1

It is convenient to represent *UITL* a formula D by its consider the syntax tree of a formula D, where each interior node n is labelled by an operator and the subtree rooted at n represents a subformula of D, denoted by $Subf(n)$. For the root of the tree, $Subf(root) = D$. For example, a node n with $Subf(n) = D_1 F_a D_2$ has two children, say n_1 and n_2 with $Subf(n_1) = D_1$ and $Subf(n_2) = D_2$. We will say n matches $n_1 F_a n_2$. A leaf node n is labelled by one of $\lceil\lceil A\rceil\rceil, \lceil\lceil A\rceil, \lceil A\rceil\rceil, \lceil A\rceil$.

Fix a formula D and let *Nodes* be the set of nodes in its syntax tree with *root* being the root node. Let $MNodes \subset Nodes$ be the subset of nodes whose operator has the form F_a, L_a, $\oplus$, or $\ominus$. For any node n, let $Ancestry(n)$ be the sequence of *Nodes* encountered on the unique path from n to the root node. For technical convenience we will append two fresh nodes $n_\triangleright$ followed by $n_\triangleleft$ to the ancestry. Formally, $Ancestry(root) = n_\triangleright.n_\triangleleft$. Also if n matches $n_1 \; op \; n_2$, then $Ancestry(n_1) = Ancestry(n_2) = n.Ancestry(n)$. We will follow the convention that $n_\triangleright$ is an $L_\triangleright$ operator and $n_\triangleleft$ is an $F_\triangleleft$ operator. Let $\ell Ancestry(n)$ be the subsequence of nodes from n to the root which are labelled with marked chops or $\ominus$ such that n is in their right subtree; $rAncestry(n)$ is similarly defined with left subtrees, marked chops and $\ominus$.

Example 3. Consider the formula D in Example 1. Figure 1 gives the syntax tree of D. At n_4, we have $Subf(n_4) = \neg(\top F_b \top)$. It is easy to see that $Ancestry(n_7) = n_6 n_4 n_3 n_1 n_\triangleright n_\triangleleft$ with $rAncestry(n_7) = n_6 n_3 n_\triangleleft$ and $\ell Ancestry(n_7) = n_1 n_\triangleright$. □

Next we consider the evaluation of D over a word w. For any word w and any subformula of a formula D we can associate a unique interval (or none) where the formula must be evaluated. This interval is fixed by the context in which the subformula occurs and does not depend upon the subformula itself. For example, the subformula $D_1 = \neg(\top F_b \top)$ of D in Example 1 is associated with the interval which begins with the last occurrence of a in w and it ends at the first subsequent occurrence of d. We call this property *unique parsability*. Formally, given word w, we can associate with each $n \in Nodes$ either a unique interval $[i,j]$ where $Subf(n)$ needs to be evaluated, or $\mathbf{u}$ denoting that the subformula of the node need not be evaluated. This is denoted by $Intv_w(n)$. Moreover, for each $n \in MNodes$ (which corresponds to a marked chop operator) we associate a chopping position $cPos_w(n)$.

Definition 1. $Intv_w : Nodes \rightarrow INTV(w) \cup \{\mathbf{u}\}$ and $cPos_w : MNodes \rightarrow pos(w) \cup \{\mathbf{u}\}$ are defined by induction on the depth of the node (from root) as follows.

- $Intv_w(root) = [1, \#w]$.
- If n matches $n_1 \vee n_2$ then $Intv_w(n_1) = Intv_w(n_2) = Intv_w(n)$. Similarly, If n matches $\neg n_1$ then $Intv_w(n_1) = Intv_w(n)$.
- If n matches $n_1 F_a n_2$ or $n_1 L_a n_2$ or $\ominus n_1$ or $\oplus n_1$ and $Intv_w(n) = \mathbf{u}$ then $Intv_w(n_1) = Intv_w(n_2) = \mathbf{u}$ and $cPos_w(n) = \mathbf{u}$.
- If n matches $n_1 F_a n_2$ or $n_1 L_a n_2$, $Intv_w(n) = [i,j]$ and if for all $k : i \leq k \leq j$ we have $w[k] \neq a$ then $Intv_w(n_1) = Intv_w(n_2) = \mathbf{u}$ and $cPos_w(n) = \mathbf{u}$.
- Let n match $n_1 F_a n_2$ with $Intv_w(n) = [i,j]$. Let $k : i \leq k \leq j$ be such that $w[k] = a$ and for all $m : i \leq m < k$ we have $w[m] \neq a$. Then, $Intv_w(n_1) = [i,k]$ and $Intv_w(n_2) = [k,j]$. Also, $cPos_w(n) = k$.
- n matches $n_1 L_a n_2$ with $Intv_w(n) = [i,j]$. Let Let $k : i \leq k \leq j$ be such that $w[k] = a$ and for all $m : k < m \leq j$ we have $w[m] \neq a$. Then, $Intv_w(n_1) = [i,k]$ and $Intv_w(n_2) = [k,j]$. Also, $cPos_w(n) = k$.
- If n matches $\oplus n_1$ or $\ominus n_1$ and $Intv_w(n) = [i,i]$ then $intv(n_1) = \mathbf{u}$ and $cPos_w(n) = \mathbf{u}$.
- If n matches $\oplus n_1$ and $Intv_w(n) = [i,j]$ with $i < j$ then $intv(n_1) = [i+1, j]$ and $cPos_w(n) = i+1$.

– If n matches $\ominus n_1$ and $Intv_w(n) = [i,j]$ with $i < j$ then $intv(n_1) = [i, j-1]$ and $cPos_w(n) = j-1$.

These definitions are extended to endmarker nodes $n_\triangleright$ and $n_\triangleleft$ as follows.
$Intv_w(n_\triangleright) = Intv_w(n_\triangleleft) = [1, \#w]$. Also, $cPos_w(n_\triangleright) = 1$ and $cPos_w(n_\triangleleft) = \#w$.

Proposition 1. *Let $Ancestry(n) = n_1, n_2, \ldots, n_k$ for a node n. For all i, j such that $1 \leq i \leq j \leq k$, $Intv_w(n_i)$ is $\mathbf{u}$ or included in $Intv_w(n_j)$. Also, if n_i is labelled F_a or L_a and $cPos_w(n_i) \neq \mathbf{u}$ then $w[cPos_w(n_i)] = a$.* □

Using the notion of unique interval associated with a node, we can define the truth value of a node n in word w as follows.

Definition 2. Define $Val_w : Nodes \to \{\mathbf{t}, \mathbf{f}, \mathbf{u}\}$ as follows. (Observe that $Val_w(root)$ is never $\mathbf{u}$ since $Intv_w(root)$ is always $[1, \#w]$.)

$$\begin{array}{l} Val_w(n) = \mathbf{u} \text{ iff } Intv_w(n) = \mathbf{u} \\ Val_w(n) = \mathbf{t} \text{ iff } Intv_w(n) = [i,j] \text{ and } w,[i,j] \models Subf(n) \\ Val_w(n) = \mathbf{f} \text{ iff } Intv_w(n) = [i,j] \text{ and } w,[i,j] \not\models Subf(n) \end{array}$$

Example 4. Consider the formula D with syntax tree as given in Figure 1. Consider the word $w = acdabacbcdbcbd$ with $pos(w) = \{1, \ldots, 14\}$. Then, we have $Intv_w(n_1) = [1, 14]$. As n_1 is labelled L_a we have $cPos_w(n_1) = 6$ and $Intv_w(n_2) = [1, 6]$ and $Intv_w(n_3) = [6, 14]$. Also, n_3 is labelled F_d and we get $cPos_w(n_3) = 10$. This gives us $Intv_w(n_4) = [6, 10]$ and $Intv_w(n_5) = [10, 14]$. Then, $Intv_w(n_6) = [10, 14]$ and as n_6 is labelled F_b we have $cPos_w(n_6) = 6$ and $Intv_w(n_7) = [6, 8]$ and $Intv_w(n_8) = [8, 10]$. Note that $n_4 = \neg n_6$ and $n_6 = n_7 F_b n_8$ and $n_7 = \top$ and $n_8 = \top$. Hence, $Val_w(n_7) = \mathbf{t}$, $Val_w(n_7) = \mathbf{t}$ giving $Val_w(n_6) = \mathbf{t}$ and $Val_w(n_4) = \mathbf{f}$. Similarly we can compute that $Val_w(n_1) = \mathbf{f}$. □

Theorem 1. *Membership of a word w in the language of a formula D is* NC^1*-hard and in the class* LOGDCFL. *Nonemptiness of the language of a formula D is* NP*-hard and in* NP *if the size of the alphabet Σ is fixed.*

Proof. After pushing negations inward and constructing the syntax tree, the Val_w function can be evaluated by a 2DPDA with auxiliary storage $O(\log(|D| + |w|))$ in time $O(poly(|D| + |w|))$. This yields a LOGDCFL procedure. If a formula D is satisfiable, we can translate it to a sentence of FO^2 of quadratic size [15] and use Weis and Immerman's result [20] to show the existence of a model of size $O((|D|^2)^{|\Sigma|})$. For a fixed size alphabet Σ, guessing the model and verifying its truth value is an NP procedure.

For the lower bounds, a Boolean assignment over n variables can be coded as a word of length n over a two-letter alphabet. The truth value of the i'th variable can be accessed using the $\oplus$ modality, which is also used to say that a model is of size n. These formulas are linear in n. Hence, any Boolean formula can be encoded as a UITL formula by replacing each variable by its corresponding UITL formula. Now we use the standard results for Boolean formulas. □

2.2 Handling context

We can further refine the characterisation of the intervals of a node. The following lemma relates intervals of nodes in an ancestry to their chopping positions in the same ancestry. Figure 2 depicts some of these relationships.

Let w be a word and let $i, j \in pos(w)$ with $i \leq j$. Then, $a \notin w[i : j)$ will abbreviate $\forall k : i \leq k < j.\ a \neq w[k]$. Similarly, we can define $a \notin w(i : j]$.

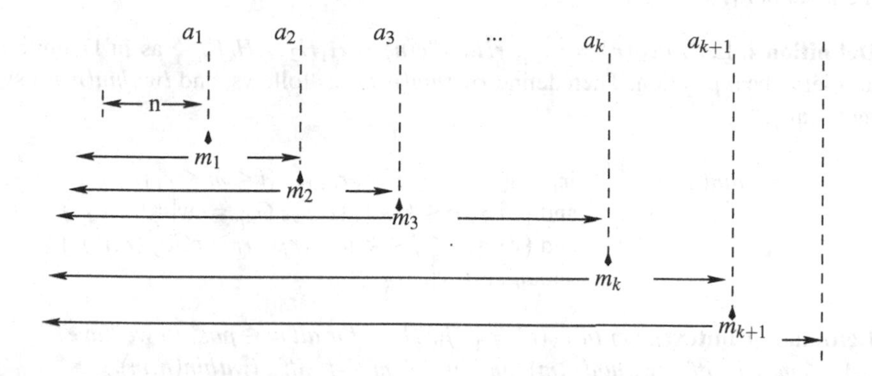

Fig. 2 Right handle and its intervals

Lemma 1. *For a node n, let $\ell Ancestry(n) = n_1, n_2, \ldots, n_p, n_{p+1}$ and $rAncestry(n) = m_1, m_2, \ldots, m_r, m_{r+1}$. (So $n_{p+1} = n_{\triangleright}$ and $m_{r+1} = n_{\triangleleft}$.) Then,*

- *$Intv_w(n) = [cPos_w(n_1), cPos_w(m_1)]$. If either of these chopping positions is **u** then $Intv_w(n) = \mathbf{u}$.*
- *$cPos_w(n_1) \geq cPos_w(n_2) \geq \ldots \geq cPos_w(n_p) \geq cPos_w(n_{p+1}) = 1$, and $cPos_w(m_1) \leq cPos_w(m_2) \leq \ldots \leq cPos_w(m_r) \leq cPos_w(m_{r+1}) = \#w$.*
- *$Intv_w(n_i) = [cPos_w(n_{i+1}), cPos_w(m_k)]$ for some $m_k \in rAncestry(n)$. Also, $Intv_w(m_i) = [cPos_w(n_k), cPos_w(m_{i+1})]$ for some $m_k \in \ell Ancestry(n)$.*
- *If m_i is a L_a node then $w[cPos_w(m_i)] = a$ and $a \notin w(cPos_w(m_i) : cPos_w(m_{i+1})]$.*
- *If m_i is a F_a node then $w[cPos_w(m_i)] = a$ and $a \notin w[cPos_w(n_1) : cPos_w(m_i))$.*
- *If m_i is $\ominus$ node then $cPos_w(m_{i+1}) = cPos_w(m_i) + 1$.*
- *If n_i is $\oplus$ node then $cPos_w(n_{i+1}) = cPos_w(n_i) - 1$.*

Proof. By induction on depth of n from the root. □

As opposed to the bottom-up evaluation of truth value, the identification of chopping positions and subintervals is defined top-down. This enables us to find the context necessary for checking whether a position m is within $Intv_w(n)$.

Definition 3. Let $\ell Handle(n)$ be the smallest prefix of $\ell Ancestry(n)$ ending with an L operator. Symmetrically let $rHandle(n)$ smallest prefix of $rAncestry(n)$ ending with an

F operator. The sequence of labels of $rHandle(n)$ will have the form $H_1H_2 \ldots H_kF_{a_{k+1}}$ where H_i is either L_{a_i} or $\ominus$. When clear from context we will often directly refer to such a sequence of labels as $rHandle(n)$. Given $rHandle(n)$, and indices i, j such that $1 \leq i \leq j \leq k+1$ let $rGap(n,i,j)$ be the count of $\ominus$ labels occurring within labels H_i to H_{j-1}. For example, given $rHandle(n) = L_{a_1} \ominus \ominus \ominus F_{a_5}$ we have $rGap(n,1,4) = 2$. Symmetrically, we can define $lGap(n,i,j)$ for $\ell Handle(n)$. □

In our running example, $rHandle(n_7) = F_b$ (the label of n_6) and $\ell Handle(n_7) = L_a$, the label of n_1.

Definition 4. Let $Intv_w(n) = [i, j]$, $rHandle(n) = H_1H_2 \ldots H_kF_{a_{k+1}}$ as in Definition 3 and let m be a position. Then define $rwithin(n,m)$ as follows, and $lwithin(n,m)$ symmetrically.

$$\begin{aligned} rwithin(n,m) \stackrel{\text{def}}{=} \; & \exists r_1 \leq r_2 \leq \ldots \leq r_k \leq r_{k+1}. \quad (i \leq m \leq r_1) \\ & \text{and } (\forall 1 \leq p \leq k+1. \, (H_p = G_{a_p} \Rightarrow w[r_p] = a_p)) \\ & \text{and } (\forall 1 \leq i \leq j \leq k+1. \, (r_j - r_i \geq rGap(n,i,j)) \\ & \text{and } a_{k+1} \notin w[i : r_{k+1}) \end{aligned}$$

Lemma 2 (Context). *Let $Intv_w(n) = [i, j]$. Then, for all $m \in pos(w)$ we have (a) $i \leq m \leq j$ iff $rwithin(n,m)$, and (b) $i \leq m \leq j$ iff $lwithin(n,m)$.*

Proof. We prove (a). Let $Intv_w(n) = [i, j]$ and $rHandle(n) = n_1 \ldots n_{k+1}$ with labels $H_1H_2 \ldots H_kF_{a_{k+1}}$. Let $j_p = cPos_w(n_p)$ for $1 \leq p \leq k+1$. As $Intv_w(n) \neq \mathbf{u}$ we also have that $cPos_w(n_p) \neq \mathbf{u}$ as n_p is ancestor of n.

For the forward direction, suppose $i \leq m \leq j$. Take $r_p = j_p$. Then, by Lemma 1, we have $r_1 \leq r_2 \leq \ldots \leq r_{k+1}$ and $w[r_p] = a_p$, for all p with $H_p = G_{a_p}$. Also, by Lemma 1, when n_p has label $\ominus$, then $j_{p+1} = j_p + 1$. Hence, for any $1 \leq p \leq q \leq n$, we have $r_q - r_p \geq rGap(n,p,q)$. Also denote $Intv_w(n_{k+1}) = [b_{k+1}, e_{k+1}]$ then $b_{k+1} \leq i$. Since n_{k+1} is labelled $F_{a_{k+1}}$, from its semantics we have that $a_{k+1} \notin w[i : j_{k+1})$. Hence the result follows.

Conversely, suppose there exist $r_1 \leq r_2 \leq \ldots \leq r_{k+1}$ such that $i \leq m \leq r_1$ and for all $1 \leq p \leq k+1$ if $H_p = G_{a_p}$ then $w[r_p] = a_p$ and for all $1 \leq p \leq q \leq k+1$ we have $r_q - r_p \geq rGap(n.p,q)$. We have to show that $m \leq j$. Assume to the contrary that $m > j$. By Lemma 1, $j_1 = j$ and $r_1 \geq m$. Hence, $r_1 > j_1$.

Consider any $1 \leq p \leq k$ such that $r_p > j_p$. There are two cases. In case 1, if p is labelled L_{a_p} then by its semantics $a_p \notin w(j_p, j_{p+1}]$. Hence, as $r_p > j_p$ and $w[r_p] = a_p$, it follows that $r_p > j_{p+1}$ which implies that $r_{p+1} > j_{p+1}$. In the second case, if p is labelled with $\ominus$ then $rGap(n,p,p+1) = 1$ and hence $r_{p+1} - r_p \geq 1$. Also, by Lemma 1, we have $j_{p+1} = j_p + 1$. Hence, as $r_p > j_p$, it follows that $r_{p+1} > j_{p+1}$.

We already have that $r_1 > j_1$. Hence by induction and using the previous step we can prove that $r_{k+1} > j_{k+1}$. But by the condition that $w[r_{k+1}] = a_{k+1}$ and $a_{k+1} \notin w[i : r_{k+1})$ we have that $j_{k+1} = r_{k+1}$, which is a contradiction. Hence we conclude that $m \leq j$. □

3 Partially ordered two-way deterministic finite automata

Partially ordered two-way DFA were introduced by Schwentick, Thérien and Vollmer [14] to characterize the unambiguous languages. We present a variant of their definition and propose a set of operators to compose these automata. Let $\Sigma' = \Sigma \cup \{\rhd, \lhd\}$ include two endmarkers. Given $w \in \Sigma^*$, the two way automaton actually scans string $w' = \rhd w \lhd$ with letters $\rhd$ and $\lhd$ at positions 0 and $|w|+1$ respectively.

Definition 5. A *po2dfa* over Σ' is a tuple $M = (Q, \leq, \delta, s, t, r)$ where $(Q, \leq)$ is a poset of states such that r,t are the only minimal elements. s is the initial state, t is the accept state and r is the rejecting state. The set $Q \setminus \{t,r\}$ is partitioned into Q_l and Q_r (the states reached from the left and the right) with $s \in Q_l$. $\delta : ((Q_l \cup Q_r) \times \Sigma) \to Q) \cup ((Q_l \times \{\lhd\}) \to Q \setminus Q_r) \cup ((Q_r \times \{\rhd\}) \to Q \setminus Q_l)$ is a transition function satisfying $\delta(q,a) \leq q$. □

If M is in a state q, reading a symbol a, it enters a state $\delta(q,a)$, and moves its head to the right if $\delta(q,a) \in Q_l$, left if $\delta(q,a) \in Q_r$, and stays in the same position if $\delta(q,a) \in \{t,r\}$. The transition function is designed to ensure that the automaton does not "fall off" either end of the input. A transition with $\delta(q,a) < q$ is said to make progress.

A po2dfa M running over word w is said to be in a configuration (q,p) if it is in a state q and head reading the position p in word. The run of a po2dfa M on an input word w starting with input head position p_0 is a sequence $(q_0,p_0),(q_1,p_1),...(q_f,p_f)$ of configurations such that:

- $q_0 = s$ and $q_f \in \{t,r\}$, for all $i(1 \leq i < l)$, $\delta(q_i, w(p_i)) = q_{i+1}$, and
- $p_{i+1} = p_i + 1$ if $q_{i+1} \in Q_l$ or $p_{i+1} = p_i - 1$ if $q_{i+1} \in Q_r$.

We abbreviate such a run by writing $M(w,p_0) = (q_f,p_f)$. The run is accepting if $q_f = t$; rejecting if $q_f = r$. A pass is a contiguous partial run where the automaton moves in one direction. An n-pass automaton is one which makes at most n passes on any input before accepting or rejecting. The automaton M is said to be start-free if for any w, M accepts w from some position iff M accepts w starting from any position.

3.1 Composition of automata

For the description of *po2dfa* we will use turtle expressions, which are extensions of the turtle programs introduced by Schwentick, Thérien and Vollmer [14]. The syntax follows and we explain the semantics below. Let A, B range over subsets of Σ'.

$$E ::= Acc \mid Rej \mid A \xrightarrow{1} \mid A \xleftarrow{1} \mid A \xrightarrow{B} \mid A \xleftarrow{B} \mid E_1 ? E_2, E_3$$

Automaton Acc accepts immediately without moving the head. Similarly, Rej rejects immediately. $A \xrightarrow{B}$ accepts at the next occurrence of a letter from B to the right, maintaining the constraint that the intervening letters are from $A \setminus B$. If no such occurrence exists the automaton rejects at the right end-marker or if a letter outside A intervenes, the automaton rejects at its position. Automaton $A \xrightarrow{1}$ accepts one position to the right if the current letter is from A, else rejects at the current position. $A \xleftarrow{B}$ and $A \xleftarrow{1}$ are symmetric in the leftward direction. The conditional construct $E_1 ?E_2, E_3$ first executes E_1 on w. On its accepting w at position j it continues with execution of E_2 from j. On E_1 rejecting w at position j it continues with E_3 from position j.

Here are some abbreviations which illustrate the power of the notation: $E_1;E_2 = E_1?E_2, Rej$, $\neg E_1 = E_1?Rej, Acc$. Moreover, if E_2 is start-free then $E_1 \vee E_2 = E_1?Acc, E_2$ and $E_1 \wedge E_2 = E_1?E2, Rej$. Notice that automata for these expressions are start-free if E_1 is start-free. We will use $A \xrightarrow{a}$ for $A \xrightarrow{\{a\}}$, $\xrightarrow{a}$ for $(\Sigma' \xrightarrow{a})$ and $\xrightarrow{1}$ for $(\Sigma' \xrightarrow{1})$. Similarly define $\xleftarrow{a}$ and $\xleftarrow{1}$. We will use the convention that $\overline{a_1, \ldots, a_k}$ denotes $\Sigma' \setminus \{a_1, \ldots, a_k\}$.

Proposition 2. *Given turtle expression E we can construct a po2dfa accepting the same language with number of states linear in* $|E|$.

We have to resort to Section 2 for the correctness of the next construction.

Definition 6. Consider a node n with $rHandle(n) = H_1 H2 \ldots H_k F_{a_{k+1}}$ as in Definition 3 and let $A \subseteq \Sigma'$. Define the one-pass automata $\mathscr{C}^+(n, A)$ and $\mathscr{C}^+(n, \xrightarrow{1})$ as follows, and symmetrically also $\mathscr{C}^-(n, A)$ and $\mathscr{C}^-(n, \xleftarrow{1})$. Let $perf(H_i)$ be $(\overline{a_{k+1}} \xrightarrow{a_i})$ if $H_i = L_{a_i}$ and $\overline{a_{k+1}} \xrightarrow{1}$ if $H_i = \ominus$.

$$\begin{aligned}
\mathscr{C}^+(n, A) &= (\overline{a_{k+1}} \xrightarrow{A}); perf(H_1); \ldots; perf(H_k); (\overline{a_{k+1}} \xrightarrow{a_{k+1}}) \\
\mathscr{C}^+(n, \xrightarrow{1}) &= (\overline{a_{k+1}} \xrightarrow{1}); perf(H_1); \ldots; perf(H_k); (\overline{a_{k+1}} \xrightarrow{a_{k+1}})
\end{aligned}$$

Since $rHandle(n)$ and $\ell Handle(n)$ are linear in the depth of n it follows that the sizes of the $\mathscr{C}^-(n), \mathscr{C}^+(n)$ automata are also linear in the depth of n.

Lemma 3. *Let* $Intv_w(n) = [i, j]$.

- *Started at position i,* $\mathscr{C}^+(n, A)$ *accepts iff* $\exists k.\ rwithin(n, k)$ *and* $w[k] \in A$ *and* $w[i : k) \notin A$.
- *Started at position i,* $\mathscr{C}^+(n, \xrightarrow{1})$ *accepts iff* $i + 1 \leq j$.

Symmetric properties hold for $\mathscr{C}^-$.

Proof. The context lemma (Lemma 2) proved the required "within" property. That the automata check this "within" is easy to see. □

4 From formulae to automata

Now we are all set to construct a *po2dfa* $\mathscr{M}(D)$ which precisely accepts the word models of a given formula D. Our turtle expressions are a convenient syntax for the two-way movement of *po2dfa*. For example, expression $\xrightarrow{\triangleleft};\xleftarrow{a};\xrightarrow{d}$ denotes an automaton which first finds the endpoint of the word (by looking for endmarker $\triangleleft$) it then reverses its direction and searches for the first a in backward direction and then searches in the forward direction for the first subsequent d. Clearly, such an automaton locates the right endpoint of the interval of the subformula $D_1 = \neg(\top F_b \top)$ of D in Example 1. In general, for each subformula D_1 we can construct automata $\mathscr{L}(D_1)$ and $\mathscr{R}(D_1)$ which locate the left and right endpoints of the unique interval associated with D_1. Now it remains to check that the subword of this interval satisfies the subformula D_1. D_1 evaluates to true iff there is no (first) occurrence of letter b within its unique interval. While turtle expressions lack a simple way of checking a property within a specific subinterval,Lemma 3 shows how we can use "handles" to code this checking. Putting all this together, we give a construction of a language equivalent *po2dfa* of size d^2 for a formula of size d.

Definition 7. By induction on depth of a node n, define automata $\mathscr{L}(n)$ and $\mathscr{R}(n)$.

- $\mathscr{L}(root) = \xleftarrow{\triangleright};\xrightarrow{1}$ and $\mathscr{R}(root) = \xrightarrow{\triangleleft};\xleftarrow{1}$.
- Let n match $\neg n_1$. Then $\mathscr{L}(n_1) = \mathscr{L}(n)$ and $\mathscr{R}(n_1) = \mathscr{R}(n)$.
- Let n match $n_1 \vee n_2$. Then $\mathscr{L}(n_1) = \mathscr{L}(n_2) = \mathscr{L}(n)$ and $\mathscr{R}(n_1) = \mathscr{R}(n_2) = \mathscr{R}(n)$.
- Let n match $n_1 F_a n_2$. Then $\mathscr{L}(n_1) = \mathscr{L}(n)$ and $\mathscr{R}(n_2) = \mathscr{R}(n)$. Also, $\mathscr{R}(n_1) = \mathscr{L}(n);\xrightarrow{a}$ and $\mathscr{L}(n_2) = \mathscr{R}(n_1)$.
- Let n match $n_1 L_a n_2$. Then $\mathscr{L}(n_1) = \mathscr{L}(n)$ and $\mathscr{R}(n_2) = \mathscr{R}(n)$. Also, $\mathscr{R}(n_1) = \mathscr{R}(n);\xleftarrow{a}$ and $\mathscr{L}(n_2) = \mathscr{R}(n_1)$.
- Let n match $\oplus n_1$. Then, $\mathscr{L}(n_1) = \mathscr{L}(n);\xrightarrow{1}$ and $\mathscr{R}(n_1) = \mathscr{R}(n)$.
- Let n match $\ominus n_1$. Then, $\mathscr{L}(n_1) = \mathscr{L}(n)$ and $\mathscr{R}(n_1) = \mathscr{R}(n);\xleftarrow{1}$.

Lemma 4. *As the inductive automaton construction follows the inductive definition of $Intv_w(n)$, it is immediate that for any node n with $Intv_w(n) = [i,j]$ (not* **u***), for any position k in w, $\mathscr{L}(n)(w,k) = (t,i)$ and $\mathscr{R}(n)(w,k) = (t,j)$. Thus, $\mathscr{L}(n)$ and $\mathscr{R}(n)$ are start-free. Note that $\mathscr{L}(n)$ and $\mathscr{R}(n)$ grow linearly with the depth of n.* □

Definition 8. We define $\mathscr{M}(n)$ for each node n by induction on the height of n.

- If n is labelled $\lceil\lceil A \rceil\rceil$ then $\mathscr{M}(n) = \mathscr{L}(n);\mathscr{C}^+(n,\overline{A})?Rej,Acc$. The translations for $\lceil\lceil A\rceil$, $\lceil A\rceil\rceil$ and $\lceil A\rceil$ are similar.
- For a Boolean expression, $\mathscr{M}(n)$ is defined by the corresponding turtle expression. E.g. if n matches $n_1 \vee n_2$ then $\mathscr{M}(n) = \mathscr{M}(n_1) \vee \mathscr{M}(n_2)$.
- Let n match $n_1 F_a n_2$. Let $\mathscr{M}(n) = \mathscr{L}(n);\mathscr{C}^+(n,a);\mathscr{M}(n_1);\mathscr{M}(n_2)$.
- Let n match $n_1 L_a n_2$. Let $\mathscr{M}(n) = \mathscr{R}(n);\mathscr{C}^-(n,a);\mathscr{M}(n_1);\mathscr{M}(n_2)$.
- Let n match $\oplus n_1$. Let $\mathscr{M}(n) = \mathscr{L}(n);\mathscr{C}^+(n,\xrightarrow{1});\mathscr{M}(n_1)$.
- Let n match $\ominus n_1$. Let $\mathscr{M}(n) = \mathscr{R}(n);\mathscr{C}^-(n,\xleftarrow{1});\mathscr{M}(n_1)$.

Example 5. For the formula $D \stackrel{\text{def}}{=} \top L_a((\neg(\top F_b \top))F_d \top)$ of Example 1 we give the construction of *po2dfa*. The formula is represented as syntax tree in Figure 1.

- The root n_1 matches $n_2 L_a n_3$. Hence, $\mathscr{M}(n_1) = \mathscr{R}(n_1); \mathscr{C}^-(n_1, a); \mathscr{M}(\top); \mathscr{M}(n_3)$. Also $\mathscr{L}(n_1) = \overset{\triangleright}{\leftarrow}; \overset{1}{\rightarrow}$ and $\mathscr{R}(n_1) = \overset{\triangleleft}{\rightarrow}; \overset{1}{\leftarrow}$ giving $\mathscr{C}^-(n_1, a) = \overset{a}{\leftarrow}$. Since $n_2 = \top$, we have $\mathscr{M}(n_2) = Acc$.
- n_3 matches $n_4 F_d n_4$ with $n_5 = \top$. Hence $\mathscr{M}(n_3) = \mathscr{L}(n_3); \mathscr{C}^+(n_3, d); \mathscr{M}(n_4); Acc$. Now, $\mathscr{L}(n_3) = \mathscr{R}(n_1); \overset{a}{\leftarrow}$ and $\mathscr{R}(n_3) = \mathscr{R}(n_1)$. Also $rHandle(n_3) = F_{\triangleleft}$. Hence, $\mathscr{C}^+(n_3, F_d) = \overset{d}{\rightarrow}$.
- n_4 matches $\neg n_6$. Hence $\mathscr{M}(n_4) = \mathscr{M}(n_6)?Rej, Acc$.
- $Subf(n_6) = \top F_b \top$. Hence, $\mathscr{M}(n_6) = \mathscr{L}(n_6); \mathscr{C}^+(n_6, b); Acc; Acc$. We have $rHandle(n_6) = F_d$. Hence, $\mathscr{C}^+(n_6, F_b) = (\overline{d} \overset{b}{\rightarrow}); \overset{d}{\rightarrow}$ and $\mathscr{L}(n_6) = \mathscr{L}(n_3) = \overset{\triangleleft}{\rightarrow}; \overset{1}{\leftarrow}; \overset{a}{\leftarrow}$. □

Theorem 2. *Given a formula D, the language L(D) is accepted by the po2dfa automaton $\mathscr{M}(root)$ of Definition 8 where root is the root node of parse tree of D. Moreover, $\mathscr{M}(root)$ has $O(|D|^2)$ states.*

Proof. Construct the syntax tree and let $Intv_w(n) = [i, j]$ for any node n. By induction on the height of node n, for any word w, we prove that $\mathscr{M}(n)$ accepts w iff $Val_w(n) = \mathbf{t}$. Note that $\mathscr{M}(n)$ is start-free since each $\mathscr{M}(n)$ is either Acc or it begins with $\mathscr{L}(n)$ or $\mathscr{R}(n)$, which are start-free by the previous lemma. Below are the proofs of three cases, the rest are similar.

- Let $n = \lceil[A]\rceil$. Let $Intv_w(n) = [i, j]$. Then,
  $Val_w(n) = \mathbf{t}$
  iff $\forall k : i \leq k \leq j : w[k] \in A$
  iff $\forall k : i \leq k : w[k] \notin A$ implies $k \notin Intv_w(n)$
  iff not ($\exists k : i \leq k : w[k] \notin A$ and $rwithin(n, k)$)
  iff the $\mathscr{C}^+(n, \overline{A})$ automaton rejects starting from (w, i). (by Lemma 3)
  iff $\mathscr{M}(n)$ accepts w.
- Let n match a Boolean expression. The result holds since the smaller automata are start-free.
- Let n match $n_1 F_a n_2$. Let $Intv_w(n) = [i, j]$. Then,
  $Val_w(n) = \mathbf{t}$
  iff $w, [i, j] \models (Subf(n_1)) F_a (Subf(n_2))$
  iff $\exists k : i \leq k \leq j$ s.t. $w[k] = a$ and $a \notin w[i : k)$ and $Val_w(n_1) = \mathbf{t}$
  and $Val_w(n_2) = \mathbf{t}$ (giving $Intv_w(n_1) = [i, k]$ and $Intv_w(n_2) = [k, j]$)
  iff $\exists k$ s.t. $rwithin(n, k)$ and $w[k] = a$ and $a \notin w[i : k)$ and
  $\mathscr{M}(n_1)$ accepts w and $\mathscr{M}(n_2)$ accepts w (by induction hypothesis)
  iff $\mathscr{C}^+(n, a)$ accepts (w, i) (by Lemma 3) and $\mathscr{M}(n_1), \mathscr{M}(n_2)$ accept w.
  iff $\mathscr{M}(n)$ accepts w
- Let n match $\oplus n_1$. Let $intv(n) = [i, j]$ Then,
  $Val_w(n) = \mathbf{t}$
  iff $i + 1 \leq j$ and $w, [i + 1, j] \models (Subf(n_1))$
  iff $i + 1 \leq j$ and $\mathscr{M}(n_1)$ accepts w (By induction hypothesis)
  iff $rwithin(n, i + 1)$ and $\mathscr{M}(n_1)$ accepts w

iff $\mathscr{C}^+(n, \stackrel{1}{\rightarrow})$ accepts (w,i) and $\mathscr{M}(n_1)$ accepts w
iff $\mathscr{M}(n)$ accepts w

The number of nodes in the syntax tree of a formula is linear in its size $|D|$. At each node, at most $O(|D|)$ states to the automaton are added before recursively translating sub-nodes. Hence, the the number of states of $\mathscr{M}(root)$ is $O(|D|^2)$. □

The complexities of membership and satisfiability problem for the logic UITL were analysed in Theorem 1. Here we give alternate upperbounds on these complexities which are obtained using the formula automaton construction. Note that even when automaton based procedures have higher complexities, in practice, they are very amenable to implementation.

Corollary 1. *Membership of a word w in the language a formula D is in* DTIME$(|w| \times |D|^2)$. *The satisfiability of D can be checked in* NSPACE$(|D|^2 \log |D|)$.

Proof. Since the number of states of $\mathscr{M}(root)$ in the theorem above is $O(|D|^2)$, whether a word model w satisfies D can be checked in time $O(|w| \times |D|^2)$ by simulating the *po2dfa*.

We can also check satisfiability of D by reducing the *po2dfa* of size $O(|D|^2)$ to a one-way DFA of size $O((|D|^2)^{|D|^2})$ using the standard 2DFA to 1DFA reduction. The emptiness of this one-way DFA is contained in nondeterministic $\log((|D|^2)^{|D|^2})$ space, i.e.NSPACE$(|D|^2 \log |D|)$. □

5 From automata to formulae

Fix a *po2dfa* M. We give the construction of a formula exactly specifying the language of M. Consider a progress transition e of M. For simplicity, we assume that e is not labelled by the endmarkers $\lhd$ or $\rhd$. We construct a formula $\psi(e)$ such that the following lemma holds. Its proof is by induction on the partial order.

Lemma 5. $w \models \psi(e)$ *iff there exists a partial run of M on w (starting at position* 1*) and ending with the e transition.* □

The formula $\psi(e) = \bigvee_{i \in I(e)} \xi_i$ consists of finitely many disjoint disjuncts where each ξ_i is a *pointed simple formula*. Such a formula does not use boolean operators, has a pointer to one of its sub formulas. For example, see $\psi(e_b)$ in Example 6 Such $\psi(e)$ defines a class of words with a unique factorization [13]. For convenience a pointed simple formula is represented as (T,n) with syntax tree T and pointer node n.

Fix a progress transition e with $\delta(p,c) = q$ such that the incoming progress transitions into p are $e_1, \ldots, e_k$. Also assume that $A = \{a \in \Sigma \mid \delta(p,a) = p\}$ are the letters on which the automaton loops in state p. Inductively assume that $\psi(e_i)$ has been constructed. Then, we define $\psi(e) = \vee\{Extend(\xi, e) \mid \xi \in \psi(e_i), 1 \leq i \leq k\}$. Here $Extend(\xi, e)$ extends the partial runs satisfying ξ to their extensions ending with e. We now define $Extend(\xi, e)$.

An execution ending with one of the e_i can be extended with finitely many steps involving letters of A and then taking the transition e. Moreover the head moves backwards iff $p \in Q_r$ (except at the last step where it moves in the direction of q). To take the direction into account, let $Extend(\xi,e) = rExtend(\xi,e)$ if $p \in Q_l$ and $Extend(\xi,e) = \ell Extend(\xi,e)$ if $p \in Q_r$. We define these below.

Let the inorder traversal of T be $n_1,n_2,\ldots,n_r$ (thus n_1 is the leftmost leaf and n_r is the rightmost leaf) and $n = n_i$. It is easy to see that nodes n_1 and n_2 in inorder traversal nodes are adjacent iff $Intv_w(n_1)$ and $Intv(n_2)$ are adjacent. Then, $rExtend(T,n_i) \stackrel{\text{def}}{=} rChange(T,n_i)$ if $n_i = n_r$ (last node in inorder), and $rExtend(T,n_i) \stackrel{\text{def}}{=} rChange(T,n_i) \lor rExtend(rSkip(T,n_i),n_{i+1})$ otherwise.

The function $rChange(T,n_i)$ modifies (T,n_i) by propagating a subalphabet (corresponding to the selfloop of the state). If n_i is labelled with any of the four atomic formulas $\lceil\lceil B\rceil\rceil$, or $\lceil B\rceil\rceil$, (or $\lceil\lceil B\rceil$ or $\lceil B\rceil$), and $c \in B$, then the corresponding leaf node n_i is replaced by the subtree corresponding to $\oplus(\lceil\lceil A\cap B\rceil F_a\lceil B\rceil\rceil)$ (or $\oplus(\lceil\lceil A\cap B\rceil F_a\lceil B\rceil)$, respectively). The new pointer points to the subformula to the left of F_a if $q \in Q_r$ and to the right of F_a otherwise." If $c \notin B$, then $rChange(T,n_i) = \bot$. If n_i is an F_c or L_c node, then the parse tree remains unchanged. However, if n_i is labelled F_b or L_b, for some $b \neq c$, then $rChange(T,n_i) = \bot$.

The function $rSkip(T,n_i)$ alters (T,n_i) as follows. If n_i is labelled with any of the atomic formulas $\lceil\lceil B\rceil\rceil$, $\lceil B\rceil\rceil$, $\lceil\lceil B\rceil$ or $\lceil B\rceil$, then this node is replaced by $\lceil A\cap B\rceil$. If n_i is a F_b or L_b node, then $rSkip(T,n_i) = T$ if $b \in A$ and $rSkip(T,n_i) = \bot$ if $b \notin A$.

The base case of finding formula $\psi(e)$ for an outgoing transition e from the initial state $s \in Q_l$ is $\psi(e) \stackrel{\text{def}}{=} rExtend((\lceil\lceil\Sigma\rceil,1),e)$.

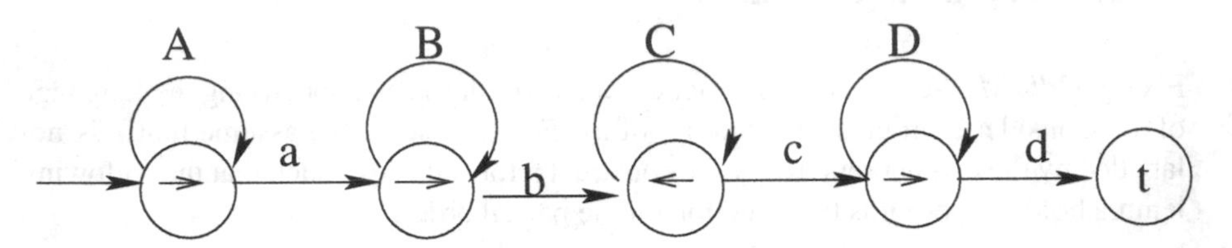

Fig. 3 A simple automaton with reversals

Example 6. Consider the automaton given in Figure 3. In this automaton, we have the conditions $a \notin A, b \notin B, c \notin C, d \notin D$, required for determinism, and we assume that $c \in B\cap A$ and $a,b \in C$. Let e_a,e_b,e_c,e_d be the edges labelled with a,b,c,d, respectively.

For convenience, a pointed simple formula T,n is denoted by underlining the subformula of node n.

- $\psi(e_a) = rExtend(\underline{\lceil\lceil\Sigma\rceil},e_a) = \lceil\lceil A\rceil F_a\underline{\lceil\Sigma\rceil\rceil}$.
- $\psi(e_b) = rExtend(\overline{\psi(e_a)},e_b) = \lceil\lceil A\rceil F_a(\oplus(\overline{\lceil\lceil B\rceil} F_b\lceil\Sigma\rceil\rceil))$.
- $\psi(e_c) = \ell Extend(\psi(e_b),e_c) = \ell Extend(\lceil\lceil\overline{A\rceil F_a}(\oplus(\underline{\lceil\lceil B\rceil} F_b\lceil\Sigma\rceil\rceil)),e_c)$
  $= \lceil\lceil A\rceil F_a(\oplus(\ominus(\lceil\lceil B\rceil L_c\underline{\lceil B\cap C\rceil\rceil})F_b\lceil\Sigma\rceil\rceil))$

$$
\begin{aligned}
& \vee\ \ell Extend(\lceil\lceil A\rceil \underline{F_a}(\oplus(\ominus(\lceil\lceil B\cap C\rceil)F_b\lceil\Sigma\rceil\rceil)),e_c) \\
= \ & \lceil\lceil A\rceil F_a(\oplus(\ominus(\lceil\overline{\lceil B\rceil L_c \underline{\lceil B\cap C\rceil\rceil}})F_b\lceil\Sigma\rceil\rceil)) \\
& \vee\ lChange(\lceil\lceil A\rceil \underline{F_a}(\oplus(\overline{\ominus(\lceil\lceil B\cap C\rceil)F_b\lceil\Sigma\rceil\rceil})))
\\
& \vee\ \ell Extend(\underline{\lceil\lceil A\rceil} F_a(\oplus(\ominus(\lceil\lceil B\cap C\rceil)F_b\lceil\Sigma\rceil\rceil)),e_c)
\end{aligned}
$$

(The above step follows from the assumption that $a \in C$)

$$
\begin{aligned}
= \ & \lceil\lceil A\rceil F_a(\oplus(\ominus(\lceil\lceil B\rceil L_c \underline{\lceil B\cap C\rceil\rceil})F_b\lceil\Sigma\rceil\rceil)) \ \vee \ false \\
& \vee\ \ominus(\lceil\lceil A\rceil L_c\lceil A\cap C\rceil\rceil)\overline{F_a(\oplus(\ominus(}\lceil\lceil B\cap C\rceil)F_b\lceil\Sigma\rceil\rceil))
\end{aligned}
$$

(The above step follows from the assumptions that $a \neq c$ and $c \in A$)

- $\psi(e_d)$ may be similarly worked out.

Theorem 3. *Given an n-pass po2dfa M, there is a formula in $R_n \cup L_n$ of size exponential in the number of its transitions, which defines the language accepted by M.*

Proof. Let E be the set of transitions leading into the accepting state t. Define the formula $F(M) \stackrel{\text{def}}{=} \bigvee_{e\in E} \psi(e)$. Then by the previous lemma M accepts the language $L(F(M))$. From the definition of *Extend* we see that the alternation between F_a and L_a modalities takes place only when the automaton changes direction, hence the constructed formula is in $R_n \cup L_n$.

For any transition e let $depth(e)$ denote the length of the longest progress path from the start state to e. By examining the construction, it can be seen that in $\psi(e) = \bigvee_{i\in I(e)} \xi_i$ the size of each ξ_i is linear in $depth(e)$. However, each $Extend(T,e)$ gives rise to up to $|depth(e)|$ disjuncts. Hence the number of disjuncts $|I(e)|$ can be exponential in $depth(e)$ as also the size of the $F(M)$. □

This also shows that unambiguous polynomials over the piecewise testable languages [16] are matched by deterministic and co-deterministic products, a result independently obtained by Kufleitner and Weil [8]. From the main theorem of [8], we get the corollary that there is an $FO^2[<]$ formula with n quantifier alternations for the language of M above.

References

1. M. Bojańczyk, C. David, A. Muscholl, T. Schwentick and L. Segoufin. Two-variable logic on words with data, *Proc. LICS*, Seattle, 2006, 7–16.
2. M. Bojańczyk. Two-way unary temporal logic over trees, *Proc. LICS*, Wrocław, 2007.
3. B. Borchert and P. Tesson. The atnext/atprevious hierarchy on the starfree languages, *Report* WSI-2004-11 (U. Tübingen, 2004).
4. V. Diekert, P. Gastin and M. Kufleitner. A survey on small fragments of first-order logic over finite words, *Int. J. Found. Comp. Sci.*, to appear.
5. K. Etessami, M.Y. Vardi and T. Wilke. First-order logic with two variables and unary temporal logic, *Inf. Comput.* 179, 2002, 279–295.
6. S.N. Krishna and P.K. Pandya. Modal strength reduction in quantified discrete duration calculus, *Proc. FSTTCS*, Hyderabad (R. Ramanujam and S. Sen, eds.), *LNCS* 3821, 2005, 444–456.

7. M. Kufleitner. Polynomials, fragments of temporal logic and the variety DA over traces, *Theoret. Comp. Sci.* 376, 2007, 89–100.
8. M. Kufleitner and P. Weil. On FO^2 quantifier alternation over words, Manuscript, , Jan 2008.
9. C. Löding and W. Thomas. Alternating automata and logics over infinite words, *Proc. IFIP TCS*, Sendai (J. van Leeuwen, O. Watanabe, M. Hagiya, P.D. Mosses, T. Ito, eds.), *LNCS* 1872, 2000, 521–535.
10. M. Mortimer. On language with two variables, *Zeit. Math. Log. Grund. Math.* 21, 1975, 135–140.
11. B.C. Moszkowski and Z. Manna. Reasoning in interval temporal logic, *Proc. Logics of programs*, Pittsburgh (E.M. Clarke and D. Kozen, eds.), *LNCS* 164, 1983, 371–382.
12. J.-E. Pin and P. Weil. Polynomial closure and unambiguous products, *Theory Comput. Syst.*30, 1997, 383–422.
13. M.-P. Schützenberger. Sur le produit de concaténation non ambigu, *Semigroup Forum* 13, 1976, 47–75.
14. T. Schwentick, D. Thérien and H. Vollmer. Partially-ordered two-way automata: a new characterization of DA, *Proc. DLT '01*, Vienna (W. Kuich, G. Rozenberg and A. Salomaa, eds.), *LNCS* 2295, 2002, 239–250.
15. S.S. Shah. FO^2 and related logics, Master's thesis (TIFR, 2007).
16. I. Simon. Piecewise testable events, *Proc. GI Conf. Autom. Theory and Formal Lang.*, Kaiserslautern (H. Barkhage, ed.), *LNCS* 33, 1975, 214–222.
17. L.J. Stockmeyer and A.R. Meyer. Word problems requiring exponential time, *Proc. STOC*, Austin, 1973, 1–9.
18. P. Tesson and D. Thérien. Diamonds are forever: the variety DA, *Semigroups, algorithms, automata and languages* (G.M.S. Gomes, P.V. Silva and J.-E. Pin, eds.), (World Scientific, 2002), 475–500.
19. D. Thérien and T. Wilke. Over words, two variables are as powerful as one quantifier alternation: $FO^2 = \Sigma_2 \cap \Pi_2$, *Proc. STOC*, Dallas, 1998, 41–47.
20. P. Weis and N. Immerman. Structure theorem and strict alternation hierarchy for FO^2 on words, *Proc. CSL*, Lausanne (J. Duparc and T. Henzinger, eds.), *LNCS* 4646, 2007, 343–357.

On Boundedness in Depth in the π-Calculus*

Roland Meyer

University of Oldenburg
Roland.Meyer@informatik.uni-oldenburg.de

Abstract. We investigate the class $\mathcal{P}_{BD}$ of π-Calculus processes that are bounded in the function *depth*. First, we show that boundedness in depth has an intuitive characterisation when we understand processes as graphs: a process is bounded in depth if and only if the length of the simple paths is bounded. The proof is based on a new normal form for the π-Calculus called *anchored fragments*. Using this concept, we then show that processes of bounded depth have well-structured transition systems (WSTS). As a consequence, the termination problem is decidable for this class of processes. The instantiation of the WSTS framework employs a new well-quasi-ordering for processes in $\mathcal{P}_{BD}$.

1 Introduction

Concurrent systems are known to be hard to design correctly. Dynamically reconfigurable systems add to concurrency the problem of changing connection structures between system components. To ensure the correct behaviour of systems, automatic verification techniques have proven useful. This automation comes with a tradeoff. To automate the analysis requires a decidable class of models, but to model the systems of interest requires an expressive class. We use the π-Calculus to model dynamically reconfigurable systems [17, 18]. The contribution of this paper is the up-to-now most expressive subclass of π-Calculus for which termination is decidable. The importance of termination for the π-Calculus has been recognised in [19, 5].

The class $\mathcal{P}_{BD}$ we propose contains the processes that are *bounded in depth.* The function *depth* measures the interdependence of restricted names in process terms. Boundedness in depth is a very liberal requirement as it turns out that all decidable subclasses of π-Calculus known so far are subclasses of $\mathcal{P}_{BD}$: finitary agents [9], finite control processes [4], bounded processes [3], unique receiver and bounded input systems (up to bisimilarity) [2], finite handler processes [14], structurally stationary processes [14], and restriction-free processes [2].

But the definition of depth is difficult to grasp as the function refers to all processes in a congruence class. To provide an intuition to $\mathcal{P}_{BD}$, we make use of the standard graph-theoretic interpretation of the π-Calculus [17, 18]. Our first main result states that boundedness in depth is equivalent to boundedness in the

* This work was supported by the German Research Council (DFG) as part of the Graduate School "TrustSoft" (GRK 1076/1).

Please use the following format when citing this chapter:

Meyer, R., 2008, in IFIP International Federation for Information Processing, Volume 273; *Fifth IFIP International Conference on Theoretical Computer Science*; Giorgio Ausiello, Juhani Karhumäki, Giancarlo Mauri, Luke Ong; (Boston: Springer), pp. 477–489.

length of the simple paths (i.e., without repetition of edges) in the graphs. The proof is based on a new normal form for processes called *anchored fragments.*

The decidability result for $\mathcal{P}_{BD}$ is obtained by viewing this class as an instance of *well-structured transition systems (WSTS)* [10, 1, 11]. WSTS are a framework for infinite state systems that generalises decidability results for particular models. Technically, a WSTS is a transition system with an ordering relation on the states which is compatible with the transition relation. Depending on the ordering, the compatibility, and decidability properties the framework yields decision procedures, e.g., for termination [10, 11] or simulation [1].

Our second main result is the instantiation of the WSTS framework for processes of bounded depth. As a consequence, we inherit the decision procedure for termination in [10, 11]. The technical contribution is a new ordering $\preceq_{\mathcal{P}_{BD}}$ on processes which we show to be a *well-quasi-ordering (wqo)* (i.e., in every infinite sequence of processes two comparable processes can be found) for processes of bounded depth. In the proof, anchored fragments again play a vital role. Since the ordering $\preceq_{\mathcal{P}_{BD}}$ is a simulation relation it is compatible with the reaction relation of the π-Calculus in a strong sense.

2 Preliminaries

The π-Calculus We use a π-Calculus with parameterised recursion as proposed in [18]. Let the set $(a, b \in)$ $\mathcal{N}$ of *names* contain the channels and messages that occur in communications. A process consumes *prefixes* π to communicate with other processes or to perform silent actions. The prefixes are

$$\pi ::= \overline{a}\langle b\rangle \mid a(x) \mid \tau.$$

The *output action* $\overline{a}\langle b\rangle$ sends the name b on channel a. The *input action* $a(x)$ receives a name that replaces x on a. The τ symbol stands for a *silent action.*

To denote recursive processes we use *process identifiers* K, each defined by an equation $K(\tilde{x}) := P$. When the identifier is *called,* $K\lfloor\tilde{a}\rfloor$, it is replaced by the process P where the names $\tilde{x}$ are replaced by $\tilde{a}$. More precisely, a *substitution* $\sigma = \{\tilde{a}/\tilde{x}\}$ is a function that maps the names in $\tilde{x}$ to $\tilde{a}$ and is the identity on all names not in $\tilde{x}$. The *application of a substitution,* $P\{\tilde{a}/\tilde{x}\}$, is defined in the standard way [18]. A π-Calculus process is a call to an identifier, $K\lfloor\tilde{a}\rfloor$, a *choice process* deciding between prefixes, $\Sigma_{i\in I}\pi_i.P_i$, a *parallel composition* of processes, $P_1 \mid P_2$, or the *restriction* of a name in a process, $\nu a.P$:

$$P ::= K\lfloor\tilde{a}\rfloor \mid \Sigma_{i\in I}\pi_i.P_i \mid P_1|P_2 \mid \nu a.P.$$

The *set of all processes* is $\mathcal{P}$. We abbreviate empty sums (with $I = \emptyset$) by $\mathbf{0}$ and arbitrary sums by M or N. By $\Pi_{i\in I}P_i$ we denote the parallel composition of several processes P_i with $i \in I$. The processes $K\lfloor\tilde{a}\rfloor$ and $\Sigma_{i\in I\neq\emptyset}\pi_i.P_i$ are called *sequential.* By $\mathcal{S}(P)$ we refer to the *set of sequential processes* in P. The function is defined inductively by $\mathcal{S}(\mathbf{0}) := \emptyset$, $\mathcal{S}(K\lfloor\tilde{a}\rfloor) := \{K\lfloor\tilde{a}\rfloor\}$, $\mathcal{S}(\Sigma_{i\in I\neq\emptyset}\pi_i.P_i) := \{\Sigma_{i\in I\neq\emptyset}\pi_i.P_i\}$, $\mathcal{S}(P_1 \mid P_2) := \mathcal{S}(P_1) \cup \mathcal{S}(P_2)$, and $\mathcal{S}(\nu a.P) := \mathcal{S}(P)$.

The input action $a(b)$ and the restriction $\nu c.P$ *bind* b and c, respectively. The *set of bound names* in P is $bn\,(P)$. If we refer to the *set of restricted names* in P, $rn\,(P) \subseteq bn\,(P)$, we mean the restricted names that are not covered by prefixes. A name which occurs not bound in P is *free* and the *set of free names* in P is $fn\,(P)$. We permit α-conversion of bound names. Therefore, wlog. we assume $bn\,(P) \cap fn\,(P) = \emptyset$. Unless otherwise stated, we assume that a name is bound at most once in a process. In a defining equation $K(\tilde{x}) := P$ we require $fn\,(P) \subseteq \tilde{x}$. If a substitution $\{\tilde{a}/\tilde{x}\}$ is applied to a process P, we assume $bn\,(P) \cap (\tilde{a} \cup \tilde{x}) = \emptyset$.

The results achieved in this paper make heavy use of the *structural congruence* relation $\equiv$ of processes. It is the smallest congruence where α-conversion of bound names is allowed, $+$ and $\mid$ are commutative and associative and have $\mathbf{0}$ as neutral element, and the following laws for restriction hold:

$$\nu x.\nu y.P \equiv \nu y.\nu x.P \qquad\qquad \nu x.\mathbf{0} \equiv \mathbf{0}$$
$$\nu x.(P \mid Q) \equiv P \mid (\nu x.Q), \text{ if } x \notin fn\,(P).$$

The latter law is called *scope extrusion*.

We distinguish two normal forms for processes. A process $\nu\tilde{a}.(P_1 \mid \ldots \mid P_n)$ where $\tilde{a} \subseteq fn\,(P_1 \mid \ldots \mid P_n)$ and all P_i are sequential is in *standard form* [17]. Via structural congruence every process P can be rewritten as a process P_{sf} in standard form as follows. First, the scope of every restricted name not under a prefix is extruded over all processes composed in parallel. Then unused restricted names and empty sums are removed. Since all bound names are different and disjoint with the free names, α-conversion is not required. Thus, the rewriting does not change the sequential processes, $\mathcal{S}(P) = \mathcal{S}(P_{sf})$.

The *restricted form* [14] is based on the notion of *fragments*, i.e., processes where the scopes of restricted names are minimal:

$$F ::= K\lfloor\tilde{a}\rfloor \mid \Sigma_{i\in I\neq\emptyset}\pi_i.P_i \mid \nu a.(F_1 \mid \ldots \mid F_n),$$

where $a \in fn\,(F_i)$ for all i. The *set of all fragments* is $(F, G \in)\ \mathcal{P}_{\mathcal{F}}$. Fragments that are sequential processes, $K\lfloor\tilde{a}\rfloor$ or $\Sigma_{i\in I\neq\emptyset}\pi_i.P_i$, are *elementary* and referred to by F_e. A process P_ν is in *restricted form*, if it is a parallel composition of fragments, $P_\nu = \Pi_{i\in I}F_i$. The *set of fragments in* P_ν is $Frag\,(P_\nu) := \bigcup_{i\in I}\{F_i\}$. The *set of all processes in restricted form* is $\mathcal{P}_\nu$.

To compute the restricted form $P_\nu \in \mathcal{P}_\nu$ of a process $P \in \mathcal{P}$, we minimise the scopes of all restricted names not under a prefix and remove processes congruent with $\mathbf{0}$, in particular unused restricted names. Again, this does not change the sequential processes, $\mathcal{S}(P) = \mathcal{S}(P_\nu)$. The restricted form of a process is invariant under structural congruence up to rewriting of fragments: $P \equiv Q$ iff $P_\nu \mathrel{\hat{\equiv}} Q_\nu$, where $\hat{\equiv}$ is the smallest equivalence on processes in restricted form that permits

(1) associativity and commutativity with regard to $|$ and (2) replacing fragments by structurally congruent ones, i.e., $F \mid P_\nu \hat{\equiv} G \mid P_\nu$ if $F \equiv G$.

The behaviour of π-Calculus processes is determined by the *reaction relation* $\rightarrow \subseteq \mathcal{P} \times \mathcal{P}$ defined by the following rules:

$$(\text{Tau})\ \tau.P + M \rightarrow P$$

$$(\text{React})\ (x(y).P + M) \mid (\overline{x}\langle z\rangle.Q + N) \rightarrow P\{z/y\} \mid Q$$

$$(\text{Const})\ K\lfloor\tilde{a}\rfloor \rightarrow P\{\tilde{a}/\tilde{x}\}, \text{ if } K(\tilde{x}) := P$$

$$(\text{Par})\ \frac{P \rightarrow P'}{P \mid Q \rightarrow P' \mid Q} \qquad (\text{Res})\ \frac{P \rightarrow P'}{\nu a.P \rightarrow \nu a.P'}$$

$$(\text{Struct})\ \frac{P \rightarrow P'}{Q \rightarrow Q'}, \text{ if } P \equiv Q \text{ and } P' \equiv Q'.$$

By $Reach\,(P)$ we denote the *set of all processes reachable from* P with the reaction relation. The reaction relation is *image finite*, i.e., for every process P there are up to structural congruence only finitely many Q with $P \rightarrow Q$.

To relate a reachable fragment $F \in Frag\,(Reach\,(P))$ with the initial process P, we recall that F consists of *derivatives* of P [14]. Derivatives are sequential subprocesses of P gained by removing prefixes as if they were communicated. Let P use $n \in \mathbb{N} = \{0, 1, 2, \ldots\}$ recursive definitions $K_i(\tilde{x}_i) := P_i$. We define $derivatives(P) := der(P) \cup \bigcup_{i=1}^{n} der(P_i)$, where $der(\mathbf{0}) := \emptyset$, $der(K\lfloor\tilde{a}\rfloor) := \{K\lfloor\tilde{a}\rfloor\}$, $der(\Sigma_{i\in I\neq\emptyset}\pi_i.P_i) := \{\Sigma_{i\in I\neq\emptyset}\pi_i.P_i\} \cup \bigcup_{i\in I} der(P_i)$, $der(P_1|P_2) := der(P_1) \cup der(P_2)$, and $der(\nu a.P) := der(P)$. Then every $F \in Frag\,(Reach\,(P))$ is structurally congruent with $\nu\tilde{a}.(\Pi_{i\in I\neq\emptyset}Q_i\sigma_i)$, where $Q_i \in derivatives(P)$ and $\sigma_i : fn\,(Q_i) \rightarrow fn\,(P) \cup \tilde{a}$.

To define the function *depth*, we require the *nesting of restrictions* measured as follows: $nest_\nu\,(K\lfloor\tilde{a}\rfloor) := 0$, $nest_\nu\,(\Sigma_{i\in I}\pi_i.P_i) := 0$, $nest_\nu\,(P_1 \mid P_2) := max\{nest_\nu\,(P_1), nest_\nu\,(P_2)\}$, and $nest_\nu\,(\nu a.P) := 1 + nest_\nu\,(P)$.

Definition 1. The *depth* of $F \in \mathcal{P}_\mathcal{F}$ is the minimal nesting of restrictions in all fragments in the congruence class: $depth(F) := min\{nest_\nu\,(F') \mid F' \equiv F\}$. A process $P \in \mathcal{P}$ is *bounded in depth*, iff there is $k_D \in \mathbb{N}$ such that the depth of all reachable fragments is less or equal to k_D, i.e.,

$$\exists k_D \in \mathbb{N} : \forall Q \in Reach\,(P) : \forall F \in Frag\,(Q_\nu) : depth(F) \leq k_D.$$

The *set of all processes that are bounded in depth* is $\mathcal{P}_{BD}$. ◊

Well-Quasi-Orderings A *quasi-ordering (qo)* on a set of elements A is a reflexive and transitive relation $\preceq_A \subseteq A \times A$. We also call $(A, \preceq_A)$ a qo. The qo $(A, \preceq_A)$ is a *well-quasi-ordering (wqo)*, iff in every infinite sequence $(a_i)_{i\in\mathbb{N}}$ in A there are two comparable elements, i.e., there are indices $i < j$ with $a_i \preceq_A a_j$.

A result by Higman [13] lifts a wqo $\preceq_A$ on a set of elements A to a wqo $\preceq_A^H$ on the set of finite sequences A^*. The ordering $u \preceq_A^H v$ demands u to be a subsequence of v which is dominated elementwise wrt. $\preceq_A$, i.e., $u = (u_1, \ldots, u_m)$

and $v = (v_1, \ldots, v_n)$ and there are $1 \leq i_1 < \ldots < i_m \leq n$ such that $u_k \preceq_A v_{i_k}$ for all $1 \leq k \leq m$.

In Section 4 we define a qo on fragments. To prove it is a wqo, we relate it with a wqo on trees. Consider a qo $(A, \preceq_A)$. The *trees over* A are defined by

$$T ::= a \mid (a, (T_1, \ldots, T_n)),$$

where $a \in A$. The *set of all trees over* A is $\mathcal{T}(A)$. The *height* of a tree is measured similar to the nesting of restrictions in fragments, $height(a) := 0$ and $height((a, (T_1, \ldots, T_n))) := 1 + max\{height(T_i) \mid 1 \leq i \leq n\}$. For $n \in \mathbb{N}$ we denote by $\mathcal{T}(A)_n$ the trees of height less or equal to n.

We use the *rooted tree embedding* $\preceq_{\mathcal{T}(A)}$ as qo on the trees in $\mathcal{T}(A)$. Intuitively, $T_1 \preceq_{\mathcal{T}(A)} T_2$ if T_1 is a subtree of T_2 so that the levels of T_1 are preserved in T_2. In particular, the root of T_1 is mapped to the root of T_2 and the leaves in T_1 are leaves in T_2. Technically, the rooted tree embedding is defined by two rules. If $a \preceq_A a'$ then $a \preceq_{\mathcal{T}(A)} a'$ (Elem) and if $a \preceq_A a'$ and $(T_1, \ldots, T_m) \preceq^H_{\mathcal{T}(A)} (T'_1, \ldots, T'_n)$ then $(a, (T_1, \ldots, T_m)) \preceq_{\mathcal{T}(A)} (a', (T'_1, \ldots, T'_n))$ (Comp). It is not hard to see that the relation $\preceq_{\mathcal{T}(A)} \subseteq \mathcal{T}(A) \times \mathcal{T}(A)$ is a qo. It is a wqo on trees of bounded height.

Lemma 1. *If* $(A, \preceq_A)$ *is a wqo then* $(\mathcal{T}(A)_n, \preceq_{\mathcal{T}(A)})$ *is a wqo for all* $n \in \mathbb{N}$.

3 A Characterisation of Boundedness in Depth

In this section, we interpret fragments F as hypergraphs $\mathcal{G}[\![F]\!]$. With this interpretation we call the process $P \in \mathcal{P}$ *bounded in the simple paths*, iff there is $k_{Sim} \in \mathbb{N}$ such that the length of the longest simple path in the hypergraphs of all reachable fragments is less or equal to k_{Sim}, i.e.,

$$\exists k_{Sim} \in \mathbb{N} : \forall Q \in Reach\,(P) : \forall F \in Frag\,(Q_\nu) : lsp(\mathcal{G}[\![F]\!]) \leq k_{Sim},$$

where $lsp(\mathcal{G}[\![F]\!])$ denotes the length of the longest simple path in $\mathcal{G}[\![F]\!]$. We prove that a process is bounded in depth if and only if it is bounded in the simple paths. Thus, processes in $\mathcal{P}_{BD}$ can be intuitively understood as hypergraphs where the length of the simple paths is bounded.

The main technical contribution is the definition of *anchored fragments*. In this section, we use them to derive boundedness in depth from boundedness in the simple paths (Lemma 3). In Section 4 they help us prove that the given qo is a wqo. In particular we need that the nesting of restrictions in anchored fragments is bounded if the depth is (Corollary 1). Before we turn to anchored fragments, we make the interpretation of processes as hypergraphs precise.

3.1 The Graph-theoretic Interpretation of the π-Calculus

A *hypergraph* [12] is a graph where several vertices may be connected with one hyperedge, i.e., it is a tuple $\mathcal{G} = (V, E, l, inc)$, where V is a finite set of *vertices*, E is a finite set of *hyperedges*, $l : V \to \mathcal{P}$ is a *vertex labelling function*, and $inc : E \to \mathbb{P}(V)$ is an *incidence function*. In the graphical representation we draw a dot labelled by $l(v)$ for each $v \in V$ and a box labelled by e for every $e \in E$. There is an arc between v and e, if $v \in inc(e)$. In our setting edges are names, $E \subseteq \mathcal{N}$. We also call hypergraphs *graphs* and hyperedges *edges*.

Two graphs $\mathcal{G}_1$ and $\mathcal{G}_2$ are *equal*, $\mathcal{G}_1 = \mathcal{G}_2$, if $E_1 = E_2$ and there is a bijection $f : V_1 \to V_2$ that is compatible with the labelling and the incidence functions. Hence, the identity of elements $v \in V$ is not important and we can always assume $V_1 \cap V_2 = \emptyset$.

A *path* in $\mathcal{G}$ is a finite sequence $p = (v_1, e_1, \ldots, v_n, e_n, v_{n+1})$ such that the edges e_i connect v_i and v_{i+1}, i.e., $v_i, v_{i+1} \in inc(e_i)$ for all i. The *length* of p, *length*(p), is the number of edges in p. By $fe(p)$ we refer to the *first element in* p, v_1. A path is *simple*, if $e_i \neq e_j$ for all $i \neq j$. By $lsp(\mathcal{G})$ we denote the *length of the longest simple path* in $\mathcal{G}$. The *set of all paths* in $\mathcal{G}$ is $Paths(\mathcal{G})$.

We require three operations on graphs. The *disjoint union* of $\mathcal{G}_1$ and $\mathcal{G}_2$, where $E_1 \cap E_2 = \emptyset$, puts both graphs side by side. Formally, it is the graph $\mathcal{G}_1 \uplus \mathcal{G}_2 := (V_1 \uplus V_2, E_1 \uplus E_2, l_1 \uplus l_2, inc_1 \uplus inc_2)$. The *connect operator* takes a graph $\mathcal{G}$ and a name $a \notin E$. The result is the graph $\mathcal{G} \otimes a$, where a is added to E. The new edge connects the processes that have a as a free name, i.e., $\mathcal{G} \otimes a := (V, E \uplus \{a\}, l, inc \uplus \{(a, V_a)\})$, where $V_a \subseteq V$ with $v \in V_a$ iff $a \in fn\,(l(v))$. We define the *application of a substitution* $\{a/x\}$ to $\mathcal{G}$ by $\mathcal{G}\{a/x\} := (V, E, l', inc)$, where $l'(v) := l(v)\{a/x\}$ for all $v \in V$.

The *graph-theoretic interpretation* (1) creates a vertex for every sequential process, (2) takes the restricted names not under prefixes as the edges, and (3) inserts an arc where a name is free in a process. Technically, it is the function $\mathcal{G}[\![-]\!]$ defined by $\mathcal{G}[\![\mathbf{0}]\!] := (\emptyset, \emptyset, \emptyset, \emptyset)$, $\mathcal{G}[\![K\lfloor\tilde{a}\rfloor]\!] := (\{v\}, \emptyset, \{(v, K\lfloor\tilde{a}\rfloor)\}, \emptyset)$, $\mathcal{G}[\![\Sigma_{i \in I \neq \emptyset}\pi_i.P_i]\!] := (\{v\}, \emptyset, \{(v, \Sigma_{i \in I \neq \emptyset}\pi_i.P_i)\}, \emptyset)$, $\mathcal{G}[\![P \mid Q]\!] := \mathcal{G}[\![P]\!] \uplus \mathcal{G}[\![Q]\!]$, and $\mathcal{G}[\![\nu a.P]\!] := \mathcal{G}[\![P]\!] \otimes a$ if $a \in fn\,(P)$, $\mathcal{G}[\![P]\!]$ otherwise.

Structurally congruent processes $P_1 \equiv P_2$ are mapped to equivalent hypergraphs $\mathcal{G}[\![P_1]\!] \approx \mathcal{G}[\![P_2]\!]$. The relation $\approx$ is the smallest equivalence on hypergraphs where replacement of vertex labels by structurally congruent processes is allowed, $(V \uplus \{v\}, E, l \uplus \{(v, P)\}, inc) \approx (V \uplus \{v\}, E, l \uplus \{(v, Q)\}, inc)$, if $P \equiv Q$, and renaming of edges together with the attached processes is possible, $\mathcal{G} \otimes a \approx (\mathcal{G}\{b/a\}) \otimes b$, if $b \notin fn\,(l(v))$ for all $v \in V$. The equivalence $\approx$ preserves the length of the longest simple path, $\mathcal{G}_1 \approx \mathcal{G}_2$ implies $lsp(\mathcal{G}_1) = lsp(\mathcal{G}_2)$.

3.2 Anchored Fragments

By definition, all fragments under a restriction νa share the name a. In *anchored fragments*, we demand that distinguished processes inside the fragments, the *anchors*, share the name a. The corresponding function *anc* gives for a fragment $F_{\mathcal{A}i}$ in $\nu a.(F_{\mathcal{A}1} \mid \ldots \mid F_{\mathcal{A}n})$ the process $anc(F_{\mathcal{A}i}) = P \in \mathcal{S}(F_{\mathcal{A}i})$ that knows the name, i.e., $a \in fn\,(P)$. When descending an anchored fragment $F_{\mathcal{A}} = \nu a.(F_{\mathcal{A}1} \mid \ldots \mid F_{\mathcal{A}n})$ using the function $nest_\nu$, this guarantees that the vertices labelled by the anchors $anc(F_{\mathcal{A}i})$ are connected via a in $\mathcal{G}[\![F_{\mathcal{A}}]\!]$.

Definition 2. The *set of anchored fragments* $(F_{\mathcal{A}}, G_{\mathcal{A}} \in)\ \mathcal{P}_{\mathcal{A}}$ is defined by

$$F_{\mathcal{A}} ::= K\lfloor\tilde{a}\rfloor \mid \Sigma_{i\in I\neq\emptyset}\pi_i.P_i \mid \nu a.(F_{\mathcal{A}1} \mid \ldots \mid F_{\mathcal{A}n}),$$

where $a \in fn\,(anc(F_{\mathcal{A}i}))$ for all i, with $anc(K\lfloor\tilde{a}\rfloor) := K\lfloor\tilde{a}\rfloor$, $anc(\Sigma_{i\in I\neq\emptyset}\pi_i.P_i) := \Sigma_{i\in I\neq\emptyset}\pi_i.P_i$, and $anc(\nu a.(F_{\mathcal{A}1} \mid \ldots \mid F_{\mathcal{A}n})) := anc(F_{\mathcal{A}1})$. ◊

Of course, anchored fragments are fragments. We now show that every fragment can be rewritten as an anchored fragment using structural congruence. In the proof, it is important that every sequential process inside a fragment can be chosen as the anchor.

Lemma 2. *Consider $F \in \mathcal{P}_{\mathcal{F}}$ and a process $P \in \mathcal{S}(F)$. Then there is an anchored fragment $F_{\mathcal{A}} \in \mathcal{P}_{\mathcal{A}}$ such that $F_{\mathcal{A}} \equiv F$, $\mathcal{S}(F_{\mathcal{A}}) = \mathcal{S}(F)$, and $anc(F_{\mathcal{A}}) = P$.*

We explain the induction step in the proof of Lemma 2. Given fragment F we compute the standard form $\nu\tilde{a}.(P_1 \mid \ldots \mid P_n)$. Since this does not change the sequential processes, one process P_i is the given process P, wlog. P_1. We split the set of names $\tilde{a}$ into three subsets $\tilde{a}_1, \tilde{a}_2, \tilde{a}_3$ as follows. A name a that is shared by P and $P_2 \mid \ldots \mid P_n$, i.e., $a \in fn\,(P) \cap fn\,(P_2 \mid \ldots \mid P_n)$, is in the set $\tilde{a}_1$. A name which is only in the free names of P is in $\tilde{a}_2$. The remaining names are in $\tilde{a}_3$. Shrinking the scopes yields $\nu\tilde{a}_1.(\nu\tilde{a}_2.P \mid \nu\tilde{a}_3.(P_2 \mid \ldots \mid P_n))$. To transform $\nu\tilde{a}_3.(P_2 \mid \ldots \mid P_n)$ into a parallel composition of anchored fragments, we compute the restricted form. It consists of several fragments, $(\nu\tilde{a}_3.(P_2 \mid \ldots \mid P_n))_\nu = G_1 \mid \ldots \mid G_m$. By construction, every G_i contains a process $P_{\mathcal{A}i}$ sharing a name with P. Since each G_i contains less processes than F we can apply the induction hypothesis. This yields anchored fragments $G_{\mathcal{A}i}$ where $anc(G_{\mathcal{A}i}) = P_{\mathcal{A}i}$ shares a name with P. We now have $\nu\tilde{a}_1.(\nu\tilde{a}_2.P \mid G_{\mathcal{A}1} \mid \ldots \mid G_{\mathcal{A}m})$. As the names in $\tilde{a}_1$ are shared by different $G_{\mathcal{A}i}$, we minimise their scopes to get the required anchored fragment.

Example 1. Let $F = \nu b_1, b_2, b_3, a.(K\lfloor a, b_1\rfloor \mid L\lfloor a, b_2\rfloor \mid L\lfloor a, b_3\rfloor)$. We construct the anchored fragment $F_{\mathcal{A}}$ that has $K\lfloor a, b_1\rfloor$ as the anchor, $anc(F_{\mathcal{A}}) = K\lfloor a, b_1\rfloor$. The fragment F already is in standard form. We split the set of names $\{a, b_1, b_2, b_3\}$ into $\tilde{a}_1 = \{a\}$, $\tilde{a}_2 = \{b_1\}$, and $\tilde{a}_3 = \{b_2, b_3\}$. We shrink the scopes of all $\tilde{a}_i$ which gives $\nu a.(\nu b_1.K\lfloor a, b_1\rfloor \mid \nu b_2, b_3.(L\lfloor a, b_2\rfloor \mid L\lfloor a, b_3\rfloor))$. The restricted form of $\nu b_2, b_3.(L\lfloor a, b_2\rfloor \mid L\lfloor a, b_3\rfloor)$ is $\nu b_2.L\lfloor a, b_2\rfloor \mid \nu b_3.L\lfloor a, b_3\rfloor$. Both fragments, $\nu b_2.L\lfloor a, b_2\rfloor$ and $\nu b_3.L\lfloor a, b_3\rfloor$, are also anchored fragments where the anchors share the name a with $K\lfloor a, b_1\rfloor$. The scope of a is minimal. Our computation returns $\nu a.(\nu b_1.K\lfloor a, b_1\rfloor \mid \nu b_2.L\lfloor a, b_2\rfloor \mid \nu b_3.L\lfloor a, b_3\rfloor)$. $\Diamond$

For anchored fragments $F_{\mathcal{A}}$, the nesting of restrictions corresponds to the length of a simple path p in the graph $\mathcal{G}[\![F_{\mathcal{A}}]\!]$, $nest_\nu(F_{\mathcal{A}}) = length(p)$ for some simple path $p \in Paths(\mathcal{G}[\![F_{\mathcal{A}}]\!])$. In the proof, we need that the first element of p is labelled by the anchor of $F_{\mathcal{A}}$, $l(fe(p)) = anc(F_{\mathcal{A}})$. We illustrate the construction of a suitable path p in the induction step. The idea is to extend a path p' that exists by the hypothesis by an edge and a vertex.

Example 2. Consider $F_{\mathcal{A}} = \nu a.(\nu b_1.K\lfloor a, b_1\rfloor \mid \nu b_2.L\lfloor a, b_2\rfloor \mid \nu b_3.L\lfloor a, b_3\rfloor)$. The figure to the left shows a simple path p in $\mathcal{G}[\![F_{\mathcal{A}}]\!]$ with $length(p) = 2 = nest_\nu(F_{\mathcal{A}})$ and $l(fe(p)) = K\lfloor a, b_1\rfloor = anc(F_{\mathcal{A}})$. By the hypothesis, there is a simple path p' in $\mathcal{G}[\![\nu b_3.L\lfloor a, b_3\rfloor]\!]$ with $length(p') = 1 = nest_\nu(\nu b_3.L\lfloor a, b_3\rfloor)$ and $l(fe(p')) = L\lfloor a, b_3\rfloor = anc(\nu b_3.L\lfloor a, b_3\rfloor)$. This path is $p' = (L\lfloor a, b_3\rfloor, b_3, L\lfloor a, b_3\rfloor)$, depicted by dashed lines. As $\mathcal{G}[\![\nu b_3.L\lfloor a, b_3\rfloor]\!]$ is embedded in $\mathcal{G}[\![F_{\mathcal{A}}]\!]$ (dotted line), p' is a path in $\mathcal{G}[\![F_{\mathcal{A}}]\!]$. The anchor $L\lfloor a, b_3\rfloor$ and the anchor of $F_{\mathcal{A}}$, $K\lfloor a, b_1\rfloor$, are connected with a. We define $p = (anc(F_{\mathcal{A}}), a, p') = (K\lfloor a, b_1\rfloor, a, L\lfloor a, b_3\rfloor, b_3, L\lfloor a, b_3\rfloor)$. It extends p' by the bold lines. $\Diamond$

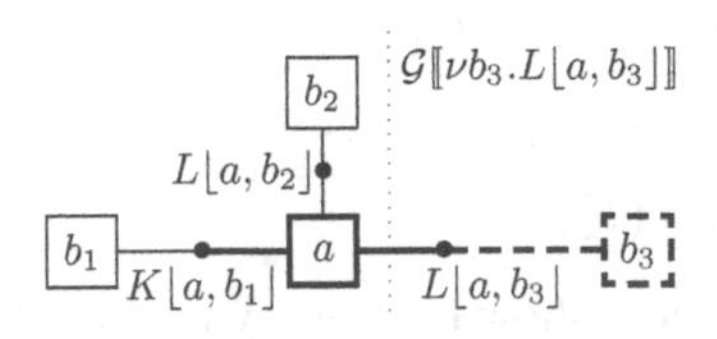

3.3 The Characterisation of Boundedness in Depth

Fragment F is structurally congruent with an anchored fragment $F_{\mathcal{A}}$ (Lemma 2). As $depth(F) \leq nest_\nu(F_{\mathcal{A}}) = length(p)$ for some simple path p in $\mathcal{G}[\![F_{\mathcal{A}}]\!]$ and as the length of the simple paths is bounded, the depth is bounded as well.

Lemma 3. *If $P \in \mathcal{P}$ is bounded in the simple paths by k_{Sim} then P is bounded in depth by k_{Sim} as well.*

It is easy to check that the length of the longest simple path in $\mathcal{G}[\![F]\!]$ is bounded by the nesting of restrictions in F as follows: $lsp(\mathcal{G}[\![F]\!]) \leq 2^{nest_\nu(F)} - 1$. Let F be bounded in depth. There is a fragment $F_D \equiv F$ where the nesting of restrictions is minimal, $nest_\nu(F_D) = min\{nest_\nu(F') \mid F' \equiv F\} = depth(F)$. Since the graphs of F and F_D are equivalent, $lsp(\mathcal{G}[\![F]\!]) = lsp(\mathcal{G}[\![F_D]\!])$ holds. The mentioned inequality and the choice of F_D yield the following lemma.

Lemma 4. *If $P \in \mathcal{P}$ is bounded in depth by k_D then P is bounded in the simple paths by $2^{k_D} - 1$.*

Combined, Lemma 3 and Lemma 4 prove our first main theorem.

Theorem 1. *A process $P \in \mathcal{P}$ is bounded in depth if and only if it is bounded in the simple paths.*

In the following section we understand anchored fragments as trees of bounded height. The boundedness is justified by the following corollary of Lemma 4.

Corollary 1. *Let P be bounded in depth by k_D and $F_{\mathcal{A}} \in$ Frag (Reach (P)) then* $nest_\nu(F_{\mathcal{A}}) \leq 2^{k_D} - 1$.

4 The Transition Systems of $\mathcal{P}_{BD}$ are Well-structured

Our second main result states that processes of bounded depth have *well-structured transition systems (WSTS)* [10, 1, 11]. A WSTS is a tuple $(S, \rightsquigarrow, \preceq)$, where $(S, \rightsquigarrow)$ is an image finite *transition system* and $\preceq \subseteq S \times S$ is a wqo on the *states* $(s, t \in)$ S which is required to be a *simulation*. By definition, the relation $s \preceq t$ is a simulation if state t imitates the transition behaviour of s, i.e., $s \preceq t$ and $s \rightsquigarrow s'$ implies there is t' with $t \rightsquigarrow t'$ and $s' \preceq t'$. To instantiate the framework, we define a qo $\preceq_{\mathcal{P}_{BD}}$ on processes and prove it (1) to be a wqo on *Reach* (P) where P is bounded in depth (Section 4.1) and (2) to be a simulation (Section 4.2). We conclude with a decision procedure for termination.

4.1 A Well-Quasi-Ordering for $\mathcal{P}_{BD}$

Our wqo $\preceq_{\mathcal{P}_{BD}}$ on processes is derived from a wqo on fragments. The idea of the *fragment ordering* $\preceq_{\mathcal{F}}$ is to use the rooted tree embedding and close it under structural congruence. The leafs in Rule (Elem) correspond to elementary fragments: $F_e \preceq_{\mathcal{F}} F_e$ (Rule (1)). Fragment $\nu a.(\Pi_{i \in I} F_i)$ is dominated by $\nu a.(\Pi_{i \in I} G_i \mid \Pi_{j \in J} G_j)$ if the G_i dominate the F_i. This mimics Rule (Comp). If F' is smaller than G' then every $F \equiv F'$ is smaller than $G \equiv G'$ (Rule (3)).

Definition 3. The *fragment ordering* $\preceq_{\mathcal{F}} \subseteq \mathcal{P}_{\mathcal{F}} \times \mathcal{P}_{\mathcal{F}}$ is defined by:

$$(1)\ \frac{}{F_e \preceq_{\mathcal{F}} F_e} \quad (2)\ \frac{F_i \preceq_{\mathcal{F}} G_i \text{ for all } i \in I}{\nu a.(\Pi_{i \in I} F_i) \preceq_{\mathcal{F}} \nu a.(\Pi_{i \in I} G_i \mid \Pi_{j \in J} G_j)}$$

$$(3)\ \frac{F \equiv F' \preceq_{\mathcal{F}} G' \equiv G}{F \preceq_{\mathcal{F}} G}. \qquad \diamond$$

Reflexivity of $\preceq_{\mathcal{F}}$ is immediate, transitivity follows from Lemma 8. To relate the fragment ordering $\preceq_{\mathcal{F}}$ with the rooted tree embedding $\preceq_{\mathcal{T}(A)}$, we interpret fragments F as (syntax) trees $\mathcal{T}[\![F]\!]$ as follows: an elementary fragment is a single leaf, $\mathcal{T}[\![F_e]\!] := F_e$, a fragment $\nu a.(F_1 \mid \ldots \mid F_n)$ is the tree $\mathcal{T}[\![\nu a.(F_1 \mid \ldots \mid F_n)]\!] := (a, (\mathcal{T}[\![F_1]\!], \ldots, \mathcal{T}[\![F_n]\!]))$. If we assume that the set A contains the sequential processes and the restricted names in F that are not under prefixes, i.e., $\mathcal{S}(F) \cup rn(F) \subseteq A$, then $\mathcal{T}[\![F]\!]$ is a tree over A, $\mathcal{T}[\![F]\!] \in \mathcal{T}(A)$.

If we furthermore assume that A is ordered by the identity, i.e., we consider the qo (A, id), then the rooted tree embedding implies the fragment ordering.

Lemma 5. *Consider the qo (A, id). If $\mathcal{T}[\![F]\!] \preceq_{\mathcal{T}(A)} \mathcal{T}[\![G]\!]$ then $F \preceq_{\mathcal{F}} G$ for all fragments $F, G \in \mathcal{P}_{\mathcal{F}}$.*

To conclude $\preceq_{\mathcal{F}}$ is a wqo from the fact that $\preceq_{\mathcal{T}(A)}$ is a wqo with Lemma 5, (A, id) needs to be a wqo (cf. Lemma 1). This is the case if A is finite. Thus, we need fragments that consist of a finite set of sequential processes and a finite set of restricted names. The idea is to reuse restricted names in parallel compositions, i.e., here we relax the requirement that a name is bound at most once. For every $i \in \mathbb{N}$ we define $con_i : \mathcal{P}_{\mathcal{F}} \to \mathcal{P}_{\mathcal{F}}$ by $con_i(F_e) := F_e$ and $con_i(\nu a.(F_1 \mid \ldots \mid F_n)) := \nu u_i.(con_{i+1}(F_1)\{u_i/a\} \mid \ldots \mid con_{i+1}(F_n)\{u_i/a\})$, where wlog. u_i is fresh for $F_1, \ldots, F_n$. Of course, $F \equiv con_i(F)$ and the restricted names are determined by $nest_\nu(F)$ since $rn(con_i(F)) \subseteq \{u_i, \ldots, u_{i+nest_\nu(F)}\}$.

Example 3. Consider $F_{\mathcal{A}} = \nu a.(\nu b_1.K\lfloor a, b_1\rfloor \mid \nu b_2.L\lfloor a, b_2\rfloor \mid \nu b_3.L\lfloor a, b_3\rfloor)$. We compute $con_0(F_{\mathcal{A}}) = \nu u_0.(\nu u_1.K\lfloor u_0, u_1\rfloor \mid \nu u_1.L\lfloor u_0, u_1\rfloor \mid \nu u_1.L\lfloor u_0, u_1\rfloor)$. ◊

Following the argumentation above, we now build particular anchored fragments $F_{\mathcal{A}}$ that consist of derivatives where the restricted names are changed by con_0.

Lemma 6. *Let $F \in Frag(Reach(P))$ for some $P \in \mathcal{P}$. There is an anchored fragment $F_{\mathcal{A}} \equiv F$ with $rn(F_{\mathcal{A}}) \subseteq \{u_0, \ldots, u_{nest_\nu(F_{\mathcal{A}})}\}$ and $\mathcal{S}(F_{\mathcal{A}}) \subseteq \{Q\sigma \mid Q \in derivatives(P)$ and $\sigma : fn(Q) \to fn(P) \cup \{u_0, \ldots, u_{nest_\nu(F_{\mathcal{A}})}\}\}$.*

Proof. Let $F \in Frag(Reach(P))$. We recalled that $F \equiv \nu\tilde{a}.(Q_1\sigma_1 \mid \ldots \mid Q_n\sigma_n)$ where $Q_i \in derivatives(P)$ and $\sigma_i : fn(Q_i) \to \tilde{a} \cup fn(P)$ in Section 2. We compute the restricted form, $(\nu\tilde{a}.(Q_1\sigma_1 \mid \ldots \mid Q_n\sigma_n))_\nu =: F'$. It is a fragment F' according to $\hat{\equiv}$. For F' we compute $F_{\mathcal{A}}'$ with Lemma 2. We now have $F \equiv F_{\mathcal{A}}'$ and $\mathcal{S}(F_{\mathcal{A}}') \subseteq \{Q\sigma \mid Q \in derivatives(P)$ and $\sigma : fn(Q) \to \tilde{a} \cup fn(P)\}$.

With the function con_0 we change the restricted names: $con_0(F_{\mathcal{A}}') \equiv F_{\mathcal{A}}'$ and $rn(con_0(F_{\mathcal{A}}')) \subseteq \{u_0, \ldots, u_{nest_\nu(F_{\mathcal{A}}')}\}$. The renaming changes the set of sequential processes. They are now derivatives where the substitutions map into $fn(P) \cup \{u_0, \ldots, u_{nest_\nu(F_{\mathcal{A}}')}\}$. Thus, $con_0(F_{\mathcal{A}}')$ satisfies the requirements. □

To see that $\preceq_{\mathcal{F}}$ is a wqo on the reachable fragments of $P \in \mathcal{P}_{BD}$, let k_D be a bound on the depth. We define the set $A := \{u_0, \ldots, u_{2^{k_D}-1}\} \cup \{Q\sigma \mid Q \in derivatives(P)$ and $\sigma : fn(Q) \to fn(P) \cup \{u_0, \ldots, u_{2^{k_D}-1}\}\}$. Obviously, A is finite and thus (A, id) is a wqo.

Let $(F_i)_{i\in\mathbb{N}}$ be a sequence in $Frag(Reach(P))$. Every F_i is structurally congruent with an anchored fragment $F_{\mathcal{A}i}$ in Lemma 6. Corollary 1 yields $nest_\nu(F_{\mathcal{A}i}) \leq 2^{k_D} - 1$. Thus $\mathcal{T}[\![F_{\mathcal{A}i}]\!] \in \mathcal{T}(A)$ with the set A we just defined. The height of $\mathcal{T}[\![F_{\mathcal{A}i}]\!]$ is equal to the nesting of restrictions in $F_{\mathcal{A}i}$. Thus, we have a sequence $(\mathcal{T}[\![F_{\mathcal{A}i}]\!])_{i\in\mathbb{N}}$ of trees in $\mathcal{T}(A)_{2^{k_D}-1}$. According to Lemma 1, $(\mathcal{T}(A)_{2^{k_D}-1}, \preceq_{\mathcal{T}(A)})$ is a wqo and so there are $i < j$ with $\mathcal{T}[\![F_{\mathcal{A}i}]\!] \preceq_{\mathcal{T}(A)} \mathcal{T}[\![F_{\mathcal{A}j}]\!]$. Since A is ordered by the identity, $F_{\mathcal{A}i} \preceq_{\mathcal{F}} F_{\mathcal{A}j}$ with Lemma 5. With Rule (3) we conclude $F_i \preceq_{\mathcal{F}} F_j$. The following lemma holds.

Lemma 7. *Let $P \in \mathcal{P}_{BD}$. Then $(Frag\,(Reach\,(P)), \preceq_{\mathcal{F}})$ is a wqo.*

We define the qo $\preceq_{\mathcal{P}_{BD}}$ on $Reach\,(P)/_{\equiv}$ by $[\Pi_{i\in I}F_i] \preceq_{\mathcal{P}_{BD}} [\Pi_{i\in I}G_i \mid \Pi_{j\in J}G_j]$ if $F_i \preceq_{\mathcal{F}} G_i$ for all $i \in I$. Wqo follows from Lemma 7 and Higman's result.

Proposition 1. *Let $P \in \mathcal{P}_{BD}$. Then $(Reach\,(P)/_{\equiv}, \preceq_{\mathcal{P}_{BD}})$ is a wqo.*

4.2 The Relation $\preceq_{\mathcal{P}_{BD}}$ is a Simulation

In the proof that $\preceq_{\mathcal{P}_{BD}}$ is a simulation, the following Lemma 8 is crucial. It relates the fragment ordering $F \preceq_{\mathcal{F}} G$ with the standard form of F. This standard form is covered by G in a way that reveals $\preceq_{\mathcal{F}}$ is a simulation.

Lemma 8. *For all $F, G \in \mathcal{P}_{\mathcal{F}}$: $F \preceq_{\mathcal{F}} G$ if and only if $F \equiv \nu\tilde{a}.(P_1 \mid \ldots \mid P_n)$ in standard form and $G \equiv \nu\tilde{a}.(P_1 \mid \ldots \mid P_n \mid R)$ for some $R \in \mathcal{P}$.*

Let $[P] = [\Pi_{i\in I}F_i] \preceq_{\mathcal{P}_{BD}} [\Pi_{i\in I}G_i \mid \Pi_{j\in J}G_j] = [Q]$, which means $F_i \preceq_{\mathcal{F}} G_i$ for all $i \in I$. With Lemma 8 we get $\Pi_{i\in I}F_i \equiv \Pi_{i\in I}\nu a_i.(P_{1_i} \mid \ldots \mid P_{n_i})$. We extrude the names νa_i and check that $[P] \to [P']$ implies $[Q] \to [Q']$ with a case distinction. The direction from right to left in Lemma 8 yields $[P'] \preceq_{\mathcal{P}_{BD}} [Q']$.

Proposition 2. *The relation $\preceq_{\mathcal{P}_{BD}}$ is a simulation on $\mathcal{P}/_{\equiv}$.*

With Proposition 1, Proposition 2, and the fact that $\to$ is image finite up to $\equiv$, we conclude that processes of bounded depth have WSTS.

Theorem 2. *Let $P \in \mathcal{P}_{BD}$. Then $(Reach\,(P)/_{\equiv}, \to, \preceq_{\mathcal{P}_{BD}})$ is a WSTS.*

4.3 Decidability of Termination for $\mathcal{P}_{BD}$

The WSTS $(S, \rightsquigarrow, \preceq)$ *has a non-terminating computation from $s_0 \in S$* iff an infinite sequence $s_0 \rightsquigarrow s_1 \rightsquigarrow \ldots$ exists. If $\rightsquigarrow$ is effectively computable and $\preceq$ is decidable the following algorithm decides the termination problem [10, 11].

Let $s_0 \in S$. We construct the *finite reachability tree* $FRT(s_0)$. The root is labelled by s_0. For every node labelled by s in the tree, we create a new node for every successor t of s. We connect the node labelled by s and the new node. If there is a node labelled by s' on the path from the root to the new node with $s' \preceq t$, we label the new node by t_+. Otherwise we label it by t. We do not create successors for nodes t_+. The idea is that t with $s' \preceq t$ can simulate the behaviour of s' and thus repeat $s' \rightsquigarrow \ldots \rightsquigarrow t$.

Proposition 3 ([10, 11]). *A WSTS $(S, \rightsquigarrow, \preceq)$ has a non-terminating computation from $s_0 \in S$ if and only if $FRT(s_0)$ contains a node t_+. As $\preceq$ is a wqo, the tree $FRT(s_0)$ is finite and containment of t_+ is decidable.*

The reaction relation is effectively computable and $\preceq_{\mathcal{P}_{BD}}$ is decidable.

Corollary 2. *For $P \in \mathcal{P}_{BD}$ it is decidable whether there is a non-terminating computation starting from $[P]$.*

Example 4. Let $P_0 = \nu u_0.(\nu u_1.K\lfloor u_0, u_1\rfloor \mid \nu u_1.L\lfloor u_0, u_1\rfloor \mid \nu u_1.L\lfloor u_0, u_1\rfloor)$ with $K(x,y) := K\lfloor x,y\rfloor \mid \nu z.\overline{y}\langle z\rangle$ and $L(x,y) := \overline{x}\langle y\rangle$. Then $FRT([P_0])$ is

$[P_0]$ → $[P_1]_+$
$[P_2]$ → $[P_3]_+$
$[P_4]$ → $[P_5]_+$

$$P_1 = \nu u_0.(\nu u_1.(K\lfloor u_0, u_1\rfloor \mid \nu u_2.\overline{u_1}\langle u_2\rangle) \mid \nu u_1.L\lfloor u_0, u_1\rfloor \mid \nu u_1.L\lfloor u_0, u_1\rfloor)$$
$$P_2 = \nu u_0.(\nu u_1.K\lfloor u_0, u_1\rfloor \mid \nu u_1.\overline{u_0}\langle u_1\rangle \mid \nu u_1.L\lfloor u_0, u_1\rfloor)$$
$$P_3 = \nu u_0.(\nu u_1.(K\lfloor u_0, u_1\rfloor \mid \nu u_2.\overline{u_1}\langle u_2\rangle) \mid \nu u_1.\overline{u_0}\langle u_1\rangle \mid \nu u_1.L\lfloor u_0, u_1\rfloor)$$
$$P_4 = \nu u_0.(\nu u_1.K\lfloor u_0, u_1\rfloor \mid \nu u_1.\overline{u_0}\langle u_1\rangle \mid \nu u_1.\overline{u_0}\langle u_1\rangle)$$
$$P_5 = \nu u_0.(\nu u_1.(K\lfloor u_0, u_1\rfloor \mid \nu u_2.\overline{u_1}\langle u_2\rangle) \mid \nu u_1.\overline{u_0}\langle u_1\rangle \mid \nu u_1.\overline{u_0}\langle u_1\rangle).$$

The root of $FRT([P_0])$ is labelled by $[P_0]$. We have $[P_0] \rightarrow [P_1]$ and $[P_0] \rightarrow [P_2]$. Thus, we insert two new nodes. For the first node, $[P_0] \preceq_{\mathcal{P}_{BD}} [P_1]$ holds, so we label it by $[P_1]_+$. Since $[P_0] \preceq_{\mathcal{P}_{BD}} [P_2]$ does not hold, the second node is labelled by $[P_2]$. With $[P_2] \rightarrow [P_3]$ we construct a new node. As $[P_0] \not\preceq_{\mathcal{P}_{BD}} [P_3]$ but $[P_2] \preceq_{\mathcal{P}_{BD}} [P_3]$, we label it by $[P_3]_+$. The remaining nodes are constructed similarly with $[P_4] \preceq_{\mathcal{P}_{BD}} [P_5]$. The existence of $[P_1]_+$ implies the system has a non-terminating computation from $[P_0]$. ◊

5 Related Work and Conclusion

The interpretation of processes as graphs was proposed in [15, 16] and has been recalled in [17, 18] for the π-Calculus. We related the depth of a process P with a function on the graph $\mathcal{G}[\![P]\!]$. We are not aware of similar results in the literature. The proof required an intricate normal form called anchored fragments.

In [6, 8] decidability of structural congruence relations was investigated. The authors proposed normal forms related with the restricted form in [14]. The standard form of processes is due to [17]. Anchored fragments are more stringent than the normal forms above, and thus reveal more information about the connection structure of process terms.

Finkel generalised the coverability graph procedure for Petri nets to what he called WSTS [10]. He presented algorithms to decide termination and boundedness problems in the general setting. Abdulla et. al. generalised decidability results of temporal properties and simulation relations for lossy channel systems to their notion of WSTS [1]. Both definitions were unified in [11]. This paper is the first to instantiate the WSTS framework for the π-Calculus. Compatibility with the reaction relation required a non-trivial ordering $\preceq_{\mathcal{P}_{BD}}$.

Based on a translation of π-Calculus into multisets, orderings on processes defined by multiset containment relations were studied in [7]. We considered the more intricate wqos, i.e., $\preceq_{\mathcal{P}_{BD}}$ needed to be well-behaved under reaction.

In [19, 5] type systems for the π-Calculus were presented that ensure termination of well-typed processes. We observe that terminating processes are always bounded in depth due to the finite number of reachable processes. Furthermore, our result is more general in that we instantiate the WSTS framework for $\mathcal{P}_{BD}$ and then derive decidability of termination as a corollary. To turn our decidability result into a practical procedure, approximations on $\preceq_{\mathcal{P}_{BD}}$ should be developed to prune the finite reachability tree.

References

1. P. A. Abdulla, K. Čerans, B. Jonsson, and Y.-K. Tsay. Algorithmic analysis of programs with well quasi-ordered domains. *Information and Computation*, 160(1–2):109–127, 2000.
2. R. M. Amadio and C. Meyssonnier. On decidability of the control reachability problem in the asynchronous π-calculus. *Nordic Journal of Computing*, 9(1):70–101, 2002.
3. L. Caires. Behavioural and spatial observations in a logic for the π-Calculus. In *FOSSACS 2004*, volume 2987 of *LNCS*, pages 72–89. Springer-Verlag, 2004.
4. M. Dam. Model checking mobile processes. *Information and Computation*, 129(1):35–51, 1996.
5. Y. Deng and D. Sangiorgi. Ensuring termination by typability. *Information and Computation*, 204(7):1045–1082, 2006.
6. J. Engelfriet and T. Gelsema. Multisets and structural congruence of the pi-calculus with replication. *Theoretical Computer Science*, 211(1–2):311–337, 1999.
7. J. Engelfriet and T. Gelsema. Structural inclusion in the pi-calculus with replication. *Theoretical Computer Science*, 258(1–2):131–168, 2001.
8. J. Engelfriet and T. Gelsema. A new natural structural congruence in the pi-calculus with replication. *Acta Informatica*, 40(6):385–430, 2004.
9. G.-L. Ferrari, S. Gnesi, U. Montanari, and M. Pistore. A model-checking verification environment for mobile processes. *ACM Transactions on Software Engineering and Methodology*, 12(4):440–473, 2003.
10. A. Finkel. Reduction and covering of infinite reachability trees. *Information and Computation*, 89(2):144–179, 1990.
11. A. Finkel and Ph. Schnoebelen. Well-structured transition systems everywhere! *Theoretical Computer Science*, 256(1–2):63–92, 2001.
12. A. Habel. *Hyperedge Replacement: Grammars and Languages*, volume 643 of *LNCS*. Springer-Verlag, 1992.
13. G. Higman. Ordering by divisibility in abstract algebras. *Proc. London Math. Soc.* (3), 2(7):326–336, 1952.
14. R. Meyer. A theory of structural stationarity in the π-Calculus. *Under revision*, 2008.
15. G. Milne and R. Milner. Concurrent processes and their syntax. *JACM*, 26(2):302–321, 1979.
16. R. Milner. Flowgraphs and flow algebras. *JACM*, 26(4):794–818, 1979.
17. R. Milner. *Communicating and Mobile Systems: the π-Calculus.* Cambridge University Press, 1999.
18. D. Sangiorgi and D. Walker. *The π-calculus: a Theory of Mobile Processes.* Cambridge University Press, 2001.
19. N. Yoshida, M. Berger, and K. Honda. Strong normalisation in the π-Calculus. *Information and Computation*, 191(2):145–202, 2004.

A Unified View of Tree Automata and Term Schematisations

Nicolas Peltier

LIG, CNRS
46, avenue Félix Viallet
38031 Grenoble Cedex, France
Nicolas.Peltier@imag.fr

Abstract. We propose an extension of tree automata, called N-automata, which captures some of the features of term schematisation languages, for instance the use of counter variables and parameters. We show that the satisfiability problem is decidable for positive, purely existential, membership formulae which permits to include the proposed formalism into most existing symbolic computation procedures (such as SLD-resolution).

1 Introduction

Formalisms able to handle infinite sets of terms (and manipulate them) are useful in various domains of computer science, for instance for preventing divergence of symbolic computation procedures (such as resolution, superposition, etc.). Among these formalisms, *tree automata* (TA) play a central rôle, mainly due to their nice computational properties [4]: the set of regular term languages (i.e. the languages representable by a tree automaton) is closed under all boolean operations (intersection, union and complement) and the emptiness problem (i.e. the problem of deciding whether a given automaton denotes an empty set of terms) is decidable. TA have many applications, for instance in rewriting [7, 13] or constraint solving [12]. As for word automata, a TA can be defined by a set of states and by a transition function, and the set of recognized terms is specified by a final state. Alternatively, it can be seen as a set of (Horn) monadic clauses satisfying some additional properties, where each predicate corresponds to a state. The recognized language is simply the interpretation of the final predicate in the minimal model of this set of clauses. Using this view, TA can be easily extended by considering non-monadic predicate symbols [1, 9, 10], representing (synchronized) *term tuple languages*.

Other formalisms, called *term schematisations* (TS), have been proposed during the 90's to denote infinite sequences of structurally similar terms. The idea is to denote infinite sequences of terms obtained by starting from a given base term s and by iterating from s a particular "context" $C[\diamond]$, where $\diamond$ is a distinguished subterm in C (denoting a "hole"). If $C[x]$ denotes the term obtained from C by replacing $\diamond$ in C by x, we get the sequence s, $C[s]$, $C[C[s]]$, $\ldots C^n[s]$. For instance, the set of terms $x, g(g(x)), g(g(g(g(x)))), \ldots, g^{2m}(x)$ is

Please use the following format when citing this chapter:

Peltier, N., 2008, in IFIP International Federation for Information Processing, Volume 273; *Fifth IFIP International Conference on Theoretical Computer Science*; Giorgio Ausiello, Juhani Karhumäki, Giancarlo Mauri, Luke Ong; (Boston: Springer), pp. 491–505.

obtained by iterating m times the context $C = g(g(\diamond))$ on the base term x (m denotes an arithmetic variable).

There exist several classes of term schematisation languages corresponding to different classes of contexts: the *recurrent terms* [2] (unique context with only one hole), the *terms with integer exponents* [3] (arbitrary contexts with one hole), the *R-terms* [14] (contexts containing several holes) and the most expressive language of *primal grammars* [8], in which the contexts can depend on the rank in the iteration. Unification is decidable for all these languages.

There are important differences between TA and TS and the representable languages are not comparable. TS allow one to denote terms containing several occurrences of the same (non ground) term, which is not possible using TA[1]. For instance, one can denote using a TS a list of the form $[x, x, x, x, \ldots, x]$, where x is an arbitrary term. The list is obtained by iterating the context $cons(x, \diamond)$ on the base term *nil*. This set of terms cannot be denoted by a TA because this would require an arbitrary number of equality tests. Moreover, TS use arithmetic variables to count the number of iterations in the sequences. This feature can be used for instance to denote the sequence $(f^n(g^n(a)))_{n \in \mathbb{N}}$ which is well known to be non regular, i.e. not representable be a (tree) automaton.

On the other hand, TA can denote many sets of terms that are not representable by a TS. Indeed, a TS cannot denote a term containing an arbitrary number of variables: for instance it is not possible to denote the sequence $a, f(x_1, a), f(x_2, f(x_1, a)), f(x_3, f(x_2, f(x_1, a))), \ldots$, because the variables cannot depend on the rank of the iteration[2]. Moreover, more flexible iterations can be denoted using TA, with non-unique contexts, for instance one can denote the term $f(t_1, f(t_2, f(t_3, \ldots (f(t_n, x)) \ldots)))$ where for all $i \in [1..n]$, $t_i \in \{b, c\}$. Such a term cannot be described using existing TS. More generally, one can denote iterations combining different contexts.

A very natural question arises: is it possible to *unify* these two approaches? The goal is to define a formalism that *combines* all the above features: use of *counter variables*, *indexed* and *non-indexed* variables and *non-unique contexts*. Ideally, it should be *strictly more expressive* than both approaches, and hopefully also more expressive than the union of the two languages, because some "hybrid" terms representable neither by TA nor by TS could be denoted by combining both approaches. Of course, a basic requirement is that both emptiness *and* unification problems should remain decidable.

The present paper is a first answer to this problem. More precisely, we propose a (strict) extension of tree automata, called N**-automata**. We shall prove that this formalism strictly subsumes the terms with integer exponents of [3]. Other, more expressive term schematisation languages are non comparable with N-automata.

---

[1] Some limited equality tests can be safely considered [4].

[2] The formalism of *non flat* primal grammar does offer the possibility of considering "indexed" (or *marked*) variables but unification is decidable only for flat primal grammars, i.e. for primal grammars without indexed variables.

As in [1, 9], we extend TA by adding additional parameters to the states. Some of these parameters are arithmetic variables allowing one to count the number of times the automaton enters some specific states. The other ones denote standard terms, that are allowed to occur several times into the terms recognized by the automaton. They play the same rôle as (non-indexed) variables in TS. The language recognized by the N-automaton depends on the value of the above parameters.

We shall show that the emptiness problem is decidable for N-automata. Moreover, the set of recognized languages is stable under intersection. More generally, we define a notion of N^+*-formulae*, which are positive and purely existential logical formulae combining arithmetic (linear) equality with atoms of the form $p(t_1, \ldots, t_n, s)$, meaning that s occurs in the language recognized by the N-automaton at state p, using $t_1, \ldots, t_n$ as parameters (one can view N^+-formulae as existential, positive, "membership" formulae [5]). With these semantics, we show that the satisfiability problem is decidable by providing an algorithm transforming any (closed) N^+-formula into a purely arithmetic formula. N^+-formulae subsume both emptiness and unification problems. To the best of our knowledge there is no formalism sharing these features[3]. Our results do not follow from the ones in [15] since the iterations we consider cannot be expressed using positive formulae built on equality and subterm ordering, nor from the ones in [11], because the (ground) rewrite rules in [11] are not comparable with the iterations we use in the present paper, and also because the considered problems are different.

Due to space restriction the proofs are not included.

2 Preliminaries

We denote by T_Σ the set of terms constructed as usual on a set of *function symbols* Σ and on a set of *ordinary variables* $\mathcal{X}$ and by T_N the set of arithmetic terms built on the function symbols $0, succ, +$ and on a set of *arithmetic variables* $\mathcal{X}_N$ disjoint from $\mathcal{X}, \Sigma$. As usual the term $succ^n(0)$ is simply denoted by n. A term (arithmetic or standard) is said to be *ground* iff it contains no variable.

We shall consider predicate symbols whose arguments will be either natural numbers or standard terms. Thus, we assume a set of *predicate symbols* Ω is given with a function pr mapping each symbol $p \in \Omega$ to a *profile* $pr(p)$, which is a finite sequence $\tau_1 \times \ldots \times \tau_n$ where n denotes the arity of p and where for every $i \in [1..n]$, τ_i is either `int` (natural numbers) or `t` (standard terms). A predicate is said to be *monadic* if its arity is 1. If O is a subset of Ω then $Atom(O)$ denotes the set of *atoms* of the form $p(t_1, \ldots, t_n)$ where p is a predicate symbol of profile

[3] For instance the languages in [1] or [9] are very expressive, but lack the same decidability results.

$(\tau_1, \ldots, \tau_n)$ and for all $i \in [1..n]$, if $\tau_i = \mathtt{t}$ then $t_i \in T_\Sigma$, and if $\tau_i = \mathtt{int}$ then $t_i \in T_N$.

A *substitution* is a function mapping each ordinary variable in $\mathcal{X}$ to a term in T_Σ and each arithmetic variable in $\mathcal{X}_N$ to an arithmetic term in T_N. As usual a substitution can be extended to a homomorphism of T_Σ, T_N and $Atom(\Omega)$. The image of a term t by a substitution σ is denoted by $t\sigma$. Two terms t, s are said to be *unifiable* iff there exists a substitution σ s.t. $t\sigma = s\sigma$. As usual two unifiable terms have a most general unifier. A substitution is *ground* if for all variables x, $x\sigma$ is ground.

A *rule* is a formula of the form $H_1 \wedge \ldots \wedge H_n \Rightarrow C$, where $H_1, \ldots, H_n, C$ are atoms such that all the variables occurring in $H_1, \ldots, H_n$ also occur in C. C is called the *head* of the rule and $H_1, \ldots, H_n$ are the *premises*. We may have $n = 0$, in this case $H_1 \wedge \ldots \wedge H_n \Rightarrow C$ is to be read as C.

The notions of interpretations, models etc. are defined as usual. It is well known that any set of rules S has a minimal model, denoted by $Mod(S)$.

3 N-Automata

For technical convenience we use a clausal view of tree automata. A tree automaton (in the usual sense) can be seen as a set of Horn monadic clauses. In this section we extend the definition to handle (some classes of) non-monadic predicate symbols.

3.1 Rules and Automata

We assume that the profile of every predicate symbol p (corresponding to a *state*) is of the form $\tau_1 \times \ldots \times \tau_n \times \mathtt{t}$. The last argument can be seen as the term to be recognized by the automaton, and the first ones correspond to parameters. We denote by $A_{\mathtt{int}}(p)$ the set of indices $i \in [1..n]$ s.t. $\tau_i = \mathtt{int}$ and by $A_{\mathtt{t}}(p)$ the set of indices s.t. $\tau_i = \mathtt{t}$.

We associate to every predicate p:

- a unique natural number *level*(p) used to control the "dependencies" between the predicates (recursive calls): if a predicate symbol p depends on another predicate symbol q, then the level of q must be lower or equal to the one of p.
- two disjoint sets $A_c(p) \subseteq A_{\mathtt{int}}(p)$ and $A_=(p) \subseteq A_{\mathtt{t}}(p)$. The elements of $A_c(p)$ are called the *counters* of p. Intuitively, $A_=(p)$ denotes the set of non arithmetic parameters that must be equal to the terms accepted by p (this corresponds to a kind of equality test: at any state p one can test that the

consider term is equal to a non arithmetic parameter, see Condition 1 below) and $A_c(p)$ denotes the arithmetic parameters used by p.

Definition 1. An N-*rule* is a rule $H \Rightarrow p(t_1, \ldots, t_n, s)$ satisfying the following conditions.

1. For all $i \in A_{\mathtt{t}}(p)$, if $i \in A_{=}(p)$ then $t_i = s$, otherwise t_i is a variable occurring only once in the head.
2. For all $i \in A_{\mathtt{int}}(p)$, t_i is either n_i or 0 or $succ(n_i)$, where n_i is a variable occurring only once in the head, and $t_i \neq n_i$ then $i \in A_c(p)$.
3. s is either a term of the form $f(x_1, \ldots, x_k)$ where $x_1, \ldots, x_k$ are distinct variables, or a variable and in this case H is empty.
4. If s is of the form $f(x_1, \ldots, x_k)$ then $H = \bigwedge_{i=1}^{k} q_i(s_1^i, \ldots, s_n^i, x_i)$ where:

   a. For all $i \in [1..k]$, *level*$(q_i) \leq$ *level*(p) and q_i has the same profile as p.
   b. For any $i \in [1..k]$, $j \in A_{\mathtt{int}}(p)$, if $t_j = succ(n_j)$, then $s_j^i = n_j$ or $s_j^i = 0$. Otherwise we have either $s_j^i = t_j$ or $s_j^i = 0$. Moreover, if $s_j^i \neq t_j$ then $j \in A_c(p)$.
   c. For any $i \in [1..k]$, $j \in A_{\mathtt{t}}(p)$, $s_j^i = t_j$.
   d. There exists *at most* one $i \in [1..k]$ s.t. the two following conditions hold: *level*$(q_i) =$ *level*(p) **and** there exists $j \in A_c(q_i)$ s.t. $s_j^i \neq 0$.
   Moreover, for all $j \in A_c(p)$ and for all $l \neq i$, we must have $s_j^l = 0$. A rule containing such a literal is called *inductive* and in this case the literal $q_i(s_1^i, \ldots, s_n^i, x_i)$ satisfying the above conditions is called the *principal* literal[4].
   e. If *level*$(q_i) =$ *level*(p) then $A_c(q_i) = A_c(p)$ and $A_{=}(q_i) = A_{=}(p) = \emptyset$.

In the particular case where $n = 0$, our definition coincides with the standard rules of tree automata (all the predicate symbols have the same level and there is no inductive rule).

Example 1. Here are examples of N-rules:

$$\begin{array}{ll}
p(x, n, m, y_1) \wedge r(x, 0, m, y_2) \wedge p(x, 0, m, y_3) & \Rightarrow p(x, succ(n), m, f(y_1, y_2, y_3)) \\
q(x, 0, m, y) & \Rightarrow p(x, 0, m, g(y)) \\
r(h(y), n, m, y) & \Rightarrow q(h(y), n, m, h(y)) \\
r(x, n, m, y) & \Rightarrow r(x, n, succ(m), i(y)) \\
 & r(x, n, 0, a)
\end{array}$$

We have
level$(p) = 2$, *level*$(q) =$ *level*$(r) = 1$. $A_{\mathtt{int}}(p) = A_{\mathtt{int}}(q) = A_{\mathtt{int}}(r) = \{2, 3\}$, $A_{\mathtt{t}}(p) = A_{\mathtt{t}}(q) = A_{\mathtt{t}}(r) = \{1\}$. $A_c(p) = \{2\}$, $A_c(q) = \emptyset$, $A_c(r) = \{3\}$, $A_{=}(p) = A_{=}(r) = \emptyset$, $A_{=}(q) = \{1\}$. The first and fourth clauses are inductive and the first literal is principal in these clauses.

[4] This condition is the most complex and non-intuitive one. Roughly speaking, it states that the counter variables can only be used along **one** position in the term. The remaining subterms should not depend on the variables in $A_c(p)$.

The reader can check that the minimal model of the above set of rules is the set of terms of the form $p(h(v), n, m, u), q(h(v), n, m, h(v)), r(t, n, m, v)$ where $t \in T_\Sigma$, $n, m \in \mathbb{N}$, $v = i^m(a)$ and $u = f(f(\dots(f(f(g(h(v)), v, g(h(v)), v, g(h(v))), \dots, v, g(h(v))\dots), v, g(h(v)))))$.

On the other hand the following rules are *not* N-rules:

$q(x, n, m, y) \Rightarrow p(g(x), n, m, f(y))$	Condition 1 is violated
$p(x, n, n, y) \Rightarrow p(x, n, n, g(y))$	Condition 2
$p(x, n, m, y) \Rightarrow p(x, n, m, f(y, y))$	Condition 3, $f(y, y)$ non linear
$p(x, n, m, y) \wedge q(x, n, m, y) \Rightarrow p(x, n, m, g(y))$	Condition 4, y occurs twice
$p(a, n, m, y) \Rightarrow p(x, n, m, g(y))$	Condition 4.*c*, a should be x
$p(x, n, m, y_1) \wedge r(x, n, m, y_2) \Rightarrow p(x, s(n), m, j(y_1, y_2))$	Cond. 4.*d*, arg. 2 of r is not 0

A rule is called a *p-rule* if its head is of the form $p(\mathbf{t})$ for some vector of terms $\mathbf{t}$.

Definition 2. An *N-automaton* $\mathcal{A}$ is a pair $(S_\mathcal{A}, \rho_\mathcal{A})$, where $S_\mathcal{A}$ is a set of predicate symbols (of arity > 0) and $\rho_\mathcal{A}$ is a set of N-rules built on the set of predicates $S_\mathcal{A}$ s.t. for every $p \in S_\mathcal{A}$:

- $\rho_\mathcal{A}$ contains at most one inductive p-rule.
- There exists no pair of distinct rules with unifiable heads (in the usual setting this means that the automaton is deterministic).

For any $n+1$-ary predicate symbol $p \in S_\mathcal{A}$, $(t_1, \dots, t_m)$ is said to be a *p-vector* iff $m = n$ and for every $i \in [1..n]$, $t_i \in T_\Sigma$ if $i \in A_\mathbf{t}(p)$ and $t_i \in T_N$ if $i \in A_{\mathtt{int}}(p)$. This implies that $p(t_1, \dots, t_m, s)$ is an atom (where s denotes an arbitrary term in T_Σ).

Definition 3. (Accepted Language) Let $\mathcal{A}$ be an automaton and $p \in S_\mathcal{A}$. For any p-vector $\mathbf{t}$, we denote by $p_\mathcal{A}(\mathbf{t})$ the set of terms s s.t. $Mod(\rho_\mathcal{A}) \models p(\mathbf{t}, s)$. $p_\mathcal{A}(\mathbf{t})$ is *the language recognized by* $\mathcal{A}$ *at state* p *with parameters* $\mathbf{t}$.

Note that by definition, if $s \in p_\mathcal{A}(t_1, \dots, t_n)$ then for every $i \in A_=(p)$, we have $t_i = s$.

We need to introduce some additional notations. Let $\mathcal{A}$ be an N-automaton. We write $p \geq_\mathcal{A} q$ iff there exists a p-rule $H \Rightarrow p(\mathbf{t})$ s.t. q occurs in H. $\geq_\mathcal{A}^*$ denotes the reflexive and transitive closure of $\geq_\mathcal{A}$. An index i is said to be an *inductive counter* for p if there exists a predicate symbol q s.t. $p \geq_\mathcal{A}^* q$ and i is a counter for q. The set of inductive counters of a predicate p is denoted by $IC_\mathcal{A}(p)$.

A natural number i is said to be *active* for a predicate symbol p if $i \in A_c(p) \cup A_=(p)$. It is said to be *inductively active* if there exists a predicate symbol q s.t. $p \geq_\mathcal{A}^* q$ and i is active for q. The set of inductively active arguments of a predicate p is denoted by $IA_\mathcal{A}(p)$. An essential property of $IA_\mathcal{A}(p)$ is that if $i \notin IA_\mathcal{A}(p)$ (i.e. if i is not inductively active for p) then the language $p_\mathcal{A}(\mathbf{t})$ does not depend on the i-th component of the vector $\mathbf{t}$.

Lemma 1. *Let $\mathcal{A}$ be an N-automaton. Let p be a $n+1$-ary predicate symbol in $S_{\mathcal{A}}$ and let $(s_1, \dots, s_n), (s'_1, \dots, s'_n)$ be two p-vectors s.t. for all $i \in [1..n]$, if $s_i \neq s'_i$ then $i \notin IA_{\mathcal{A}}(p)$.*

We have $p_{\mathcal{A}}(s_1, \dots, s_n) = p_{\mathcal{A}}(s'_1, \dots, s'_n)$.

A N-automaton is said to be *normal* iff all its rules are of the form $H \Rightarrow p(\mathbf{t}, f(x_1, \dots, x_k))$ for some function symbol f (with possibly $k = 0$). It is easy to see that any N-automaton can be transformed into an equivalent normal automaton.

Lemma 2. *For any N-automata $\mathcal{A}$ one can construct a normal $\mathcal{A}$-automaton $\mathcal{A}'$ s.t. for all $p \in S_{\mathcal{A}}$ and for all ground p-vectors $\mathbf{t}$: $p_{\mathcal{A}}(\mathbf{t}) = p_{\mathcal{A}'}(\mathbf{t})$.*

3.2 N^+-Formulae

Sometimes N-automata alone are not expressive enough and one has to add conditions on the parameters, in particular arithmetic conditions. Rather than including them into the rules, it is more convenient to put them outside the automaton, yielding the following definition:

Definition 4. The set of N^+*-formulae* for an N-automaton $\mathcal{A}$ is the smallest set of formulae satisfying the following properties:

- *true*, *false* are N^+-formulae.
- Any atom $p(t_1, \dots, t_n)$ in $Atom(S_{\mathcal{A}})$ is an N^+-formula.
- If $t, s \in T_{\Sigma}$ or $t, s \in T_N$ then $t = s$ is an N^+-formula.
- If ϕ, ψ are N^+-formulae, then $\phi \vee \psi$ and $\phi \wedge \psi$ are N^+-formulae.
- If ϕ is an N^+-formula and x is a variable (occurring either in $\mathcal{X}$ or in $\mathcal{X}_N$) then $(\exists x)\phi$ is an N^+-formula.

Definition 5. A ground substitution σ is said to be a *solution* of an N^+-formula ϕ w.r.t. an N-automaton $\mathcal{A}$ iff one of the following condition holds:

- ϕ is $t = s$, $t, s \in T_{\Sigma}$ and $t\sigma = s\sigma$.
- ϕ is $t = s$, $t, s \in T_N$ and $t\sigma$ and $s\sigma$ can be reduced to the same natural number by the usual rules of Presburger arithmetic: $0 + x \to x$ and $succ(x) + y \to succ(x + y)$.
- ϕ is $p(t_1, \dots, t_n, s)$ and $s\sigma \in p_{\mathcal{A}}(t_1\sigma, \dots, t_n\sigma)$.
- ϕ is $\phi_1 \vee \phi_2$ (resp. $\phi_1 \wedge \phi_2$) and σ is a solution of ϕ_1 or ϕ_2 (resp. ϕ_1 and ϕ_2).
- ϕ is $(\exists x)\phi$ and there exists a term t s.t. σ is a solution of $\phi\{x \to t\}$.

We denote by $sol_{\mathcal{A}}(\phi)$ the set of solutions of ϕ w.r.t. $\mathcal{A}$ and we write $\phi \equiv_{\mathcal{A}} \psi$ iff $sol_{\mathcal{A}}(\phi) = sol_{\mathcal{A}}(\psi)$.

3.3 Examples and Comparisons

Example 2. Let $\Sigma = \{a, f, g\}$. Let $\mathcal{A}$ be the N-automaton defined as follows: $S_{\mathcal{A}} \stackrel{\text{def}}{=} \{p, q, r, s\}$, $A_{\texttt{int}}(u) = \{1\}$, for all $u \in S_{\mathcal{A}}$.

$$\rho_{\mathcal{A}} \stackrel{\text{def}}{=} \begin{cases} q(0,x,y) \wedge p(n,x,l) & \Rightarrow p(succ(n), x, cons(y,l)) \\ & \Rightarrow p(0,x,nil) \\ r(0,x,y_1) \wedge s(0,x,y_2) & \Rightarrow q(0,x,f(y_1,y_2)) \\ & \Rightarrow r(0,x,x) \\ & \Rightarrow s(0,x,y) \end{cases}$$

We have $A_c(p) = \{1\}$, $A_c(q) = A_c(r) = A_c(s) = \emptyset$, $A_=(p) = A_=(q) = A_=(s) = \emptyset$, $A_=(r) = 2$. $p_{\mathcal{A}}(n,x)$ denotes the set of terms of the form $\{cons(f(x,y_1), cons(f(x,y_2), \ldots, cons(f(x,y_n), nil) \ldots))\}$ where $x, y_1, \ldots, y_n$ are arbitrary terms, i.e. the lists of the form $[f(x,y_1), \ldots, f(x,y_n)]$. Notice that this set of terms cannot be denoted by a standard tree automaton (due to the several occurrences of x), nor by any known term schematisation for which unification is decidable (due to the indexed variables $y_1, \ldots, y_n$).

The N^+-formula $(\exists m, x)[n = m + m \wedge r(0,x,a) \wedge p(n,x,y)]$ has the following set of solutions: $\{x \to a, n \to 2m, y \to [f(a,y_1), \ldots, f(a,y_{2m})]\}$, where $m \in \mathbb{N}$.

As already seen, N-automata are strict extensions of usual TA. Some of the existing extensions of TA could be included into N-automata, for instance we could add equality or disequality tests between brothers (i.e. between the variables $x_1, \ldots, x_k$ in Definition 1). We did not consider these additional possibilities in the present paper for the sake of simplicity and conciseness. We now compare N-automata and I-terms.

***N*-automata and *I*-Terms** The *terms with integer exponents* (or I-terms [3]) are a particular class of term schematisations. Formally speaking, the set of *I-terms* T_I and the set of *contexts* (terms with one hole) $T_\diamond$ are the least sets that satisfies the following conditions:

- $\mathcal{X} \subseteq T_I$ and $\diamond \in T_\diamond$.
- If $t_1, \ldots, t_n \in T_I^n$, and f is a function of arity n in Σ, then $f(t_1, \ldots, t_n) \in T_I$.
- If $t_1, \ldots, t_{i-1}, t_{i+1}, \ldots, t_n \in T_I^{n-1}$, f is a function of arity n in Σ and $t_i \in T_\diamond$, then $f(t_1, \ldots, t_n) \in T_\diamond$.
- If $t \in T_\diamond$, $t \neq \diamond$, $n \in T_N$ and $s \in T_I$, then $t^n.s \in T_I$.

If t is a term in $T_\diamond$ and $s \in T_I$ then $t[s]$ denotes the term of T_I obtained by replacing $\diamond$ with s, formally defined as follows: $\diamond[s] \stackrel{\text{def}}{=} s$, $f(t_1, \ldots, t_n)[s] \stackrel{\text{def}}{=} f(t_1[s], \ldots, t_n[s])$ and $(t^n.u)[s] \stackrel{\text{def}}{=} t^n.u$. Then the semantics of (ground) I-terms is given by the following rewriting rules: $t^0.s \to s$ and $t^{n+1}.s \to t[t^n.s]$. Using these two rules (and the usual arithmetic rules), any ground I-term t can be transformed into a standard term $t{\downarrow}$. For instance, the I-term $f(x,\diamond)^n.a$ denotes the term $f(x, f(x, \ldots, f(x,a) \ldots))$. The next lemma shows that I-terms can be denoted by N^+-formulae.

Lemma 3. *Let t be an I-term of free variables $x_1, \ldots, x_n$. There exist an N-automaton $\mathcal{A}_t$ and an N^+-formula $\phi_t(y)$ of free variables $x_1, \ldots, x_n, y$ (where y does not occur in $x_1, \ldots, x_n$) s.t. for every ground substitution σ, we have: $\sigma \in sol_{\mathcal{A}_t}(\phi_t(y))$ iff $y\sigma = t\sigma\downarrow$. Moreover, $\phi_t(y)$ contains no equation between non-arithmetic terms.*

Remark 1. Consequently, any unification problem $t = s$ between I-terms can be associated to an N^+-formula $\psi = (\exists x)[\phi_t(x) \wedge \phi_s(x)]$ s.t. $(t\sigma)\downarrow = (s\sigma)\downarrow$ iff σ is a solution of ψ w.r.t. the union of the automata $\mathcal{A}_t$ and $\mathcal{A}_s$.

Example 3. The equation $y = f(g(\diamond), x)^n.a$ is equivalent to the N^+-formula: $p(x, n, y)$ where p is defined by the rules:

$$q(x, n, y_1) \wedge r(x, 0, y_2) \Rightarrow p(x, succ(n), f(y_1, y_2)) \qquad p(x, 0, a)$$
$$p(x, n, y) \Rightarrow q(x, n, g(y)) \qquad r(x, 0, x)$$

We have $A_{\mathtt{int}}(u) = \{2\}$ and $A_{\mathtt{t}}(u) = \{1\}$, for every $u \in \{p, q, r\}$, $A_c(p) = A_c(q) = \{2\}$, $A_c(r) = A_=(p) = A_=(q) = \emptyset$ and $A_=(r) = \{1\}$.

Unfortunately, the previous result does not extend to other more expressive term schematisation languages such as primal grammars. This is mainly due to the possibility of "diagonalisation" i.e. inductive contexts depending on the rank of the iteration, as in the term $f(g^n(x), f(g^{n-1}(\ldots, f(g(x), x))))$. Such a term can be expressed easily by a primal grammar, but it cannot be denoted by an N-automaton. Thus N-automata do not subsume primal grammars and the two formalisms are not comparable.

4 Intersection

In this section, we show how to compute the intersection of two languages denoted by N-automata, which is the first step toward solving N^+-formulae. The obtained language can itself be denoted by an N-automaton. This is more complicated than in the case of standard tree automata, because one has to handle the additional parameters, but the procedure is similar.

Two predicate symbols p, q are said to be *disjoint* in an N-automaton $\mathcal{A}$ if $pr(p) = pr(q)$ and $IC_{\mathcal{A}}(p) \cap IC_{\mathcal{A}}(q) = \emptyset$. We first show how to handle this particular case.

Let $\mathcal{A}$ be an N-automaton. We denote by $S^\star_{\mathcal{A}}$ the set of predicates $[p, q]^I$ where p, q are disjoint symbols of arity $n + 1$ in $S_{\mathcal{A}}$ and $I \subseteq [1..n]$. Intuitively, we will have $[p, q]^I{}_{\mathcal{A}}(\mathbf{t}) = p_{\mathcal{A}}(\mathbf{t}) \cap p_{\mathcal{A}}(\mathbf{t})$, if the I-components of $\mathbf{t}$ are 0 (I is useful mainly to ensure that the level decreases). We construct an automaton $\hat{\mathcal{A}}$ defined on the set of predicate symbols $S_{\mathcal{A}} \cup S^\star_{\mathcal{A}}$ as follows.

We first define: $level([p, q]^I) \stackrel{\text{def}}{=} level(p) + level(q) + arity(p) - |I|$, $pr([p, q]^I) \stackrel{\text{def}}{=} pr(p) = pr(q)$, $A_c([p, q]^I) \stackrel{\text{def}}{=} A_c(p) \cup A_c(q)$ and $A_=([p, q]^I) \stackrel{\text{def}}{=} A_=(p) \cup A_=(q)$.

A substitution θ is said to be a *I-unifier* of two vectors $(t_1, \ldots, t_n)$ and $(s_1, \ldots, s_n)$ iff for every $i \in [1..n]$ we have $t_i\theta = s_i\theta$ and if $i \in I$ then $t_i\theta = s_i\theta = 0$.

We denote by $R_{\mathcal{A}}$ the set of rules of the form $H\theta \wedge H'\theta \rightarrow [p,q]^I(\mathbf{t}\theta)$ s.t. p, q are two $n+1$-ary predicate symbols in $S_{\mathcal{A}}$, I is a subset of $[1..n]$, $H \Rightarrow p(\mathbf{t}), H' \rightarrow q(\mathbf{s})$ are two rules in $\rho_{\mathcal{A}}$ and θ is the most general I-unifier of $\mathbf{t}$ and $\mathbf{s}$.

Lemma 4. *Let $\mathcal{A}$ be an N-automaton. Let $\mathcal{I} = Mod(\rho_{\mathcal{A}})$ and $\mathcal{J} = Mod(R_{\mathcal{A}} \cup \rho_{\mathcal{A}})$. For any pair of disjoint predicate symbols (p,q) of arity n, for every $I \subseteq [1..n]$ for every term s and for every ground p-vector $(t_1, \ldots, t_n)$ we have $\mathcal{J} \models [p,q]^I(t_1, \ldots, t_n, s)$ iff $\forall i \in I, t_i = 0$ and $\mathcal{I} \models p(t_1, \ldots, t_n, s) \wedge q(t_1, \ldots, t_n, s)$.*

In particular, Lemma 4 shows that the language denoted by the predicate $[p,q]^{\emptyset}$ is the intersection of the languages denoted by p and q, which is the desired result, but of course the rules in $R_{\mathcal{A}}$ are not N-rules. In order to transform them into N-rules with the same minimal model, we introduce the following rewrite rules (operating on rules):

Merging: $[p(t_1, \ldots, t_n, x) \wedge q(t_1, \ldots, t_n, x) \wedge H \Rightarrow C] \longrightarrow$
$[p,q]^I(t_1, \ldots, t_n, x) \wedge H \Rightarrow C$
if p, q are disjoint and I is the set of indices $i \in [1..n]$ s.t. $t_i = 0$.
Agreement: $[p(t_1, \ldots, t_n, x) \wedge q(s_1, \ldots, s_n, x) \wedge H \Rightarrow C] \longrightarrow$
$[p(t_1, \ldots, t_{i-1}, s_i, t_{i+1}, \ldots, t_n, x) \wedge q(s_1, \ldots, s_n, x) \wedge H \Rightarrow C]$
if $p \in S_{\mathcal{A}}$ and $i \notin IA_{\mathcal{A}}(p)$.

It is clear that these rules terminate on any set of rules. We denote by $R_{\mathcal{A}}\!\downarrow$ an arbitrarily chosen normal form of $R_{\mathcal{A}}$ w.r.t. the two rules above. The two following lemmata show in some sense the soundness and completeness of the above rules.

Lemma 5. *Let $\mathcal{A}$ be an N-automaton. $Mod(R_{\mathcal{A}} \cup \rho_{\mathcal{A}}) = Mod(R_{\mathcal{A}}\!\downarrow \cup \rho_{\mathcal{A}})$.*

Lemma 6. *Let $\mathcal{A}$ be an N-automaton. $(S^{\star}_{\mathcal{A}} \cup S_{\mathcal{A}}, R_{\mathcal{A}}\!\downarrow \cup \rho_{\mathcal{A}})$ is an N-automaton.*

We take $\hat{\mathcal{A}} \stackrel{\text{def}}{=} (S^{\star}_{\mathcal{A}} \cup S_{\mathcal{A}}, R_{\mathcal{A}}\!\downarrow \cup \rho_{\mathcal{A}})$. By the above lemmata, for every pair of disjoint predicates p, q and for every ground p-vector $\mathbf{s}$, we have $t \in [p,q]^{\emptyset}{}_{\hat{\mathcal{A}}}(\mathbf{s})$ iff $t \in p_{\mathcal{A}}(\mathbf{s}) \cap q_{\mathcal{A}}(\mathbf{s})$.

We need the following:

Lemma 7. *Let $\mathcal{A}$ be an automaton. Let $p \in S_{\mathcal{A}}$ and let $l \in A_{\mathbf{t}}(p)$. For all ground terms $t_1, \ldots, t_n, t$, $Mod(\rho_{\mathcal{A}}) \models p(t_1, \ldots, t_{l-1}, \bot, t_{l+1}, \ldots, t_n, t)$ iff $Mod(\rho_{\mathcal{A}}) \models p(t_1, \ldots, t_n, t)$ and t_l does not occur in t.*

The next lemma handles the more general case of non-disjoint intersection.

Lemma 8. *Let $\mathcal{A}$ be an N-automaton. For any N^+-formula $\phi = p(\mathbf{t}, x) \wedge q(\mathbf{t}', x)$, one can compute an extension $\mathcal{A}'$ of $\mathcal{A}$ and an N^+-formula $\Lambda(\phi)$ of the form $r(\mathbf{s}, x)$ s.t. all the components of $\mathbf{s}$ are components of $\mathbf{t}$ or $\mathbf{t}'$ and $sol_{\mathcal{A}}(\phi) = sol_{\mathcal{A}'}(\Lambda(\phi))$.*

5 Solving N^+-Formulae

In this section, we show (constructively) that there exists an algorithm checking whether a given (closed) N^+-formula has solutions or not. This entails in particular that emptiness problems or unification problems are decidable since they can be easily encoded into N^+-formulae. According to Lemma 3, any equation $t = s$ between terms in T_I (hence also between terms in T_Σ) can be eliminated and replaced by an equivalent N^+-formula ϕ not containing any such equations (see Remark 1). Moreover, we assume, w.l.o.g., that for all non-arithmetic atoms $p(t_1, \dots, t_n)$ occurring in the formula, $t_1, \dots, t_n$ are either variables, or 0 or $\perp$, where $\perp$ is a special constant symbol not occurring in the considered automaton.

5.1 Emptiness Problems

We first consider a particular case. An N^+-formula ϕ of the form $(\exists x)p(t_1, \dots, t_n, x)$ where x does not occur in $t_1, \dots, t_n$ is called an *emptiness problem*. ϕ is said to be *simple* if $A_=(p) = \emptyset$, and for all $i \in A_c(p)$, $t_i = 0$.

If S is a set of rules, and $p(\mathbf{t})$ an atom (where $\mathbf{t}$ denotes a vector of terms), we denote by $S_{[p(\mathbf{t})]}$ the set of rules $(H \Rightarrow p(\mathbf{s}))\theta$ s.t. $H \Rightarrow p(\mathbf{s}) \in S$ and θ is a most general unifier of $\mathbf{t}$ and $\mathbf{s}$. Note that the heads of the rules in $S_{[p(\mathbf{t})]}$ are instances of $p(\mathbf{t})$.

Let $\phi = (\exists x)p(\mathbf{t}, x)$ be an emptiness problem (e.p. for short) and let $\mathcal{A}$ be an N-automaton. We denote by $D_\mathcal{A}(\phi)$ the set of formulae of the form $(\exists \mathbf{z})[\mathbf{t} = \mathbf{s} \wedge \bigwedge_{i=1}^{k} (\exists x_i) q_i(\mathbf{v_i}, x_i)]$, where $\bigwedge_{i=1}^{k} q_i(\mathbf{u_i}, x_i) \Rightarrow p(\mathbf{s}, f(x_1, \dots, x_k))$ is a rule occurring in $\rho_{\mathcal{A}[p(\mathbf{t},x)]}$, $\mathbf{z}$ denotes the variables in $\mathbf{s}$ and $\mathbf{v_i}$ is obtained from $\mathbf{u_i}$ by replacing any occurrence of $f(x_1, \dots, x_k)$ by $\perp$. We denote by $U_\mathcal{A}(\phi)$ the disjunction of all the formulae occurring in $D_\mathcal{A}(\phi)$.

Proposition 1. *For any e.p. ϕ and for every N-automaton $\mathcal{A}$, $\phi \equiv_\mathcal{A} U_\mathcal{A}(\phi)$.*

For any e.p. ϕ, we shall define an equivalent N^+-formula $\Gamma_\mathcal{A}(\phi)$ containing no existential non-arithmetic variable. To this purpose, we need to introduce some additional definitions. If p is a predicate symbol, then we denote by $n(p)$ the (necessarily unique) predicate q s.t. the principal atom of the inductive p-rule is of the form $q(\dots)$ (if there is no inductive p-rule then $n(p)$ is defined arbitrarily), and by $m(p)$ the smallest integer k s.t. there exists l s.t. $n^l(n^k(p)) = n^k(p)$ (k, l exist since the number of predicate symbols is finite).

Let $\mathcal{A}$ be an N-automaton and let $\phi = (\exists x)p(\mathbf{t}, x)$ and $\psi = (\exists y)q(\mathbf{s}, y)$ be two e.p.'s. We write $\phi > \psi$ iff:

- Either *level*(p) > *level*(q), or *level*(p) = *level*(q) and $|var(\mathbf{s})| < |var(\mathbf{t})|$.
- Or *level*(p) = *level*(q), $|var(\mathbf{s})| = |var(\mathbf{t})|$, ϕ is simple and ψ is not.

- Or $level(p) = level(q)$, $|var(\mathbf{s})| = |var(\mathbf{t})|$, neither ϕ nor ψ is simple and $m(p) > m(q)$.

$\Gamma_{\mathcal{A}}(\phi)$ is constructed by induction on the ordering $>$. If ξ is a complex formula then we shall denote by $\Gamma_{\mathcal{A}}(\xi)$ the formula obtained by replacing each e.p. ψ occurring in ξ by $\Gamma_{\mathcal{A}}(\psi)$. Of course this definition makes sense only if $\Gamma_{\mathcal{A}}(\psi)$ has been defined, i.e. if $\phi > \psi$ for every e.p. ψ occurring in ξ.

Let $\phi = (\exists x)p(\mathbf{t}, x)$ where $\mathbf{t} = (t_1, \dots, t_n)$. $\Gamma_{\mathcal{A}}(\phi)$ is defined as follows:

1. **If $A_=(p)$ is non empty**, then $\Gamma_{\mathcal{A}}(\phi) \stackrel{\text{def}}{=} p(t_1, \dots, t_n, t_i)$ where i is an arbitrarily chosen index in $A_=(p)$.
2. **If there exists $j \in A_c(p)$ s.t. $t_j \neq 0$ and if either $\rho_{\mathcal{A}}$ contains no inductive p-rule or $m(p) > 0$**, then $\Gamma_{\mathcal{A}}(\phi) \stackrel{\text{def}}{=} \Gamma_{\mathcal{A}}(U_{\mathcal{A}}(\phi))$.
3. **If for all $j \in A_c(p)$, $t_j = 0$:**
   We denote by E the smallest set of conjunctions of e.p. s.t. $\phi \in E$ and if $\phi \wedge \psi \in E$, ϕ is simple, and $(\exists \mathbf{u})[\mathbf{t} = \mathbf{s} \wedge \gamma]$ is in $D_{\mathcal{A}}(\phi)$ then $\gamma \wedge \psi \in E$.
   E must be finite. Indeed, since the head of the rules contains all the variables, all the variables in E must occur in ϕ, hence the number of possible e.p. is finite (up to equivalence). Thus the number of distinct disjunctions is finite. Let ξ be the disjunction of conjunctions $\psi \in E$ that contain no simple e.p. We define $\Gamma_{\mathcal{A}}(\phi) \stackrel{\text{def}}{=} \Gamma_{\mathcal{A}}(\xi)$.
4. **If $\rho_{\mathcal{A}}$ contains an inductive p-rule and $m(p) = 0$:**
   Let $\{i_1, \dots, i_m\}$ be the elements in $A_c(p)$ s.t. t_{i_j} is a variable. Starting from the formula ϕ we repeatedly replace any e.p. of the form $(\exists x)n^i(p)(\mathbf{v}_i, x)$ (initially we have $i = 0$) by $U_{\mathcal{A}}((\exists x)n^i(p)(\mathbf{v}_i, x_i))$, until we obtain another e.p. of head p (which is possible since $m(p) = 0$). The obtained formula can be reduced (by miniscoping and distributivity) to a formula of the form $\psi \vee [(\exists \mathbf{z}) \bigwedge_{j=1}^m t_{i_j} = succ^{k_j}(z_j) \wedge \gamma \wedge (\exists x)p(\mathbf{v_k}, x)]$, where $k_1, \dots, k_m \in \mathbb{N}$, $z_1, \dots, z_m$ are either 0 or variables not occurring in γ, $\mathbf{z}$ denotes the vector of variables in $z_1, \dots, z_m$ and $\mathbf{v_k}$ is obtained from $\mathbf{t}$ by replacing each component t_{i_j} by z_j.
   We define: $\Gamma_{\mathcal{A}}(\phi) \stackrel{\text{def}}{=} (\exists l, \mathbf{z})[\bigwedge_{j=1}^m (t_{i_j} = l \times k_i + z_j) \wedge \Gamma_{\mathcal{A}}(\gamma) \wedge \Gamma_{\mathcal{A}}(\psi')]$, where ψ' is obtained from ψ by replacing t_{i_j} by z_j ($l \times k_i$ denotes the term $l + l + \ldots + l$ (k_i times)).

Lemma 9. *$\Gamma_{\mathcal{A}}(\phi)$ is well defined, for every emptiness problem ϕ and for every automaton $\mathcal{A}$. Moreover, $\phi \equiv_{\mathcal{A}} \Gamma_{\mathcal{A}}(\phi)$ and the quantified variables in $\Gamma_{\mathcal{A}}(\phi)$ are arithmetic.*

5.2 Reduction to Presburger Arithmetic

By distributivity and miniscoping, any N^+-formula ϕ can be reduced into a formula of the form $\bigvee_{i=1}^n (\exists \mathbf{x_i})\psi_i$ where $\psi_i = \bigwedge_{j=1}^{k_i} \gamma_{ij}$ and where the γ_{ij}'s are atoms. The algorithm is defined by the following rules, applied in the specified order, on the disjuncts ψ_i (and not on the formulae occurring in them). The formulae are normalized after each rule application.

$$
\begin{array}{lll}
(r_0) & (\exists \mathbf{x})\ [\psi \wedge p(\mathbf{t}, \bot)] & \rightarrow \mathit{false} \\
(r_1) & (\exists \mathbf{x})\ [\psi \wedge p(\mathbf{t}, x) \wedge q(\mathbf{t}', x)] & \rightarrow (\exists \mathbf{x})[\psi \wedge \Lambda(p(\mathbf{t}, x) \wedge q(\mathbf{t}', x))] \\
(r_2) & (\exists x_1, \ldots, x_k)\ (\exists \mathbf{n})\ \phi & \rightarrow \bigvee_{i=1}^{k} (\exists x_1, \ldots, x_k)\ \exists \mathbf{n}\ rm_{x_i}(\phi) \\
& \multicolumn{2}{l}{\text{If for all } i \in [1..k] \text{ there exists an atom } p(\mathbf{t}, x_j) \text{ in } \phi \text{ s.t. } j \neq i} \\
& \multicolumn{2}{l}{\text{and } x_i \text{ occurs in } \mathbf{t},\ x_1, \ldots, x_k \in \mathcal{X} \text{ and } \mathbf{n} \text{ is a vector of variables in } \mathcal{X}_N.} \\
& \multicolumn{2}{l}{rm_x(\phi) \text{ is defined below.}} \\
(r_3) & (\exists x_1, \ldots, x_k)\ (\psi \wedge p(\mathbf{t}, x_k)) & \rightarrow (\exists x_1, \ldots, x_{k-1})\ [\psi \wedge p(\mathbf{t}, x_k)\{x_k \rightarrow t_i\}] \\
& \multicolumn{2}{l}{\text{If } x_k \text{ does not occur in } \psi \text{ and } i \text{ is an index in } A_{=}(p) \text{ s.t. } t_i \neq x_k.} \\
(r_4) & (\exists x_1, \ldots, x_k)\ [\psi \wedge p(\mathbf{t}, x_k)] & \rightarrow (\exists x_1, \ldots, x_{k-1})\ [\psi \wedge U_{\mathcal{A}}((\exists x_k)\ p(\mathbf{t}, x_k))] \\
& \multicolumn{2}{l}{\text{If } x_k \text{ occurs in } \mathbf{t} \text{ but not in } \psi.} \\
(r_5) & (\exists x_1, \ldots, x_k)\ [\psi \wedge p(\mathbf{t}, x_k)] & \rightarrow (\exists x_1, \ldots, x_{k-1})\ [\psi \wedge \Gamma_{\mathcal{A}}((\exists x_k) p(\mathbf{t}, x_k))] \\
& \multicolumn{2}{l}{\text{If } A_{=}(p) = \emptyset,\ x_k \text{ does not occur in } \mathbf{t} \text{ nor in } \psi.}
\end{array}
$$

$rm_x(\phi)$ is defined by the following (auxiliary) rule:

$$
\begin{array}{ll}
(rm_x)\ (\exists x_1, \ldots, x_n, y)\ (p(\mathbf{t}, y) \wedge \psi) \rightarrow & (\exists x_1, \ldots, x_n)\ (p(\mathbf{t}, y) \wedge \psi)\{y \rightarrow x\} \\
& \vee (\exists x_1, \ldots, x_n, y)\ (p(\mathbf{t}\{x \rightarrow \bot\}, y) \wedge \psi)
\end{array}
$$

If x occurs in $\mathbf{t}$ and $y \neq x$.

Lemma 10. *Let ϕ be a N^+-formula. The rules $r_0, \ldots, r_5$ (with the above strategy) terminate on ϕ and preserve equivalence. Moreover, any irreducible formula is purely arithmetic.*

Since Presburger arithmetic is known to be decidable, Lemma 10 provides an algorithm for checking whether a given closed N^+-formula is satisfiable or not.

6 A Simple Application

We show a simple example of application in the context of Logic Programming. If the satisfiability problem is decidable for N^+-formulae (as shown by Lemma 10), then N-automata can be integrated in Logic Programs. The corresponding unification problems can be solved by our constraint solving algorithm.

Assume that one wants to define a predicate $last(l, x)$ which is true iff x is the last element of the list l. Using standard Horn clauses, $last(l, x)$ is defined as follows: $\{last(cons(x, nil), x), last(cons(y, l), x) \Leftarrow last(l, x)\}$.

Using N-automata we obtain: $last(l, x) \Leftarrow p(x, l)$, where p is defined by the following N-rules:

$$\{q(x, y) \Rightarrow p(x, cons(y, nil)), p(x, l) \Rightarrow p(x, cons(y, l)), q(x, x)\}.$$

Both techniques yield exactly the same result (from a semantic point of view). But the use of N-automata allows one to compute the solution of a request $last(l, x)$ in a symbolic way, rather than enumerating all possible lists.

Thus a request of the form $last(l, 1) \wedge last(l, 2)$ *diverges* with the first approach and simply *terminates* and *fails* using our technique. Of course, the programmer does not need to write the N-automaton explicitly: it could be compiled automatically from the set of Horn clauses (in case they are of the required form).

7 Conclusion

We have presented a framework unifying tree automata [4] with some term schematisation languages [3]. By combining the features of both approaches, we obtained a formalism which is strictly more expressive than the original ones. We provided an algorithm to check the satisfiability of positive, purely existential membership formulae, which allows one to include N-automaton into most existing symbolic computation procedures (such as SLD-resolution in Logic Programming). Our work extends the power of tree automata by showing how to include integer counters and parameters. It also strictly enhances the expressive power of term schematisations by using more general contexts.

Future works include the extension of the presented approach in order to capture more expressive term schematisation languages such as R-terms or primal grammars, and the extension of the class of considered formulae in order to handle formulae with negations and universal quantifiers. In the context of tree automata this corresponds to the complement problem (i.e. compute the complement of the set of terms recognized by a tree automaton) and in the context of term schematisations, this corresponds to disunification problems [6].

References

1. J. Chabin, J. Chen, and P. Réty. Synchronized-context free tree-tuple languages. Technical Report RR-2006-13, LIFO, 4 rue Léonard de Vinci, BP 6759, F-45067 Orléans Cedex 2 FRANCE, 2006.
2. H. Chen and J. Hsiang. Logic programming with recurrence domains. In *Automata, Languages and Programming (ICALP'91)*, pages 20–34. Springer, LNCS 510, 1991.
3. H. Comon. On unification of terms with integer exponents. *Mathematical System Theory*, 28:67–88, 1995.
4. H. Comon, M. Dauchet, R. Gilleron, F. Jacquemard, D. Lugiez, S. Tison, and M. Tommasi. Tree automata techniques and applications. Available on: `http://www.grappa.univ-lille3.fr/tata`, 1997.
5. H. Comon and C. Delor. Equational formulae with membership constraints. *Information and Computation*, 112(2):167–216, August 1994.
6. H. Comon and P. Lescanne. Equational problems and disunification. *Journal of Symbolic Computation*, 7:371–475, 1989.

7. G. Feuillade, T. Genet, and V. Viet Triem Tong. Reachability Analysis over Term Rewriting Systems. *J. Automated Reasoning*, 33 (3-4):341–383, 2004.
8. M. Hermann and R. Galbavý. Unification of Infinite Sets of Terms schematized by Primal Grammars. *Theoretical Computer Science*, 176(1–2):111–158, 1997.
9. S. Limet and G. Salzer. Manipulating tree tuple languages by transforming logic programs. *Electr. Notes Theor. Comput. Sci.*, 86(1), 2003.
10. S. Limet and G. Salzer. Proving properties of term rewrite systems via logic programs. In *RTA*, volume 3091 of *Lecture Notes in Computer Science*, pages 170–184. Springer, 2004.
11. C. Löding. Model-checking infinite systems generated by ground tree rewriting. In M. Nielsen and U. Engberg, editors, *Foundations of Software Science and Computation Structures, 5th International Conference, FOSSACS 2002.*, volume 2303 of *Lecture Notes in Computer Science*, pages 280–294. Springer, 2002.
12. D. Lugiez and J. L. Moysset. Tree automata help one to solve equational formulae in AC-theories. *Journal of Symbolic Computation*, 18(4):297–318, 1994.
13. P. Réty and J. Vuotto. Tree automata for rewrite strategies. *J. Symb. Comput.*, 40(1): 749–794, 2005.
14. G. Salzer. The unification of infinite sets of terms and its applications. In *Logic Programming and Automated Reasoning (LPAR'92)*, pages 409–429. Springer, LNAI 624, July 1992.
15. K. N. Venkataraman. Decidability of the purely existential fragment of the theory of term algebras. *J. ACM*, 34(2):492–510, 1987.

Deconstructing behavioural theories of mobility*

Julian Rathke and Paweł Sobociński

ECS, University of Southampton

Abstract. We re-examine the standard structural operational semantics of the π-calculus with the view that both process structure and contextual observational power should play roles in describing the behavioural theory. To that end we provide a decomposition of the operational semantics of π which allows for a systematic definition of labelled transitions. These are derived from the calculus' underlying reduction rules by following the contexts-as-labels philosophy while being presented using the structural approach. Our novel transition system refines to a composite description of the standard early LTS. We generalise our technique to higher-order and asynchronous variants.

Introduction

The π-calculus [6,14] is a foundational model for the study of mobile processes. It has become one of the most well-known and widely studied process calculi and extensions of it are now beginning to be used in a variety of application areas. Each of these applications is typically based on specialising the π-calculus to the particular domain; usually by extending one or more features of the language. However, a weak point of this approach is that with each change, the behavioural theory of the language must be reworked in order to accommodate the new language features. This can be a non-trivial task and often leads to ad hoc solutions based upon a tailor-made LTS. This is an undesirable situation and our goal is thus to develop methods by which suitable labelled transition systems for general process languages can be systematically defined based upon *both* structural rules [15] *and* contextual observable power [12]. This will be difficult to achieve in general but in this paper we take our first step by showing our approach to the problem as it applies to some π-calculus variants.

It may seem churlish of us to re-examine the well-established and finely crafted labelled transition semantics of the π-calculus [14] but we believe a deconstructive reading of these will deepen the understanding of LTSs more generally. In particular, we challenge the notion that structure is paramount in giving operational semantics (cf. Plotkin's SOS [15]) and argue that if observational power is also taken in to consideration when building labelled transition systems then the transition systems can be defined in a systematic manner such that the resulting bisimulation equivalences are better suited for characterising

* Research partially supported by EPSRC grant EP/D066565/1.

Please use the following format when citing this chapter:

Rathke, J. and Sobociński, P., 2008, in IFIP International Federation for Information Processing, Volume 273; *Fifth IFIP International Conference on Theoretical Computer Science*; Giorgio Ausiello, Juhani Karhumäki, Giancarlo Mauri, Luke Ong; (Boston: Springer), pp. 507–520.

contextually defined equivalences. In this paper we support this argument by first considering the π-calculus without matching.

Traditional presentations of the LTS of the π-calculus emphasise the structural approach while neglecting observational power. To exemplify this, consider the standard early labelled transition rule for output actions:

$$a!b.P \xrightarrow{a!b} P$$

From a purely structural point of view this rule is perfectly sensible. But, if we consider the underlying reduction rule of π-calculus:

$$a!b.P \parallel a?x.Q \rightarrow P \parallel Q[b/x] \qquad (1)$$

we see that the specific name b is parametric in this rule and does not genuinely play a structural role in the transition labelled $a!b$ – any context that provides an input on a enables the passing of any other name. Indeed, in a π-calculus without name-equality tests there actually is no context that justifies the observation of the specific name b being communicated and the standard 'structural' rule is inappropriate. This makes it untenable for us to accept the standard label above as a canonical labelled transition of the π-calculus. However, we can resolve this by focusing on the structure of the underlying reduction rule: we decompose the rule in to structure provided by the process and parameters to the rule provided by the context. In the case for output, the former of these provides a (partial) labelled transition of the form:

$$a!b.P \xrightarrow{a!} \lambda X.(P \parallel X(b))$$

in which the communication, but no further non-structural information, is represented. Here, the contribution from the context (process Q) is abstracted away. The complete labelled transition is then obtained by allowing the context to supply the missing parameters by applying the resulting abstraction above to an arbitrary process. Doing this may or may not subsequently ascertain the identity of b, depending on the power of the contexts of the language. Such output transitions do not rely on subtle observational powers regarding matching – they are generated from the reduction rule alone. A similar approach can be taken for input transitions and thus an entire LTS can be derived systematically thereby avoiding ad hoc solutions.

The focus on the relationship of an LTS with an underlying reduction system is clearly shared with previous work on systematically *deriving* transition systems from reductions [11, 12, 20, 21]. Indeed, the labels of our derived LTS have corresponding contexts which justify them from the point of view of the contexts-as-labels approach. This does not hold for the standard early LTS as for instance there is no context that accounts for the bound-output label.

In the above-mentioned approaches, labels are defined to be contexts which trigger reduction. However, while tied closely to observability, they suffer from a lack of a structural presentation. Our LTS enjoys the best of both approaches as it is presented structurally yet satisfies the contexts-as-labels criteria.

Related work We use a meta-syntax based on the simply-typed λ-calculus in order to denote terms that have a context-component as a result of an interaction. The technically related approaches in the literature include [4, 22] which use variants of the λ-calculus as a metasyntax. The difference in approaches arises from different underlying goals: the aforementioned works use the meta-syntax to study systems of SOS rules a posteriori, we are interested in defining new LTSs.

Milner's [13] approach to capturing the *late* semantics of π using abstractions and concretions is closer to our approach in spirit. Abstractions and concretions are syntactic entities that arise as a result of the complementary roles of inputs and outputs in π. In contrast to Milner, we do not need to define a special notion of application and substitution for abstractions and concretions because our 'concretions' retain structural information from the reduction rules which enables us to use the standard notion of capture-avoiding substitution.

Structure In §1 we introduce our base-calculus, a typed version of π without choice or matching, together with the meta-syntax for expressing partial interactions. In §2 we give the SOS rules which define our LTS and show that ordinary bisimilarity agrees with contextual equivalence. In §3 we show that the SOS methodology can be used also to define the standard early LTS. In §4 we show that the technique naturally generalises to higher-order and asynchronous settings.

1 A simply-typed π-calculus (without matching)

Here we revisit the syntax, structural congruence and the reduction semantics of π in a typed setting. We study a core language without choice and without name equality testing as this serves to demonstrate our point well.

The syntax, the types and the axioms of structural congruence are given in *Fig.* 1. A significant departure from the π-calculus is the inclusion of simply-typed λ-terms and applications in the language. These features are not to be considered as an extension of π-calculus, rather as a *meta-language*. λ-bindings are used as meta-syntax for manipulating terms.

The type system, presented in the upper section of *Fig.* 2 is very simple: there are three base types; a name type Nm and a process type Pr and a unit type $\mathbf{1}$. Our type system is used only in order to formalise the meta-syntax and is simpler than usual π-calculus type systems based on channel types [18, Part III]. We will consider only typeable terms.

A type context consist of a finite set of names Δ together with a finite map Γ from variables to types. We use the notation $\Gamma, x : \sigma$ to mean the context Γ extended with the mapping $x \mapsto \sigma$; implicitly it is always assumed that x is not

$$\sigma \ ::= \ \mathsf{Nm} \mid \mathsf{Pr} \mid \mathbf{1} \mid \sigma \to \sigma$$

$$M ::= x \mid a \mid 0 \mid M \| M \mid M!MM \mid M?xM \mid \nu aM \mid \mathrm{rp}(M) \mid 1 \mid \lambda x{:}\sigma.M \mid M(M)$$

$$(P\|Q)\|R \equiv P\|(Q\|R) \qquad P\|Q \equiv Q\|P \qquad P\|0 \equiv P$$

$$\nu a \nu b P \equiv \nu b \nu a P \qquad \nu a 0 \equiv 0 \qquad \nu a(P\|Q) \equiv P\|\nu a Q \ (a \notin P) \qquad \nu a P \equiv \nu b P[b/a] \ (b \notin P)$$

$$\mathrm{rp}(P) \equiv P\|\mathrm{rp}(P) \qquad \mathrm{rp}(P\|Q) \equiv \mathrm{rp}(P)\|\mathrm{rp}(Q) \qquad \mathrm{rp}(0) \equiv 0$$

$$k?xP \equiv k?yP[y/x] \ (y \notin fr(P)) \qquad (\lambda x{:}\sigma.M)(N) \equiv M[N/x] \qquad \lambda x{:}\sigma.M \equiv \lambda y{:}\sigma.M[y/x] \ (y \notin fr(M))$$

Fig. 1 Types, syntax and structural congruence.

$$\frac{a \in \Delta}{\Delta;\Gamma \vdash a{:}\mathsf{Nm}}\ (\text{:Name}) \qquad \frac{\Gamma(x)=\sigma}{\Delta;\Gamma \vdash x{:}\sigma}\ (\text{:Var}) \qquad \frac{}{\Delta;\Gamma \vdash 0{:}\mathsf{Pr}}\ (\text{:Null}) \qquad \frac{}{\Delta;\Gamma \vdash 1{:}1}\ (\text{:Unit})$$

$$\frac{\Delta;\Gamma \vdash k{:}\mathsf{Nm} \quad \Delta;\Gamma \vdash l{:}\mathsf{Nm} \quad \Delta;\Gamma \vdash P{:}\mathsf{Pr}}{\Delta;\Gamma \vdash k!lP{:}\mathsf{Pr}}\ (\text{:OutPref}) \qquad \frac{\Delta;\Gamma \vdash k{:}\mathsf{Nm} \quad \Delta;\Gamma,x \vdash P{:}\mathsf{Pr}}{\Delta;\Gamma \vdash k?xP{:}\mathsf{Pr}}\ (\text{:InPref})$$

$$\frac{\Delta,a;\Gamma \vdash P{:}\mathsf{Pr}}{\Delta;\Gamma \vdash \nu aP{:}\mathsf{Pr}}\ (\text{:Nu}) \qquad \frac{\Delta;\Gamma \vdash P{:}\mathsf{Pr} \quad \Delta;\Gamma \vdash Q{:}\mathsf{Pr}}{\Delta;\Gamma \vdash P\|Q{:}\mathsf{Pr}}\ (\text{:Par}) \qquad \frac{\Delta;\Gamma \vdash P{:}\mathsf{Pr}}{\Delta;\Gamma \vdash \mathrm{rp}(P){:}\mathsf{Pr}}\ (\text{:Rep})$$

$$\frac{\Delta;\Gamma,X{:}\sigma \vdash M{:}\sigma'}{\Delta;\Gamma \vdash \lambda X{:}\sigma.M{:}\sigma \to \sigma'}\ (:\lambda) \qquad \frac{\Delta;\Gamma \vdash M{:}\sigma \to \sigma' \quad \Delta;\Gamma \vdash N{:}\sigma}{\Delta;\Gamma \vdash M(N){:}\sigma'}\ (\text{:App})$$

Fig. 2 Type rules of first-order π.

already in the domain of Γ. Similarly, for names, $\Delta, a = \Delta + \{a\}$. We assume a countable supply of variables of each type in addition to a separate countable supply of name constants. We shall use the syntactic convention of a, b for name constants, k, l for terms of name type (either constants or variables), x, y for variables of name type, X, Y for variables generally, P, Q for terms of process type and M, N for general terms.

A *closed* term V is a typeable term that does not contain free variables – *i.e.* there exist Δ, σ such that $\Delta;\varnothing \vdash V{:}\sigma$ (we will often write just $\Delta \vdash V{:}\sigma$).

Structural congruence $\equiv$ is the smallest relation over terms that satisfies the axioms and is closed under all the syntactic features of the calculus: the output prefix, the input binder, the ν binder, the λ-binder and the parallel composition. Exhibiting the non-computational role of the meta-language, β-reduction is part of structural congruence. Our language contains three binders. Substitution within the β-rule is the usual capture-avoiding notion *with respect to all three.* Structurally congruent terms have the same type.

Definition 1 (Indexed transition system) An indexed transition system $\mathcal{T}$ has states comprising pairs of a set of names Δ and a closed term V which has its free names in Δ; *i.e.* the states are contained in the set $\{(\Delta, V) \mid \exists\sigma.\ \Delta \vdash V{:}\sigma\}$. We shall use the notation $\langle \Delta \rhd V \rangle$ to refer to a state. Our transition systems are presented in the structural style. We make one non-standard assumption: we work with abstract syntax and thus assume the implicit presence of the rule:

$$\frac{P' \equiv P \quad \langle \Delta \rhd P \rangle \xrightarrow{\alpha} \langle \Delta' \rhd Q \rangle \quad Q \equiv Q'}{\langle \Delta \rhd P' \rangle \xrightarrow{\alpha} \langle \Delta' \rhd Q' \rangle} \text{(StrCng)}.$$

The choice of including the rule (StrCng) in our transition system for the full structural congruence is a technical convenience rather than necessity and greatly reduces the number of rules required. This allows us to concentrate on the more interesting cases and not on the rather standard "structural" rules. The price is that proofs based on structural induction over terms become less straightforward. The bare minimum we require in (StrCng) is a congruence which contains the axioms for alpha- and beta-equivalence. To implement our LTS in the absence of the full $\equiv$ relation one would need to include symmetric rules for parallel composition and suitable versions of the exisiting rules for replicated processes. Notably, though, no extra rules for scoping are required as no scope extrusion is performed in our LTS.

Our first transition system is the reduction semantics for the π-calculus: rule

$$\frac{}{\langle \Delta \rhd a!bP \| a?xQ \rangle \rightarrow \langle \Delta \rhd P \| Q[b/x] \rangle}$$

and rules which close the relation under parallel composition and the ν binder. Subject reduction is easily shown using a straightforward induction on the derivation of the transition.

We can give an alternative definition of the reduction relation as the reduction relation of a *reactive system* [11, 12]; this style of definition makes the parametric nature of π-reductions more explicit.

We shall first need to define a general notion of *context*. Contexts are constructed in two stages. Firstly, for each type σ, we add *σ-typed holes* $-_\sigma$ and n-tuples (for any $n \in \mathbb{N}$) to the syntax, together with two additional type rules, given below. We refer to a term $(V_1, \ldots, V_n)$ as a *pre-context.*

$$M ::= \ldots \mid -_\sigma \mid (M,\ldots,M) \qquad \frac{}{\Delta;\Gamma \vdash -_\sigma{:}\sigma} \text{(:Hole)} \qquad \frac{\Delta \vdash V_1{:}\sigma_1 \ \ldots\ \Delta \vdash V_n{:}\sigma_n \ (n \in \mathbb{N})}{\Delta \vdash (V_1,\ldots,V_n){:}[\sigma_1 \ldots \sigma_n]} \text{(:Tup)}.$$

Secondly, given a pre-context of type $[\sigma_1 \ldots \sigma_n]$ which contains m holes, we replace each hole symbol with a unique integer from 1 to m. Such a numbering uniquely determines a word over the set of types; the ith-letter being the type σ'_i of the ith-numbered hole. We say that the resulting tuple $(V'_1, \ldots, V'_n)$ is a $[\sigma'_1 \ldots \sigma'_m] \rightarrow [\sigma_1 \ldots \sigma_n]$ *context.* Each hole appears exactly once, thus the contexts are linear.

Note that ordinary closed terms of type σ are in 1-1 correspondence (and can be identified with) the $[] \rightarrow [\sigma]$ contexts. A *π-context* is a $[\mathsf{Pr}] \rightarrow [\mathsf{Pr}]$ context that does not contain elements of the meta-language – *i.e.* does not contain λ-terms or applications. An *evaluation context* is a π-context in which the process hole does not appear within the scope of a prefix. Substitution is syntactic: given a $[\overrightarrow{\sigma_1}] \rightarrow [\overrightarrow{\sigma_2}]$ context g and a $[\overrightarrow{\sigma_2}] \rightarrow [\overrightarrow{\sigma_3}]$ context f, $f \circ g$ is the $[\sigma_1] \rightarrow [\sigma_3]$ context obtained by substituting the ith component of g for the ith hole of

f. Context substitution may involve free names of g being captured by name restrictions of f. The input-binder cannot capture because, by construction, contexts do not contain free variables. A λ-term of type $\sigma' \to \sigma$ is inherently different from a context $[\sigma'] \to [\sigma]$, since the former is possibly non-linear and in the latter the substitution is not capture-avoiding.

In order to consider π as a reactive system, we start with a set of name-indexed reduction rules: pairs l_a, r_a of $[\mathsf{Nm}, \mathsf{Pr}, \mathsf{Nm}{\to}\mathsf{Pr}]{\to}[\mathsf{Pr}]$ contexts defined

$$l_a \stackrel{\text{def}}{=} a!1_{\mathsf{Nm}}.2_{\mathsf{Pr}} \parallel a?y.3_{\mathsf{Nm}\to\mathsf{Pr}}(y) \quad \text{and} \quad r_a \stackrel{\text{def}}{=} 2_{\mathsf{Pr}} \parallel 3_{\mathsf{Nm}\to\mathsf{Pr}}(1_{\mathsf{Nm}}).$$

We construct the reduction relation as follows: $\langle\, \Delta \rhd P \,\rangle \to \langle\, \Delta \rhd P' \,\rangle$ iff there exist a name a, a $[] \to [\mathsf{Nm}, \mathsf{Pr}\,\mathsf{Nm} \to \mathsf{Pr}]$ context p (the parameters) and an evaluation context d such that $P \equiv d \circ l_a \circ p$ and $P' \equiv d \circ r_a \circ p$. The transition system defined is the same relation as given by the inductive, structural presentation.

The reduction semantics naturally leads to a notion of contextually-defined equivalence, the barb-congruence; defined here in the dynamic style [9]. Although the results of [19] suggest that this rendering of contextual equivalence does not coincides with that in [18] say, it is useful to point out that the results in [19] depend crucially upon the blurring of the distinction between names and variables and hence, as suggested in [7], we believe the two approaches to barbed congruence to coincide in our setting.

We use the notion of strong barb: the ability to immediately input or output on a particular channel a, denoted $\downarrow_a$. The natural notion of equivalence relation on states of a typed transition system is an indexed relation.

Definition 2 (Indexed relation) A name-indexed relation R is a set of triples (Δ, P, Q) where Δ is a finite set of names and P and Q are closed terms typeable in Δ ($\Delta;\varnothing \vdash P,Q{:}\sigma$). We write $PR_\Delta Q$ for $(\Delta, P, Q) \in R$.

Definition 3 (Reduction barbed congruence) *Barb-congruence, denoted* $\simeq$, *is the largest symmetric relation that, for any* $P \simeq_\Delta Q$ *is:*

- *(i)* Reduction-preserving: *if* $\langle\, \Delta \rhd P \,\rangle \to \langle\, \Delta \rhd P' \,\rangle$ *then there exists a reduction* $\langle\, \Delta \rhd Q \,\rangle \to \langle\, \Delta \rhd Q' \,\rangle$ *such that* $P' \simeq_\Delta Q'$;
- *(ii)* Barb-preserving: *if* $\langle\, \Delta \rhd P \,\rangle \downarrow_a$ *then* $\langle\, \Delta \rhd Q \,\rangle \downarrow_a$;
- *(iii)* A congruence: *for all* π*-contexts* $\Delta' \vdash C{:}[\mathsf{Pr}]{\to}[\mathsf{Pr}]$ *we have* $C \circ P \simeq_{\Delta \cup \Delta'} C \circ Q$.

2 A structured LTS

In this section we shall describe our approach to endowing the π-calculus with an LTS. We split a labelled transition which corresponds to an interaction of a term with a context into a process-view of the interaction (*Fig.* 3) and a context-view (*Fig.* 4). The complete LTS is obtained by combining the two (*Fig.* 5). We emphasise that this LTS is obtained systematically from the underlying

$$\frac{}{\langle\, \Delta \rhd a?xP \,\rangle \xrightarrow{a?} \langle\, \Delta \rhd \lambda x{:}\mathsf{Nm}.\, P \,\rangle}\ (\textsc{In}) \qquad \frac{}{\langle\, \Delta \rhd a!MP \,\rangle \xrightarrow{a!} \langle\, \Delta \rhd \lambda X{:}\mathsf{Nm}{\to}\mathsf{Pr}.\, P \| X(M) \,\rangle}\ (\textsc{Out})$$

$$\frac{\langle\, \Delta \rhd P \,\rangle \xrightarrow{a!} \langle\, \Delta \rhd T \,\rangle \quad \langle\, \Delta \rhd Q \,\rangle \xrightarrow{a?} \langle\, \Delta \rhd U \,\rangle}{\langle\, \Delta \rhd P \| Q \,\rangle \xrightarrow{\tau} \langle\, \Delta \rhd \lambda{\cdot}{:}1.\, T(U) \,\rangle}\ (\textsc{Tau})$$

$$\frac{\langle\, \Delta \rhd P \,\rangle \xrightarrow{\alpha} \langle\, \Delta \rhd V \,\rangle}{\langle\, \Delta \rhd P \| Q \,\rangle \xrightarrow{\alpha} \langle\, \Delta \rhd \lambda X.\, V(X) \| Q \,\rangle}\ (\textsc{Par}) \qquad \frac{\langle\, \Delta, a \rhd P \,\rangle \xrightarrow{\alpha} \langle\, \Delta, a \rhd V \,\rangle \quad a \notin \alpha}{\langle\, \Delta \rhd \nu a P \,\rangle \xrightarrow{\alpha} \langle\, \Delta \rhd \lambda X.\, \nu a V(X) \,\rangle}\ (\textsc{Res})$$

Fig. 3 Process-view fragment ($\mathcal{C}$).

$$\frac{\Delta \subseteq \Delta' \quad \Delta' \vdash N{:}\sigma}{\langle\, \Delta \rhd \lambda x{:}\sigma.M \,\rangle \xrightarrow{N} \langle\, \Delta' \rhd (\lambda x{:}\sigma.M)(N) \,\rangle}\ (\textsc{Inst})$$

Fig. 4 Canonical context actions ($\mathcal{A}$).

$$\frac{\langle\, \Delta \rhd P \,\rangle \xrightarrow{\alpha}_{\mathcal{C}} \langle\, \Delta \rhd V \,\rangle \quad \langle\, \Delta \rhd V \,\rangle \xrightarrow{\beta}_{\mathcal{A}} \langle\, \Delta' \rhd P' \,\rangle}{\langle\, \Delta \rhd P \,\rangle \xrightarrow{\alpha\beta} \langle\, \Delta' \rhd P' \,\rangle}\ (\textsc{Comb})$$

Fig. 5 Combined system of complete actions ($\mathcal{CA}$).

reduction rule of the π-calculus and, as such, may appear less elegant than an optimised or ad hoc system for the same language.

We begin with the process-view LTS $\mathcal{C}$, with its structural rules given in *Fig. 3*. Here and henceforward we shall use the syntactic convention of writing T for terms of type $(\mathsf{Nm}{\to}\mathsf{Pr}){\to}\mathsf{Pr}$ and U for terms of type $\mathsf{Nm}{\to}\mathsf{Pr}$ and omit the types of variables where they are clear from context. First, we focus on output transitions: a process $a!MP$ offering an output on a channel a, matches a subterm of the source of the single π-calculus reduction rule. Thus, in some context, it can engage in an interaction to evolve into (according to the target of the same reduction rule) a process consisting of its continuation P, in parallel with the continuation of the interacting context, Q, say. Furthermore, Q, has been passed the communicated name M. In the process-view, the interacting context is left unspecified and the target of the transition is a λ-abstraction that binds a variable of type $\mathsf{Nm} \to \mathsf{Pr}$ (cf (Out)).

On the other hand, a process offering an input on a channel a can interact and obtain some unspecified name – the result is a λ-abstraction that binds a variable of type Nm (cf (In)). A process with both capabilities can perform the synchronisation by itself – the abstractions are combined via an application (cf (Tau)). Note that the subtleties of scope extrusion are dealt with cleanly by leveraging the capture-free substitution of the λ-calculus metalanguage. The remaining rules ((Par) and (Res)) account for interaction within evaluation contexts and are purely structural.

Transitions which arise from process terms, as presented in *Fig.* 3, represent the part of the interaction which is controlled by the process. In fact, if the sole purpose of the LTS were to structurally define the reduction relation we could stop here as *Fig.* 3 fulfills this role. However, in order to characterise a contextually defined equivalence as a bisimilarity we need to account for the interactions with arbitrary contexts. The transitions which arise from λ-abstracted terms, presented in *Fig.* 4, represent these parts of interactions controlled by the context. Combining the process and context views allows us to completely describe behaviour in π. We do this in *Fig.* 5 by a simple conjunction of the two views of the interaction. The labels γ here are composite actions $(\alpha\beta)$ consisting of both the structural process contribution α and contextual contribution β. The context-view transitions take a very simple form, that is, applicative labels carrying *any* well-typed process. This quantification may appear rather crude but it does at least provide a robust means of defining the context contribution of parameters to the completed labelled transition which is insensitive to subtleties in observational power of the underlying calculus. The same cannot be said of the standard semantics of π-calculus in which the identities of transmitted names are directly observed. We return to this point in §3 where we refine $\mathcal{CA}$ by limiting the contextual contributions β to a more tractable class of terms.

Owing to the construction of the LTS, there are contexts which witness the labels of $\mathcal{CA}$ in the sense that they induce a reduction with the same result; moreover they are parametric in the context component provided in $\mathcal{A}$. Let $\mu_{a?} \stackrel{\text{def}}{=} 1_{\mathsf{Pr}} \parallel a!2_{\mathsf{Nm}}$ and $\mu_{a!} \stackrel{\text{def}}{=} 1_{\mathsf{Pr}} \parallel a?y.2_{\mathsf{Nm}\to\mathsf{Pr}}(y)$.

Lemma 4 (Witness contexts)

(i) If $\langle \Delta \rhd P \rangle \xrightarrow{\alpha\beta}_{\mathcal{CA}} \langle \Delta' \rhd P' \rangle$ *then* $\langle \Delta' \rhd \mu_\alpha \circ (1_{\mathsf{Pr}}, \beta) \circ P \rangle \to \langle \Delta' \rhd P' \rangle$ $(\alpha \in \{a?, a!\})$*;*
(ii) if $\langle \Delta \rhd P \rangle \xrightarrow{\tau 1}_{\mathcal{CA}} \langle \Delta \rhd P' \rangle$ *then* $\langle \Delta \rhd P \rangle \to \langle \Delta \rhd P' \rangle$. □

The converse of the the first part of above lemma does not hold; the context which triggers a reduction may provide redundant parts not vital to the actual interaction with the term. However, a converse relationship between reductions and labels from $\mathcal{CA}$ can be established provided the context can be forced to interact with a term; this is done with the aid of a fresh auxiliary name constant to provide a barb. This relationship is used in order to prove that contextual equivalence is contained in bisimilarity (completeness) and in general does not follow from our systematic derivation.

Lemma 5 (Characterising contexts) *Given a name constant u fresh for P, let* $\chi_u(a?b) \stackrel{\text{def}}{=} a!b.u!$, $\chi_u(a!U) \stackrel{\text{def}}{=} a?x.u!.U(x)$ *and* $\chi_u(\tau 1) \stackrel{\text{def}}{=} u!$. *If*

$$\langle \Delta \rhd P \| \chi_u(\gamma) \rangle \to \langle \Delta \rhd P'' \rangle \text{ with } \langle \Delta \rhd P'' \rangle \downarrow_u$$

and

$$\langle \Delta \rhd P'' \| u? \rangle \to \langle \Delta \rhd P' \rangle \text{ with } \langle \Delta \rhd P' \rangle \not\downarrow_u$$

$$M ::= \ldots \mid M\uparrow \qquad \frac{\Delta;\Gamma \vdash k{:}\mathrm{Nm}}{\Delta;\Gamma \vdash k\uparrow{:}\mathrm{Pr}}\ (:\textsc{Lk})$$

$$\frac{\Delta \subseteq \Delta' \quad a \in \Delta'}{\langle \Delta \rhd \lambda x.M \rangle \xrightarrow{a} \langle \Delta' \rhd (\lambda x.M)(a) \rangle}\ (\textsc{InstNm}) \qquad \frac{\langle \Delta \rhd \lambda X.M(\lambda x.x\uparrow) \rangle \xrightarrow{\alpha\uparrow} \langle \Delta' \rhd P \rangle}{\langle \Delta \rhd \lambda X.M \rangle \xrightarrow{\alpha} \langle \Delta' \rhd P \rangle}\ (\textsc{InstPr})$$

$$\frac{}{\langle \Delta \rhd \lambda.\,.P \rangle \xrightarrow{1} \langle \Delta \rhd P \rangle}\ (\textsc{InstUn}) \qquad \frac{}{\langle \Delta \rhd a\uparrow \rangle \xrightarrow{a\uparrow} \langle \Delta \rhd 0 \rangle}\ (\textsc{Lk}) \qquad \frac{\langle \Delta, a \rhd P \rangle \xrightarrow{a\uparrow} \langle \Delta, a \rhd P' \rangle}{\langle \Delta \rhd \nu a.P \rangle \xrightarrow{(a)\uparrow} \langle \Delta, a \rhd P' \rangle}\ (\textsc{Fr})$$

$$\frac{\langle \Delta \rhd P \rangle \xrightarrow{\alpha\uparrow} \langle \Delta' \rhd P' \rangle}{\langle \Delta \rhd P \| Q \rangle \xrightarrow{\alpha\uparrow} \langle \Delta' \rhd P' \| Q \rangle}\ (\|\textsc{Lk}) \qquad \frac{\langle \Delta, b \rhd P \rangle \xrightarrow{\alpha\uparrow} \langle \Delta', b \rhd P' \rangle}{\langle \Delta \rhd \nu b.P \rangle \xrightarrow{\alpha\uparrow} \langle \Delta' \rhd \nu b.P' \rangle}\ (\nu\textsc{Lk})$$

Fig. 6 Refined context actions: syntax and rules ($\mathcal{R}$).

then $\langle \Delta \rhd P \rangle \xrightarrow{\gamma} \langle \Delta \rhd P' \rangle$. *The converse of this also holds.* □

Given the LTS generated by $\mathcal{CA}$, we can make use of the standard notion of (strong) bisimilarity which we denote by $\sim_{\mathcal{CA}}$. The close relationship of $\mathcal{CA}$ with the underlying reduction system means that it is straightforward to prove that bisimilarity is a congruence.

Proposition 6 $\sim_{\mathcal{CA}}$ *is preserved by all* π*-contexts.* □

These propositions and the fact that bisimilarity is barb-preserving are sufficient to establish soundness ($\sim_{\mathcal{CA}}$ implies $\simeq$). The conclusion of Lemma 5 also lets us demonstrate completeness ($\simeq$ implies $\sim_{\mathcal{CA}}$) because every label of $\mathcal{CA}$ can be simulated as a barb-sensitive reduction.

Theorem 7 $\sim_{\mathcal{CA}} = \simeq$. □

The reader may like to compare this with a related result in [2] in which sophisticated mappings of extruded names are required in the definition of the LTS. Although the result there captures the equivalence using a more tractable LTS, our approach captures the same equivalence in a systematic way which extends immediately to π-calculus with matching and further variants. It is the robustness of our technique which is of value here.

3 Refining context actions

In the calculus above we did not include a choice operator and test for name equality. It can be shown that in this setting standard early bisimilarity does not agree with barbed congruence [2]. However, bisimilarity on our novel LTS does (*cf.* Theorem 7). More significantly though, the proof of this theorem does not rely on the ability to compare names and remains true in the presence of matching.

As is well known, when name equality *is* present in the π-calculus, standard early bisimilarity does happen to provide a labelled characterisation of barbed congruence. In this section we shall show that the early LTS for π can itself be decomposed into our process and context views analogously to the presentation of $\mathcal{CA}$. Although we do not derive this refinement in a systematic manner, it turns out that the standard LTS can still usefully be construed as combination of the base structural rules $\mathcal{C}$ and a refinement of the context-view transitions $\mathcal{A}$. We include the standard early LTS $\mathcal{S}$ in Appendix 5 for reference.

Technically the refinement is achieved by introducing a new part of the meta-language – a term $k\uparrow$ of process type where k is a term of name type; the syntax and the additional type rule are presented in the upper section of *Fig.* 6. This meta-term interacts with the syntactic features of π as shown by rules (LK), (‖LK), (νLK) and (FR) in the lower part of *Fig.* 6. The label $k\uparrow$ is an abbreviation for the observation of a successful interaction of a context that tests the identity of the name k, while the label $(x)\uparrow$ is an abbreviation for the observation of a fresh name. Thus, the rule (FR) is related to the (OPEN) rule in $\mathcal{S}$ but, unlike (OPEN), this rule is divorced from the underlying communication rules. Instead of allowing an instantiation by an arbitrary process as we find in $\mathcal{A}$, (INSTPR) simply relays the observation of the meta-process $\lambda x.x\uparrow$; name bindings are dealt canonically, as shown by (INSTNM).

We shall consider the system $\mathcal{CR}$ that is obtained by combining the $\mathcal{C}$ system with the refined system $\mathcal{R}$, analogously to how the system $\mathcal{CA}$ was obtained via the rules presented in *Fig.* 5. Using structural analysis it is not difficult to prove that $\mathcal{CR}$ and $\mathcal{S}$ are equal in the sense that (up to minor relabeling on τ transitions) exactly the same transitions are derived.

Theorem 8 $\mathcal{CR} = \mathcal{S}$. $\square$

As an immediate corollary of Theorem 7 (with matching), Theorem 8 and the known result that standard early bisimilarity $\sim_{\mathcal{S}}$ coincides with $\simeq$ we obtain:

Corollary 9 $\sim_{\mathcal{S}} = \sim_{\mathcal{CR}} = \sim_{\mathcal{CA}} = \simeq$. $\square$

4 Modular variants of the π-calculus

We believe that splitting an LTS into a process-view and a context-view, based on the underlying reductions, naturally leads to more modular and robust theories. In order to justify this belief, we now apply these ideas to two variants of the π-calculus: the higher-order π-calculus and the asynchronous π-calculus. For the former, it should be of no surprise that this can be done as the original LTSs for the higher-order π-calculus are presented using concretions and abstractions and so avoid difficulties with scope extrusion [17]. For the asynchronous language only the communication fragment differs and thus we expect to isolate any changes to the LTS to that for the process-view $\mathcal{C}$.

$$\frac{\Delta;\Gamma \vdash M{:}\mathsf{Pr} \quad \Delta;\Gamma \vdash k{:}\mathsf{Nm} \quad \Delta;\Gamma \vdash R{:}\mathsf{Pr}}{\Delta;\Gamma \vdash k!RM{:}\mathsf{Pr}}\ (\text{:OutPref}) \qquad \frac{\Delta;\Gamma,x{:}\mathsf{Pr} \vdash M{:}\mathsf{Pr} \quad \Delta;\Gamma \vdash k{:}\mathsf{Nm}}{\Delta;\Gamma \vdash k?xM{:}\mathsf{Pr}}\ (\text{:InPref})$$

Fig. 7 Type rules for second-order.

$$\frac{}{\langle \Delta \triangleright \lambda x.M \rangle \xrightarrow{\mathsf{Tr}} \langle \Delta,a \triangleright (\lambda x.M)(a\uparrow) \rangle}\ (\text{InstTr})$$

$$\frac{}{\langle \Delta \triangleright \lambda X.M \rangle \xrightarrow{\mathsf{Ab}} \langle \Delta,a \triangleright (\lambda X.M)(\lambda y.a\downarrow(y)) \rangle}\ (\text{InstAb})$$

$$\frac{}{\langle \Delta \triangleright a\uparrow \rangle \xrightarrow{a\uparrow} \langle \Delta \triangleright 0 \rangle}\ (\text{TrOut}) \qquad \frac{}{\langle \Delta \triangleright a\downarrow(P) \rangle \xrightarrow{a\uparrow} \langle \Delta \triangleright P \| a\downarrow(P) \rangle}\ (\text{TrIn})$$

$$\frac{\langle \Delta \triangleright P \rangle \xrightarrow{a\uparrow} \langle \Delta \triangleright P' \rangle}{\langle \Delta \triangleright P\|Q \rangle \xrightarrow{a\uparrow} \langle \Delta \triangleright P'\|Q \rangle}\ (\|\text{Tr}) \qquad \frac{\langle \Delta,b \triangleright P \rangle \xrightarrow{a\uparrow} \langle \Delta,b \triangleright P' \rangle}{\langle \Gamma \triangleright \nu bP \rangle \xrightarrow{a\uparrow} \langle \Gamma \triangleright \nu bP' \rangle}\ (\nu\text{Tr})$$

Fig. 8 Refined context actions: second-order ($\mathcal{R}^{so}$).

4.1 The higher-order π-calculus

Following [17], to simplify the presentation we will actually present the LTS for the second-order π-calculus In order to define the second-order π-calculus we simply need to modify the type system in *Fig.* 2 to allow the communication of process terms rather than names. The new type rules for input and output are given in *Fig.* 7.

Now, remarkably, modulo types, the rules of *Fig.* 3 need no modifications whatsoever. That is, up to typing, the CCS style communication core is identical for both first and higher-order languages. Perhaps more remarkably, the rules in *Fig.* 4 and *Fig.* 5 need no modifications either. The differences between the first- and second-order languages are dealt with using types alone.

In essence, the notion of bisimilarity yielded by $\mathcal{CA}$ for the second-order language is Sangiorgi's context bisimilarity [17]. It is therefore interesting to note by analogy that $\sim_{\mathcal{CA}}$ provides a definition of context bisimulation for the first-order π-calculus. As for the first-order case though, context bisimulation is unattractive due to its reliance upon context actions containing arbitrary (typed) process terms. It is known [10, 17] that context bisimulation can be refined to so-called 'normal' bisimulation, much in the same way as the $\mathcal{CA}$ system is refined to $\mathcal{R}$ by using a limited form of context action. We give the rules for the second-order refined system $\mathcal{R}^{so}$, in *Fig.* 8. In this case however, we need to adjust the completed actions system, $\mathcal{CA}$, to include the rule

$$\frac{\langle \Delta \triangleright P \rangle \xrightarrow{a\uparrow}_{\mathcal{R}^{so}} \langle \Delta \triangleright P' \rangle}{\langle \Delta \triangleright P \rangle \xrightarrow{a\uparrow} \langle \Delta \triangleright P' \rangle}\ (\text{CTr})$$

and by combining $\mathcal{C}$ and $\mathcal{R}^{so}$ using this augmented $\mathcal{CA}$, we obtain the refined system $\mathcal{CR}^{so}$ for second-order π-calculus.

In order to present the refined system, again we introduce an augmented meta-language. This takes the form of processes $a \uparrow : \mathsf{Pr}$ and process abstractions $a \downarrow : \mathsf{Pr} \to \mathsf{Pr}$. These are entirely analogous to the triggers and abstractions of [17]. The behaviour of the 'trigger' process $a \uparrow$ is simply to announce itself whenever it is executed and the 'abstraction' $a \downarrow (P)$ will repeatedly allow copies of P to be accessed by calling on the name a. We know from [10, 17] that these syntactic gadgets actually have real syntactic counterparts in the second-order language and are in effect just macros. We also know from [10, 17] that bisimulation equivalence generated by this refined LTS ($\sim_{\mathcal{CR}^{so}}$) coincides with context bisimilarity over second-order processes ($\sim_{\mathcal{CA}}$).

4.2 The asynchronous π-calculus

There is an established [3, 8] presentation of the variant of the π-calculus in which all communication is done asynchronously. This involves simply restricting the syntax of the language such that the residual of any output prefix is the nil process. The obvious effect of this is such that no process can be blocked waiting on a send of data. A less obvious effect is that this language restriction actually impacts upon the behavioural theory of the language considerably and makes inputs unobservable. This is well-accounted for in the literature [1], but here we show that a simple modification to the combined module $\mathcal{CA}$ only can also account for the change in behavioural theory. We think of the $\mathcal{CA}$ transitions as observations that an interacting process can make. We include an additional rule which allows one to make an artificial $a?b$ observation predicated upon some unobservable activity in P when offered $a!b$ by the context.

To obtain an LTS appropriate for the asynchronous language we define the system $\mathcal{CA}^a$ by adding the following rule to $\mathcal{CA}$:

$$\frac{\langle \Delta \rhd P \rangle \xrightarrow{\tau}_{\mathcal{C}} \langle \Delta \rhd P' \rangle \quad \langle \Delta \rhd \lambda x.\, P' \| a!x \rangle \xrightarrow{b}_{\mathcal{A}} \langle \Delta' \rhd P'' \rangle}{\langle \Delta \rhd P \rangle \xrightarrow{a?b}_{\mathcal{CA}} \langle \Delta' \rhd P'' \rangle} \text{(AIN)}$$

This $\mathcal{CA}^a$ system can also be used to combine the $\mathcal{R}$ system with $\mathcal{C}$ to yield the corresponding system $\mathcal{CR}^a$ and the techniques described above can easily be applied to the asynchronous variant of the language also to establish analogous results.

Conclusion and future work

We have provided a new way of understanding the well-trodden labelled transition semantics of the π-calculus. Our approach attempts to combine the structural approach to semantics with the contexts-as-labels approach in order to obtain systematically defined labelled transition systems for process calculi. In this paper we have shown that this approach works very well for the π-calculus and we believe that the technique is robust and widely applicable. We have applied our approach to the ambient calculus [5] for which we have obtained a new, systematically derived, labelled transition system along with a sound and complete context-bisimilarity for that language [16]. Furthermore we plan to develop a general setting for our approach to pursue the synthesis of labelled transition systems from reduction rules in the spirit of [11, 12, 20, 21].

References

1. R. Amadio, I. Castellani, and D. Sangiorgi. On bisimulations for the asynchronous pi-calculus. *Theoretical Computer Science*, 195(2):291–324, 1998.
2. M. Boreale and D. Sangiorgi. Bisimulation in name-passing calculi without matching. In *IEEE Symposium on Logic in Computer Science LiCS*, 1998.
3. G. Boudol. Asynchrony and the π-calculus. Technical Report 1702, INRIA, Sophia-Antipolis, 1991.
4. R. Bruni and U. Montanari. Cartesian closed double categories, their lambda-notation, and the Pi-calculus. In *IEEE Symposium on Logic in Computer Science LiCS*, 1999.
5. L. Cardelli and A. Gordon. Mobile ambients. In *Foundations of Software Science and Computation Structures, FoSSaCS*, LNCS. Springer-Verlag, 1998.
6. U. Engberg and M. Nielsen. A calculus of communicating systems with label passing. Technical Report DAIMI PB-208, University of Aarhus, May 1986.
7. C. Fournet and G. Gonthier. A hierarchy of equivalences for asynchronous calculi. In *International Colloquium on Automata, Languages and Programming ICALP*, volume 1443 of *LNCS*. Springer-Verlag, 1998.
8. K. Honda and M. Tokoro. An object calculus for asynchronous communication. In *European Conference on Object-Oriented Programming ECOOP*, 1991.
9. K. Honda and N. Yoshida. On reduction-based process semantics. *Theoretical Computer Science*, 152(2):437–486, 1995.
10. A.S.A. Jeffrey and J. Rathke. Contextual equivalence for higher-order pi-calculus revisited. *Logical Methods in Computer Science*, 1(1:4), 2005.
11. B. Klin, V. Sassone, and P. Sobociński. Labels from reductions: towards a general theory. In *Conference on Algebra and Coalgebra in Computer Science Calco*, volume 3629 of *LNCS*, pages 30–50. Springer, 2005.
12. J. Leifer and R. Milner. Deriving bisimulation congruences for reactive systems. In *International Conference on Concurrency Theory Concur*, volume 1877 of *LNCS*, pages 243–258. Springer, 2000.
13. R. Milner. *Communicating and Mobile Systems: the π-calculus.* CUP, 1999.
14. R. Milner, J. Parrow, and D. Walker. A calculus of mobile processes, II. *Information and Computation*, 100(1):41–77, 1992.
15. G. D. Plotkin. A structural approach to operational semantics. Technical Report FN-19, DAIMI, Computer Science Department, Aarhus University, 1981.

16. J. Rathke and P. Sobociński. Deriving structural labelled transitions for mobile ambients, 2008. Submitted for publication.
17. D. Sangiorgi. Bisimulation for higher-order process calculi. *Information and Computation*, 131(2):141–178, 1996.
18. D. Sangiorgi and D. Walker. *The π-calculus*. CUP, 2001.
19. D. Sangiorgi and D. Walker. Some results on barbed equivalences in pi-calculus. In *International Conference on Concurrency Theory CONCUR*, volume 2154 of *LNCS*. Springer-Verlag, 2001.
20. V. Sassone and P. Sobociński. Locating reaction with 2-categories. *Theoretical Computer Science*, 333(1-2):297–327, 2005.
21. Peter Sewell. From rewrite rules to bisimulation congruences. *Theoretical Computer Science*, 274(1–2):183–230, 2002.
22. A. Ziegler, D. Miller, and C. Palamidessi. A congruence format for name-passing calculi. In *Workshop on Structural Operational Semantics SOS*, volume 156 of *ENTCS*, pages 169–189, 2006.

5 Standard early labelled transitions, typed ($\mathcal{S}$)

For reference purposes, we list below the standard early labelled transition system semantics for the π-calculus. They have been adapted to our typed setting but remain substantially the same as can be found in say, [18].

$$\frac{\Delta \subseteq \Delta' \quad b \in \Delta'}{\langle \Delta \rhd a?xP \rangle \xrightarrow{a?b} \langle \Delta' \rhd P[b/x] \rangle}\ (\textsc{In}) \qquad \frac{}{\langle \Delta \rhd a!bP \rangle \xrightarrow{a!b} \langle \Delta \rhd P \rangle}\ (\textsc{Out})$$

$$\frac{\langle \Delta \rhd P \rangle \xrightarrow{a!b} \langle \Delta \rhd P' \rangle \quad \langle \Delta \rhd Q \rangle \xrightarrow{a?b} \langle \Delta \rhd Q' \rangle}{\langle \Delta \rhd P \| Q \rangle \xrightarrow{\tau} \langle \Delta \rhd P' \| Q' \rangle}\ (\textsc{Comm})$$

$$\frac{\langle \Delta, b \rhd P \rangle \xrightarrow{a!b} \langle \Delta, b \rhd P' \rangle \ (a \neq b)}{\langle \Delta \rhd \nu bP \rangle \xrightarrow{a!(b)} \langle \Delta, b \rhd P' \rangle}\ (\textsc{Open}) \frac{\langle \Delta \rhd P \rangle \xrightarrow{a!(b)} \langle \Delta, b \rhd P' \rangle \quad \langle \Delta \rhd Q \rangle \xrightarrow{a?b} \langle \Delta, b \rhd Q' \rangle}{\langle \Delta, b \rhd P \| Q \rangle \xrightarrow{\tau} \langle \Delta \rhd \nu b(P' \| Q') \rangle}\ (\textsc{Close})$$

$$\frac{\langle \Delta \rhd P \rangle \xrightarrow{\alpha} \langle \Delta' \rhd P' \rangle}{\langle \Delta \rhd P \| Q \rangle \xrightarrow{\alpha} \langle \Delta \rhd P' \| Q \rangle}\ (\textsc{Par}) \qquad \frac{\langle \Delta, a \rhd P \rangle \xrightarrow{\alpha} \langle \Delta', a \rhd P' \rangle \quad a \notin \alpha}{\langle \Delta \rhd \nu aP \rangle \xrightarrow{\alpha} \langle \Delta' \rhd \nu aP' \rangle}\ (\textsc{Res})$$

Adequacy of Compositional Translations for Observational Semantics

Manfred Schmidt-Schauß[1], Joachim Niehren[2], Jan Schwinghammer[3], and David Sabel[4]

[1] J. W. Goethe-Universität, Frankfurt, Germany,
schauss@ki.informatik.uni-frankfurt.de
[2] INRIA, Lille, France, Mostrare Project
[3] Saarland University, Saarbrücken, Germany
[4] J. W. Goethe-Universität, Frankfurt, Germany,
sabel@ki.informatik.uni-frankfurt.de

Abstract. We investigate methods and tools for analysing translations between programming languages with respect to observational semantics. The behaviour of programs is observed in terms of may- and must-convergence in arbitrary contexts, and *adequacy* of translations, i.e., the reflection of program equivalence, is taken to be the fundamental correctness condition. For compositional translations we propose a notion of *convergence equivalence* as a means for proving adequacy. This technique avoids explicit reasoning about contexts, and is able to deal with the subtle role of typing in implementations of language extensions.

1 Introduction

Proving correctness of program translations on the basis of operational semantics is an ongoing research topic (see e.g. the recent [7, 18]) that is still poorly understood when it comes to concurrency and mutable state. We are motivated by implementations of language extensions that are often packaged into the language's library. Typical examples are implementations of channels, buffers, or semaphores using mutable reference cells and futures in Alice ML [1, 12], or using MVars in Concurrent Haskell [13]. Ensuring the correctness of such implementations of higher-level constructs is obviously important.

In this paper we adopt an *observational semantics* based on may- and must-convergence. Two programs are considered equivalent if they exhibit the same may- and must-convergence behaviour in all contexts. This definition is flexible and has been applied to a wide variety of programming languages and calculi in the past. The observation of may- and must-convergence is particularly well-suited for dealing with nondeterminism as it arises in concurrent programming [2, 17, 11].

We study implementations of language extensions in the compilation paradigm, i.e., by viewing them as translations $T : L \to L'$ from a language L into another language L'. Such translations are usually *compositional* in that $T(C[t]) = T(C)[T(t)]$ for all contexts C and programs t of L. In a naive approach, one might even want to assume

---

Please use the following format when citing this chapter:

Schmidt-Schauß, M., et al., 2008, in IFIP International Federation for Information Processing, Volume 273; *Fifth IFIP International Conference on Theoretical Computer Science*; Giorgio Ausiello, Juhani Karhumäki, Giancarlo Mauri, Luke Ong; (Boston: Springer), pp. 521–535.

that L is a conservative extension of L' so that (non-)equivalences of L' continue to hold in L. However, this fails in many cases (see below) due to subtle typing problems.

A translation $T : L \to L'$ is *adequate* if $T(s) \sim_{L'} T(t)$ implies $s \sim_L t$ for all programs s and t of L, where $\sim_L$ and $\sim_{L'}$ are the program equivalences of the respective languages. Adequacy is the basic correctness requirement to ensure that program transformations of the target language L' can be soundly applied with respect to observations made in the source language L.

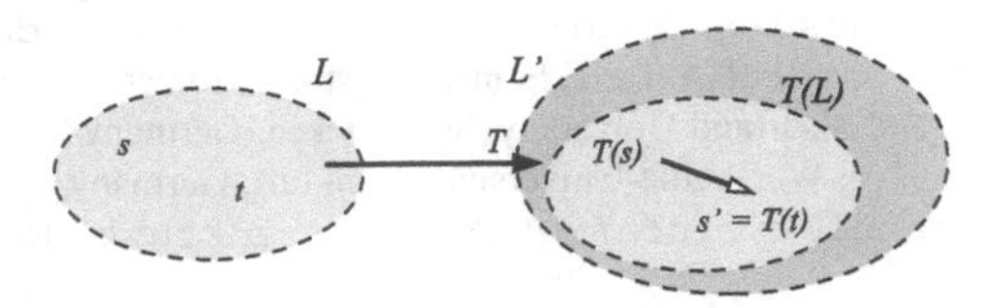

Suppose a translation $T(s)$ is optimized to an equivalent program $s' \sim_{L'} T(s)$ and that s' is the translation of some t, i.e. $T(t) = s'$. Any useful notion of correctness must enforce that s and t are indistinguishable, i.e. $s \sim_L t$. This is precisely what adequacy of T guarantees. With respect to implementations, adequacy opens the possibility of transferring contextual equivalences from the target language L' to the source language L. For non-deterministic and concurrent languages, such equivalences have been established for instance by inductive reasoning using diagram-based methods directly on an underlying small-step operational semantics [6, 11].

Full abstraction extends adequacy by the inverse property, i.e., that program equivalence is also preserved by the translation. In the general situation, however, the language L' may be more expressive than L and allows us to make more distinctions, also on the image $T(L)$. Thus we can have $T(s) \not\sim_{L'} T(t)$ for some expressions s, t with $s \sim_L t$.

In denotational semantics, adequacy and full abstraction are well-studied concepts. In contrast, in this paper we provide a general criterion for proving adequacy of translations that is not tied to specific models. More precisely, we show that convergence equivalence implies adequacy of compositional translations, meaning it is enough to establish that all convergence tests yield the same results before and after the translation. We also provide a criterion for the full abstractness of compositional translations for which the target language is a conservative extension of the source language.

In order to demonstrate these tools, we consider the standard Church encoding of pairs in a call-by-value lambda calculus with a fixed point operator and nondeterministic choice. In order to reason that the encoding of pairs is adequate, one needs to check, for all lambda terms t with pairs and projections, that reduction from t may-converges (must-converges, respectively) if and only if reduction from its encoding $T(t)$ may-converges (must-converges, respectively). However, even in this seemingly well-understood example, this condition *fails* if the lambda calculus is untyped, since the implementation may remove errors, i.e., $T(t)$ terminates more often than t. If the source-language is typed so that stuck expressions are excluded, then our tools apply in a smooth way and show the adequacy of the standard translation, even for differently typed versions of the lambda calculus that is used as target language. Since neither simple typing nor Hindley-Milner polymorphic typing are sufficient to make the source language an extension of the target language, we cannot expect to have an extension situation under type systems that are commonly used in programming languages.

Related work. Various proof methods have been developed for establishing contextual equivalences. These include context lemmas (e.g., [9]), bisimulation methods (for instance, [5]), diagram-based methods (e.g., [6, 11]), and characterizations of contextual equivalence in terms of logical relations (e.g. [14]). In most cases, language extensions and their effect on equivalences are not discussed. There are some notable exceptions: a translation from the core of Standard ML into a typed lambda calculus is given in [16], and full abstraction is shown by exhibiting an inverse mapping, up to contextual equivalence. Adequate translations (with certain additional constraints) between call-by-name and call-by-value versions of PCF are considered in [15], via fully abstract models (necessitating the addition of parallel constructs to the languages) and domain-theoretic techniques. The fact that adequate (and fully abstract) translations compose is exploited in [8], where a syntactic translation is used to lift semantic models for FPC to ones for the lazy lambda calculus. In a similar vein, the recent [18] develops a translation from an aspect-oriented language to an ML-like language, to obtain a model for the former. The adequacy proof follows a similar pattern to ours, but does not abstract away from the particularities of the concrete languages.

Shapiro [20] categorizes implementations and embeddings in concurrent scenarios, but does not provide concrete proof methods based on contextual equivalence. For deterministic languages (where may- and must-convergence agree), frameworks similar to our proposal were considered by Felleisen [4] and Mitchell [10]. Their focus is on comparing languages with respect to their expressive power; the non-deterministic case is only briefly mentioned by Mitchell. Mitchell's work is concerned with (the impossibility of) translations that additionally preserve representation independence of ADTs, and consequently assumes, for the most part, source languages with expressive type systems. Felleisen's work is set in the context of a Scheme-like untyped language. Although the paper discusses the possibility of adding types to get stronger expressiveness statements, the theory of expressiveness is developed by abandoning principles similar to adequacy.

Outline. Section 2 recalls the encoding of pairs in the non-deterministic lambda calculus, introduces rigorous notions of observables, and illustrates the need for types. In Section 3 a general framework for proving observational correctness as well as adequacy of translations is introduced. Section 4 shows the adequacy of the pair encoding using a simple type system and discusses two extensions.

2 Non-deterministic Call-by-Value Lambda Calculi

In this section, we recall the call-by-value lambda calculus with a fixed point operator and nondeterministic choice, and present its observational semantics on the basis of may- and must-convergence. We illustrate why Church's encoding of pairs in this calculus fails to be observationally correct in the untyped case.

$$\begin{aligned} x,y &\in Var \\ r,s,t &\in Exp_{cp} ::= w \mid t_1\, t_2 \mid t_1 \oplus t_2 \\ v,w &\in Val_{cp} ::= x \mid \lambda x.t \mid \mathbf{unit} \mid \mathbf{fix} \\ & \qquad \mid (w_1,w_2) \mid \mathbf{fst} \mid \mathbf{snd} \end{aligned}$$

Fig. 1 Syntax of λ_{cp}

$$\mathbb{E} ::= [\,] \mid \mathbb{E}t \mid w\mathbb{E} \mid \mid \mathbb{E}\oplus t \mid w \oplus \mathbb{E}$$

Fig. 2 Evaluation Contexts $\mathbb{E}$

$$\begin{array}{ll} (\beta\text{-CBV}) & \mathbb{E}[(\lambda x.t)\, w] \to \mathbb{E}[t[w/x]] \\ (\text{FIX}) & \mathbb{E}[\mathbf{fix}\, \lambda x.t] \to \mathbb{E}[t[(\lambda y.(\mathbf{fix}\, \lambda x.t)y)/x]] \\ (\oplus\text{L}) & \mathbb{E}[w_1 \oplus w_2] \to \mathbb{E}[w_1] \\ (\oplus\text{R}) & \mathbb{E}[w_1 \oplus w_2] \to \mathbb{E}[w_2] \\ (\text{SEL-F}) & \mathbb{E}[\mathbf{fst}\,(w_1,w_2)] \to \mathbb{E}[w_1] \\ (\text{SEL-S}) & \mathbb{E}[\mathbf{snd}\,(w_1,w_2)] \to \mathbb{E}[w_2] \end{array}$$

Fig. 3 Small-Step Reduction

2.1 Languages

The calculus λ_{cp} is the usual call-by-value lambda calculus extended by a (demonic, see [21]) choice operator, a call-by-value fixed point operator for recursion, pairs (w_1,w_2) and selectors **fst** and **snd** as data structure, and a constant **unit**. Fixing a set of variables *Var*, the syntax of expressions Exp_{cp} and values Val_{cp} is shown in Fig. 1. The subcalculus λ_c is the calculus without pairs and selectors and will be used as target language. We use Exp_c (Val_c, resp.) for the set of λ_c-expressions (λ_c-values, resp.).

A *context* C is an expression with a hole denoted with $[\,]$, $C[s]$ is the result of placing the expression s in the hole of C. For both calculi we require call-by-value evaluation contexts $\mathbb{E}$ which are introduced in Fig. 2. With $s_1[s_2/x]$ we denote the capture-free substitution of variable x with s_2 for all free occurrences of x in s_1. To ease reasoning we assume that the distinct variable convention holds for all expressions, i.e. that the bound variables of an expression are all distinct and free variables are distinct from bound variables.

The reduction rules for both calculi are defined in Fig. 3. Small step reduction $\to_{cp}$ of λ_{cp} is the union of all six rules, and small step reduction $\to_c$ of λ_c is the union of the first four rules. We assume that reduction preserves the distinct variable convention by implicitly performing α-renaming if necessary.

2.2 Contextual Equivalence

Let *Exp* be a language, let $Val \subseteq Exp$ be a set of values and $\to$ be a reduction relation. Then *may-convergence* for expressions $s \in Exp$ is defined as $s\downarrow$ iff $\exists v \in Val : s \xrightarrow{*} v$, and *must-convergence* is defined as $s\Downarrow$ iff $\forall s' : s \xrightarrow{*} s' \implies s'\downarrow$. For a discussion and motivations for the latter notion see [2, 17, 11]. Note that there is also another notion of must-convergence found in the literature (e.g. [3]), which holds if an expression has only evaluations to values, in particular, if the expression has no infinite evaluations (i.e. if $s \not\to^{\omega}$).

For an expression s we also write $s\Uparrow$ if $s\downarrow$ does not hold, and say that s is *must-divergent*. We write $s\uparrow$ if s is not must-convergent and then say s is *may-divergent*.

$enc(x)$	$= x$	$enc(\mathbf{fix})$	$= \mathbf{fix}$
$enc(\mathbf{unit})$	$= \mathbf{unit}$	$enc((w_1,w_2))$	$= \lambda s.\,(s\; enc(w_1)\; enc(w_2))$
$enc(\lambda x.t)$	$= \lambda x.enc(t)$	$enc(\mathbf{fst})$	$= \lambda p.\,(p\; \lambda x.\lambda y.x)$
$enc(t_1\; t_2)$	$= enc(t_1)\; enc(t_2)$	$enc(\mathbf{snd})$	$= \lambda p.\,(p\; \lambda x.\lambda y.y)$
$enc(t_1 \oplus t_2)$	$= enc(t_1) \oplus enc(t_2)$		

Fig. 4 Translation of λ_{cp} into λ_c

Note that may-divergence can equivalently be defined as $s\uparrow$ iff $\exists s' \in Exp : s \xrightarrow{*} s'$ and $s' \Uparrow$. This view allows us to use inductive proofs for showing may-divergences. For Exp_c, Val_c, and $\rightarrow_c$ we use $\downarrow_c$ for may-convergence and $\Downarrow_c$ for must-convergence. Accordingly for Exp_{cp}, Val_{cp}, and $\rightarrow_{cp}$ we use $\downarrow_{cp}$ and $\Downarrow_{cp}$ for the predicates.

Contextual equivalence for a (non-deterministic) calculus $(Exp, Val, \rightarrow)$ is defined by observing may- and must-convergence in all contexts. We first define two preorders for both predicates:

$$s_1 \leq_{\downarrow} s_2 \text{ iff } \forall C : C[s_1]\downarrow \implies C[s_2]\downarrow \qquad s_1 \leq_{\Downarrow} s_2 \text{ iff } \forall C : C[s_1]\Downarrow \implies C[s_2]\Downarrow$$

These are combined to obtain the contextual preorder $\leq$ as their intersection $\leq_{\downarrow} \cap \leq_{\Downarrow}$, and the contextual equivalence $\sim$ as $\leq \cap \geq$. To distinguish between the relations for λ_c and λ_{cp}, we index the symbols for the preorders and equivalence with c or cp, respectively, e.g. contextual equivalence in λ_c is $\sim_c$, and contextual preorder in λ_{cp} is $\leq_{cp}$.

2.3 Implementation of Pairs

We will mainly investigate the translation enc of λ_{cp} into λ_c as defined in Fig. 4 under different restrictions. Conversely, it is trivial to encode λ_c into λ_{cp} via the identity $inc(s) = s$ (which is more an *embedding* than a translation).

The following counter example shows that the implementation of pairs is not correct in the untyped setting.

Example 2.1. Let $t := \mathbf{fst}(\lambda z.z)$. Then $t\Uparrow_{cp}$, since t is irreducible and not a value. However, the translation $enc(t)$ results in the expression $t' := (\lambda p.p\; (\lambda x.\lambda y.x))\; (\lambda z.z)$, which deterministically reduces by some (β-CBV)-reductions to $\lambda x.\lambda y.x$, hence $enc(t)\Downarrow_c$. This is clearly not a correct translation, since it removes an error. Therefore, the observations are not preserved by this translation. This example also invalidates the implication $T(p_1) \leq_c T(p_2) \implies p_1 \leq_{cp} p_2$, since $enc(t') = t'$, and hence $enc(t') = t' \leq_c t' = enc(t)$, but $t' \not\leq_{cp} t$ by the arguments above. In the terminology of Definition 3.2 below, the translation enc is not adequate.

This counter example is also valid for deterministic calculi, where may- and must-convergence coincide. There, it is possible to circumvent the problem by weakening the definition of correctness to only one direction of the logical equivalence,

$s\downarrow \implies T(s)\downarrow$, but this results in weaker properties and is not the appropriate notion for compilations. In particular, this notion of correctness of a translation (which is called *weak expressibility* in [4]) implies the correctness of a trivial translation that maps all expressions to a (may-) convergent expression.

One potential remedy to the failure of the untyped approach to correctness of translations is to distinguish divergence from typing errors. From a different point of view, this simply means that only correctly typed programs should be considered by a translation: in Section 4.1 we will obtain adequacy after adding a type system to λ_{cp}.

3 Adequacy of Translations

We present a general framework for reasoning about different notions of language translations which are related to correctness.

We assume that languages come equipped with a small-step operational semantics and a notion of observables, expressed through convergence tests, with respect to which contextual equivalence can be defined. Since we are interested in concurrent calculi, a typical case will be the observations of may- and must-termination behavior, as introduced in the previous section. In the following we generalize slightly and, instead of contexts, speak of observers: this makes it easier to fit formalisms without an obvious notion of context into the framework, like abstract machines.

Definition 3.1. A *program calculus with observational semantics (OSP-calculus)* consists of the following components:

- A set $\mathcal{T}$ of *types*, ranged over by τ.
- For every type τ, a set $\mathcal{P}_\tau$ of *programs*, ranged over by p.
- For every pair τ_1, τ_2 of types, a set of functions $\mathcal{O}_{\tau_1,\tau_2}$ with $O : \mathcal{P}_{\tau_1} \to \mathcal{P}_{\tau_2}$ for $O \in \mathcal{O}_{\tau_1,\tau_2}$, called *observers*, such that also the identity function Id_τ is included in $\mathcal{O}_{\tau,\tau}$ for every type τ, and such that $\bigcup_{\tau_1,\tau_2 \in \mathcal{T}} \mathcal{O}_{\tau_1,\tau_2}$ is closed under function composition whenever the types are appropriate.
- A set $\{\Downarrow_1, \ldots, \Downarrow_n\}$ of *convergence tests* with $\Downarrow_i$: $\bigcup_{\tau \in \mathcal{T}} \mathcal{P}_\tau \to \{\text{true}, \text{false}\}$ for all $i = 1, \ldots, n$.

This definition is also applicable to the special case of deterministic calculi, where usually only a single termination predicate is considered. Moreover, it allows for untyped calculi like λ_{cp} by considering a single, 'universal' type. The calculus λ_{cp} then fits this definition of OSP-calculus, after identifying a context C with the map $t \mapsto C[t]$, and taking $\{\Downarrow_1, \Downarrow_2\} = \{\downarrow_{cp}, \Downarrow_{cp}\}$.

Since this framework has arbitrary observers (not only contexts) and there are types, the observational preorders at type τ are defined as follows, where $p_1, p_2 \in \mathcal{P}_\tau$:

- $p_1 \leq_{\Downarrow_i,\tau} p_2$ iff for all $\tau' \in \mathcal{T}$ and all $O : \tau \to \tau'$, $O(p_1)\Downarrow_i$ implies $O(p_2)\Downarrow_i$.
- $p_1 \leq_\tau p_2$ iff $\forall i : p_1 \leq_{\Downarrow_i,\tau} p_2$.
- $p_1 \sim_\tau p_2$ iff $p_1 \leq_\tau p_2$ and $p_2 \leq_\tau p_1$.

The relations $\leq_{\Downarrow_i,\tau}$ and $\leq_\tau$ are *precongruences*, i.e. they are preorders, and $p_1 \leq_{\Downarrow_i,\tau} p_2$ implies $\forall O : \tau \to \tau' : O(p_1) \leq_{\Downarrow_i,\tau'} O(p_2)$. For proving the latter implication let O' be an observer with $O'(O(p_1)) \Downarrow_i$. Then $O' \circ O$ is also an observer, hence $O' \circ O(p_2) \Downarrow_i$. Obviously, the same holds for $\leq_\tau$. The relation $\sim_\tau$ is a *congruence*, i.e. it is a precongruence and an equivalence relation.

In the following we only consider translations between OSP-calculi that have the same number n of convergence tests $\{\Downarrow_1,\ldots,\Downarrow_n\}$, in a fixed ordering. We define some characterizing notions of translations. In the remainder of this section we exhibit their dependencies and prove some consequences.

Definition 3.2. A translation $T : \mathscr{C} \to \mathscr{C}'$ between two calculi $\mathscr{C} = (\mathscr{T},\mathscr{P},\mathscr{O},\leq)$ and $\mathscr{C}' = (\mathscr{T}',\mathscr{P}',\mathscr{O}',\leq')$ maps types to types $T : \mathscr{T} \to \mathscr{T}'$, programs to programs $T : \mathscr{P}_\tau \to \mathscr{P}'_{T(\tau)}$, and observers to observers $T : \mathscr{O}_{\tau,\tau'} \to \mathscr{O}'_{T(\tau),T(\tau')}$ such that their types correspond for all $\tau,\tau' \in \mathscr{T}$ and such that $T(Id_\tau) = Id_{T(\tau)}$ for all τ.

Adequacy. A translation T is *adequate* iff for all τ, and $p_1,p_2 \in \mathscr{P}_\tau$, $T(p_1) \leq'_{T(\tau)} T(p_2) \implies p_1 \leq_\tau p_2$.

Full abstraction. A translation T is *fully abstract* iff for all τ, and $p_1,p_2 \in \mathscr{P}_\tau$, $p_1 \leq_\tau p_2 \iff T(p_1) \leq'_{T(\tau)} T(p_2)$.

Observational correctness. A translation T is *observationally correct* iff for all τ, $p \in \mathscr{P}_\tau, O \in \mathscr{O}_{\tau,\tau'}$ and all i: $O(p) \Downarrow_i$ if and only if $T(O)(T(p)) \Downarrow'_i$.

Convergence equivalence. A translation T is *convergence equivalent* (i.e. *preserves and reflects convergence*) iff for all p and convergence tests $\Downarrow_i$: $p \Downarrow_i$ if and only if $T(p) \Downarrow'_i$.

Compositionality. A translation T is *compositional* iff for all types $\tau,\tau' \in \mathscr{T}$, for all observers $O \in \mathscr{O}_{\tau,\tau'}$ and all programs $p \in \mathscr{P}_\tau$ we have $T(O(p)) = T(O)(T(p))$.

If in the following types are omitted, we implicitly assume that type information follows from the context.

As motivated in the Introduction, we consider adequacy as the right notion of correctness. Observational correctness is a sufficient criterion for adequacy (see Proposition 3.3). Convergence equivalence is implied by observational correctness, since T preserves identity observers. For compositional translations, the converse is true, i.e., it is sufficient to prove convergence equivalence in order to prove observational correctness. Full abstraction is not necessary for the adequacy of translations. If it holds in addition, for surjective translations it means that both program calculi are identical w.r.t. $\leq$.

Note that Definition 3.2 is stated only in terms of convergence tests and sets of observers, and hence only relying on the syntax and the operational semantics. Thus it can be used in all calculi with such a description. In the case of two calculi with convergence tests defined in terms of a small-step semantics, the definition also allows for reduction sequences in the translation that may lead outside of the image of the translation, i.e., that may not be retranslatable.

Proposition 3.3. *For a translation T the following hold:*

1. *If T is compositional, then T is convergence equivalent if and only if T is observationally correct.*
2. *If T is observationally correct, then T is adequate.*

Proof. 1. The only if direction holds, since T preserves identity observers: $Id_\tau(p) \Downarrow_i \iff T(Id_\tau)T(p) \Downarrow_i' \iff Id_{T(\tau)}T(p) \Downarrow_i' \iff T(p) \Downarrow_i'$.
For the if-direction let us assume that T is compositional and convergence equivalent. If $O(p) \Downarrow_i$, then preservation of convergence yields $T(O(p)) \Downarrow_i'$. Compositionality implies $T(O(p)) = T(O)(T(p))$, hence $T(O)(T(p)) \Downarrow_i'$. If $T(O)(T(p)) \Downarrow_i'$ then compositionality implies $T(O(p)) \Downarrow_i'$ so that reflection of convergence yields $O(p) \Downarrow_i$.
2. To show adequacy, let us assume that $T(p_1) \leq_{T(\tau)} T(p_2)$. We must prove that $p_1 \leq_\tau p_2$. Thus let O be such that $O(p_1) \Downarrow_i$. By observational correctness this implies $T(O)(T(p_1)) \Downarrow_i'$. From $T(p_1) \leq_{T(\tau)} T(p_2)$, we obtain $T(O)(T(p_2)) \Downarrow_i'$, since $T(O)$ is an admissible observer. Observational correctness in the other direction implies $O(p_2) \Downarrow_i$. This proves $p_1 \leq_\tau p_2$. □

As the following counter examples show, convergence equivalence is in general not sufficient for adequacy, and full abstraction is not implied by observational correctness. Similarly, convergence equivalence is not even implied by full abstraction (and thus neither by adequacy):

Example 3.4 (Convergence equivalence does not imply adequacy). Let the OSP-calculus L have three programs: a,b,c with $a\uparrow$, $b\downarrow$ and $c\downarrow$. Assume there are two observers O_1, O_2 with $O_1(x) = x$ and $O_2(a) = a, O_2(b) = a, O_2(c) = c$. Then $b \not\sim_L c$. The language L' has three programs A,B,C with $A\uparrow$, $B\downarrow$ and $C\downarrow$. There is only the identity observer O in L'. Then $B \sim_{L'} C$. Let the translation be defined as $T : L \to L'$ with $T(a) = A, T(b) = B, T(c) = C$, and $T(O_1) = T(O_2) = O$. Then convergence equivalence holds, but neither equational adequacy nor observational correctness. Note that T is not compositional, since $T(O_2(b)) = A$ while $T(O_2)(T(b)) = O(B) = B$.

Example 3.5 (Observational correctness does not imply full abstraction). A simple example taken from [10] is the identity encoding from the OSP-calculus λ_{cp} without the projections **fst** and **snd** into full λ_{cp}. Then, in the restricted OSP-calculus, all pairs are indistinguishable but the presence of the observers (here simply taken as contexts) **fst** $[\cdot]$ and **snd** $[\cdot]$ in λ_{cp} permits more distinctions to be made.

Example 3.6 (Convergence equivalence is not implied by full abstraction). A trivial example is given by two calculi $\mathscr{C}$ with $p \Downarrow$ for all p, and $\mathscr{C}'$ with the same programs and $\neg p \Downarrow'$ for all p. For the translation $T(p) = p$ for all p it is clear that $\forall p_1, p_2 : p_1 \leq p_2 \iff T(p_1) \leq' T(p_2)$ holds, but T does not preserve convergence.

By standard arguments it can be shown that translations compose:

Proposition 3.7. *Let $\mathscr{C}, \mathscr{C}', \mathscr{C}''$ be program calculi, and $T : \mathscr{C} \to \mathscr{C}'$, $T' : \mathscr{C}' \to \mathscr{C}''$ be translations. Then $T' \circ T : \mathscr{C} \to \mathscr{C}''$ is also a translation, and for every property P from Definition 3.2, if T, T' have property P, then also the composition $T' \circ T$.*

We now consider the case that only new language primitives are added to a language, together with their operational semantics, which are then encoded by the translation. This is usually known as removing 'syntactic sugar'.

Definition 3.8. An OSP-calculus $\mathscr{C}$ is an *extension* of the OSP-calculus $\mathscr{C}'$ iff there is a compositional translation $\iota : \mathscr{C}' \to \mathscr{C}$, called an *embedding*, which is injective on the expressions, types and observers, and is convergence equivalent.

Informally, this can be described (after identifying $\mathscr{C}'$-programs with their image under ι) as follows: every $\mathscr{C}'$-type is also a $\mathscr{C}$-type, $\mathscr{P}'_\tau \subseteq \mathscr{P}_\tau$, and $\mathscr{O}'_{\tau,\tau'}$ is a subset of $\mathscr{O}_{\tau,\tau'}$, and the test-predicates coincide on $\mathscr{C}'$-programs. The embedding of $\mathscr{O}'_{\tau,\tau'}$ into $\mathscr{O}_{\tau,\tau'}$ is slightly more involved, since the $\mathscr{C}'$-observers are restrictions (as functions) of $\mathscr{C}$-observers. Note that for the case of contexts as observers, the embedding of $\mathscr{O}'_{\tau,\tau'}$ into $\mathscr{O}_{\tau,\tau'}$ is unique. The conditions imply that an embedding ι is adequate, but not necessarily fully abstract.

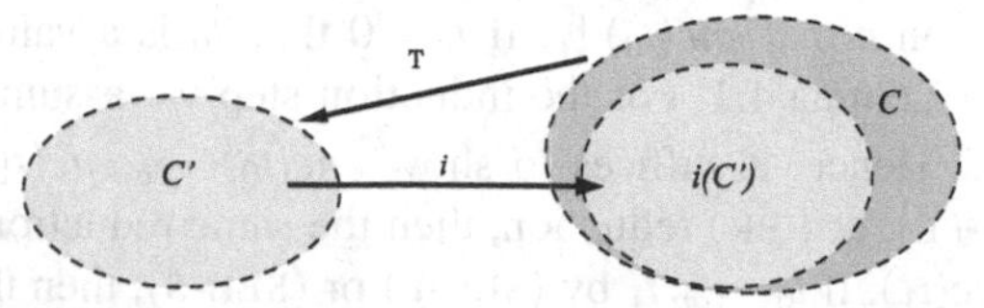

If $\mathscr{C}$ is an extension of $\mathscr{C}'$, then an observationally correct translation $T : \mathscr{C} \to \mathscr{C}'$ (plus some obvious conditions) has the nice consequence of T and ι being fully abstract.

An example for an embedding is the trivial embedding $inc : \lambda_c \to \lambda_{cp}$, which is adequate by Proposition 3.3, since the embedding inc is compositional and convergence equivalent. This allows us to reason about contextual equivalence in λ_{cp} and transfer this result to λ_c, i.e. a proof of $t_1 \sim_{cp} t_2$ where t_1, t_2 are also expressions of λ_c directly shows $t_1 \sim_c t_2$. Disproving an equivalence in λ_{cp}, however, does *not* imply that this equivalence is false in λ_c.

Proposition 3.9 (Full Abstraction for Extensions). *Let $\mathscr{C}$ be an extension of $\mathscr{C}'$, and let $T : \mathscr{C} \to \mathscr{C}'$ be an observationally correct translation, such that $T \circ \iota$ is the identity on $\mathscr{C}'$-programs, on $\mathscr{C}'$-observers, and on $\mathscr{C}'$-types. Then the translation T as well as the embedding ι are fully abstract.*

Proof. First we show that T is fully abstract. Adequacy follows from Proposition 3.3. It remains to show the converse condition for full abstraction. Let p_1, p_2 be $\mathscr{C}$-programs of type τ, and assume $p_1 \leq_{\Downarrow_i,\tau} p_2$. We have to show that $T(p_1) \leq'_{\Downarrow_i,T(\tau)} T(p_2)$. Let O' be a $\mathscr{C}'$-observer with $O'(T(p_1)) \Downarrow'_i$. Then by definition of ι there exists an observer O of $\mathscr{C}$ with $O := \iota(O')$. Since $T \circ \iota$ is the identity, we have $T(O) = O'$ and thus we obtain $T(O)(T(p_1)) \Downarrow'_i$. Observational correctness implies that $O(p_1) \Downarrow_i$. From $p_1 \leq_{\Downarrow_i,\tau} p_2$ we now derive $O(p_2) \Downarrow_i$. Again observational correctness can be applied and shows that $T(O)(T(p_2)) \Downarrow'_i$. This is equivalent to $O'(T(p_2)) \Downarrow'_i$. Since the observer O' was chosen arbitrarily, we have $T(p_1) \leq'_{\Downarrow_i,T(\tau)} T(p_2)$.

The embedding ι is already shown to be adequate. The missing direction, i.e. that $\iota(p_1) \leq'_{\Downarrow_i,T(\tau)} \iota(p_2)$ implies $p_1 \leq'_{\Downarrow_i,\tau} p_2$ follows from full abstraction of T and the assumption that $T \circ \iota$ is the identity. $\square$

$(.,.) :: \forall\alpha,\beta.\alpha \to \beta \to (\alpha,\beta)$	**unit** :: unit
fst $:: \forall\alpha,\beta.(\alpha,\beta) \to \alpha$	$\oplus :: \forall\alpha.\alpha \to \alpha \to \alpha$
snd $:: \forall\alpha,\beta.(\alpha,\beta) \to \beta$	**fix** $:: \forall\alpha,\beta.((\alpha \to \beta) \to (\alpha \to \beta)) \to (\alpha \to \beta)$

Fig. 5 Types of constants

4 Adequacy of Pair Encoding

We analyze the translation *enc* on the untyped language λ_c. Inspecting the definition of *enc* the following lemma is easy to verify:

Lemma 4.1. *For all* $s \in \lambda_{cp}$*: s is a* λ_{cp}*-value iff enc(s) is a* λ_c*-value.*

Lemma 4.2. *Let* $t \in \lambda_{cp}$ *with* $t\downarrow_{cp}$*, then* $enc(t)\downarrow_c$*.*

Proof. Let $t_0 \in \lambda_{cp}$ with $t\downarrow_{cp}$, so $t_0 \to_{cp} t_1 \to_{cp} \cdots \to_{cp} t_n$ where t_n is a value. We show by induction on n that $enc(t_0)\downarrow_c$. If $n = 0$ then t_0 is a value and $enc(t_0)$ must be a value, too, by Lemma 4.1. For the induction step we assume the induction hypothesis $enc(t_1)\downarrow_c$. Hence, it suffices to show $enc(t_0) \xrightarrow{*}_c enc(t_1)$. If $t_0 \to_{cp} t_1$ is a (β-CBV), (FIX), ($\oplus$L), or ($\oplus$R) reduction, then the same reduction can be used in λ_c, and $enc(t_0) \to_c enc(t_1)$. If $t_0 \to_{cp} t_1$ by (SEL-F) or (SEL-S), then three (β-CBV) steps are necessary in λ_c, i.e., $enc(t_0) \xrightarrow{3}_c enc(t_1)$. □

For the other direction, i.e., for proving the claim $enc(t)\downarrow_c \implies t\downarrow_{cp}$ the counter example 2.1 shows that the translation *enc* is not adequate and not observationally correct. Moreover, this example shows that an untyped language does in general not permit an adequate – and hence also not an observationally correct – translation into a subset of itself.

4.1 Typing λ_{cp}

One solution to prevent the counter example 2.1 is to consider a simply typed variant λ_{cp}^T of λ_{cp} as follows. The types are given by $\tau ::= \mathsf{unit} \mid \tau \to \tau \mid (\tau,\tau)$, and only typed expressions and typed contexts are in the language λ_{cp}^T, where we assume a hole $[\cdot]_\tau$ for every type τ. For typing, we treat pairs, projections, the unit value, and the operators $\oplus$ and **fix** as a family of constants with the types given in Fig. 5. Type safety can be stated by a preservation theorem for all expressions and a progress theorem for closed expressions. The framework now permits to prove adequacy via observational correctness of the translations.

Proposition 4.3. *For* λ_{cp}^T*, the (correspondingly restricted) translation* $enc : \lambda_{cp}^T \to \lambda_c$ *is compositional and convergence equivalent, and hence adequate.*

Proof. Compositionality follows from the definition of *enc* (see Fig. 4). Lemma 4.1 also holds if *enc* is restricted to λ_{cp}^T. We split the proof into four parts:

1. $t \downarrow_{cp} \implies enc(t) \downarrow_c$: Follows from Lemma 4.2.
2. $enc(t) \downarrow_c \implies t \downarrow_{cp}$: An inspection of the reductions shows that if t_1 is reducible, then for every reduction *Red* of $enc(t_1)$ to a value, there is some t_2 with $t_1 \rightarrow_{cp} t_2$ and $enc(t_1) \xrightarrow{+}_c enc(t_2)$ is a prefix of *Red*. We use induction on the length of a reduction *Red* of $enc(t)$ to a value to show that a corresponding reduction can be constructed. The base case is proved in Lemma 4.1. If t is an irreducible non-value, then due to typing it is an open expression of one of the forms $\mathbb{E}[(x\ r)], \mathbb{E}[\mathbf{fix}\ x], \mathbb{E}[\mathbf{fst}\ x], \mathbb{E}[\mathbf{snd}\ x]$, where x is a free variable. But the cases are not possible, since $enc(t)$ is either an irreducible non-value, or $enc(t)$ reduces in one step to an irreducible non-value.
3. $enc(t) \Downarrow_c \implies t \Downarrow_{cp}$: We prove that $t \uparrow_{cp} \implies enc(t) \uparrow_c$ by induction on the length of a reduction $t \xrightarrow{*}_{cp} t'$, where $t' \Uparrow_{cp}$. For the base case $t \Uparrow_{cp}$ and (2) show that $enc(t) \Uparrow_c$. The induction consists in computing a reduction sequence $enc(t) \xrightarrow{*}_c r$ where $r \Uparrow_{cp}$ and the correspondence is as in the proof of Lemma 4.2, such that $t \xrightarrow{*}_{cp} t'$ and $r = enc(t')$. By type preservation, t' is well-typed and now the base-case reasoning applies.
4. $t \Downarrow_{cp} \implies enc(t) \Downarrow_c$: Proving $enc(t) \uparrow_c \implies t \uparrow_{cp}$ can be done using the same technique as in the previous parts. □

Note that Proposition 3.9 cannot be applied since λ^T_{cp} is not an extension of untyped λ_c. As expected, full abstraction does not hold. For instance, let $s = \lambda p.((\lambda y.\lambda z.(y,z))\ (\mathbf{fst}\ p)\ (\mathbf{snd}\ p))$, and $t = \lambda p.p$. Then the equation $s \sim_{cp,(\mathsf{unit},\mathsf{unit})\to(\mathsf{unit},\mathsf{unit})} t$ holds in λ^T_{cp} by standard reasoning, but after translation to λ_c, we have $enc(s) \not\sim_c enc(t)$. The latter can be seen with the context $C = ([\cdot]\ \mathbf{unit})$, since $C[enc(s)]$ is must-divergent while $C[enc(t)]$ must-converges.

The extension situation could perhaps be regained by a System F-like type system, which we leave for future research. Here we just observe that the use of a simple type system for λ_c is insufficient since the encoding of pairs with components of different types cannot be simply typed. The same holds for Hindley-Milner polymorphic typing: to see this, let $s, r \in \lambda_{cp}$ where s is defined as before and $r = s\ (\mathbf{unit}, \lambda x.x)$. The most general type of $enc(s)$ in a Hindley-Milner system is $((\alpha \to \alpha \to \alpha) \to \beta) \to (\beta \to \beta \to \gamma) \to \gamma$, which essentially means that the encoding requires the components of a pair to have equal type. The reason for the insufficient type is the monomorphic use of the argument variable p of $enc(s)$. Hence, $enc(r)$ is not typeable using a Hindley-Milner type system.

One can establish a fully-abstract translation between λ^T_{cp} and a variant of λ_c by using a 'virtual typing' in λ_c which, intuitively, restricts λ_c to the image of the translation (see [19, Appendix C]).

4.2 Modifying Reduction Strategies

As a final example we extend λ^T_{cp} in two steps. First, in λ_{cpg}, we allow pairs with arbitrary expressions as components (see Fig. 6). Second, in λ_{cpig}, we relax the reduction

$w \in Val_{cpig}$ iff $w \in Val_{cpg}$ iff $w \in Val_{cp}$
$t \in Exp_{cp(i)g} ::= w \mid t_1\, t_2 \mid t_1 \oplus t_2 \mid (t_1,t_2)$

Fig. 6 Syntax of λ_{cpg} and λ_{cpig}

$$\mathbb{E}_{cpig} ::= [\,] \mid \mathbb{E}_{cpig}\, t \mid w\mathbb{E}_{cpig} \mid \mid \mathbb{E}_{cpig} \oplus t \mid t \oplus \mathbb{E}_{cpig} \mid (\mathbb{E}_{cpig},t) \mid (t,\mathbb{E}_{cpig})$$

Fig. 7 Evaluation Contexts $\mathbb{E}_{cpig}$ for λ_{cpig}

$$\mathbb{E}_{cpg} ::= [\,] \mid \mathbb{E}_{cpg}\, t \mid w\mathbb{E}_{cpg} \mid \mathbb{E}_{cpg} \oplus t \mid w \oplus \mathbb{E}_{cpg} \mid (\mathbb{E}_{cpg},t) \mid (w,\mathbb{E}_{cpg})$$

Fig. 8 Evaluation Contexts for λ_{cpg}

$$\begin{array}{lll}
enc_i : \lambda_{cpg} \to \lambda^T_{cp} : & enc_i(t) & = t \\
enc_g : \lambda_{cpig} \to \lambda_{cpg} : & enc_g((t_1,t_2)) & = (\lambda x\, y.(x,y))\ enc_g(t_1)\ enc_g(t_2) \quad \text{if } \{t_1,t_2\} \not\subseteq Var \\
: & enc_g((x,y)) & = (x,y) \\
: & enc_g(t) & = \text{descending, not changing the structure} \quad \text{otherwise;} \\
enc_{ig} : \lambda_{cpig} \to \lambda^T_{cp} : & enc_{ig} & = enc_g \circ enc_i
\end{array}$$

Fig. 9 Translations between λ_{cpig}, λ_{cpg} and λ^T_{cp}

strategy by allowing interleaving evaluation of pair components and of the arguments of the choice-operator. The corresponding evaluation contexts $\mathbb{E}_{cpig}$ for the calculus λ_{cpig} are in Fig. 7.

4.2.1 Permitting General Pairs We consider the extension λ_{cpg} of the language λ^T_{cp} where λ_{cpg} is simply typed, and where pairs are not restricted to values. The syntax is shown in Fig. 6, the evaluation contexts in λ_{cpg} are introduced in Fig. 8. The reductions are as in λ^T_{cp}. We show that $enc_g : \lambda_{cpg} \to \lambda^T_{cp}$ is a fully abstract translation and hence nothing is lost by restricting pairs to values. Type preservation and progress also hold for λ_{cpg}. Moreover, enc_g is compositional and is easily seen to map well-typed terms of λ_{cpg} to well-typed terms of λ^T_{cp}.

Lemma 4.4. *For the translation enc_g the following holds: For all s, if s is a λ_{cpg}-value, then $enc_g(s)$ is must-convergent and has a deterministic reduction to a value. Moreover, for all s, if $enc_g(s)$ is a value, then s is a λ_{cpg}-value.*

Proof. By induction on the size of expressions and inspection of all cases. This holds also for the case $(w_1,w_2) \mapsto (\lambda x\, y.(x,y))\ enc_g(w_1)\ enc_g(w_2)$, since $enc_g(w_1)$, $enc_g(w_2)$ are must-convergent and independently reduce to values, and then two deterministic beta-reductions reduce the resulting expression to a value. □

Proposition 4.5. *The translation enc_g is fully abstract.*

Proof. By Proposition 3.9, and since the identity: $\lambda^T_{cp} \to \lambda_{cpg}$ is an embedding (see Definition 3.8), it suffices to prove observational correctness of the translation. Note that $enc_g(t) = t$, for all λ^T_{cp}-terms t, which makes Proposition 3.9 applicable. We have to show four implications.

1. $t\downarrow_{cpg} \Longrightarrow enc_g(t)\downarrow_{cp}$: This follows by a straightforward translation from the $t\downarrow_{cpg}$-reduction into a reduction of $enc_g(t)$. In the case of non-value pairs, (β-CBV)-reductions have to be added to produce pairs in λ^T_{cp}.

2. $enc_g(t)\downarrow_{cp} \Longrightarrow t\downarrow_{cpg}$: A reduction $enc_g(t)\downarrow_{cp}$ can be re-translated into one of t, by observing that (t_1,t_2) on the λ_{cpg}-side may correspond to three different possibilities on the λ_{cp}-side: it may be (t_1',t_2'), $(\lambda xy.(x,y))\ t_1'\ t_2'$ or $(\lambda y.(t_1',y))\ t_2'$.
3. $t\Downarrow_{cpg} \Longrightarrow enc_g(t)\Downarrow_{cp}$: We show $enc_g(t)\uparrow_{cp} \Longrightarrow t\uparrow_{cpg}$. Again the reductions correspond, up to the (β-CBV)-reductions for the pair-encoding. The base case is that $enc_g(t)\Uparrow_{cp} \Longrightarrow t\Uparrow_{cpg}$, which follows from (1).
4. $enc_g(t)\Downarrow_{cp} \Longrightarrow t\Downarrow_{cpg}$: We show $t\uparrow_{cpg} \Longrightarrow enc_g(t)\uparrow_{cp}$. As above, the reductions correspond up to the (β-CBV)-reductions for the pair-encoding. The base case is $t\Uparrow_{cpg} \Longrightarrow enc_g(t)\Uparrow_{cp}$, and follows from (2). □

Remark 4.6. The combined translation from λ_{cpg} to λ_c is $enc_{gc} := enc \circ enc_g$. It operates on pairs of non-variables s,t as follows: $enc_{gc}((s,t)) = enc(\lambda xy.(x,y))\ enc_{gc}(s)\ enc_{gc}(t) = (\lambda xy.(\lambda p.p\,x\,y))\ enc_{gc}(s)\ enc_{gc}(t)$. The naive translation $T'((s,t)) = (\lambda p.p\ T'(s)\ T'(t)))$ is not convergence equivalent, since for example $T'((\Omega,\Omega)) = \lambda p.p\ \Omega\ \Omega$. However, (Ω,Ω) must-diverges, whereas $\lambda p.p\ \Omega\ \Omega$ is a value and thus converges.

4.2.2 Permitting Interleaved Reductions In this subsection we will show that it is also correct to modify the reduction strategy in the OSP-calculus λ_{cpg}, where we allow that the arguments of choice and of pairs may be evaluated independently (i.e. interleaved, in any order). The OSP-calculus λ_{cpig}, i.e. its syntax and the evaluation contexts $\mathbb{E}_{cpig}$ used for reduction have been introduced in Fig. 6 and Fig. 7. The translation $enc_i : \lambda_{cpig} \to \lambda_{cpg}$ is just the identity (see Fig. 9). However, it is not immediately obvious that the convergence predicates of λ_{cpig} and λ_{cpg} are the same, due to the independent reduction possibilities in λ_{cpig}. We denote the reduction in λ_{cpig} with $\to_{cpig}$ and the reduction in λ_{cpg} with $\to_{cpg}$.

Proposition 4.7. *The identity translation enc_i from λ_{cpig} into λ_{cpg} is fully abstract.*

Proof. Obviously enc_i (and its inverse) are compositional. Thus, to prove observational correctness it suffices to establish convergence equivalence. We have to show four implications:

1. $enc_i(t)\downarrow_{cpg} \Longrightarrow t\downarrow_{cpig}$: This follows by using the same reduction sequence.
2. $t\downarrow_{cpig} \Longrightarrow enc_i(t)\downarrow_{cpg}$: A reduction corresponding to $t\downarrow_{cpig}$ can be rearranged until it is a reduction w.r.t. λ_{cpg}, since the reductions are at independent positions, and the final result is a value without any reductions.
3. $enc_i(t)\Downarrow_{cpg} \Longrightarrow t\Downarrow_{cpig}$: We show the equivalent $t\uparrow_{cpig} \Longrightarrow enc_i(t)\uparrow_{cpg}$. Let *Red* be a λ_{cpig}-reduction of $enc_i(t)$ to a must-divergent expression. We use induction on the measure (l,n), where l is the number of reductions and n is the number of non-value surface positions of $enc_i(t)$, i.e. positions not within abstractions. Now consider the λ_{cpg}-redex in $enc_i(t)$. If the reduction of the redex is contained in *Red*, then we can shift it to the start, and we obtain a shorter reduction, i.e. l is decreased. Otherwise, if the reduction of the redex is not contained in *Red*, there are two possibilities. If the redex is must-divergent, then we are finished, since then $enc_i(t)$ is also must-divergent. Otherwise, if the redex is not must-divergent, then we simply select a converging reduction of the redex to a value. This reduction can be integrated into

Red. In this case the number of reductions does not change, but the number n of the measure will be reduced. In any case, we can use induction. The base case follows from (1).

4. $t{\Downarrow_{cpig}} \Longrightarrow enc_i(t){\Downarrow_{cpg}}$: We show $enc_i(t){\uparrow_{cpg}} \Longrightarrow t{\uparrow_{cpig}}$. We can leave the reduction unchanged. The base case is $enc_i(t){\Uparrow_{cpg}} \Longrightarrow t{\Uparrow_{cpig}}$, which follows from (2).

Finally, full abstraction follows from Proposition 3.9, since the proof also shows that the inverse of enc_i is convergence equivalent. □

Remark 4.8. Note that in languages with shared variable concurrency (for instance, extensions of λ_{cp} with reference cells) the modification of the reduction strategy given in this subsection is no longer correct: permitting interleaving reductions of the arguments can be observed through their read and write effects on shared variables.

Using Proposition 3.7 we have:

Theorem 4.9. *The translation enc_{ig} is fully abstract. For $enc : \lambda_{cp}^{T} \to \lambda_c$ the combined translation $enc \circ enc_{ig} : \lambda_{cpig} \to \lambda_c$ is adequate.*

Conclusions and Outlook

Motivated by translation problems between concurrent programming languages, this paper succeeded in clarifying the methods, and providing tools, to assess the correctness of translations. The framework is general enough to apply directly to an operational semantics and the derived contextual equivalences, without relying on the availability of models.

In future research we want to exploit these results, to prove the correctness of various implementations of synchronization constructs in concurrent languages.

Acknowledgements We thank the anonymous referees for their valuable comments.

References

1. *The Alice Project* (2007). Saarland University, `http://www.ps.uni-sb.de/alice`
2. Carayol, A., Hirschkoff, D., Sangiorgi, D.: On the representation of McCarthy's amb in the pi-calculus. Theoret. Comput. Sci. **330**(3), 439–473 (2005)
3. De Nicola, R., Hennessy, M.: Testing equivalences for processes. Theoret. Comput. Sci. **34**, 83–133 (1984)
4. Felleisen, M.: On the expressive power of programming languages. Sci. Comput. Programming **17**(1–3), 35–75 (1991)
5. Gordon, A.D.: Bisimilarity as a theory of functional programming. Theoret. Comput. Sci. **228**(1–2), 5–47 (1999)
6. Kutzner, A., Schmidt-Schauß, M.: A nondeterministic call-by-need lambda calculus. In: Proc. ICFP, pp. 324–335. ACM (1998)

7. Matthews, J., Findler, R.B.: Operational semantics for multi-language programs. In: 34th ACM POPL, pp. 3–10. ACM (2007)
8. McCusker, G.: Full abstraction by translation. In: Advances in Theory and Formal Methods of Computing. IC Press (1996)
9. Milner, R.: Fully abstract models of typed lambda calculi. Theoret. Comput. Sci. **4**(1), 1–22 (1977)
10. Mitchell, J.C.: On abstraction and the expressive power of programming languages. Sci. Comput. Programming **21**(2), 141–163 (1993)
11. Niehren, J., Sabel, D., Schmidt-Schauß, M., Schwinghammer, J.: Observational semantics for a concurrent lambda calculus with reference cells and futures. Electron. Notes Theor. Comput. Sci. **173**, 313–337 (2007)
12. Niehren, J., Schwinghammer, J., Smolka, G.: A concurrent lambda calculus with futures. Theoret. Comput. Sci. **364**(3), 338–356 (2006)
13. Peyton Jones, S., Gordon, A., Finne, S.: Concurrent Haskell. In: 23rd ACM POPL, pp. 295–308. ACM (1996)
14. Pitts, A.D.: Parametric polymorphism and operational equivalence. Math. Structures Comput. Sci. **10**, 321–359 (2000)
15. Riecke, J.G.: Fully abstract translations between functional languages. In: 18th ACM POPL, pp. 245–254. ACM (1991)
16. Ritter, E., Pitts, A.M.: A fully abstract translation between a lambda-calculus with reference types and Standard ML. In: Proc. 2nd TLCA, pp. 397–413. Springer (1995)
17. Sabel, D., Schmidt-Schauß, M.: A call-by-need lambda-calculus with locally bottom-avoiding choice: Context lemma and correctness of transformations. Math. Structures Comput. Sci. (2008). Accepted for publication
18. Sanjabi, S.B., Ong, C.H.L.: Fully abstract semantics of additive aspects by translation. In: Proc. 6th OASD, pp. 135–148. ACM (2007)
19. Schmidt-Schauß, M., Niehren, J., Schwinghammer, J., Sabel, D.: Adequacy of compositional translations for observational semantics. Technical report Frank-33, J.W.Goethe Universität, Frankfurt (2008)
20. Shapiro, E.: Separating concurrent languages with categories of language embeddings. In: 23rd ACM STOC, pp. 198–208. ACM (1991)
21. Søndergaard, H., Sestoft, P.: Non-determinism in functional languages. Comput. J. **35**(5), 514–523 (1992)

The Surprising Robustness of (Closed) Timed Automata against Clock-Drift

Mani Swaminathan[1,2], Martin Fränzle[1], and Joost-Pieter Katoen[3]

[1] University of Oldenburg {mani.swaminathan,fraenzle}@informatik.uni-oldenburg.de
[2] ASKON Consulting Group GmbH
[3] RWTH Aachen University katoen@cs.rwth-aachen.de

Abstract. We investigate reachability (or equivalently, safety) for timed systems modelled as Timed Automata (TA) under notions of "robustness", i.e., when the clocks of the TA may drift by small amounts. Our contributions are two-fold: (1) We first consider the model of clock-drift introduced by Puri [1] and subsequently studied in other works [2, 3, 4]. We show that the standard zone-based forward reachability analysis performed by tools such as UPPAAL is in fact exact for TA with closed guards, invariants, and targets, when testing robust safety of timed systems having an arbitrary, but finite lifetime. (2) Next, we consider a more realistic model of drifting clocks that takes into account the regular resynchronization performed in most practical systems. We then show that the standard reachability analysis of tools like UPPAAL again suffices to test for robust safety in this model of clock-drift, for TA with closed guards, invariants, and targets, but now without any restrictions on system life-time.

1 Introduction

Real-time systems, which have strict timing requirements, have emerged as an enabling technology for several important application domains such as air traffic control, telecommunications, and medicine, to name a few. Such systems are becoming increasingly pervasive, and hence rigorous methods and techniques to ensure their correct functioning are of utmost importance. Timed Automata (TA) [5] have been extensively studied as a formalism for modelling real-time systems. TA extend ω-automata by augmenting them with "clock" variables based on a dense-time model, which quantitatively capture the behaviour of the system with time. TA model checkers such as UPPAAL [10] and KRONOS [7] are now available and have been successfully used in several industrial case studies, such as [8].

A key result for the decidability properties of TA is the region-automaton construction [5], which partitions the inherently infinite state space of the TA into finitely many equivalence classes or "regions". The number of such regions is, however, exponential in the number of clocks, and the region construction is therefore not suited in practice for model checking TA when the number of clocks is large. Most available tools for model checking TA (such as UPPAAL) instead use on-the-fly algorithms that dynamically search through the

---

Please use the following format when citing this chapter:

Swaminathan, M., Fränzle, M. and Katoen, J.-P., 2008, in IFIP International Federation for Information Processing, Volume 273; *Fifth IFIP International Conference on Theoretical Computer Science*; Giorgio Ausiello, Juhani Karhumäki, Giancarlo Mauri, Luke Ong; (Boston: Springer), pp. 537–553.

state space of the TA, which is partitioned into "zones" [10]. Associated data structures such as Difference Bound Matrices (DBMs) [10] are used to represent zones in TA-based verification. Reachability analysis forms the core of such verification tools [9] and is implemented by a Forward Reachability Analysis (FRA) algorithm that computes the set of successors of a zone, with termination being enforced by zone-widening using k-normalization [10].

However, such analyses, whether region- or zone-based, assume that the clocks of the TA are perfectly synchronous, which is not the case in practice, where the clocks could drift by small amounts. It is shown in [1] that the usual region-based analysis is not correct w.r.t. reachability when considering perturbations in the clocks, in the sense that an unsafe state, reported as unreachable for perfect clocks, might well be reachable by iterating often enough through a cycle in the TA, even when the clocks drift by infinitesimally small amounts, and such a TA is therefore not "robustly safe". This insight leads to the definition of robust reachability, where a reachability property is considered to be "robustly (in-)valid" iff it does not change its validity for some small relative drift between clocks.

"Robust" reachability analysis [1, 2] therefore computes the set of states that are reachable for *every* (i.e., even the slightest) drift, reporting the TA as not being robustly safe iff that enlarged reach-set contains an unsafe state (where the guards and invariants of the TA, and the unsafe target state, are all assumed to be closed). Robust reach-set computation in [1, 2] is based on searching the strongly connected components of the region-graph, thus suffering from the exponential size of the region-graph in the number of clocks. Zone-based algorithms that compute this reach-set more efficiently are presented in [3, 4]. For a given TA with maximum clock-drift parameterized by $\varepsilon > 0$, with the corresponding reachable state-space being $Reach^{\varepsilon}$, the algorithms in [1, 2, 3] compute the set $\cap_{\varepsilon>0} Reach^{\varepsilon}$ and test it for empty intersection with the (closed) target. It is shown in [1] that $\cap_{\varepsilon>0} Reach^{\varepsilon}$ has an empty intersection with the closed target state iff there exists some $\varepsilon > 0$ such that the intersection of $Reach^{\varepsilon}$ with the (closed) target state is again empty. The algorithms in all these works however alternate between forward and backward analysis, and thus induce a performance overhead compared to the standard FRA algorithm used within tools like UPPAAL. All the above works (except [4]) assume that the guards, invariants, and targets of the TA are closed. Furthermore, all of them assume that each cycle of the TA is a *progress cycle*, wherein every clock is reset at-least once per cycle. The unsafe states that become reachable with drifting clocks (but which are unreachable with perfect clocks) are added to such robust reach-sets only by iterating an *unbounded* number of times through the (progress) cycles of the TA, thereby requiring that the *life-time of the systems be infinite*. Moreover, the model of clock-drift considered in these works is one of *unbounded* relative drift between the clocks, which does not take into account the *regular resynchronization* of clocks that is performed in practical real-time systems. This paper addresses these two issues, with two main contributions:

1. We first consider the model of clock-drift introduced by Puri [1] and studied subsequently by others [2, 3, 4]. We show that, under the assumption of closed guards, invariants, and targets, the standard zone-based FRA of TA performed by tools such as UPPAAL is indeed exact when testing for robust safety of timed systems having an *arbitrary, but finite* life-time. We test here whether the TA can robustly avoid the target arbitrarily long, in the following sense: for any given number i of iterations of the transition relation, there is $\varepsilon_i > 0$ such that $Reach_i^{\varepsilon_i}$ has an empty intersection with the target state, where $Reach_i^{\varepsilon_i}$ is the reachable state space after i iterations of the transition relation under maximum perturbation ε_i of the clocks. Note that ε_i may depend on the number i of executed iterations, with ε_i decreasing (not necessarily strictly) with i, and potentially tending to 0 as i tends to ∞. Thus, robust safety under our notion does not imply the existence of a homogeneous $\varepsilon > 0$ that is independent of the number of iterations and such that $Reach^{\varepsilon}$ has an empty intersection with the target state (which is the notion considered in previous works [1, 2, 3, 4]). However, our notion of robust safety implies avoidance of the target state by some strictly positive value of the perturbation for any *arbitrary, but finite* number of iterations. This is applicable to all systems having a finite life-time.
2. Next, we introduce a more realistic model of clock-drift that takes into account the *regular resynchronization* performed in practical real-time systems (such as bit-stuffing in communication protocols), which results in a *bounded* relative clock-drift. Under the assumption of closed guards, invariants, and targets, we show that the standard zone-based FRA of TA is again exact when testing for robust safety of such timed systems with clock resynchronization. In this case, a certification of robust safety imposes no restriction on the life-time of the system — it implies avoidance of the (closed) target by all $0 < \varepsilon < 1$ (where the ε now parameterizes the maximum relative bounded clock-drift subject to periodic resynchronization) independent of the number of iterations.

The rest of the paper is organized as follows: Section 2 briefly reviews TA definitions and semantics, along with our assumptions. It also presents the standard algorithm for zone-based FRA. Section 3 describes the robustness problem for TA in the context of the model of clock-drift considered by Puri and others, and shows the exactness of the standard zone-based FRA algorithm w.r.t robust safety for systems having a finite life-time. Section 4 then introduces our model of bounded clock-drift that accounts for regular clock resynchronization, and shows the exactness of the standard zone-based FRA algorithm w.r.t robust safety, but now without any restrictions on the life-time of the system. Section 5 concludes the paper and sketches future research directions.

2 Timed Automata (TA)

Given a finite set C of *clocks*, a *clock valuation* over C is a map $v : C \to \mathbb{R}_{\geq 0}$ that assigns a non-negative real value to each clock in C. If n is the number of clocks, a clock valuation is basically a point in $\mathbb{R}^n_{\geq 0}$, which we henceforth denote by $\mathbf{u}, \mathbf{v}$ etc.

Definition 1. A *zone* over a set of clocks C is a system of constraints defined by the grammar $g ::= x \rhd d \mid x - y \rhd d \mid g \wedge g$, where $x, y \in C$, $d \in \mathbb{N}$, and $\rhd \in \{<, \leq, >, \geq\}$. The set of zones over C is denoted $Z(C)$.

A *closed zone* is one in which $\rhd \in \{\leq, \geq\}$, and we denote the set of closed zones over C by $Z_c(C)$. A zone with no bounds on clock differences (i.e., with no constraint of the form $x - y \rhd d$) is said to be *diagonal-free*, and we denote the corresponding set of zones by $Z_d(C)$. The set $Z_{cd}(C)$ denotes zones that are *both closed and diagonal-free*. The set $Z_{cdU}(C)$ denotes the set of *closed, diagonal-free zones having no lower bounds on the clocks.*

Definition 2. A TA is a tuple $A = (L, C, (l_0, \mathbf{0}), T, Inv)$, with

- a finite set L of *locations* and a finite set C of *clocks*, with $|C| = n$.
- An *initial* location $l_0 \in L$ together with the initial clock-valuation $\mathbf{0}$ where all clocks are set to 0 [1]
- a set $T \subseteq L \times Z_{cd}(C) \times 2^C \times L$ of possible *transitions* between locations. A transition t between two locations (l, l') is denoted $l \xrightarrow{t} l'$, and involves a *guard* $G(t) \in Z_{cd}(C)$ and a *reset set* $Res_t \subseteq C$.
- $Inv : L \to Z_{cdU}(C)$ assigns *invariants* to locations

In the sequel, we will denote by k the *clock ceiling* of the TA A under investigation, which is the largest constant among the constraints of A (including the predicate defining the unsafe state). Note that we assume that the guards of the automaton are *closed and diagonal-free zones.* Invariants in addition have only *upper-bounds* on clocks. Diagonal constraints of the form $x - y \rhd d$ thus are not part of the TA syntax, but are of relevance, since they occur during the course of forward reachability analysis as a result of the *time-passage* operation defined as follows:

Definition 3. For a clock valuation $\mathbf{x}$, its *time-passage* is $timepass(\mathbf{x}) = \{\mathbf{x} + d \mid d > 0\}$, where $\mathbf{x} + d$ denotes the addition of a strictly positive scalar d to each component of $\mathbf{x}$. This is canonically lifted to clock-zones Z as $timepass(Z) = \bigcup_{\mathbf{x} \in Z} \; timepass(\mathbf{x})$.

Definition 4. $\lfloor \mathbf{x} \rfloor_k$ denotes the k-region containing $\mathbf{x}$, which is the equivalence class induced by the *k-region-equivalence relation* $\approx_k$. For two clock valuations $\mathbf{x}$ and $\mathbf{y}$, $\mathbf{x} \approx_k \mathbf{y}$ iff

[1] We assume without loss of generality that all clocks are initially set to 0.

$$\forall i \leq n : \left(\begin{array}{c} (x_i > k) \wedge (y_i > k) \\ \vee\ ((int(x_i) = int(y_i)) \wedge (fr(x_i) = 0 \Leftrightarrow fr(y_i) = 0) \wedge \\ \forall j \leq n :\ (fr(x_i) \leq fr(x_j) \Leftrightarrow fr(y_i) \leq fr(y_j))) \end{array} \right)$$

Here, for a clock valuation $\mathbf{x} \in \mathbb{R}^n_{\geq 0}$, x_i denotes its i-th component, i.e., the value of the i-th clock, and $int(x_i)$ and $fr(x_i)$ respectively denote the integer and fractional parts of x_i.

Definition 5. [11] A *k-bounded zone (k-zone)* has no constant exceeding k among its constraints. For a zone Z, its *k-normalization*, denoted $norm_k(Z)$, is the smallest k-bounded zone containing Z.

If Z is a k-zone, $norm_k(Z) = Z$. k is taken to be the largest constant appearing in the constraints (including the unsafe state) of the TA.

Definition 6. $Reach \subseteq L \times (C \to \mathbb{R}_{\geq 0})$ is the reach-set of the TA A, consisting of an infinite set of (concrete) states of the TA of the form $(l, \mathbf{x})$, where $l \in L$ and $\mathbf{x} \in \mathbb{R}^n_{\geq 0}$. It is defined inductively as follows, with $Reach_i$ denoting the reach-set under $i \in \mathbb{N}$ steps, starting from the initial state $(l_0, \mathbf{0})$ and alternating between time-passage and discrete-location transitions:[2]

- $Reach_0 = \{(l_0, \mathbf{0})\}$.
- if i even $Succ(Reach_i) = \left\{ (l, \mathbf{x}) \,\middle|\, \begin{array}{l} \exists \mathbf{u} \in Inv(l) : (l, \mathbf{u}) \in Reach_i \\ \wedge\ \mathbf{x} \in timepass(\mathbf{u}) \cap Inv(l) \end{array} \right\}$
- if i odd $Succ(Reach_i) = \left\{ (l, \mathbf{x}) \,\middle|\, \begin{array}{c} \exists t \in T, l' \in L, \mathbf{u} \in Inv(l') \cap G(t) : \\ l' \xrightarrow{t} l \wedge (l', \mathbf{u}) \in Reach_i \\ \wedge\ \mathbf{x} \in Inv(l) \cap Res_t(\mathbf{u}) \end{array} \right\}$,
  where $Res_t(\mathbf{u})(c) = \mathbf{u}(c)$ iff $c \notin Res_t$, else $Res_t(\mathbf{u})(c) = 0$.
- $\forall i \geq 0, Reach_{i+1} = Reach_i \ \cup\ Succ(Reach_i)$.
- $Reach = \bigcup_{i \in \mathbb{N}} Reach_i$.

Reach is computed in tools like UPPAAL by the following zone-based forward reachability algorithm. Given a timed automaton A with the target (l, B), it decides whether $Reach \cap (l, B) \neq \emptyset$. Reachable state sets are represented by lists $\langle (l_1, Z_1), \ldots, (l_m, Z_m) \rangle$ of location-zone pairs. Let R_i denote the (symbolic) reachable state-space at the i-th ($i \geq 0$) iteration.

1. Start with the state-set $R_0 = \{(l_0, \mathbf{0})\}$, or equivalently, in DBM form, $R_0 = l_0 \times \{\bigwedge_{x \in C} x - x_0 \leq 0)\}$, where $x_0 \notin C$ is a pseudo-clock used to represent the constant 0.
2. For $i \geq 0$, compute the symbolic successors of R_i, denoted $Post(R_i)$, separately for even and odd values of i, as follows:
   - If i even, $Post(R_i) = \{(l, Z) \mid \exists (l, Z') \in R_i$ $: Z = norm_k(timepass(Z')) \wedge Inv(l)\}$

[2] To simplify the proofs, we use even- and odd-numbered steps to distinguish between time-passage (of possibly zero duration) and transitions between discrete locations.

- If i odd, $Post(R_i) = \{(l, Z) \mid \exists (l', Z') \in R_i, t \in T : l' \xrightarrow{t} l$ $\wedge Z = Res_t(Z' \wedge G(t)) \wedge Inv(l)\}$

3. Build $R_{i+1} = R_i \cdot Post(R_i)$, where $\cdot$ denotes conditional concatenation that suppresses subsumed zones, i.e., removes (l, Z) if there is another (l, Z') with Z implying Z'.
4. Repeat steps (2) and (3) until $R_{i+1} = R_i$. Denote the last set R_i thus computed as R. Termination is guaranteed by the use of k-normalization, as there are only finitely many different k-zones such that only subsumed zones arise eventually.
5. Test whether $Z \wedge B$ is satisfiable for some $(l, Z) \in R$. If so then report "(l, B) is reachable", otherwise report "(l, B) is un-reachable".

It has been shown that this algorithm is sound and complete w.r.t. reachability [10] in the sense that $Reach \cap (l, B) = \emptyset$ iff $R \cap (l, B) = \emptyset$.

3 Robustness w.r.t. Clock-Drift

We have hitherto considered perfectly synchronous clocks. We now consider drifting clocks that could occur in practice, as introduced in [1] and studied subsequently by others [2, 3, 4]. This phenomenon is modelled by introducing a parameter $\varepsilon > 0$ that characterizes the relative drift between the clocks. The slopes of the clocks are assumed to be within the range $\left[\frac{1}{1+\varepsilon}, 1+\varepsilon\right]$. This is equivalent to a relative drift in the range $\left[(\frac{1}{1+\varepsilon})^2, (1+\varepsilon)^2\right]$ between the clocks. We could alternatively consider the slopes to be in the range $[1-\varepsilon, 1+\varepsilon]$. The behaviour of both models w.r.t. infinitesimally small values of ε is identical, only that in our case, the slope of a clock never becomes negative no matter how large ε is. We then have a modification of the time-passage operation as follows:

Definition 7. For a clock valuation $\mathbf{x}$, its *time-passage under perturbation of* ε is: $timepass^{\varepsilon}(\mathbf{x}) = \left\{\mathbf{x} + d \cdot \mathbf{e} \,\middle|\, d > 0,\ \mathbf{e} \in \left[\frac{1}{1+\varepsilon}, 1+\varepsilon\right]^n\right\}$.
For a Zone Z, $timepass^{\varepsilon}(Z) = \bigcup_{\mathbf{x} \in Z} timepass^{\varepsilon}(\mathbf{x})$

While this model restricts the slopes of the clocks based on the value of parameter ε, the actual relative drift between the clocks increases without bound with increasing delay $d > 0$. The reachable state space also gets enlarged. For a given perturbation of ε, the corresponding *perturbed reach-set* $Reach^{\varepsilon}$ is defined inductively, similar to the non-perturbed case, by accounting for drifting clocks through the replacement of the deterministic $timepass()$ by an appropriate non-deterministic $timepass^{\varepsilon}()$ for steps corresponding to time-passage.

We now consider the effect of clock-drift on deciding whether some location-zone pair (l, B) is reachable. As an example (cf. Fig. 1), consider a timed automaton A, consisting of a single location l_0, two clocks x, y, the invariant of l_0

being $x \leq 2$, and a self-looping transition t consisting of a guard $x = 2?$, with the associated resets $x := 0, y := 0$. Let the unsafe state of A be characterized by $(l_0, B) = (l_0, y > 2)$. Assuming perfect clocks, the state-space of A is given by $Reach = (l_0, Z)$, where $Z \equiv (x \leq 2 \wedge y = x)$, and A is clearly safe, as $Reach \cap (l_0, B) = \emptyset$. For drift characterized by a given $\varepsilon > 0$, the corresponding state space is $Reach^{\varepsilon} = (l_0, Z^{\varepsilon})$, where $Z^{\varepsilon} \equiv x \leq 2 \wedge \frac{x}{(1+\varepsilon)^2} \leq y \leq x(1+\varepsilon)^2$. Thus, $\forall \varepsilon > 0 : Reach^{\varepsilon} \cap (l_0, B) \neq \emptyset$ and A is therefore not "robustly" safe. The automaton along with the associated state-space for each case is illustrated in Fig. 1.

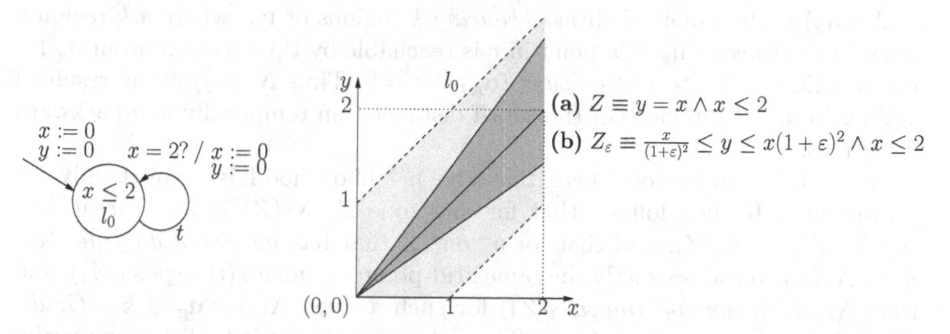

Fig. 1 A timed automaton A along with its state-spaces (a) without drift: (l_0, Z), (b) for a drift of $\varepsilon : (l_0, Z^{\varepsilon})$.

Related work on robust reachability of TA [1, 2, 3] compute the set $\cap_{\varepsilon>0} Reach^{\varepsilon}$. For this example, $\cap_{\varepsilon>0} Reach^{\varepsilon} = Reach$. This is because, for a zone Z, $\cap_{\varepsilon>0}\ timepass^{\varepsilon}(Z) = timepass(cl(Z))$, where $cl(Z)$ is the closure of Z, obtained by relaxing each strict inequality of Z to the corresponding non-strict one. In the present case, $Z \equiv \mathbf{0}$ is closed, as is $(Z \cup timepass(Z)) \cap Inv(l_0) \equiv y = x \wedge x \leq 2$, so $\cap_{\varepsilon>0} Reach^{\varepsilon} \ \cap\ (l_0, B) = \emptyset$. Hence, if open target states were allowed, the algorithms in [1, 2, 3] would all report this automaton as being robustly safe, while even the slightest perturbation would actually make the unsafe state reachable. However, if $B \equiv y \geq 2$, we see that the automaton of Fig. 1 is unsafe even with perfect clocks, while for $B \equiv y \geq 3$, the automaton is now safe even for drifting clocks, for all $0 < \varepsilon < \sqrt{1.5} - 1$.

We thus observe that *closed constraints* give consistent results while testing the automaton of Fig. 1 for safety, both with perfect clocks and under drift. Note also that this automaton has a single *progress cycle*, which additionally *resets all clocks simultaneously in a single transition.* The remit of this paper is to formulate conditions under which tests on TA for robust safety give identical results for both perfect and drifting clocks. We define for this purpose a *grid-point* and its associated *neighbourhood* as follows:

Definition 8. *Grid* denotes the set-of all *grid-points* in $\mathbb{R}^n_{\geq 0}$, i.e., $Grid = \{\mathbf{x_g} \in \mathbb{R}^n_{\geq 0} \mid \forall 1 \leq i \leq n : fract(x_{gi}) = 0\}$. For $\mathbf{x} \in \mathbb{R}^n_{\geq 0}$,

$grid(\mathbf{x}) = \{\mathbf{x_g} \in Grid \mid dist(\mathbf{x}, \mathbf{x_g}) < 1\}$, where $dist(\mathbf{x}, \mathbf{x_g}) = max_{1 \leq i \leq n}|x_i - x_{gi}|$. The subset of *Grid* that contains only those grid-points bounded by k is denoted $k - Grid$.

Thus, $\forall \mathbf{x_{kg}} \in k - Grid$: $\lfloor \mathbf{x_{kg}} \rfloor_k = \mathbf{x_{kg}}$. We will henceforth denote points in *Grid* by the suffix g ($\mathbf{x_g}$ etc.) and points in $k - Grid$ by the suffix kg ($\mathbf{x_{kg}}$ etc.).

Definition 9. For $\mathbf{u_g} \in Grid$, we define its *neighbourhood* $N_k(\mathbf{u_g}) = \bigcap_{\varepsilon > 0} \lfloor timepass^{\varepsilon}(\mathbf{u_g}) \rfloor_k$. For a zone Z, its *neighbourhood* is defined as: $N_k(Z) = \bigcup_{\mathbf{u_g} \in Z \cap Grid} N_k(\mathbf{u_g})$

$N_k(\mathbf{u_g})$ is the union of all *neighbouring* k-regions of $\mathbf{u_g}$, where a k-region r is said to *neighbour* $\mathbf{u_g}$ iff a point in r is reachable by time-passage from $\mathbf{u_g}$ for *every* drift, i.e., $\forall \varepsilon > 0 : timepass^{\varepsilon}(\mathbf{u_g}) \cap r \neq \emptyset$. Thus $N_k(\mathbf{u_g})$ is the result of adding to $\mathbf{u_g}$ all k-regions of Hausdorff distance 0 in temporally non-backward directions.

It must be understood here that the neighbourhood is defined only for grid-points[3]. It then follows that for any zone Z, $N_k(Z)$ is *idempotent*, i.e., $N_k(N_k(Z)) = N_k(Z)$, and that for a zone Z that has *no closed diagonal borders*, $N_k(Z)$ contains exactly the same grid-points as $norm_k(timepass(Z))$, and thus $N_k(Z) = norm_k(timepass(Z))$ for such a zone. Also, $\forall \mathbf{u_g} \notin k - Grid$: $N_k(\mathbf{u_g}) = norm_k(timepass(\mathbf{u_g}))$. The following lemmas establish some useful properties of the neighbourhood operator.

Lemma 1. $\forall \mathbf{x} \in \mathbb{R}^n_{\geq 0}, \forall \mathbf{u_g} \in Grid :$
$\mathbf{x} \in N_k(\mathbf{u_g}) \Leftrightarrow \forall \varepsilon > 0 \ \exists \mathbf{y} \in \lfloor \mathbf{x} \rfloor_k : \mathbf{y} \in timepass^{\varepsilon}(\mathbf{u_g})$

The proof is immediate from the definition of $N_k(\mathbf{u_g})$.

Lemma 2. *For any* $\mathbf{u_g} \in Grid$, $N_k(\mathbf{u_g})$ *is given by:*

$$N_k(\mathbf{u_g}) = norm_k \left\{ \mathbf{u_g} + d + \sum_{i=1}^{n} a_i \cdot \mathbf{e_i} \mid d > 0, \ a_i \in [0,1) \right\},$$

where $\mathbf{e_i}$ *is the* i*-th unit vector.*

Here $\mathbf{u_g} + d$ denotes the addition of d to each component of $\mathbf{u_g}$. The proof follows from Lemma 1 and the definition of $\lfloor \mathbf{x} \rfloor_k$ for $\mathbf{x} \in \mathbb{R}^n_{\geq 0}$. This means that for any zone Z, $N_k(Z)$ is obtained as follows: First apply the standard unperturbed time-passage operator on Z, and then widen the *diagonal constraints which are non-strict inequalities* of the resulting conjunctive system by 1, to the *next higher strict inequalities*, i.e., $x - y \leq c$ is widened to $x - y < c + 1$, followed by standard k-normalization.

[3] By considering only closed guards and invariants for the automaton, we ensure that all the zones we encounter during FRA contain at least one grid-point.

Lemma 3. *Given any (diagonal-free) k-zone Z, $\mathbf{x} \in \mathbb{R}^n_{\geq 0}$, $\mathbf{u_g} \in Grid$,*
$\mathbf{x} \in N_k(\mathbf{u_g}) \cap Z \Leftrightarrow \forall \varepsilon > 0\ \exists \mathbf{y} \in \lfloor \mathbf{x} \rfloor_k : \mathbf{y} \in timepass^{\varepsilon}(\mathbf{u_g}) \cap Z$

The proof follows from Lemma (1) and the following property of any diagonal-free k-zone B [11]:

Property 1. $\forall \mathbf{x}, \forall \mathbf{y} \in \lfloor \mathbf{x} \rfloor_k : \mathbf{x} \in B \Leftrightarrow \mathbf{y} \in B$.

Lemma 4. *For any two* closed zones Z_1 *and* Z_2,

$$Z_1 \cap N_k(Z_2) = \emptyset \Leftrightarrow Z_1 \cap norm_k(timepass(Z_2)) = \emptyset$$

The proof of "$\Rightarrow$" is immediate, as $N_k(Z_2) \supseteq norm_k(timepass(Z_2))$. The proof of "$\Leftarrow$" is also obvious if $N_k(Z_2) = norm_k(timepass(Z_2))$.
When $N_k(Z_2) \supset norm_k(timepass(Z_2))$, we prove "$\Leftarrow$" as follows:
Z_1, Z_2 (and thus $timepass(Z_2)$, except for its "bottom") are closed. For $N_k(Z_2) \supset norm_k(timepass(Z_2))$, it must be the case that Z_2 is a k-zone and so $norm_k(timepass(Z_2)) = timepass(Z_2)$ is also closed (except for its "bottom"). Thus, in order for Z_1 to have an empty intersection with $norm_k(timepass(Z_2))$, the two must be separated by a (max. norm) distance of at least 1. It also follows that the only additions to $norm_k(timepass(Z_2))$ to form $N_k(Z_2)$ are the open diagonal borders (obtained by relaxing the diagonal constraints of Z_2 by 1). These borders thus added being open, can at most *touch, but not intersect* Z_1, which entails our result.

Lemma 5. *For any* closed k-zone Z, *for any* $\mathbf{u_g} \in Grid$, *any* $\mathbf{v} \in \mathbb{R}^n_{\geq 0}$

$$\mathbf{v} \in Z \cap N_k(\mathbf{u_g}) \Rightarrow \exists \mathbf{v_g} \in (grid(\mathbf{v}) \cap Z \cap norm_k(timepass(\mathbf{u_g})))$$

The proof follows as a consequence of Lemma 4 and the definition of $grid(\mathbf{v})$. This means that any closed guard (Z, referring to Lemma 5) that is enabled by a point ($\mathbf{v}$) obtained by time-passage from a grid-point ($\mathbf{u_g}$) under the smallest of drifts (and thus included into that point's ($\mathbf{u_g}$'s) neighbourhood) is also enabled by a different grid-point ($\mathbf{v_g}$) obtained by time-passage (without drift) from that grid-point ($\mathbf{u_g}$).

Lemma 6. *For any* closed, diagonal-free k-zone Z, *any* $\mathbf{x}, \mathbf{u} \in \mathbb{R}^n_{\geq 0}$,

$$\begin{aligned} &\mathbf{u} \in Z \ \wedge\ \forall \varepsilon > 0 : (\lfloor \mathbf{x} \rfloor_k \cap timepass^{\varepsilon}(\mathbf{u}) \cap Z) \neq \emptyset \\ &\Rightarrow \exists \mathbf{u_g} \in grid(\mathbf{u}) \cap Z \ \wedge\ \mathbf{x} \in N_k(\mathbf{u_g}) \cap Z \end{aligned}$$

The proof is immediate from Lemmas 3 and 5, and the definition of $grid(\mathbf{u})$.

Definition 10. Let R_i^* be the reach-set at the i-th iteration, computed by modifying the time-passage steps of the standard FRA algorithm as follows: the $norm_k(timepass())$ operator is replaced by its neighbourhood $N_k()$. R_i^* is termed the corresponding *robust reach-set.*

Let R^* be the robust reach-set that is ultimately computed by the FRA algorithm by using $N_k()$ instead of $norm_k(timepass())$, while computing the time-passage successors of zones [4], and R be the reach-set that is computed by the standard zone-based FRA (cf. Definition 6).
From Lemma 4, we get $R^* = \{(l, Z \cup (N_k(Z) \wedge Inv(l))) \mid (l, Z) \in R\}$, thereby resulting in the following corollary:

Corollary 1. *For any closed zone B and any $l \in L$,*
$R^* \cap (l, B) = \emptyset \Leftrightarrow R \cap (l, B) = \emptyset$.

We now establish useful properties of the sets R_i^* through the following lemmas.

Lemma 7. *Given any $i \in \mathbb{N}$, any $l \in L$, and any $\mathbf{x} \in \mathbb{R}^n_{\geq 0}$,*

$$(l, \mathbf{x}) \models R_i^* \Rightarrow \forall \varepsilon > 0 \ \exists \mathbf{y} \in \lfloor \mathbf{x} \rfloor_k : (l, \mathbf{y}) \in Reach_i^\varepsilon$$

Here, by $(l, \mathbf{x}) \models R_i^*$, we mean that there exists a zone $Z \in R_i^*$ such that $\mathbf{x} \in Z$. This lemma shows that the set R_i^* collects the regions that can be "touched" in the sense of some (but not necessarily all) points within being reachable for *every* perturbation. The proof is by induction over the number i of iterations, separately for even and odd values of i, using Lemma 3, Property 1, and the definitions of R_i^* and $Reach_i^\varepsilon$.

Lemma 8. *For any $l \in L$, any diagonal-free k-zone B, any $i \in \mathbb{N}$,*
$R_i^* \cap (l, B) \neq \emptyset \Rightarrow \forall \varepsilon > 0 : \ Reach_i^\varepsilon \cap (l, B) \neq \emptyset$. [5]

This lemma implies that at any iteration depth i, if the set R_i^* intersects with a target state, then the corresponding perturbed reach-set under even the smallest of perturbations likewise intersects with the target state. The proof follows from Lemma 7 and Property 1.

Lemma 9. *For any* even i, $l \in L$, $\mathbf{u_g} \in Grid \cap Inv(l)$, $\mathbf{v} \in \mathbb{R}^n_{\geq 0}$,

$$\begin{aligned} &(l, \mathbf{u_g}) \models R_i^* \wedge \exists l' \in L \ \exists t \in T : l \xrightarrow{t} l' \wedge \ \mathbf{v} \in N_k(\mathbf{u_g}) \cap Inv(l) \cap G(t) \\ &\Rightarrow \exists \mathbf{v_g} \in norm_k(timepass(\mathbf{u_g})) \cap grid(\mathbf{v}) \cap Inv(l) \cap G(t) : \\ &\qquad \exists \mathbf{w_g} \in (Inv(l') \cap Res_t(\mathbf{v_g}) : (l', \mathbf{w_g}) \models R_{i+2}^* \end{aligned}$$

The proof is immediate from Lemma 5 and the definition of R_i^*. Here we assume, in addition to the guards and invariants being closed and diagonal-free, the following condition of *admissible target locations*, which ensures consistency between the invariants of a location and the guards of the transitions entering and leaving that location:
For any locations l and l', and any transition t with $l \xrightarrow{t} l'$:
$Inv(l) \cap G(t) \neq \emptyset \ \wedge \ Inv(l') \cap G(t) \neq \emptyset$.

[4] Termination is guaranteed for such an algorithm by the use of k-normalization in the computation of the neighbourhood $N_k()$ of zones encountered during the FRA.

[5] Here $R_i^* \cap (l, B) \neq \emptyset$ denotes $Z \wedge B$ being satisfiable for some $(l, Z) \in R_i^*$.

Lemma 10. *For any* even i, *any* $l \in L$, *any* $\mathbf{x} \in \mathbb{R}^n_{\geq 0}$,
$\forall \mathbf{u_g} \in Grid \cap Inv(l) : (l, \mathbf{u_g}) \models R^*_i,\ \ \mathbf{x} \notin N_k(\mathbf{u_g}) \cap \bar{I}nv(l)$
$\Rightarrow \exists \varepsilon_i > 0\ \forall \mathbf{y} \in \lfloor \mathbf{x} \rfloor_k, \forall \mathbf{u} \in Inv(l) : (l, \mathbf{u}) \in Reach^{\varepsilon_i}_i :$
$\mathbf{y} \notin timepass^{\varepsilon_i}(\mathbf{u}) \cap Inv(l)$

The proof follows from Lemma 6 and Lemma 9, by induction over even i. A consequence is that the following converse of Lemma 7 also holds:

Lemma 11. *Given any* $i \in \mathbb{N}$, *any* $l \in L$, *and any* $\mathbf{x} \in \mathbb{R}^n_{\geq 0}$,

$$(l, \mathbf{x}) \not\models R^*_i \Rightarrow \exists \varepsilon_i > 0\ \forall \mathbf{y} \in \lfloor \mathbf{x} \rfloor_k :\ (l, \mathbf{y}) \notin Reach^{\varepsilon_i}_i$$

The proof is by induction over the number i of iterations, separately for even and odd values of i, using Lemmas 3 and 10, Property 1, and the definitions of R^*_i and $Reach^{\varepsilon_i}_i$.

Lemma 12. *For any* $l \in L$, $i \in \mathbb{N}$, *any diagonal-free k-zone* B,

$$R^*_i \cap (l, B) = \emptyset \Rightarrow \exists \varepsilon_i > 0 : Reach^{\varepsilon_i}_i \cap (l, B) = \emptyset.$$

The above lemma implies that at any iteration depth i, the set R^*_i does not intersect with a target state iff there exists a strictly positive value of the perturbation, such that the corresponding perturbed reach-set at that iteration depth likewise avoids the target state. The proof follows from Lemma 11 and Property 1. The following corollary is then a direct consequence of Lemmas 8 and 12.

Corollary 2. *Given any* $l \in L$, *any diagonal-free k-zone* B,
$R^* \cap (l, B) = \emptyset \Leftrightarrow \forall i \in \mathbb{N}\ \exists \varepsilon_i > 0 : Reach^{\varepsilon_i}_i \cap (l, B) = \emptyset$

Corollaries 1 and 2 lead us to the following theorem, which is a main result of this paper.

Theorem 1. *Let* R *be the final reach-set computed by the standard zone-based FRA, for a TA with closed and diagonal-free guards and invariants. Then for any closed and diagonal-free k-zone* B *and any* $l \in L$, $R \cap (l, B) = \emptyset \Leftrightarrow \forall i \in \mathbb{N}\ \exists \varepsilon_i > 0 : Reach^{\varepsilon_i}_i \cap (l, B) = \emptyset$

It follows from this theorem that the standard zone-based FRA used in tools like UPPAAL is exact (sound and complete) while testing TA with closed guards and invariants for robust safety against closed targets.

The "$\Leftarrow$" part of Theorem 1 states that a closed target is reported as reachable by standard zone-based FRA only if it is also reachable in a finite number of iterations of the transition relation of the TA, under even the slightest of perturbations. This result is intuitively obvious, because even the smallest perturbed reach-set is a strict superset of its non-perturbed version. The "$\Rightarrow$" part of Theorem 1 states that a closed target is reported as unreachable by zone-based FRA only if for any given number of iterations i of the transition relation, there exists a strictly positive value of the perturbation ε_i that the automaton can tolerate and yet remains safe, in the sense that the corresponding perturbed reach-set $Reach_i^{\varepsilon_i}$ has an empty intersection with the (closed) target state. It must be noted here that this does not mean the existence of a homogeneous $\varepsilon > 0$ independent of the number of iterations, for which the unsafe state can be avoided, which is the notion considered in related works [1, 2, 3, 4]. Rather, as mentioned in the introduction, the magnitude of the tolerated perturbation ε_i could (but not necessarily) decrease with the number i of iterations, with ε_i potentially tending to 0 as i tends to ∞[6]. However, so long as we execute an *arbitrary, but finite number* of iterations, we are guaranteed a positive value of the tolerable perturbation for robust safety.

The analyses in [1, 2, 3, 4], on the other hand, add states that can be reached in any (unbounded) number of iterations through the (progress) cycles of the automaton[7], for even the slightest perturbation. Therefore, a state $(l, \mathbf{x})$ is considered to be *robustly unreachable* in our sense (i.e., not included in R^*), but reachable in the sense of the works in [1, 2, 3, 4] iff $\lim_{\varepsilon \to 0} \min\{i \in \mathbb{N} \mid (l, \mathbf{x}) \in Reach_i^{\varepsilon}\} = \infty$.

4 Robustness w.r.t. Imperfect Synchronization

In the previous section, we considered a model of drifting clocks where the relative drift between the clocks increases without bound with the passage of time, although the clock-slopes are themselves bounded according to the parameter ε. This is, however, rarely the case in practice, where the clocks, though subject to drift, are *regularly resynchronized* by diverse means, ranging from bit-stuffing in communication protocols to high-level clock synchronisation schemes. A parameter Δ characterizes the *post-synchronization-gap* and a parameter μ the longest possible gap between synchronizations. If the slopes of the clocks (w.r.t absolute time) are in the range $\left[\frac{1}{1+\theta}, 1+\theta\right]$ between synchronizations, such a resynchronization enforces a *uniform bound* given by

$$\varepsilon = \max\left(\Delta + \mu\left(\frac{1}{1+\theta}\right)^2, \Delta + \mu(1+\theta)^2\right) = \Delta + \mu(1+\theta)^2$$

6 For closed TA in which each cycle has at-least one transition that resets all clocks simultaneously, the robust reach-sets computed by the algorithms in [2, 3, 4] coincide with the standard reach-set computed by UPPAAL, as seen in automaton of Fig. 1. Thus, a certificate of safety by standard UPPAAL for such TA w.r.t. closed targets implies a robust safety margin independent of iteration depth.

7 We make no assumption on the cycles of the automaton.

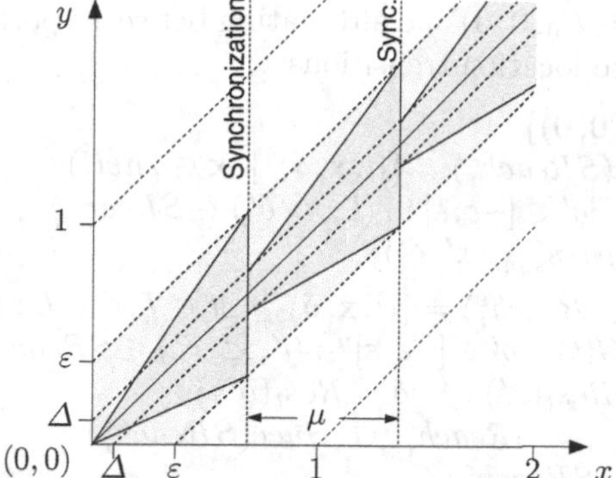

Fig. 2 Periodic resynchronization resulting in a bound ε on relative drift between clocks

on the relative drift between the clocks, irrespective of the extent of time-passage. The phenomenon is illustrated for two clocks x and y in Fig. 2. Throughout this section, we assume $0 < \varepsilon < 1$.

We incorporate such a resynchronization into TA by associating a *drift-offset* $\boldsymbol{\delta} \in [-\varepsilon, \varepsilon]^n$ for each clock valuation $\mathbf{x} \in \mathbb{R}^n_{\geq 0}$. This drift-offset keeps track of the extent to which the individual clocks in $\mathbf{x}$ have deviated from an implicit reference clock maintained by the synchronization scheme. The states of a TA in this semantics are thus tuples $(l, \mathbf{x}, \boldsymbol{\delta}) \in L \times \mathbb{R}^n_{\geq 0} \times [-\varepsilon, \varepsilon]^n$. As the deviation $\boldsymbol{\delta}$ is controlled by the synchronization scheme such that it always remains below ε, the (perturbed) time-passage under synchronization is as follows:

Definition 11. Given any $\mathbf{x} \in \mathbb{R}^n_{\geq 0}$, any $\boldsymbol{\delta} \in [-\varepsilon, \varepsilon]^n$,

$$timepass^{\varepsilon}_{sync}(\mathbf{x}, \boldsymbol{\delta}) = \{(\mathbf{x}', \boldsymbol{\delta}') \mid \boldsymbol{\delta}' \in [-\varepsilon, \varepsilon]^n \wedge \exists d > 0 : \mathbf{x}' = \mathbf{x} - \boldsymbol{\delta} + d + \boldsymbol{\delta}'\}$$

A *run* of a perturbed TA subject to clock synchronization with accuracy ε is a sequence $\langle (l_0, \mathbf{x}_0, \boldsymbol{\delta}_0), (l_1, \mathbf{x}_1, \boldsymbol{\delta}_1), \ldots \rangle$ of states such that

1. l_0 is the initial location and $\mathbf{x}_0 = \boldsymbol{\delta}_0 = \mathbf{0}$,
2. For even i, $l_{i+1} = l_i$, $\mathbf{x}_{i+1} \in Inv(l_i)$
   $\wedge\ (\mathbf{x}_{i+1}, \boldsymbol{\delta}_{i+1}) \in \{(\mathbf{x}_i, \boldsymbol{\delta}_i)\} \cup timepass^{\varepsilon}_{sync}(\mathbf{x}_i, \boldsymbol{\delta}_i)$ [8]
3. For odd i, $\exists t_i \in T : l_i \xrightarrow{t_i} l_{i+1} : \mathbf{x}_i \in Inv(l_i) \cap G(t_i)$,
   $\mathbf{x}_{i+1} \in Inv(l_{i+1}) \cap Res_{t_i}(\mathbf{x}_i)$, $\boldsymbol{\delta}_{i+1} = Res_{t_i}(\boldsymbol{\delta}_i)$.

Due to memorizing the current deviation $\boldsymbol{\delta}$ and adjusting it consistently to the constraint that the overall accuracy is better than ε, this semantics is subtly more constrained than the —superficially similar— semantics permitting an arbitrarily directed ε-deviation upon every time passage.

SReach$^{\varepsilon}$ is the corresponding perturbed reach-set, defined inductively as follows, with *SReach*$^{\varepsilon}_i$ denoting the perturbed reach-set in $i \in \mathbb{N}$ steps, starting

[8] By abuse of notation, the subscripts i here denote the sequence of tuples in a run, and not individual vector components.

from the initial state $(l_0, \mathbf{0}, \mathbf{0})$ and alternating between (perturbed) time-passage and (exact) discrete-location transitions:

- $SReach_0^\varepsilon \equiv \{(l_0, \mathbf{0}, \mathbf{0})\}$
- For i even, $Succ(SReach_i^\varepsilon) = \{(l, \mathbf{x}, \boldsymbol{\delta}) \mid \mathbf{x} \in Inv(l)$
  $\wedge\ \exists \mathbf{x'} \in Inv(l), \exists \boldsymbol{\delta'} \in [-\varepsilon, \varepsilon]^n : (l', \mathbf{x'}, \boldsymbol{\delta'}) \in SReach_i^\varepsilon$
  $\wedge\ (\mathbf{x}, \boldsymbol{\delta}) \in timepass_{sync}^\varepsilon(\mathbf{x'}, \boldsymbol{\delta'})\}$
- For i odd, $Succ(SReach_i^\varepsilon) = \{(l, \mathbf{x}, \boldsymbol{\delta}) \mid \exists t \in T, l' \in L : l' \xrightarrow{t} l,$
  $\exists \mathbf{x'} \in Inv(l') \cap G(t)\ \exists \boldsymbol{\delta'} \in [-\varepsilon, \varepsilon]^n : (l', \mathbf{x'}, \boldsymbol{\delta'}) \in SReach_i^\varepsilon :$
  $\wedge\ \mathbf{x} \in Inv(l) \cap Res_t(\mathbf{x'})\ \wedge\ \boldsymbol{\delta} = Res_t(\boldsymbol{\delta'})\}$
- $\forall i \geq 0, SReach_{i+1}^\varepsilon = SReach_i^\varepsilon\ \cup\ Succ(SReach_i^\varepsilon)$
- $SReach^\varepsilon = \bigcup_{i \in \mathbb{N}} SReach_i^\varepsilon$

As before, we assume that all guards and invariants are closed and diagonal-free. Let *Reach* denote the reach-set obtained by considering perfectly synchronous clocks ($\varepsilon = 0$), where $Reach_i$ denotes the reach-set at step i, as defined previously (cf. Definition 6). We establish the relationship between the sets $SReach^\varepsilon$ and *Reach* through the following lemmas.

Lemma 13. *For any* $i \in \mathbb{N}$, $l \in L$, $\mathbf{x} \in \mathbb{R}^n_{\geq 0}$, $\boldsymbol{\delta} \in [-\varepsilon, \varepsilon]^n$,

$$(l, \mathbf{x}, \boldsymbol{\delta}) \in SReach_i^\varepsilon \;\Rightarrow\; \exists \mathbf{x_g} \in grid(\mathbf{x}) : (l, \mathbf{x_g}) \in Reach_i$$

The proof is by induction over i, from the definitions of $SReach_i^\varepsilon$ and $Reach_i$. The following corollary is an immediate consequence.

Corollary 3. *For any* $i \in \mathbb{N}$, *it holds that:*

$$sup_{s \in SReach_i^\varepsilon}\ dist(s, Reach_i) < 1\ ,$$

where for $s = (l, \mathbf{x}, \boldsymbol{\delta}) \in SReach_i^\varepsilon$, $dist(s, Reach_i) = \inf_{(l,\mathbf{x'}) \in Reach_i} dist(\mathbf{x}, \mathbf{x'})$.

Corollary 3 intuitively means that irrespective of the iteration depth i, the perturbed reach-set $SReach_i^\varepsilon$ stays "close-enough" to the standard reach-set $Reach_i$, in the sense that even the "farthest" point in the perturbed reach-set is less than unit distance away from the standard reach-set.

Lemma 14. *For a TA with only closed and diagonal-free guards and invariants, and any closed target location-zone pair of the form* (l, B)*:*

$$Reach \cap (l, B) = \emptyset \Leftrightarrow \forall 0 < \varepsilon < 1 : SReach^\varepsilon \cap (l, B) = \emptyset\ ,$$

where $SReach^\varepsilon \cap (l, B) = \emptyset$ *denotes* $\forall (l, \mathbf{x}, \boldsymbol{\delta}) \in SReach^\varepsilon : \mathbf{x} \notin B$.

The proof of "$\Leftarrow$" is obvious as $\forall \varepsilon > 0 : SReach^\varepsilon \supset Reach$, in the following sense: $\forall (l, \mathbf{x}) \in Reach : (l, \mathbf{x}, \mathbf{0}) \in SReach^\varepsilon$. The proof of "$\Rightarrow$" follows from Corollary 3, in conjuction with the fact that B is a closed zone, as are all the guards and invariants of the TA, and $0 < \varepsilon < 1$. This lemma, together with the soundness and completeness result for standard zone-based FRA [10], leads us to the following theorem, which is the second main result of this paper:

Theorem 2. *For a TA with only closed and diagonal-free guards and invariants, any location l, and any closed, diagonal-free k-zone B :*

$$R \cap (l, B) = \emptyset \Leftrightarrow \forall 0 < \varepsilon < 1 : SReach^{\varepsilon} \cap (l, B) = \emptyset \ ,$$

where R is the symbolic reachable state-space that is ultimately computed by the standard zone-based FRA.

Theorem 2 thus establishes the exactness of standard zone-based forward analysis using a tool like UPPAAL for TA with closed guards and invariants, when testing for robust safety against closed targets, with drifting clocks subject to periodic resynchronizations that enforce accuracy better than 1. A certification of robust safety in this case implies that the target state could be avoided by all values of the perturbation ε that are strictly less than 1, independent of the depth of iteration, unlike the case for unbounded relative clock-drift that was considered in the previous section.

Theorem 2 may also be proven using the neighbourhood construction for grid-points, as was previously done for Theorem 1.

5 Conclusion

We have investigated reachability (and thus, safety) of TA subject to drifting clocks – a phenomenon that occurs in practical implementations of timed systems. We first considered the model of clock-drift introduced in [1] and studied in [2, 3, 4], and analyzed the reachability for TA with closed guards, invariants, and targets, but without the assumption of progress cycles, as was made in [1, 2, 3, 4]. We showed the exactness of the standard zone-based FRA of UPPAAL for such TA, under a notion of robustness weaker than that in [1, 2, 3, 4], in the sense that we do not add states that require an unbounded number of iterations in order to be reached, under infinitesimally small clock-drift (cf. Theorem 1). Our notion is applicable to all systems having a finite life-time, where for any particular projected life-time, an appropriate worst-case clock drift enforcing behavior indistinguishable from the ideal can be chosen. For long life-times, the permissible clock drift may become extremely small. As technical realizations in many systems (like, e.g., bit-stuffing in communication protocols or the central-master synchronization incorporated in GPS-controlled systems) address this problem by regular clock resynchronization, thus bounding the relative drift within an set of clocks even over arbitrarily long life-times, we have also modelled and analyzed such synchronization schemes. We have shown that the standard zone-based analysis of UPPAAL is again exact while testing such models for robust safety, but now with the assertion of a uniform strictly positive robustness margin of 1, independent of system life-time.
Note that our definition of TA admits only *diagonal-free constraints* for the guards, invariants, and targets. This is because TA with diagonal constraints

of the form $x - y \triangleright c$ have been shown to be incompatible with forward reachability analysis that employs standard k-normalization for termination, and a modified normalization that takes into account the diagonal constraints of the TA is in fact necessary for dealing with such cases [11, 10]. However, the techniques of this paper extend quite naturally to TA with diagonal constraints and a suitably modified normalization operation. An extension of these techniques to Probabilistic TA [12] (TA with discrete probability distributions annotating transitions between locations) also appears straight-forward.
We finally wish to mention the following alternate notions of robustness for TA: [13] imposes a topological closure on timed traces, which has been shown in [14] to affect digitization of TA. [15] considers robust model-checking of LTL properties, while [16] considers robustness analysis via channel machines.

Acknowledgements This work was supported mainly by the Deutsche Forschungsgemeinschaft (DFG) under grant GRK 1076/1 for the Graduate School Trustsoft `http://www.uni-oldenburg.de/trustsoft/en/`. We also acknowledge support of the DFG under the AVACS project `www.avacs.org`, of the EU under the Quasimodo Project
`http://www.quasimodo.aau.dk/`, and of ASKON Consulting Group GmBH `www.askon.de`.
The comments of two of the reviewers were most helpful in refining this paper, as were discussions with Alexandre David, Kim Larsen, Laurent Doyen, Jean-Francois Raskin, Olaf Owe, Wang Yi, Stefan Leue, Michael Adelaide, Piotr Kordy, Christian Herde, and Abhishek Dhama.

References

1. A. Puri, "Dynamical Properties of Timed Automata", *Discrete Event Dynamic Systems* **10**, Kluwer (2000), 87–113
2. M. de Wulf, L. Doyen, N. Markey, and J.-F. Raskin, "Robustness and Implementability of Timed Automata", *Proc. of FORMATS-FTRTFT'04*, LNCS **3253**, Springer (2004), 118-133
3. C. Daws and P. Kordy, "Symbolic Robustness Analysis of Timed Automata", *Proc. of FORMATS'06*, LNCS **4202**, Springer (2006), 143-155
4. C. Dima, "Dynamical Properties of Timed Automata Revisited", *Proc. of FORMATS'07*, LNCS **4763**, Springer (2007), 130–146
5. R. Alur and D. Dill, "A Theory of Timed Automata", *Theoretical Computer Science* **126**, Elsevier (1994), 183–235
6. G. Behrmann, A. David, and K. Larsen, "A Tutorial on Uppaal", *Formal Methods for the Design of Real-Time Systems*, LNCS **3185**, Springer-Verlag (2004), 200–236
7. M. Bozga, C. Daws, O. Maler, A. Olivero, S. Tripakis, S. Yovine, "KRONOS: A Model-Checking Tool for Real-Time Systems", *Proc. of FTRTFT'98*, LNCS **1486**, Springer (1998), 298–302
8. M. Lindahl, P. Pettersson, and W. Yi, "Formal Design and Analysis of a Gear Controller", *International Journal on Software Tools for Technology Transfer* **3**, Springer-Verlag (2001), 353–368
9. L. Aceto, P. Bouyer, A. Burgueno, and K. Larsen, "The Power of Reachability Testing for Timed Automata", *Theoretical Computer Science* **300**, Elsevier (2003), 411-475

10. J. Bengtsson and W. Yi, "Timed Automata: Semantics, Algorithms, and Tools", *Lectures on Concurrency and Petri Nets*, LNCS **3098**, Springer (2004), 87–124
11. P. Bouyer, "Forward Analysis of Updatable Timed Automata", *Formal Methods in System Design* **24**, Kluwer (2004), 281–320
12. M. Kwiatkowska, G. Norman, R. Segala, and J. Sproston, "Automatic Verification of Real-time Systems with Discrete Probability Distributions", *Theoretical Computer Science* **282**, Elsevier (2002), 101-150
13. V. Gupta, T. A. Henzinger, and R. Jagadeesan, "Robust Timed Automata", *Proc. of HART'97*, LNCS **1201**, Springer (1997), 331–345
14. J. Ouaknine and J. Worrell, "Revisiting Digitization, Robustness, and Decidability for Timed Automata", *Proc. of LICS'03*, IEEE CS Press (2003), 198-207
15. P. Bouyer, N. Markey, and P.-A. Reynier, "Robust Model-Checking of Linear-Time Properties in Timed Automata", *Proc. of LATIN'06*, LNCS **3887**, Springer (2006), 238–249.
16. P. Bouyer, N. Markey, and P.-A. Reynier, "Robust Analysis of Timed Automata via Channel Machines", *Proc. of FoSSaCS'08*, LNCS **4962**, Springer (2008), 157-171